# Advances in Reconfigurable Mechanisms and Robots I

Jian S. Dai · Matteo Zoppi
Xianwen Kong
Editors

# Advances in Reconfigurable Mechanisms and Robots I

Springer

*Editors*
Jian S. Dai
Chair of Mechanisms and Robotics
School of Natural and Mathematical Sciences
King's College London
University of London
London
UK

and

Centre for Advanced Mechanisms and Robotics
Key Laboratory for Mechanism Theory and Equipment Design of Ministry of Education
School of Mechanical Engineering
Tianjin University
Tianjin
China

Matteo Zoppi
Department of Mechanics and Machine Design
University of Genova
Via Opera Pia 15A
Genova
Italy

Xianwen Kong
School of Engineering and Physical Sciences
Heriot-Watt University
Edinburgh EH14 4AS
UK

ISBN 978-1-4471-4140-2 ISBN 978-1-4471-4141-9 (eBook)
DOI 10.1007/978-1-4471-4141-9
Springer London Heidelberg New York Dordrecht

Library of Congress Control Number: 2012938714



Printed on acid-free paper

Springer is part of Springer Science+Business Media (www.springer.com)

# Preface

Human history elucidates the spiral development of production and its corresponding technologies when they evolve back into the tailor-made and small batch production as that in the nineteenth century before the second industrial revolution for mass production. The encounters in the twenty-first century with the overwhelming demands for variety and the fierce competition in the consumer markets require the tailor-made and small batch production with additional dimensions in rapidness and adaptability. This needs a quick changeover and reconfiguration for machines and production lines. Further to this, the changing climates along with natural catastrophes, the hazardous environments, and the space exploration need devices and robots with variable capacity and changeable structures with certain versatility and flexibility.

As a backbone and an essential part of machines and robots, mechanisms and also exoskeletons of robots need to have additional abilities to vary mobility and/or topology to adapt for the demands. These types of mechanisms with variable mobility which emerged in the 1990s raised much interest in the first decade of this century in their topology, kinematics, dynamics and stiffness, and evolve into reconfigurable robots in many fields and applications.

The philosophy underlying this development is adaptability and reconfigurability, which present the challenges in mechanism design and robots development that the kinematicians and roboticists are to face in the new decade with the increasing production demand of small batch, short run, and quick changeover, with the worsening environments and climate, and with the increasing space exploration.

To meet the above trend and with an overwhelming support, we have successfully held the First ASME/IFToMM International Conference on Reconfigurable Mechanisms and Robots (ReMAR 2009). The conference was a great success with excellent presentations and papers on seven topic areas in reconfigurable mechanisms and robots. The collection of the papers presented in the First ASME/IFToMM International Conference on Reconfigurable Mechanisms and Robots can be found in *Reconfigurable Mechanisms and Robots* (ed. Dai, Zoppi and Kong), KC Edizioni, IEEE Catalog No. CFP0943G-PRT, 2009.

Following this success, we are holding this Second ASME/IFToMM International Conference on Reconfigurable Mechanisms and Robots (ReMAR 2012) to provide an international forum for presenting and exploring the reconfigurable mechanisms and robots with their properties of changing the mobility and topological structure, and for presenting discussions of their uses in domestic, hazardous, outer-space, and manufacturing environments. With over 103 submissions, the rigorous review process was held with detailed and in-depth comments on each paper. This resulted in the acceptance of 78 papers with an additional month for authors to follow the reviewers' comment to revise the papers. We are very pleased with the overwhelming support from authors and from the communities and with the excellent work from the authors and their serious effort to improve and finalize the papers.

This book is the result of the successful international conference for all 78 papers on seven areas that are compiled into seven parts as reconfiguration theory, topology, kinematics, and design of reconfigurable mechanisms, reconfigurable parallel mechanisms, bio-reconfiguration techniques and biomedical devices, analysis and design of reconfigurable robots, control of reconfigurable robots, deployable mechanisms, and application of reconfigurable mechanisms.

This book starts from the reconfiguration theory part that presents the representation theory, topology analysis, and type synthesis of reconfigurable mechanisms. The matrix and configuration representations are studied and the multiple operation forms are investigated. The geometric constraints are elaborated and the discontinuously movable constant velocity is scrutinized. This leads to the resistance-gradient model for development of metamorphic mechanisms and reconfigurable mechanisms.

Part II of the book is dedicated to the topology, kinematics, and design of reconfigurable mechanisms on structure synthesis, topology analysis, and mobility analysis. The biological modeling and configuration evolution are initiated, leading to mechanism evolution to various mobility configurations with adaptability and reconfigurability. The variable graph is proposed and a compliant reconfigurable mechanism is presented utilizing the transformation.

Part III presents the recent study of reconfigurable parallel mechanisms which evolves into multiple operational modes and utilizes the actuation scheme. The development of variable-axis revolute joint and variable Hooke joint enables the type change of a parallel mechanism during operation and raised much interest in the further direction of parallel mechanism development, leading to type-variable parallel mechanisms that experience the metamorphosis during the operation. Experimental modal is analyzed and the principle of parallel mechanisms is utilized to generate a new concept of self-reconfigurable intelligent swarm fixtures.

Part IV is dedicated to bio-inspired knowledge in reconfiguration and in biodevices. This part first presents the work on applying the metamorphic-compliant mechanism to the bi-behavioral prosthetic knee and the advances on reconfigurable stiffness for a knee exoskeleton. This is followed by the development of a dexterous surgical instrument with reconfigurable units and a novel MR-compatible robot. The minimally invasive surgical robot is introduced with the portable reconfigurable unit and with a passive robotic platform. In this part, new reconfigurable robots are presented and origami mechanisms are displayed.

Part V focuses on the analysis and design of reconfigurable robots with design and test on modular self-reconfigurable robots, structures and characteristics exploration on self-reconfigurable robots, gait analysis on six-legged reconfigurable robots, and architecture synthesis on modular robots. It is interesting to readers that a massage robot is developed with a metamorphic mechanism and a novel quadruped robot is investigated with the extendable body. This leads to dynamics analysis based on Lagrange's equation and contact information identification with intelligent fingertip sensors. This part ends with an interesting robotic methodology proposed for system reconfiguration of waste collections.

Part VI continues the topics of the reconfigurable robots and focuses on the control with the decentralized control scheme, a back-stepping terminal sliding mode algorithm, and the centralized receding horizon coverage control framework. The transparency analysis and friction compensation are implemented to control the master–slave robots, the cable-driven manipulators, and the reconfigurable manipulators. This part concludes with a review and prospects of robotics studies in Europe for an interesting map of the robotics development.

Part VII is dedicated to deployable mechanisms and applications of the reconfigurable mechanisms. The section starts from the deployable masts based on the Bennett linkage and continues with an interesting study of structural synthesis of an ancient Chinese Chu-state repeating crossbow. The duality of the platonic polyhedrons and the isomorphism of regular deployable polyhedral mechanisms are presented and the survey on inflatable deployable space structures is made. The applications of reconfigurable mechanisms to a modular deployable antenna, to lift mechanisms, and to box-taking mechanisms are presented. The interesting foldable parallel mechanisms and the color-flipping ball mechanism are investigated. More application is presented in the cutting system.

This book presents a vivid development of reconfigurable mechanisms and robots and is a collection of the work in design and development across the disciplines. With the contributions from the authors, this book delivers a lasting impact on the study and development of reconfigurable mechanisms and robots.

In compiling this book, we are pleased to see the variety of the topics, the depth of the study, and the wide range of the applications of reconfigurable mechanisms. The development of reconfigurable mechanisms is seen in the book contributing to development of mechanisms and robotics and presents a strong part for advancing the knowledge in the field and for economical development. We thank all the authors for their contributions and meticulous manner in preparing their manuscripts and thank all the reviewers for their rigorous review and detailed comments to help authors to improve their papers. Further, we thank Helge Wurdemann and Ketao Zhang for their dedication and persistence in checking every manuscript and contacting the authors for their revision of the manuscripts and for finalizing this book.

We have received immense support from Tianjin University, King's College London, the Ministry of Education of China, the National Natural Science

Foundation of China (NSFC), the Chinese Mechanical Engineering Society (CMES), and the European Commission Seventh Framework under grant agreements 231143 and 270436. We have the overwhelming support from the Design Division and the Mechanisms and Robotics Committee of ASME, the ASME UK and Ireland Section, the Automation and Robotics Society of IEEE, and the International Federation for the Promotion of Mechanism and Machine Science (IFToMM). We acknowledge the financial support from Tianjin University, the NSFC Key Research grant 51135008 and grant 51175366, the NSFC conference grant, and the Key Laboratory of Ministry of Education of China for Mechanism Theory and Equipment Design.

In organizing this second international conference on reconfigurable mechanisms and robots, we are grateful to members of the scientific committee, the program committee, and the track chairs/co-chairs for the rigorous peer review of papers. We thank the organizing teams in both Tianjin University and King's College London, University of London who worked so hard to put together exciting sessions and presentations. We particularly thank Helge Wurdemann for conference preparation and website editing, Ketao Zhang for paper handling and author contacting, Guowu Wei for financial arrangement, Jianmin Li for conference planning, and Yuhu Yang and Zhongxia Xiang for local arrangement. We also extend our thanks to Just Herder, Anurag Purwar, Chao Chen, and Yan Chen for their continuing support and kind help.

We are grateful to Tianjin University which is the first Chinese university established in 1895 for hosting this conference and for their generous financial support. We particularly thank Professor Shuxin Wang in his capacity as the Dean of the School of Mechanical Engineering and Co-Chair of the conference and Professor Tian Huang in his capacity as the Chairman of the Academic Board of the School for their strong support and thank Professor Guobiao Wang of the National Natural Science Foundation of China (NSFC) for his invaluable support.

Tianjin, China, July 2012

Jian Sheng Dai
Matteo Zoppi
Xianwen Kong

# Contents

## Part V Analysis and Design of Reconfigurable Robots

# Part I
# Reconfiguration Theory

# Chapter 1
# Metamorphic Structure Representation: Designing and Evaluating Anatomies of Metamorphic Manipulators

**Charalampos Valsamos, Vassilis C. Moulianitis and Nikos Aspragathos**

**Abstract** In this work the metamorphic structure representation (MSR) is proposed for the systematic development and evaluation of structures of metamorphic serial manipulators. The basic elements of MSR are presented, in terms of module and connection type. The conceptual design of metamorphic structures using MSR is presented. The criteria for evaluating emerging structures are introduced in terms of simplicity and solvability of inverse kinematics. Finally, a case study of two different structures is presented along with their subsequent evaluation.

**Keywords** Open chain metamorphic manipulators · Reconfigurable robots · Structure representation

## 1.1 Introduction

During the last decades, it was revealed that the use of modular reconfigurable manipulators the flexibility and adaptability of industrial robotic systems could be enhanced significantly. Such robotic manipulators incorporate a modular architecture, while at the same time efforts were made in the direction of reconfigurable

C. Valsamos (✉) · N. Aspragathos
Department of Mechanical and Aeronautics Engineering,
University of Patras, 26500 Rio, Achaia, Greece
e-mail: balsamos@mech.upatras.gr

N. Aspragathos
e-mail: asprag@mech.upatras.gr

V. C. Moulianitis
Department of Product and Systems Design, University of the Aegean,
84100 Ermoupolis, Syros, Greece
e-mail: moulianitis@syros.aegean.gr

J. S. Dai et al. (eds.), *Advances in Reconfigurable Mechanisms and Robots I*,
DOI: 10.1007/978-1-4471-4141-9_1,  

control. The envisaged outcome of current reconfigurable robotics research is the development of a system able to address a multitude of various tasks, in a rapid, optimal and seamless fashion.

The reconfigurable modular manipulator system (RMMS) developed at Carnegie Mellon [1] was composed by the reconfigurable manipulator, a modular and reusable control software and a novel agent based approach for the task—based design of modular manipulators, using four kinds of self contained mechatronic modules. The rapidly reconfigurable robotic work cell [2] utilized standard and inter—operable components, which could be easily assembled into robots with the chosen anatomy and degrees of freedom of either parallel or serial architecture. A kinematic graph based representation, termed assembly incidence matrix (AIM) was introduced to mathematically describe modular robot configurations [3].

Metamorphic robots were presented as a class of serial reconfigurable manipulators allowing the creation of structures presenting a large number of different anatomies each, many of which are outside the current line of design practice for robots [4]. The general structure of the proposed class of serial manipulators is modular, but the transition from one anatomy to a new one is conducted rapidly and without any need for dismantling the initial robot structure. Under specific perquisites it is possible to construct a parametric analytical solution to the kinematics of a given structure and all of its emerging anatomies, providing additional advantages in the reduction of required time for changes in control and task planning.

A metamorphic manipulator can achieve a considerable range of different anatomies, by exploiting the off-line variation of its passive modules. Although this is their main advantage, it greatly increases the search space for design processes, since both the best structure and the best anatomy of the corresponding structure for a given task are sought. Therefore, a systematic representation of different metamorphic structures is required in order to allow their rapid development and evaluation.

The problem of designing a metamorphic robot can be viewed at two levels. In the upper level, metamorphic structures are created and evaluated depending on their structural properties to determine the best one(s). In the lower level, these structures are evaluated in order to select the overall best one under a given performance type [5] and the method presented in [4, 6] is used in order to determine both the optimal anatomy and task location in the vicinity of the optimal anatomy's workspace.

In this paper, the metamorphic structure representation (MSR) is introduced as a tool for supporting the rapid and systematic conceptual design of open chain metamorphic manipulators. In addition, two criteria are presented, simplicity of the structure and analytical kinematic solvability in order to evaluate the emerging structures. It will be discussed that the introduced representation can be used with A.I. techniques, towards systematic automated conceptual design of metamorphic robotic workcells.

## 1.2 Basic Structural Definitions

In order to establish the MSR, the following definitions are introduced:

- **Active joint**. Self contained rotational active module (joint). Provides a rotational degree of freedom for a structure.
- **Pseudo joint** Passive 1 rotational d.o.f. connector. Changing its angle off-line allows the transition of a structure to a new anatomy. For any given anatomy, the pseudo joint setting is locked.
- **Metamorphic Structure**. A metamorphic structure is an open chain composed of the three main types of modules, connected in a specific way to each other, so as to form a serial manipulator. It is characterized by the presented degrees of freedom, the number and types of modules, their connections and their relative sequence in the created body.
- **Metamorphic anatomy**. A metamorphic anatomy is defined as the current posture of a metamorphic structure. This posture is dependant of the setting of the pseudo joints contained within the structure. As pseudo joints settings are altered, new anatomies emerge for the given structure.
- **Configuration** is defined by the values of the variables of the active joints (angles) for a given anatomy**.** The **Reference configuration** is defined as the configuration for a given anatomy where all the kinematic variables of the manipulator are set to zero.

## 1.3 Metamorphic Structure Representation

The representation of emerging serial metamorphic robotic structures is of outmost importance so as to provide a means for conceptualization of structures as well as to allow such representations to be used in A.I. methods used for the determination of the best possible structure. It seems that the binary representation is very convenient for the description of the topology and geometry of metamorphic mechanisms [7].

The introduced serial metamorphic manipulators utilize 3 main type of modules for their structures, active joints, pseudo joints and rigid connectors [5] (Appendix A). Rigid connectors are incorporated as lengths to active joints and pseudo joints and therefore not needed to be modeled separately. Thus, the module set cardinality is limited in two types. Connections are modeled examining the relative configuration of selected geometrical properties for adjacent modules which are the joint twists both for active and pseudo joints (see Appendix A). There are three discrete connection possibilities: perpendicular, parallel or skew (Fig. 1.1). Following the above presentation, the MSR is defined by the following elements:

- A single digit is appointed to represent each type of available modules.
- A single digit is appointed to represent the connection of two adjacent modules.

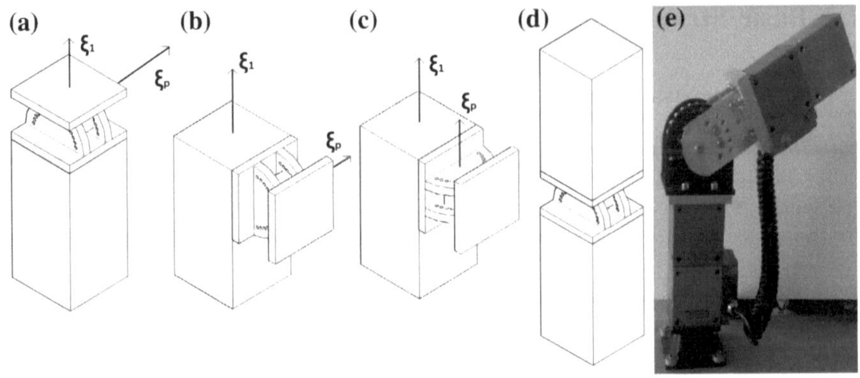


**Fig. 1.1** Drawing examples of **a** string 001#, **b** string 021#, **c** string 011#, **d** string 00100#, **e** Physical implementation of string 00100#

**Table 1.1** Representation of module and connection types for metamorphic structures

| | Module type | | |
|---|---|---|---|
| | Active joint | Pseudo joint | |
| Digit | 0 | 1 | |
| | Connection | | |
| | Perpendicular | Parallel | Skew |
| Digit | 0 | 1 | 2 |

- A 2 digit basic string in the form **module-connection** uniquely describes the type of module and its connective relation to its adjacent one within a structure.

According to these assignments, Table 1.1 presents the encoding for module and connection types. Metamorphic structures can be described by string arrays of the 2 digit basic string.

## 1.4 Conceptual Design Using the MSR

Using the notations established, metamorphic structures can be developed by string arrays composed of any given number of basic strings. If two basic strings are considered **module$_1$-connection$_1$** and **module$_2$-connection$_2$** and are combined to form a structure then it will be **module$_1$-connection$_1$-module$_2$-connection$_2$**.

The following apply for every metamorphic structure represented using this notation:

(a) The modules and the connections between two successive active joints define a link.
(b) The end of the string describing a metamorphic structure depicts the connection of the last module of the kinematic chain with the tool frame's Z axis.

**Table 1.2** Links created using one pseudo joint. Possible relation of active joint twists for a metamorphic link created for various setting of the pseudo joint displacement

| MSR Notation | Active Joint$_1$-connection$_1$ | Pseudo joint-connection$_2$ | Active Joint$_2$-connection$_3$ | Active joint twists relation relative to pseudo joint setting angle ($\theta_p$) | | |
|---|---|---|---|---|---|---|
| | | | | Intersecting | Parallel | Skew |
| 00 10 0# | 00 | 10 | 0# | $\theta_p \neq 0^0$ | N.A. | N.A. |
| 00 11 0# | 00 | 11 | | $\theta_p = 0^0$ | $\theta_p \neq 0^0$ | N.A. |
| 00 12 0# | 00 | 12 | | $\theta_p \neq \pm 90^0$ | N.A. | $\theta_p = \pm 90^0$ |
| 01 11 0# | 01 | 11 | | N.A. | N.A. | Always |
| 01 12 0# | 01 | 12 | | N.A. | Always | |
| 02 12 0# | 02 | 12 | | $\theta_p \neq 0^0$ | N.A. | $\theta_p = 0^0$ |

Symbol # is a wildcard for the connection type of the remaining structure. (N.A. = Not Available)

Table 1.2 illustrates the links created using one pseudo joint, also depicting the resulting possible relation of the two active joint twists thus connected relative to the pseudo joint parameter $\theta_p$. Fig. 1.1 illustrates a physical example of strings **001#, 011#**, **021#**, string **00100#** presented in Table 1.2 and it's physical implementation. The presented link leads in intersecting active joint twists except the case showed in Fig. 1.1d ($\theta_p = 0^0$).

## 1.5 Emerging Open Chain Structure Evaluation

Using the MSR, a wide variety of serial metamorphic structures can be defined. In order to proceed with the selection of the best possible for the given requirements, a method for their initial evaluation is. Therefore, evaluation criteria need to be established. The criteria selected are the simplicity of a structure and the kinematic solvability in an analytical form of a structure.

### *1.5.1 Simplicity*

The simplicity of a structure is subject to the number of modules used to from it. Most importantly the minimum number of pseudo joints is sought to be placed in a link, so as the maximum number of possible relations for the active joint twists connected will be presented.

Table 1.3 presents the 24 different feasible types of links constructed using one or two pseudo joints. The types are categorized in groups based on the available relation achieved by the twists of the active joints. By examination of the resulting link properties, it becomes evident that in the case of using group one types to form a link, the active joint twists achieve all possible relations (intersecting, parallel or

**Table 1.3** Metamorphic links created using one or two pseudo joints and relations between the connected active joints as pseudo joint parameters are altered

| Group of metamorphic links | MSR notation | Available relations of active joint twists in achieved links | | |
|---|---|---|---|---|
| | | Intersecting | Parallel | Skew |
| 1 | 00 10 10 0# | Yes | Yes | Yes |
| | 00 10 12 0# | Yes | Yes | Yes |
| | 00 12 10 0# | Yes | Yes | Yes |
| | 00 12 11 0# | Yes | Yes | Yes |
| | 00 12 12 0# | Yes | Yes | Yes |
| | 01 12 12 0# | Yes | Yes | Yes |
| | 02 10 12 0# | Yes | Yes | Yes |
| 2 | 00 11 10 0# | Yes | Yes | No |
| | 00 11 12 0# | Yes | Yes | No |
| | 02 11 12 0# | Yes | Yes | No |
| | 00 12 0# | Yes | Yes | No |
| | 02 12 0# | Yes | Yes | No |
| 3 | 00 10 12 0# | No | Yes | Yes |
| | 01 10 12 0# | No | Yes | Yes |
| | 02 12 12 0# | No | Yes | Yes |
| 4 | 00 11 11 0# | Yes | No | Yes |
| | 01 10 11 0# | Yes | No | Yes |
| | 01 11 12 0# | Yes | No | Yes |
| | 00 11 0# | Yes | No | Yes |
| 5 | 00 10 0# | Yes | No | No |
| 6 | 01 11 11 0# | No | Yes | No |
| | 01 11 0# | No | Yes | No |
| 7 | 01 12 11 0# | No | No | Yes |
| | 01 12 0# | No | No | Yes |

skew). Any further addition of pseudo joints will not result in active joint twists achieving new types of configurations. Such being the case, links in metamorphic structures are to be constructed using either one or two pseudo joints.

Additionally, it can be seen that groups 2, 4, 6 and 7 can be constructed with either one or two pseudo joints. In such cases, structures with one pseudo joint are to be favored against the others.

### *1.5.2 Solvability of Kinematics*

One of the main reasons that reconfigurable robots proposed in the relevant literature present similarities to current fixed robot anatomies is the problem of solving their inverse kinematics. The inverse kinematics problem is non-linear and directly dependant on both a robot's anatomy and configuration making it quite

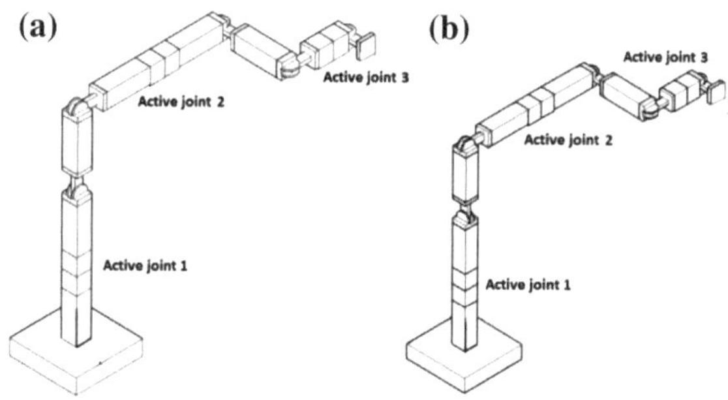


**Fig. 1.2** Visual representation of two MSR notations

difficult to solve analytically for reconfigurable robots and therefore, numerical methods are preferred [8, 9].

The same problem stands for metamorphic robots. However, utilizing the way in which their anatomy is changed via the off-line pseudo joint movement, it is possible to develop a parametric analytical solution for all emerging anatomies of a structure, having as parameters the pseudo joint angles [6]. The solvability of inverse kinematics is depended on the relation between the active joints of the structure. According to Gao [10] a non redundant serial manipulator allows the derivation of analytical solution to its kinematics if one pair of consecutive active joint twists is either parallel or intersects at a point. Additionally, Pieper's theorem also applies to ensure analytical kinematic solvability. Such being the case pseudo joints in a structure should be inserted in such a way so that these conditions are fulfilled. Using the data from Table 1.3, it can be seen that groups 2, 5 and 6, if used in a metamorphic structure's first two links will always allow the derivation of an analytical parametric solution.

## 1.6 Structuring and Evaluating Metamorphic Structures Using the MSR

Having established the basics of the MSR metamorphic structures can be designed, and evaluated using the selected criteria. Figure 1.2 illustrates two different 3 d.o.f. metamorphic structures. Fig. 1.2a structure MSR notation is **0011100011100010**, while the notation for structure shown in 2 (b) is **0012100011100010**. Their subsequent evaluation based on the selected criteria is presented in Table 1.4.

Based on the evaluation of the two structures, structure **0012100011100010** fares better than **0011100011100010** and therefore is most likely to be selected for the next stage in the design of a metamorphic workcell.

As seen the MSR helps both in the representation as well as the subsequent evaluation of the created anatomies. In addition to this, this form of representation allows its usage to A.I. techniques, especially Genetic Algorithms, so as to develop a method of automatic and rapid creation, evaluation and selection of metamorphic structures, as a tool for the rapid, systematic conceptual design of metamorphic robotic workcells.

**Table 1.4** Evaluation of the two presented structures according to the selected criteria

| Criteria | Evaluation |
|---|---|
| Feasibility | • Both structures contain the same number of pseudo joints<br>• Structure (a) contains two Type 2 links which can also be achieved using a single pseudo joint, while structure (b) contains a single Type 2 link. Therefore, the latter achieves a better score regarding simplicity.<br>• Additionally, structure (b) also contains a Type 1 link, which allows the first two active joints to achieve all possible configurations, and therefore, can achieve a wider range of different anatomies. |
| Solvability | • Both structures utilize a Type 2 link (**0011100#**), and therefore they will both always present maximum solvability for every emerging anatomy achieved |

## 1.7 Conclusion: Future Work

A MSRwas proposed for the rapid systematic creation of metamorphic structures for serial metamorphic manipulators. The representation of the modules available for the representation of the different types of modules and their interconnections was presented. The support for conceptual design of metamorphic structures was also presented in detail. The criteria for the evaluation of developed structures were identified and presented. A case study of two different structures was presented along with their subsequent evaluation.

Future work includes the development of a multi criteria index to be used along with an automated system for the generation and evaluation of different metamorphic structures to be used in the design of metamorphic robotic workcells using AI techniques.

## A.1 Appendix A

Modules types for structuring metamorphic robots

| | |
|---|---|
| | **Pseudo joint**<br>• 1 D.O.F. passive connector<br>• Twist can be set by the user Two options, as shown or opposite direction)<br>• Reference configuration $\theta_p = 0^o$ set by designer<br>• Range of displacement for configuration shown<br>$[-90^0, +90^0]$ (adjusted according to reference configuration set by the designer)<br>• 2 Connecting Surfaces (as shown)<br>• Not scalable, Connective surfaces set to match all types of other modules<br>**Active Joint**<br>• 1 D.O.F. rotating active joint<br>• Self contained active module<br>• Twist and displacement as shown<br>• Scalable in three different sizes |

(continued)

(continued)

| |
|---|
| **Rigid Connector** |
| • Passive connector |
| • Provides length for a metamorphic structure |
| • For connection encoding the direction of the length is used (as shown) |
| • 6 connective surfaces (as shown) |
| • Also scalable to required length |

## References


1. Paredis C, Brown B, Khosla P (1996) A rapidly deployable manipulator system, In: Proceedings of the International Conference on Robotics and Automation, pp 1434–1439
2. Chen I-M (2001) Rapid response manufacturing through a rapidly reconfigurable robotic workcell. ROBOT CIM-INT MANUF 17:199–213
3. Chen I-M, Yang G, Yeo S-H (2006) Automatic modeling for modular reconfigurable robotic systems, theory and practice, industrial-robotics-theory-modelling-control. ARS/plV, Germany, ISBN 3-86611-285-8, p 964
4. Charalampos D Valsamos CD, Moulianitis V, Aspragathos N (2011) Index based optimal anatomy of a metamorphic manipulator for a given task, Accepted for publication to ROBOT CIM-INT MANUF
5. Valsamos H, Moulianitis V, Aspragathos N, Rapid evaluation of reconfigurable robots anatomies using computational intelligence, LECT NOTES COMPUT SC 6277 LNAI (PART 2), pp 341–350
6. Valsamos H, Aspragathos NA (2009) Determination of anatomy and configuration of a reconfigurable manipulator for the optimal manipulability, IEEE/ASME ReMAR'09 Conference, pp 497–504
7. Zhang T Dai JS Fang YF, Zeng Q (2011) String matrix based geometrical and topological representation of mechanisms, the 13th world congress in mechanism and machine science, Guanajuato, 19–25 June2011 No.A23_493
8. Chen I-M, Gao Y (1998) Inverse kinematics for modular robots. In: Proceeding Of the 1998 IEEE International Conference on Robotics and Automation, Leuven, Belgium
9. Aghili F, Parsa K, (2007) Configuration control and recalibration of a new reconfigurable robot. 2007 IEEE Conference on Robotics and Automation, Roma 10–14 April 2007, pp 4047–4083
10. Gao Y (2000) Decomposible closed-form inverse kinematics for reconfigurable robots using product of exponential formula. Master Thesis, School of Mechanical and Production Engineering Nanyang Technological University, Singapore

# Chapter 2
# A Method for Configuration Representation of Metamorphic Mechanisms with Information of Component Variation

**Wuxiang Zhang and Xilun Ding**

**Abstract** The metamorphic mechanism has the characteristics of multi-configuration, variable constraints, and multi-function. For achieving the variation characteristics of the topological structures in different configurations and the coupling relations intuitively during the process of configuration transformation, a novel comprehensive symbolic matrix for representing the topological structures of the metamorphic mechanism is proposed. The information including variation of components, the relative joint orientations is involved in this new adjacency matrix. The variation characteristics and coupling features of the metamorphic mechanism can be obtained by the generalized operations including intersection and difference on the corresponding symbolic matrices.

**Keywords** Configuration representation · Component variation · Metamorphic mechanisms

## 2.1 Introduction

In contrast to a traditional mechanism, the metamorphic mechanism has the characteristics of multi-configuration, variable constraints, and multi-function. It forms a class of mechanisms that have the ability to change configuration sequentially from one to another as a resultant change of the number of effective links and topological structure to achieve different tasks following its working

W. Zhang (✉) · X. Ding
Robotics Institute, Beijing University of Aeronautics and Astronautics, Beijing, P.R. China
e-mail: zhangwuxiang@me.buaa.edu.cn

X. Ding
e-mail: xlding@buaa.edu.cn

J. S. Dai et al. (eds.), *Advances in Reconfigurable Mechanisms and Robots I*,
DOI: 10.1007/978-1-4471-4141-9_2, 

conditions and requirements. Thus a metamorphic mechanism can be considered as a mechanism set composed by multiple kinematic chains which have the ability of reconfiguration [1].

In the practical process of the transformation between these kinematic chains, there exist three basic metamorphic characteristics including variation of components, adjacent relations and properties of kinematic joints. In most cases, these three ways are occurred simultaneously. On the other hand, some basic constituent elements (component and its connectivity relationship) of these mechanisms keep unchanged to make the mechanisms in adjacent configurations have complex coupling features [2]. These two aspects are key factors affecting the study on the method for configuration synthesis of metamorphic mechanisms [3].

Therefore, for achieving the variation characteristics and the coupling relations of the topological structures in different configurations, a description approach which is appropriate to the metamorphic mechanism should be researched firstly.

Mechanism diagram and topological graph are two simple and intuitive methods for describing the structure of a mechanism, but it can not be represented in a mathematical form. Therefore, adjacency matrix method was used to describe the topological structure of the mechanism in a single configuration. Further, elementary transformation of the matrices was proposed to represent the corresponding variation of mechanisms [4–7]. Wang and Dai [1] introduced the joint symbols into the adjacency matrix for expressing the variation of kinematic joints. In this method, all the components are numbered sequentially and placed in the principal diagonal position. And the off-diagonal elements are expressed by using the joint symbols, such as revolute joint (R), prismatic joint (P) and higher joint (G), etc. representing the connectivity relationship of the corresponding rows and columns. So the matrix is symmetrical, and the number of its rows (columns) is equal to the number of equivalent components. But the joint axis information is not be expressed by using this method, so planar mechanisms and spatial mechanisms cannot be distinguished.

Because the axis orientation of the kinematic joints are time-varying, it is infeasible to express the corresponding information by taking advantage of eulerian angles relative to a fix coordinate in adjacency matrix. The relative axis orientation of the kinematic joints located at two ends of the same component is invariable while in motion so that we can utilize use it. Yang [8] introduced the concept of geometric constraint for expressing the relative position and orientation of the joint axes and generalized it into six types: parallelism, coincidence, intersection, perpendicularity and randomicity. Li, Wang and Dai [9, 10] developed a topological representation matrix with information of loops, types of links and joints, and orientations of joints.

As stated previously, the components have their own unified numbers in these methods, and the variation information of these components can not be showed up with producing confusion of components and their connectivity relationships when comparing the adjacency matrices. In addition, different description sequence to the same mechanism will lead to different adjacency matrices which need to perform additional isomorphism identification.

Based on the above-mentioned research achievements and insufficiencies, a symbolic matrix which can represent the topological structures of the mechanism in all configurations with information of component variation is proposed.

## 2.2 The Method for Configuration Representation of Metamorphic Mechanisms

The basic characteristic of the method is to construct a symbolic adjacency matrix by introducing the information of component variation and joint orientation. The variation characteristics and coupling features of the mechanisms in adjacent configurations can be obtained by the generalized operations including intersecting and difference on the corresponding symbolic matrices.

The symbolic matrix $A$ is constructed to describe the topological structures of the metamorphic mechanism completely as

$$A = \begin{pmatrix} L_1 & J_{1,2} & \cdots & J_{1,i} & \cdots & J_{1,n-1} & J_{1,n} \\ J_{1,2} & L_2 & \cdots & J_{2,i} & \cdots & J_{2,n-1} & J_{2,n} \\ \vdots & \vdots & \ddots & \vdots & \vdots & \vdots & \vdots \\ J_{1,i} & \cdots & \cdots & L_i^K & & \cdots & J_{i,n} \\ \vdots & \vdots & \vdots & \vdots & \ddots & \vdots & \vdots \\ J_{1,n-1} & J_{2,n-1} & \cdots & J_{i,n-1} & \cdots & L_{n-1} & J_{n-1,n} \\ J_{1,n} & J_{2,n} & \cdots & J_{i,n} & \cdots & J_{n-1,n} & L_n \end{pmatrix} \tag{2.1}$$

where the principal diagonal element $L_i^K$ represent the component of the mechanism as well as the off-diagonal element $J_{i,j}$ represent the connectivity relationship of components $L_i$ and $L_j$. The specific explanations are shown as follows.

(1) Principal diagonal element $L_i^K$ represent the component whose sequence number is $i$ in the mechanism. The subscript $i$ represents the order of the corresponding component in the matrix and the superscript $K$ represents the driving component or the frame component as symbols $D$ or $F$ respectively. In general conditions, the component whose connectivity characteristic keeps unchanged during the course of configuration transformation is prior to be selected as the first principal diagonal element. Then other components should be arranged in sequence.

In the symbolic matrix, for expressing the component variation including combination and decomposition distinctly, 'direct sum' (operator '$\oplus$') is used here. The combination of more than two components can be represented as $L_i \oplus L_j \oplus \cdots$. It is different with the former representation methods. In those methods, several components fixed together was treated as one of them on the assumption of the components are coincident without considering the influence of their relative position and orientation constraints at the moment of configuration transformation. But in fact, some newly components are usually generated. The geometric

**Table 2.1** Expression of the geometric constraint

| Geometric constraint relations | $W$ |
|---|---|
| Parallelism | $//R,\ //P,\ //H, \ldots$ |
| Intersection | $\overset{\frown}{R},\ \overset{\frown}{RR},\ \overset{\frown}{RRR}, \ldots$ |
| Coincidence | $/R,\ /H$ |
| Perpendicularity | $\perp R,\ \perp P,\ \perp H, \ldots$ |
| Randomicity | $-R,\ -P,\ -H, \ldots$ |

parameter of them will be variable with the moment of configuration transformation. So 'direct sum' is introduced into the matrix for expressing the component combination at a certain position. This can provide the basis for the research on sequent dimensional synthesis of metamorphic mechanisms.

(2) Off-diagonal element $J_{i,j}$ ($j = 2, \ldots, nj\ 2, \cdots, n$) shows the type of joint and the relative axis orientation of the joints located at two ends of the component $L_i$ as

$$_{i,j}=J_{j,i}=\begin{cases} U_W & \begin{array}{c} U\,represent\,the\,joint\,symbols, for\,example-\mathrm{R\,(revolute\,joint)},\\ Prismatic\,joint, Spherical\,joint,\cdots \\ The\,subscript\,W\,represent\,the\,relative\,geometric\,constraint\,relation\,of \\ the\,join\,t\,axes\,located\,at\,two\,ends\,of\,the\,component\,L_i \end{array} \\ 0 & Components\,L_i\,and\,L_j\,are\,not\,connected \end{cases} \tag{2.2}$$

In Eq. (2.2), the joint axis between the first component and the second component can be treated as the reference axis, and the other joint axis of the second component can be expressed as the relative orientation of the reference axis. By inference, all the joint axis of the mechanism can be represented relatively by geometric constraint relations of the two ends of the same component successively. The expressions of the geometric constraint are illustrated in Table 2.1 [8–10].

*Note*. For describing the intersection of more than two joint axes in the same point, the corresponding $W$ should be expressed as $\overset{\frown}{RR}\ldots$.

So the matrix $A$ has the following characteristics as follows:

(1) The matrix is symmetrical, and the principal diagonal elements represent the sequential connected components. The quantity of its rows (columns) is equal to the quantity equivalent components.
(2) Nonzero elements in the lower-left/upper-right part of the matrix are correspondent with the types of joints. And the subscript of these nonzero elements can represent the relative axis orientation of all joints in the mechanism for distinguishing planar mechanisms from spatial mechanisms clearly.
(3) The quantity of the nonzero elements in the $i$th row (the principal diagonal element $L_i$ is not included) $N_i$ represents the amount of the kinematic joints connected with the component $L_i$. If there exist $N_i = 1$, the mechanism has at least one open chain; if all the amount of $N_i$ are equal to 2, the mechanism is a

single closed chain; If there exist $N_i > 2$, the mechanism has multi-closed chains.

(4) The joint axis orientation between the last component and the first component in the closed kinematic chain should be represented as the geometric constraint relative to the reference axis. To the open chain, the case does not exist.

The symbolic matrix can describe the topological structure of the mechanism in a single configuration. Based on the matrix, the variation characteristics and coupling features of the metamorphic mechanism in adjacent configurations can be obtained by the corresponding operations.

## 2.3 Topological Variation Characteristics of the Metamorphic Mechanism

For achieving the topological variation characteristics of the metamorphic mechanism, it is necessary to apply generalized difference operation on the corresponding configuration representation matrices as

$$A_{i+1,i}^{sub} = A_{i+1} - \mathrm{A_i} \overset{\text{Equivalence}}{\Leftrightarrow} \mathrm{A}'_{\mathrm{i+1}} - \mathrm{A}'_{\mathrm{i}} \tag{2.3}$$

where $A_{i+1,i}^{sub}$ represent the result of the generalized difference operation, $A'_{i+1}$ and $A'_i$ representing the increased order matrices are equivalent with the original configuration representation matrices in configurations $i + 1$ and $i$ respectively. The orders of the two involved matrices are probably different, so it is necessary to make their orders have the same value by the operation of increasing order. The operator"−"represent the generalized difference operation for achieving the different parts. In set theory, the matrix $A_{i+1,i}^{sub}$ is the relative complement set of $A'_i$ to $A'_{i+1}$. Its physical meaning is that after comparing the two topological representation matrices $A_{i+1}$ 和 $A_i$, the different parts of the two matrices are extracted and reserved. The order of the result matrix is correspondent with the quantity of all related components. Its operation steps are described as follows.

Firstly, if the orders of the two matrices involved in the operation are different, the operation of increasing order should be done according to the number of all related components. Components fixed together should be resolved to several single components, and the corresponding $J_{i,j}$ in Eq. (2.1) is given as the number '1'. The other connectivity relations should be transferred to the corresponding components.

The operation starts with the first line and column, and the principal diagonal elements and their connectivity relationship should be compared gradually. The same elements should be assigned '0' while the different elements in the minuend matrix should be reserved. Variation of the driving components and components fixed on the ground which are important factors influencing the topological structures of the metamorphic mechanism should be included. In addition, the same joint axis may have different representations in two matrices because of component variation, so it is necessary to identify the variation in combination with the metamorphic mechanism.

## 2.4 Coupling Features of the Metamorphic Mechanism

There exist some basic constituent elements (component and its connectivity relations) keep unchanged to make the mechanisms in adjacent configurations have complex coupling features. For achieving these coupling features, it is necessary to apply generalized intersection operation (operator '∩') on the corresponding configuration representation matrices as

$$A_{i+1,i}^{int} = A_{i+1} \cap A_i \overset{\text{Equivalence}}{\Leftrightarrow} A_{i+1}^{'} \cap A_i^{'} \tag{2.4}$$

where $A_{i+1,i}^{int}$ represent the result of the generalized intersection operation whose physical meaning is that after comparing the two topological representation matrices $A_{i+1}$ 和 $A_i$, the same parts of the matrix are extracted and reserved. And the different elements should be assigned '0'.

The operation steps are coincident with the above expressed difference operation. The first step is making the order of the involved matrices unified. Then the principal diagonal elements and their connectivity relations should be compared gradually, and the different elements should be assigned '0' while the same elements should be reserved. Therefore, the nonzero elements in the result matrix manifest the coupling features.

## 2.5 Relation Between Variation Characteristics and Coupling Features

Because the variation characteristics and coupling features of the metamorphic mechanism in adjacent configurations can be obtained by the generalized operations including intersecting and differentiating on the corresponding symbolic matrices, it is obvious that there exist the following relation as

$$\begin{aligned} A_{i+1,i}^{sub} \cup A_{i+1,i}^{int} &= (A_{i+1} - -A_i) \cup (A_{i+1} \cap A_i) \\ &= [(A_{i+1} - -A_i) \cup A_{i+1}] \cap [(A_{i+1} - -A_i) \cup A_i] \\ &= A_{i+1} \cap (A_{i+1} \cup A_i) \\ &= A_{i+1} \end{aligned} \tag{2.5}$$

In Eq. (2.5), symbol "∪" represent the generalized union operation. Equation (2.5) shows that the configuration representation matrix in configuration $i$ can be

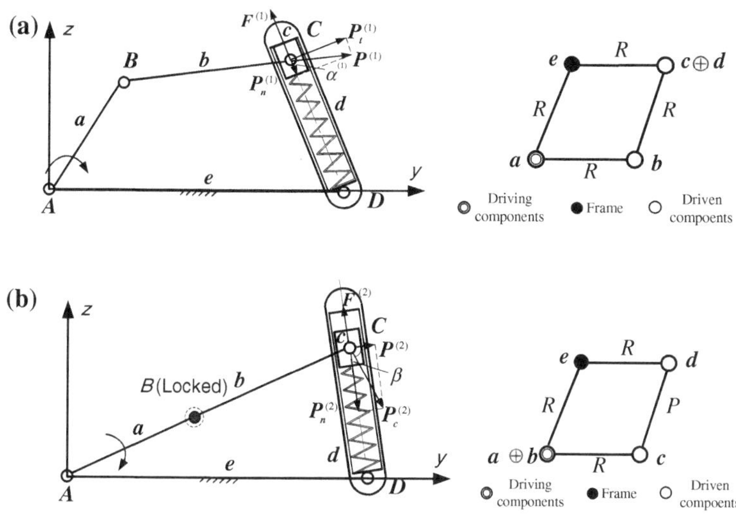


**Fig. 2.1** A planar five bar force-limit metamorphic mechanism **a** Mechanism schematic in configuration 1 and its topological structure. **b** Mechanism schematic in configuration 2 and its topological structure

obtained by uniting the topological variation characteristics matrix and coupling characteristics matrix.

## 2.6 Case Study

### *2.6.1 Case 1*

A planar five bar force-limit metamorphic mechanism is shown in Fig. 2.1 [3, 11]. When the mechanism is in configuration 1 as in Fig. 2.1a, slider $c$ locates at the top end of the slot in component $d$ under the action of spring force $F^{(1)}$ which satisfies $F^{(1)} \geq P_n^{(1)} = P^{(1)} \cos \alpha^{(1)}$. Force $P_n^{(1)}$ (the force direction is along the link $d$) is the component force of the transmission force $P^{(1)}$ generated by link $b$, $\alpha^{(1)}$ is the transmission angle. The mechanism in this configuration can be treated as a four bar mechanism, in this case, spring force $F^{(1)}$ guarantees the slider $c$ stays static. Without such a force $F^{(1)}$, slider $c$ inclines to move down along the slot under the action of $P_n^{(1)}$, and the mechanism leads to uncertain motion. When the mechanism is in configuration 2, see Fig. 2.1b, links $a$ and $b$ are fixed together by locking revolute joint $B$ using geometric limit, and the mechanism is transformed to a slider mechanism. Under this condition, spring force $F^{(2)} < P_n^{(2)} = P_c^{(2)} \cos \beta$ makes the prismatic joint $C$ active. $P_n^{(2)}$ is the component force of $P_c^{(2)}$ (generated by link $b$, and its direction is perpendicular to link $b$). Angle $\beta$ represents the angle between the directions of $P_n^{(2)}$ and $P_c^{(2)}$.

Applying the proposed method, the symbolic matrices of the mechanisms in its two configurations can be expressed as follows:

$$A_1 = \begin{pmatrix} e^F & R & 0 & R_{\|R} \\ R & a^D & R_{\|R} & 0 \\ 0 & R_{\|R} & b & R_{\|R} \\ R_{\|R} & 0 & R_{\|R} & c \oplus d \end{pmatrix} \tag{2.6}$$

$$A_2 = \begin{pmatrix} e^F & R & 0 & R_{\|R} \\ R & a^D \oplus b & R_{\|R} & 0 \\ 0 & R_{\|R} & c & P_{\|R} \\ R_{\|R} & 0 & P_{\|R} & d \end{pmatrix} \tag{2.7}$$

The matrices show that all the joint axes are parallel to each other, indicating that it is a planar mechanism. And components $a$ and $e$ are the driving component and the frame respectively.

By applying the generalized difference operation, the topological structure variation characteristics can be achieved as

$$\begin{aligned} A_{2,1}^{sub} &= A_2 - \mathrm{A}_1 \overset{\text{Equivalence}}{\Leftrightarrow} \mathrm{A}_2^{'} - \mathrm{A}_1^{'} \\ &= \begin{bmatrix} \mathrm{e^F} & \mathrm{R} & 0 & 0 & \mathrm{R_{\|R}} \\ \mathrm{R} & \mathrm{a^D} & 1 & 0 & 0 \\ 0 & 1 & \mathrm{b} & \mathrm{R_{\|R}} & 0 \\ 0 & 0 & \mathrm{R_{\|R}} & \mathrm{c} & \mathrm{P_{\|R}} \\ \mathrm{R_{\|R}} & 0 & 0 & \mathrm{P_{\|R}} & \mathrm{d} \end{bmatrix} - \begin{bmatrix} \mathrm{e^F} & \mathrm{R} & 0 & 0 & \mathrm{R_{\|R}} \\ \mathrm{R} & \mathrm{a^D} & \mathrm{R_{\|R}} & 0 & 0 \\ 0 & \mathrm{R_{\|R}} & \mathrm{b} & \mathrm{R_{\|R}} & 0 \\ 0 & 0 & \mathrm{R_{\|R}} & \mathrm{c} & 1 \\ \mathrm{R_{\|R}} & 0 & 0 & 1 & \mathrm{d} \end{bmatrix} \\ &= \begin{bmatrix} \mathrm{e^F} & 0 & 0 & 0 & 0 \\ 0 & \mathrm{a^D} & 1 & 0 & 0 \\ 0 & 1 & \mathrm{b} & 0 & 0 \\ 0 & 0 & 0 & \mathrm{c} & \mathrm{P_{\|R}} \\ 0 & 0 & 0 & \mathrm{P_{\|R}} & \mathrm{d} \end{bmatrix} \end{aligned} \tag{2.8}$$

According to Eq. (2.8), the element indicating the connectivity relation of components $a$ and $b$ is '1', it means the two components are fixed together to generate a new link $a \oplus b$. Component $d$ is separated from the component $c \oplus d$ and their connectivity joint with component $c$ is changed to a prismatic joint.

The coupling features of the mechanisms in two configurations can be obtained as

$$A_{2,1}^{int} = A_2 \cap A_1 \overset{\text{Equivalence}}{\Leftrightarrow} \begin{bmatrix} e^F & R & 0 & 0 & R_{\|R} \\ R & a^D & 0 & 0 & 0 \\ 0 & 0 & b & R_{\|R} & 0 \\ 0 & 0 & R_{\|R} & c & 0 \\ R_{\|R} & 0 & 0 & 0 & d \end{bmatrix} \tag{2.9}$$

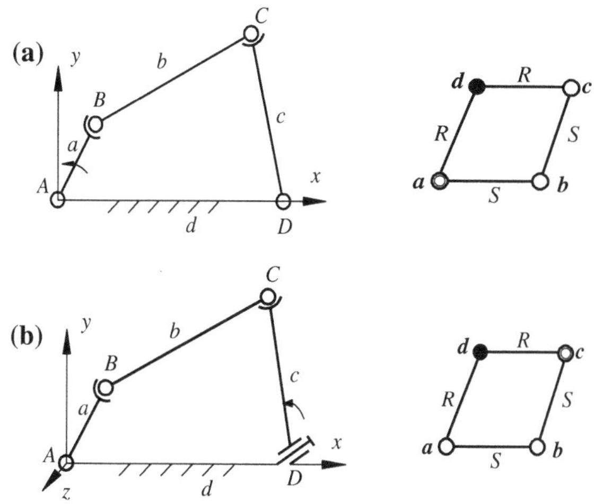


**Fig. 2.2** A planar-spatial four bar metamorphic mechanism **a** Mechanism schematic in configuration 1 and its topological structure **b** Mechanism schematic in configuration 2 and its topological structure

by applying the generalized intersection operation. According to Eq. (2.9), the connectivity relationship between component *e* and the related components such as links *a* and *d* keep unchanged.

### 2.6.2 Case 2

A planar-spatial four bar metamorphic mechanism is shown in Fig. 2.2. When the mechanism is in configuration 1 as in Fig. 2.2a, it can be treated as a planar RSSR four bar mechanism. Links *a* and *d* are the driving component and frame respectively. At the moment of configuration transformation,the axis of kinematic joint between components *c* and *d* is changed to make the mechanism transformed to a spatial mechanism as shown in Fig. 2.2b as well as the driving component becomes link *a*. The special metamorphic mechanism transforms its configuration by changing joint orientation [10].

The topological structures in these two configurations can be expressed in matrix form as follows:ą

$$A_1 = \begin{pmatrix} d^F & R & 0 & R_{\parallel R} \\ R & a^D & S_{-R} & 0 \\ 0 & S_{-R} & b & S_{-s} \\ R_{\parallel R} & 0 & S_{-s} & c \end{pmatrix} \tag{2.10}$$

$$A_2 = \begin{pmatrix} d^F & R & 0 & R_{\perp R} \\ R & a & S_{-R} & 0 \\ 0 & S_{-R} & b & S_{-s} \\ R_{\perp R} & 0 & S_{-s} & c^D \end{pmatrix} \tag{2.11}$$

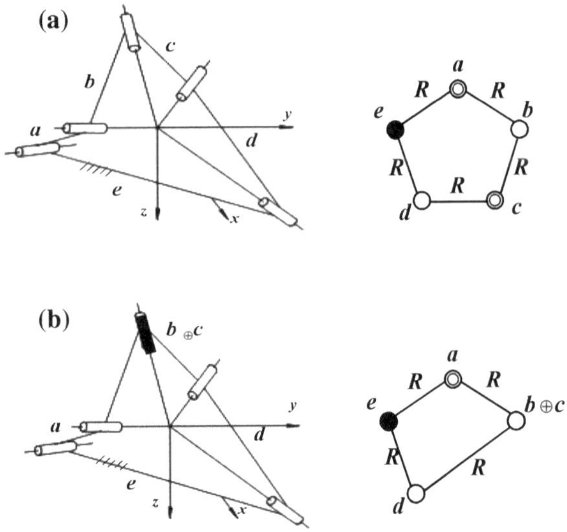


**Fig. 2.3** A spherical 5R–4R metamorphic mechanism
**a** Mechanism schematic in configuration 1 and its topological structure.
**b** Mechanism schematic in configuration 2 and its topological structure

The topological structure variation characteristics and coupling features can be achieved as

$$A_{2,1}^{sub} = A_2 - \mathrm{A}_1 = \begin{pmatrix} \mathrm{d^F} & 0 & 0 & \mathrm{R_{\perp R}} \\ 0 & \mathrm{a^D} & 0 & 0 \\ 0 & 0 & \mathrm{b} & 0 \\ \mathrm{R_{\perp R}} & 0 & 0 & \mathrm{c^D} \end{pmatrix} \tag{2.12}$$

$$A_{2,1}^{int} = A_2 \cap A_1 = \begin{pmatrix} d^F & R & 0 & 0 \\ R & a^D & S_{-R} & 0 \\ 0 & S_{-R} & b & S_{-S} \\ 0 & 0 & S_{-S} & c^D \end{pmatrix} \tag{2.13}$$

Equations (2.12) and (2.13) show that the driving components becomes link *c* and the axis geometric constraint relationship between two ends of component *d* are changed from parallelism to perpendicularity as shown in Fig. 2.2.

### 2.6.3 Case 3

A spherical 5R–4Rmetamorphic mechanism and its topological structures are shown in Fig. 2.3. All the components are connected by revolute joints whose joint axes intersect in the origin of coordinates. When the mechanism is transformed to configuration 2, components *b* and *c* are fixed together to realize configuration transformation.

The topological structures of the mechanisms in two configurations can be expressed in matrix form as follows:

$$A_1 = \begin{pmatrix} e^F & R & 0 & 0 & R_{\widehat{R}} \\ R & a^D & R_{\widehat{R}} & 0 & 0 \\ 0 & R_{\widehat{R}} & b & R_{\widehat{RR}} & 0 \\ 0 & 0 & R_{\widehat{RR}} & c & R_{\widehat{RRR}} \\ R_{\widehat{R}} & 0 & 0 & R_{\widehat{RRR}} & d \end{pmatrix} \tag{2.14}$$

$$A_2 = \begin{pmatrix} d^F & R & 0 & R_{\widehat{R}} \\ R & a^D & R_{\widehat{R}} & 0 \\ 0 & R_{\widehat{R}} & b \oplus c & R_{\widehat{RR}} \\ R_{\widehat{R}} & 0 & R_{\widehat{RR}} & d \end{pmatrix} \tag{2.15}$$

By applying the generalized difference and intersection, topological structure variation characteristics of the adjacent matrices can be obtained as

$$\begin{aligned} A_{2,1}^{sub} &= A_2 - \mathrm{A}_1 \overset{\text{Equivalence}}{\Leftrightarrow} \mathrm{A}_2' \cap \mathrm{A}_1' \\ &= \begin{pmatrix} \mathrm{e}^{\mathrm{F}} & \mathrm{R} & 0 & 0 & \mathrm{R}_{\widehat{\mathrm{R}}} \\ \mathrm{R} & \mathrm{a}^{\mathrm{D}} & \mathrm{R}_{\widehat{\mathrm{R}}} & 0 & 0 \\ 0 & \mathrm{R}_{\widehat{\mathrm{R}}} & \mathrm{b} & 1 & 0 \\ 0 & 0 & 1 & \mathrm{c} & \mathrm{R}_{\widehat{\mathrm{RR}}} \\ \mathrm{R}_{\widehat{\mathrm{R}}} & 0 & 0 & \mathrm{R}_{\widehat{\mathrm{RR}}} & \mathrm{d} \end{pmatrix} - \begin{pmatrix} \mathrm{e}^{\mathrm{F}} & \mathrm{R} & 0 & 0 & \mathrm{R}_{\widehat{\mathrm{R}}} \\ \mathrm{R} & \mathrm{a}^{\mathrm{D}} & \mathrm{R}_{\widehat{\mathrm{R}}} & 0 & 0 \\ 0 & \mathrm{R}_{\widehat{\mathrm{R}}} & \mathrm{b} & \mathrm{R}_{\widehat{\mathrm{RR}}} & 0 \\ 0 & 0 & \mathrm{R}_{\widehat{\mathrm{RR}}} & \mathrm{c} & \mathrm{R}_{\widehat{\mathrm{RRR}}} \\ \mathrm{R}_{\widehat{\mathrm{R}}} & 0 & 0 & \mathrm{R}_{\widehat{\mathrm{RRR}}} & \mathrm{d} \end{pmatrix} \\ &= \begin{pmatrix} \mathrm{e}^{\mathrm{F}} & 0 & 0 & 0 & 0 \\ 0 & \mathrm{a}^{\mathrm{D}} & 0 & 0 & 0 \\ 0 & 0 & \mathrm{b} & 1 & 0 \\ 0 & 0 & 1 & \mathrm{c} & 0 \\ 0 & 0 & 0 & 0 & \mathrm{d} \end{pmatrix} \end{aligned} \tag{2.16}$$

$$A_{2,1}^{int} = A_2 \cap A_1 = \begin{pmatrix} e^F & R & 0 & 0 & R_{\widehat{R}} \\ R & a^D & R_{\widehat{R}} & 0 & 0 \\ 0 & R_{\widehat{R}} & b & 0 & 0 \\ 0 & 0 & 0 & c & R_{\widehat{RR}} \\ R_{\widehat{R}} & 0 & 0 & R_{\widehat{RR}} & d \end{pmatrix} \tag{2.17}$$

As shown in Eqs. (2.16) and (2.17), components $b$ and $c$ are fixed together to produce a new component $b \oplus c$.

## 2.7 Conclusion

A novel comprehensive symbolic matrix for representing the topological structures of the metamorphic mechanism is proposed. The information including variation of components, relative orientation of the kinematic joints are involved in this new adjacency matrix. The variation characteristics and coupling features of the metamorphic mechanism in adjacent configurations can be obtained by the generalized operations including intersection and difference on the corresponding symbolic matrices.

**Acknowledgments** The authors are thankful for the fundamental support of the Natural Science Foundation of China (NSFC) under grant number 51125020&51105013 and the Research Fund for the Doctoral Program of Higher Education of China (RFDP) under grant number 200800060009.

## References

1. Wang DL, Dai JS (2007) Theoretical foundation of metamorphic mechanism and its synthesis. Chin Mech Eng 43(8):32–42
2. National natural science foundation of China (2010) Development strategy for mechanical engineering. Science Press, Beijing
3. Zhang WX, Ding XL, Dai JS (2011) Morphological synthesis of metamorphic mechanisms based on constraint variation. Proc IME C J Mech Eng Sci 225(1):2997–3010
4. Dai JS, Ding XL, Wang DL (2005) Topological changes and the corresponding matrix operations of a spatial metamorphic mechanism. Chin Mech Eng 41(8):30–35
5. Dai JS, Rees JJ (2005) Matrix representation of topological changes in metamorphic mechanisms. ASME Trans J Mech Des 127(4):610–619
6. Wu YR, Jin GG, Li DF et al (2007) Adjacent matrix method describing the structure changing of metamorphic mechanisms. Chin Mech Eng 43(7):23–26
7. Lan ZH, Du R (2008) Representation of topological changes in metamorphic mechanisms with matrices of the same dimension. J Mech Des 130:074501(1–4)
8. Yang TL (2004) Topology structure design of robot mechanisms. China Machine Press, Beijing
9. Li SJ, Wang DL, Dai JS (2009) Topology of kinematic chains with loops and orientation of joints axes. Chin Mech Eng 25(6):34–40
10. Li SJ, Dai SJ (2010) Configuration transformation matrix of metamorphic mechanisms and joint-orientation change metamorphic method. Chin Mech Eng 21(14):1698–1703
11. Li SJ, Dai JS (2010) Structure of metamorphic mechanisms based on augmented assur groups. Chin Mech Eng 46(13):22–41

# Chapter 3
# Enumeration Problems: A Bridge Between Planar Metamorphic Robots in Engineering and Polyforms in Mathematics

**Anelize Zomkowski Salvi, Roberto Simoni and Daniel Martins**

**Abstract** This paper relates two problems: enumeration of metamorphic robots in mechanical engineering and enumeration of polyforms in mathematics. First, a review of the two problems is presented. Some particularities of the enumeration of metamorphic robots and theoretical results about the enumeration of polyforms are described in order to create a bridge between these problems. Then, based on the results and the complexity of the enumeration of polyforms, some directions for further works on the planar enumeration of metamorphic robots are proposed.

## 3.1 Introduction

A metamorphic robotic system is a collection of mechatronic modules that can dynamically self-reconfigure [1]. A change in the macroscopic morphology results from the locomotion of each module over its neighbors. Chirikjian [1, 2] presents some applications of metamorphic robotic systems. One cited application in particular, civil structures in times of emergency, evinces the importance of previously knowing the possible configurations that a predetermined finite number of modules can assume.

A. Z. Salvi (✉) · R. Simoni · D. Martins
Federal University of Santa Catarina, Campus Universitário,
Trindade Florianópolis, SC 88040-900, Brazil
e-mail: anezomkosalvi@gmail.com

R. Simoni
e-mail: roberto.emc@gmail.com

D. Martins
e-mail: daniel@emc.ufsc.br

J. S. Dai et al. (eds.), *Advances in Reconfigurable Mechanisms and Robots I*,
DOI: 10.1007/978-1-4471-4141-9_3, 

A frequent problem found in the literature on metamorphic robots is: how to enumerate all possible configurations that a metamorphic robot system can assume from a finite set of modules. This problem has a counterpart in the context of mathematics: enumeration of polyforms.

In set theory, enumeration means listing the elements of a given set. In combinatorics, it relates to counting, i.e. determining the exact number of elements of finite sets. In mechanical engineering, in some situations, it is necessary to present all of the configurations that a metamorphic robot can assume. Thus, in this context, throughout this work, enumeration means listing all of these configurations.

For polyforms, the term enumeration can refer to both counting or listing the configurations. The problem of listing polyforms is a problem of exponential complexity; therefore, the number of configurations grows very fast, as will be discussed in Sect. 3.1. Consequently, the problem of enumerating (listing) all possible configurations of a planar metamorphic robot for a given set of modules presents a significant challenge.

This paper presents a review of the enumeration for both metamorphic robot configurations and polyforms, with the aim of applying several results obtained for polyforms to metamorphic robots. This approach is particularly useful to understand the growth of the problem, the complexity and the limitations of the brute force generation (generation of all configurations). Based on this review and on the particularities of each problem, the future of the enumeration of metamorphic robots is discussed, considering the growing number of configurations, mechanical constraints of the modules and tasks to be executed.

The remainder of this paper is structured as follows: Sect. 3.2 presents a brief review of the enumeration of metamorphic robots. Section 3.3 presents a brief review of the enumeration of polyforms and discusses the computational complexity of the problem. Section 3.4 describes a bridge between metamorphic robots and polyforms. Section 3.4.1 discusses further advances in the enumeration of metamorphic robot configurations including the mechanical constraints on modules and task execution for reducing the number of configurations. Section 3.5 presents the conclusions.

## 3.2 Enumeration of Metamorphic Robots

For a given number of modules, a metamorphic robot made up of these modules can attain different configurations based on their relative coupling. The problem of enumerating all possible distinct arrangements of the metamorphic robotic systems given a number of modules is called *enumeration of metamorphic robot configurations* or, in short, *enumeration of metamorphic robots*.

Configurations are considered as distinct (nonisomorphic) when none of them is a rigid transformation (translation, rotation or reflection) of another. For example, the four configurations shown in Fig. 3.1a are considered isomorphic, while the two configurations shown in Fig. 3.1b are considered distinct.

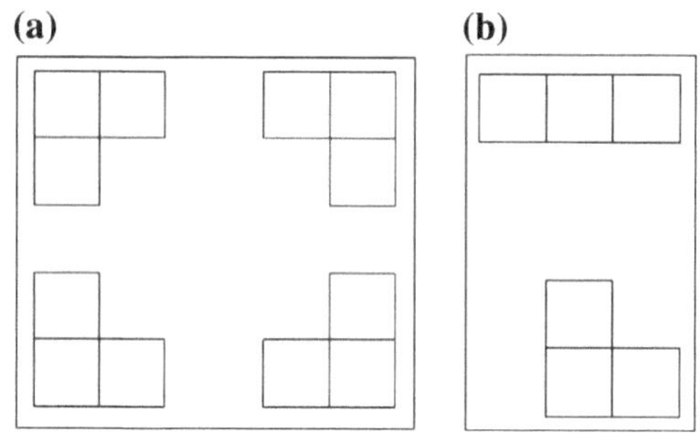


**Fig. 3.1** A set of four isomorphic configurations (**a**), and a set of two distinct configurations (**b**)

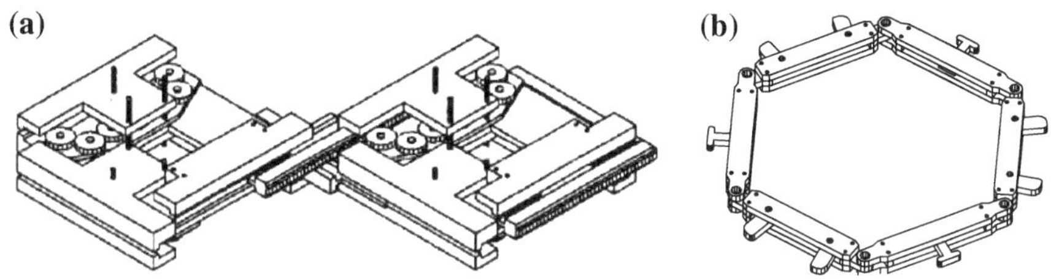


**Fig. 3.2** Two standard modules of planar metamorphic robots

Two common module designs applied to planar metamorphic robots are: square modules [3–5], Fig. 3.2a, and hexagonal modules [3, 4, 6–8], Fig. 3.2b. Common spatial metamorphic systems have cubic [9, 10] and dodecahedral [11, 12] modules which are similarly symmetrical; however, spatial systems will not be further discussed in this paper.

The problem of the enumeration of planar metamorphic robots has been addressed in the literature by Chen and Burdick [13]. The focus of their work was the kinematics of metamorphic robots, so their algorithm enumerates kinematically distinct modular robot assembly configurations. They introduced a representation of a modular robot assembly configuration as an assembly incidence matrix and defined equivalence relations based on symmetries in module geometry and graph isomorphisms on the assembly incidence matrix. They also presented an algorithm to identify kinematically equivalent robots.

Chitta and Ostrowski [14] also focused on the enumeration of kinematically distinct configurations of a modular robot. They proposed an algorithm based on testing two conditions given a pair of modular robots: the equivalence of individual link assembly patterns and the isomorphism of the underlying graph structure.

Martins and Simoni [15] considered the enumeration of all non-isomorphic planar metamorphic robots from the topologically point of view, i.e. based on the relative arrangement of the modules. They use group and graph theory tools in order to avoid repetitions in the process of enumeration. They developed a constructive technique to determine all distinct configurations and presented the configurations for a small quantity of square and hexagonal modules.

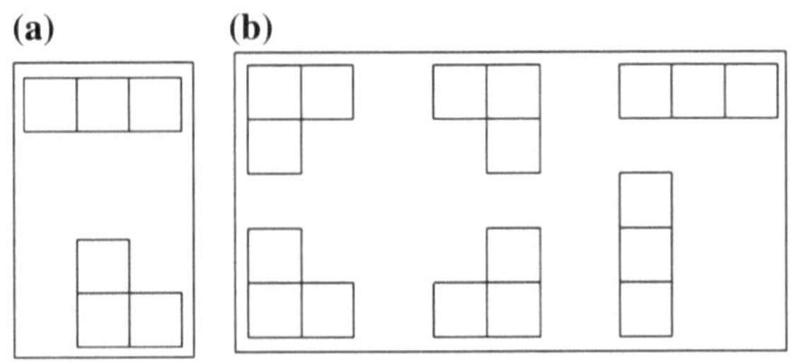


**Fig. 3.3** All free triominoes (**a**), all fixed triominoes (**b**)

In mathematics, enumeration of metamorphic robots has a counterpart, enumeration of polyforms. The next section introduces the enumeration of polyforms and presents a bibliography review. Some particularities of the enumeration of polyforms and theoretical results are described in order to create a bridge with the enumeration of metamorphic robots in Sect. 3.4.

## 3.3 Enumeration of Polyforms

A polyomino is a collection of equal-sized squares in a plane, joined along complete edges [16]. In the same way, a polyform is a plane figure constructed by joining together identical basic polygons. More specific names have been given to polyforms resulting from specific basic polygons, such as: polyomino (square), polyhex (regular hexagon), and polydrafter (drafting triangle with angles of 30°, 60° and 90°).

For polyominoes, the problem of enumeration can be formulated as: *how many different polyominoes* of area $n$ (number of pieces) can be constructed on a square grid? The same formulation can be applied to other polyforms. Most of the mathematical approaches focus on counting and not on listing configurations, mainly because the growth in the number of distinct configurations is a considerable limitation of the brute force generation approach.

Before discussing this last point, it is important to understand the concept of different polyforms from the mathematical point of view.

There are two common ways of defining distinct sets of polyominoes: fixed polyominoes are considered distinct if they have different shapes or orientations, while free polyominoes are considered distinct if they have different shapes, while their orientation and location in the plane is not important [16]. This concept can be extended to other types of polyforms.

Figure 3.3 shows all polyominoes with three pieces called triominoes; however, while Fig. 3.3a both free triominoes, Fig. 3.3b shows all of the six fixed triominoes (of which the two free triominoes are naturally a subset).

Tables 3.1 and 3.2 give the number of all possible free configurations for polyominoes and polyhexes. These results were obtained through indirect counting techniques [16, 17].

**Table 3.1** Number of free configurations of polyominoes with $n$ pieces

| $n$ | Configurations | $n$ | Configurations | $n$ | Configurations | $n$ | Configurations |
|---|---|---|---|---|---|---|---|
| 1 | 1 | 8 | 369 | 15 | 3426576 | 22 | 43191857688 |
| 2 | 1 | 9 | 1285 | 16 | 13079255 | 23 | 168047007728 |
| 3 | 2 | 10 | 4655 | 17 | 50107909 | 24 | 654999700403 |
| 4 | 5 | 11 | 17073 | 18 | 192622052 | 25 | 2557227044764 |
| 5 | 12 | 12 | 63600 | 19 | 742624232 | 26 | 9999088822075 |
| 6 | 35 | 13 | 238591 | 20 | 2870671950 | 27 | 39153010938487 |
| 7 | 108 | 14 | 901971 | 21 | 11123060678 | 28 | 153511100594603 |

**Table 3.2** Number of free configurations of polyhexes with $n$ pieces

| $n$ | Configurations | $n$ | Configurations | $n$ | Configurations |
|---|---|---|---|---|---|
| 1 | 1 | 8 | 1448 | 15 | 76581875 |
| 2 | 1 | 9 | 6572 | 16 | 372868101 |
| 3 | 3 | 10 | 30490 | 17 | 1822236628 |
| 4 | 7 | 11 | 143552 | 18 | 8934910362 |
| 5 | 22 | 12 | 683101 | 19 | 43939164263 |
| 6 | 82 | 13 | 3274826 | 20 | 216651036012 |
| 7 | 333 | 14 | 15796897 | 21 | 1070793308942 |

The next section analyzes the complexity of the enumeration of polyforms to provide a more theoretical basis for understanding the problem.

### 3.3.1 On the Complexity of the Enumeration of Polyforms

Tables 3.1 and 3.2 indicate the manner in which the number of configurations increases when the number of pieces increases. The following results are important for the comprehension of the problem complexity. Let $a_n$ be the number of $n$-celled fixed polyforms

$$\lambda = \lim_{n \to \infty} a_n^{\frac{1}{n}} = \mu \tag{3.1}$$

which exists and is non-zero [18].

Equation 3.1 implies that,

$$a_n > \mu^n \tag{3.2}$$

in other words, $a_n$, the number of fixed configurations of area $n$, grows exponentially with $n$. Also the time required for an algorithm that lists the $a_n$ configurations necessarily grows exponentially.

Due to the exponential growth, mathematics has focused on indirect counting techniques or on listing specific classes of polyominoes such as the convex, ECO [19], convex and near convex, Feretic [20] or even plane fitting polyominoes, Fukuda [21].

For fixed polyominoes it can be estimated that

$$a_n \sim \frac{c\lambda^n}{n} \tag{3.3}$$

where the best known lower and upper bounds for $\lambda$ are [22–24].

$$3.980137 < \lambda < 4.65 \tag{3.4}$$

and

$$c = 0.3169 \tag{3.5}$$

In the case of fixed polyhexes, Voge and Guttmann [25] gave a rigorous proof that the growth constant $\lambda$ exists and satisfies the inequality

$$4.8049 \leq \lambda \leq 5.9047. \tag{3.6}$$

Up to this point, the discussion has focused solely on the enumeration of fixed polyforms. The problem becomes even harder for free polyforms. However, some important results on the enumeration of fixed polyforms can be extended to free polyforms:

- Equation 3.1 also holds for free polyominoes [18];
- The number of free polyominoes (square) is approximately $\frac{1}{8}$ of the corresponding fixed polyominoes [26].

An important consideration is that polyhexes have a counterpart in chemistry, the polyhex hydrocarbons. Polyhex hydrocarbons are organic compounds which contain only carbon and hydrogen atoms and whose graph is embedded in a hexagonal lattice. Many techniques for enumerating this class of polyhexes have been developed; a few examples are show in Caporossi and Hansen [27] and in Müller et al. [28] and references therein.

The next section discusses how to apply the presented mathematical results to the enumeration of metamorphic robots, especially how to deal with the exponential growth of the problem considering the mechanical constraints of the modules.

## 3.4 A Bridge Between Metamorphic Robots and Polyforms

The mathematical analysis in Sect. 3.1 showed that the growth of the enumeration problem is exponential, thus, when the number of modules increases the problem of enumeration becomes difficult or even impractical.

As in many situations it is still important to know the configurations for a given number of modules, the issue to be addressed in this regard is *how to deal with the growth in the number of configurations.*

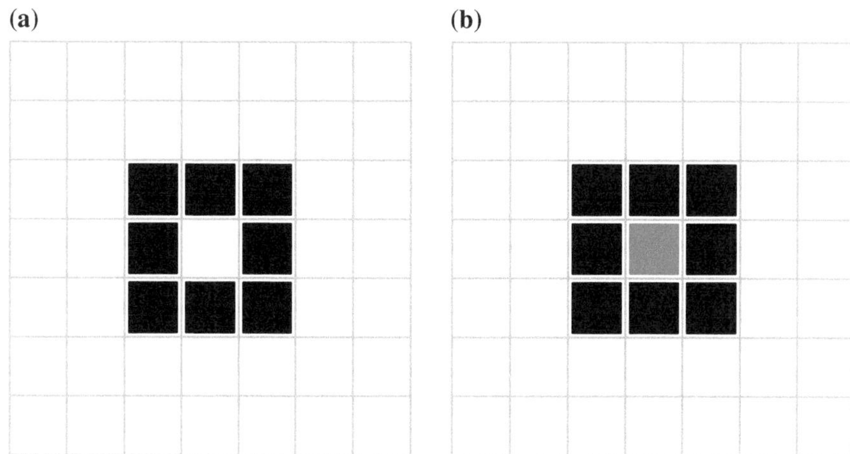


**Fig. 3.4** Configuration (**b**) cannot be achieved from configuration (**a**)

The answer to this problem provides a bridge between the mathematical approach and the particularities of the enumeration problem in relation to metamorphic robots.

The central point is: when dealing with metamorphic robots some mechanical constraints cannot be neglected. For example, the configuration shown in Fig. 3.4b cannot be achieved directly from the configuration shown in Fig. 3.4a via a new planar module without disengaging some of the existing planar modules, i.e. without a change in the robot configuration. Mechanically, this is not possible since planar modules cannot flip over other existing planar modules. On the other hand, adding an extra module to the robot in order to fill this gap has arguably no advantages at all. Furthermore, even if this regions is reachable, this "internal assembly" poses many coupling problems. Finally, if the configuration in Fig. 3.4b is desirable, it is possibly to arrive at it *before* creating the hole in Fig. 3.4a through a different assembly route.

Such constraints discussed above simply do not apply to polyforms. Therefore, enumeration algorithms for polyforms, in both contexts (counting and listing) often consider the configuration where the position marked in gray is occupied by a piece (Fig. 3.4b) for instance, one of the natural sequences from the configuration where this position is empty (Fig. 3.4a), or even do not consider the configuration shown in Fig. 3.4a as a polyform.

On the other hand, these constraints have to be considered in the enumeration of metamorphic robots. Thus, a certain degree of care has to be taken especially when adopting a technique originally used for polyforms in the context of metamorphic robots.

These constraints also play an important role in the enumeration process, because they decrease the number of methamorphic robots configurations to be generated by cutting some branches of the enumeration process. This is an important feature because the enumeration (listing) is a non polynomial problem.

The next section discusses the role of these constraints and suggests further advances in the enumeration of metamorphic robots.

### 3.4.1 Further Advances in Enumeration of Metamorphic Robots

As discussed at the beginning of Sect. 3.4 and shown in Fig. 3.4 in the enumeration of metamorphic robot mechanical constraints cannot be neglected, especially because with a relatively large number of modules this problem becomes intractable. An important question at this point is: *what other constraints should be considered?*

Certainly, the answer is strongly related to the type of information to be extracted from the enumeration and which variables are relevant to the enumeration process.

Working with modular robotic systems that can be arranged in a tree-like structure and knowing that for this type of arrangement the number of configurations is also typically exponential in terms of the number $n$ of modules, Yim [29] suggest that several factors contribute to the number of distinct configurations: (a) the number of connection ports per module, (b) the number of ways that two connection ports can be attached, (c) symmetries in the connection port, and (d) symmetries in the module.

In this context, there is even another way to reduce the number of configurations, which needs to be included during the first step of enumeration: a series of task characteristics as constraints, i.e. a heuristic can be addressed to reduce the number of configurations and provides only configurations that achieve the task requirements.

In chemistry, this type of enumeration process with appropriate constraints is widely used, and certainly the number of configurations is substantially decreased, and the information extracted from the process is more accurate. Some examples are [30]:

- generation of structures based on a molecular formula,
- generation based on a given set of reactions and reactants,
- generation based on a generic structural formula.

In this context, to decrease the number of configurations, the enumeration of metamorphic robots can explore two (broad) directions:

1. consider mechanical constraints of the modules and
2. consider the task requirements.

Thus, as further advances required in the enumeration of metamorphic robots can be pointed out the need to (i) develop a heuristic related to the task variables, (ii) study the mechanical constraints inherent to the problem and, finally, (iii) develop an algorithm that integrates both of these advances.

## 3.5 Conclusions

This paper presents a bibliography review of the enumeration of metamorphic robots and polyforms showing how these two problems are related. Only a few studies have addressed the enumeration of metamorphic robots. On the other hand, the enumeration of polyforms is a well known mathematical problem. This paper discussed the main connections between these two problems, forming a bridge between them, especially concerning the growth of the enumeration problem. Based on the results and complexity of the enumeration of polyforms and particularities of the modules of metamorphic robots some guidelines are provided for further work on the enumeration of metamorphic robots. Mathematical analysis showed that the enumeration of polyforms and metamorphic robots are problems of exponential complexity and thus, to decrease the number of configurations it is suggested two research areas that need to be explored in the enumeration of metamorphic robots: (1) consideration of the mechanical constraints of the modules and (2) consideration of the task requirements.

**Acknowledgments** This work was partially supported by two Brazilian Government Agencies: CNPq and CAPES.

## References

1. Chirikjian G (1994) Kinematics of a metamorphic robotic system. In: IEEE international conference on robotics and automation, pp 449–455
2. Chirikjian G, Pamecha A (1996) Bounds for self-reconfiguration of metamorphic robots. In: IEEE international conference on robotics and automation, vol. 2, pp 1452–1457
3. Pamecha A, Chiang C, Stein D, Chirikjian G (1996) Design and implementation of metamorphic robots. In: ASME design engineering technical conference and computers in engineering conference, pp 1–10
4. Dumitrescu A, Suzuki I, Yamashita M (2002) High speed formations of reconfigurable modular robotic systems. In: IEEE international conference on robotics and automation, vol 1, pp 123–128
5. Chiang C, Chirikjian G (2001) Modular robot motion planning using similarity metrics. Auton Robot 10(1):91–106
6. Abrams A, Ghrist R (2004) State complexes for metamorphic robots. Int J Robot Res 23(7):811–830
7. Walter J, Welch J, Amato N (2004) Distributed reconfiguration of metamorphic robot chains. Distrib Comput 17(2):171–189
8. Walter J, Welch J, Amato N (2002) Concurrent metamorphosis of hexagonal robot chains into simple connected configurations. IEEE Trans Robot Autom 18(6):945–956
9. Rus D, Vona M (2001) Crystalline robots: self-reconfiguration with compressible unit modules. Auton Robot 10(1):107–124
10. Yoshida E, Murata S, Kurokawa H, Tomita K, Kokaji S (1998) A distributed reconfiguration method for 3D homogeneous structure. In: IEEE/RSJ international conference on intelligent robots and systems, vol 2, pp 852–859
11. Yim M, Lamping J, Mao E, Chase J (1997) Rhombic dodecahedron shape for self-assembling robots. Xerox PARC, SPL TechReport P9710777

12. Yim M, Zhang Y, Lamping J, Mao E (2001) Distributed control for 3D metamorphosis. Auton Robot 10(1):41–56
13. Chen I, Burdick J (1998) Enumerating the non-isomorphic assembly configurations of modular robotic systems. Int J Robot Res 17(7):702–719
14. Chitta S, Ostrowski J (2006) Enumeration and motion planning for modular mobile robots. Department of Computer and Information Science, University of Pennsylvania, Technical report no. MS-CIS-01-08
15. Martins D, Simoni R (2009) Enumeration of planar metamorphic robots configurations. In: ASME/IFToMM international conference on ReMAR2009, pp 610–618
16. Redelmeier D (1981) Counting polyominoes: yet another attack. Discrete Math 36(3):191–203
17. I. Jensen (2003) Counting polyominoes: a parallel implementation for cluster computing. In: Computational science—ICCS 2003, pp 698–698
18. Klarxer D (1967) Cell growth problems. Can J Math 19(4):851
19. Del Lungo A, Frosini A, Rinaldi S (2003) Eco method and the exhaustive generation of convex polyominoes. In: DMTCS, Springer, pp 129–140
20. Feretic S (2009) Polyominoes with nearly convex columns: a semidirected model. Arxiv preprint arXiv:0910.4573
21. Fukuda H, Mutoh N, Nakamura G, Schattschneider D (2008) Enumeration of polyominoes, polyiamonds and polyhexes for isohedral tilings with rotational symmetry. Lect Notes Comput Sci 4535:68–78
22. Barequet G, Moffie M, Ribó A, Rote G (2005) Counting polyominoes on twisted cylinders. In: Felsner S (ed) EuroComb2005. Discrete mathematics and theoretical computer science proceedings AE, Citeseer
23. Klarner D, Rivest R (1973) A procedure for improving the upper bound for the number of n-ominoes. Can J Math 25(3), 585–602
24. Jensen I, Guttmann A (2000) Statistics of lattice animals (polyominoes) and polygons. J Phys A Math Gen 33:L257
25. Voge M, Guttmann A (2003) On the number of hexagonal polyominoes. Theor Comput Sci 307(2):433–453
26. Lunnon W (1971) Counting polyominoes. In: Atkin AOL, Birch BJ (eds) Computers in number theory, Academic Press, London, pp 347–372
27. Caporossi G, Hansen P (1998) Enumeration of polyhex hydrocarbons to h = 21. J Chem Inf Comp Sci 38(4):610–619
28. Müller W, Szymanski K, Knop J, Nikolić S, Trinajstić N (1990) On the enumeration and generation of polyhex hydrocarbons. J Comput Chem 11(2):223–235
29. Yim M (1994) Locomotion with a unit-modular reconfigurable robot. PhD dissertation, Stanford University
30. Kerber A, Laue R, Meringer M, Rücker C (2004) Molecules in silico: the generation of structural formulae and its applications. J Comput Chem Jpn 3(3):85–96

# Chapter 4
# A Discontinuously Movable Constant Velocity Shaft Coupling of Koenigs Joint Type

**Chung-Ching Lee and Jacques M. Hervé**

**Abstract** Based on the original Koenigs joint, we present a novel discontinuously movable (DM) RC-//-RC parallel mechanism having a bifurcation of motion. At a singular posture of bifurcation, the end effector has locally two degrees of freedom of infinitesimal translation and two working modes happen at the positions of double points on intersections of two congruent revolute cylinders, which are generated by the two RC open sub-chains. In the mode I, two axes of rotations are transmitted by the ratio of minus one and two shafts rotate in an opposite direction. In the model II, angular velocity ratio between two intersecting shafts is positive one and two cranks rotate in the same direction. Moreover, the same type of phenomenon occurs when two RC open chains are displaced by an offset of translation. Last but not least, a further investigation on a general DM chain of RC-//-RC or HC-//-HC type is expected.

**Keywords** Koenigs joint · A bifurcation · Double point · Discontinuously movable mechanism · Revolute cylinder

## 4.1 Introduction

The joint of Koenigs invented by Gabriel Koenigs [1], a French mathematician, is a constant velocity shaft coupling (CVSC), which can transmit rotation with the angular velocity ratio 1:1 between two intersecting shafts. This original mechanism

C.-C. Lee (✉)
Department Mold and Die Engineering, National Kaohsiung University of Applied Sciences, 415 Chien Kung Road, Kaohsiung, 80782, Taiwan ROC
e-mail: cclee@cc.kuas.edu.tw

J. M. Hervé
Ecole Centrale des Arts et Manufactures, Grande Voie des Vignes, Chatenay-Malabry F-92295, France
e-mail: jacques.herve07@orange.fr

J. S. Dai et al. (eds.), *Advances in Reconfigurable Mechanisms and Robots I*, 
DOI: 10.1007/978-1-4471-4141-9_4, 

is plane symmetric and has a singular phenomenon when the angle between two axes of rotation is zero. To overcome a drawback of singular self-motion when the shaft axes are aligned, a new patented version of the joint with equal offsets between the R and C axes was proposed by Hervé [2], referring to the figures in the appendix. This kind of CVSC for a variable angle between the input and the output shafts is also called a homokinetic joint. It is different from the Oldham-style homokinetic shaft-couplings [3, 4], which only transmit the uniform motion between the two axes that can undergo relative motions while keeping parallel.

The mobility of mechanisms depends not only on the topological graph of the considered mechanisms but also on the geometrical positions in space of their kinematic pairs. The kinematotropy [5] results in the changes of mobility due to a variation of joint variables. It has a phenomenon of discontinuous mobility. In fact, there are different kinds of discontinuity [6]. A bifurcation is a kind of discontinuity. In general, a kinematic bifurcation may just happen at some transition postures of mechanism. In this article, we focus on the distinct singular posture of discontinuity mobility called a bifurcation occurring in the original Koenigs joint and propose a discontinuously movable (DM) RC-//-RC mechanism. This DM mechanism has a finite bifurcation at two double points located at the intersection of two revolute cylinders, which is a special bicylindric curve.

## 4.2 Exceptional Mobility of Koenigs Joint

The RC-//-RC parallel mechanism shown in Fig. 4.1 is movable with one degree of freedom (DoF) of finite motion if the two axes in each of the two RC open chains are parallel. Here, R denotes the 1-DoF revolute pair and C the 2-DoF cylindrical pair. The underline and double underline indicate the parallelism of pair axes. Hence, a general RC-//-RC linkage has two couples of parallel joints, one cylinder and one revolute in each couple. The 1-DoF mobility of RC-//-RC mechanism is overconstrained. The geometric property of parallelism is sufficient to prove its mobility. That category of overconstrained mobility is qualified as exceptional [7]. The exceptional mobility can be established by the intersection of two 3D submanifolds of Schoenflies (or *X*-) motion 4D groups as explained in the following. When the mechanism is plane-symmetric and is used as a constant velocity shaft coupling, it is the joint of Koenigs as described in R. Bricard's book [1]. Koenigs joint is a special RC-//-RC mechanism with the equal distances between two parallel axes as shown in Fig. 4.2 which can uniformly transmit motion between two intersecting axes; the proper name Koenigs is often misspelled König or Königs in the literature on shaft couplings. In its general version that is not always homokinetic, the RC-//-RC mechanism was used to move the railway signs as mentioned in [8], in 1931. In general, the RC-//-RC parallel chain can move with one finite DoF when the mechanism is not plane symmetric. The mobility of RC-//-RC chain is based on the group dependence of two Lie

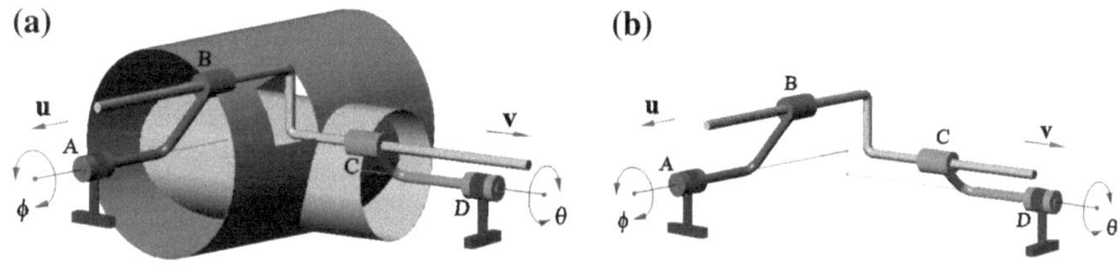


**Fig. 4.1** A general RC-//-RC chain. **a** With two different cylindric surfaces; **b** mechanism structure

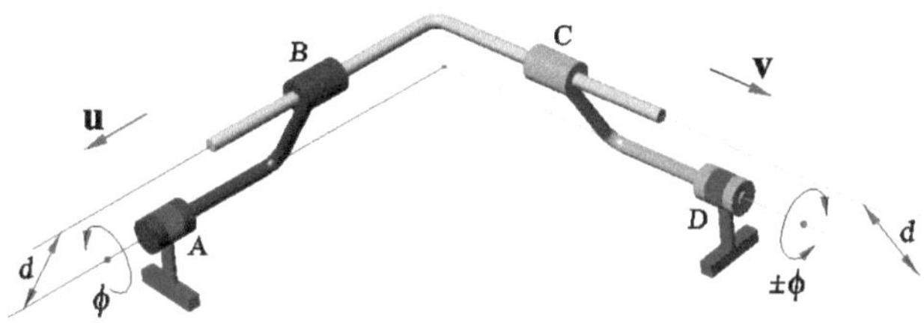


**Fig. 4.2** A Koenigs-joint chain

subgroups of the general Lie group of rigid-body displacements. It is ready for verification as follows.

Figure 4.1 shows a general RC-//-RC chain with offsets and distinct distances between two parallel axes. A RC subchain is an open chain, which generates a 3D submanifold of the $X$-motion 4D group, $\chi(\mathbf{u})$, whose axes of rotation are parallel to the unit vector $\mathbf{u}$. From the theory of displacement group, the intersection of two 3D submanifolds of two distinct 4D $X$-groups is a 1D submanifold of the 3D group $\mathcal{T}$ of spatial translations; it is a set of curvilinear translations. In form of a mathematical equation, the above statement can be generally expressed as follows,

$$\chi(\mathbf{u}) \cap \chi(\mathbf{v}) = \mathcal{T} \Rightarrow \chi_{-1}(\mathbf{u}) \cap \chi_{-1}(\mathbf{v}) = \mathcal{T}_{-2} \tag{4.1}$$

in which the notation $\chi_{-1}(\mathbf{u})$ designates any one of the 3D subsets of $\chi(\mathbf{u})$. A special 3D subsets of $\chi(\mathbf{u})$ is $\mathcal{T}$, which is the intersection $\chi(\mathbf{u}) \cap \chi(\mathbf{v})$. If $\chi_{-1}(\mathbf{u}) = \mathcal{T}$ then $\chi_{-1}(\mathbf{u}) \cap \chi_{-1}(\mathbf{v}) = \mathcal{T} \cap \chi_{-1}(\mathbf{v}) = \mathcal{T}_{-1}$. In a general case, assuming $\chi_{-1}(\mathbf{u}) \neq \mathcal{T}$ and $\chi_{-1}(\mathbf{v}) \neq \mathcal{T}$. $\mathcal{T}_{-2}$ is a 1D submanifold of the 3D group $\mathcal{T}$ of spatial translations. A RC open chain with the R and C axes parallel to $\mathbf{u}$ generates a special $\chi_{-1}(\mathbf{u})$, which is not equal to $\mathcal{T}$. The other RC open chain with R and C axes parallel to $\mathbf{v}$, ($\mathbf{v} \neq \mathbf{u}$), generates a special $\chi_{-1}(\mathbf{v})$, which is also not equal to $\mathcal{T}$. Therefore, it is proven that a general form of RC-//-RC chain has one finite DoF. However, under more special geometric conditions, the RC-//-RC mechanism has a discontinuous mobility and there is a bifurcation [6] of motion, which will be explored in the next section.

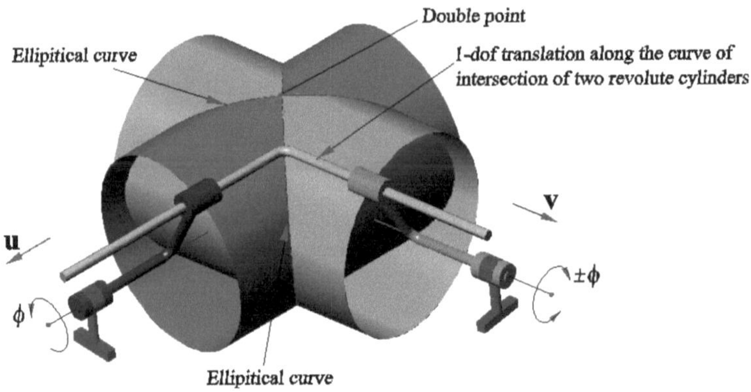


**Fig. 4.3** Discontinuously movable RC-//-RC mechanism

## 4.3 Configuration Analysis of Bifurcation

The RC-//-RC parallel mechanism is movable with one DoF if the two axes in each of the two RC open chains are parallel as aforesaid. When the mechanism is plane-symmetric and is used as a constant velocity shaft coupling, it is the joint of Koenigs [1]. We focus on the special case of a RC-//-RC mechanism which has a plane of symmetry at its home posture because very special properties appear in that case. The end body in RC-//-RC parallel chain is the coupler of the corresponding RC CR four-bar linkage. The coupler moves with 1-DoF translation. The trajectory of each point belonging to the coupler is a curve. Because of the coupler translation, all the trajectories are congruent curves. When there is plane symmetry, the two C axes intersect at a point lying in the plane of symmetry. Let us consider the trajectory of the coupler point that is common to the C axes. In an RC open chain, a point located on the C axis obviously moves on a revolute cylinder. Therefore, the foregoing curve is the intersection of two revolute cylinders and is called a bicylindric curve, which, in the general case, may have various shapes. When the two revolute cylinders are congruent, the bicylindric curve is degenerated into the union of two ellipses, which intersect at two double points as shown in Fig. 4.3. We focus on that special case.

The set of the feasible motions of the end body has two connected parts or branches; each branch is constituted by 1-dof translational motion along one elliptical curve. In two singular configurations, a bifurcation of translational motion can occur. A bifurcation belonging to one of these two singular configurations is illustrated in Fig. 4.4. In one mode (branch 1), Fig. 4.4c, there is a plane of symmetry; the plane is the plane of one ellipse. Two cranks rotate with opposite angles, which are denoted $\phi$ and $-\phi$ in figures; these angles are oriented by the unit vectors **u** and **v**, which are homologous by plane symmetry. In the other mode (branch 2), Fig. 4.4d, there is no plane symmetry. The two RC open chains are

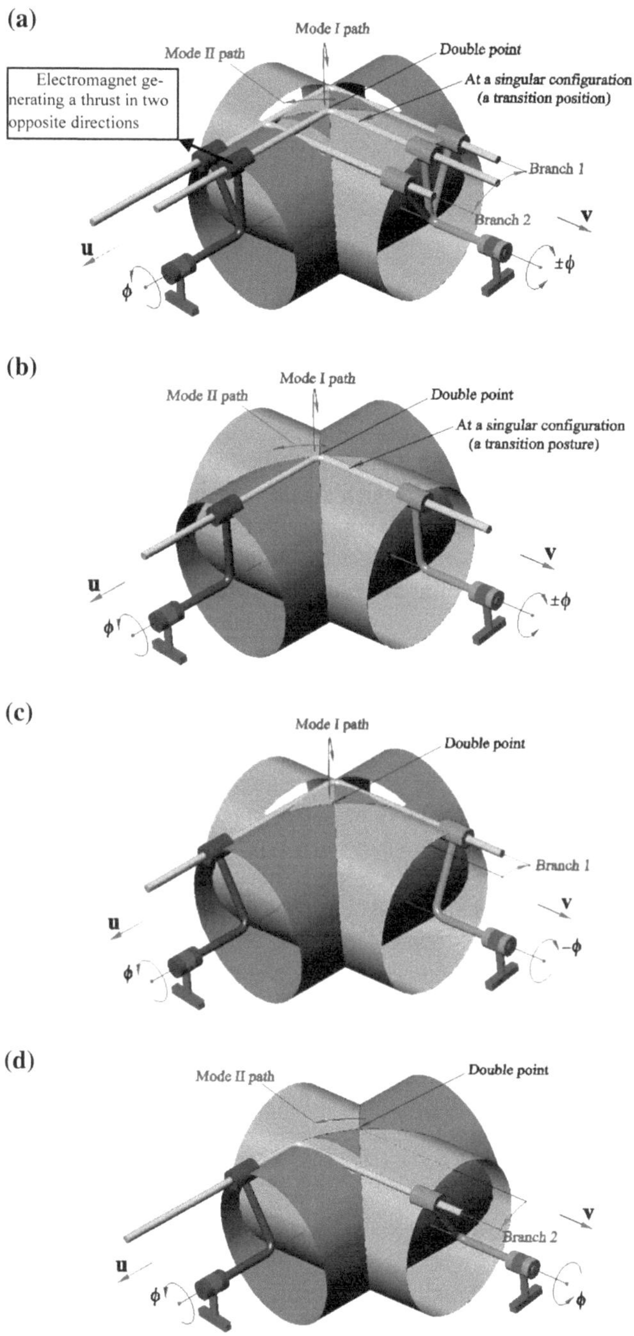


**Fig. 4.4** One bifurcation configuration of DM RC-//-RC CVSC. **a** A bifurcation at the double point of top surface; **b** at a transition posture; **c** symmetric motion at a mode I; **d** motion at a mode II

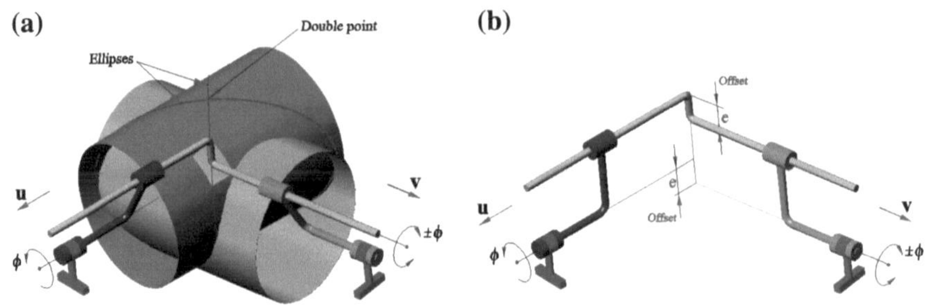


**Fig. 4.5** A slightly general DM RC-//-RC CVSC. **a** RC-//-RC chain with the offset; **b** a formation

homologous by a rotation with a given specified angle ($5\pi/9$ in figures) about the axis of the two double points, which is also the intersection of the two ellipse planes, Fig. 4.4b. Then, the cranks rotate with the same angle $\phi$. Each mode cited above corresponds to the end-body translation along one elliptical curve. At a singular posture of bifurcation, the end-body or coupler gains one DoF of infinitesimal mobility and has locally two DoFs of infinitesimal translation. Moreover, the singular posture is a dead-point position; a crank does not transmit its rotation to the other crank. At any one of its two singular postures of bifurcation, the RCCR mechanism gains locally and transitorily one degree of freedom of infinitesimal mobility. Hence, the control of the change of working modes demands the addition of a second motor acting at the singular postures. We notice that the axial motion in the C pair of the input crank can have two opposite directions depending on the output crank motion. An electromagnet of type solenoid (or coil-winding) can be integrated in the C pair. The electromagnet can generate an axial force in two opposite directions in function of its electric current. Switching the current allows the selection of the motion branch.

The same properties are possessed by a slightly general mechanism, Fig. 4.5a, when one between the two RC open chains is displaced by a given translation, which is parallel to the axis of the two double points as shown in Fig. 4.5b. Two offsets appear and both are equal to the translation amplitude. Figure 4.6 illustrates the bifurcation configurations similar to the previous special case.

## 4.4 Conclusions

A constant velocity shaft coupling of Koenigs joint type, a special RC-//-RC chain, has two bifurcations located at the intersection of two ellipses, which are the degenerate one of bicylindric curve. The configuration analysis of this kind of DM mechanism is elucidated in detail. One motion mode of bifurcation has the rotations in the same direction but the other one has the rotation motion in the opposite

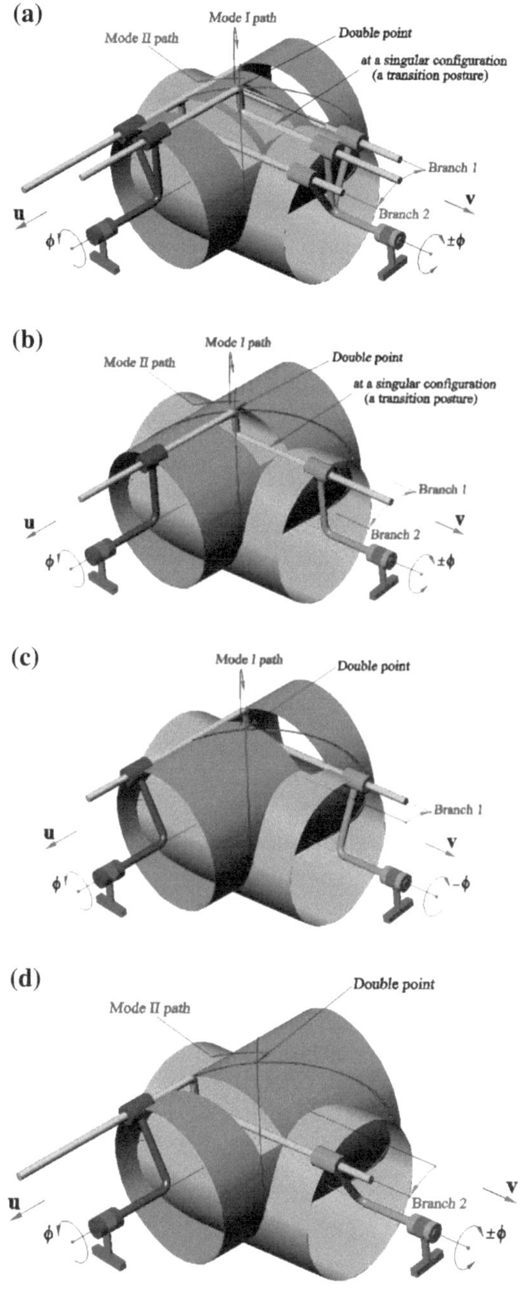


**Fig. 4.6** One bifurcation configuration of DM RC-//-RC CVSC with the offsets. **a** Bifurcation at the double point of top surface; **b** at a transition posture; **c** motion at mode I; **d** motion at mode II

direction. In practical applications, the C pair of the input crank may include an auxiliary device with an electromagnet controlled by a switch. A more general case of RC-//-RC parallel DM chain with unequal revolute cylinders has no CV

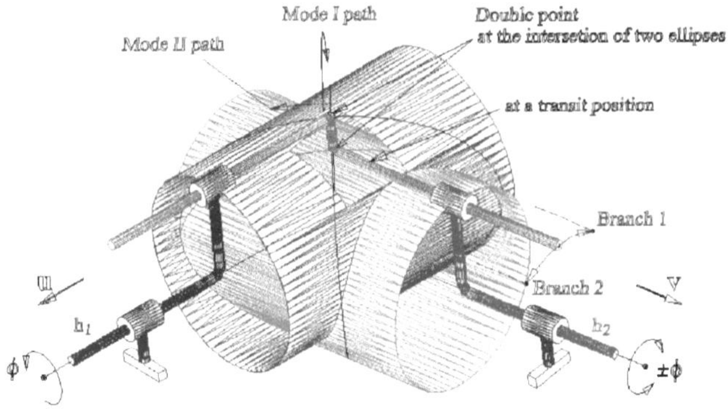


**Fig. 4.7** a HC-//-HC DM parallel chain

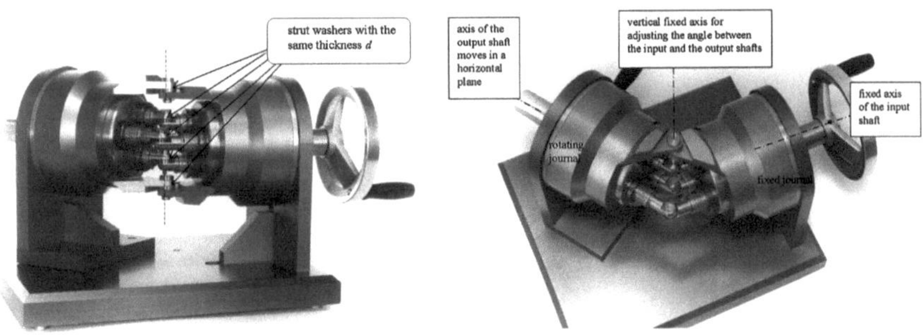


(a) (b)

**Fig. A.1** A new patented homokinetic joint

shaft coupling but it is expected to have only one bifurcation at the self intersection of bicylindrical curve, which will be a broad topic in the future. It is obvious that a most general case having HC-//-HC type shown in Fig. 4.7 produces the same type of phenomenon of bifurcations but it is beyond the scope of the paper.

**Acknowledgments** The authors are very thankful to the National Science Council for supporting this research under grants NSC 99-2221-E-151-016 and NSC 100-2221-E-151-022.

# Appendix

The figures show a new patented version of the homokinetic joint proposed by Hervé [2]. Fig. A.1.

## References

1. Bricard R, de Cinématique L (1927) Tome II. Gauthier-Villars, Pairs, pp 203–204
2. Hervé JM (1986) Le joint de Koenigs, ses variantes, son application possible en robotique. Entrainements et systemes 19(6):4–6
3. Lee, C-C, Hervé, JM (2009) "Synthesize oldham-style constant-velocity shaft couplings via Schoenflies displacement group." Proc. JSME Intl Conf Motion Power Transm, Sendai, 13–15 May 2009, pp 584–589
4. Lee C-C, Hervé JM (2009) Uncoupled actuation of overconstrained 3T-1R hybrid parallel manipulators. Robotica 27(1):103–117
5. Wohlhart K (1996) Kinematotropic linkages. Recent advances in robot kinematics. Kluwer, Dordrecht, pp 359–368
6. Lee, C-C, Hervé, JM (2005) Discontinuously movable seven-link mechanisms via group-algebraic approach. Proc. IMechE, Vol.219, Part C: J. Mechanical Engineering Science, No.6, 2005, pp 577–587
7. Hervé JM (1978) Analyse structurelle des mécanismes par groupe des déplacements. Mech Mach Theory 13(4):437–450
8. Nicaise M (1931) Les mouvements mécaniques. Librairie Polytechnique CH. Béranger, Paris

# Chapter 5
# A Special Wohlhart's Double-Goldberg 6*R* Linkage and its Multiple Operation Forms Among 4*R* and 6*R* Linkages

**Chaoyang Song and Yan Chen**

**Abstract** For Wohlhart's double-Goldberg 6*R* linkage constructed from two original Goldberg 5*R* linkages, there are two constructive forms and one non-constructive form, which can be reconfigured through bifurcation points. A special Wohlhart's double-Goldberg 6*R* linkage is proposed by introducing a special geometry constraint. One of the constructive forms of the resultant linkage is degenerated into a pseudo 4*R* linkage, which is essentially a Bennett linkage. Therefore, the special Wohlhart's double-Goldberg 6*R* linkage achieves multiple forms among 4*R* and 6*R* linkages. Physical models are also made to validate the special Wohlhart's double-Goldberg 6*R* linkage in different operation forms.

**Keywords** Wohlhart's double-Goldberg linkage · Overconstrained linkage · Multiple operation modes

## 5.1 Introduction

The concept and design of reconfigurable mechanisms have been developed to meet the need of the emerging new frontiers of multi-functional machines and robots since 1990s. One effective method of reconfiguration is through the bifurcation points on kinematic paths of different linkages. The kinematotropic linkage proposed by Wohlhart [1] can even change its global mobility with positional parameter actuations at the bifurcation points. Galletti and Fanghella [2]

C. Song · Y. Chen (✉)
School of Mechanical and Aerospace Engineering, Nanyang Technological University, Singapore, Singapore
e-mail: chenyan@ntu.edu.sg

J. S. Dai et al. (eds.), *Advances in Reconfigurable Mechanisms and Robots I*,
DOI: 10.1007/978-1-4471-4141-9_5, © Springer-Verlag London 2012

used the displacement groups method and designed a series of kinematotropy mechanisms. The metamorphic mechanisms were proposed by Dai and Rees [3] with variable topology and mobility. Recently, Kong and Huang [4] proposed a number of one degree-of-freedom single-loop overconstrained linkages with two operation modes by using type synthesis method. Wohlhart [5] also proposed a series of multifunctional 7*R* linkages by inserting an overconstrained 4*R*, 5*R* or 6*R* mobile chain into a closed-loop 7*R* linkage. In this paper, a special geometry condition is introduced to Wohlhart's double-Goldberg 6*R* linkage to make it have multiple operation forms among 4*R* and 6*R* linkages.

This paper is organized as follows. Sect. 5.2 introduces three forms of Wohlhart's double-Goldberg 6*R* linkage. In Sect. 5.3, the original form of Wohlhart's double-Goldberg 6*R* linkage is degenerated into a Bennett linkage. Section 5.4 describes the multiple operation forms and their transitions. The conclusion is enclosed in the last section. In this paper, $a_{(i-1)i}$ is length of link $(i-1)i$, $\alpha_{(i-1)i}$ is twist of link $(i-1)i$, $R_i$ is offset of joint $i$ and $\theta_i$ is revolute variable of joint $i$. $a/\alpha$, $b/\beta$, $c/\gamma$ and $d/\delta$ are the lengths and twists of the link, e.g., $a/\alpha$ is a link with length $a$ and twist $\alpha$.

## 5.2 Wohlhart's Double-Goldberg 6*R* Linkage

Recently, Wohlhart's double-Goldberg 6*R* linkage [6] was re-examined by Song and Chen [7]. Shown in Fig. 5.1 are the two constructive forms of Wohlhart's double-Goldberg 6*R* linkage. Both linkages are obtained by merging two Goldberg 5*R* linkages on the commonly shared link-pair $a/\alpha \sim c/\gamma$. After removing the common links and joint in dash lines, the rest part will form the Wohlhart's double-Goldberg 6*R* linkage. The one in Fig. 5.1a was originally formed by Wohlhart and named as Form I linkage. The one in Fig. 5.1b is the newly found constructive form [7] and named as Form II linkage. During bifurcation analysis, the third linkage form was detected and named as Form III linkage. The Form III linkage cannot be decomposed into the combination between two Goldberg 5*R* linkages as Forms I and II, and therefore it is non-constructive. All three forms of Wohlhart's double-Goldberg 6*R* linkage share the identical geometry conditions as Eqs. (5.1a), (5.1b), (5.1c) and (5.1d) yet have different closure equations.

$$a_{12} = a_{45} = a + c,\ a_{23} = a_{61} = b,\ a_{34} = a_{56} = d, \tag{5.1a}$$

$$\alpha_{12} = \alpha_{45} = \alpha + \gamma,\ \alpha_{23} = \alpha_{61} = \beta,\ \alpha_{34} = \alpha_{56} = \delta, \tag{5.1b}$$

$$\frac{\sin\alpha}{a} = \frac{\sin\beta}{b} = \frac{\sin\gamma}{c} = \frac{\sin\delta}{d}, \tag{5.1c}$$

$$R_i = 0\ (i = 1,\ 2,\ \ldots,\ 6). \tag{5.1d}$$

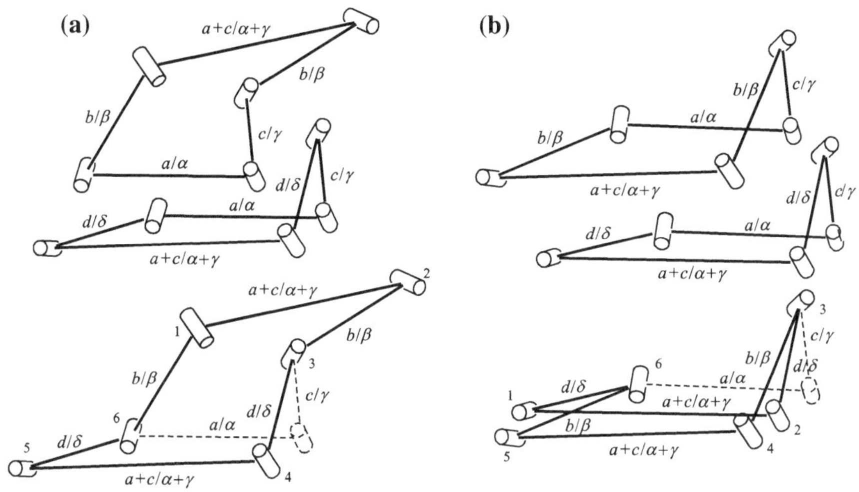


**Fig. 5.1** Two constructive forms of Wohlhart's double-Goldberg 6*R* linkage: **a** original Form I linkage reported by Wohlhart [6] and **b** new constructive Form II linkage reported by Song and Chen [7]

The closure equations of Forms I and II linkages can be derived from their construction process [7], which are

$$\begin{array}{c} \tan\frac{\theta_2}{2} = \dfrac{m_1 m_2}{\tan\frac{\theta_1}{2}},\ \theta_3 = 2\tan^{-1}\left(\dfrac{m_1}{\tan\frac{\theta_1}{2}}\right) + 2\tan^{-1} P_{\theta_1} - \pi \\ \tan\dfrac{\theta_4}{2} = m_4 P_{\theta_1},\ \tan\dfrac{\theta_5}{2} = \dfrac{m_3}{P_{\theta_1}},\ \theta_6 = -\theta_3, \end{array}, \tag{5.2}$$

in which

$$m_1 = \frac{\sin\dfrac{\beta+\alpha}{2}}{\sin\dfrac{\beta-\alpha}{2}},\ m_2 = \frac{\sin\dfrac{\beta+\gamma}{2}}{\sin\dfrac{\beta-\gamma}{2}},\ m_3 = \frac{\sin\dfrac{\delta+\alpha}{2}}{\sin\dfrac{\delta-\alpha}{2}},\ m_4 = \frac{\sin\dfrac{\delta+\gamma}{2}}{\sin\dfrac{\delta-\gamma}{2}} \tag{5.3}$$

For Form I linkage

$$\begin{array}{c} q = |m_1|\sqrt{\dfrac{(m_2+m_3) - m_2 m_3 (m_1+m_4)}{m_1 m_4 (m_2+m_3) - (m_1+m_4)}},\ Q = \dfrac{(m_1 m_2 - 1)\tan\frac{\theta_1}{2}}{m_1 m_2 + \tan^2\frac{\theta_1}{2}} \\ P_{\theta_1} = \begin{cases} \dfrac{(1-m_3 m_4) + \sqrt{(1-m_3 m_4)^2 - 4 m_3 m_4 Q^2}}{2 m_4 Q} & \begin{pmatrix} -\pi \le \theta_1 \le -2\tan^{-1} q \\ 2\tan^{-1} q \le \theta_1 \le \pi \end{pmatrix} \\ \dfrac{(1-m_3 m_4) - \sqrt{(1-m_3 m_4)^2 - 4 m_3 m_4 Q^2}}{2 m_4 Q} & (-2\tan^{-1} q \le \theta_1 \le 2\tan^{-1} q) \end{cases}, \end{array} \tag{5.4}$$

and for Form II linkage, change $P_{\theta_1}$ into the following,

$$P_{\theta_1} = \begin{cases} \dfrac{(1 - m_3m_4) + \sqrt{(1 - m_3m_4)^2 - 4m_3m_4Q^2}}{2m_4Q} & (-2\tan^{-1} q \leq \theta_1 \leq 2\tan^{-1} q) \\ \dfrac{(1 - m_3m_4) - \sqrt{(1 - m_3m_4)^2 - 4m_3m_4Q^2}}{2m_4Q} & \begin{pmatrix} -\pi \leq \theta_1 \leq -2\tan^{-1} q \\ 2\tan^{-1} q \leq \theta_1 \leq \pi \end{pmatrix} \end{cases} \tag{5.5}$$

As the Form III linkage is non-constructive, its closure equations cannot be derived using the same method as Forms I and II. However, we can use the singular value decomposition method [8] to obtain its kinematic paths numerically.

## 5.3 A Special Wohlhart's Double-Goldberg 6*R* Linkage

From Eqs. (5.2), (5.4) and (5.5), $\theta_3$ and $\theta_6$ are not equal to zero generally during the motion. If constraining $\theta_3 = \theta_6 = 0$, we will get

$$\frac{(1 - m_3m_4) \pm \sqrt{(1 - m_3m_4)^2 - 4m_3m_4\left[\dfrac{(m_1m_2 - 1)\tan\frac{\theta_1}{2}}{m_1m_2 + \tan^2\frac{\theta_1}{2}}\right]^2}}{2m_4\dfrac{(m_1m_2 - 1)\tan\frac{\theta_1}{2}}{m_1m_2 + \tan^2\frac{\theta_1}{2}}} = \frac{\tan\frac{\theta_1}{2}}{m_1}, \tag{5.6}$$

i.e.,

$$[m_4(m_1m_2 - 1) - m_1(1 - m_3m_4)]\tan^2\frac{\theta_1}{2} = m_1^2[m_2(1 - m_3m_4) - m_3(m_1m_2 - 1)] \tag{5.7}$$

Here, $m_i(i = 1, 2, 3, 4)$ in Eq. (5.3) are fixed for a given linkage. Thus, in order to make $\theta_1$ free, Eq. (5.7) can be held only when

$$\begin{cases} m_4(m_1m_2 - 1) - m_1(1 - m_3m_4) = 0 \\ m_2(1 - m_3m_4) - m_3(m_1m_2 - 1) = 0 \end{cases} \tag{5.8}$$

One solution of Eq. (5.8) is

$$\begin{cases} m_4 = -m_1 \\ m_3 = -m_2 \end{cases}, \tag{5.9}$$

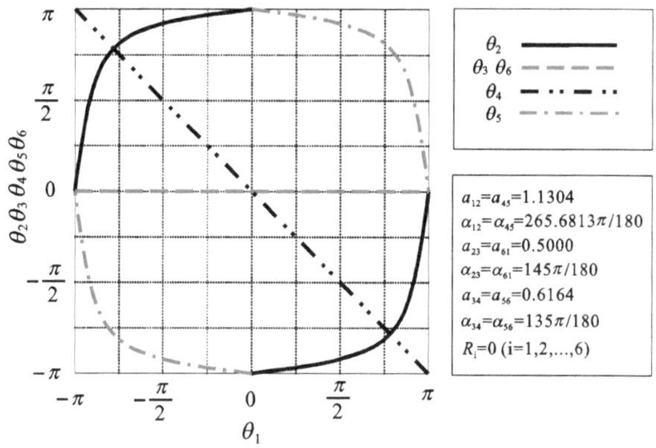


**Fig. 5.2** Kinematic paths of the special Wohlhart's double-Goldberg 6*R* linkage in Bennett form

which is

$$\tan\frac{\alpha}{2}\tan\frac{\gamma}{2}=\tan\frac{\beta}{2}\tan\frac{\delta}{2}, \tag{5.10}$$

in terms of link twists. Furthermore, the following relationship can be obtained by considering Eqs. (5.1c) and (5.10) that

$$\frac{\sin(\alpha+\gamma)}{a+c}=\frac{\sin(\beta+\delta)}{b+d}. \tag{5.11}$$

Therefore, when the condition shown in Eq. (5.11) is introduced to Eq. (5.1), a special Wohlhart's double-Goldberg 6*R* linkage will be formed, whose revolute variables $\theta_3$ and $\theta_6$ are expected to be zero during full-circle motion. In other words, $\theta_3$ and $\theta_6$ are kinematically constrained to zero during the full-circle motion. The closure equations of Forms I and II of the special Wohlhart's double-Goldberg 6*R* linkage can be obtained as below by substituting Eq. (5.11) into Eqs. (5.2), (5.4) and (5.5) that

$$\tan\frac{\theta_2}{2}=\frac{m_1m_2}{\tan\frac{\theta_1}{2}},\ \tan\frac{\theta_4}{2}=-m_1P_{\theta_1},\ \tan\frac{\theta_5}{2}=-\frac{m_2}{P_{\theta_1}}, \tag{5.12a}$$

$$\theta_3=2\tan^{-1}\left(\frac{m_1}{\tan\frac{\theta_1}{2}}\right)+2\tan^{-1}P_{\theta_1}-\pi,\ \theta_6=-\theta_3, \tag{5.12b}$$

in which

$$P_{\theta_1}=\begin{cases}\dfrac{\tan\frac{\theta_1}{2}}{m_1} & \text{for Form I linkage,}\\[2ex] \dfrac{m_2}{\tan\frac{\theta_1}{2}} & \text{for Form II linkage.}\end{cases} \tag{5.13}$$

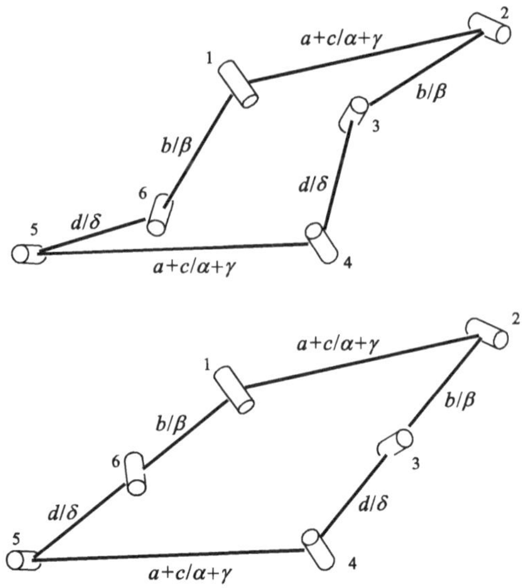


**Fig. 5.3** Reconfiguration of Form I of Wohlhart's double-Goldberg 6*R* linkage into a Bennett 4*R* linkage

Only for Form I linkage, $\theta_3 = \theta_6 = 0$ after introducing the constraint in Eq. (5.11) to the 6*R* linkage. Thus, this 6*R* linkage is degenerated into a pseudo 4*R* linkage, whose kinematic paths are shown in Fig. 5.2.

After careful examining the geometry conditions of the special Wohlhart's double-Goldberg 6*R* linkage and its corresponding kinematic closure equations with $\theta_3 = \theta_6 = 0$, it is interesting to find that the pseudo 4*R* linkage is essentially a Bennett linkage connected by joints 1, 2, 4 and 5 with links $b/\beta$ and $d/\delta$ adjacent to joints 3 and 6 rigidified into a composite link $b + d/\beta + \delta$, see Fig. 5.3.

## 5.4 Multiple Operation Forms of the Special Wohlhart's Double-Goldberg 6*R* Linkage

Generally, Wohlhart's double-Goldberg 6*R* linkage has three forms: two constructive 6*R* forms and one non-constructive 6*R* form with different kinematic paths, which can transform into each other through bifurcation points [7]. After introducing the special geometry constraint in Eq. (5.11), the Form I linkage, degenerates to a Bennett form with only four active revolute variables. Thus, the operation form of a 4*R* linkage has been successfully introduced to Wohlhart's double-Goldberg 6*R* linkage.

The transition among different forms of the special Wohlhart's double-Goldberg 6*R* linkage is shown in Fig. 5.4. Only the relationship between $\theta_1$ and $\theta_5$ is used for the ease of representation. The black solid lines correspond to the Form II of Wohlhart's double-Goldberg linkage in constructive 6*R* form. The grey solid lines correspond to the Form I of Wohlhart's linkage in Bennett 4*R* form.

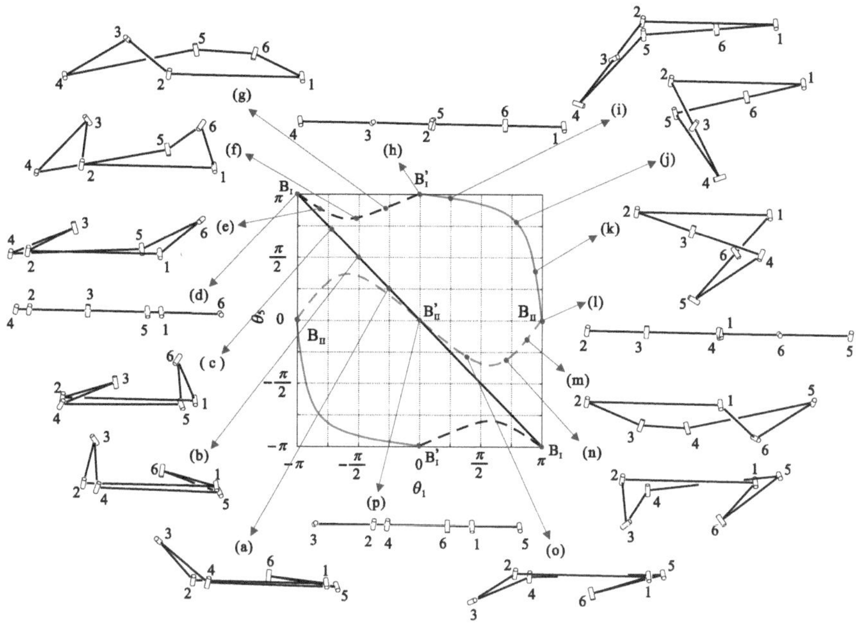


Fig. 5.4 Transitions of the special Wohlhart's double-Goldberg 6*R* linkage in different operation forms: (*a*) ~ (*c*) are in constructive 6*R* form; (*e*) ~ (*g*) and (*m*) ~ (*o*) are in different configurations of non-constructive 6*R* form; (*i*) ~ (*k*) are in Bennett 4*R* form. (*d*), (*h*), (*l*) and (*p*) are the transition configuration at bifurcation points

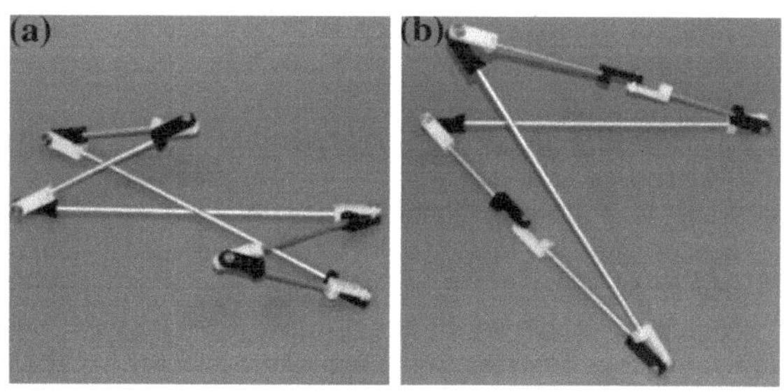


**Fig. 5.5** Physical models of the special Wohlhart's double-Goldberg 6*R* linkage. **a** Constuctive 6*R* form **b** Bennett 4*R* form

The black and grey dash lines represent the Form III of Wohlhart's double-Goldberg linkage in different configurations, which are in non-constructive 6*R* forms [7]. $B'_I$, $B_{II}$ and $B'_{II}$ are the bifurcation points between different forms. Physical models of the special Wohlhart's double-Goldberg 6*R* linkage are also presented in Fig. 5.5.

## 5.5 Conclusion

In this paper, a special Wohlhart's double-Goldberg 6*R* linkage is presented to have multiple operation forms among 4*R* and 6*R* linkages. Originally, the Wohlhart's double-Goldberg 6*R* linkage has three operation forms: Forms I and II are in constructive 6*R* forms; and Form III is in non-constructive 6*R* form. By investigating the closure equations, a special geometry condition that $\sin(\alpha+\gamma)/(a+c) = \sin(\beta+\delta)/(b+d)$ is introduced, resulting two revolute variables to be kinematically inactive in Form I linkage. The Form I linkage degenerates into a Bennett linkage. Therefore, the special Wohlhart's double-Goldberg 6*R* linkage has three operation forms: a constructive 6*R* form, a non-constructive 6*R* form and a Bennett 4*R* form. Different forms of the linkage can be transited into each other through bifurcation points.

The work in this paper demonstrates the possibility and approach to propose reconfigurable mechanisms among kinematic loops with different number of links. Such mechanisms will play a fundamental role in the multifunctional robots and manufacture system of new generation.

**Acknowledgments** Song C.Y. would like to thank NTU for providing the University Graduate Scholarship during his PhD study.

## References

1. Wohlhart, K (1996) Kinematotropic linkages. In: The fifth international symposium on advances in robots kinematics, Kulwer Academic Publishers, Portoroz, 1996, pp 359–368
2. Galletti C, Fanghella P (2001) Single-loop kinematotropic mechanisms. Mech Mach Theory 36(6):743–761
3. Dai JS, Rees JJ (1999) Mobility in metamorphic mechanisms of foldable/erectable kinds. J Mech Des 121(3):375–382
4. Kong, X, Huang, C (2009) Type synthesis of single-DOF single-loop mechanisms with two operation modes. In: ASME/IFToMM international conference on reconfigurable mechanisms and robots, IEEE, 2009, pp. 136–141
5. Wohlhart, K (2010) Multifunctional 7R Linkages. In: Proceedings of the international symposium of mechanism and machine science, AzCIFToMM, Izmir, 2010, pp 85–91
6. Wohlhart K (1991) Merging two general goldberg 5*R* linkages to obtain a new 6*R* space mechanism. Mech Mach Theory 26(7):659–668
7. Song CY, Chen Y (2011) Bifurcation analysis of the original Double-Goldberg 6*R* linkage. International symposium on multibody systems and mechatronics, Valencia
8. Gan WW, Pellegrino S (2006) A numerical approach to the kinematic analysis of deployable structures forming a closed loop. Proc Inst Mech Eng C J Mech Eng Sci 220(7):1045–1056

# Chapter 6
# The Equivalent Resistance Gradient Model of Metamorphic Mechanisms and the Design Approach

**Shujun Li and Jian S. Dai**

**Abstract** The practical metamorphic process, which include configuration transforming, configuration keeping and working sequence problems, is introduced. The concept of equivalent resistance gradient of metamorphic joints is proposed to investigate a new way for regulating the constrained metamorphic process and/or design of metamorphic joints. A kinds of metamorphic cyclogram is introduced, and the resistance characteristics of typical metamorphic joints are studied. A non-dimension equivalent resistance coefficient of metamorphic joints and the equivalent resistance gradient model of metamorphic mechanism are defined. The connection ships of equivalent resistance, constraint characteristics and forms, and the structure of metamorphic joints in the working stages are revealed. A design approach is demonstrated based on the metamorphic cyclogram and the equivalent resistance gradient model.

**Keywords** Metamorphic mechanism · Configuration keeping · Working sequence · Equivalent resistance gradient · Metamorphic cyclogram

---

S. Li (✉)
College of Mechanical Engineering and Automation, Northeastern University, Shenyang, 110004, People's Republic of China
e-mail: sjli_cn@yahoo.com.cn

J. S. Dai
Centre for Advanced Mechanisms and Robotics, MOE Key Laboratory for Mechanism Theory and Equipment Design, Tianjin University, Tianjin, China
e-mail: jian.dai@kcl.ac.uk

J. S. Dai
Centre for Robotics Research, King's College, University of London, Strand, London, WC2R 2LS, UK

J. S. Dai et al. (eds.), *Advances in Reconfigurable Mechanisms and Robots I*,
DOI: 10.1007/978-1-4471-4141-9_6, © Springer-Verlag London 2012

## 6.1 Introduction

Since the metamorphic mechanisms were proposed a decade ago [1], the researches on metamorphic ways have been contributed over the world. Parise, Howell and Magleby introduced metamorphic ortho-planar mechanisms [2], and Li et al. presented a method for structural synthesis of metamorphic mechanisms based on the configuration transformations [3]. Liu and Wang studied the characteristics of metamorphic mechanisms and ways of metamorphic [4, 5]. Ding and Yang proposed a reconfiguration theory for a new kind of metamorphic mechanisms with assembly-circles artifact [6], and Li et al. studied joint-gene based topological representation and configuration transformation of metamorphic mechanisms [7] in 2009. Li and Dai introduced a new idea of Augmented Assur Group, and be used in the structural theory of metamorphic mechanisms similar to Assur Group been used in general planar mechanisms [8], and Zhang and Dai classified the metamorphic techniques and analyzed the characteristics of geometric reconfiguration in the same year [9]. Lan discussed the structure and evolution of planar metamorphic mechanisms in 2010 [10]. Generally, the topological structure of a source metamorphic mechanism was determined by the configuration transmission matrix, which it only characterize the connection ship of the links according to the kinematic requirements of the working stages. In fact, it is very important for design practical metamorphic mechanisms that the forms and structures of constraint metamorphic joints, and as well as the kinematic parameters to satisfy the configuration transforming, configuration keeping and working sequence problems. The idea of metamorphic cyclogram and equivalent resistance gradient are introduced to investigate the relationship between the kinematic status, resistant forces and constraint forms of metamorphic joints, and to find a new design approach for practical constrained metamorphic mechanisms.

## 6.2 Complete Metamorphic Process of Constrained Metamorphic Mechanisms

The general constrained metamorphic processes are: firstly, degenerating the multi-DOF source metamorphic mechanism into single-DOF mechanism by constraints of metamorphic joints to form one of the working stages and keeping the working configuration. Then, releasing the constrained metamorphic joint free and constraining other metamorphic joint to change the working configuration and forming another working stage and keeping the working configuration. The working sequences of required working stages are controlled by constraining the different metamorphic joints in the sequences. The sketch map of a complete metamorphic process of constrained metamorphic mechanisms is shown as in Fig. 6.1.

To obtain all metamorphic working configurations of the mechanism, we can constrain the joints of the mechanism alternately according to the required working stages and working sequences by kinematic geometry arrangements, geometric and force constraints, designated profiles of links and joints, etc.

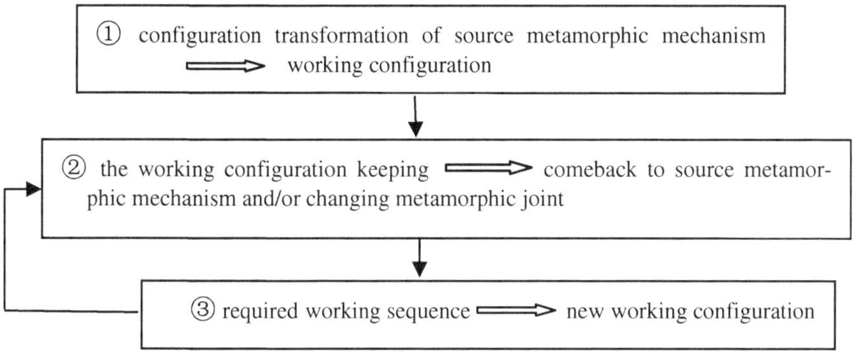


**Fig. 6.1** Sketch map of a complete metamorphic process of constrained metamorphic mechanisms

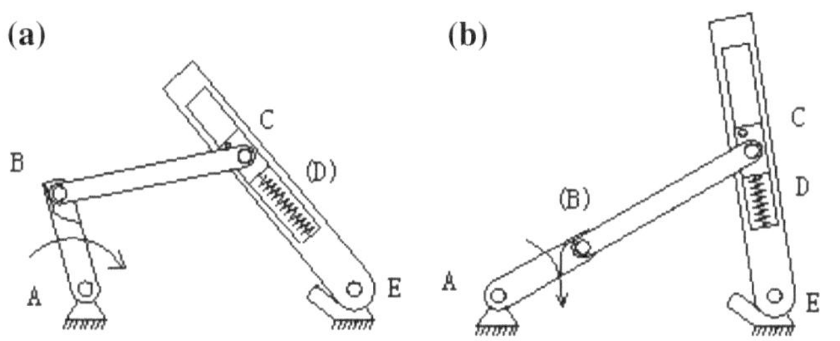


**Fig. 6.2** Two metamorphic working-stage configurations of 2 DOF metamorphic mechanism

The mechanism in Fig. 6.2, for examples, the metamorphic configurations can be obtained by using geometric and force constraints of metamorphic joints to constrain one mobility one at a time. Figure 6.2a is a configuration with mobility one and the movement of slider D is constrained by spring force with spring forces being larger than the joint frictions. In this case, the mechanism is a four-bar linkage with only four revolute joints while the slider D does not move. When the mechanism moves to another configuration in Fig. 6.2b while the movement of joint B is constrained by a geometric constraint to make two links to become one. Therefore, the driving force presses the spring to move the slider D. The mechanism becomes a crank-slider mechanism.

## 6.3 Constrained Metamorphic Joints and Force Characteristics Sketches

Generally constrained metamorphic operations are implemented by using geometric constraints and/or force constraints of metamorphic joints to overlap two links to become one, or to make the metamorphic joints to be locked. Proposed

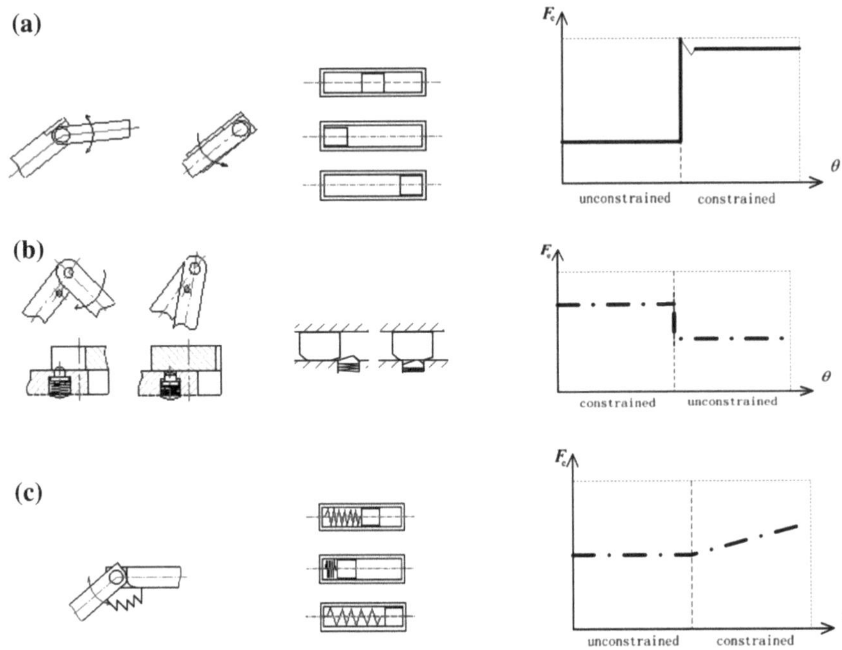


**Fig. 6.3** Constrained metamorphic joints and force characteristics sketches

typical structures of constrained metamorphic joints are shown as in the Fig. 6.3. Figure 6.3a are turning joint and prismatic joint with geometric constraint to overlap two links to become one and provide infinite constraint force at the point, and Fig. 6.3b are turning joint and prismatic joint with geometric constraint controlled by spring to provide limited constraint force at the point. Figure 6.3c are turning joint and prismatic joint with geometric constraint and spring force constraint respectively. The right parts of the Fig. 6.3 are the corresponding force characteristics sketches.

## 6.4 The Equivalent Resistance Gradient Model

### *6.4.1 Metamorphic Cyclogram of Metamorphic Mechanisms*

The working statuses of constrained metamorphic joints will transfer between moving and static depend on the changing of constrained forces to form the corresponding working stages, while the diver joint and general joints are keeping working. The displacement (turning angle or moving distance) $\theta$ of the driver is proposed in horizontal axis, and the working conditions of joints $J$ ($J = R$ for revolute joint, $J = P$ for prismatic joint etc.) in corresponding working stages (or

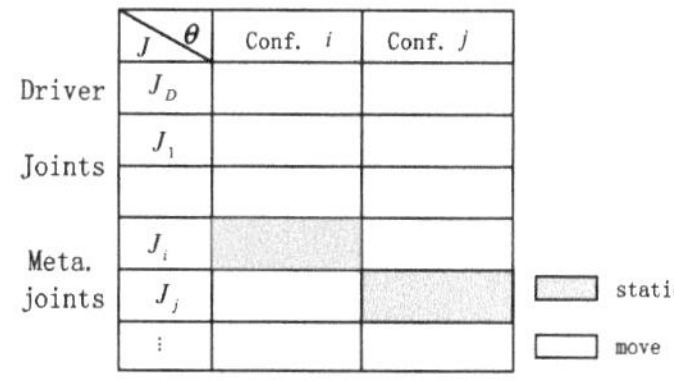


**Fig. 6.4** Metamorphic cyclogram of 2 DOF metamorphic mechanism

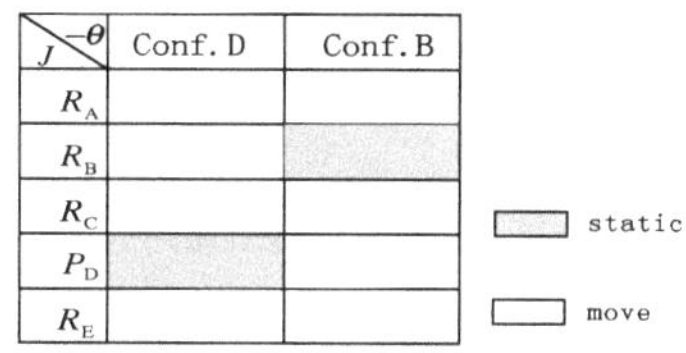


**Fig 6.5** Metamorphic cyclogram of metamorphic mechanism shown in Fig. 6.2

metamorphic configurations) is proposed in vertical axis, so that the working configuration transferring and the working statuses of metamorphic joints can be described in following ***metamorphic cyclogram*** as shown in Fig. 6.4, which shows the working statuses of the joints with the movement of driver in the corresponding working stages. The metamorphic cyclogram of metamorphic mechanism of Fig. 6.2 can be formed as in the Fig. 6.5.

## *6.4.2 Equivalent Resistance Gradient Model of Metamorphic Mechanisms*

There is only one metamorphic joint moving (working) at a time in corresponding working stage while the other(s) keeping static in the metamorphic process. The interchanges from moving to static of metamorphic joints are controlled by the constraints or constrained forces by the designed structure of metamorphic joints. A new idea of equivalent resistance coefficient is proposed to describe the working statuses and constraint resistance characteristics of the joints and in order to compare the constrained forces between revolute and prismatic metamorphic joints. It is defined as: the ratio of the force and/or torque in moving direction provided by the constraint of the metamorphic joint to the force and/or torque in the moving direction acted on the metamorphic joint in the working process.

$$f_e(\theta_i) = \frac{\mathbf{F}_c(\theta_i)}{\mathbf{F}(\theta_i)} = \frac{\mathbf{T}_c(\theta_i)}{\mathbf{T}(\theta_i)} \quad i = 1, 2, \ldots, m \tag{6.1}$$

where, $f_e(\theta_i)$ is equivalent resistance coefficient of metamorphic joint, and $\theta_i$ is displacement of the driver in corresponding working stages, and $m$ is the number of the working stages. $\mathbf{F}_c(\theta_i)$ and $\mathbf{T}_c(\theta_i)$ are resistant force and resistant torque in moving direction provided by the constraint of metamorphic joint. $\mathbf{F}(\theta_i)$ and $\mathbf{T}(\theta_i)$ are actual force and torque in the moving direction acted of on the metamorphic

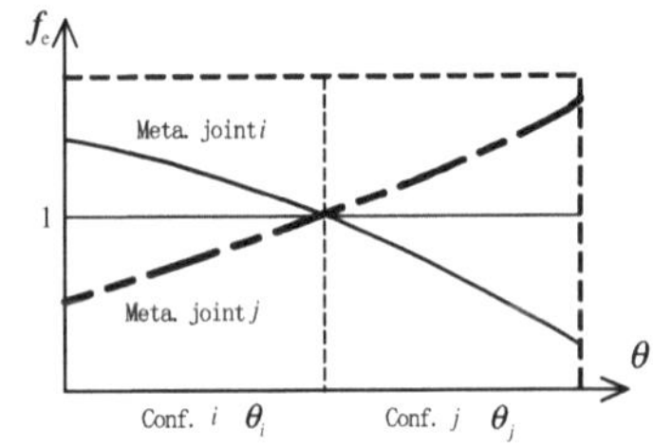


**Fig. 6.6** Equivalent resistance gradient sketch of 2 DOF metamorphic mechanism

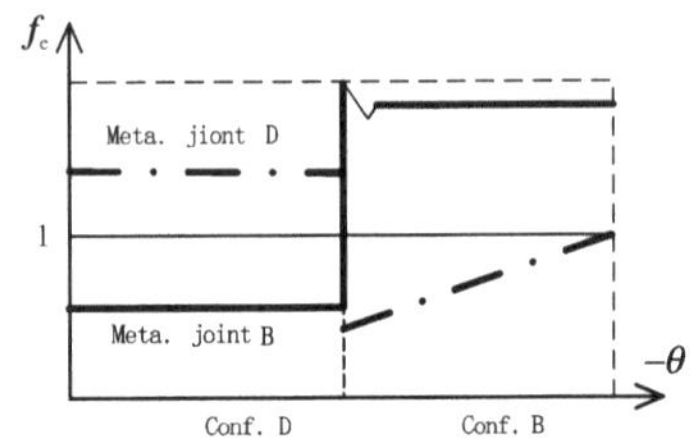


**Fig. 6.7** Equivalent resistance gradient sketch that of Fig. 6.2

joint in the working process. In order to show the resistant characteristics of metamorphic joint such as in the Fig. 6.3 to get the constraint form and structure information of metamorphic joints, the formula (6.1) be modified as

$$f_{em}(\theta_i)=\frac{\mathbf{F}_c(\theta_i)}{F_{min}}=\frac{\mathbf{T}_c(\theta_i)}{T_{min}} \tag{6.2}$$

$$f_{es}(\theta_i)=\frac{\mathbf{F}_c(\theta_i)}{F_{max}}=\frac{\mathbf{T}_c(\theta_i)}{T_{max}} \tag{6.3}$$

where $f_{em}(\theta_i)$ is the equivalent resistance coefficient of ***moving metamorphic joint*** in the working stages. $F_{min}$ and $T_{min}$ are the minimum actual force and torque acted on moving metamorphic joint in the corresponding working stages. $f_{es}(\theta_i)$ is the equivalent resistance coefficient of ***static metamorphic joint*** in the working stages. $F_{max}$ and $T_{max}$ are the maximum actual force and torque acted on static metamorphic joint in the corresponding working stages. The $f_{em}(\theta_i)$ and $f_{es}(\theta_i)$ become in proportion with $\mathbf{F}_c(\theta_i)$ and/or $\mathbf{T}_c(\theta_i)$ by modified formulas (6.3)–(6.4), and lead the equivalent resistant force sketches of metamorphic joint are similar with the force characteristics sketches of metamorphic joints of the Fig. 6.3.

The moving sequences of metamorphic joints should be proportional to the equivalent resistant forces according to the law of minimum resistance of kinematics, the *equivalent resistance gradient* of the metamorphic joints in the working stages of constrained metamorphic process should be

$$f_{em}(\theta_i)\leq 1 \quad \text{and} \quad f_{es}(\theta_i)\geq 1 \tag{6.4}$$

The equivalent resistance gradient sketch of 2 DOF metamorphic mechanism can be formed as shown in Fig. 6.6 according to formula (6.4). The equivalent resistance gradient sketch of the mechanism of the Fig. 6.2 can be obtained as shown in Fig. 6.7 according to formulas (6.1)–(6.4) and Fig. 6.5 and Fig. 6.6.

To compare the equivalent resistance gradient sketch with the metamorphic cyclogram, one can find out the relationships between kinematic statuses and constraint forces of metamorphic joints in corresponding working stages.

## 6.5 Metamorphic Process Regulating and Design Approach

### 6.5.1 The Characteristics of Equivalent Resistance Gradient

There are following characteristics of equivalent resistance gradient by analyzing formulas (6.1)–(6.4) and Fig. 6.6:

(1) The intersection point of two equivalent forces sketch is the dividing point of two working stages. The configuration keeping period is depends on the position of the intersection point.
(2) The equivalent resistance coefficient of static metamorphic joint is larger than one, and that moving metamorphic joint is less than one in the same working stage, so that regulating the values of two equivalent resistant forces (or the relative positions of two equivalent resistant force sketches) would change the working sequences (configurations) of metamorphic joints.
(3) The equivalent resistant force sketch by modified formulas (6.3)–(6.4) is proportional to the force characteristics sketch of the constrained joint, so that the form and constraint parameters of the metamorphic joints can design and/ or chosen based on the sketch.
(4) The metamorphic cyclogram and equivalent resistance gradient sketch are interrelated from Fig. 6.5 and Fig. 6.6. The corresponding equivalent resistance gradient can be obtained by referencing the metamorphic cyclogram sketch formed by a task based metamorphic process, and this can be used in the design of metamorphic mechanisms and regulating metamorphic process.

### 6.5.2 Metamorphic Process Regulating and Design Approach

If the metamorphic cyclogram of Fig. 6.5, for example, were to be changed into Fig. 6.8 according to working process, the mechanism can be designed as follow.

The Fig. 6.8 shows that driver is counter-clock-wise run and the metamorphic joint C and E are constrained in configuration $i$ and $j$ respectively, so that the corresponding equivalent resistance gradient sketch of the two configurations can be formed as shown in the Fig. 6.9 according to formula (6.4). The metamorphic

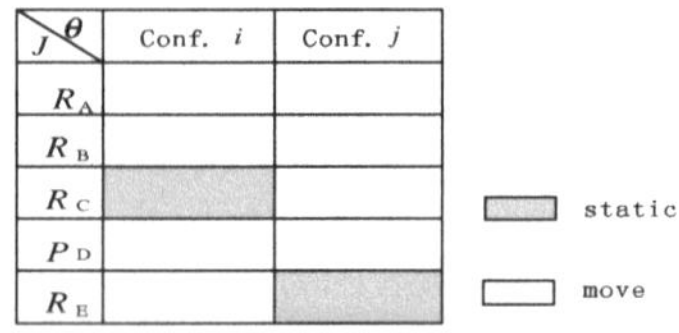


**Fig. 6.8** Metamorphic cyclogram of 2 DOF metamorphic mechanism

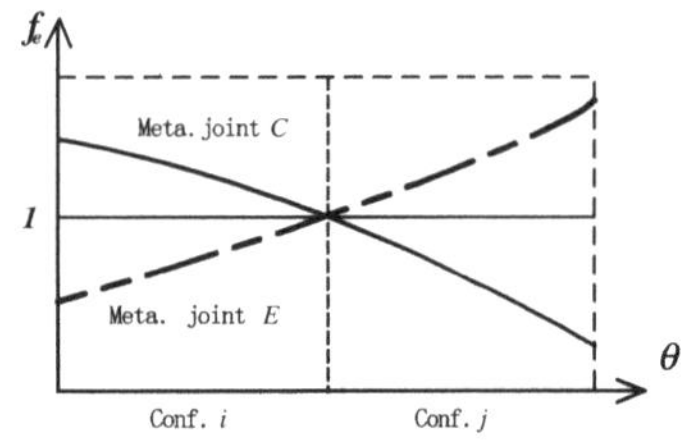


**Fig. 6.9** Equivalent resistance gradient sketch of the two configurations

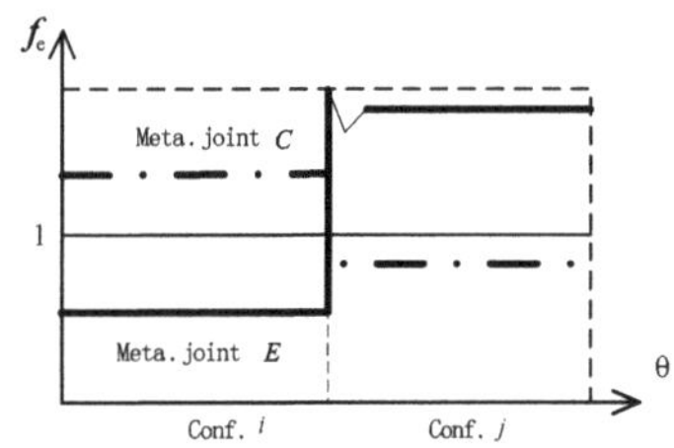


**Fig. 6.10** Equivalent resistance gradient sketch with the forms of the constrained metamorphic joints

joint C can be kept static by the spring force controlled constraint and joint E works in configuration *i*; metamorphic joint E can be kept static by geometric constraint to provide a lager force then that spring force of the joint C and joint C works in configuration *j*. The equivalent resistance gradient sketch with the forms of the constrained metamorphic joints can be formed as shown in Fig. 6.10 according Fig. 6.3 and Fig. 6.9. It forms one of the constrained metamorphic mechanisms which meet the kinematic requirements of the Fig. 6.8, and the corresponding structural sketch of the mechanism is shown in Fig. 6.11. By actuating joint A counter-clock-wise as that in Fig. 6.11a while joint C is constrained by a peg that the mechanism is a crank-slider mechanism with joints A, B and E and slider D movements. When the slider D moves to a certain position in the housing when the housing link is restricted by a bolt at joint E in Fig. 6.11b, the peg is released that joint C is free to move. The mechanism becomes another crank-slider mechanism with the housing link becoming the ground link.

The example shows that the new metamorphic cyclogram can be formed with the change of the working process, and the corresponding equivalent resistance gradient sketch can be formed. The connection-ship of metamorphic processes are set up to form the bases of regulation the configuration keeping period and working sequence of the mechanism, as well as the design and/or choosing of constrained metamorphic joints.

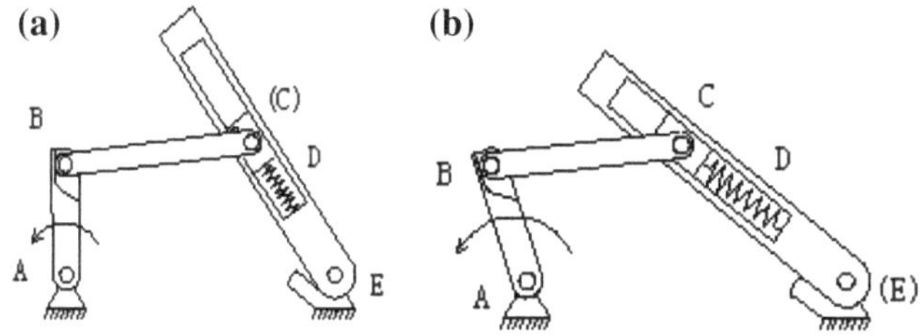


**Fig. 6.11** Structural sketches of 2 DOF metamorphic mechanism

## 6.6 Conclusions

The concept of equivalent resistance gradient of metamorphic joints is introduced to describe the metamorphic process and provide a new approach to regulate the working stages and design practical metamorphic joints to satisfy the configuration transforming, configuration keeping and working sequence requirements. A metamorphic cyclogram is proposed, and a kind of non-dimension equivalent resistance gradient model of metamorphic joints is defined to form the equivalent resistance gradient sketch of metamorphic mechanisms, so that the working status and the resistant forces of metamorphic mechanism in corresponding working stages are obtained and the inter-relationship between equivalent resistance, constraint form and characteristics, and structure of metamorphic joints are connected. The metamorphic cyclogram and the equivalent resistance gradient model established the theoretical base of a new design approach of practical metamorphic mechanism. The corresponding design method will be studied in the following works.

**Acknowledgments** The authors would like to acknowledge the financial support of the National Natural Science Foundation of China (NSFC) (Grant NO. 50875038 and 51175069).

## References

1. Dai JS, Rees JJ (1999) Mobility in metamorphic mechanisms of foldable/erectable kinds. J MECH DES-T ASME 121(3):375–382
2. Parise, JJ, Howell, LL, Magleby, SP (2000) Ortho-planar mechanisms. In: Proceedings of .26th biennial mechanisms and robotics conference, Baltimore, 2000
3. Li DL, Dai JS, Zhang QX, Jin GG (2002) Structure synthesis of metamorphic mechanisms based on the configuration transformations. Chinese J Mech Eng 38(7):12–16
4. Liu, CH, Yang TL (2004) Essence and characteristics of metamorphic mechanisms and their metamorphic ways. In: Proceedings of. 11th world congress in mechanism and machine science, Mech Eng Press, Tianjing 2004, pp 1285–1288
5. Wang DL, Dai JS (2007) Theoretical foundation of metamorphic mechanisms and its synthesis. Chinese J Mech Eng 43(8):32–42
6. Ding, XL, Yang, Y (2009) Investigation of reconfiguration theory based on an assembly-circles artifact. In: ASME/IFToMM international conference on reconfigurable mechanisms and robots, London, 2009, pp 456–463

7. Li, SJ, Zhang, YL, Yang, S, Wang, HG (2009) Joint-gene based variable topological representations and configuration transformations. In: ASME/IFToMM international conference on reconfigurable mechanisms and robots, London, 2009, pp 348–354
8. Li SJ, Dai JS (2009) On the structure of metamorphic mechanisms based on augmented Assur Groups. Chinese J Mech Eng 46(10):19–28
9. Zhang, LP, Dai, JS (2009) Metamorphic techniques and geometric reconfiguration principles. In: ASME/IFToMM international conference on reconfigurable mechanisms and robots, London, (2009), pp 32–40
10. Lan, ZH (2010) Structure and evolution of planar metamorphic mechanisms. In: CCMMS, Shanghai, 2010, pp 104–106

# Chapter 7
# Biological Modeling Representations and Configuration Evolution Analysis of a Novel Metamorphic Loading Mechanism

**Ganwei Cai, Yuchen Pan, Hongzhou Wang and Jian S. Dai**

**Abstract** A novel electric loading mechanism is introduced in this paper by applying metamorphic mechanisms to the innovative design of loaders, which can change its topological structure to satisfy various kinds of demands during the working process. One driving motor can finish multiple tasks such as shoveling, rolling, lifting, and unloading. This paper introduces the characteristics during the evolutionary process first, based on which, two 0-DOF (Degree-of-Freedom) directional Assur groups are developed to be the new metamorphic cells as well as their representations. Configuration transformation and its degeneration ways during the genetic growth are studied. Biological modeling for working-phase mechanisms is proposed. The evolutionary synthesis process for generating the source-metamorphic loading mechanism and its representation is introduced through evolutionary operations.

**Keywords** Metamorphic mechanisms · Loader · Metamorphic cell · Evolutionary Operation

---

G. Cai (✉) · H. Wang
College of Mechanical Engineering, Guangxi University,
Nanning 530004, China
e-mail: caiganwei@163.com

Y. Pan
College of Civil Engineering and Architecture, Guangxi University,
Nanning 530004, China
e-mail: gxu_panyuchen@163.com

J. S. Dai
King's College London, University of London, Strand, London,
WC2R 2LS, UK
e-mail: jian.dai@kcl.ac.uk

J. S. Dai et al. (eds.), *Advances in Reconfigurable Mechanisms and Robots I*,
DOI: 10.1007/978-1-4471-4141-9_7, © Springer-Verlag London 2012

## 7.1 Introduction

A loader, as shown in Fig. 7.1, is a heavy equipment machine often used in construction, primarily used to load material such as asphalt, dirt, snow, rock, sand, and woodchips into a dump truck. But the hydraulic system of a traditional loader uses fluid for operating media with a large number of hydraulic components, resulting in high cost of manufacturing and maintaining. Without proper maintenance, hydraulic leaks are inevitable.

Metamorphic mechanisms were proposed during the study of reconfigurable packaging, which are the mechanisms with the ability to change their topological structures [1, 2]. The approaches to achieve metamorphic function can be divided into two categories: one is to change the number of links by link coincidence and self locking [3]. The other is to apply geometric constraint to joints to change the joint property as that introduced in Ref. [4]. Parise et al. [5] presented new ortho-planar metamorphic mechanisms in 2000. Liu and Yang [6] investigated the essence and characteristics of metamorphic mechanisms in 2004,. Dai and Rees Jones [7] presented the topology of variable mobility-configurations of metamorphic mechanisms and described transformation using matrix operations in 2005. Yan and Kuo [8, 9] investigated variable kinematic joints based on topological representations in the form of graphs and emphasized the meaning of variable topologies in 2006 and 2007, respectively. Zhang et al. [10] introduced the theories and methods of synthesis of metamorphic mechanisms based on biological modeling and evolution.

The potential use of this type of mechanisms is promising due to their multiple working phases, variable topology, and the changeable mobility. A novel type of electric loading mechanism, whose virtual prototype is shown in Fig. 7.1, is proposed in this paper by applying metamorphic mechanisms to the innovative design of loaders, which can change its topological structure to satisfy various kinds of demands during the working process. One driving motor can finish multiple tasks such as shoveling, loading, lifting, unloading, lowing and resetting. The motor is placed on the frame, problems such as low stiffness, high inertia of the linkages of traditional serial manipulators can be avoided. This paper introduces the characteristics during the evolutionary process first, based on which, two 0-DOF (Degree-of-Freedom) directional Assur groups are developed to be the new metamorphic cells as well as their representations. Configuration transformation and its degeneration ways during the genetic growth are studied. Biological modeling for working-phase mechanisms is proposed. The evolutionary synthesis process for generating the source-metamorphic loading mechanism and its representation is introduced after evolutionary operations.

## 7.2 Source-Metamorphic Mechanism

A source-metamorphic mechanism $^{S}M$ possesses the complete topological elements of all working-phases involved in a loading cycle of the loader, and has the highest level of mobility than each of its sub-working-phase mechanisms [10].

**Fig. 7.1** The traditional hydraulic loader

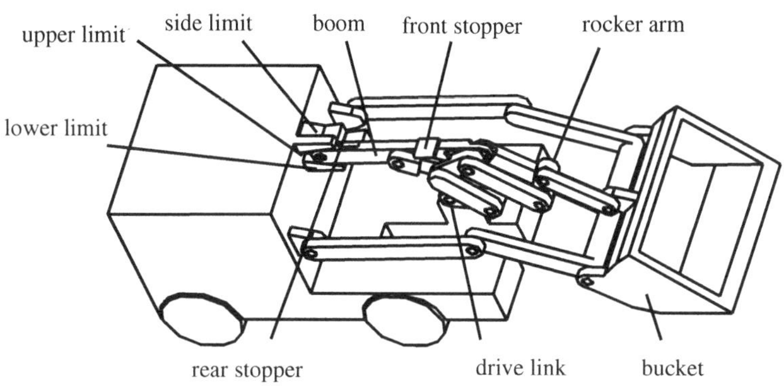


**Fig. 7.2** Virtual prototype of the novel loader

As is shown in Fig. 7.2 and Fig. 7.3, the motion of the boom (link 6) of the mechanism is limited by upper limit and lower limit which are both attached to the frame. The limits stop the boom and cause topological changes when the boom rotates to the limiting positions during the working process of the loader. There is a spindle with two stoppers inside the boom where a groove for the slider 9 is at the forepart. A compound revolute joint connects link 3, link 4 and link 5 together so that the drive link 2 is able to drive the boom and the slider via link 4 and link 5 during different working phases, respectively. Link 10 links the boom, the bucket (link 14), link 7 and link 8, of which the middle part connects to the rocker arm 12.

## 7.3 Working-Phase Mechanisms

Loading process of the loader consists of seven parts: shoveling, bucket rolling out, raising, dumping, bucket rolling in, lowing and bucket resetting. As introduced in another paper, the topological structures are identical in raising and

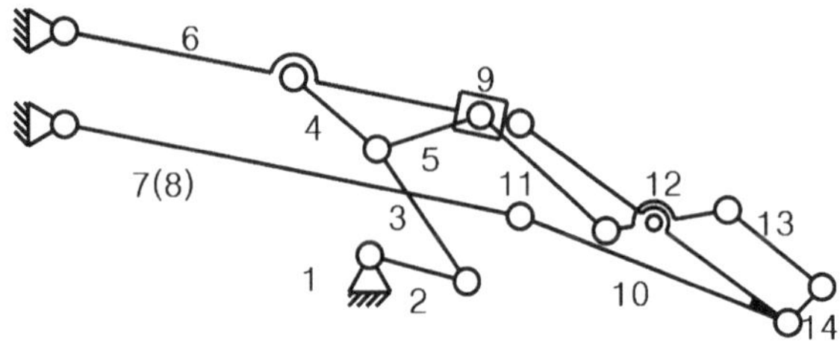


**Fig. 7.3** Sketch of the source-metamorphic mechanism

lowing phases when the other five share another topological structure. The sketches of working-phase mechanisms are depicted in Fig. 7.4, where Fig. 7.4a represents the structure of shoveling, bucket rolling out, dumping, bucket rolling in and bucket resetting phases, Fig. 7.4b stands for the raising and lowing process.

## 7.4 Novel Metamorphic Cells

### *7.4.1 Definition of Metamorphic Cell*

According to the definition given by Zhang, et al. a metamorphic cell is the simplest group constructed by successive links and connected joints with zero number of the degrees of freedom and cannot be divided further, which was introduced in Ref. [10]. For planar mechanisms, the cell is the variational Assur group with the orientation characteristic what exactly are: i) the metamorphic cell has the directionality. This can be deduced from the direction of successive links of a mechanism in the determined motion transformation. ii) the metamorphic cell has the calculability, metamorphic cells with the same pattern but different directions may have distinct calculation results. iii) the cell can be used as an independent and basic unit chain, which can be operated with aggregation and segregation evolution to adapt to metamorphic growth and variation Table 7.1.

### *7.4.2 Two Novel Metamorphic Cells*

Zhang et al. also introduced the representation for cell with the combination of kinematic pair symbols and geometric symbols with "$-$" representing a link and "$\prod$" the main part of a slider, where the direction of metamorphic cells represents the motion transformation sequence, and gave 17 examples of metamorphic cells. In this paper, we proposed two new expressions for the novel mechanism, where "$\perp$" stands for a link with three revolute pairs, "$+$" represents a bar that has four revolute pairs, "$\coprod$" means a linkage with two revolute pairs and one prismatic pair, and "$\pm$" is the expression of a link with three revolute pairs and a slider.

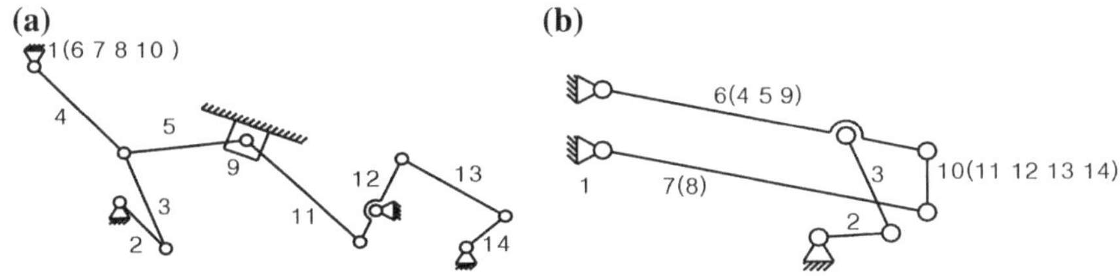


**Fig. 7.4** Sketches of working-phase mechanisms

**Table 7.1** The novel metamorphic cells

| Type | Expression | Scheme |
|---|---|---|
| $A_{18}$ | $(R-R\overset{R}{\perp}R-R-R)$ | 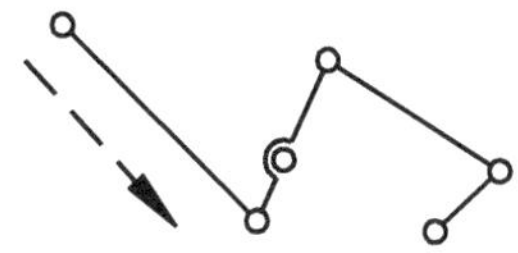 |
| $A_{19}$ | $R-R\left\langle\begin{matrix} R\overset{R}{\pm}R\overset{R}{+}R-R \\ P\ {}^{R}\!\searrow R \\ R\amalg R-R \end{matrix}\right.$ | 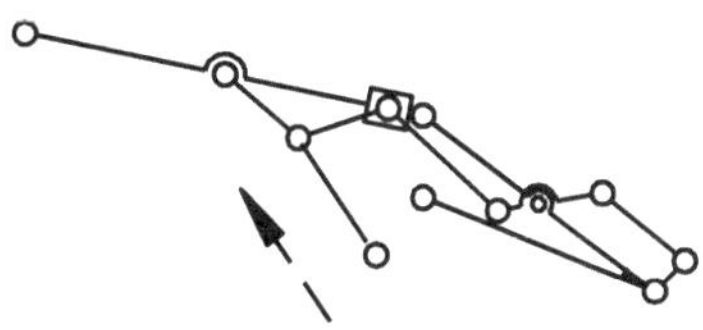 |

## 7.5 Biological Modeling for Source-Metamorphic Mechanism

The source-metamorphic mechanism with multiple degrees of freedom can be modeled biologically as an organism, say:

$$^{S}M=\left[a_i;\ A_i;\ a_f\right] \tag{7.1}$$

in which, $a_i$ represents the active gene, $A_i$ is the expression of metamorphic cell, and $a_f$ stands for the frame gene. From the proposed novel loading mechanism point of view, it can be depicted as follows, shown in Fig. 7.5.

## 7.6 Configuration Evolution of the Source-Metamorphic Mechanism

Working-phases of the loading mechanism include: shoveling, bucket rolling out, raising, dumping, bucket rolling in, lowing, bucket resetting. As introduced in another paper, the topological structures are identical in raising and lowing phases when the other five share another topological structure.

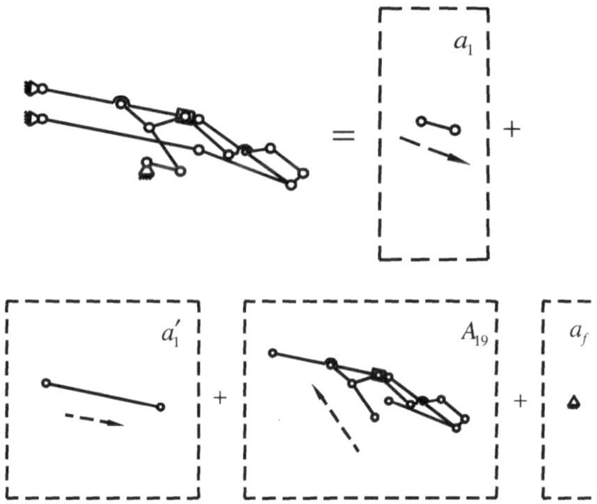


**Fig. 7.5** Biological building blocks for source-metamorphic mechanism

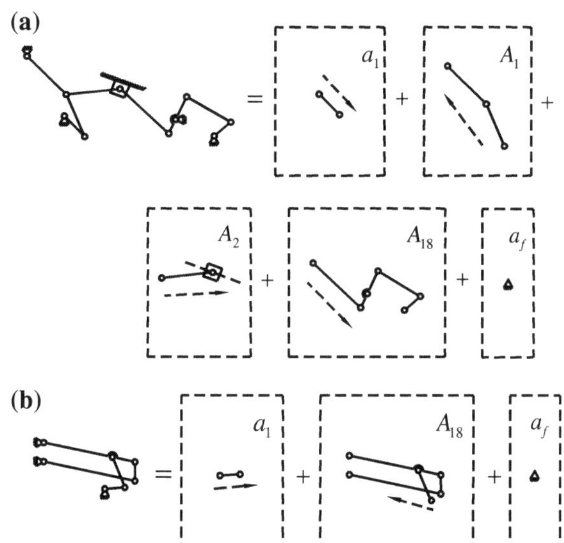


**Fig. 7.6** Biological modeling for the two working-phases mechanisms

Then the biological decomposition can be expressed as follows:

$$^1M = {}^2M = {}^4M = {}^5M = {}^7M = \left[a_1; A_1 \cup A_2 \cup A_{18}; a_f\right] \tag{7.2}$$

$$^3M = {}^6M = \left[a_1; A_{18}; a_f\right] \tag{7.3}$$

Generation of the source-metamorphic mechanism between two working-phase mechanisms is depicted in Fig. 7.6 where the two share the same active metamorphic gene but nonidentical metamorphic cells. The active metamorphic gene

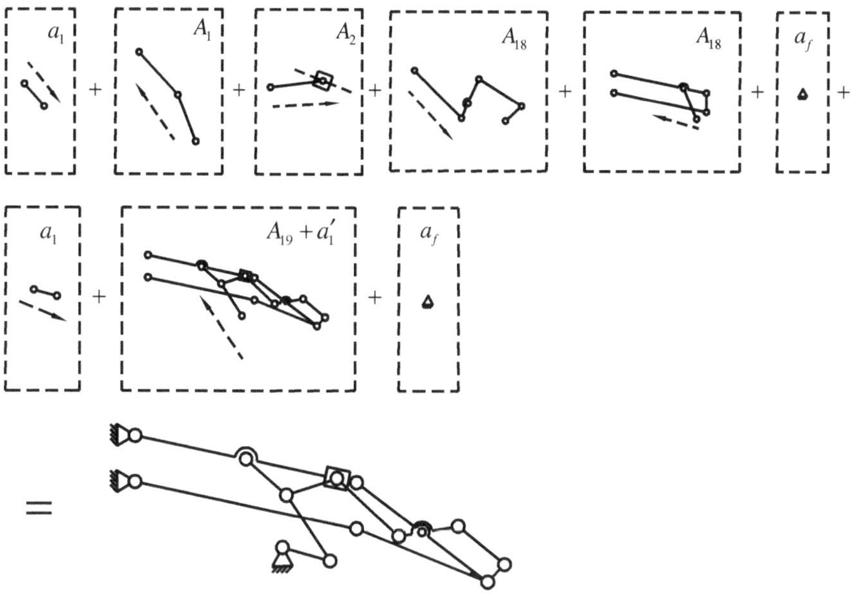


**Fig. 7.7** Evolutionary synthesis for generating the source-metamorphic mechanism

$a_1$, will be retained as the active gene of the source-metamorphic mechanism. So the operation of cell aggregation can be written as:

$$\begin{aligned} & A_1 \cup A_2 \cup A_{18} \cup A_{18} \\ & = (R-R-R) \cup (R-R \sqcap P) \\ & \cup (R-R \overset{R}{\perp} R-R-R) \cup (R-R \overset{R}{\perp} R-R-R) \\ & = R-R \left\langle \begin{matrix} R \overset{R}{\pm} R \overset{R-R}{+} R-R \\ P \quad {}^{R}\!\searrow R \nearrow \\ R \amalg R - \overset{|}{R} \end{matrix} \right. = A_{19} a_1 \end{aligned} \tag{7.4}$$

resulting in a residual single gene $a_1$ which will be put forward as an additional active gene. As a matter of fact, the annexed gene $a_1$ is not the same as the $a_1$ in the working-phases. For clearance, it will be written as $a_1'$in the following operations. In this specific case, the gene is provided by the gravity and the lower limit. Then we have

$$(A_1 \cup A_2 \cup A_{18}) \cup A_{18} = A_{19} a_1' \tag{7.5}$$

Consequently, the evolution process of source-metamorphic mechanism can be obtained, reads:

$$\begin{aligned} {}^{S}M &= {}^{1}M \cup {}^{2}M \cup {}^{3}M \cup {}^{4}M \cup {}^{5}M \cup {}^{6}M \cup {}^{7}M \\ &= {}^{1}M \cup {}^{1}M \cup {}^{3}M \cup {}^{1}M \cup {}^{1}M \cup {}^{3}M \cup {}^{1}M \\ &= {}^{1}M \cup {}^{3}M \\ &= \left[a_1; A_1 \cup A_2 \cup A_{18}; a_f\right] \cup \left[a_1; A_{18}; a_f\right] \\ &= \left\{(a_1 \cup a_1); [(A_1 \cup A_2 \cup A_{18}) \cup A_{18}]; \left(a_f \cup a_f\right)\right\} \\ &= \left[a_1; \left(A_{19}a_1'\right); a_f\right] = \left[a_1, a_1'; A_{19}; a_f\right] \end{aligned} \tag{7.6}$$

The evolutionary synthesis for generating the source-metamorphic mechanism is depicted in Fig. 7.7.

## 7.7 Conclusions

i) The characteristics during the evolutionary process was introduced first, based on which, two 0-DOF directional Assur groups were proposed to be the new metamorphic cells as well as their representations. ii) Configuration transformation and its degeneration ways during the genetic growth were studied. Biological modeling for working-phase mechanisms was proposed. After evolutionary operations, the evolutionary synthesis process for generating the source-metamorphic loading mechanism and its representation were introduced

**Acknowledgments** This work was jointly supported by the National Natural Science Foundation of China (No. 51075077), Guangxi Key Laboratory of Manufacturing System & Advanced Manufacturing Technology Project (09-007-05S013) and Teams for Innovation in the Construction of Talent Highlands in Guangxi Institutions of Higher Learning.

## References

1. Dai JS, Rees JJ (1999) Mobility in metamorphic mechanisms of foldable/erectable kinds. J Trans ASME Mech Des 121(3):375–382
2. Dai JS (1996) Conceptual study of the dexterous reconfigurable assembly and packaging system. Science and technology report. Unilever Research, Port Sunlight
3. Leonesio, M, Bianchi, G, Manara, P (2007) A general approach for self-locking analysis in closed kinematic chains. Proceedings of the 12th world congress in mechanism and machine theory, Besancon, 2007, pp. 141–147 )
4. Gan DM, Dai JS, Liao QZ (2010) Constraint analysis on mobility change of a novel metamorphic parallel mechanism. J Mech Mach Theory 45:1864–1876
5. Parise, JJ, Howell, LL, Magleby, SP (2000) Ortho-planar mechanisms. Proceedings 26th biennial mechanisms and robotics conference, Baltimore, 2000
6. Liu, C, Yang, T (2004) Essence and characteristics of metamorphic mechanisms and their metamorphic ways. Proceedings 11th world congress in mechanism and machine science, Tianjing, China, 2004, pp 1285–1288

7. Dai JS, Rees JJ (2005) Matrix representation of topological changes in metamorphic mechanisms. J ASME Mech Des 127(4):837–840
8. Yan HS, Kuo CH (2006) Topological representations and characteristics of variable kinematic joints. J ASME Mech Des 128(2):384–391
9. Yan HS, Kuo CH (2007) On the mobility and configuration singularity of mechanisms with variable topologies. J ASME Mech Des 129:617–624
10. Zhang LP, Wang DL, Dai JS (2008) Biological modeling and evolution based synthesis of metamorphic mechanisms. J ASME Mech Des 130:1–11

# Chapter 8
# On the Matrix Representation Methods for Variable Topology Mechanisms

**Lung-Yu Chang and Chin-Hsing Kuo**

**Abstract** Mechanism with variable topologies (MVTs) can change its topological structure because the kinematic types of certain joints are changeable during operation process. This work reviews and discusses the matrix methods for the representation and analysis of the topological structures of MVTs. First, we study several matrix representation methods for MVTs and compare their characteristics. Accordingly, these methods are classified into two different types based on the existence of joint types recorded. Furthermore, we discuss the applications of these methods for the structural decomposition and the topological homomorphism identification of MVTs. Last, we summarize the current limitations of the matrix representation methods for MVTs for future studies.

**Keywords** Reconfigurable mechanisms · Mechanisms with variable topologies · Matrix representation · Configuration decomposition · Topological homomorphism

## 8.1 Introduction

For presenting the topological structure [1] of a mechanism, the most common way is the use of graph theory. However, since the topological graphs cannot be directly applied for computer programming, matrix representations are in general more popular than the graph representations for the purpose of topological representation and analysis of mechanisms.

L.-Y. Chang · C.-H. Kuo (✉)
Department of Mechanical Engineering, Nation Taiwan University of Science and Technology, 43, Sect. 4, Keelung road, Taipei, 106, Taiwan
e-mail: chkuo717@mail.ntust.edu.tw

J. S. Dai et al. (eds.), *Advances in Reconfigurable Mechanisms and Robots I*,
DOI: 10.1007/978-1-4471-4141-9_8, © Springer-Verlag London 2012

During the past decade, several research groups have paid significant attention to the topological representations of the mechanisms with variable topologies (MVTs) [2]. They began to represent the topological structures of MVTs by using matrix representations in oder to anaylze their topologies computationally. For example, in 2005, Dai and Rees Jones proposed EU-matrix transformations [3] to formulate the topological changes of MVTs. In 2006, Yan and Kuo provided the directionality topology matrix [4, 5] which is particularly suitable for representing the topological structures of MVTs. In 2011, Slaboch and Voglewede presented the mechanism state matrix [6] which can record the mobilities and topological characteristics of variable topology mechanisms. By overview, Slaboch and Voglewede's method should suffice for identifing the characteristics of variable topology mechanisms. However, since MVTs have multiple topological structures, it is not easy to decompose all the sub-phase working mechanisms [7] under the circumstances when only a source mechanism is known. Therefore, it will be appreciated if the topology of the source mechanism of an MVT can be identified, following which all the sub-phase mechanisms can be computed accordingly.

The topological structures of MVTs are in general dominated by the changes of joint types. In this paper, we will revisit the present matrix representation methods for MVTs and classify them based on the existence of joint types recorded. Next, we will discuss the differences between these methods. Finally, we summarize the advantages of these methods and conclude their limits for the applications for topological analysis of MVTs.

## 8.2 Topological Characteristics of the Variable Topology Mechanisms

The topological structure of a mechanism stands for the numbers of links and joints, the joint types and the adjency and incidence between links and joints. A mechanism is said with variable topology or with changeable configuration if at least one of following conditions is met during operation [8]:

a. The numbers of effective links and/or joints are changeable.
b. The kinematic types of certain joints are changeable.
c. The adjacency and incidence among links and joints are changeable.
d. The relative arrangement between joints is changeable.

## 8.3 Matrix Representations for Variable Topology Mechanisms

Perhaps, the most computationally effective method for analyzing the topology of a mechanism is via the matrix methods. Since the key point of the change of topology of an MVT is due to the variable joints [5], we classfy the matrix representation methods based on the existence of joint type recorded.

## 8.3.1 Matrix Representations Without Joint Type Recorded

There are three ways for representing the topological structures of MVTs as matrices, that are, adjacency matrix, incidence matrix and EU-matrix transformation.

### 8.3.1.1 Adjacency Matrix

The *adjacency matrix* [9] is used to represent the adjacency relationship among links in a mechanism. The links in a mechanism are labeled 1 to $n$, and the adjacency matrix is an $n$ by $n$ symmetric matrix whose diagonal elements are all denoted by zeros. The element in an adjacency matrix **A** can be defined as follows:

$$a_{ij} = \begin{cases} 1, & \text{If link } i \text{ is connected to link } j \text{ by a joint} \\ 0, & \text{otherwise. (including } i = j) \end{cases} \tag{8.1}$$

In 2008, Lan and Du [10] modified adjacency matrix by introducing "$-1$" to describe a fixed kinematic pair [5]. By using this method, the dimension of the link adjacency matrix of the MVT can remain the same even if the number of effective links is changed during operation.

### 8.3.1.2 Incidence Matrix

The *incidence matrix* [9] is used to represent the incidence relationship among links and joints in a mechanism. First, the links and joints of a mechanism are numbered sequentially from 1 to $n$ and 1 to $m$, respectively. Then, each row will represent a link and each column will represent a joint. The element of an incidence matrix **B** can be defined as follows:

$$b_{ij} = \begin{cases} 1, & \text{If link } i \text{ is connected to joint } j \\ 0, & \text{otherwise.} \end{cases} \tag{8.2}$$

### 8.3.1.3 EU-Matrix Transformation

*EU-matrix transformation* [3] is a method that can manipulate its adjacency matrix when the topology of the mechanism changes. This method uses the concept of E-elementary matrix and U-elementary matrix in linear algebra to demonstrate the transformation of mechanism configurations. The method is the first attempt that uses matrix operation to algebraically manipulate the topological structure of MVTs. Suppose that link $i$ contacts with link $j$ ($i > j$) in an MVT. Then, the matrix $\mathbf{E}_j$ is a unit matrix $\mathbf{I}_{i*i}$ add the $j$th column with 0 and matrix $\mathbf{U}_{i,j}$ is a unit matrix $\mathbf{I}_{j*j}$ which element $a_{ij}$ be replaced to 1.

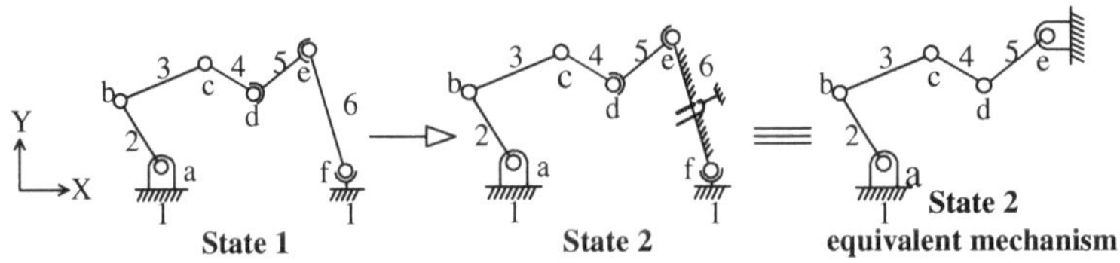


**Fig. 8.1** Two states of a six-bar linkage with variable topology

## 8.3.2 Matrix Representations with Joint Types Recorded

### 8.3.2.1 Directionality Topology Matrix

The *directionality topology matrix* [4] not only represents the adjacency among the links but also indicates the joint types in an MVT in every stage. The links of the MVT are labeled 1 to $n$ and the joints are $a$ to $z$. The numbers of the rows and columns of the matrix are equal to the number of links. The diagonal elements represent the labels of links. The upper triangle matrix represents the adjacency between links $i$ and $j$ which is expressed by *joint code* [5]. The lower triangle matrix represents the labels of the joints. If link $i$ is not connected to link $j$ by a joint, the corresponding matrix element $(i, j)$ and $(j, i)$ are written as zeros. Accordingly, the *joint sequences* [5] can be employed when a mechanism has multiple topology states.

For example, Eq. (8.5) is the directionality topology matrix of the two-stage six-bar linkage shown in Fig. 8.1:

$$\mathrm{M}_{DT} = \begin{bmatrix} 1 & J_{z,z}^{R,R} & 0 & 0 & 0 & J_{v,v}^{S,X} \\ a & 2 & J_{z,z}^{R,R} & 0 & 0 & 0 \\ 0 & b & 3 & J_{z,z}^{R,R} & 0 & 0 \\ 0 & 0 & c & 4 & J_{v,z}^{S,R} & 0 \\ 0 & 0 & 0 & d & 5 & J_{v,z}^{S,R} \\ f & 0 & 0 & 0 & e & 6 \end{bmatrix} \tag{8.3}$$

### 8.3.2.2 Mechanism State Matrix

*Mechanism state matrix* [5] perhaps is the most complete method so far for recording the topological characteristics of MVTs. The links of an MVT are numbered sequentially from 1 to $n$. Each row represents the state of mechanism and each column represents a link. If an MVT has $n$ links and $m$ states, the mechanism state matrix is an $m \times n$ matrix. The element in a mechanism state matrix $\mathbf{M}_{\mathrm{SM}}$ is defined as follows:

$$m_{sm} = \kappa_{(i,j)}\gamma_{1(i,j)\mu_{1(i,j)}}^{\lambda_{1(i,j)}}, \gamma_{2(i,j)\mu_{2(i,j)}}^{\lambda_{2(i,j)}} \cdots \gamma_{r(i,j)\mu_{r(i,j)}}^{\lambda_{r(i,j)}} \tag{8.4}$$

where $\kappa_{(i,j)}$ : in state i$\begin{cases} \alpha, \text{if link } j \text{ is a fixed link} \\ \beta, \text{if link } j \text{ is not a fixed link} \end{cases}$

$\gamma_{x(i,\ j)}$: a link number (1 to $n$) corresponding to a link connected to link $j$ by a joint, where $\gamma_{x+1(i,\ j)} > \gamma_{x(i,\ j)} > j$.

$x$: An integer from 1 to $r$.

$r$: The number of links connected to link $j$ for $\gamma_{x(i,\ j)} > j$.

$\lambda_{x(i,\ j)}$: The type of kinematic pair connecting link $j$ to link $\gamma_{x(i,\ j)}$.

$\mu_{x(i,\ j)}$: The orientation of the kinematic pair connecting link $j$ to link $\gamma_{x(i,\ j)}$.

For example, Eq. (8.5) is the mechanism state matrix for the six bar linkage in Fig. 8.1.

$$\mathbf{M}_{\mathrm{SM}} = \begin{bmatrix} \alpha 2_z^R,\ 6_v^S & \beta 3_z^R & \beta 4_z^R & \beta 5_v^S & \beta 6_v^S & \beta \\ \alpha 2_z^R,\ 6_v^X & \beta 3_z^R & \beta 4_z^R & \beta 5_z^R & \beta 6_z^R & \alpha \end{bmatrix} \tag{8.5}$$

where $R$ represents the revolve joint, $S$ indicates the spherical joint, $z$ is defined as the direction of rotation axis and $v$ is defined as the spherical motion.

## 8.4 Topological Homomorphism

A very familiar problem revealed in the structural analysis of mechanisms is the isomorphism identification between two mechanisms. Two mechanisms are said to be isomorphic if they have the same topological structure. Similarly, it is possible that two MVTs can have same topological structure(s) at one or more topology states. In such a situation, these two MVTs are said *topological homomorphic* [4]. More precisely, if MVT $m_a$ has one or more topological structures identical to that of another MVT $m_b$, we said that $m_a$ is topological homomorphic to $m_b$. Furthermore, if there is a sequentially one-to-one correspondence between the topological structures of $m_a$ and $m_b$, then $m_a$ and $m_b$ are said *mutually topological homomorphic.*

Clearly, if we want to represent and identify the topological homomorphism between two MVTs, we may decompose all the topological structures of the MVTs and identify their isomorphism in each topological structure step by step. For example, Fig. 8.2a–c shown three simple mechanisms (with two links and one joint only). It shows that Fig. 8.2c has five topology states whereas Fig. 8.2a and b both have only four. By inspecting their joint sequences [5], the MVTs in Fig. 8.2a–c possess the joint topologies with $J_{x,x,x,x}^{P,R,P,R}$, $J_{x,x,x,x}^{R,P,R,P}$, and $J_{x,x,x,x,x}^{R,P,R,P,R}$, respectively. It is obvious that there exists a sequentially one-to-one correspondence to the joint code of Fig. 8.2a and b, i.e., a cyclic $R_x$–$P_x$–$R_x$–$P_x$ period. So the two mechanisms shown in Fig. 8.2a and b are mutually topological homomorphic. On the other hand, we identify that Fig. 8.2a and b have four identical joint codes (mechanism topologies) which are corresponding to the parts of that in Fig. 8.2c. So the mechanisms shown in Fig. 8.2a and b are topological homomorphic to that in Fig. 8.2c.

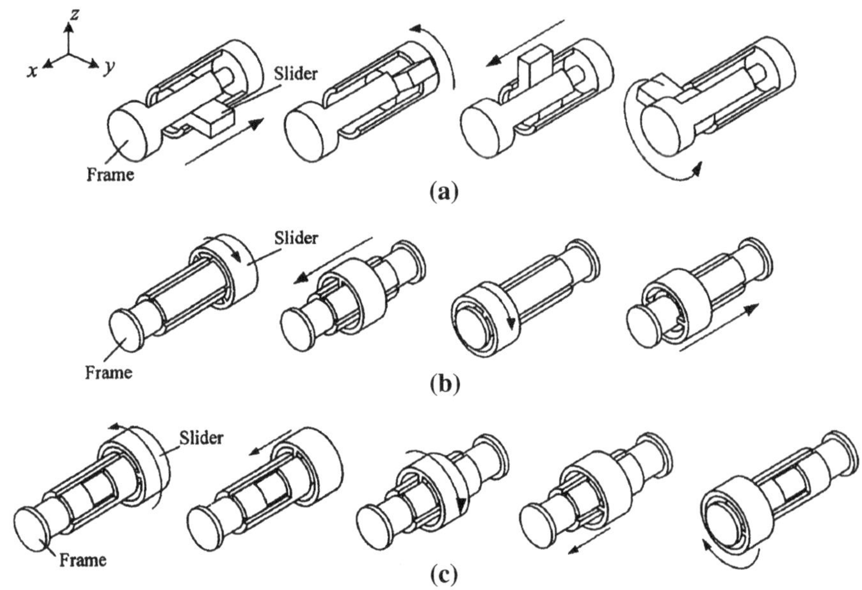


**Fig. 8.2** Three topological homomorphic mechanisms

**Fig. 8.3** Variable joint with two types [11]

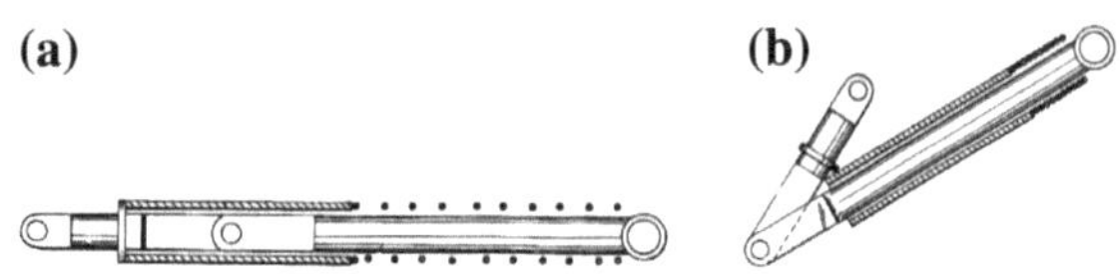


Let us examine the topological homomorphism by another case. Figure 8.3 shows a variable joint with two states, one with two links connected by a prismatic joint, Fig. 8.3a, and the other with three links connected by one prismatic joint and one revolute joint, Fig. 8.3b. Based on using this joint, Fig. 8.4 shows a mechanism for folding bicycle with two topology states. Figure 8.4a is state one when bicycle is in use. When the bike is to-be-folded, its topologies will be changed to that as shown in Fig. 8.4b. The corresponding kinematic chains of both states for the bicycle folding mechanism are depicted in Fig. 8.5. On the other hand, Fig. 8.6 illustrates an airplane loading gear mechanism. Figure 8.6a is the folding state, while Fig. 8.6b is a dead point configuration for the mechanism. The corresponding kinematic chains of both states for the airplane loading gear mechanism are depicted in Fig. 8.7. By comparing the kinematic chains in Figs. 8.5 and 8.7, it shows that the chains in Figs. 8.5a and 8.7a are identical, which implies that these two mechanisms are topological homomorphic.

Table 8.1 provides a comparison for the matrix representation methods aforementioned. It shows that only EU-matrix transformation method can be used for structural decomposition, but it cannot indicate the changes of the kinematic types

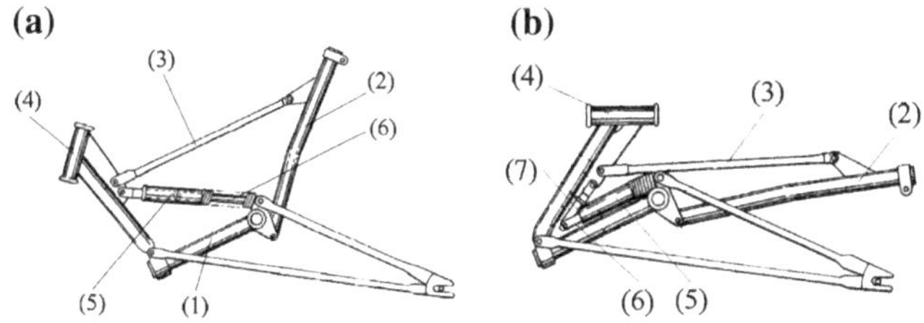


Fig. 8.4 A bicycle folding mechanism [11]

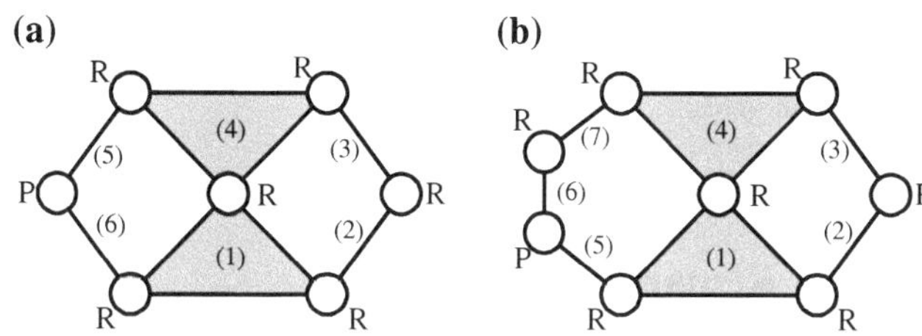


Fig. 8.5 The kinematic chains of the bicycle folding mechanism

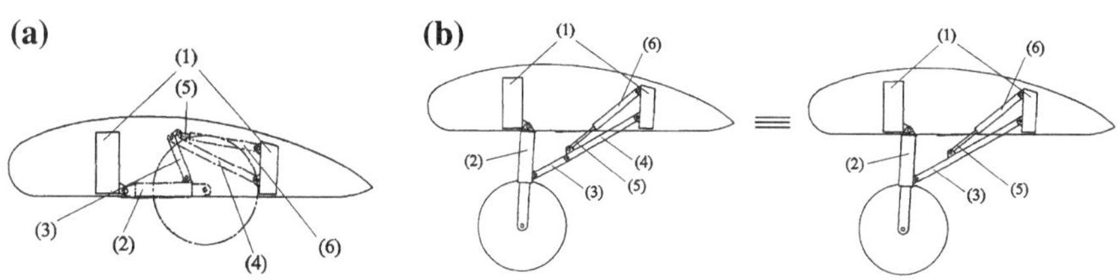


Fig. 8.6 An airplane loading gear mechanism [12]

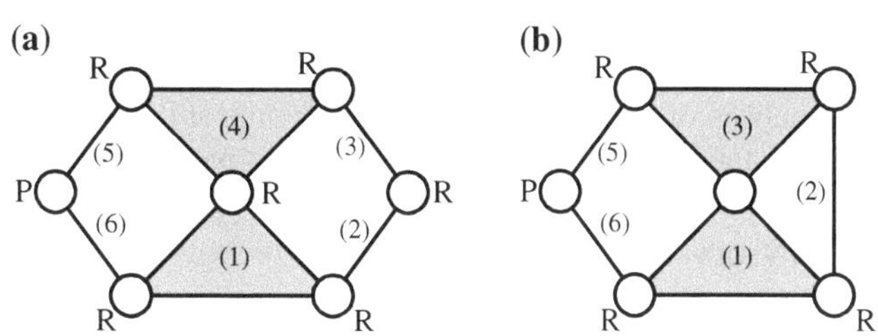


Fig. 8.7 The kinematic chains of the airplane loading gear mechanism

of joints. Besides, only directionality topology matrix can be applied for analyze the topological homomorphism of MVTs, while other methods do not consider the properties of topological homomorphism among MVTs.

**Table 8.1** Information provided by the matrix representation methods for MVTs

| Matrix representations | (i) | (ii) | (iii) | (iv) | (v) | (vi) |
|---|---|---|---|---|---|---|
| Adjacency matrix | ○ | | | | | |
| Improve adjacency matrix | ○ | ○ | | | | |
| Incidence matrix | ○ | | | | | |
| EU-matrix transformation | ○ | | | | * | |
| Directionality topology matrix | ○ | | ○ | ○ | | • |
| Mechanism state matrix | ○ | ○ | ○ | ○ | | |

(i) The connectivity of the links and joints
(ii) The fixed links
(iii) The types of joints
(iv) The kinematic orientations of the joints
(v) Mechanism topological structures decomposition
(vi) Topological homomorphism
(*) Cannot record joint type
(•) Can only be used on simplest mechanisms

## 8.5 Concluding Remarks

This paper compares all the existing matrix representation methods for MVTs, including adjacency matrix, incidence matrix, EU-matrix transformations, directionality topology matrix and mechanism state matrix. It shows that although EU-matrix transformation method can consider multiple topological structures when the number of effective links is changed, this method is still not perfect for representing the topological structures of an MVT since the joint types are not recorded by this method. On the other hand, although directionality topology matrix can record the joint types in each state, this method cannot be applied for the topological decomposition of MVTs. Mechanism state matrix seems to be the most complete method so far for recording the topological characteristics of an MVTs. However, it still cannot be utilized for completely decomposing the topological structures in each state of the variable topology mechanisms. Topological homomorphism is a newly arising issue for MVTs, and we need to know all topology structures of an MVT in order to identify its topological homomorphism. However, the analysis of topological homomorphism is still not considered by all the matrix representation methods we discussed in this paper. Hence, a new matrix method that can be used to recognize all the topological structures of MVTs including the topological homomorphism is worth to be developed.

## References

1. Yan H-S (2006) Creative design of mechanical devices. Springer, Singapore
2. Dai JS, Jones RJ (1999) Mobility in metamorphic mechanisms of foldable/erectable kinds. ASME J Mech Des 121:375–382
3. Dai JS, Jones RJ (2005) Matrix representation of topological changes in metamorphic mechanisms. ASME J Mech Des 127:837–840

4. Yan HS, Kuo C-H (2006) Representations and identifications of structural and motion state characteristics of mechanisms with variable topologies. Trans Can Soc Mech Eng 30:19–40
5. Yan H-S, Kuo C-H (2006) Topological representations and characteristics of variable kinematic joints. ASME J Mech Des 128:384–391
6. Slaboch BJ, Voglewede PA (2011) Mechanism state matrices for planar reconfigurable mechanisms. ASME J Mech Robotics 3:011012(1-7)
7. Zhang L, Dai JS (2009) Reconfiguration of spatial metamorphic mechanisms. ASME J Mech Des 1:011012
8. Kuo C-H, Dai JS, Yan H-S (2009) Reconfiguration principles and strategies for reconfigurable mechanisms. In: ASME/IFToMM international conference on reconfigurable mechanisms and robots (ReMAR2009), pp 1–7, London
9. Tasi L-W (2001) Mechanism design: enumeration of kinematic structures according to function. CRC Press, Boca Raton
10. Lan ZH, Du R (2008) Representation of topological change in metamorphic mechanisms with matrices of the same dimension. ASME J Mech Des 130:074501(1-4)
11. Takamiya K, Ishibashi H (1984) Frame for folding bicycle. United States Patent No. US4460192
12. Onions JH (1940) Shock absorber for aircraft. United States Patent No. 2186266

# Chapter 9
# Geometric Constraints Resulting From Puzzles

**Liping Zhang and Jian S. Dai**

**Abstract** Wooden puzzles consisting of interlocking assembly of notched sticks are often referred to as bar-puzzles, sometime known as the Chinese Puzzles or Chinese Cross. By far the more familiar of three-dimensional puzzles in general is the multi-piece bar. But there does not appear to be any evidence that the idea of its mechanism property and any configuration analysis originated. To this purpose, this paper proposes a static and discrete reconfiguration theory of geometric puzzles for modeling the topology changement as Put Together, Take Apart, Sequential Movement and various others. The partition and assembly process analysis aims to extract the kinematic chains as links and joints. The puzzle unlocking leads to configuration constraints rearrangement problems which accompanying pieces of bars self-grouped as defined reconfiguration links and joints. The puzzle solutions are first described as reconfigurable topology mechanism and the constrained mobility is analyzed based on an ingenious and distinctive reconfiguration property.

**Keywords** Interlocking · Mobility · Reconfiguration · Constraint · Mechanism

L. Zhang (✉)
Department of Mechanical Engineering, Dalian Polytechnic University,
116034 Dalian, China
e-mail: lipingzhang3@163.com

J. S. Dai
Tianjin University, Tianjin, People's Republic of China
e-mail: jian.dai@kcl.ac.uk

J. S. Dai
King's College London, University of London, Strand,
London, WC2R 2LS, UK

J. S. Dai et al. (eds.), *Advances in Reconfigurable Mechanisms and Robots I*,
DOI: 10.1007/978-1-4471-4141-9_9, © Springer-Verlag London 2012

## 9.1 Introduction

Nearly everyone may have a few amusements among the childhood treasures based on the simple principle of taking things apart and fitting them back together again [1]. Construction pastimes in the form of geometrical assembled puzzles have a universal appeal that transcends all cultural boundaries and all age levels. The most geometrical puzzles are based on mathematical principles known. The three-dimensional puzzles in the so-called bar or notched bars are certainly the most popular.

Although instances of puzzles have constant sizes and structures, the natural generalization of many of them to sizes based on some parameter are NP-complete [2, 3]. Some are presented as Instant Insanity [4] and sliding block puzzles like the 15-puzzle [5]. Different puzzles and magics are described by Demaine [6, 7]. The design methodology is discussed on polyform [8]. Hyparhedra and origami folding for the mathematical surface approximation are introduced in [9]. Puzzles lead to research problems while puzzle solving and how to compute their algorithmically are arising the fun and interesting problems.

In stead of the mathematical recreations in general, the shadow of every geometrical models looms the mechanism structure and its configuration relations. Within every tortuous puzzle solution lurks the mobility regeneration under configuration variation or internal constraints changed.

This paper proposes a static reconfiguration theory of geometric puzzles for modeling the topology changement as Put Together, Take Apart, Sequential Movement and various others. The partition and assembly process aim to extract kinematic sequence with links and joints. By movement analysis, the kinematic chain is specified according the partition pattern.

The reason we called the wooden bar structure as puzzle is in its complexity of the interlocking properties. This paper reveals its interlocking property as configuration constraints including many passive constraints. The configuration constraints are not only determined by the size and dimension, but also have relations with the link and joint reconfigurations.

Then the paper discloses the mechanism modeling by two partition methods. A simple partition method utilizes bar pieces being removed one by one in order. An improved partition method allows bar pieces being taken apart in a bar-group with self-rearrangement and regeneration [10, 11]. The first is according to conventional mechanism and the second is a meta-source for generation of reconfigurable mechanisms [12, 13] with regenerated links and joints developed by bar pieces reunited. In particular, the puzzles are first described as reconfigurable mechanism and mobility is analyzed based on its ingenious metamorphic configuration property.

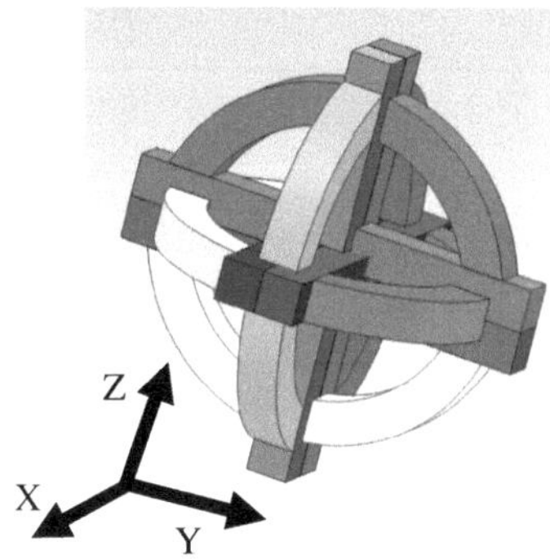


**Fig. 9.1** A 12-piece puzzle

## 9.2 The Complexity of 12-Piece Puzzles

A proper analysis on the geometrical puzzles should probably begin with partition or assembly. A proper analysis on the geometrical puzzles should probably begin with partition or assembly process.

### *9.2.1 Structure*

This 12-piece puzzle consists of six arcs and six stick-bars as illustrated in Fig. 9.1. The green bar is a proper cuboid without any spur-like appearance and other five bars are notched sticks. The 12 bars are labeled according to different structures.

The bar pieces can be divided into two groups as notchable and unnotchable pieces. Piece 1 is in cubic block (unnotchable) without producing any voids. It is important in configuration selflocking and topology closing. The duplication of identical pieces leads to pieces having an axis if symmetry. Six pieces with same arc structure are identical with stick notched structure.

The 12 pieces are arranged symmetrically in three mutually perpendicular intersecting pairs. The structure can be chosen in set of 12 pieces.

### *9.2.2 Kinematic Links*

The wooden puzzling has its own configuration while fixed together. This structure can be seen as a static mechanism with mobility 0. The constructive bars can be disconnected as mechanism links and joints being interlocked.

To analysis kinematic chain, its 12 constructive pieces is illustrated in Fig. 9.2.

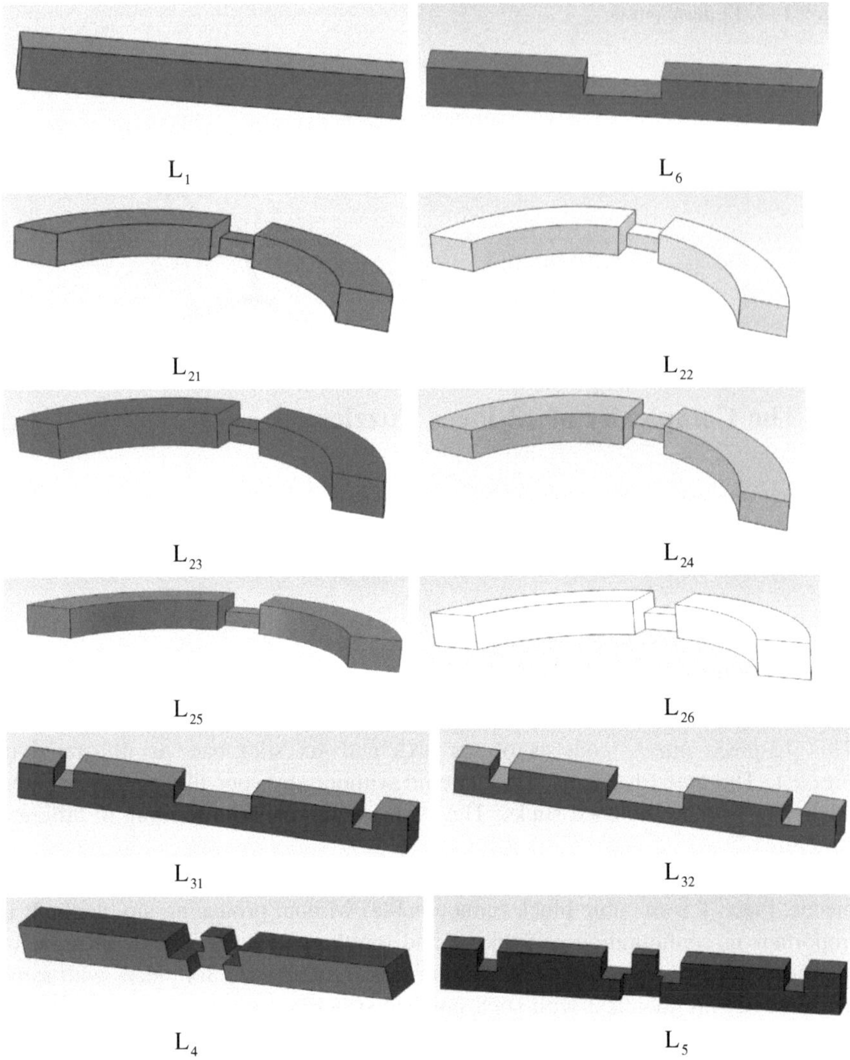


**Fig. 9.2** 12-piece link family

## 9.3 Partition Scheme and Mechanism Configuration

Solving puzzles based on partition problem is challenging. Partition solution of a puzzle is not unique in practice. Actually, most good puzzles have only few pieces but are still very hard. By tracing the partition process, the modeling of mechanism configuration can be achieved.

**Fig. 9.3** A prismatic joint generated by bar piece splicing

### *9.3.1 A Simple Partition Scheme and Kinematic Joint*

**A simple scheme**: basically, the simple partition can be interpreted as one piece bar being taken away from the puzzle at one partition operation.

**Kinematic joint**: this operation generates a relative motion of the adjacent pieces and forms a static kinematic joint here. As illustrated in Fig. 9.3, if the red piece is removed from intersected, then a translation motion is achieved to generate a prismatic joint.

### *9.3.2 An Improved Scheme and a Reunited Link*

**An improved scheme**: in some special cases, an improved partition can be interpreted as multi-piece bars being reorganized together and taken away in a group to move a slight space temporarily to permit the following single piece out. This improved scheme is to ensure the previous simple partition scheme could be well executed.

The improved scheme is aim to find a flexible route to assist puzzles partition. This additional understanding gives us a lot of insight into puzzle design, which leading to interesting puzzles.

**An intermediate reconfiguration link** $\Delta L$: in this operation, multiple bars are gathered as one intermediate reunited link to move together as one link.

It is very important to find the available and proper internal link to solve the partition problem. Solving a puzzle is like solving a research problem. Both require the right cleverness to see the problem from the right angle and then the careful execution of that idea until you find a solution. By this self-rearrangement and self-organization method, the formed intermediate link can sever as a container for topology changement and geometric constraint regeneration. To some extent, it presents the difficulty and essence of a puzzle solution.

As illustrated in Fig. 9.4, after the bars $L_1$, $L_{21}$, $L_{22}$ and $L_{31}$ being removed from structure one by one, then the partition of the rest part of bar pieces (as shown

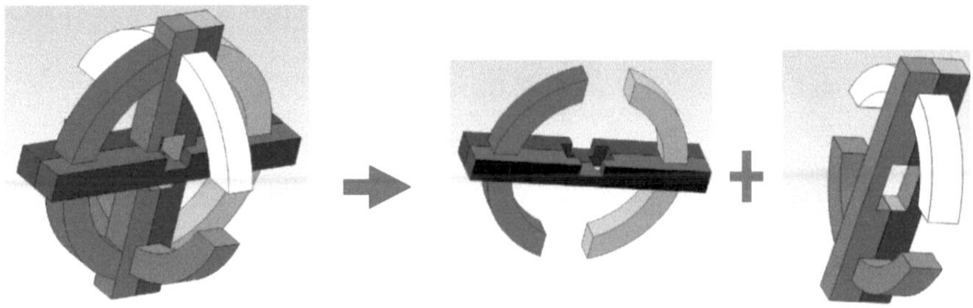

$\Delta L$ ($L_4L_{23}L_{24}$ $L_5$) $L_6$,$L_{25}$,$L_{26}$,$L_{32}$

**Fig. 9.4** An intermediate reconfiguration link $\Delta L$ ($L_4L_5L_{23}L_{24}$)

in Fig. 9.4) needs to utilize an improved partition scheme. Since we cannot find any piece bar can be taken way anymore at this time, but if we move out the links $L_4$, $L_{23}$, $L_{24}$ and $L_5$ as one piece $\Delta$L, a vacant space is released, then piece $L_4$ can have an out way.

### 9.3.3 Partition Scheme for 12-Piece Bars

The partition scheme for this 12-piece bar has multiple ways. Herein, we take a more simple way to illustrate the sequential movements with bars decomposition process.

Combining the previous scheme, the static kinematic chain according to the partition scheme for 12-piece bar is set up as shown in Table 9.1.

The first row of Table 9.1 indicates the move out order of bars (links). This means that the *i*th column presents *i*th bar being taken away. The first column in Table 9.1 is the transposed of the first row elements.

The intersection zone element *ij* of row and column presents the movements generated by the *j*th column bar being taken away. In this bar mechanism, the kinematic joints are all translational joints in three-dimension which denoted as 'T'. To distinguish the directions and motions, the prismatic joint is presented by subscript '*x*, *y*, or *z*,' according to the coordinate. 'T*x*' means this partition operation is a translation along the *x* axis.

Take the second column as an example, when the link $L_1$ is moved out, this arises eight movements. Since link one is connected with eight links (as $L_{21}$, $L_{22}$, $L_{31}$, $L_4$, $L_5$, $L_6$, $L_{25}$ and $L_{32}$), when $L_1$ is taken apart, this can be seen as eight pair movements as ($L_1L_{21}$), ($L_1L_{22}$), ($L_1L_{31}$), ($L_1L_4$), ($L_1L_5$), ($L_1L_6$), ($L_1L_{25}$) and ($L_1L_{32}$). At the same time, these total nine links are connected by the kinematic joints as a sub-chain which incurred by the sequential movements.

**Table 9.1** Sequential mechanism movements for a 12-piece puzzle

| Links | $L_1$ | $L_{21}$ | $L_{22}$ | $L_{31}$ | $L_{4,\ 2\text{-}3,\ 2\text{-}4,\ 5}$ | $L_4$ | $L_{23}$ | $L_{24}$ | $L_5$ | $L_6$ | $L_{25}$ | $L_{26}$ | $L_{32}$ |
|---|---|---|---|---|---|---|---|---|---|---|---|---|---|
| $L_1$ | | | | | | | | | | | | | |
| $L_{21}$ | Ty | | | | | | | | | | | | |
| $L_{22}$ | Ty | | | | | | | | | | | | |
| $L_{31}$ | Ty | Tz | Tz | | Tz | | | | | | | | |
| $L_{4,\ 2\text{-}3,\ 2\text{-}4,\ 5}$ | | | | | | | | | | | | | |
| $L_4$ | Ty | Tz | | | | | | | | | | | |
| $L_{23}$ | | | | | | Ty | | | | | | | |
| $L_{24}$ | | | | | | Ty | | | | | | | |
| $L_5$ | Ty | | Tz | | | | Ty | Ty | | | | | |
| $L_6$ | Ty | | | Ty | Tz | Ty | | Ty | Ty | | | | |
| $L_{25}$ | Ty | | | Ty | | | | | | Tz | | | |
| $L_{26}$ | | | | | | | | | | Tz | | | |
| $L_{32}$ | Ty | | | Ty | Tz | Ty | Ty | | Ty | Tz | Tx | Tx | |

## 9.4 Kinematic Property of Reconfigurable Link $\Delta L$

Known this bar puzzling has 12 pieces, by combining the kinematic constraints during the partition, so rules for modeling chain of this bar mechanism is available.

A more unique rule for the puzzle mechanism generation is that, in some cases, the mechanism links or joints are not equal to the pieces of the puzzles. The flexibility and complexity of puzzle partition is reflected by the bar piece combinations or joint regeneration.

Just as summarized in the Table 9.1, to be noted the $\Delta L$ ($L_{4,\ 2\text{-}3,\ 2\text{-}4,\ 5}$), is 4 bar pieces ($L_4$, $L_{23}$, $L_{24}$ and $L_5$) combined as one piece to move out a bit space together. This internal generated link can be identified as a reconfigurable link with self-reconnected and self-reunited bars.

It is absolutely necessary for the puzzles have some intrinsic reconfigurable links or joints to change the current static topology and put forth a fresh way to generate kinematic chain. By this reconfigurable link, the whole mechanism constraint properties can change totally to realize interlocking.

For the self-interlocking property analysis of puzzle, the partition is an unlocking process to disconnect the bars. The conventional idea is to remove the bars one by one. Since the partition is to change a structure with mobility zero, i.e. an interlocking solid, to different open chains with variable mobility. The reconfigurable link plays an important role in this configuration rearrangement.

As illustrated in Table 9.2, when link $\Delta L(L_{4,\ 2\text{-}3,\ 2\text{-}4,\ 5})$ is moved out, three translational motions are incurred by $\Delta L - L_{31}$ (Tz), $\Delta L - L_6$ (Tz) and $\Delta L - L_{32}$ (Tz).

From the Table 9.2, we can see that a internal produced link $\Delta L$, accompanying three fresh joints as $Tz$ ($\Delta L\ L_{31}$), $Tz$ ($\Delta L\ L_6$), $Tz$ ($\Delta L\ L_{32}$). The regenerated link and joints totally changed the puzzle topology and inside constraints.

**Table 9.2** A reconfigurable link with its movements

| Links | $L_1$ | $L_{21}$ | $L_{22}$ | $L_{31}$ | $L_{4, 2\text{-}3, 2\text{-}4, 5}$ | $L_4$ | $L_{23}$ | $L_{24}$ | $L_5$ | $L_6$ | $L_{25}$ | $L_{26}$ | $L_{32}$ |
|---|---|---|---|---|---|---|---|---|---|---|---|---|---|
| $L_1$ | | | | | | | | | | | | | |
| $L_{21}$ | | | | | | | | | | | | | |
| $L_{22}$ | | | | | | | | | | | | | |
| $L_{31}$ | | | | | Tz | | | | | | | | |
| $L_{4, 2\text{-}3, 2\text{-}4, 5}$ | | | | | | | | | | | | | |
| $L_4$ | | | | | | | | | | | | | |
| $L_{23}$ | | | | | | | | | | | | | |
| $L_{24}$ | | | | | | | | | | | | | |
| $L_5$ | | | | | | | | | | | | | |
| $L_6$ | | | | | Tz | | | | | | | | |
| $L_{25}$ | | | | | | | | | | | | | |
| $L_{26}$ | | | | | | | | | | | | | |
| $L_{32}$ | | | | | Tz | | | | | | | | |

## 9.5 Configuration Constraint Analysis

By the interlocking configuration and reconfigurable link analysis, the constraints of the kinematic chain are derived from the movements based on their partition sequence. Take the 12-piece puzzle as an example.

### *9.5.1 Simplify Spatial Mobility to Planar Mobility*

As indicated in the Table 9.1, there are 13 links in this bar mechanism model. However, the kinematic joints are all prismatic joints along $x$–$y$–$z$ directions.

Based on the mobility calculation, iff the prismatic joints exist in the chain, then the three dimensional rotations are restricted. This means six dimensional motions (three dimensional rotations and three dimensional translations) of spatial rigid body are reduced to three dimensional translations only. The complete rotation constraints lead to the spatial mobility formula converts to the planar mobility formula. It has

$$\mathrm{F} = 3n - 2P_L \tag{9.1}$$

In Eq. (9.1), mobility F equals to the number of the degrees of freedom for the active links (denoted as $n$) deducted the number of the constraints (indicated as $P_L$).

A rigid body with three dimensional translation motions has 3 degrees of freedom. Since the prismatic joint has only one degree of freedom, i.e., a joint can only slide along one direction $x$ (or $y$ or $z$), this leads to the other two direction motions are constrained.

**Table 9.3** The passive constraints

| Links | $L_1$ | $L_{21}$ | $L_{22}$ | $L_{31}$ | $L_{4,\ 2\text{-}3,\ 2\text{-}4,\ 5}$ | $L_4$ | $L_{23}$ | $L_{24}$ | $L_5$ | $L_6$ | $L_{25}$ | $L_{26}$ | $L_{32}$ |
|---|---|---|---|---|---|---|---|---|---|---|---|---|---|
| $L_1$ | | | | | | | | | | | | | |
| $L_{21}$ | T*y* | | | | | | | | | | | | |
| $L_{22}$ | T*y* | | | | | | | | | | | | |
| $L_{31}$ | T*y* | T*z* | T*z* | | Tz | | | | | | | | |
| $L_{4,\ 2\text{-}3,\ 2\text{-}4,\ 5}$ | | | | | | | | | | | | | |
| $L_4$ | T*y* | T*z* | | | | | | | | | | | |
| $L_{23}$ | | | | | | T*y* | | | | | | | |
| $L_{24}$ | | | | | | T*y* | | | | | | | |
| $L_5$ | T*y* | | T*z* | | | | T*y* | T*y* | | | | | |
| $L_6$ | T*y* | | | T*y* | Tz | T*y* | | T*y* | T*y* | | | | |
| $L_{25}$ | T*y* | | | T*y* | | | | | | T*z* | | | |
| $L_{26}$ | | | | | | | | | | T*z* | | | |
| $L_{32}$ | T*y* | | | T*y* | Tz | T*y* | T*y* | | T*y* | T*z* | T*x* | T*x* | |

## *9.5.2 Passive Constraints and Mobility*

By added the reunited link $\Delta L$, there are 13 links in this 12-piece bar mechanism. Pick up one link as frame, then we have 12 active links finally. It has

$$n = \left( \sum_{i=1}^{12} L_i + \Delta L \right) - 1 = 13 - 1 = 12 \tag{9.2}$$

As for the kinematic joint motions, we can see there are multiple passive constraints exist. Take row '$L_{31}$' as an example, it has four movements T*y*, T*z*, T*z* and T*z*. There are triplicate prismatic motions 'Tz', while this goes with two passive constraints in the same motion type and direction. So the four movements T*y*, T*z*, T*z* and T*z* can be seen as two dependent motions T*y* and T*z*, the other T*z* and T*z* are redundant.

These passive constraints are all highlighted in Table 9.3 in orange.

Sum up the passive constraints $C$ (=15) and deduct them from the existing motions $M$ (=33), it has

$$P_L = C - M = 33 - 15 = 18 \tag{9.3}$$

Further for the mobility analysis, considering the partition process, its final bur mechanism mobility can be calculated as the following

$$\mathrm{F} = 3n - 2P_L = 3 * 12 - 2 * 18 = 0 \tag{9.4}$$

This mobility is calculated under partition constraints and configuration constraints of the interlocking bars. The process shows the interesting puzzling may have multiple ways to change because of its self-rearrangement while moving with more and more passive constraints.

## 9.6 Conclusions

A static constraint based reconfiguration analysis method of a geometric puzzle is developed in this paper. Discrete topology are easy to combine describing the kinematic chain while puzzle partition. Simple partition way may lead to conventional mechanism by bar piece take apart or put together one by one in sequence. Improved partition way may bring about variable configuration by bar piece take apart or put together in self-regroup. The second way is aim to change the current interlocking condition and put forth an novel method to unlocking the trouble constraints. This leads to an available topology condition of an interlocking structure to mechanism reconfiguration analysis.

The discrete movements of bar pieces are employed for the generation checking of kinematic joints. Links or reconfigurable links of every possible connection in a topology is combined with its puzzle model partition method directly. This means that for the same puzzle, topology can be varied corresponding to different partition results. In view of this, a puzzle is a source of topology reconfiguration.

Apart from the arisen of novel reconfigurable links, the ingenious way to unlock the self-locking structure is to employ the passive or redundant constraints flexibly. This makes such kinds of constraint checking and also improves the mobility efficiency. The topology analysis results this method are well defined. These are practical for further shape and size or for direct design and fabrication.

## References

1. Coffin S (2007) Geometric puzzle design. A K Peters Ltd., Wellesley
2. Garey MR, Johnson DS, Tarjan RE (1976) The Planar Hamiltonian circuit problem is NP-complete. SIAM J Comput 5(4):704–714
3. Robertson E, Munro I (1978) NP-completeness, puzzles, and games. Utilitas Math 13:99–116
4. Garey MR, Johnson DS (1979) Computers and intractability: a guide to the theory of NP-completeness. W. H Freeman, New York
5. Ratner D, Warmuth M (1990) Finding a shortest solution for the N_N-extension of the 15-puzzle is intractable. J Symb Comp 10:111–137
6. Demaine ED (2001) Playing games with algorithms: algorithmic combinatorial game theory. In: Proceedings of the mathematical foundations of computer science, pp 18–32
7. Demaine ED, Demaine ML (2006) Puzzles, art, and magic with algorithms. Theory Comput Syst 39(3):473–481
8. Demaine ED, Demaine ML, Eppstein D, Frederickson DG, Friedman E (2005) Hinged dissection of polyominoes and polyforms. Comput Geom Theory Appl 31(3):237–262
9. Jackson P (1989) Origami: a complete step-by-step guide. Hamlyn Publishing Group Ltd, London
10. Dai JS, Jones RJ (1998) Mobility in metamorphic mechanisms of foldable/erectable kinds. 25th ASME biennial mechanisms and robotics conference, Atlanta, Sept 1998
11. Wohlhart K (1996) Kinematotropic linkages. In: Lenarcic J, Parenti-Castelli V (eds) Recent advances in robot kinematics, Kluwer Academic, Dordrecht, pp 359–368

12. Zhang LP, Wang DL, Dai JS (2008) Biological modeling and evolution based synthesis of metamorphic mechanisms. Trans ASME J Mech Des 130(7):072303-1-11
13. Zhang LP, Dai JS (2009) Spatial metamorphic mechanisms reconfiguration. Trans ASME J Mech Robotics 1:011012-1-8

# Chapter 10
# Reconfigurable Assembly Approach for Wind Turbines Using Multiple Intelligent Agents

**C. Deters, H. A. Wurdemann, Jian S. Dai, L. D. Seneviratne and K. Althoefer**

**Abstract** This paper investigates the use of an agent based assembly strategy for a wind turbine hub. The manual assembly procedure for a wind turbine is presented. The hub parts are constantly optimised and therefore a fully automated assembly line requires continously reprogramming. Thus, a new reconfigurable assembly system is introduced which is flexible and self-adaptive. The methodology of implementing an intelligent agent for designing the assembly strategy for the wind generator hub and the algorithm for the optimal task sequence are described. This reconfigurable automation runs a Partial Order Planning algorithm in real-time using Beckhoff TwinCAT® 3.

**Keywords** Reconfigurable assembly · Intelligent agent · Partial order planning · Wind turbine · Beckhoff TwinCAT® 3

## 10.1 Introduction

Power generation using wind has certain advantages due to its technology maturity, good infrastructure and relative cost competitiveness [1]. The wind turbine industry is growing and plays a major role in the future international scene [2]. It is predicted that this renewable energy technology will develop to about 12 % of the world's electrical supply by 2020 [3, 4]. Thus, wind turbines have become more and more of a mass product.

C. Deters (✉) · H. A. Wurdemann · J. S. Dai · L. D. Seneviratne · K. Althoefer
School of Natural and Mathematical Sciences, Centre for Robotics Research,
Strand, London WC2R 2LS, UK
e-mail: christian.deters@kcl.ac.uk

J. S. Dai et al. (eds.), *Advances in Reconfigurable Mechanisms and Robots I*,
DOI: 10.1007/978-1-4471-4141-9_10, © Springer-Verlag London 2012 

In the wind turbine manufacturing industry, assembly is one of the core processes in the full production chain. Most of these assembly operations are executed manually [5] the main disadvantages of a fully automated assembly are a relatively high degree of variability, low process stability, the necessity to repeat certain tasks that are difficult to perform correctly first time and the generation of a high rework rate. Further, adaptations within the assembly process can easily be integrated when they are applied in a manual production line. Despite the difficulties to implement a lean manufacturing strategy in these factories, there is an enormous potential for improvement in terms of productivity, quality costs and delivery times [6, 7].

Every control or assembly strategy targets specific goals [8], which can be a successfully assembled turbine, for instance. In order to react to changes the system needs to be reconfigurable and self-adaptive, which requires sensors and actuators for identification. Further, performing a reconfiguration, mechanisms need to be defined as type, autonomy, organization, scope and duration [8], which are described as a set of system reactions. Continuous changes in systems may be caused by continuous changes in the environment, which influence the control system. These changes will be handled with the capability of the system to adjust its behaviour in order to provide the desired system output. A self-adaptive system should not only be able to adapt itself to changes in the environment, but also optimise itself. Moreover, a self-adaptation is classified into two main groups: open- and closed-adaptive [9]. Open-adaptive means that the system self-adapts to new system behaviours, such as new applications. On the other hand, a closed-adaptive system is not able to self-adapt to unknown applications but it is self-contained.

Self-Adaptive software is normally implemented in an agent based context [9]. An agent is an individual software program which runs in a specific environment. This can be a mobile robot in wich case the agent decides on where to go. An agent can have the following attributes: autonomy [10], social ability [11], reactivity [12], and pro-activeness. Agents in artificial intelligence applications are usually specified as a computer system [12] with the additional previously discussed attributes. The systems are both designed and implemented based on concepts which are normally based on human behaviour. An agent based on artificial intelligence is normally characterised using mentalistic notations [13] which are subdivided into knowledge, belief, intention and obligations.

In general, a modern factory is systemised so that each part is prepared in a specific workstation and then forwarded. The assembly line needs to be set up in a way that it also considers assembly times. This approach can be extended to an agent based concept [14], with each part considered by individual agents who communicate with each other. This considers a virtual assembly line to start up [15–17], which describes the commonality of an assembly process for an overall product series. The structure should outline an overview of all possible configurations and also about possible reconfigurations due to product changes or other aspects. Furthermore, it should provide a general structure for the assembly strategy of an individual out of a product family and a high abstraction level of the assembly characteristics for the overall product family [18]. A product can be made more individual and customised according to specific customer requirements and wishes as more variants can be introduced. This leads to flexibility in low or fluctuating volumes. The requirements on volume might vary due to limited markets or due to a short life cycle, or, as with time competitors may join the market, the overall demand for the product may decrease [19].

These features of an intelligent agent based reconfigurable system are now considered for a wind turbine assembly. An overview of the manual hub bearing assembly process is given in Sect. 10.2. The methodology of implementing an agent for designing the assembly strategy for the wind generator hub and the algorithm for the optimal task sequence are described in Sects. 10.3 and 10.4. This self-adaptive automation runs in real-time using Beckhoff TwinCAT® 3 (Sect. 10.5).

## 10.2 Manual Hub Bearing Operation

As part of this paper, the reconfigurable process is introduced to the hub bearing assembly which is an essential part of the wind turbine manufacturing process. In the most frequently used topology of wind turbines, three fiberglass blades are bolted to the hub housing at the front of the nacelle. The example of a wind turbine hub considered for this paper consists of the main parts [20] listed below.

### *10.2.1 Wind Turbine Hub Body*

The hub body is manufactured upfront and is normally delivered to the assembly line. This hub body is designed to use 3 rotor blades. Further, there are also hubs designed for one or two rotor blades. The standard rotor hub has a diameter of more than 3 m and a weight of 13 tons. In preparation for the bearing assembly, additional elements will be attached to the hub increasing its weight to 25 tons.

### *10.2.2 Rotational Movement Unit*

The pitch rotation of each blade requires the use of a bearing between the hub and the blade. The bearing weights 6 tons. The wind turbine hub uses up to three bearings depending on the type. A rotor blade is mounted on each bearing when the turbine is installed in the field. The bearing can contain up to 128 stud bolts on each ring, the outside ring is used for mounting the blade.

### *10.2.3 Integration of Electrical Components*

The pitch system are electrical components which are normally prepared and wired so that it can be directly installed in the turbine. Therefore, it can be treated

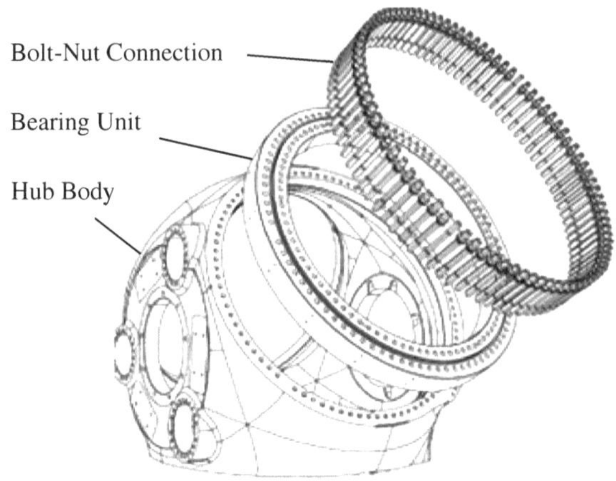


**Fig. 10.1** Final hub bearing assembly [20]: the hub has a diameter of more than 3 m and a weight of 13 tons. 128 bolts of type M36 with a length of 450 mm are used in order to fix the bearing and pitch system to the hub

as a single item. It is installed in an electrical cabinet which is then mounted inside the turbine hub. However, there are also hydraulic pitch systems which do not have an electrical cabinet.

### *10.2.4 Bolt-Nut Connection*

The stud bolts are used to mount the bearing onto the hub. 128 bolts of type M36 with a length of 450 mm are used. Most critically, the assembly process consists of a sequence of iterative operations of screw tensioning (forcefully elongating the screw) and bolt tightening (leading to a torque on the bolt and forces on the nut). The process is repeated as many times as needed to ensure the correct bolt tension; the number of interactions is currently based on the experience of technicians conducting these tasks manually. Automation is only possible, when reconfiguration is used replacing the worker's experience. The complete assembly process in Fig. 10.1 is particularly complex due to the size and weight of the parts involved.

## 10.3 Reconfigurable Assembly Strategy Using Intelligent Agents

Since this wind turbine hub assembly has many flexible contraints that cannot be met by obedient automation, a new system is required. This new system should be able to decide itself what it needs to do in order to satisfy its objectives. Thus, an assembly strategy based on intelligent agents is introduced which can operate in rapidly changing and open environments. Figure 10.2 shows the layout of the reconfigurable assembly strategy.

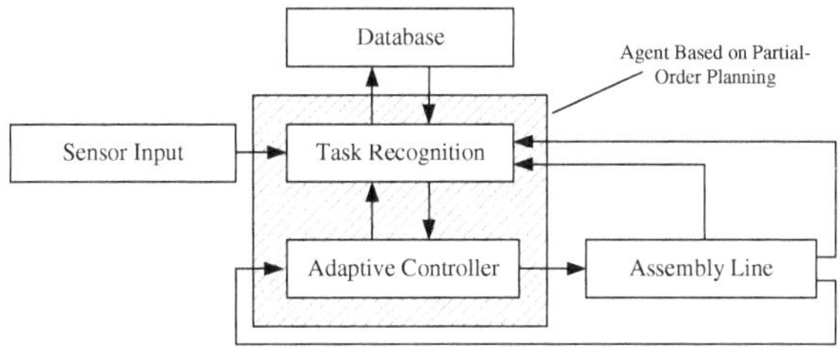


**Fig. 10.2** Reconfigurable assembly using intelligent agent strategy

Pre-defined patterns are stored in the database. These pre-defined patterns are actions or assembly steps which include pre and post conditions and are based on sensor inputs. In addition, an initial and goal state is defined for each assembly part. The task recognition block runs the actual intelligent agent. It contains two possible operational modes: online and offline mode. When the assembly line is activated, the agent will start in the online mode. It analyses the initial state and a target state and creates a suitable assembly strategy. In fact, when the assembly line is running and the strategy requires a change due to a new or changed item, the agent switches to the online mode and updates the assembly strategy. The agent also transfers the assembly strategy to the controller during the online mode which can either be a pre-defined strategy or a strategy derived from pre-defined patterns. The intelligent agent turns to the offline mode once a strategy is found and transferred to the controller. At this point the controller will only switch online again if it detects a new item to be assembled. The adaptive controller will receive its strategy from the agent. It will run as a step chain. The actual plant is the assembly line for a wind turbine.

Further, the database may store a pre-defined assembly strategy or, moreover, it will store any found assembly strategy while the controller runs in the online mode. If a new item is introduced to the assembly line, the database needs to be updated with a new item. The task recognition will then start online and run the assembly strategy algorithm, which is also introduced in the next section.

## 10.4 Optimisation Using Partial Order Planning

The Partial-Order Planning algorithm [21] is of special interest in this research case since it can be used and extended for automation approaches in the assembly industry. The reason for choosing this algorithm is that a strategy which the algorithm returns is always correct and complete. If the algorithm cannot return a strategy, the assembly actions need to be revised as there is either an incorrect or missing action. The Partial-Order Planning algorithm consists of three main parts, which are a set of ordering constraints describing the pre and post conditions of each step. The following actions can be defined for assembly process:

1. Hub placement at the assembly station
2. Hub preparation (Bolt hole check, installation)
3. Bearing installation
4. Bolt nut tightening

### *10.4.1 Conditions for Assembly Actions*

Connections between the actions are given by pre and post conditions, which are defined. For Action 1, the hub needs to be available, in a correct size and ready to be placed in the assembly station. Action 2 includes two internal actions, where the hub is prepared. The overall action requires the hub to be ready in the assembly line which is a requirement for the 'bolt hole check' action. This is followed by the bolt installation, where the bolts are placed in the turbine hub. Post conditions are defined by finishing the pre conditions. Bearing installation (Action 3) requires that the bolts are installed. Post conditions are set by an installed bearing and the bolts are ready for nut tightening. Action 4 is the final action for finalising the hub, the nuts are tightened on the bolts. The conditions for executing this action expect an installed bearing as well as installed nuts. These actions can be summarised using first order logic:

**Action 1:** Place_Hub(Astation, Warehouse):
**Require** ¬At(Astation, Hub) ∧ At(Hub, Warehouse)
**Return** At(Astation, Hub) ∧ ¬At(Hub, Warehouse)
**Action 2:** Prep_Hub(astation, Hub):
**Require** At(Astation, Hub)
**Return** At(Hub, holes) ∧ At(Hub, bolts)
**Action 3:** Bearing_Inst(bearing, Hub):
**Require** At(bolts, Hub) ∧ At(Astation, Hub)
**Return** At(Bearing, Hub) ∧ ¬At(Bearing, Nuts)
**Action 4:** Nut_Tighten(Nuts, Hub):
**Require** At(Hub, Bearing) ∧ At(Hub, Bolts)
**Return** At(Nuts, Tighten)

### *10.4.2 Partial-Order Planning and Assembly Strategy*

This leads to the Partial-Order Planning algorithm which has not been implemented in the automation area on industrial hardware so far. The overall strategy is shown in Fig. 10.3. This defined step chain can be stored in the database and transferred to the controller. Any changes can be introduced and the chain order redesigned. In the turbine industry, there are many different turbine hubs with various sizes of bolts, bearings etc. As wind turbines are constantly developed, a change can be introduced by using this algorithm. In fact, the advantage of using an intelligent agent is that it can check whether there is a new or updated action in the database which is linked to

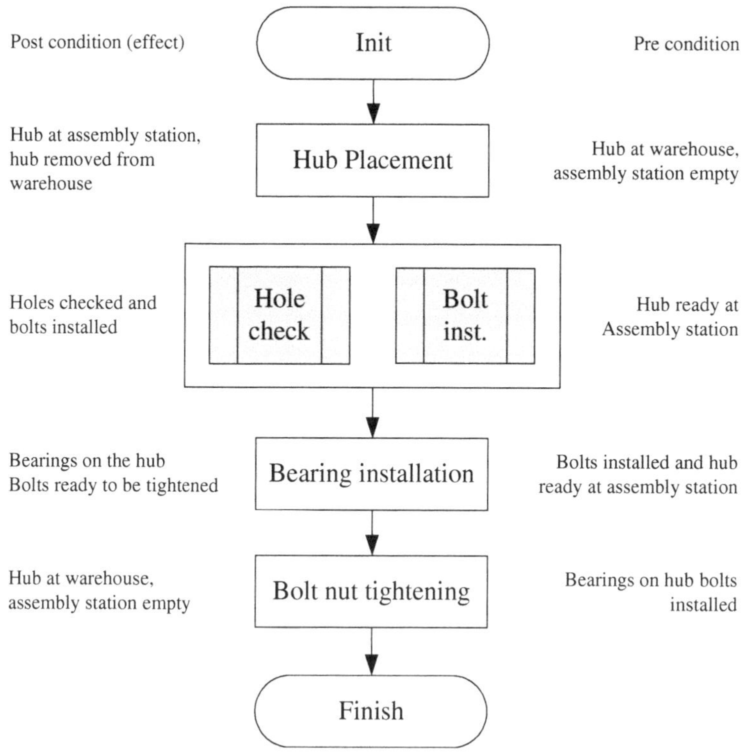


**Fig. 10.3** Self-adaptive assembly using intelligent agent strategy

the assembly process. It can then automatically re-design the assembly chain and, since the Partial-Order Planning algorithm is always correct and complete, it can automatically generate a new step chain which is then executed in the next PLC cycle. So, neither re-programming nor a shutdown of the machinery is required.

## 10.5 Flexible Hub Assembly with Industrial Standard Hardware

The presented industrial control is implemented on a standard industrial controller programming tool called TwinCAT® 3 provided by Beckhoff Automation GmbH. TwinCAT® is an automation software environment. It integrates into Visual Studio 2010 and builds a bridge between classical automation and computer science. The system architecture is illustrated in Fig. 10.4. In order to set up the controller based on Partial-Order Planning, MATLAB® is chosen. Standard PLCs run normally based on a sequential programming language, such as Structured Text (ST), which is an intuitive programming language. MathWorks® introduced recently the PLC

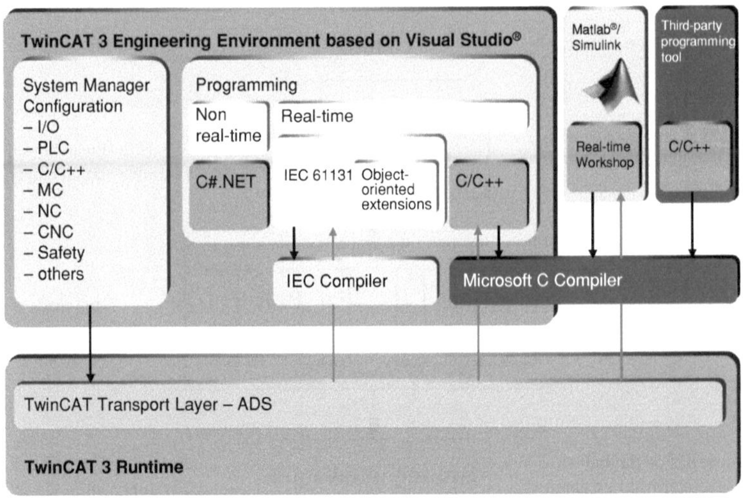


**Fig. 10.4** TwinCAT® 3 by Beckhoff automation GmbH system architecture [22]

coder, which allows code conversion from Simulink® models to ST code, which is limited basic Simulink® functions as well as Stateflow® models. Beckhoff TwinCAT® 3 uses integrated C++ code from MATLAB® Windows® real-time target. The agent is defined as a Simulink® model, converted into C++ code, and then integrated into a PLC cycle.

## 10.6 Conclusions and Future Works

This paper proposes the conversion approach of a manual wind turbine hub assembly into a reconfigurable semi-automated process. Thus, an intelligent agent based strategy is presented in order to fix the pitch and bearing system to the hub body using a bolt-nut connection. The general advantage of the used strategy is the higher flexibility of the assembly line. The reconfigurable strategy is optimised using a Partial-Order Planning algorithm. An assembly line is normally pre-programmed to specific products, an assembly strategy for each available product needs to be selected before the production can start. An update of the assembly program requires normally a redesign of the PLC program, which is rather time consuming. Further, Beckhoff TwinCAT® 3 is introduced as an industrial standard hardware.

**Acknowledgments** The work described in this paper is supported by COSMOS project grant from the European Communities Seventh Framework Programme (FP7-NMP-2009-SMALL-3, NMP- 2009-3.2-2) under grant agreement 246371-2.

## References


1. Joselin Herbert GM, Iniyan S, Sreevalsan E, Rajapandian S (2005) A review of wind energy technologies. Renew Substainable Energy Rev 11((2007):1117–1145
2. Ezio S, Claudio C (1998) Exploitation of wind as an energy source to meet the world's electricity demand. Wind Eng 74(76):375–87
3. Yao F, Bansal RC, Dong ZY, Saket RK, Shakya JS (2011) Wind energy resources: theory, design and applications. Handbook of renewable energy technology, pp 3–20. World Scientific Publishing Co. Pte. Ltd., Singapore
4. Saidur R, Islam MR, Rahim NA, Solangi KH (2010) A review on global wind energy policy. Renew Substainable Energy Rev 14:1744–1762
5. Sharpe M (2009) Robotic fabrication of wind turbine power generators. Weld J 8:40–44
6. Li H, Chen Z (2007) Overview of different wind generator systems and their comparisons. IET Renew Power Gener 2(2):123–138
7. Li H, Chen Z (2009) Design optimization and site matching of direct-drive permanent magnet wind power generator systems. Renew Energy 34:1175–1184
8. Cheng BHC, de Lemos R, Giese H, Inverardi P, Magee J (2009) Software engineering for self-adaptive systems: a research roadmap. LNCS 5525. Springer, Berlin, pp 1–26
9. Oreizy P, Gorlick MM, Taylor RN, Heimhigner D, Johnson G, Medvidovic N, Quilici A, Rosenblum DS, Wolf AL (1999) An achitecture-based approach to self-adaptive software. IEEE Trans Intell Syst Appl 14(3):54–62
10. Castelfranchi C (1995) Guarantees for autonomy in cognitive agent architecture. In: Wooldridge M, Jennings N (eds) Intelligent agents, volume 890 of lecture notes in computer science, Springer, Berlin, pp 56–70
11. Bradshaw JM (1997) Software agents. MIT, Cambridge
12. Wooldridge M, Jennings NR (1995) Intelligent agents: theory and practice. Knowl Eng Rev 10(02):115–152
13. Shoham Y (1993) Agent-orientated programming. Artif Intell 60(1):51–92
14. Vbra P (2003) Manufacturing agent simulation tool. In IEEE conference of emerging technologies and factory automation, pp 282–287
15. Xuemei H (2007) Implementing manufacturing manufacturing reconfiguration methodolgy in multi agent system of reconfigurable assembly line. In: International conference on mechatronics and automation, Harbin, China, 2007
16. Heisel U, Jovane F, Moriwaki T, Pritschow G, Ulsoy G, Koren Y, Van Brussel H (1999) Reconfigurable manufacturing systems. CIRP Ann Manuf Technol 48(2):527–540
17. Koren Y, Shpitalni M (2011) Design of reconfigurable manufacturing systems. J Manuf Syst 29(4):130–141
18. Huang X (2004) Research on methodology of intelligent and reconfigurable assembly lionde and simulation plantform. Chinese Academny of Science, Shenyang
19. Lang SYT, Shen W, Bi ZM, Wang L (2007) Reconfigurable manufacturing systems: the state of the art. Int J Prod Res 46(4):967–992
20. Cad drawings of a wind turbine hub (2011) Technical report, Gamesa Corporacin Tecnolgica
21. Russell SJ, Norvig P (2002) Artificial intelligence: a modern approach (International edn.). Pearson US Imports & PHIPEs, Upper Saddle River
22. Beckhoff Automation GmbH (2010) TwinCAT® 3: extended automation

# Chapter 11
# Development of a Reconfigurable Compliant Education Kit for Undergraduate Mechanical Engineering Education

**Shouzhong Li, Jingjun Yu, Yue Wu and Guanghua Zong**

**Abstract** In the curriculum of mechanism design, degree of freedom (DOF) is by far one of the most important issues, but it is not easy to whether impart from teachers or learn for students, especially it needs strong space imagination to determine the freedoms and constraints. To make this teaching or learning more convenient, an education kit is no doubt helpful to understand the complicated concept and theory via demonstration or DIY. For this purpose, a design philosophy for a reconfigurable compliant education kit is proposed, and a prototype is provided correspondingly. In fact, this education kit can be assembled into a variety of different parallel flexure mechanisms, which can be derived by Blanding's rule. The education kit shows its advantages in the following applications: (1) demonstrate the complementary rule of freedom and constraint proposed by Maxwell; (2) identify the redundant constraints; (3) analyze degree of freedom in both a qualitative and quantitative way; (4) develop the design of a specific compliant mechanism with multiple topologies.

**Keywords** Compliant mechanism · Education kit · Reconfigurable · Redundant constraint

## 11.1 Introduction

Education plays an important role in creating and inheriting human civilization. With the development of education, education kits emerged. These education kits have some inherent advantages in teaching, such as intuitional, convenient demonstration, and long reserving time. However, there is little research on education kits [1–3] compared

S. Li · J. Yu (✉) · Y. Wu · G. Zong
Robotics Institute, Beihang University, Beijing 100191, China
e-mail: jjyu@buaa.edu.cn

J. S. Dai et al. (eds.), *Advances in Reconfigurable Mechanisms and Robots I*,
DOI: 10.1007/978-1-4471-4141-9_11, 

to the instruments for the scientific purpose. In the curriculum of mechanism design, many concepts are not intuitional, such as degree of freedom and constraint. It needs strong space imagination for students, so it is hard to both teach and learn. Considering the advantages of education kits, it is urgent to design a new education kit to assist teaching mechanism design. This new education kit should have some features, such as simple structure, convenient for demonstration and manufacturing. In view of this, a reconfigurable compliant education kit is developed in this paper.

In fact, there are many traditional commercial education kits relating to mechanisms produced via industrialization. However, most of them are commonly less variable and used only for demonstration. Recently, some new-type education kits are more concerned due to their practice and innovation [2, 3]. As far as mechanisms are concerned, there are many methods to design these new-type education kits. One attractive method is based on the complementary rule of freedom and constraint which proposed by Blanding [4] and developed by Hopkins [5, 6] as the FACT method. Shivers at MIT [7] designed a compliant education kit. This compliant education kit can realize a helical motion and enable to measure some motion parameters and relationship between load and displacement experimentally.

Enlightened by education kit in [7] with the latest study on flexure mechanisms and compliant mechanisms [8–14], a reconfigurable compliant education kit is presented in this paper. Idea of modularization is introduced in designing this education kit, which makes types of education kits more abundant and cost down. On the other hand, the proposed education kit is a flexure mechanism essentially which is different from the traditional rigid education kits. In teaching, this education kit has three main applications. One is to assist teaching some non-intuitional concepts, such as degrees of freedom, constraints, redundant constraints, etc. Another application is to demonstrate the complementary rule of freedom and constraint as an object. The third application of this education kit is to arouse students' inspiration to construct more novel types of flexure mechanisms, which is the issue on design domain. In a word, as teaching tools this compliant education kit is very useful in teaching mechanism design.

The rest of this paper is organized as follows. Section 11.2 derives the generalized Blanding rule by introducing freedom and constraint patterns. Section 11.3 presents the specific procedure to design the compliant education kit. Section 11.4 presents some main applications with case studies which illustrate how this education kit to be used in teaching. At last Sect. 11.5 presents conclusions.

## 11.2 Fundamental Concept and Theory

### *11.2.1 Freedom and Constraint Patterns*

As well known, a free spatial body has six degrees of freedom, three independent rotations and three independent translations. Generally, a rotation can be represented as a line coinciding with the rotational axis called a freedom line, and a pure

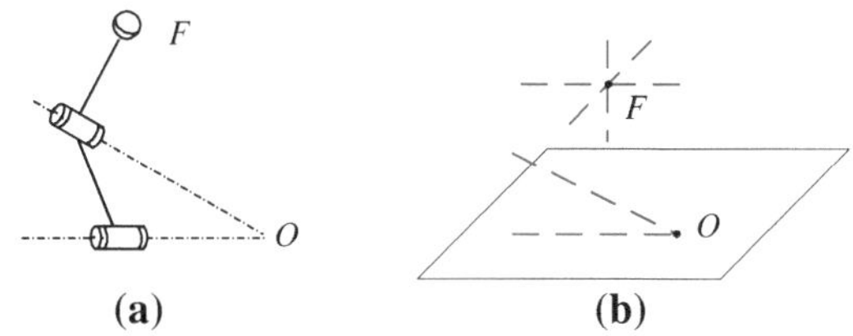


**Fig. 11.1** RRS kinematic chain and its corresponding freedom pattern

translation can be represented as a line with arrows on both ends called a freedom couple which along the direction of translation. In fact, the translational freedom line is a free vector which depends on its orientation rather than the location.

Similarly, a constraint force can be represented as a line coinciding with its action line called a constraint line, and a constraint torque can be represented as a line with arrows on both ends called a constraint couple whose direction is parallel to the normal line of the constraint torque. In this paper, freedom lines are represented by dot lines and constraint lines are represented by solid lines.

A series of freedom lines and freedom couples can be formed a freedom pattern. As an example, Fig. 11.1a shows a RRS chain, where the axes of two revolute joints intersect at point O. Its corresponding freedom pattern is illustrated in Fig. 11.1b.

Also, a constraint pattern contains a series of constraint lines and constraint couples. Figure 11.2 illustrates the constraint pattern corresponding to the case shown in Fig. 11.1. Section 11.2.2 will explain how to obtain this specified constraint pattern.

### *11.2.2 Generalized Blanding Rules*

In fact, there always exists a reciprocal relationship between the freedoms that a mechanism possesses and constraints acting on the mechanism. The complementary rule is first proposed by Maxwell [15], and he indicates that the sum of independent DOFs and independent degrees of constraints (DOCs) on a body is six. By means of freedom pattern and constraint pattern, the reciprocity of freedoms and constraints can be also visualized, and it is called the freedom and complementary constraint patterns by Blanding [4].

Blanding [4] has also addressed *the Rule of Complementary Patterns* which states that ***every freedom line intersects all constraint lines***. In fact, this rule can be explained and deduced easily through screw theory [15, 16]. In order to make the rule more convenient, this rule is modified to treat lines and couples differently [14]. Three conclusions are revealed in the following section, which can be used to guide mechanism analysis and design.

a. *Every rotational freedom line intersects every constraint force line.* (According to the projective geometry, all parallel lines intersect each other at a single point at infinity. Then two lines on the same plane must intersect.)

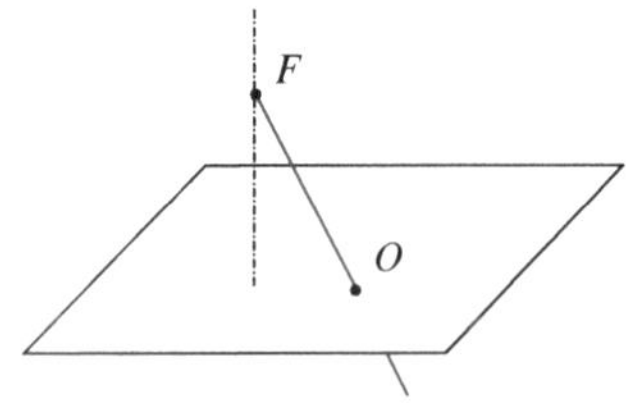


**Fig. 11.2** Constraint pattern of RRS kinematic chain

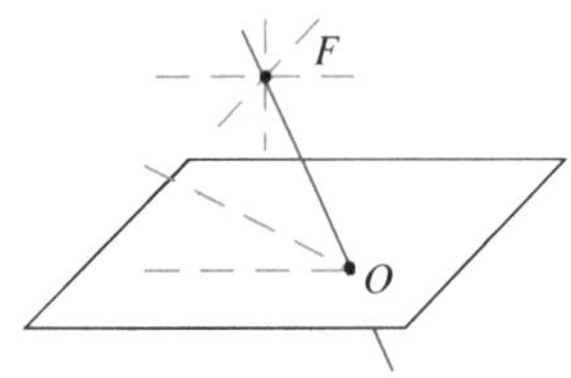


**Fig. 11.3** Freedom pattern of RRS chain and its constraint pattern

b. *The direction of every translational freedom couple is orthogonal to every constraint force line, and every rotational freedom line is orthogonal to the direction of every constraint torque.*
c. *The direction of translational freedom couple is free with its complementary constraint torque.*

The *Rule of Complementary Patterns* is completely covered within these three rules. Therefore, here we called all these rules *the generalized Blanding rules*.

According to *the generalized Blanding rules*, constraint pattern of RRS chain in Fig. 11.1 can be easily found. The freedom pattern and its complementary constraint pattern are shown in Fig. 11.3.

Generally, there are many lines which satisfy the generalized Blanding rules when finding freedoms or constraints of a specific mechanism. However, these lines are not all independent to each other. In order to determine the number of independent freedoms or constraints, we need to primarily judge the independence of these lines. Some rules are proposed in paper [17] to help us judge this independence.

## 11.3 Design of a Compliant Education Kit

### *11.3.1 Typical Flexure Primitives*

As a mechanical generator of some specified mechanism, an ideal flexure intends to be as flexible as possible in DOF directions while being as stiff as possible in other directions in order to obtain a desired motion or constraint. However, any real flexure has no infinite stiffness or infinite compliance. In the case of flexures

**Table 11.1** Equivalent constraint model of typical flexures

| Flexures | Equivalent constraint model | |
|---|---|---|
| | Pattern | Dimension |
|  |   | 1 |
|  |  | 2 |
|  |  | 3 |

with non-ideal constraint behavior, therefore, the quality of the constraint may be regarded as better or close to ideal one when the ratio of DOF' compliance to non-DOF's compliance can be orders of magnitude.

We have derived equivalent constraint model of several typical flexure primitives [14], and presented in Table 11.1 again.

As a result, a wire flexure (length:radius $\geq$10) provides a line constraint, a narrow plate (length:width:thickness $\approx$100:10:1) provides a constraint pattern which contains constraint lines parallel to each other in the same plane, a plate (length:width:thickness $\approx$100:60:1) provides a constraint pattern which contains all constraint lines in the same plane. In this paper, these typical flexures will be used to construct compliant education kits.

### *11.3.2 Procedure of Designing Reconfigurable Compliant Education Kits*

In fact, the compliant education kit involved in this paper is indeed a reconfigurable parallel flexure mechanism. All configurations can be synthesized via applying the generalized Blanding rules mentioned above. In this section, we will illustrate the general procedure of designing this education kit by taking a 2T1R spatial mechanism for instance.

(1) *Specify the freedom pattern of desired mechanism.* In this example, the desired freedom pattern contains two independent translational couples and a rotational line which is orthogonal to both couples, as shown in Fig. 11.4a.
(2) *Determine the complementary constraint pattern.* According to the generalized Blanding rules, obtain the constraint pattern reciprocal to the desired freedom pattern, as shown in Fig. 11.4b.

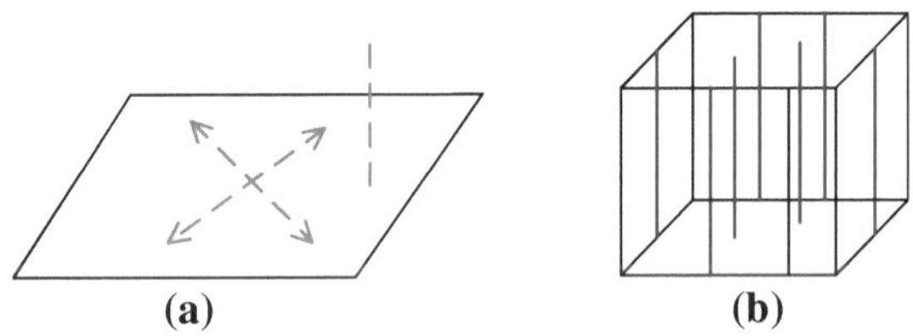


**Fig. 11.4** The desired freedom pattern of 2T1R type and its complementary constraint pattern

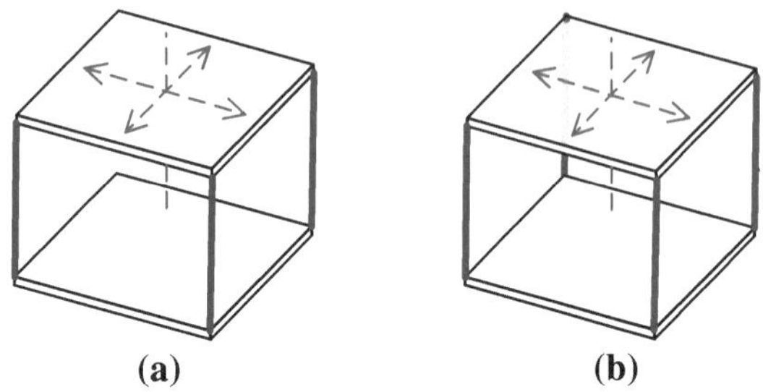


**Fig. 11.5** A 2T1R compliant education kit

**Fig. 11.6** Two kinds of wire flexures

(3) *Calculate the dimension of constraint pattern, denoted by n.* By judge the independence mentioned above, obtain the dimension of constraint pattern.
(4) *Select n independent constraint line within constraint pattern, and realize them* via *appropriate flexures,* as shown in Fig. 11.5a.
(5) *Add the redundant constraint to improve performance,* shown in Fig. 11.5b.

As a result, various configurations can be derived by implementing the synthesis procedure. However, it is only limited to a conceptual design strictly, and it is hardly to manufacture prototypes for each type. A feasible idea is that many different education kits can be constructed using same modules. In other words, these education kits should be reconfigurable. In our previous research, two kinds of wire flexures with circle cross-section are used as constraint elements. They are straight and pre-buckling ones respectively, as shown in Fig. 11.6.

In order to achieve various reconfigurable configurations, many holes are dilled in the top and bottom platform, as shown in Fig. 11.7. These flexure elements and plates are used pole-hole connection. Layout of these holes with above two types of flexure elements can satisfy most of constraint patterns. Figure 11.8 illustrates the model of a compliant mechanism with 1T DOF.

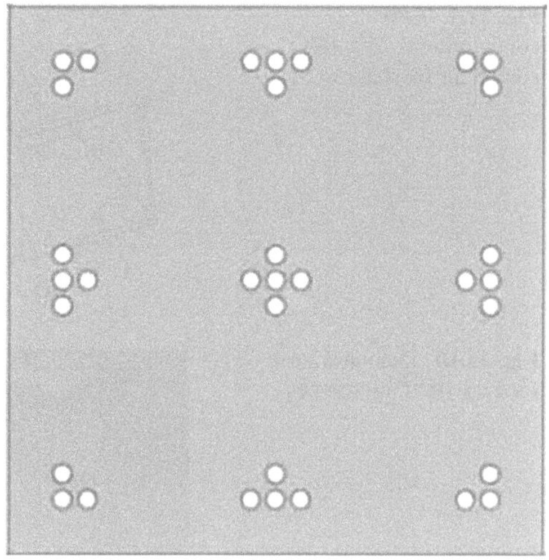

Fig. 11.7 Layout of holes in the platform

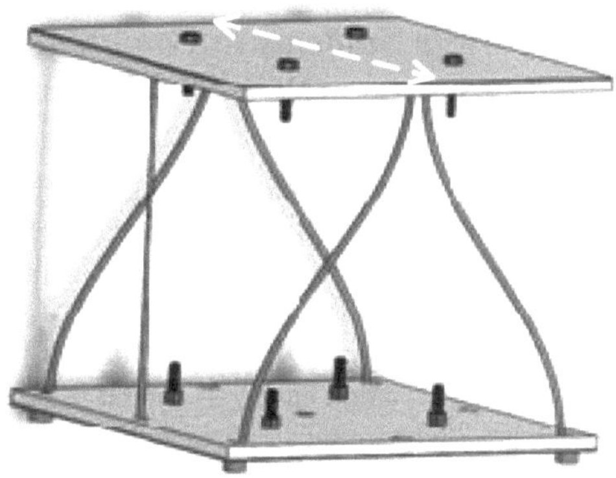

Fig. 11.8 A model of 1T education kit

## 11.4 Instructions of Education Kits Using in Teaching with Case Studies

The purpose of designing compliant education kits in this paper is to assist teaching the curriculum of mechanism design. As a main part, this section presents how to use these education kits in teaching, and how to be coupled with or assist teaching and make students learn more knowledge easily. In fact, these education kits can be applied in the following aspects: (1) help students to comprehend some concepts, such as degree of freedom, constraint, redundant constraint, etc. (2) Demonstrate how to analyze the DOF of an arbitrary mechanism by using the complementary rule of freedom and constraint. (3) Arouse students' inspiration to develop the mechanism design.

In this section, instructions of these education kits will be illuminated concretely.

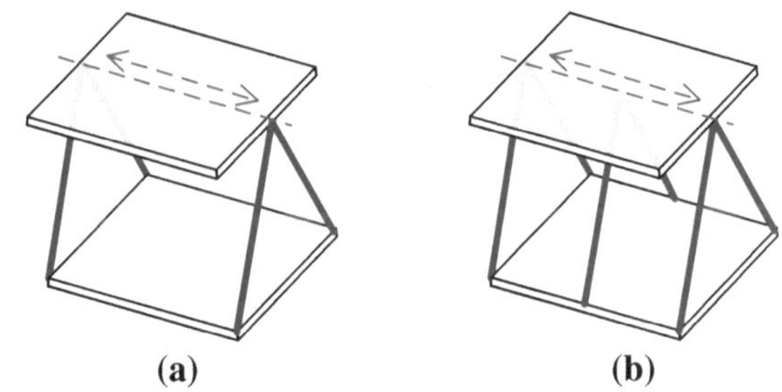


**Fig. 11.9** Two configurations of a 1R1T compliant mechanism

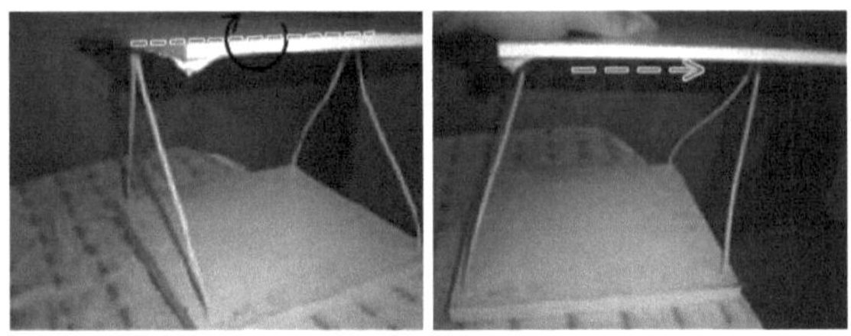

**Fig. 11.10** Demonstration for two 1R1T prototypes

## *11.4.1 Help Students to Comprehend Some Non-Intuitional Concepts (DOF, Constraint, Redundant Constraint)*

Given an education kits shown in Fig. 11.9a, it is easy to deduce that this mechanism have a rotation and a translation along the rotational axis (shown in red lines) via the generalized Blanding rules. In teaching, students can construct this type and operate it practically. Obviously, only when a force along the translational direction or a torque around the rotational axis acts on the top platform, this mechanism can generate translation or rotation easily, as shown in Fig. 11.10. However, this mechanism can hardly generate motion along the constraints, even acting loads. After demonstrating this example, students maybe comprehend these concepts more thoroughly.

For the redundant constraints, we can add another two constraints in the above education kit, as shown in Fig. 11.9b. Although adding two constraints, this mechanism has the same DOF characteristic as the mechanism shown in Fig. 11.9a. Our demonstration can verify this point, so the added two constraints prove to be the redundant constraints. In addition, students can try to find the redundant constraints of other given types.

## *11.4.2 Demonstrate the Complementary Rule of Freedom and Constraint*

The DOF number of spatial mechanism can be calculated via some formula, such as G-K criterion. However, we can hardly know DOF characteristic qualitatively only by this formula. Fortunately, the DOF type can be evaluated by the

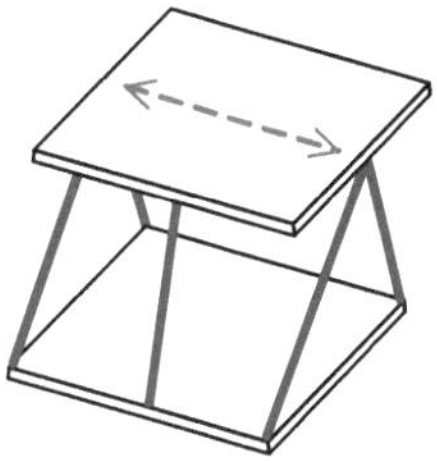

(a)

(b)

**Fig. 11.11** A 1T compliant mechanism

complementary rule of freedom and constraint. One crucial application of the proposed education kit is to demonstrate this complementary rule.

*Application*: In teaching, given a specific type, let students judge the DOF topology via applying the complementary rule of freedom and constraint. Then students can construct the education kit with this DOF topology, and check whether the result is right or not via DIY. Take the education kit shown in Fig. 11.8 for example, all five constraints in this mechanism are independent via judging independence mentioned above. According to the complementary rule of freedom and constraint, this mechanism has a translation DOF illustrated in red line, shown again in Fig. 11.11a. Figure 11.11b indicates this mechanism has a translation DOF along the red line indeed.

By experimental learning, students can understand the complementary rule of freedom and constraint more easily and thoroughly.

## *11.4.3 Arouse Inspiration of Designing the Novel Mechanisms*

It can be seen that the proposed education kits is not only modular but also reconfigurable. As a result, the modules in the kit toolbox can be assembled and derived into many different configurations. Students could construct some novel type with the aid of imagination and inspiration. Two application trials are presented in the following.

*Application* 1: Let students construct the types with one DOF using these modules as more as possible. Figures 11.12 and 11.13 give some common types with 1R and 1T respectively, where freedom topologies are denoted by red line. Remarkably, there are redundant constraints in those types shown in Figs. 11.12b, d and 11.13b.

*Application* 2: Given a known constraint, let students construct types with 1–3 DOF as more as possible. For example, given a constraint shown in Fig. 11.14, three appropriate types are illustrated in Fig. 11.15. These three types have 2R, 1R1T, 2R1T DOF topology respectively and are marked in red lines.

These experimental trainings not only make the students master the relevant knowledge, but also cultivate the practical ability and spatial imagination and inspire students' creativity.

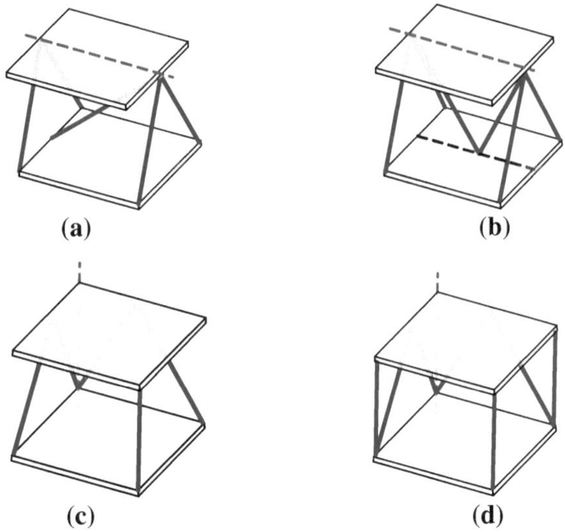

Fig. 11.12 Four types of compliant mechanisms with 1R DOF

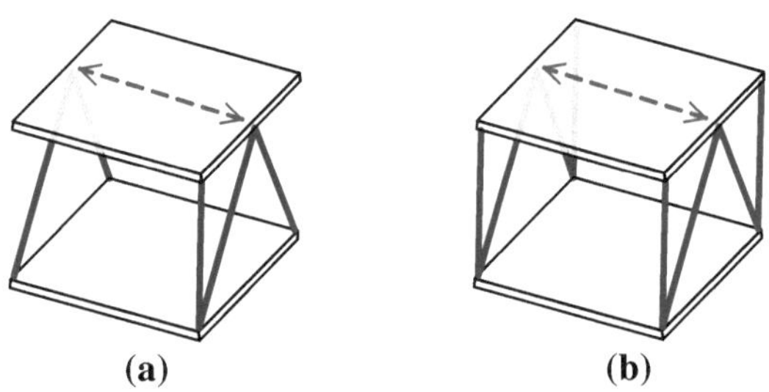

Fig. 11.13 Two types of compliant mechanisms with 1T DOF

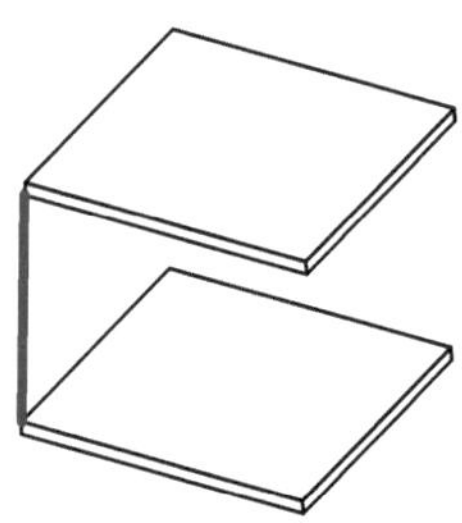

Fig. 11.14 A configuration with a given constraint

## 11.5 Conclusions

In this paper, a method for designing a reconfigurable compliant education kit is addressed based on the generalized Blanding rules and knowledge of compliant mechanisms. After analyzing the ideal constraint flexures, a specific design procedure for the education kits is presented. As a result, a reconfigurable compliant education kit is derived. Then for the perspective of teaching, instructions and potential applications of this proposed education kit are discussed. Evidently, the

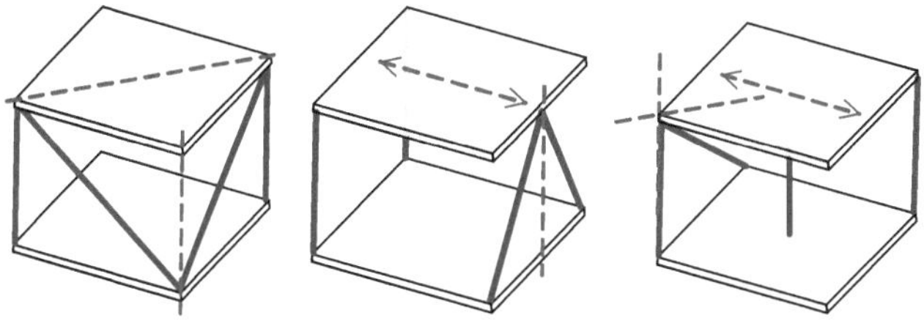

**Fig. 11.15** Three types of education kits with 2–3 DOF

usage of the education kit can not only to help engineering students learn the relevant knowledge, but also cultivate their practical ability and arouse their inspiration. In recent works, the education kit toolbox contained only by wire flexures, which limits the configuration types and rigidity of the system. Therefore, other type of flexures such as plate flexures will be added to the kit toolbox in the future work.

**Acknowledgments** The authors gratefully acknowledge the financial support of the NSFC (Grant No. 50875008, 51175010, 51175011). The authors also express their gratitude for Prof. Culpepper at MIT for his pioneering work on undergraduate education kits.

## References

1. Institute for Personal Robotics in Education. Learning computing with robots. http://wiki.roboteducation.org
2. Fiorini P (2005) LEGO kits in the lab. IEEE Trans Robotics Autom 12(4):5–12
3. Papadimitriou V, Papadopoulos E (2007) Putting low-cost commercial robotics components to the test-development of an educational mechatronics/robotics platform using LEGO components. IEEE Trans Robotics Autom 14(4):99–110
4. Blanding DL (1999) Exact constraint: machine design using kinematic principle. ASME Press, New York
5. Hopkins JB, Culpepper ML (2010) Synthesis of multi-degree of freedom, parallel flexure mechanism concepts via freedom and constraint topology (FACT). Part I Princ Precis Eng 34(2):259–270
6. Hopkins JB, Culpepper ML (2010) Synthesis of multi-degree of freedom, parallel flexure mechanism concepts via freedom and constraint topology (FACT). Part II Pract Precis Eng 34(2):271–278
7. Shivers SE (2007) Design modeling and fabrication of experimental apparatus compliant mechanism education kit. Massachusetts Institute of Technology, Cambridge
8. Howell LL (2001) Compliant mechanisms. Wiley, New York
9. Smith ST (2000) Flexures: elements of elastic mechanisms. Gordon and Breach Science Publishers, New York
10. Hale LC (1999) Principles and techniques for designing precision machines. Massachusetts Institute of Technology, Cambridge
11. Awtar S, Slocum AH (2007) Characteristics of beam-based flexure modules. J Mech Des 129:625–638
12. Su H-J, Dorozhkin DV, Vance JM (2009) A screw theory approach for the conceptual design of flexible joints for compliant mechanisms. J Mech Robotics 1(4):041009

13. Hao GB, Kong XW (2009) A 3-DOF translational compliant parallel manipulator based on flexure motion. ASME/DETC2009-86075, San Diego, CA
14. Yu JJ, Li SZ et al (2011) Screw theory based methodology for the deterministic type synthesis of flexure mechanisms. J Mech Robotics 3(3):031008
15. Maxwell JC, Niven WD (1890) General considerations concerning scientific apparatus. Courier Dover Publications, New York
16. Ball RS (1900) A treatise on the theory of screws. Cambridge University Press, Cambridge
17. Yu JJ, Dong X et al (2011) Mobility and singularity analysis of a class of 2-DOF rotational parallel mechanisms using a visual graphic approach. ASME/DETC2011-48274, Washington, DC, USA

# Part II
# Topology, Kinematics and Design of Reconfigurable Mechanisms

# Chapter 12
# Structural Synthesis of a Class of Metamorphic Parallel Mechanisms with Variable Mobility

**Wei Ye, Yuefa Fang and Sheng Guo**

**Abstract** In this paper, a systematic approach is presented for the structural synthesis of a class of metamorphic parallel mechanisms which possess variable mobility. The metamorphic parallel mechanisms are constructed using limbs with lockable joints. When different joints are locked, the mechanisms will have different mobility from single degree of freedom to full six degrees of freedom. First, the feasible limbs are enumerated based on constraint analysis. Then a metamorphic parallel mechanism constructed with feasible limbs is presented as example and its ability to perform variable mobility is demonstrated. The metamorphic parallel mechanisms presented in this paper will be perfect suitable in the complex process of industrial production.

**Keywords** Metamorphic parallel mechanisms · Variable mobility · Lockable joints

## 12.1 Introduction

With the development of science and technology, mechanisms which have variable mobility are expected in certain process of industrial production. Metamorphic mechanisms [1, 2] are a type of mechanisms with changeable mobility because their

W. Ye (✉) · Y. Fang · S. Guo
Beijing Jiaotong University, Beijing, China
e-mail: 10116304@bjtu.edu.cn

Y. Fang
e-mail: yffang@bjtu.edu.cn

S. Guo
e-mail: shguo@bjtu.edu.cn

J. S. Dai et al. (eds.), *Advances in Reconfigurable Mechanisms and Robots I*,
DOI: 10.1007/978-1-4471-4141-9_12, © Springer-Verlag London 2012 

topological structures are changeable. Since proposed in 1998 [3], metamorphic mechanisms have attracted many researchers' interest.

There are many literatures focused on the study of metamorphic mechanisms in the past decade. Parise et al. [4] presented ortho-planar metamorphic mechanisms in 2000, its can perform topological structural change in two orthogonal planes. Liu and Yang [5] did a research focus on the essence and characteristics of metamorphic mechanisms, presented three metamorphic ways in changing the topological structures of a mechanism. Yan and Kuo [6] presented the topological representations and characteristic analysis of variable joints. Guo [7] investigated the mobility change of planar metamorphic and presented five kind of metamorphic kinematic pairs. Zhang and Ding [8] proposed a method for designing new metamorphic mechanisms by using variable kinematic pairs to replace traditional common kinematic pairs. Zhang et al. [9] presented topological analysis for metamorphic mechanisms and investigated a new metamorphic kinematic pair and its evolved metamorphic chain and metamorphic parallel mechanism. Gan et al. [10] presented a new joint coined as the rT joint and investigated the mobility change of metamorphic parallel mechanisms assembled with this rT joint. Dai and Rees [11] presented a five-bar spherical linkage which has two degrees of freedom and can change its topological configurations to one degree of freedom. Using this mechanism as a metamorphic palm, Dai et al. [12–14] presented a novel metamorphic robotic hand which has high dexterity and versatility.

Noticed that the available mobility of the presented metamorphic mechanisms are limited to some particular motion, the application of these mechanisms are limited. While in the process of industrial production, the mobility required are diversity, varying from single degree of freedom to fully six degrees of freedom, thus, if we can design mechanisms with all these mobility, its will be perfect suitable in the process of industrial production.

In this paper, we will investigate the structural synthesis of a class of metamorphic parallel mechanisms which possess variable mobility. These mechanisms are constructed using limbs with lockable joints. When different joints are locked, the mechanisms will perform different mobility and motion characteristics.

## 12.2 Enumeration of Feasible Limb Structures

As we know, the motion characteristics of moving platform are determined by the nature of the platform constraint system which is formed by all the constraints provided by the limbs, we assume that all the constraints are linearly independent, such that the number of the total constraints provided by the limbs is equal to the rank of the platform constraint system, denoted as C, the number of degrees of freedom of the moving platform F is given by

$$F = 6 - C \tag{12.1}$$

Hence, if we want to design mechanisms which can perform variable mobility from single degree of freedom to full six degrees of freedom, the total constraints provided by all the limbs should be changeable as follow

$$C = \begin{cases} 0 & \text{for 6-DOFmechanisms} \\ 1 & \text{for 5-DOF mechanisms} \\ 2 & \text{for4-DOF mechanisms} \\ 3 & \text{for 3-DOF mechanisms} \\ 4 & \text{for 2-DOF mechanisms} \\ 5 & \text{for 1-DOFmechanisms} \\ 6 & \text{for immobile structures} \end{cases} \tag{12.2}$$

To achieve this aim, a simple way is to design limbs with lockable joints.

Next we will investigate the structures of feasible limbs, let the limb number to be three without loss of generality. According to the constraint analysis, some preliminary principles are obtained as follow:

1. Since multi-DOF joints can be regard as the combination of single-DOF revolute joint and prismatic joint, we will only consider revolute joints and prismatic joints for simplicity.
2. The total joints number of a limb should be six in order to make sure that the limb provides no constraint when all the joints are released.
3. Each limb can provide one constraint force or one constraint couple when a particular joint is locked.
4. Each limb can provide one constraint force and one constraint couple when some two particular joints are locked.
5. In order to make sure that the moving platform has a clear rotation center, the constraint forces provided by the limbs should intersect at a common point and this point should keep immobile with the moving platform.

Take the above preliminary principles into account, three more rules should be satisfied when we investigate the structures of feasible limbs:

(1) To make sure that the constraint force always passing through a common point, at least two intersecting revolute joints are required in each limb. Considering this common point should keep immobile with the moving platform, the top one of the intersecting revolute joints of each limb must connect to the moving platform directly. As the maximum number of independent revolute joints intersecting at a common point is three, the maximum number of revolute joints of each limb intersecting at a common point is three.

(2) In order to make sure that each limb can provide a constraint couple when one particular joint is locked, the direction of the axis of unlock revolute joints should be two, as a result, when the number of revolute joints intersecting at one point is three, one of the three joints must be locked. Except these three revolute joints, the axis of the remaining revolute joints should parallel with the axis of the lowest revolute joint of the three revolute joints, as the maximum number of independent revolute joints parallel with each other is three,

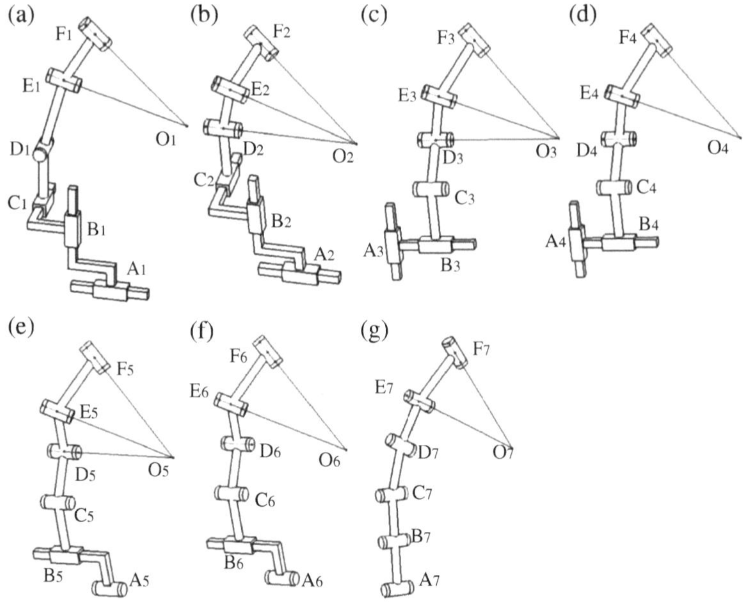


**Fig. 12.1** The feasible limbs

the maximum number of the remaining revolute joints is two, so the maximum number of the total revolute joints is five. But when the number of revolute joints intersecting at one point is two, the maximum number of revolute joints can be six, in this condition, the direction of the axis of these revolute joints should be three and the number of joints in each direction is one, two and three respectively.

(3) To make sure that the limbs provide no constraint when all the joints are released, namely no constraint force and couple, the minimum number of revolute joint in each limb is three.

According to the above considerations, the feasible limb types can be 3R3P, 4R2P, 5R1P and 6R, in which R represented revolute joint and P represented prismatic joint. As Fang presented in [15], if a limb provides a constraint force, all the revolute joints must be intersecting or parallel to the constraint force, all the prismatic joint axis are perpendicular to the constraint force. If a limb provides a constraint couple, all the revolute joint axis must be perpendicular to the constraint couple and prismatic joint axis can be oriented arbitrarily as long as they are linearly independent. It is also need to notice that we must lock some joints in order to ensure the limb provides some constraints. Based on these conditions, seven feasible limbs are constructed as shown in Fig. 12.1.

## 12.3 Constraint Analysis of Feasible Limbs

According to above analysis, the constraint provided by feasible limbs should be changeable when different joints are locked, namely, each limb provides no constraint when all the joints are released, each limb can provide one constraint force or one constraint couple when a particular joint is locked, each limb can provide one constraint force and one constraint couple when some two particular joints are locked. So when we construct a parallel mechanism using three identical feasible limbs, the rank of the platform constraint system can be changeable from 0 to 6, in the follow part, we will testify the feasibility of the above seven limbs.

In a-type configuration, there are two revolute joints intersecting at a common point $O_1$, the axis of the revolute joint $D_1$ is parallel with the axis of the prismatic joint $C_1$ and perpendicular to the axis of the prismatic joints $A_1$ and $B_1$. When all the joints are released, it provides no constraint, when we lock the prismatic joint $C_1$, it provides a constraint force passing through point $O_1$ and parallel with the axis of the locked prismatic joint, while when we lock one of the three revolute joints, it provides a constraint couple perpendicular to the plane decided by the axis of the remaining two revolute joints.

In b-type configuration, there are three revolute joints intersecting at a common point $O_2$, the three prismatic joints axis can be oriented arbitrarily as long as they are linearly independent. When all the joints are released, it provides no constraint. When we lock one of the three prismatic joints, it provides a constraint force passing through point $O_2$ and perpendicular to the plane decided by the axis of the remaining two prismatic joints. When we lock one of the three revolute joints, it provides a constraint couple perpendicular to the plane decided by the axis of the remaining two revolute joints.

In c-type configuration, there are three revolute joints intersecting at a common point $O_3$, the axis of the revolute joint $C_3$ and prismatic joint $B_3$ are both parallel with the revolute joint $D_3$ and perpendicular to the axis of the prismatic joints $A_3$. When all the joints are released, it provides no constraint, when we lock the prismatic joint $B_3$, it provides a constraint force passing through point $O_3$ and parallel with the axis of the locked prismatic joint, when we lock the revolute joint $E_3$ or $F_3$, it provides a constraint couple perpendicular to all the axis of the remaining revolute joints.

In d-type configuration, there are two revolute joints intersecting at a common point $O_4$, axis of prismatic joint $B_4$ and revolute joints $C_4$ and $D_4$ are parallel with each other and perpendicular to the axis of the prismatic joints $A_4$. When all the joints are released, it provides no constraint, when we lock the prismatic joint $B_4$, it provides a constraint force passing through point $O_4$ and parallel with the axis of the locked prismatic joint, when we lock the revolute joint $E_4$ or $F_4$, it provides a constraint couple perpendicular to all the axis of the remaining revolute joints.

In e-type configuration, there are three revolute joints intersecting at a common point $O_5$, axis of revolute joints $A_5$ and $C_5$ and prismatic joint $B_5$ are all parallel with the axis of the revolute joint $D_5$. When all the joints are released, it provides no constraint, when we lock the prismatic joint $B_5$, it provides a constraint force passing through point $O_5$ and parallel with the axis of the locked prismatic joint, when we lock the revolute joint $E_5$ or $F_5$, it provides a constraint couple perpendicular to all the axis of the remaining revolute joints.

In f-type configuration, there are two revolute joints intersecting at a common point $O_6$, axis of revolute joints $A_6$, $C_6$ and $D_6$ and prismatic joint $B_6$ are parallel with each other. When all the joints are released, it provides no constraint, when we lock the prismatic joint $B_6$, it provides a constraint force passing through point $O_6$ and parallel with the axis of the locked prismatic joint, when we lock the revolute joint $E_6$ or $F_6$, it provides a constraint couple perpendicular to all the axis of the remaining revolute joints.

In g-type configuration, there are two revolute joints intersecting at a common point $O_7$, axis of revolute joints $D_7$ is parallel with axis of revolute joints $E_7$, axis of revolute joints $A_7$, $B_7$ and $C_7$ are parallel with each other. When all the joints are released, it provides no constraint, when we lock the revolute joint $D_7$, it provides a constraint force passing through point $O_7$ and parallel with the axis of the revolute joint $A_7$, $B_7$ and $C_7$, when we lock the revolute joint $F_7$, it provides a constraint couple perpendicular to all the axis of the remaining revolute joints.

It is obviously that in all above configurations, if we lock some two particular joints at the same time, the limbs could provide one constraint force and one constraint couple on the moving platform. For example, in a-type configuration, if we lock joint $C_1$ and $D_1$ at the same time, the limb will provide one constraint force passing through point $O_1$ and parallel with the axis of joint $C_1$ and one constraint couple perpendicular to the plane decided by the axis of joint $E_1$ and $F_1$, moreover, there exist many other combinations, we won't enumerate its one by one in detail for simplicity.

## 12.4 Mobility Analysis of a Metamorphic Parallel Mechanism Constructed with Feasible Limbs

A parallel mechanism constructed with three b-type limbs is given as example as shown in Fig. 12.2. These limbs will provide different constraints on the moving platform when different joints are locked, such as no constraint, one force constraint, one couple constraint, two force constraints, two couple constraints, three force constraints, three couple constraints or the combinations of two kinds of the above constraints. Let R denote the rotational motion and T denote the translational motion, this parallel mechanism will possess diversity mobility varying from 1R, 1T, 2R, 2T, 3R, 3T, 1R1T, 1R2T, 1R3T, 2R1T, 2R2T, 2R3T, 3R1T, 3R2T, 3R3T. Different combinations of locked joints and its corresponding constraints and mobility are listed in Table 12.1, it is necessary to make sure that the constraints provided by the limbs are linearly independent when the locked joints are selected in each condition.

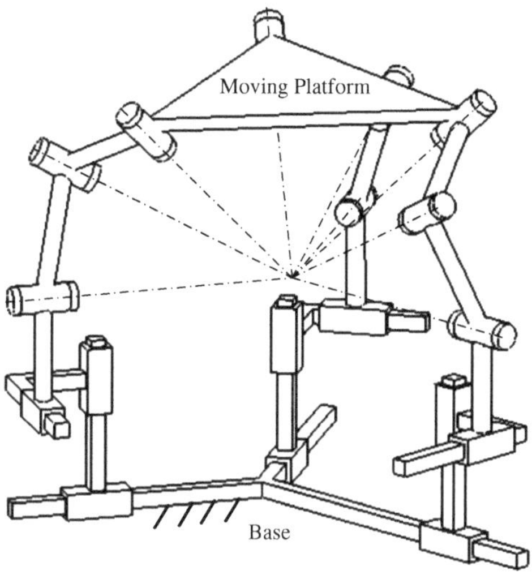


**Fig. 12.2** A metamorphic parallel mechanism constructed with three b-type limbs

**Table 12.1** Different combinations of locked joints and its corresponding constraints and mobility

| Mobility | Locked joints | Platform constraint system |
|---|---|---|
| 0 | One prismatic joint and one revolute joint in each limb | Three constraint forces and three constraint couples |
| 1R | One prismatic joint in each limb and one revolute joint in two of the three limbs | Three constraint forces and two constraint couples |
| 1T | One prismatic joint in two of the three limbs and one revolute joint in each limb | Two constraint forces and three constraint couples |
| 2R | One prismatic joint in each limb and one revolute joint in one of the three limbs | Three constraint forces and one constraint couple |
| 2T | One prismatic joint in one of the three limbs and one revolute joint in each limb | One constraint force and three constraint couples |
| 3R | One prismatic joint in each limb | Three constraint forces |
| 3T | One revolute joint in each limb | Three constraint couples |
| 1R1T | One prismatic joint in two of the three limbs and one revolute joint in two of the three limbs | Two constraint forces and two constraint couples |
| 1R2T | One prismatic joint in one of the three limbs and one revolute joint in two of the three limbs | One constraint force and two constraint couples |
| 1R3T | One revolute joint in two of the three limbs | Two constraint couples |
| 2R1T | One prismatic joint in two of the three limbs and one revolute joint in one of the three limbs | Two constraint forces and one constraint couple |
| 2R2T | One prismatic joint in one of the three limbs and one revolute joint in one of the three limbs | One constraint force and one constraint couple |
| 2R3T | One revolute joint in one of the three limbs | One constraint couple |
| 3R1T | One prismatic joint in two of the three limbs | Two constraint forces |
| 3R2T | One prismatic joint in one of the three limbs | One constraint force |
| 3R3T | No joint | No constraint |

## 12.5 Conclusion

This paper focus on the structural synthesis of a class of metamorphic parallel mechanisms which possess variable mobility. The primary feature of the metamorphic parallel mechanisms is that they are constructed with limbs with lockable joints. The mechanisms have the ability to perform variable mobility from single degree of freedom to full six degrees of freedom with different joints locked. The metamorphic parallel mechanisms presented in this paper will be suitable in some complex process of industrial production where the mobility of the mechanisms is expected to be changed during operation.

**Acknowledgments** This research work was sponsored by National Science Foundation of China under grant No.51175029, No.51075025.

## References

1. Zhang LP, Wang DL, Dai JS (2008) Biological modeling and evolution based synthesis of metamorphic mechanisms. ASME J Mech Des 130:072303
2. Zhang LP, Dai JS (2009) Reconfiguration of spatial metamorphic mechanisms. ASME J Mech Robotics 1(1):011012
3. Dai JS, Rees JJ (1998) Mobility in metamorphic mechanisms of flodable/erectable kinds. Proceedings of the 25th ASME biennial mechanisms and robotics conference, Atlanta, Sept 1998, DETC98/MECH-5902
4. Parise JJ, Howell LL, Magleby SP (2000) Ortho-planar mechanisms. Proceedings of the 26th biennial mechanisms and robotics conference, Baltimore, Paper No. DETC2000/MECH-14193
5. Liu CH, Yang TL (2004) Essence and characteristics of metamorphic mechanisms and their metamorphic ways. Proceedings of the 11th world congress in mechanism and machine science, Tianjin, pp 1285–1288
6. Yan HS, Kuo CH (2006) Topological representations and characteristics of variable kinematic joints. ASME J Mech Des 128(2):384–391
7. Guo ZH (2007) Planar metamorphic mechanism and its structural analysis. J Shandong Univ Technol Sci Technol 21(1):1–4
8. Zhang WX, Ding XL (2009) A method for designing metamorphic mechanisms and its application. ASME/IFToMM international conference on reconfigurable mechanisms and robots, 22–24 June, London, pp 171–174
9. Zhang KT, Dai JS, Fang YF (2010) Topology and constraint analysis of phase change in the metamorphic chain and its evolved mechanism. ASME J Mech Des 132(12):121001
10. Gan DM, Dai JS, Liao QZ (2009) Mobility change in two types of metamorphic parallel mechanisms. ASME J Mech Robotics 1:041007-1-9
11. Dai JS, Rees JJ (2005) Matrix representation of topological changes in metamorphic mechanisms. ASME J Mech Des 127(7):837–840
12. Dai JS (2004) Robotics hand with palm section comprising several parts able to move relative to each other. U.K. Patent No. GB20040009548
13. Dai JS, Wang DL (2007) Geometric analysis and synthesis of the metamorphic robotic hand. ASME J Mech Des 129(11):1191–1197
14. Dai JS, Wang DL, Cui L (2009) Orientation and workspace analysis of the multifingered metamorphic hand—metahand. IEEE Trans Robotics 25(4):942–947
15. Fang YF, Tsai LW (2002) Structure synthesis of 4-DoF and 5-DoF parallel manipulators with Identical Limb. Int J Robotics Res 21(9):799–810

# Chapter 13
# Structural Representation of Reconfigurable Linkages

**Ketao Zhang, Yuefa Fang, Guowu Wei and Jian S. Dai**

**Abstract** This paper presents adjacency matrix originated three-dimensional matrix for structural representation of mechanisms by integrating binary string in the sense of displacement subgroup theory. The improved elementary matrix operation for expressing of topological transformation of metamorphic mechanism is integrated in the three-dimensional matrix representation. A novel reconfigurable eight-bar linkage employing variable-axis revolute joints is proposed and the three-dimensional matrix for each working phase is expressed to reveal the distinct geometry. The improved elementary matrix operation is used to express the topological transformation of a metamorphic eight-bar linkage and to identify the effectiveness of the operation.

**Keywords** Variable-axis revolute joint · Three-dimensional matrix · Binary string · Reconfigurable linkage

---

K. Zhang (✉) · G. Wei · J. S. Dai
Centre for Robotics Research, King's College London, University of London, Strand, London WC2R 2LS, UK
e-mail: ketao.zhang@kcl.ac.uk

K. Zhang · Y. Fang
Department of Mechanical Engineering, Beijing Jiaotong University, Beijing 100044, People's Republic of China
e-mail: yff@bjtu.edu.cn

G. Wei
e-mail: guowu.wei@kcl.ac.uk

J. S. Dai
e-mail: jian.dai@kcl.ac.uk

J. S. Dai et al. (eds.), *Advances in Reconfigurable Mechanisms and Robots I*,
DOI: 10.1007/978-1-4471-4141-9_13, © Springer-Verlag London 2012

## 13.1 Introduction

The study of metamorphic mechanisms [1] and kinematotropic mechanisms [2] and other developments in reconfigurable mechanisms started in the mid-1990 s when the reconfigurable assembly and packaging system was addressed for folding decorative gifts and boxes, and the kinematotropic linkage was presented together with the position parameter variation. Kinematotropic linkages are coined as one kind of linkages that change their full-cycle mobility at certain configuration. The mobility change of this kind of mechanisms is induced by a certain geometric constraint to limit the motion and lead to different motion branches. Galletti and Fanghella [3] presented various single and multi-loop kinematotropic mechanisms. Fanghella et al. [4] addressed the parallel robots whose platforms can change their subgroups of displacement. Gogu [5] addressed the geometric constraint induced transitory phase of bifurcated platform motion in parallel mechanisms and the mobility change of the mechanisms.

The metamorphic mechanism which takes the concept of metamorphosis has the ability to transform its structure from one kind to another. The structure variation, topological configuration transformation and subsequently mobility change of this kind of mechanism are induced by either geometrical constraints or physical constraints. Elementary matrix operations were introduced to describe the topological configuration transformation of metamorphic mechanisms, and formal matrix operations [6] were further developed to describe the distinct topology of configurations embodied in metamorphic mechanisms. Lan and Du [7] introduced element $-1$ to indicate the frozen links and represent the transformation using matrices with same dimension. However, the operations in terms of matrix form were not addressed for bringing in the element $-1$. Besides the methods based on link-to-link adjacency matrix, Wang and Dai [8] introduced the symbol of kinematic pairs in the topological representation of metamorphic mechanisms and the approach is aimed to satisfy the characteristics including multi-stage, multi-topological structure and multi-degree of freedom of metamorphic mechanisms. Tsai [9] introduce the letters into link-to-link adjacency matrix to denote types of joints which were employed in a mechanism. Yan and Kuo [10] emphasized the meaning of variable topologies of the kinematic pairs based on descriptive joint representation. For convenience of computer programming, Kuo [11] employed hexadecimal system to digitize descriptive joint code and joint sequence. Zeng and Fang [12] proposed systematic approach for representation of hybrid mechanisms base on mathematical logic. Zhang et al. [13] proposed an approach for geometrical and topological representation of mechanisms with the newly developed three-dimensional matrix by integrating 16 bits binary strings.

This paper presents the augmented three-dimensional matrix with integrated binary strings for structural representation of linkages and improved elementary matrix operation for topological transformation of metamorphic mechanism. The proposed matrix and matrix operations are utilized in the representation of new reconfigurable linkage that is capable of implementing different functions in distinct motion phases. This provides an effective approach for unraveling the underline principle of reconfiguration.

## 13.2 The Augmented Three-Dimensional Matrix and Improved Elementary Matrix Operation

Topological structures of the mechanisms are characterized by the kinematic joints and links which compose the mechanical assemblage. Since 1960 s, graph theory has been extensively used as one fundamental tool for topological representation, analysis and structure synthesis of mechanisms. The topological structure presented by a graph can be conveniently mapped into matrix form which is capable of making analytical operation. The well-known link-to-link adjacency matrix is used to identify graphs up to graph isomorphism. However, the geometric properties of kinematic joints are not counted in both the graph representation and the graph mapped adjacency matrix. In order to accurately represent the structure of mechanisms particularly the varying structure and configuration of reconfigurable mechanisms, three dimensional (3D) matrix representation originated from the link-to-link adjacency matrix and improved elementary matrix operation are introduced.

### *13.2.1 The Three-Dimensional Matrix with Binary Strings*

The three dimensional (3D) matrix based structural representation of mechanisms is originated from the general link-to-link adjacency matrix [9] by integrating 16 bits binary strings developed based on the genesis of displacement subgroups in Lie algebraic structure [14].

The links of a kinematic chain are numbered sequentially from 1 to n. The binary string based 3D adjacency matrix, **BA**, is defined as follows:

$$a_{ijk} = \begin{cases} b & \text{if link } i \text{ is connected to link } j \text{ by a joint,} \\ 0 & \text{otherwise (including } i = j). \end{cases} \tag{13.1}$$

in which $b$ is 16 bits binary string consists of four groups sub-strings, the subscript $k$ ($k$ = 1, 2, 3 and 4) corresponds to the four binary string groups.

In each 16 bits binary string, the lower 4 bits sub-string is used to represent the 16 types of universal sets of spatial displacement subsets [12, 13]. The middle 4 bits string and the higher 4 bits string are set to represent the first and second order relative relation of axial vectors which expressing the joint axes. The highest 4 bits sub-string presents the effectiveness of joints in various topological configurations of reconfigurable mechanisms. Simply, the highest 4 bits sub-string is expressed by its corresponding decimal number and the other three sub-strings are expressed by their corresponding hexadecimal numbers. Since a joint in reconfigurable mechanism may have two working stages including active stage and physically constrained stage, decimal 1 and $-1$ are used to represent these two working stages and expressed in binary strings 0001 and 1001 respectively [13]. The highest 4 bits sub-strings expressed in decimal numbers compose a matrix which represents the connectivity of a mechanism and effectiveness of joints, and the matrix is coined phase matrix of the mechanism.

### 13.2.2 Improved Elementary Matrix Operation

The inherent properties of the metamorphic mechanism are structure evolution, topology variation and subsequently mobility change resorting to link-annex with suppressed relative motion between two adjacent links. The structural transformation of metamorphic mechanism leads to variation of topology which can be represented in the matrix form. The improved two-step EU-elementary matrix operation is developed to express the transformation between two distinct phases.

The improved EU-elementary matrix operation consists of steps of applying a $\mathbf{U}_{i,j}$-elementary matrix operation which is similar as that in literature [6] and a new introduced $\mathbf{E}_{j(-1)}$-elementary matrix operation. The elementary matrix $\mathbf{E}_{j(-1)}$ is introduced to perform the function of integrating the element $-1$ to the $j$th row and column of a phase matrix when it is premultiplied to a phase matrix and its transpose is postmultiplied. The information of each annexed link is represented by an entry with element $-1$ and the connectivity of effective links in subphase is reserved by the entries with element 1 in phase matrix. The modified matrix operation uses modulu-2 arithmetic. In the process of matrix operation and the corresponding structure transformation, the dimension of the phase matrix and the identified number of the links are kept unchanged.

The phase change induced by annexing the $j$th link to $i$th link of a linkage with $n$ links, the improved elementary matrix operation is give by

$$\mathbf{P} = (\mathbf{E}_{j(-1)}\mathbf{U}_{i,j})\mathbf{P}^s(\mathbf{E}_{j(-1)}\mathbf{U}_{i,j})^T \tag{13.2}$$

where

$$\mathbf{E}_{j(-1)} = \begin{bmatrix} e_1 & & & & & & \\ & \ddots & & & & & \\ & & e_{j-1} & & & & \\ & & & -e_j & & & \\ & & & & e_{j+1} & & \\ & & & & & \ddots & \\ & & & & & & e_n \end{bmatrix}$$

in which the entry $e_k$ and $u_k$ $(k = 1,\cdots, i,\cdots, j,\cdots, n)$ are given as 1.

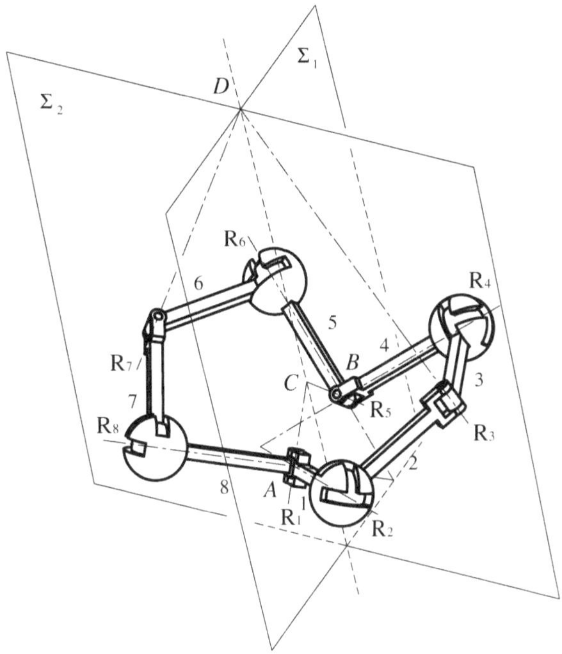


**Fig. 13.1** A reconfigurable eight-bar linkage

In the reverse transformation from a subphase configuration to the source phase of the metamorphic mechanisms, the relative annexed links are released and the connectivity of the mechanism changes in the physical process of structure evolution. This transformation is able to be represented by the elementary matrix operation since the $\mathbf{U}_{i,j}$-elementary matrix and new introduced elementary matrix $\mathbf{E}_{j(-1)}$ are invertible. The matrix operation is expressed as

$$\mathbf{P}^s = (\mathbf{U}_{i,j}^{-1}\mathbf{E}_{j(-1)}^{-1})\mathbf{P}(\mathbf{U}_{i,j}^{-1}\mathbf{E}_{j(-1)}^{-1})^T \qquad (13.3)$$

.

## 13.3 Structure Representation of a Reconfigurable Eight-Bar Linkage with Variable-Axis Revolute Joint

A new reconfigurable mechanism with symmetrical structure is illustrated in Fig. 13.1. The eight links from link 1 through to link 8 are connected by four variable-axis revolute joints [10] and the commonly used revolute joints alternately.

The joints in the mechanism are labeled $R_1$ to $R_8$ sequentially in Fig. 13.1. The mechanism is characterized by arrangement of the joints and links with equal length. Since the variable-axis revolute joint is able to undergo different rotational

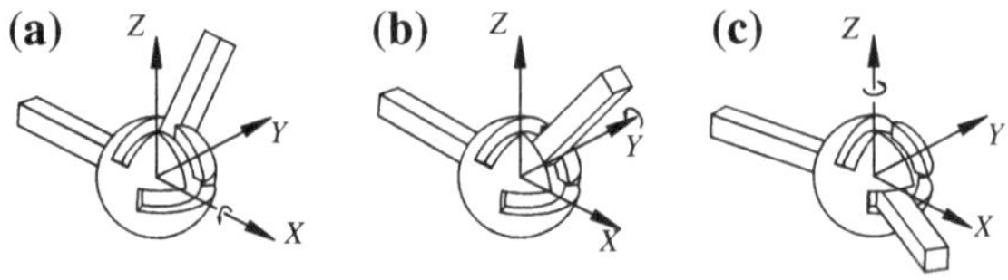


**Fig. 13.2** Three stable topological phases of the variable-axis revolute joint **a** about X-axis **b** about Y-axis **c** about Z-axis

motions about three orthogonal axes respectively, the kinematic joint has three stable working phases in Fig. 13.2 and the constraints supplied by this kinematic joint are changeable following the variation of the rotational axis.

### *13.3.1 Working Phase with Two Symmetric Planes*

As illustrated in Fig. 13.1, the variable-axis revolute joints are working in their first stable phase in Fig. 13.2a. In this working phase, the effective joint axes $R_1$, $R_2$, and $R_8$ have a common point *A* and $R_1$ is perpendicular to the plane formed by $R_2$ and $R_8$. Symmetrically, the revolute joint axes $R_4$, $R_5$, and $R_6$ have a common point *B* and $R_5$ is perpendicular to the plane formed by $R_4$ and $R_6$. The revolute joint axes $R_1$ and $R_5$ have common point *C* while $R_3$ and $R_7$ have common point *D*. The two planes $\Sigma_1$ formed by $R_1$and $R_5$ and $\Sigma_2$ by $R_3$ and $R_7$ are the symmetrical plane of the linkage. The common line *CD* is the symmetrical axis of the mechanism in this working phase.

According to the geometry of the working phase in Fig. 13.1, the augmented 3D matrix presentation of the linkage is expressed as $\mathbf{BA}_1$ in Appendix.

### *13.3.2 Working Phase with Skew Axial Motion*

When the variable-axis revolute joint connecting links 3 and 4 and that connecting links 5 and 6 change their configuration from the first stable phase to the second phase in Fig. 13.2b, the reconfigurable eight-bar linkage changes its structure to a different working phase in Fig. 13.3.

In this working phase, the effective joint axes $R_1$, $R_2$, and $R_8$ have a common point *A* and $R_1$ is perpendicular to $R_2$ and $R_8$. The revolute joint axes $R_3$ and $R_7$ have common point *B* while $R_4$ and $R_6$ have common point *C*. Since $R_3$ and $R_4$, $R_6$ and $R_7$ are parallel respectively, the plane formed by $R_3$ and $R_7$ and that by $R_4$ and $R_6$ are parallel. The mechanism is symmetrical with the plane $\Sigma_1$ in which the common points *A*, *B* and *C* located. As constrained by the geometry of the linkage, the joint axis $R_5$ locates in the symmetrical plane $\Sigma_1$ and always passes the common point *A*. Accordingly, the augmented 3D matrix of the linkage in this working phase becomes $\mathbf{BA}_2$ in Appendix.

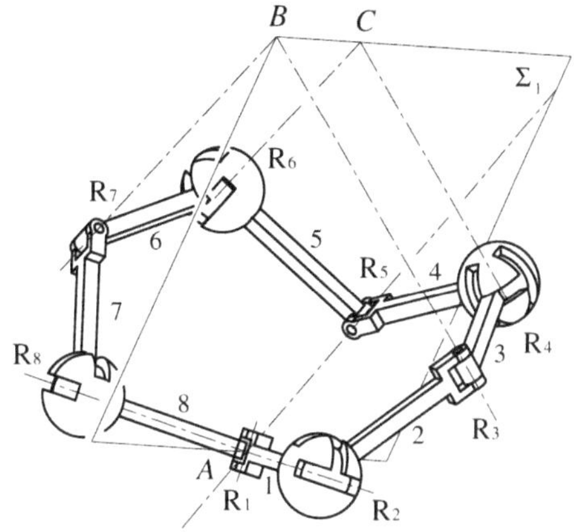


Fig. 13.3 Working phase with skew axial motion

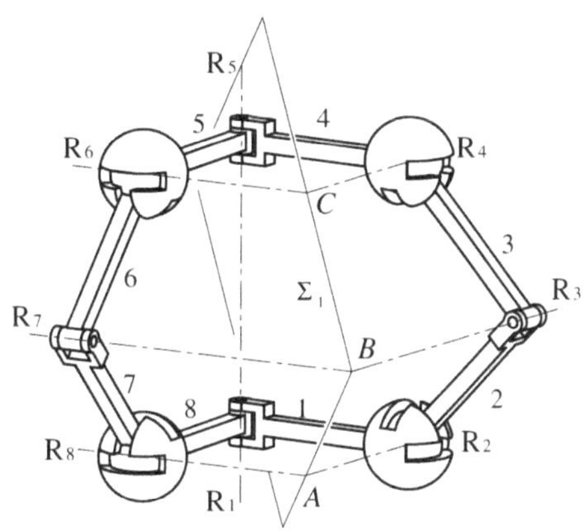


Fig. 13.4 Working phase with straight-line motion

### 13.3.3 Working Phase with Straight-Line Motion

When the four variable-axis revolute joints change their configuration to the second phase in Fig. 13.2b, the reconfigurable eight-bar linkage changes its structure to the third working phase in Fig. 13.4.

The revolute joint axes $R_2$ to $R_4$ which are connected in series are parallel, and the axes $R_6$ to $R_8$ which are connected in series are parallel and pointing in another direction. Under such a condition, the revolute joints $R_1$ and $R_5$ becomes collinear. The linkage in this working phase is symmetrical with the plane $\Sigma_1$ and the links 4 and 5 implement straight line motion with respect to links 1 and 8. The augmented 3D matrix of the linkage in Fig. 13.4 becomes $\mathbf{BA}_3$ in Appendix.

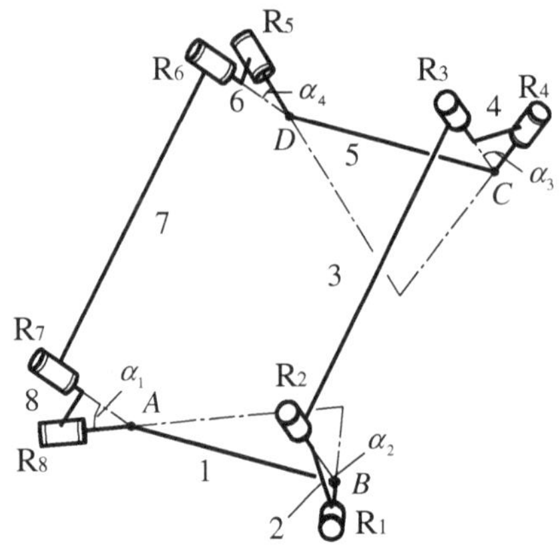


**Fig. 13.5** A metamorphic eight-bar linkage

## 13.4 Structure Representation of a Metamorphic Linkage

The linkage in Fig. 13.5 is a metamorphic linkage which is able to change its topology resorting to link annex at typical configuration [13]. The links are labeled 1 to 8 and the joint exes are labeled $R_1$ to $R_8$. The angles between each couple of revolute joints intersecting at common points *A*, *B*, *C* and *D* are determinate, that $\alpha_1 = \alpha_2 = \alpha_3 = \alpha_4 = 45°$. The axes of joints $R_1$ and $R_8$ are perpendicular, and the axes of joints $R_4$ and $R_5$ are perpendicular in the other end. The axes of revolute joints at the distal end of link 3 and 7 are parallel respectively.

The augmented 3D matrix of the linkage in the source phase with all the revolute joints are effective is expressed as $\mathbf{BA}_{m1}$ in Appendix.

One of the subphases of the metamorphic eight-bar linkage is an overconstrained six-bar linkage in Fig. 13.6. This overconstrained six-bar linkage is derived by annexing links 8 and 1 and links 6 and 5. The axes $R_7$ and $R_1$ have common point E, and the axes $R_6$ and $R_4$ have common point F. The point *A* is projection of common point *B* on axis $R_7$, and *D* is the projection of *C* on axis R6.

The improved elementary matrix operations which express the configuration transformation is given by

$$\mathbf{P} = (\mathbf{E}_{8(-1)}\mathbf{U}_{1,8})\mathbf{P}^{s}(\mathbf{E}_{8(-1)}\mathbf{U}_{1,8})^{T} \tag{13.4}$$

and

$$\mathbf{P} = (\mathbf{E}_{6(-1)}\mathbf{U}_{5,6})\mathbf{P}^{s}(\mathbf{E}_{6(-1)}\mathbf{U}_{5,6})^{T} \tag{13.5}$$

Resultantly, the augmented 3D matrix of the linkage in the subphase in Fig. 13.6 becomes $\mathbf{BA}_{m2}$ in Appendix.

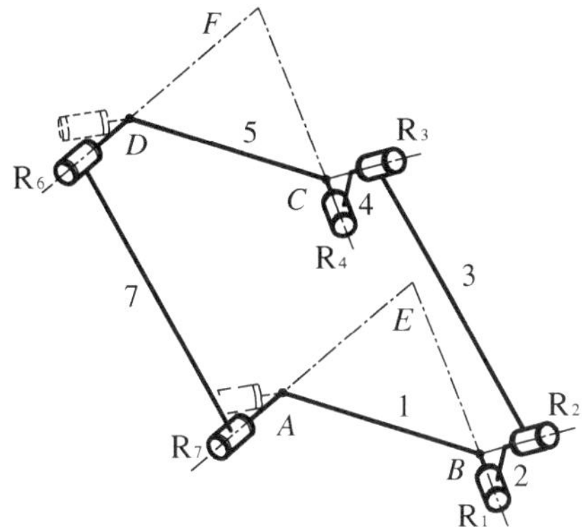


**Fig. 13.6** Overconstrained six-bar linkage

## 13.5 Conclusions

This paper presented the link-to-link adjacency matrix originated three-dimensional matrix for structural representation of mechanisms by integrating the binary strings. The improved elementary matrix operation which is integrated in the three-dimensional matrix operation is capable of characterize the topological transformation of reconfigurable linkages. The three typical working phases of a reconfigurable eight-bar linkage are obtained by altering the working phase of the variable-axis revolute joint. The distinct geometry corresponding to each working phase of the reconfigurable linkage is revealed by the three-dimensional matrix representation. The effectiveness of the improved elementary matrix operation for representing the topological transformation is verified with of metamorphic eight-bar linkage.

**Acknowledgments** The authors thank the support of European Commission—Framework 7 Programme under grant number 270436.

## Appendix

$$\mathbf{BA}_1 = \begin{bmatrix} 0000 & 1A44 & 0000 & 0000 & 0000 & 0000 & 0000 & 12C4 \\ 1A44 & 0000 & 1244 & 0000 & 0000 & 0000 & 0000 & 0000 \\ 0000 & 1244 & 0000 & 1AC4 & 0000 & 0000 & 0000 & 0000 \\ 0000 & 0000 & 1AC4 & 0000 & 12C4 & 0000 & 0000 & 0000 \\ 0000 & 0000 & 0000 & 12C4 & 0000 & 1A44 & 0000 & 0000 \\ 0000 & 0000 & 0000 & 0000 & 1A44 & 0000 & 1244 & 0000 \\ 0000 & 0000 & 0000 & 0000 & 0000 & 1244 & 0000 & 1AC4 \\ 12C4 & 0000 & 0000 & 0000 & 0000 & 0000 & 1AC4 & 0000 \end{bmatrix} \quad (A.1)$$

$$\mathbf{BA}_2 = \begin{bmatrix} 0000 & 1444 & 0000 & 0000 & 0000 & 0000 & 0000 & 12C4 \\ 1444 & 0000 & 1414 & 0000 & 0000 & 0000 & 0000 & 0000 \\ 0000 & 1414 & 0000 & 1A44 & 0000 & 0000 & 0000 & 0000 \\ 0000 & 0000 & 1A44 & 0000 & 1444 & 0000 & 0000 & 0000 \\ 0000 & 0000 & 0000 & 1444 & 0000 & 1414 & 0000 & 0000 \\ 0000 & 0000 & 0000 & 0000 & 1414 & 0000 & 1244 & 0000 \\ 0000 & 0000 & 0000 & 0000 & 0000 & 1244 & 0000 & 1AC4 \\ 12C4 & 0000 & 0000 & 0000 & 0000 & 0000 & 1AC4 & 0000 \end{bmatrix} \quad (A.2)$$

$$\mathbf{BA}_3 = \begin{bmatrix} 0000 & 1114 & 0000 & 0000 & 0000 & 0000 & 0000 & 1444 \\ 1114 & 0000 & 1414 & 0000 & 0000 & 0000 & 0000 & 0000 \\ 0000 & 1414 & 0000 & 1A44 & 0000 & 0000 & 0000 & 0000 \\ 0000 & 0000 & 1A44 & 0000 & 1144 & 0000 & 0000 & 0000 \\ 0000 & 0000 & 0000 & 1144 & 0000 & 1114 & 0000 & 0000 \\ 0000 & 0000 & 0000 & 0000 & 1114 & 0000 & 1414 & 0000 \\ 0000 & 0000 & 0000 & 0000 & 0000 & 1414 & 0000 & 1A44 \\ 1444 & 0000 & 0000 & 0000 & 0000 & 0000 & 1A44 & 0000 \end{bmatrix} \quad (A.3)$$

$$\mathbf{BA}_{m1} = \begin{bmatrix} 0000 & 12A4 & 0000 & 0000 & 0000 & 0000 & 0000 & 12C4 \\ 12A4 & 0000 & 1214 & 0000 & 0000 & 0000 & 0000 & 0000 \\ 0000 & 1214 & 0000 & 12A4 & 0000 & 0000 & 0000 & 0000 \\ 0000 & 0000 & 12A4 & 0000 & 12C4 & 0000 & 0000 & 0000 \\ 0000 & 0000 & 0000 & 12C4 & 0000 & 12A4 & 0000 & 0000 \\ 0000 & 0000 & 0000 & 0000 & 12A4 & 0000 & 1214 & 0000 \\ 0000 & 0000 & 0000 & 0000 & 0000 & 1214 & 0000 & 12A4 \\ 12C4 & 0000 & 0000 & 0000 & 0000 & 0000 & 12A4 & 0000 \end{bmatrix} \quad (A.4)$$

$$\mathbf{BA}_{m2} = \begin{bmatrix} 0000 & 12A4 & 0000 & 0000 & 0000 & 0000 & 12A4 & -12C4 \\ 12A4 & 0000 & 1214 & 0000 & 0000 & 0000 & 0000 & 0000 \\ 0000 & 1214 & 0000 & 12A4 & 0000 & 0000 & 0000 & 0000 \\ 0000 & 0000 & 12A4 & 0000 & 12A4 & 0000 & 0000 & 0000 \\ 0000 & 0000 & 0000 & 12A4 & 0000 & -12A4 & 1214 & 0000 \\ 0000 & 0000 & 0000 & 0000 & -12A4 & 0000 & -1214 & 0000 \\ 12A4 & 0000 & 0000 & 0000 & 1214 & -1214 & 0000 & -12A4 \\ -12C4 & 0000 & 0000 & 0000 & 0000 & 0000 & -12A4 & 0000 \end{bmatrix} \quad (A.5)$$

## References

1. Dai JS, Rees Jones J (1998) Mobility in metamorphic mechanisms of foldable/erectable kinds. In: Proceedings of the 25th ASME biennial mechanisms and robotics conference, DETC-5902. Atlanta, USA
2. Wohlhart K (1996) Kinematotropic Linkages Recent Advances in Robot Kinematics. Kluwer, Dordrecht, The Netherlands

3. Calletti C, Fanghella P (2001) Single-loop kinematotropic mechanism. Mech Mach Theory 36:743–761
4. Fanghella P, Galletti C, Giannotti E (2006) Parallel robots that change their group of motion. Advances in Robot Kinematics. Springer, Netherlands, 49–56
5. Gogu G (2009) Branching singularities in kinematotropic parallel mechanisms. In: Proceedings of the 5th international workshop on computational kinematics, Duisburg, Germanym, pp 341–348
6. Dai JS (2005) Rees Jones, J.: Matrix Representation of Topological Changes in Metamorphic Mechanisms. ASME. J Mech Des 127:837–840
7. Lan ZH, Du R (2008) Representation of Topological Changes in Metamorphic Mechanisms With Matrices of the Same Dimension. ASME J Mech Des 130:074501
8. Wang DL, Dai JS (2007) Theoretical Foundation of Metamorphic Mechanism and Its Synthesis. Chin J Mech Eng 43:32–42
9. Tsai LW (2001) Mechanism Design: Enumeration of Kinematic Structures According to Function. CRC Press, New York
10. Yan H-S, Kuo C-H (2006) Topological Representations and Characteristics of Variable Kinematic Joints. ASME J Mech Des 128:384–391
11. Kuo CH (2004) Structural characteristics of mechanisms with variable topologies taking into account the configuration singularity. MS thesis, National Cheng Kung University, Tainan, Taiwan
12. Zeng Q, Fang YF (2009) Structural Synthesis of Serial-Parallel Hybrid Mechanisms Based on Representation and Operation of Logical Matrix. J Mech Robotics 1(4):041003
13. Zhang K, Dai J S, Fang Y, Zeng Q (2011) String matrix based geometrical and topological representation of mechanisms. 13th World Congress in Mechanism and Machine Science, Guanajuato, Mexico, A23_493
14. Gogu G (2007) Structural synthesis of fully-isotropic parallel robots with Schönflies motions via theory of linear transformations and evolutionary morphology. Eur J Mech A/Solids 26:242–269

# Chapter 14
# Topological Analysis of Configuration Evolution of a Novel Type of Electric Loader with Metamorphic Functions

**Ganwei Cai, Yuchen Pan, Hongzhou Wang and Jian S. Dai**

**Abstract** Traditional loaders are mostly hydraulic type, of which there exists some problems like high costs of manufacturing, assembling and maintaining. A novel type of electric loader with the ability to change its topological structures to satisfy various kinds of needs in the working process is proposed. One driving motor can finish multiple tasks such as shoveling, loading, lifting, and unloading. This paper introduces the source-metamorphic mechanism of the newly invented loading mechanism, as well as its metamorphic ways to achieve variable topology. The adjacency matrix and topological graphs are used to describe topological change of each working phase, based on which the principles and characteristics of configuration changing are analyzed.

**Keywords** Metamorphic mechanisms · Loader · Topological graphs · Adjacency matrix

G. Cai (✉) · H. Wang
College of Mechanical Engineering, Guangxi University,
Nanning 530004, People's Republic of China
e-mail: caiganwei@163.com

Y. Pan
College of Civil Engineering & Architecture, Guangxi University,
Nanning 530004, People's Republic of China
e-mail: gxu_panyuchen@163.com

J. S. Dai
King's College London, University of London, Strand,
London WC2R 2LS, UK
e-mail: jian.dai@kcl.ac.uk

J. S. Dai et al. (eds.), *Advances in Reconfigurable Mechanisms and Robots I*,
DOI: 10.1007/978-1-4471-4141-9_14, © Springer-Verlag London 2012

## 14.1 Introduction

A loader is a heavy equipment machine often used in construction, primarily used to load material such as asphalt, dirt, snow, rock, sand, and woodchips into or onto another type of machinery like a dump truck, conveyor belt, or railcar. But the hydraulic system of a traditional loader uses fluid for operating media with a large number of hydraulic components, resulting in high cost of manufacturing and maintaining. Without proper maintenance, hydraulic leaks are inevitable.

Metamorphic mechanisms were proposed during the study of reconfigurable packaging, which are the mechanisms with the ability to change their topological structures [1, 2]. The approaches to achieve metamorphic function can be divided into two categories: one is to change the number of links by link coincidence and self locking [3]. The other is to apply geometric constraint to joints to change the joint property as that introduced in Ref. [4] by Dongming Gan et al. Parise et al. [5] presented new ortho-planar metamorphic mechanisms in 2000. Liu and Yang [6] investigated the essence and characteristics of metamorphic mechanisms in 2004. Dai and Rees Jones [7] presented the topology of variable mobility-configurations of metamorphic mechanisms and described transformation using matrix operations in 2005. Yan and Kuo [8, 9] investigated variable kinematic joints based on topological representations in the form of graphs and emphasized the meaning of variable topologies in 2006 and 2007, respectively. Zhang et al. [10] introduced the theories and methods of synthesis of metamorphic mechanisms based on biological modeling and evolution.

The potential use of this type of mechanisms is promising due to their multiple working phases, variable topology, and the changeable mobility. A novel type of electric loading mechanism is proposed with the ability to change its topological structures to adjust various kinds of demands during the working process. One driving motor can finish multiple tasks such as shoveling, loading, raising, rolling and dumping. The motor is placed on the frame, problems such as low stiffness, high inertia of the linkages of traditional serial manipulators can be avoided. In this paper, the source-metamorphic mechanism of the newly invented mechanism is introduced, as well as its metamorphic ways. The adjacency matrix and topological graphs are used to describe topological changes of each sub-phase of the mechanism, based on which the principles and characteristics of configuration changing are analyzed. The characteristics during the evolutionary process are obtained.

## 14.2 Configuration

As is shown in Figs. 14.1 and 14.2, the motion of the boom (link 6) of the mechanism is limited by upper limit and lower limit which are both attached to the frame. The limits stop the boom and cause topological change when the boom rotates to the limiting positions during the working process. A spindle with two

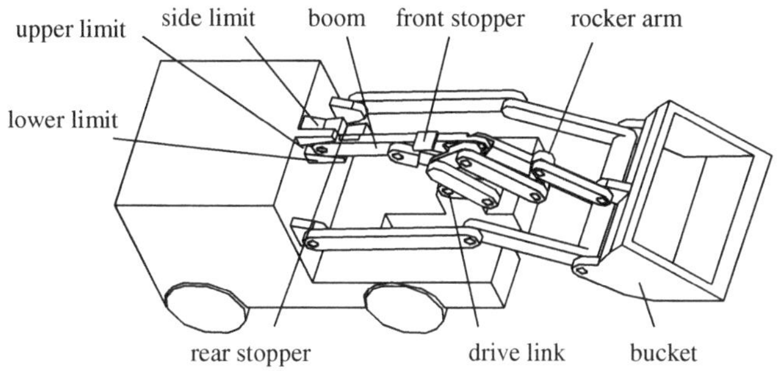


**Fig. 14.1** Virtual prototype of the novel loader

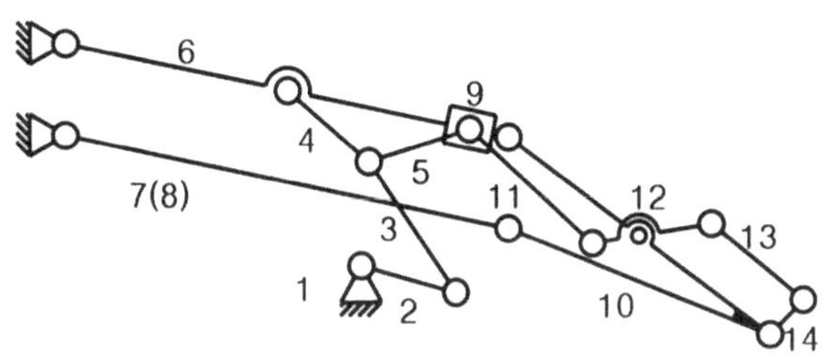


**Fig. 14.2** Sketch of the source metamorphic mechanism

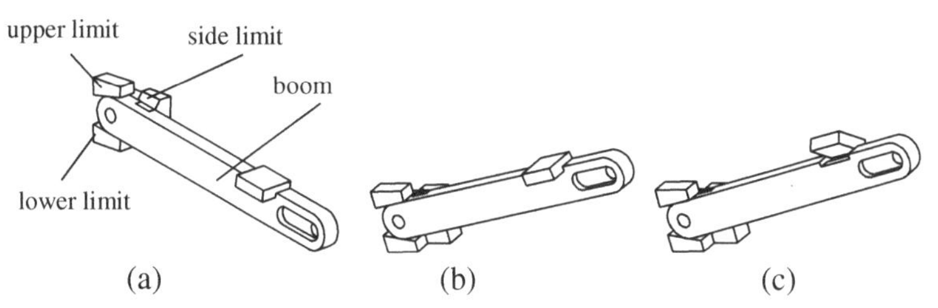


**Fig. 14.3** The boom and the spindle inside

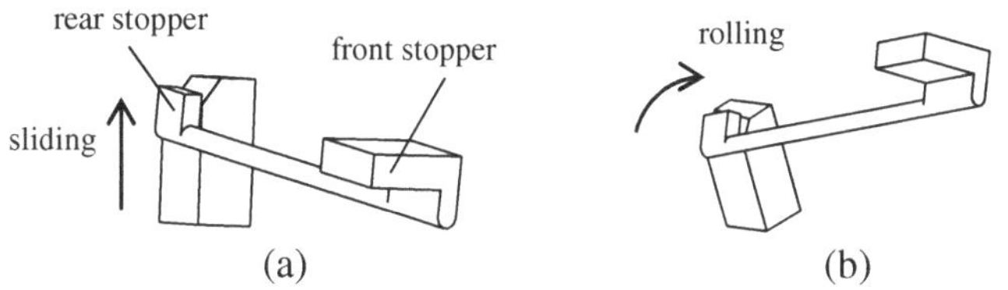


**Fig. 14.4** Two ways of motion of the spindle

stoppers is inside the boom where a groove for slider 9 is at the forepart as depicted in Figs. 14.3 and 14.4. A compound revolute joint connects link 3, link 4

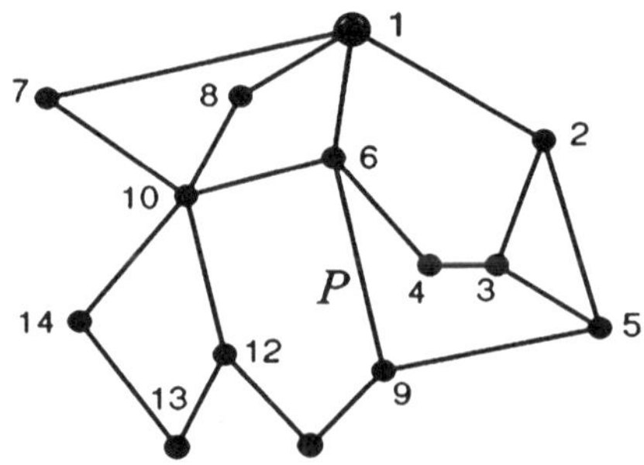


**Fig. 14.5** Topological graph of source-metamorphic mechanism ${}^{S}M$

and link 5 together so that drive link 2 is able to drive the boom and the slider via link 4 and link 5, respectively. Link 10 links the boom, the bucket (link 14), link 7 and link 8, of which the middle part connects to the rocker arm 12.

## 14.3 Topological Analysis of Working-Phase Mechanisms

### *14.3.1 Source-Metamorphic Mechanism*

Introduced by Zhang, et al., source-metamorphic mechanism ${}^{S}M$ contains the complete topological elements of all working configuration phases involved in a working cycle which can be generated from aggregationary evolution operation with the available minimum required mobility number and with elements of all working configurations [10]. Sketch and the topological graph of the source-metamorphic mechanism of the loader are shown in Figs. 14.2 and 14.5, respectively.

The adjacency matrix can be expressed as:

$$A_{0c} = \begin{array}{c} 1 \\ 2 \\ 3 \\ 4 \\ 5 \\ 6 \\ 7 \\ 8 \\ 9 \\ 10 \\ 11 \\ 12 \\ 13 \\ 14 \end{array} \begin{bmatrix} 0 & 1 & 0 & 0 & 0 & 1 & 1 & 1 & 0 & 0 & 0 & 0 & 0 & 0 \\ 1 & 0 & 1 & 0 & 0 & 0 & 0 & 0 & 0 & 0 & 0 & 0 & 0 & 0 \\ 0 & 1 & 0 & 1 & 1 & 0 & 0 & 0 & 0 & 0 & 0 & 0 & 0 & 0 \\ 0 & 0 & 1 & 0 & 1 & 1 & 0 & 0 & 0 & 0 & 0 & 0 & 0 & 0 \\ 0 & 0 & 1 & 1 & 0 & 0 & 0 & 0 & 1 & 0 & 0 & 0 & 0 & 0 \\ 1 & 0 & 0 & 1 & 0 & 0 & 0 & 0 & 1 & 1 & 0 & 0 & 0 & 0 \\ 1 & 0 & 0 & 0 & 0 & 0 & 0 & 0 & 0 & 1 & 0 & 0 & 0 & 0 \\ 1 & 0 & 0 & 0 & 0 & 0 & 0 & 0 & 0 & 1 & 0 & 0 & 0 & 0 \\ 0 & 0 & 0 & 0 & 1 & 1 & 0 & 0 & 0 & 0 & 1 & 0 & 0 & 0 \\ 0 & 0 & 0 & 0 & 0 & 1 & 1 & 1 & 0 & 0 & 0 & 1 & 0 & 1 \\ 0 & 0 & 0 & 0 & 0 & 0 & 0 & 0 & 1 & 0 & 0 & 1 & 0 & 0 \\ 0 & 0 & 0 & 0 & 0 & 0 & 0 & 0 & 0 & 1 & 1 & 0 & 1 & 0 \\ 0 & 0 & 0 & 0 & 0 & 0 & 0 & 0 & 0 & 0 & 0 & 1 & 0 & 1 \\ 0 & 0 & 0 & 0 & 0 & 0 & 0 & 0 & 0 & 1 & 0 & 0 & 1 & 0 \end{bmatrix} \tag{14.1}$$

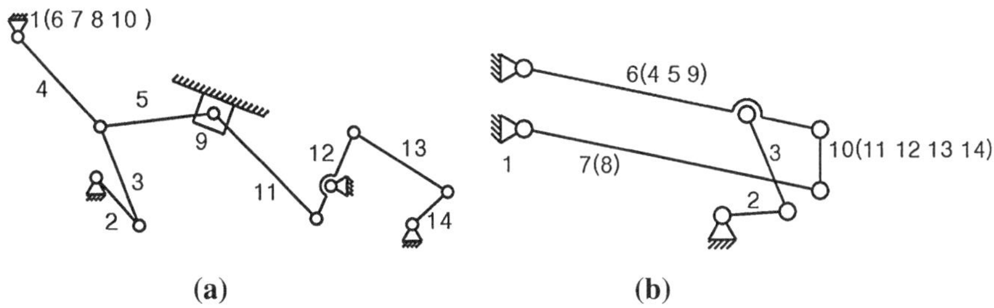


**Fig. 14.6** Sketch of the mechanisms during working phases

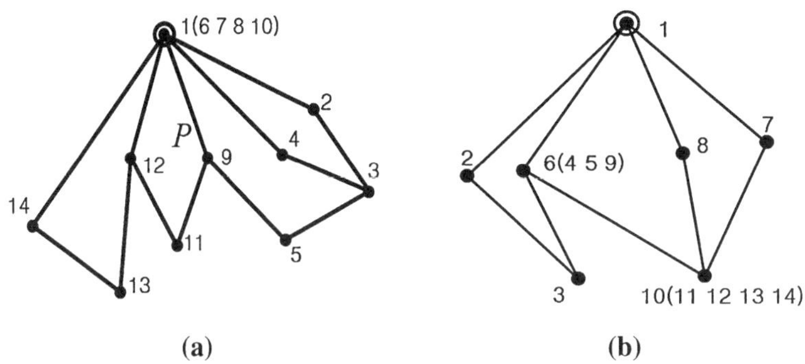


**Fig. 14.7** Topological graphs of working phases

### *14.3.2 Shoveling Phase*

The shoveling phase starts when the bucket is in the initial position. The boom stays in the lower position due to the gravity and is blocked by the lower limit, resulting in the stationary of link 7, link 8, link 10 and the bucket. The aforementioned links can be regarded as the frame during this phase shown in Fig. 14.8a. Position and posture of the spindle are shown in Figs. 14.3b and 14.4a. The sketch and the topological graph of the mechanism of are depicted in Figs. 14.6a and 14.7a, respectively. The adjacency matrix can be written as:

$$A_1 = \begin{matrix} 1 \\ 2 \\ 3 \\ 4 \\ 5 \\ 9 \\ 11 \\ 12 \\ 13 \\ 14 \end{matrix} \begin{bmatrix} 0 & 1 & 0 & 1 & 0 & 1 & 0 & 1 & 0 & 1 \\ 1 & 0 & 1 & 0 & 0 & 0 & 0 & 0 & 0 & 0 \\ 0 & 1 & 0 & 1 & 1 & 0 & 0 & 0 & 0 & 0 \\ 1 & 0 & 1 & 0 & 1 & 0 & 0 & 0 & 0 & 0 \\ 0 & 0 & 1 & 1 & 0 & 1 & 0 & 0 & 0 & 0 \\ 1 & 0 & 0 & 0 & 1 & 0 & 1 & 0 & 0 & 0 \\ 0 & 0 & 0 & 0 & 0 & 1 & 0 & 1 & 0 & 0 \\ 1 & 0 & 0 & 0 & 0 & 0 & 1 & 0 & 1 & 0 \\ 0 & 0 & 0 & 0 & 0 & 0 & 0 & 1 & 0 & 1 \\ 1 & 0 & 0 & 0 & 0 & 0 & 0 & 0 & 1 & 0 \end{bmatrix} \quad (14.2)$$

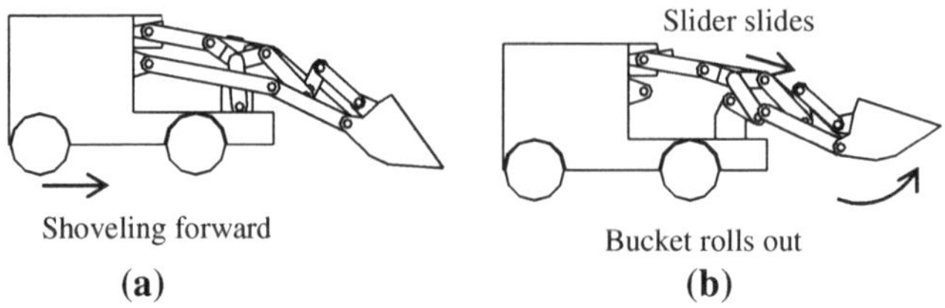


**Fig. 14.8** Shoveling and rolling out

## *14.3.3 Bucket Rolling-Out*

Referring to Fig. 14.8b, when the bucket is filled with material, it rolls back due to the rotation of drive link 2. The boom stays still as well as link 7, link 8 and link 10, in which slider 9 starts moving. Topological structure of the mechanism does not change in rolling out phase. The sketch and the topological graph of the mechanism are same to the shoveling phase, shown in Figs. 14.6a and 14.7a, respectively. The adjacency matrix can be easily obtained, yields:

$$A_2 = \begin{matrix} 1 \\ 2 \\ 3 \\ 4 \\ 5 \\ 9 \\ 11 \\ 12 \\ 13 \\ 14 \end{matrix} \begin{bmatrix} 0 & 1 & 0 & 1 & 0 & 1 & 0 & 1 & 0 & 1 \\ 1 & 0 & 1 & 0 & 0 & 0 & 0 & 0 & 0 & 0 \\ 0 & 1 & 0 & 1 & 1 & 0 & 0 & 0 & 0 & 0 \\ 1 & 0 & 1 & 0 & 1 & 0 & 0 & 0 & 0 & 0 \\ 0 & 0 & 1 & 1 & 0 & 1 & 0 & 0 & 0 & 0 \\ 1 & 0 & 0 & 0 & 1 & 0 & 1 & 0 & 0 & 0 \\ 0 & 0 & 0 & 0 & 0 & 1 & 0 & 1 & 0 & 0 \\ 1 & 0 & 0 & 0 & 0 & 0 & 1 & 0 & 1 & 0 \\ 0 & 0 & 0 & 0 & 0 & 0 & 0 & 1 & 0 & 1 \\ 1 & 0 & 0 & 0 & 0 & 0 & 0 & 0 & 1 & 0 \end{bmatrix} \tag{14.3}$$

## *14.3.4 Raising Phase*

As is depicted in Fig. 14.9a, the moment slider 9 reaches the front limit of the groove, links 4 and 5 are stopped by the front stopper. At the mean time, the slider will be locked there because the boom spindle can not rotate on this occasion due to the limit of side block, what is depicted in Figs. 14.3a, b, and 14.4a, resulting in the combination of boom 6, slider 9, link 4 and link 5. These links are lifted jointly by drive link 2.

The locked-up of the slider eliminates the relative motions between link 11, rocker arm 12, link 13 and the bucket. As a result, these four linkages can be regarded as one single linkage during the raising phase. Topological structure of the mechanism changes due to the links composition. Sketch of the mechanism during this working-phase is shown in Fig. 14.6b, when the topological graph is given in Fig. 14.7b. Consequently, the adjacency matrix for this specific sub-phase mechanism is obtained, that is:

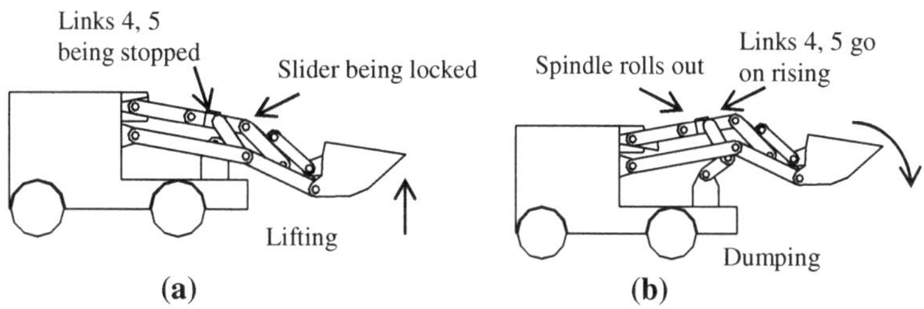


**Fig. 14.9** Raising and dumping

$$A_3 = \begin{array}{c} 1 \\ 2 \\ 3 \\ 6 \\ 7 \\ 8 \\ 10 \end{array} \begin{bmatrix} 0 & 1 & 0 & 1 & 1 & 1 & 0 \\ 1 & 0 & 1 & 0 & 0 & 0 & 0 \\ 0 & 1 & 0 & 1 & 0 & 0 & 0 \\ 1 & 0 & 1 & 0 & 0 & 0 & 1 \\ 1 & 0 & 0 & 0 & 0 & 0 & 1 \\ 1 & 0 & 0 & 0 & 0 & 0 & 1 \\ 0 & 0 & 0 & 1 & 1 & 1 & 0 \end{bmatrix} \quad (14.4)$$

### 14.3.5 Dumping Phase

The boom rotates to the upper limit position and is blocked, as well as link 7, link 8 and link 10, shown in Fig. 14.9b. These four linkages can be regarded as the frame again. The boom spindle rolls out of which the position and posture can be seen in Figs. 14.3c and 14.4b, resulting in the decomposition of boom 6, slider 9, link 4 and link 5. The last two links stop being limited by the front stopper and go on rising. The slider regains its motion in the groove and starts moving back. Topological change occurs in dumping phase. Particularly, the spindle of the boom can not roll back during the phase due to the limitation to the rear stopper provided by side limit, so that the boom can stay still in the upper limit position. Sketch and topological graph of the mechanism are exactly the same as that in the shoveling and rolling phase which are shown in Figs. 14.6a and 14.7a, as well as the adjacency matrix, reads:

$$A_4 = \begin{array}{c} 1 \\ 2 \\ 3 \\ 4 \\ 5 \\ 9 \\ 11 \\ 12 \\ 13 \\ 14 \end{array} \begin{bmatrix} 0 & 1 & 0 & 1 & 0 & 1 & 0 & 1 & 0 & 1 \\ 1 & 0 & 1 & 0 & 0 & 0 & 0 & 0 & 0 & 0 \\ 0 & 1 & 0 & 1 & 1 & 0 & 0 & 0 & 0 & 0 \\ 1 & 0 & 1 & 0 & 1 & 0 & 0 & 0 & 0 & 0 \\ 0 & 0 & 1 & 1 & 0 & 1 & 0 & 0 & 0 & 0 \\ 1 & 0 & 0 & 0 & 1 & 0 & 1 & 0 & 0 & 0 \\ 0 & 0 & 0 & 0 & 0 & 1 & 0 & 1 & 0 & 0 \\ 1 & 0 & 0 & 0 & 0 & 0 & 1 & 0 & 1 & 0 \\ 0 & 0 & 0 & 0 & 0 & 0 & 0 & 1 & 0 & 1 \\ 1 & 0 & 0 & 0 & 0 & 0 & 0 & 0 & 1 & 0 \end{bmatrix} \quad (14.5)$$

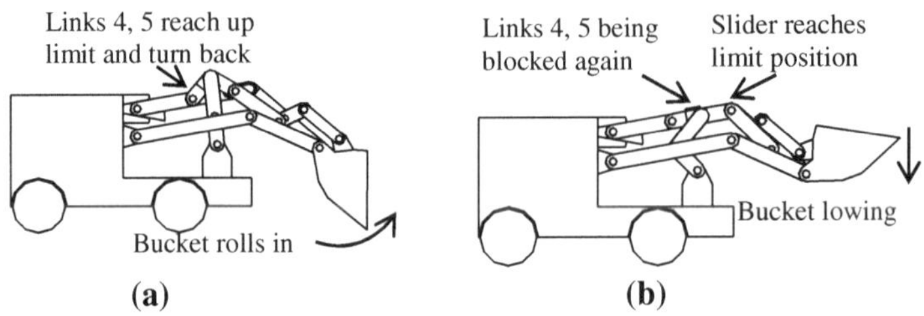


**Fig. 14.10** Bucket rolling in and lowing

## *14.3.6 Bucket Rolling-In*

As is shown in Fig. 14.10a, when dumping is finished, boom 6 stays in the upper limit position, so as link 7, link 8 and link 10, which are still treated as the frame. Drive link 2 keeps on rotating anticlockwise, resulting in the sliding of 9 and rolling-in of the bucket. Link 4 and link 5 turn back with the drive link 2 the moment they reach their up limit. Topological change does not happen in this phase. Sketch and topological graph remain the same to the previous working-phase, depicted in Figs. 14.6a and 14.7a. The adjacency matrix does not change either, writes:

$$A_5 = \begin{matrix} 1 \\ 2 \\ 3 \\ 4 \\ 5 \\ 9 \\ 11 \\ 12 \\ 13 \\ 14 \end{matrix} \begin{bmatrix} 0 & 1 & 0 & 1 & 0 & 1 & 0 & 1 & 0 & 1 \\ 1 & 0 & 1 & 0 & 0 & 0 & 0 & 0 & 0 & 0 \\ 0 & 1 & 0 & 1 & 1 & 0 & 0 & 0 & 0 & 0 \\ 1 & 0 & 1 & 0 & 1 & 0 & 0 & 0 & 0 & 0 \\ 0 & 0 & 1 & 1 & 0 & 1 & 0 & 0 & 0 & 0 \\ 1 & 0 & 0 & 0 & 1 & 0 & 1 & 0 & 0 & 0 \\ 0 & 0 & 0 & 0 & 0 & 1 & 0 & 1 & 0 & 0 \\ 1 & 0 & 0 & 0 & 0 & 0 & 1 & 0 & 1 & 0 \\ 0 & 0 & 0 & 0 & 0 & 0 & 0 & 1 & 0 & 1 \\ 1 & 0 & 0 & 0 & 0 & 0 & 0 & 0 & 1 & 0 \end{bmatrix} \tag{14.6}$$

## *14.3.7 Lowing Phase*

Referring to Fig. 14.10b, when slider 9 moves to its front limit and be locked due to the failure in rotation of the spindle, links 4 and 5 will be blocked by the front stopper again. Posture of the spindle returns to what it shows in Fig. 14.4a, which can not rotate again from this moment on. Because of the limitation of side limit on the rolling-out rear stopper, boom will not be falling in the gravity. Boom 6, slider 9, link 4 and link 5 combine together again and low down with drive link 2. Since slider 9 can not move, link 11, rocker arm 12, link 13, and the bucket lose

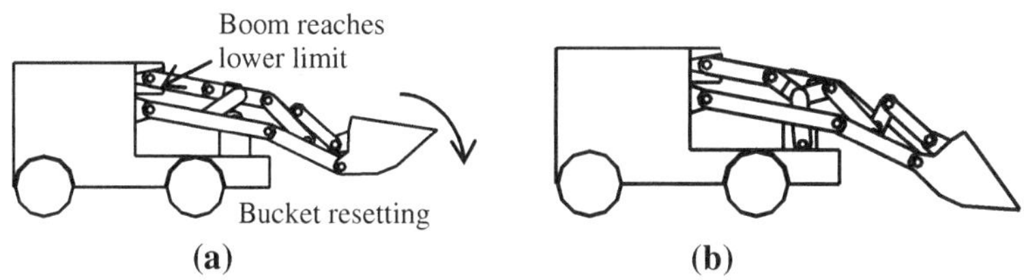


**Fig. 14.11** Bucket resetting to the initial position

the relative motions, which become one linkage once again. The topological structure of the mechanism changes again, back to that in the raising phase, to which the sketch, topological graph are identical, depicted in Figs 14.6b and 14.7b. The adjacency matrix can be written as:

$$A_6 = \begin{array}{r} 1 \\ 2 \\ 3 \\ 6 \\ 7 \\ 8 \\ 10 \end{array} \begin{bmatrix} 0 & 1 & 0 & 1 & 1 & 1 & 0 \\ 1 & 0 & 1 & 0 & 0 & 0 & 0 \\ 0 & 1 & 0 & 1 & 0 & 0 & 0 \\ 1 & 0 & 1 & 0 & 0 & 0 & 1 \\ 1 & 0 & 0 & 0 & 0 & 0 & 1 \\ 1 & 0 & 0 & 0 & 0 & 0 & 1 \\ 0 & 0 & 0 & 1 & 1 & 1 & 0 \end{bmatrix} \qquad (14.7)$$

## 14.3.8 Bucket Resetting Phase

Boom 6 rotates to the lowest position and is stopped by lower limit, as well as link7, link 8, and link 10. These four become the frame once again. The spindle turns back to the status shown in Fig. 14.4b. Slider 9 slides in the groove again. Referring to Fig. 14.11, the bucket rolls back to the initial position. Topological structure changes back to what is like in shoveling phase. Sketch and topological graph remain the same to the initial working-phase, depicted in Figs. 14.6a and 14.7a. The adjacency matrix can be obtained, say:

$$A_7 = \begin{array}{r} 1 \\ 2 \\ 3 \\ 4 \\ 5 \\ 9 \\ 11 \\ 12 \\ 13 \\ 14 \end{array} \begin{bmatrix} 0 & 1 & 0 & 1 & 0 & 1 & 0 & 1 & 0 & 1 \\ 1 & 0 & 1 & 0 & 0 & 0 & 0 & 0 & 0 & 0 \\ 0 & 1 & 0 & 1 & 1 & 0 & 0 & 0 & 0 & 0 \\ 1 & 0 & 1 & 0 & 1 & 0 & 0 & 0 & 0 & 0 \\ 0 & 0 & 1 & 1 & 0 & 1 & 0 & 0 & 0 & 0 \\ 1 & 0 & 0 & 0 & 1 & 0 & 1 & 0 & 0 & 0 \\ 0 & 0 & 0 & 0 & 0 & 1 & 0 & 1 & 0 & 0 \\ 1 & 0 & 0 & 0 & 0 & 0 & 1 & 0 & 1 & 0 \\ 0 & 0 & 0 & 0 & 0 & 0 & 0 & 1 & 0 & 1 \\ 1 & 0 & 0 & 0 & 0 & 0 & 0 & 0 & 1 & 0 \end{bmatrix} \qquad (14.8)$$

## 14.4 Conclusions

(i) A novel type of electric loading mechanism was introduced in this paper by applying metamorphic mechanisms to the innovative design of loaders, which can change its topological structures to satisfy various kinds of demands during the loading process.

(ii) The adjacency matrix and topological graphs were used to describe topological changes of each sub-phase of the working process, based on which the working principles during configuration changing was analyzed. Characteristics in the evolutionary process were obtained.

**Acknowledgments** This work was financially supported by the National Natural Science Foundation of China (No. 51075077), Guangxi Key Laboratory of Manufacturing System & Advanced Manufacturing Technology Project (09-007-05S013) and Teams for Innovation in the Construction of Talent Highlands in Guangxi Institutions of Higher Learning.

## References

1. Dai JS, Rees JJ (1999) Mobility in metamorphic mechanisms of foldable/erectable kinds. J. Trans. ASME Mech. Des. 121(3):375–382
2. Dai JS (1996) Conceptual study of the dexterous reconfigurable assembly and packaging system. Science and Technology Report, Unilever Research, Port Sunlight
3. Leonesio M, Bianchi G, Manara P (2007) A general approach for self-locking analysis in closed kinematic chains. In: Proceedings of the 12th world congress in mechanism and machine theory, Besancon, France, pp 141–147
4. Gan DM, Dai JS, Liao QZ (2010) Constraint analysis on mobility change of a novel metamorphic parallel mechanism. J. Mech. Mach. Theory. 45:1864–1876
5. Parise JJ, Howell LL, Magleby SP (2000) Ortho-planar mechanisms. In: Proceedings 26th biennial mechanisms and robotics conference, Baltimore
6. Liu C, Yang T (2004) Essence and characteristics of metamorphic mechanisms and their metamorphic ways. In: Proceedings of the 11th world congress in mechanism and machine science, Tianjing, China, pp 1285–1288
7. Dai JS, Rees JJ (2005) Matrix representation of topological changes in metamorphic mechanisms. J. ASME Mech. Des. 127(4):837–840
8. Yan HS, Kuo CH (2006) Topological representations and characteristics of variable kinematic joints. J. ASME Mech. Des. 128(2):384–391
9. Yan HS, Kuo CH (2007) On the mobility and configuration singularity of mechanisms with variable topologies. J. ASME Mech. Des. 129:617–624
10. Zhang LP, Wang DL, Dai JS (2008) Biological modeling and evolution based synthesis of metamorphic mechanisms. J. ASME Mech. Des. 130:1–11

# Chapter 15
# Analysis of Freedom Degrees of Self-Help Chair for the Old Based on the Principle of Metamorphic Mechanism

**Ju-Jiang Cao, Yu-Qi Wang and Ji-Wei Sun**

**Abstract** By analysing the work process of self-help chair for the old, combined with the topology analysis, calculated the degrees of freedom in each phase. This page elaborated on the analytical methods of composite hinge in topological diagram, and provided a design example and analytical samples for further research of metamorphic theory and analytical method.

**Keywords** Metamorphic mechanism · Topological diagram · Degrees of freedom · Self-help chair

## 15.1 Preface

Metamorphic mechanism was proposed firstly by Dai and Jones in 25th ASME Biennial Mechanisms and Robotics Conference, Atlanta, 1998 [1]. From then on, metamorphic mechanism has become a hot research spot in international academic institutions. As scholars, from various countries, researched metamorphic mechanism more in-depth, the definition and research methods of it was improved, and many metamorphic mechanism with practical value was developed.

J.-J. Cao · Y.-Q. Wang (✉) · J.-W. Sun
Shaanxi University of Science and Technology, Xi'an, 710021, China
e-mail: wangyuqi1117@126.com

J.-J. Cao
e-mail: jjcao@sust.edu.cn

J.-W. Sun
e-mail: evan2007xy@163.com

J. S. Dai et al. (eds.), *Advances in Reconfigurable Mechanisms and Robots I*,
DOI: 10.1007/978-1-4471-4141-9_15, 

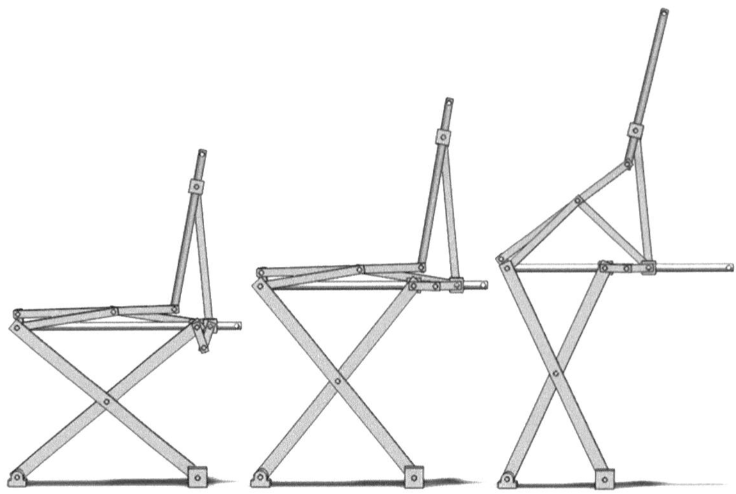

**Fig. 15.1** 3D model of self-help chair

For example: In the aviation field, metamorphic rover [2]; In the robotics field, metamorphic robotic hand [3]; In the manufacturing area, metamorphic quick-response clamp device [4]; In the field of household products, metamorphic self-help chair [5]. Metamorphic mechanism used in different phases with multiple occasions, and from one session to another session, always change the topology diagram and freedom degrees of the metamorphic mechanism [6].

This paper, researched the self-help chair for the old which based on the principle of metamorphic mechanism, by analyzing the topology diagram of self-help chair in various stages of the work process, calculated the degrees of freedom in each phase, studied of their work state at various stages in-depth, provided a design example and analytical samples for further research of metamorphic theory and analytical method.

## 15.2 Introduction to Self-Help Chair

Self-help chair is used to help the elderly to change their body posture from sitting to standing freely, and can meet their needs for leaning against the chair. The working principle of self-help chair is: using the scissor mechanism as a lifting device, and provide work platforms for the seat turning device and the backrest adjusting device on the top; using the slider-crank mechanism as the seat turning device; using the mobile guide bar dual slider mechanism as the backrest angle adjusting device; using the metamorphic components as the control device. The 3D modes of the self-help chair in working process is shown in Fig. 15.1.

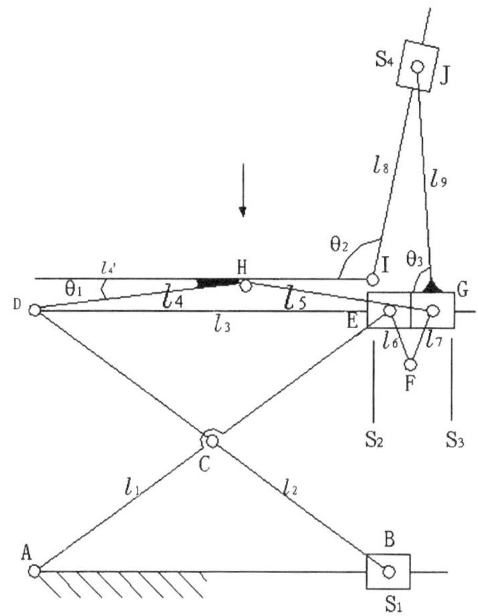


**Fig. 15.2** Sitting position of self-help chair

## 15.3 The Working Process of Self-Help Chair

The standing process of self-help chair includes four stages: sitting, rising, turning and standing, metamorphic components controlled the chair to change from rising phase into turning phase, and ultimately into standing position.

### *15.3.1 The Topology Diagram and Freedom Degrees Analysis of Sitting Stage*

The most basic function of self-help chair is meeting the elderly's needs of sitting, when the self-help chair is in the sitting stage, its mechanical diagram is shown in Fig. 15.2.

As can be seen from Fig. 15.2, in this stage, the slider S2 and the slider S3 combined together closely. The slider 3 has reached the bottom of the link L3, can not move anymore. The link L9 and the slider S3 is dead joint. The following conclusion can be drawn: L3, S2, S3, L6, L7 and L9 have been combined together, they are equivalent to a component.

The way of how to change mechanical diagram into topological graph is: change mechanical components into pitch point, change kinematic pair into line [7]. Then we can get the topological graph of the self-help chair in sitting position, shown in Fig. 15.3.

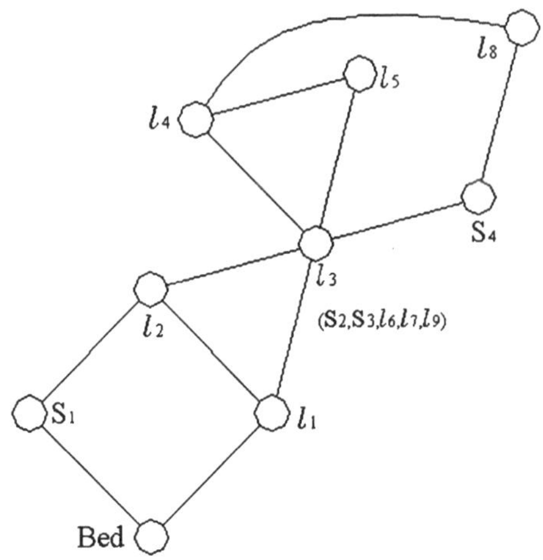


**Fig. 15.3** Topological diagram of sitting position

In the topological graph shown in Fig. 15.3, there are 9 pitch points, 12 lines. With the exception of the bed, residual 8 pitch points stands for movable components, the 12 lines stands for kinematic pair. Taking the information into the classic formula for calculating freedom degrees of planar mechanisms:

$$F = 3n - 2p_L - p_H \tag{1}$$

n The number of active components $p_L$: the number of lower pair
$p_H$ The number of higher pair

So we can obtain the freedom degrees as:

$$F_1 = 3 \times 8 - 2 \times 12 = 0$$

The degrees of freedom is 0, there is no component can move in parallel or revolve around an axis.The chair maintain a stable posture for people to use.

### *15.3.2 The Topology Diagram and Freedom Degrees Analysis of Rising Stage*

The rising stage of the self-help chair's working process is mainly to assist the body center of gravity to rise higher, its mechanical diagram is shown in Fig. 15.4.

As can be seen from Fig. 15.4, in this stage, under the external force, scissor mechanism began to run, the slider S1 and slide S2 start to run to left, the chair is in a vertical upward motion. Meanwhile, the slide S2 and slide S3 began to separate, the angle between link L6 and link L7 become larger. Change mechanical components into pitch point, and change kinematic pair into line, then we can get the topological graph of the self-help chair in rising stage, shown in Fig. 15.5.

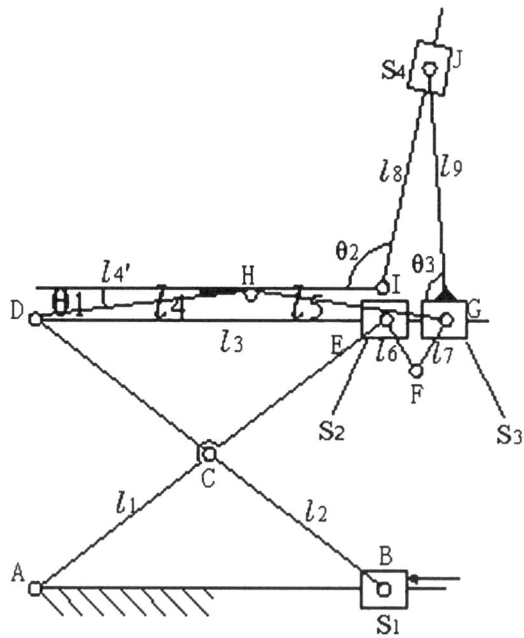


**Fig. 15.4** Rising stage of self-help chair

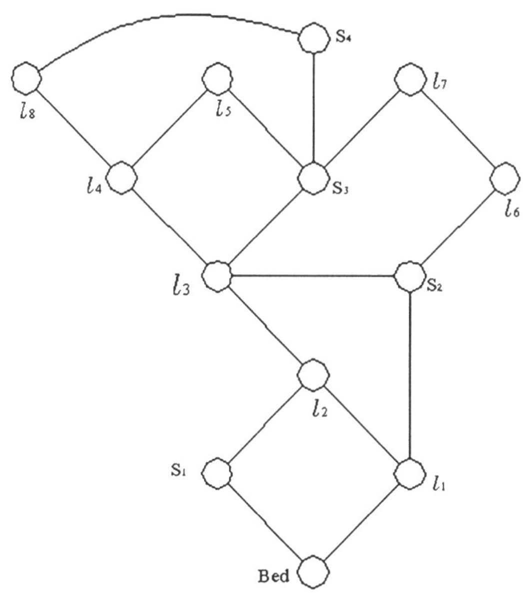


**Fig. 15.5** Topological diagram of rising stage

In the topological graph shown in Fig. 15.5, there are 13 pitch points, 17 lines. With the exception of the bed, residual 12 pitch points stands for movable

**Fig. 15.6** Turning stage of self-help chair

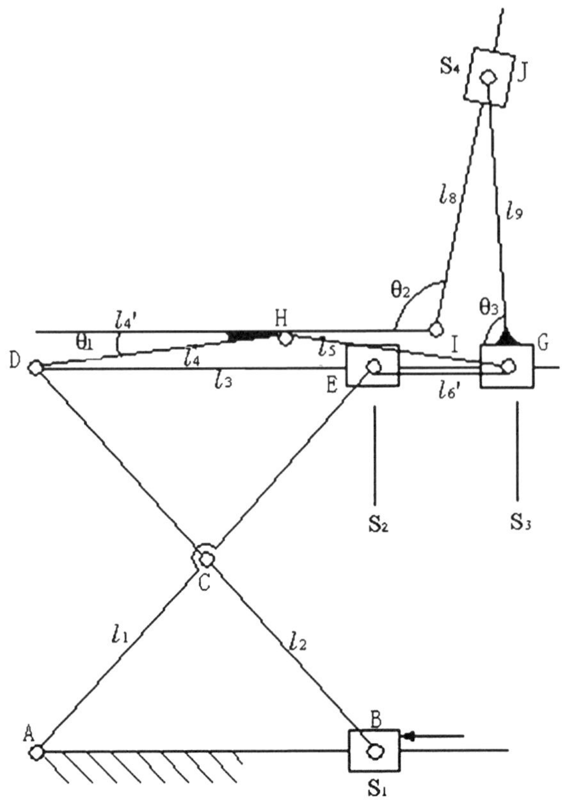


components, the 17 lines stands for kinematic pair. Taking the information into the formula (15.1), so we can obtain the freedom degrees as:

$$F_2 = 3 \times 12 - 2 \times 17 = 2$$

The degrees of freedom is 2, so the chair need dual power inputs to get a definite movemen state. The designer gave an outside force acting on the slider S1, and the other force is the gravity of the user, acting on the link L4′ which support the seat side of the chair. So the chair could do the vertical upward motion steadily.

### 15.3.3 The Topology Diagram and Freedom Degrees Analysis of Turning Stage

The turning phase of the self-help chair's working process is mainly to assist the old to change their hip position from awash to upright, its mechanical diagram is shown in Fig. 15.6.

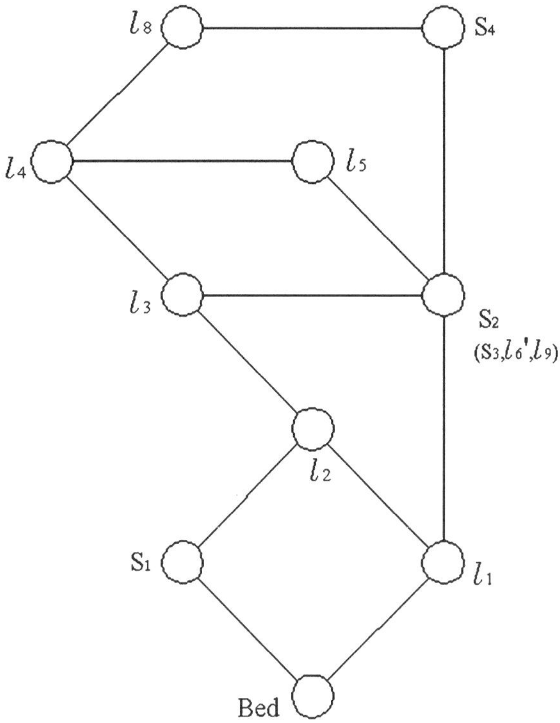


**Fig. 15.7** Topological diagram of turning stage

As can be seen from Fig. 15.6, in this stage, under the external force, scissor mechanism continue to run, the slider S1 and slide S2 run to left more far. Meanwhile, the angle between link L6 and link L7 is gradually widening to 180 °. The two links only transfer tension, they are equivalent to a link L6′. The slide S2 and slide S3 are tightly connected with link L6′, cannot move or rotate relatively. The following conclusion can be drawn: L6′, S2, S3 have been combined together, they are equivalent to a component.Change mechanical components into pitch point, and change kinematic pair into line, then we can get the topological graph of the self-help chair in turning stage, shown in Fig. 15.7.

In the topological graph shown in Fig. 15.7, there are 10 pitch points, 13 lines. With the exception of the bed, residual 9 pitch points stands for movable components, the 13 lines stands for kinematic pair. Taking the information into the formula (15.1), so we can obtain the freedom degrees as:

$$F_3 = 3 \times 9 - 2 \times 13 = 1$$

The degrees of freedom is 1, under the action of a single force, the chair has a definite state of motion: the scissor mechanism which formed by L1, L2 and L3 do the vertical upward motion steadily; the slider-crank mechanism which formed by L4, L5 and S3 do turning motion; the mobile guide bar dual slider mechanism which formed by L4, L8, S3 and S4 adjust the angle of seat back in a certain range.

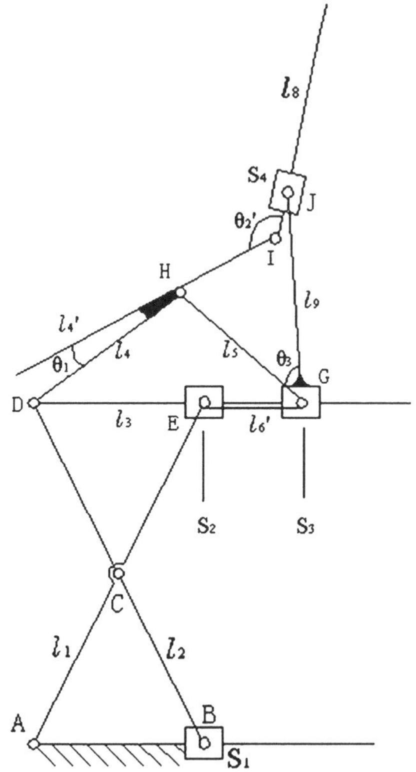


**Fig. 15.8** Standing stage of self-help chair

## 15.3.4 The Topology Diagram and Freedom Degrees Analysis of Standing Stage

The standing phase of the self-help chair's working process is the final stage when the chair have finished rising the body center of gravity to higher, turning the seat side, adjusting the angle of seat back. Its mechanical diagram is shown in Fig. 15.8.

As can be seen from Fig. 15.8, in this stage, with the seat turning device running, the link L4′ which supported seat side turned to a certain angle, the angle between link L4′ and link L8 is changed into $\theta 2'$. The chair's posture is similar with the people's posture when he leave the seat and stand up, so it can be used to lean back for the old. At the same time, the slider S1 have moved to the end of slide way on the bed, combined with the bed together, they are equivalent to a component.Change mechanical components into pitch point, and change kinematic pair into line, then we can get the topological graph of the self-help chair in standing stage, shown in Fig. 15.9.

In the topological graph shown in Fig. 15.9, there are 9 pitch points, 12 lines. With the exception of the bed, residual 8 pitch points stands for movable

Fig. 15.9 Topological diagram of standing stage

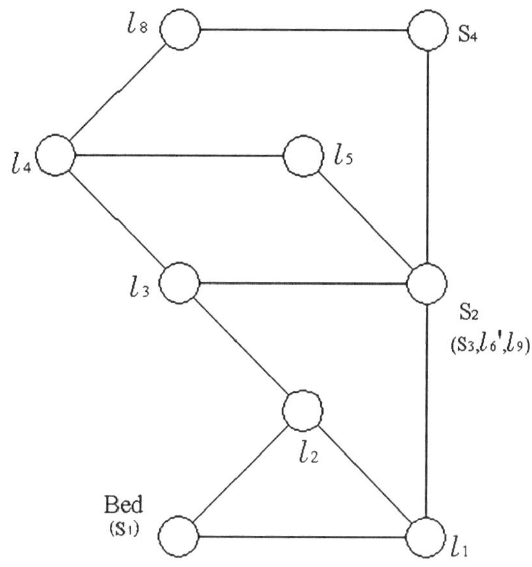


components, the 12 lines stands for kinematic pair. Taking the information into the formula (15.1), so we can obtain the freedom degrees as:

$$F_4 = 3 \times 8 - 2 \times 12 = 0$$

The degrees of freedom is zero, there is no component can move in parallel or revolve around an axis.The chair maintain a stable standing posture for people to use.

## 15.4 The Topology Diagram's Analysis of Composite Hinge

In the mechanical diagram of the chair, there are two types of composite hinges, how to properly analyze the topology of the hinges is the key point of the analysing the degree of freedom, so we need elaborate on its analysis process.

The first type of composite hinge is composed of three links L2, L3 and L4, connected by the same hinge. Using the dislocation method to separate it, we obtained the exact topological graph, as shown in Fig. 15.10.

The second type of composite hinge is composed of two links L1, L2 and a slider S2, connected by the same hinge. Using the dislocation method to separate it, we obtained the exact topological graph, as shown in Fig. 15.11.

Fig. 15.10 The first type of composite hinge

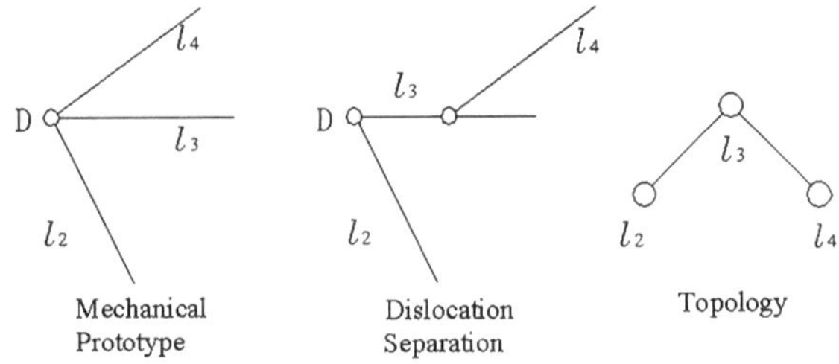


Fig. 15.11 The second type of composite hinge

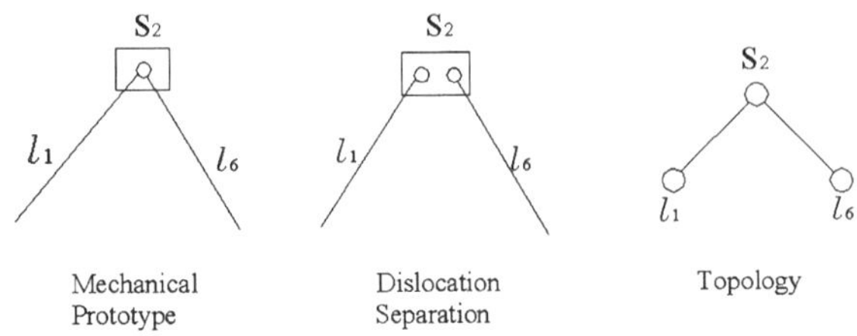


## 15.5 Conclusion

Metamorphic mechanism is a kind of special mechanism which changed its topological structure at least one more time (all structure mode to be different from each other) [8], during the operation process. By analyzing the topological graph, can obtain the number of effective links and kinematic pair intuitively, calculate the freedom degrees of the mechanism. Therefore, analysing topological graph plays a vital role on the determining and analysing metamorphic mechanism. It is certain that, with scholars from various countries researched metamorphic mechanism theory and analysis method in-depth, metamorphic mechanism will be widely used in various industries.

## References


1. Dai JS, Rees Jones J (1999) Mobility in metamorphic mechanisms of foldable/erectable kinks.R. In: 25th ASME biennial mechanisms and robotics conference, Atlanta, 1998. J Mech Des, Trans of ASME 121(3):375–385
2. Zhang K, Fang Y, Fang H (2007) Design and analysis of a rover mechanism based on the metamorphic principle. J Beijing Univ Aeronaut Astronaut 33(7):838–841
3. Dai JS (2007) WANG DL. Geometric analysis and synthesis of the metamorphic robotic hand.J.Transactions of ASME. J Mech Des 129(11):1191–1197
4. Li W, Lu J, Liu X, Wei H (2005) A New Metamorphic Quick-response Clamp Device. J. Mach Des Res 21(2):45–46
5. Sun J, Cao J (2011) Design of Self-help Chair for the Old Based on the Principle of Metamorphic Mechanism. J. Packag Eng 32(18):54–57

6. Tao X, Cao J, Liu D (2009) Some Basic Metamorphic Mechanism Theory and Its Applied Research. J. Packag Food Mach 27(4):57–60
7. Liang C, Ruan P (1986) A CAD system for linkage mechanism. Mechanical Industry Press, Beijing
8. Liu CH, Yang TL (2004) Essence and characteristics of metamorphic mechanisms and their metamorphic ways. In: Proceeding of the eleventh world congress in mechanism and machine science, April 1–4, 2004, Tianjin, China, China Machine Press, Beijing, 1285–1288

# Chapter 16
# The Variable Graph of Metamorphic Mechanisms or Variable Topology Mechanisms

**Chuan-He Liu**

**Abstract** New concepts such as the variable graph, the invariant subgraph, the variable subgraph and the cell-variator etc. of Metamorphic Mechanisms or Variable Topology Mechanisms are put forward and defined for the first time in this paper. Three new theorems on composition of the variable graph and one new theorem on the equivalent transformation of the incidence relation cell-variator are found out and their validity is strictly proved. The classifications of the cell-variators and the diagram representations of three kinds of cell-variators are studied. The diagram representation and adjacent matrix representation of the variable graph are given. The variable graph provides useful new tool for the topological structural analysis and synthesis of metamorphic mechanisms or variable topology mechanisms. It has great scientific value and engineering practicality.

**Keywords** Variable graph · Metamorphic mechanisms · Variable topology mechanisms · Cell-variator

## 16.1 Introduction

Since the concepts of the metamorphic mechanism (MM) [1, 2] and the variable topology mechanism (VTM) [3, 4] were put forward, many scholars had investigated the representation methods [5–13] of MMs or variable mechanisms. Initially, the representative tools were the schematic diagram of mechanism and the

C.-H. Liu (✉)
475 Taoshan Road, Bengbu, 233050 Anhui, People's Republic of China
e-mail: chuanheliu@sohu.com

J. S. Dai et al. (eds.), *Advances in Reconfigurable Mechanisms and Robots I*,
DOI: 10.1007/978-1-4471-4141-9_16, 

topological graph [1]. In 2000, Dai and Zhang [5] used adjacent matrix (it only includes "0" and "1") operations to represent the configuration transformation of a MM. Yan and Liu [3] used the finite state machine to represent mechanisms and chains with variable topologies. In 2003, Yan and Liu [6] used topological graph, adjacent matrix and joint-codes to represent the topological structure of mechanisms and chains with variable topologies. In 2006 and 2009, Yan and Kuo [7–9] created the unified graph that can represent and identify structural and motion state characteristics of mechanisms with variable topologies. The unified graph was used for structural analysis and configuration synthesis of mechanisms with variable topologies. This is a breakthrough in research direction on representations methods of the mechanisms with variable topologies. In 2008, Lan and Du [10] introduced "$-1$" into adjacent matrix, in order to indicate the frozen kinematic pairs. In 2009, Yang et al. [11] proposed the position and orientation characteristic matrix in order to describe reconfigurable mechanisms. Li et al. [12] proposed the concept of joint-gene based variable topological representations and configuration transformations and used the joint-gene matrix that can represent property and orientation of the joints at the same time for representation of topological structure of MMs. Liu [13] constructed the configuration-function transition digraphs of MMs or VTMs. It especially applied to the top design and analysis of MMs or VTMs. However up to now, the representation methods of MMs or VTMs had not cast off the fetter of the traditional invariant topological graph all the same. The composition and change of MMs or VTMs still cannot be represented accurately. In order to solve this problem, the fetter of the traditional invariant graph is casted off in this paper, and a completely new variable graph of MMs or VTMs (the variable graph for short hereinafter) is established. It can represent exactly the composition and change of MMs or VTMs under the condition of one graph or one adjacent matrix only. It applies to all kinds of MMs or VTMs. The definition of the variable graph is given first in this paper, next the composition of the variable graph is studied, then the diagram representation of the variable graph is discussed, and finally the adjacent matrix representation of the variable graph is investigated.

## 16.2 The Definitions on the Variable Graph

### *16.2.1 The Definition of the Invariant Topological Graph in One Phase*

**Definition 1** The $\mathbf{G}_i$, invariant topological graph of a MM or a VTM in the *i*th working-phase is defined as ordered triple as follows

$$\mathbf{G}_i = (V_i, E_i, R_i) \tag{16.1}$$

where $\boldsymbol{V}_i$ is the set of vertices that denote the links of a MM or a VTM in the $i$th working-phase. $\boldsymbol{E}_i$ is the set of edges that denote the joints of a MM or a VTM in the $i$th working-phase. $\boldsymbol{R}_i \subseteq (\boldsymbol{V}_i \& \boldsymbol{V}_i)$, the & is the unordered product and it is the function-set of the unordered pair from $\boldsymbol{E}_i$ to $\boldsymbol{V}_i$. $\boldsymbol{R}_i$ is the set of incidence functions, every element is the incidence relation between two vertices and one edge in $\boldsymbol{R}_i$. The ranges of elements which is included in $\boldsymbol{V}_i$, $\boldsymbol{E}_i$ and $\boldsymbol{R}_i$ are expanded in this paper in order to satisfy the need for research and application of MMs or VTMs.

$$V_i = \{V_{ir}, V_{ie}, V_{if}, V_{il}, V_{if}\} \neq \phi \tag{16.2}$$

where $\boldsymbol{V}_{ir}$ is the set of vertices that denote the rigid links of a MM or a VTM in the $i$th working-phase. $\boldsymbol{V}_{ie}$ is the set of vertices that denote the elastic links of a MM or a VTM in the $i$th working-phase. $\boldsymbol{V}_{if}$ is the set of vertices that denote the flexible links of a MM or a VTM in the $i$th working-phase. $\boldsymbol{V}_{il}$ is the set of vertices that denote the liquid links of a MM or a VTM in the $i$th working-phase. $\boldsymbol{V}_{i\phi}$ is the set of vertices that denote the empty (nothing) links of a MM or a VTM in the $i$th working-phase.

$$E_i = \{E_{ip}, E_{ir}, E_{i\phi}\} \tag{16.3}$$

Where $\boldsymbol{E}_{ip}$ is the set of edges that denote the kinematic pair joints of a MM or a VTM in the $i$th working-phase. $\boldsymbol{E}_{ir}$ is the set of edges that denote the rigid joints of a MM or a VTM in the $i$th working-phase. $\boldsymbol{E}_{i\phi}$ is the set of edges that denote the empty (nothing) joints of a MM or a VTM in the $i$th working-phase.

$$R_i = \{R_{ih}, R_{i\phi}\} \tag{16.4}$$

Where $\boldsymbol{R}_{ih}$ is the set of the real existent incidence relations of vertices (links) and edges (joints) of a MM or a VTM in the $i$th working-phase. $\boldsymbol{R}_{i\phi}$ is the set of the empty (inexistent) incidence relations of vertices (links) and edges (joints) of a MM or a VTM in the $i$th working-phase.

## 16.2.2 The Definition of the Variable Graph

**Definition 2** Suppose the $n$ invariant topological graphs of a MM or a VTM in $n$ (the $n$ is the finite integer and $n \geq 2$) working-phases are respectively $\mathbf{G}_1$, $\mathbf{G}_2$, …, $\mathbf{G}_n$ and they are not isomorphic pairwise, if a graph $\mathbf{G}_v$ is changed according to certain sequence or conditions among the $\mathbf{G}_1$, $\mathbf{G}_2$, …, $\mathbf{G}_n$, then

$$\mathbf{G}_v = \mathbf{G}_1 \cup \mathbf{G}_2 \cup \ldots \cup \mathbf{G}_n = \bigcup_{i=1}^{n} \mathrm{G}_i \tag{16.5}$$

is called the variable graph of the MM or the VTM.

## 16.3 The Composition of the Variable Graph

### 16.3.1 The Invariant Element and the Invariant Subgraph

**Definition 3** In a variable graph $\mathbf{G}_v$, an arbitrary vertex $v \in \boldsymbol{V}_i$, or an arbitrary edge $e \in \boldsymbol{E}_i$, or an arbitrary incidence relation $r \in \boldsymbol{R}_i$ $(i = 1, 2, \ldots, n)$ is called the element of the variable graph $\mathbf{G}_v$.

**Definition 4** In a variable graph $\mathbf{G}_v$, if an element (an arbitrary vertex $v \in \boldsymbol{V}_i$, or an arbitrary edge $e \in \boldsymbol{E}_i$, or an arbitrary incidence relation $r \in \boldsymbol{R}_i$ $(i = 1, 2, \ldots, n)$) is always unchanged during $n$ working-phases of a MM or a VTM, then the element is called the invariant element.

**Definition 5** If a graph $\mathbf{G}_{is} \subset \mathbf{G}_v$, and $\mathbf{G}_{is}$ consists of all the invariant elements in the variable graph $\mathbf{G}_v$, then $\mathbf{G}_{is}$ is called the invariant subgraph of the variable graph $\mathbf{G}_v$.

### 16.3.2 The Variable Element and the Variable Subgraph

**Definition 6** In a variable graph $\mathbf{G}_v$, if an element (an arbitrary vertex $v \in \boldsymbol{V}_i$, or an arbitrary edge $e \in \boldsymbol{E}_i$, or an arbitrary incidence relation $r \in \boldsymbol{R}_i$ $(i = 1, 2, \ldots, n)$) is to be changed during $n$ working-phases of a MM or a VTM, then the element is called the variable element.

**Definition 7** If a graph $\mathbf{G}_{vs} \subset \mathbf{G}_v$, and the $\mathbf{G}_{vs}$ consists of all the variable elements in the variable graph $\mathbf{G}_v$, then $\mathbf{G}_{vs}$ is called the variable subgraph of the variable graph $\mathbf{G}_v$.

### 16.3.3 The Composition Principle of the Variable Graph

**Theorem 1** *An arbitrary variable graph $\mathbf{G}_v$ consists of its invariant subgraph $\mathbf{G}_{is}$ and its variable subgraph $\mathbf{G}_{vs}$. Namely, if $\mathbf{G}_{is}$ and $\mathbf{G}_{vs}$ are respectively the invariant subgraph and the variable subgraph of the variable graph $\mathbf{G}_v$, then*

$$\mathbf{G}_v = \mathbf{G}_{is} \cup \mathbf{G}_{vs} \tag{16.6}$$

*Proof* Every element in the variable graph $\mathbf{G}_v$ is other invariant element, or variable element. It must be one or the other. So the union of all invariant elements is equal to the invariant subgraph $\mathbf{G}_{is}$, and the union of all variable elements is equal to the variable subgraph $\mathbf{G}_{vs}$, otherwise there is no any other element in the variable graph $\mathbf{G}_v$, and there is also no any other subgraph in it. Videlicet, there is only the invariant subgraph $\mathbf{G}_{is}$ and the variable subgraph $\mathbf{G}_{vs}$ in the variable graph $\mathbf{G}_v$. Therefore $\mathbf{G}_v = \mathbf{G}_{is} \cup \mathbf{G}_{vs}$.

**Theorem 2** *If the n invariant topological graphs of a MM or a VTM in n (n is the finite integer and n ≥ 2) working-phases are respectively* $\mathbf{G}_1$, $\mathbf{G}_2$, …, $\mathbf{G}_n$ *and they are not isomorphic pairwise, and* $\mathbf{G}_{is}$ *is the invariant subgraph of the variable graph* $\mathbf{G}_v$, *then*

$$\mathbf{G}_{is} = \mathbf{G}_1 \cap \mathbf{G}_2 \cap \ldots \cap \mathbf{G}_n = \bigcap_{i=1}^{n} \mathrm{G}_i \tag{16.7}$$

*Proof* Because $\mathbf{G}_{is}$ is the invariant subgraph of the variable graph $\mathbf{G}_v$, and $\mathbf{G}_1$, $\mathbf{G}_2$, …, $\mathbf{G}_n$ are not isomorphic pairwise, from the definition (definition 5) of the invariant subgraph we know that ∀ invariant element $g \in \mathbf{G}_{is}$, always has $g \in \mathbf{G}_1$ and $g \in \mathbf{G}_2$ and…and $g \in \mathbf{G}_n$. Namely $\mathbf{G}_{is} = \mathbf{G}_1 \cap \mathbf{G}_2 \cap \ldots \cap \mathbf{G}_n = \bigcap_{i=1}^{n} \mathrm{G}_i$.

**Theorem 3** *If the n invariant topological graphs of a MM or a VTM in n (n is the finite integer and n ≥ 2) working-phases are respectively* $\mathbf{G}_1$, $\mathbf{G}_2$, …, $\mathbf{G}_n$ *and they are not isomorphic pairwise, the number of directed edges of the configuration-function transition digraphs [13] whose n vertices are* $\mathbf{G}_1$, $\mathbf{G}_2$, …, $\mathbf{G}_n$ *is q, the symmetric difference of two invariant (in one phase) topological graphs that are incident with the ith directed edge is* $\boldsymbol{D}_i$, *and* $\mathbf{G}_{vs}$ *is the variable subgraph of the variable graph* $\mathbf{G}_v$, *then*

$$\mathbf{G}_{vs} = \bigcup_{i=1}^{q} D_i \tag{16.8}$$

Where $\boldsymbol{D}_i$ denotes the symmetric difference.

$$D_i = (\mathbf{G}_{i-1} \cup \mathbf{G}_i) - (\mathbf{G}_{i-1} \cap \mathbf{G}_i) = (\mathbf{G}_{i-1} - \mathbf{G}_i) \cup (\mathbf{G}_i - \mathbf{G}_{i-1}) \tag{16.9}$$

*Proof* Because $\mathbf{G}_{vs}$ is the variable subgraph of the variable graph $\mathbf{G}_v$, and $\mathbf{G}_1$, $\mathbf{G}_2$, …, $\mathbf{G}_n$ are not isomorphic pairwise, the number of directed edges of the configuration-function transition digraphs whose $n$ vertices are $\mathbf{G}_1$, $\mathbf{G}_2$, …, $\mathbf{G}_n$ is $q$, the symmetric difference of two invariant (in one phase) topological graphs that are incident with the $i$th directed edge is $\boldsymbol{D}_i$, therefor the union of all variable elements that exist actually in variable graph $\mathbf{G}_v$ is equal to $D_1 \cup D_2 \cup \ldots \cup D_q$. From the definition 7 we find $\mathbf{G}_{vs} = D_1 \cup D_2 \cup \ldots \cup D_q = \bigcup_{i=1}^{q} D_i$.

### *16.3.4 The Cell-Variator*

The Calculation according to the expression (16.8) can generate several cell-variators in the variable graph (accurately speaking in the variable subgraph). The cell-variators are the important components of a MM or a VTM. It is the existence and change of the cell-variators that bring on change of topological structure of the MM or the VTM. Regrettably! It remains undiscovered for more than decades. It cannot be discovered by traditional topological graph. It can be brought into sight but along with naissance of the variable graph. Well then, what is the cell-variator?

#### 16.3.4.1 Definition of the Cell-Variator

**Definition 8** A cell-variator is the variable cell that consist of $m$ $(2 \leq m \leq n)$ different configurations of one variable element (variable vertex or variable edge or variable incidence relation) in a variable graph and one configuration transition pointer, and it can bring the variable element to transit according to certain sequence or conditions among the $m$ different configurations.

#### 16.3.4.2 Classification of the Cell-Variator

The cell-variators may be divided into three classes according to its types. Namely, the vertex cell-variator, the edge cell-variator and the incidence relation cell-variator.

(1) The vertex cell-variator

A vertex cell-variator is the variable cell that consists of $m$ $(2 \leq m \leq n)$ different configurations of one variable vertex in a variable graph and one configuration transition pointer, and it can bring the variable vertex to transit according to certain sequence or conditions among the $m$ different configurations.

From the expression (16.2) in the definition we can know that one arbitrary variable vertex has five kinds of possible different configurations. 1) The vertex denoting rigid link (Fig. 16.1a); 2) The vertex denoting elastic link (Fig. 16.1b); 3) The vertex denoting flexible link (Fig. 16.1c); 4) The vertex denoting liquid link (Fig. 16.1d); 5) The vertex denoting empty (nothing) link (Fig. 16.1e).

The diagram representation of a vertex cell-variator is shown as Fig. 16.2. Figure 16.2a shows one vertex cell-variator that is 2-configurations two-way cycle transition. There are two variable configurations of one vertex inside its closed curve. The initial configuration 1 corresponds with the invariant graph $\mathbf{G}_1$ of a variable graph, and the link corresponding with it is rigid link. Configuration 2 corresponds with the invariant graph $\mathbf{G}_2$ of the variable graph, and the link corresponding with it is an empty (nothing) link, which denotes this link is inexistent. The two-way arrowhead-curve between the configuration 1 and the configuration 2 denotes that the transition is two-way cycled. The one-direction arrowhead straight line whose end connects two curves denotes the configuration transition pointer. Its end curves are the edges that are incident with the vertex which is pointed to by the pointer. The configuration transition pointer must point to the initial configuration position when a diagram of the cell-variator is drawn. Fig. 16.2a shows when the configuration transition pointer points to the vertex configuration 1, it corresponds

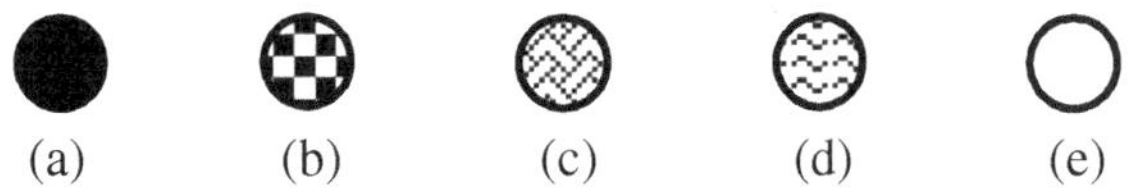


**Fig. 16.1** Possible five configurations of one variable vertex

with the invariant graph $\mathbf{G}_1$ of the variable graph, and when the configuration transition pointer points to the vertex configuration 2 the variable graph corresponding with it changes into $\mathbf{G}_2$.

Figure 16.2b shows one vertex cell-variator that is 3-configuration sequence acyclic transition. The configuration 2 is a vertex that denotes the flexible link. Figure 16.2c shows one vertex cell-variator that is 2-configuration two-way cycle transition as well. But the "3" inside closed curve is not third vertex configuration, it denotes third invariant graph $\mathbf{G}_3$ of the variable graph. The meaning is when the variable graph changes from $\mathbf{G}_2$ into $\mathbf{G}_3$ because of other cell-variator transition this vertex cell-variator keeps at the configuration 2, namely it is unchanged. Figure 16.2d shows one vertex cell-variator that is 4-configuration one-way cycle transition. The configuration 2 is a vertex that denotes the elastic link.

(2) The edge cell-variator

An edge cell-variator is the variable cell that consists of $m$ $(2 \leq m \leq n)$ different configurations of one variable edge in a variable graph and one configuration transition pointer, and it can bring the variable edge to transit according to certain sequence or conditions among the $m$ different configurations.

The diagram representation of an edge cell-variator is shown as Fig. 16.3.

Figure 16.3a shows one edge cell-variator that is 3-configuration sequence acyclic transition. The initial edge configuration 1 is $R\$_1$. $R$ denotes revolute pair, and $\$_1 = [1\ 0\ 0\ 0\ 0\ 0]$ is the Boolean motion screw [13] of $R$. It denotes that axis of the $R$ pair goes in the direction of $x$ of the right-handed coordinate system o-$x$-$y$-$z$. The o-$x$-$y$-$z$ is fixed to one link of the $R$ pair and the origin $o$ is on the axis of the revolute pair $R$. The first, second and third components of $\$_1$ are respectively the projection of the relative rotation angular velocity $\boldsymbol{\omega}$ of two links of the $R$ pair on the coordinate axis $x$, $y$ and $z$. The fourth, fifth and sixth components of $\$_1$ are respectively the projection of the relative translation velocity $\boldsymbol{v}$ of two links of the $R$ pair on the coordinate axis $x$, $y$ and $z$. If and only if one value of the projection is identically equal to 0, then the component of Boolean motion screw corresponding with it is equal to 0, otherwise equal to 1.

The edge configuration 2 is $P\$_2$, where $P$ denotes prismatic pair, and its Boolean motion screw $\$_2 = [0\ 0\ 0\ 1\ 0\ 0]$, which indicates that direction of the relative translation of two links of the $P$ pair is along the $x$ direction. The edge configuration 3 is $C\$_3$, where $C$ denotes cylindrical pair, and its Boolean motion screw $\$_3 = [1\ 0\ 0\ 1\ 0\ 0]$ indicates that direction of axis of the $C$ pair is along the $x$ direction. The basic relative motions of two links of the $C$ pair are rotation around the $x$ axis and motion of translation along the $x$ direction. The edge configuration 1, 2 and 3 correspond with $\mathbf{G}_1$, $\mathbf{G}_2$ and $\mathbf{G}_3$ of the variable graph respectively. The configuration transition pointer must point to the initial configuration position when a diagram of the cell-variator is drawn. Figure 16.3a shows when the configuration transition pointer points to the

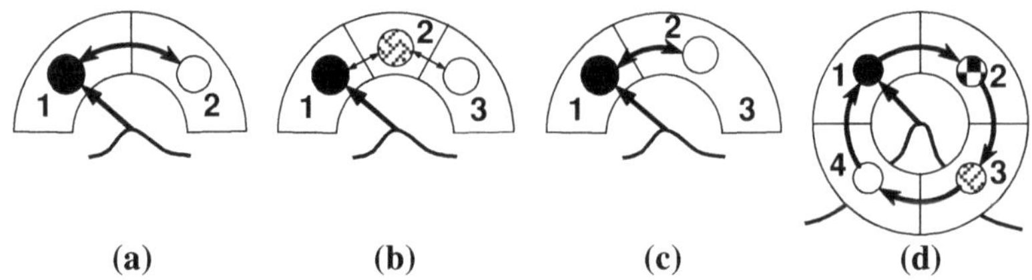


**Fig. 16.2** The diagram representation of a vertex cell-variator

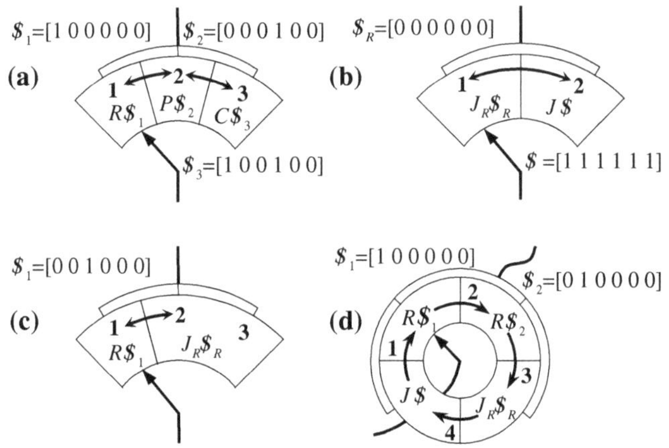


**Fig. 16.3** The diagram representation of an edge cell-variator

edge configuration 1, it corresponds with the invariant graph $\mathbf{G}_1$ of the variable graph, when the configuration transition pointer points to the edge configuration 2, the variable graph corresponding with it changes into $\mathbf{G}_2$, and when the configuration transition pointer points to the edge configuration 3, the variable graph corresponding with it again changes into $\mathbf{G}_3$. Figure 16.3b shows one edge cell-variator that is 2-configuration two-way cycle transition. Its initial edge configuration 1 is $J_R\$_R$. $J_R$ denotes rigid joint, and $\$_R = [0\,0\,0\,0\,0\,0]$ is the Boolean motion screw of $J_R$, which indicates that two links that is jointed by $J_R$ become one link and there is no any relative motion between two links. The edge configuration 2 is $J_\phi\$_\phi$. $J_\phi$ denotes empty (nothing) joint, and $\$_\phi = [1\ 1\ 1\ 1\ 1\ 1]$ is the Boolean motion screw of $J_\phi$, which indicates that there is no any direct constraint between two links that is jointed by $J_\phi$. The value of $\$_R$ and $\$_\phi$ all is constant, so they may be passed over in the diagram of an edge cell-variator. Figure 16.3b shows when the configuration transition pointer points to the edge configuration 1, the edge is rigid joint, which corresponds with the invariant graph $\mathbf{G}_1$ of the variable graph, and when the configuration transition pointer points to the edge configuration 2, the edge changes into empty (nothing) joint, the variable graph corresponding with it changes into $\mathbf{G}_2$. Figure 16.3c shows one edge cell-variator that is 2-configuration two-way cycle transition as well. But the "3" inside closed curve is not third edge configuration, it

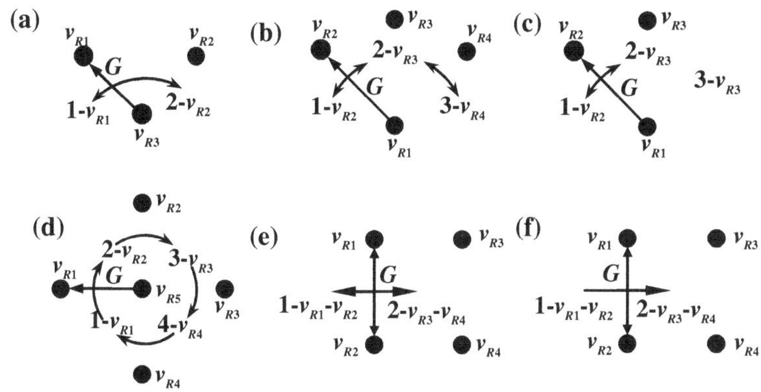


**Fig. 16.4** The diagram representation of a edge cell-variator

denotes third invariant graph $\mathbf{G}_3$ of the variable graph. The meaning is when the variable graph changes from $\mathbf{G}_2$ into $\mathbf{G}_3$ because of other cell-variator transition this edge cell-variator keeps at the configuration 2, namely it is unchanged. Figure 16.3d shows one edge cell-variator that is 4-configurations one-way cycle transition.

(3) The incidence relation cell-variator

An incidence relation cell-variator is the variable cell that consists of $m$ ($2 \leq m \leq n$) different configurations of one variable incidence relation in a variable graph and one configuration transition pointer, and it can bring the variable incidence relation to transit according to certain sequence or conditions among the $m$ different configurations.

The diagram representation of a incidence relation cell-variator is shown as Fig. 16.4. Figure 16.4a shows one incidence relation cell-variator that is 2-configuration two-way cycle transition. The initial incidence relation configuration 1 is 1-$\boldsymbol{v}_{R1}$, which shows that the incidence relation configuration transition pointer points to the $\boldsymbol{v}_{R1}$ from the $\boldsymbol{v}_{R3}$, namely the edge is incident with the $\boldsymbol{v}_{R1}$ and the $\boldsymbol{v}_{R3}$. The incidence relation configuration 2 is 2-$\boldsymbol{v}_{R2}$, which denotes that the incidence relation configuration transition pointer points to $\boldsymbol{v}_{R2}$ from $\boldsymbol{v}_{R3}$, namely the edge is incident with $\boldsymbol{v}_{R2}$ and $\boldsymbol{v}_{R3}$.

Figure 16.4a shows when the incidence relation configuration transition pointer points to $\boldsymbol{v}_{R1}$ from $\boldsymbol{v}_{R3}$, the variable graph corresponds with it is $\mathbf{G}_1$, and when the incidence relation configuration transition pointer points to $\boldsymbol{v}_{R2}$ from $\boldsymbol{v}_{R3}$, the variable graph corresponds with it changes into $\mathbf{G}_2$. This incidence relation cell-variator is just like a switch in electrocircuit. Figure 16.4b shows one incidence relation cell-variator that is 3-configuration sequence acyclic transition. Figure 16.4c shows one incidence relation cell-variator that is 2-configuration two-way cycle transition. 3-$\boldsymbol{v}_{R3}$ in the diagram denotes that this incidence relation cell-variator keeps at the configuration 2 when the variable graph changes from $\mathbf{G}_2$ into $\mathbf{G}_3$ because of other cell-variator transition, namely it is unchanged. Figure 16.4d shows one incidence relation cell-variator that is 4-configuration one-way cycle transition. In Figs. 16.4a–d, the incidence relation configuration transition pointers all have only one arrowhead in its one end. The change type of these one-arrowhead pointers is the rotation around the invariant vertex. These 4 cell-variators are all one-end incidence relation cell-variators. The cell-variators shown in Fig. 16.4e, f are all two-end incidence relation cell-variators. Namely their incidence relation configuration transition pointers all have two arrowheads in its two ends. The change type of two-arrowhead pointers is motion of translation. Figure 16.4e shows one incidence relation cell-variator that is 2-configuration two-end two-way translation cycle transition. When the incidence relation configuration transition pointer points at initial configuration 1-$\boldsymbol{v}_{R1}$-$\boldsymbol{v}_{R2}$, the edge $G$ (gear pair) is incident with $\boldsymbol{v}_{R1}$ and $\boldsymbol{v}_{R2}$, where the variable graph corresponding with it is $\mathbf{G}_1$. When the incidence relation configuration transition pointer translates and points at configuration 2-$\boldsymbol{v}_{R3}$-$\boldsymbol{v}_{R4}$, the edge $G$ changes into what is incident with $\boldsymbol{v}_{R3}$ and $\boldsymbol{v}_{R4}$, where the variable graph corresponding with it changes from $\mathbf{G}_1$ to $\mathbf{G}_2$. When the incidence relation configuration transition pointer translates and returns to configuration 1-$\boldsymbol{v}_{R1}$-$\boldsymbol{v}_{R2}$, the variable graph corresponding with it changes from $\mathbf{G}_2$ to $\mathbf{G}_1$ again. Figure 16.4f shows one incidence relation cell-variator that is 2-configuration two-end one-way translation transition.

An incidence relation cell-variator can be transformed into equivalent edge cell-variators. Here is the equivalence transformation theorem.

**Theorm 4** *Suppose $\boldsymbol{I}_v$ is one arbitrary incidence relation cell-variator that has m ($2 \leq m \leq n$) different configurations. If all vertices of $\boldsymbol{I}_v$ are invariant, and the non-empty joint corresponding with its incidence relation configuration transition pointer is $\boldsymbol{J}_I$, the empty joint is $\boldsymbol{J}_\phi$, then $\boldsymbol{I}_v$ can be transformed equivalently into m edges cell-variators (every cell-variator has 2 different configurations) $\boldsymbol{E}_{v1}$, $\boldsymbol{E}_{v2}$, ..., $\boldsymbol{E}_{vm}$, and $\boldsymbol{E}_{vi} = (\boldsymbol{J}_I - \boldsymbol{J}_\phi)_i OR(\boldsymbol{J}_\phi - \boldsymbol{J}_I)_I$, ($i = 1, 2, \ldots, m$). The m edge configuration transition pointers must be rotated simultaneously, and sequence numbers of $\boldsymbol{E}_{vi}$ must be one-to-one correspondence with sequence numbers of $\boldsymbol{I}_v$. Namely*

$$I_v = \bigcup_{i=1}^{m} E_{vi} = \bigcup_{i=1}^{m} [(J_I - J_\phi)_i OR(J_\phi - J_I)_i] \tag{16.10}$$

*Proof* Because $\boldsymbol{I}_v$ is one incidence relation cell-variator that has $m$ ($2 \leq m \leq n$) different configurations, all vertices of the $\boldsymbol{I}_v$ are invariant, and the nonempty joint corresponding with its incidence relation configuration transition pointer is $\boldsymbol{J}_I$, and the empty joint is $\boldsymbol{J}_\phi$. Therefore when the incidence relation configuration transition pointer points to the $i$th ($i = 1, 2, \ldots, m$) configuration, the empty joint of the $i$th ($i = 1, 2, \ldots, m$) configuration is $\boldsymbol{J}_I$, otherwise the joint of the $i$th ($i = 1, 2, \ldots, m$) configuration is $\boldsymbol{J}_\phi$. i.e. the $i$th configuration $\boldsymbol{I}_{vi} = (\boldsymbol{J}_I - \boldsymbol{J}_\phi)_i\ OR(\boldsymbol{J}_\varphi - \boldsymbol{J}_I)_i$, ($i = 1, 2, \ldots, m$). And $\boldsymbol{I}_{vi}$ is even the $i$th edge cell-variator. Namely $\boldsymbol{E}_{vi} = (\boldsymbol{J}_I - \boldsymbol{J}_\varnothing)_i\ OR(\boldsymbol{J}_\phi - \boldsymbol{J}_I)_i$, ($i = 1, 2, \ldots, m$). Hence $I_v = \bigcup_{i=1}^{m} E_{vi} = \bigcup_{i=1}^{m} [(J_I - J_\phi)_i\ OR(J_\phi - J_I)_i]$. The $m$ edge configuration transition pointers must be rotated simultaneously, and sequence numbers of the $\boldsymbol{E}_{vi}$ must be one-to-one corresponding with sequence numbers of the $\boldsymbol{I}_v$. Otherwise, it is impossible to realize equivalence of topological structure of the variable graph.

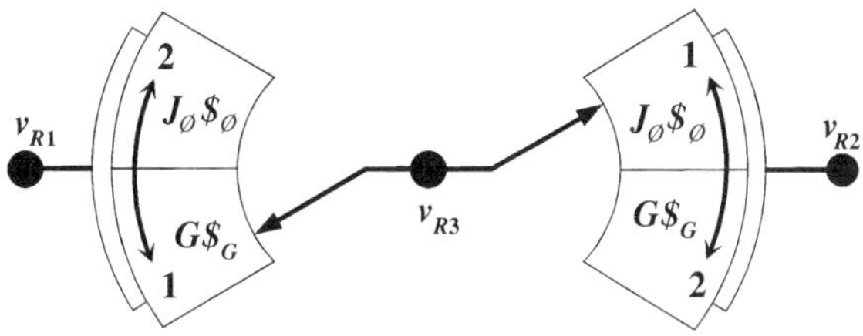


**Fig. 16.5** Verifying the theorem 4

**Fig. 16.6** The composition tree of a variable graph

- A variable graph
  - The invariant subgraph
    - The invariant vertices
    - The invariant edges
    - The invariant incidence relations
  - The variable subgraph
    - The vertex-variators
    - The edge-variators
    - The incidence relation variators

The incidence relation cell-variator in Fig. 16.4a can be transformed into two edge cell-variators in Fig. 16.5. From Fig. 16.5 we know, two edge configuration transition pointers in Fig. 16.5 must be rotated simultaneously, thus it is equivalent to the incidence relation cell-variator in Fig. 16.4a.

The cell-variators may be divided into actual cell-variators and potential cell-variators according to whether they actually exist and are applied. The actual cell-variators actually exist and they are being applied to change topological structure of a variable graph. The actual cell-variators can be determined according to the configuration-function transition digraph [13] and expression (16.8) and (16.9). The potential cell-variators are found via the calculation of symmetric difference to two invariant graphs that there is no any change path between them but they are not existent really. Any potential cell-variator cannot be drawn in diagram.

The actual cell-variators may be divided into independent cell-variators and dependent cell-variators according to whether they independently change. Change of the independent cell-variators is independent, and change of the dependent cell-variators is brought by the Change of the independent cell-variators.

### *16.3.5 The Composition Tree of a Variable Graph*

The composition tree of a variable graph is shown in Fig. 16.6.

## 16.4 The Diagram Representation of a Variable Graph

According to the composition principle of the variable graph, the diagram representation of a variable graph can be divided into the diagram representation of its invariant subgraph and the diagram representation of its variable subgraph. The diagram representation of the invariant subgraph is all the same as traditional graph. The diagram representation of the variable subgraph is the diagram representation of the actual cell-variators.

### 16.4.1 The Representation Steps of the Actual Cell-Variators

**Step 1** Determine the set of the actual vartators according to Expression (16.8) and (16.9);

**Step 2** Draw all the actual cell-variators according to 3.4.

### 16.4.2 Two Representation Examples

*Example 1* Figures 16.7a, b are the schematic diagrams of mechanism of two different configurations of extracting case accelerating mechanism of a automatic gun. Two working-phases topological graphs corresponding with them are shown in Figs. 16.7c, d.

First the invariant subgraph (Fig. 16.7e) of the variable graph is determined according to the expression (7) of the theorem 2. Then the variable subgraph (Fig. 16.7f) of the variable graph is determined according to the expression (16.8) and (16.9) of the theorem 3. Finally the variable graph (Fig. 16.7g) is determined according to the expression (16.6) of the theorem 1.

In Fig. 16.7g, the cell-variator of vertex 4 and the cell-variator of between vertex 1 and vertex 3 are all independent cell-variators. The rest two cell-variators are all dependent cell-variators.

*Example 2* Figures 16.8a, b are respectively the schematic diagrams of mechanism of two different configurations corresponding with two working-phases of a gear changement.

Two working-phases topological graphs corresponding with them are shown in Figs. 16.8c, d. First the invariant subgraph (Fig. 16.8e) of the variable graph is determined according to Expression (7) of the theorem 2. Then the variable subgraph (Fig. 16.8f) of the variable graph is determined according to Expression (8) and (9) of the theorem 3. Finally the variable graph (Fig. 16.8g) is determined according to Expression (6) of the theorem 1.

The cell-variator in Fig. 16.8g is the incidence relation cell-variator.

## 16.5 The Adjacent Matrix Representation of a Variable Graph

Definition 9 If the total number of vertices of a variable graph $\mathbf{G}_v$ is $n_v$, there are $k$ variable vertices in them. The number of configurations of the $i$th variable vertex is $m_{i,}$ then the adjacent matrix $\boldsymbol{A}_v$ of the $\mathbf{G}_v$ is a $p \times p$ square matrix.

$$A_v = \left[a_{ij}\right]_{p\prime p} \tag{16.11}$$

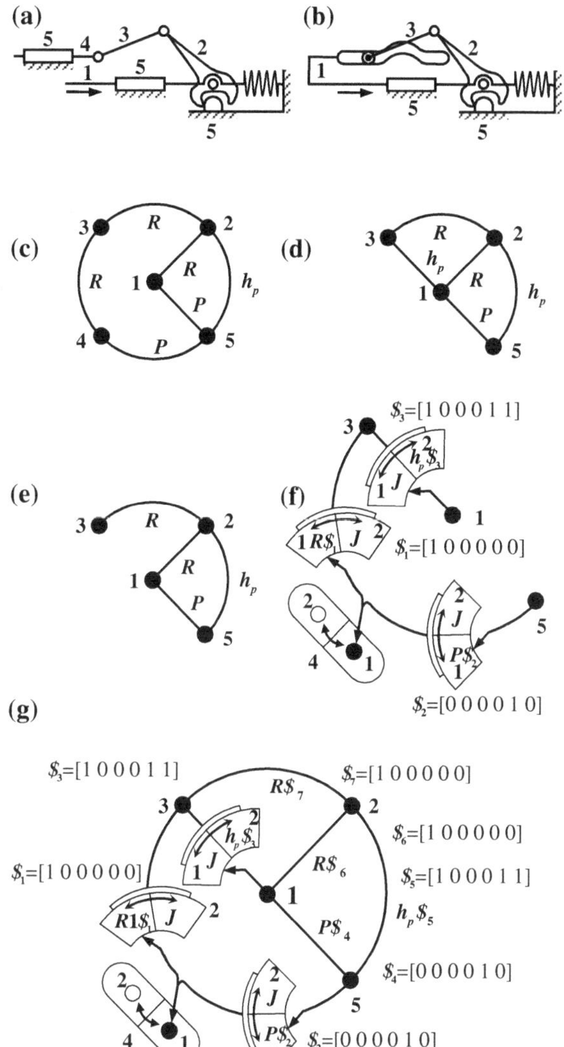


**Fig. 16.7** The variable graph of extracting case accelerating mechanism

$$p = n_v - k + \sum_{i=1}^{k} m_i \tag{16.12}$$

$$a_{ij} = \begin{cases} J_P \$_P \text{ If vertex } i \text{ and vertex } j \text{ is kinematic pair } P \text{ joint} \\ \qquad \text{and the dgree of the motion is not equal to 0} \\ J_R[000000] \text{ If vertex } i \text{ and vertex } j \text{ is rigid joint} \\ J_\phi[111111] \text{ If vertex } i \text{ or } j \text{ is empty vertex, or } i = j, \\ \qquad \text{or vertex } i \text{ and vertex } j \text{ is not adjacent} \end{cases} \tag{16.13}$$

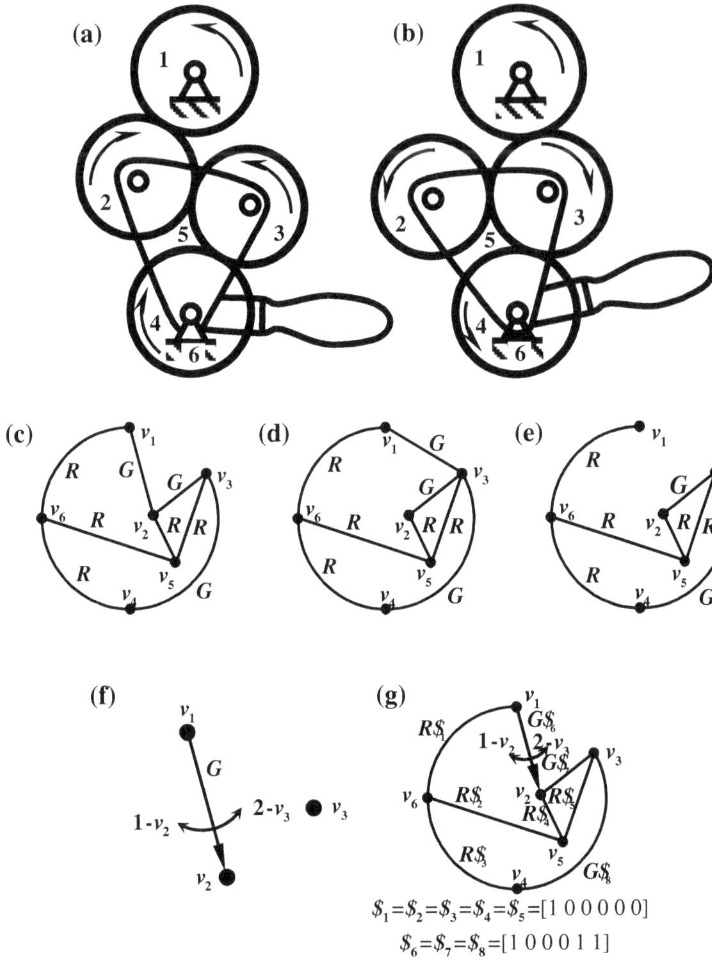


**Fig. 16.8** The variable graph of gear changement

If a variable graph contains the variable vertices, then the variable vertices may be arranged first according to their sequence number of configuration, and then the invariant vertices are arranged. Thus different configurations and their change may be identified easily.

If a variable joint that is incident with vertex *i* and vertex *j* is an edge cell-variator that has *m* configurations, then $a_{ij}$ corresponding with it is the sequence $a_{ij}^1, a_{ij}^2, \cdots, a_{ij}^m$, they should be arranged into *m* row according to their sequence numbers.

An incidence relation cell-variator should be transformed into edge cell-variators first according to the theorem 4 and expression (16.10), and then it is represented by representing method of the edge variator.

For example the adjacent matrix of the variable graph in Fig. 16.7g is

$$A_v = \begin{array}{c} \\ 4-1 \\ 4-2 \\ 1 \\ 2 \\ 3 \\ 5 \end{array} \begin{bmatrix} 4-1 & 4-2 & 1 & 2 & 3 & 5 \\ J_\phi & J_\phi & J_\phi & J_\phi & R[100000] & P[000010] \\ J_\phi & J_\phi & J_\phi & J_\phi & J_\phi & J_\phi \\ J_\phi & J_\phi & J_\phi & R[100000] & \substack{J_\phi \\ h_p[100011]} & P[000010] \\ J_\phi & J_\phi & R[100000] & J_\phi & R[100000] & h_p[100011] \\ R[100000] & J_\phi & \substack{J_\phi \\ h_p[100011]} & R[100000] & J_\phi & J_\phi \\ P[000010] & J_\phi & P[000010] & h_p[100011] & J_\phi & J_\phi \end{bmatrix}$$

According to the theorem 4 and expression (16.10), after the incidence relation cell-variator of the variable graph in Fig. 16.8g is transformed into two edge cell-variators, the adjacent matrix of the variable graph in Fig. 16.8g is as below

$$A_v = \begin{array}{c} v_1 \\ v_2 \\ v_3 \\ v_4 \\ v_5 \\ v_6 \end{array} \begin{bmatrix} v_1 & v_2 & v_3 & v_4 & v_5 & v_6 \\ J_\phi & G[100011]J_\phi & \substack{J_\phi \\ G[100011]} & J_\phi & J_\phi & R[100000] \\ G[100011]J_\phi & J_\phi & G[100011] & J_\phi & R[100000] & J_\phi \\ \substack{J_\phi \\ G[100011]} & G[100011] & J_\phi & G[100011] & R[100000] & J_\phi \\ J_\phi & J_\phi & G[100011] & J_\phi & J_\phi & R[100000] \\ J_\phi & R & R[100000] & J_\phi & J_\phi & R[100000] \\ R[100000] & J_\phi & J_\phi & R[100000] & R[100000] & J_\phi \end{bmatrix}$$

## 16.6 Conclusion

The new concepts such as the variable graph, the invariant subgraph, the variable subgraph and the cell-variator etc. of MMs or VTMs are put forward and defined for the first time in this paper. Three new theorems on composition of the variable graph and one new theorem on the equivalent transformation of the incidence relation cell-variator are found out and their validity is strictly proved. The classifications of the cell-variators and the diagram representations of three kinds of cell-variators are studied. The diagram representation and adjacent matrix representation of the variable graph are given. The variable graph provides useful new tool for the topological structural analysis and synthesis of MMs or VTMs. It has great scientific value and engineering practicality. It will certainly be applied widely in the future.

**Acknowledgments** Sincere thanks to my kind and respected teacher Yang Ting-Li for his great kindness, concern, support and instruction for many years.

Sincere thanks to professor Dai Jian-Sheng for his concern, support and help for many years.

Sincere thanks to professor Yan Hong-Sen for his support and help.

## References

1. Dai JS, Rees Jones J (1998) Mobility in metamorphic mechanisms of foldable/erectable kinds. 25th ASME Biennial mechanisms and robotics conference, Atlanta, GA, Sept 1998
2. Liu CH, Yang TL (2004) Essence and characteristics of metamorphic mechanisms and their metamorphic ways. In: Proceedings of the 11th World Congress in Mechanism and Machine Science, Tianjin, China, Beijing: China Machine Press, pp 1285-1288, April 1-4 2004
3. Yan HS, Liu NT (2000) Finite-state-machine representations for mechanisms and chains with variable topologies. In: Proceedings of ASME design engineering technical conference, Baltimore, Maryland, DETC2000/MECH-14054. 10-13 Sept 2000
4. Liu CH, Yang TL, Liu Y (2005) Basic problems of the theory for the variable topology mechanisms. Chin J Mech Eng 41(8):56–62
5. Dai JS, ZHANG QX (2000) Metamorphic mechanisms and their configuration models. Chin J Mech Eng 13(3):212–218
6. Yan HS, Liu NT (2003) Joint-codes representations for mechanisms and chains with variable topologies. T Can Soc Mech Eng 27:131–143
7. Yan HS, Kuo CH (2006) Representations and identifications of structural and motion state characteristics of mechanisms with variable topologies. T Can Soc Mech Eng 30:19–40
8. Yan HS, Kuo CH (2006) Topological representations and characteristics of variable kinematic joints. J Mech Des-T ASME 128:384–391
9. Yan HS, Kuo CH (2009) Structural analysis and configuration synthesis of mechanisms with variable topologies. In: Proceedings of ASME/IFToMM international conference on reconfigurable mechanisms and robots, London, pp 23-31, 22-24 June 2009
10. Lan ZH, Du R (2008) Representation of topological changes in metamorphic mechanisms with matrices of the same dimension. J Mech Des-T ASME 130:074501-1-074504
11. Yang TL et al (2009) Structure composition principle of reconfigurable mechanisms and basic methods for changing topological structure. In: Proceedings of ASME/IFToMM international conference on reconfigurable mechanisms and robots, London, pp 104-109, 22-24 June 2009
12. Li SJ et al (2009) Joint-gene based variable topological representations and configuration transformations. In: Proceedings of ASME/IFToMM international conference on reconfigurable mechanisms and robots, London, pp 348-354, 22-24 June 2009
13. Liu CH (2009) The configuration-function transition digraphs of metamorphic mechanisms or variable topology mechanisms. In: Proceedings of ASME/IFToMM international conference on reconfigurable mechanisms and robots, London, pp 110-117, 22-24 June, 2009

# Chapter 17
# Type Synthesis of Planar Parallel Mechanism Incorporating Actuated Limb with Zero/One Constraint with Set Conception

**Gang Dong, Yimin Song and Tao Sun**

**Abstract** This paper presents a method of type synthesis for planar parallel mechanisms (PPMs) incorporating actuated limbs with zero/one constraint with set conception. Firstly, according to their contributions to the moving platform, the limbs of PPMs can be classified as four categories: actuated limb with zero constraint, fully-constrained limb with zero actuator, fully-constrained limb with one actuator and under-constrained limb with one actuator. Followed by the descriptions of actuated space and constrained space for the end of the limbs by virtue of the screw theory, the type syntheses of the four type limbs are carried out. Then different sets and functions are defined respectively, whose elements and variables are the topologies of four type limbs. The different function values and the composition principle of PPMs lead to existence of PPMs topologies by four solutions. Finally, taking three typical PPMs with linear motors for examples, their virtual prototypes are obtained with commercial 3-dimensional CAD software.

**Keywords** Planar parallel mechanism · Type synthesis · Actuated space · Constrained space · Set

G. Dong · Y. Song (✉) · T. Sun
School of Mechanical Engineering, Tianjin University, Tianjin, People's Republic of China
e-mail: ymsong@tju.edu.cn

G. Dong
e-mail: dg_tju@tju.edu.cn

T. Sun
e-mail: suntaotju@gmail.com

J. S. Dai et al. (eds.), *Advances in Reconfigurable Mechanisms and Robots I*,
DOI: 10.1007/978-1-4471-4141-9_17, © Springer-Verlag London 2012

## 17.1 Introduction

In general, a parallel mechanism (PM) typically consists of a moving platform that is connected to a fixed base by several limbs, and the number of actuated limbs is equal to the degrees of freedom (DoFs) of a PM. Since the inception of known Stewart platform in 1965 [1], the PMs have been drawn great attentions over the academics and industries because of some advantages in payload capability, stiffness, dynamic behavior comparing with their serial counterparts. Thereafter, type syntheses or topology syntheses of PMs have become a research focus for finding new and better PMs in the subsequent decades. Early researches on type syntheses of PMs concentrated primarily on 6-DoF Stewart-type PMs, what methods can be mainly focused on the enumeration method, the method of trial and error, and the evolutionary method, such as 6-UPS PM, 6-SPS PM and so on [2–4]. The results of these type synthesis are lack of the systematism. Then the object of type synthesis of PMs turns the direction to the ones with fewer than six DoFs, called lower-DoF PMs. A French Jean-Pierre Merlet points out that “the type synthesis of lower-DoF PMs is a very difficult problem”, the difficulties focus on (1) there is none of a DoF formula suiting for use in any circumstance, (2) there is none of any mathematical tools to describe all lower-DoF PMs with different constraints in diverse layout. In recent years, the methods of type synthesis for lower-DoF PMs concentrate on Lie Group, Screw Theory, Differentiable Manifolds and so on. Many scholar and engineers both at home and abroad make their contributions to the type synthesis of lower-DoF PMs, including Clément-M Gosselin, Xian-wen Kong, Jacques Marie Hervé, Gogu, Zhen Huang, Ting-li Yang, Ze-xiang Li, Jian S. Dai, Xin-jun Liu, Qin-chuan Li, Feng Gao and so on [5–12].

On the basis of the review mentioned above, it is necessary to consider two problems as follows.

(1) The lower-DoF PMs possess some advantages as same as serial mechanisms, but suffer several disadvantages, such as worse singularity, small workspace, and poor accuracy. To overcome these shortcomings, the planar parallel mechanisms (PPMs) may be a favorable solution. The PPMs are usually a neglected domain, and the type synthesis of PPMs can not be preceded systematically.
(2) Even though the advanced mathematical tool can synthesize the topologies of lower-DoF PMs systemically to some extent, what is very difficult or unavailable to non-mathematical students, it can not be applied in the engineering extensively.

Therefore, this paper studies the type synthesis of PPMs incorporating actuated limbs with zero/one constraint using a simple method. Firstly, the actuated space and the constrained space of the moving platform of PPMs and the end of limbs are defined respectively, and the contributions to the moving platform of PPMs made by its limbs lead to that can be classified as four categories: Actuated limb with zero constraint, Fully-constrained limb with zero actuator, Fully-constrained limb with one actuator and Under-constrained limb with one actuator. Secondly, on the basis of descriptions of the actuated space and constrained space of the end for the four type limbs using the screw theory, the type synthesis of the four type limbs are carried out. Thirdly, different sets and functions are defined respectively, whose elements and variables are the topologies of four type limbs; the different function values and the composition principle of PPMs lead to existence of PPMs topologies by four solutions. Finally, taking three typical PPMs with linear motors for examples, their virtual prototypes are obtained with commercial 3-dimensional CAD software. The aforementioned contents can be seen in Sect. 17.2–17.4 in detail.

## 17.2 Type Synthesis of Limbs in PPMs

### 17.2.1 Descriptions of Limbs in PPMs

The pose space of moving platform for PPMs can be divided into two spaces, one is the actuated space and the other one is the constrained space, whose relationships can be expressed as

$$S = S^{M} \oplus S^{C},\ \lambda = \lambda^{M} + \lambda^{C},\ S^{M} \cap S^{C} = \Phi. \tag{17.1}$$

where $S$ denotes the pose space of the moving platform with its dimension is $\lambda$, $S^M$ and $S^C$ represent its actuated space and constrained space, whose dimensions are noted by $\lambda^M$ and $\lambda^C$, the notations $\cap$, $\oplus$ and $\Phi$ denote the intersection, direct sum and the empty set. If the DoFs of PPMs are defined as $f$, there is $f = \lambda^M$.

From the respective of a set, the actuated space $S^M$ and the constrained space $S^C$ of the moving platform can be expressed as follows.

$$S^M = \{\$_a \mid \$_a \in \mathbb{R}^3\},\ S^C = \{\$_c \mid \$_c \in \mathbb{R}^3\} \tag{17.2}$$

where $\$_a$ and $\$_c$ denote the elements of $S^M$ and $S^C$ respectively.

With the same principle, the actuated space and the constrained space for the end (close to the moving platform) of the limbs in PPMs can be expressed as $S_i^M$ and $S_i^C$ respectively, and then

$$S^C = \bigcup_{i=1}^{L} S_i^C,\ S^M = \overline{S^C} \tag{17.3}$$

where the notations $\cup$ and $\overline{S^C}$ represent the union set and the complementary set.

Therefore, the topology of limbs in PPMs is directly bound up with that of the PPMs. According to the contributions to the moving platform of PPMs made by its limbs, the limbs can be classified as four categories with the space definition.

(1) Type I—Actuated Limb with Zero Constraints

This limb is named as planar limb I. The actuated space $S_I^M$ of the end for planar limb I is the same with the pose space $S$ of the moving platform for PPMs, and the constrained space $S_I^C$ of the end for limb I equals $\Phi$, which can be expressed as

$$S = S_I^M,\ S_I^C = \Phi.\ \lambda_I^M = \lambda,\ \lambda_I^C = 0 \tag{17.4}$$

(2) Type II—Fully-Constrained Limb with Zero actuators

This limb is named as planar limb II, the actuated space $S_{II}^M$ and the constrained space $S_{II}^C$ of the end for planar limb II are equal to the actuated space $S^M$and the constrained space $S^C$ of the moving platform for PPMs respectively, which can be described as

$$S^M = S_{II}^M,\ S^C = S_{II}^C,\ \lambda_{II}^M = f,\ \lambda_{II}^C = \lambda - f \tag{17.5}$$

(3) Type III—Fully-Constrained Limb with One actuator

This limb is named as planar limb III, the actuated space $S_{III}^M$ and the constrained space $S_{III}^C$ of the end for planar limb III are the same with the actuated space $S^M$ and the constrained space $S^C$ of the moving platform for PPMs respectively, which can be expressed as

$$S^M = S_{III}^M,\ \lambda_{III}^M = f\ \lambda_{III}^C = \lambda - f \tag{17.6}$$

(4) Type IV—Under-Constrained Limb with One actuator

This limb is named as planar limb IV, the constrained space $S_{IV}^C$ of the end for planar limb IV is a subspace of the constrained space $S^C$ of the moving platform for PPMs, or the constrained space $S^C$ of moving platform for PPMs is the subspace of the actuated space $S_{IV}^M$ of the end of limb IV, which can be described as

$$S_{IV}^C \subset S^C,\ S_{IV}^M \subset S^M,\ f \leq \lambda_{IV}^M \leq \lambda \tag{17.7}$$

It should be noted that the descriptions of $S$, $S^M$, $S^C$, $S_i^M$, and $S_i^C$ are proceeded with help of the screw theory, and then takes limb $i$ in a general PM for an example.

As shown in Fig. 17.1, a global coordinate frame designated as the frame $O$-$xyz$ is fixed at the center of the fixed base and a moving coordinate frame $P$-$uvw$ is assigned to the center of the moving platform similarly, and the screws of the joints in limb $i$ can be formulated under the frame $P$-$uvw$ as

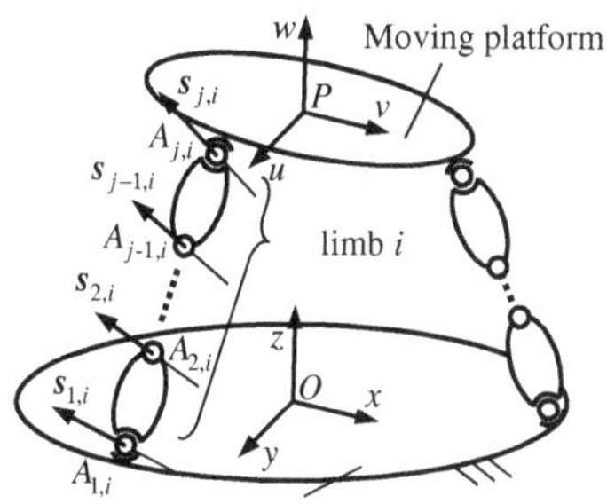


**Fig.17.1** The schematic diagram of limb $i$ in a general PM

$$\hat{\$}_{1,i} = \begin{pmatrix} s_{1,i} \\ l_{1,i} \times s_{1,i} \end{pmatrix}, \hat{\$}_{2,i} = \begin{pmatrix} s_{2,i} \\ l_{2,i} \times s_{2,i} \end{pmatrix}, \ldots, \hat{\$}_{j-1,i} = \begin{pmatrix} s_{j-1,i} \\ l_{j-1,i} \times s_{j-1,i} \end{pmatrix}, \hat{\$}_{j,i} = \begin{pmatrix} s_{j,i} \\ l_{j,i} \times s_{j,i} \end{pmatrix} \tag{17.8}$$

herein, $s_{j,i}$ denotes the axis of $j$th 1-DoF joints in $i$th limb, $l_{j,i} = \overline{PA_{j,i}}$ and $A_{j,i}$represents center of $j$th 1-DoF joints in $i$th limb, and it is suppose that the screws of joints in Eq. (17.8) are independent linearly.

Therefore, the end motions of the $i$th limb can be formulated as follows.

$$\$_i^{\boldsymbol{M}} = \sum_{j=1}^{N} C_{j,i} \hat{\$}_{j,i} \tag{17.9}$$

where $C_{j,i}$ denotes the amplitude of angular or linear velocity of $j$th 1-DoF joints in $i$th limb, $N$ is the number of 1-DoF joints.

According to the reciprocal screw theory, the reciprocal product of Eq. (17.9) can be expressed as

$$\hat{\$}_{k,i}^{r} \circ \$_i^{\boldsymbol{M}} = 0 \tag{17.10}$$

where $\hat{\$}_{k,i}^{r}$ represents the reciprocal screw of joint's screw correspondingly, and it is supposed that the reciprocal screws of joints in Eq. (17.10) are independent linearly.

In terms of definitions of the actuated and constrained space of the limb mentioned above, that of limb $i$ in PPMs can be described as follows.

$$S_i^{\boldsymbol{M}} = \left\{ \hat{\$}_{j,i} \mid \hat{\$}_{j,i} \in \mathbb{R}^3 \right\}, \ S_i^{C} = \left\{ \hat{\$}_{k,i}^{r} \mid \hat{\$}_{k,i}^{r} \in \mathbb{R}^3 \right\}, \tag{17.11}$$

In general, $(j + k)$ equals three if the PM is a PPM.

**Table 17.1** The topologies of type I-IV limb

| Limb type | Limb topology |
|---|---|
| Type I | PRR, RPR, RRP, PRP, PRP, RRR, RPR, RRP, RRP, RPR, PRR, PRR, PRP, PRPa, PPaR, PaPR, RPPa, PaRP, RPaP, RRPa, RRPa, RPaR, RPaR, PaRR, PaRR, PRPa, PPaR, PaPR, RPPa, PaRP, RPaP |
| Type II | PR, RP, RR, P, R, PPa, RPa, PaP, PaR, Pa |
| Type III | PR, RP, P, PR, RP, RR, RR, R, PPa, RPa, PaP, PaR |
| Type IV | PR, RP, RP, PR, PPa, RPa, PaP, PaR |

### 17.2.2 Type Synthesis of Limb I-IV in PPMs

As is well known, the limbs are composed of rods connected by joints. Normally, the common joints includes prismatic joint (P for short), rotary joint (R for short), helical joint (H for short), cylindrical joint (C for short), universal joint (U for short), Even joint (E for short), Spherical joint (S for short), which can be called as general joint. Besides, the joints in limbs can be some complicated joints, named as complex joint, such as 4R complex joint, 4U complex joint, 4S complex joint.

Owing to this paper focus on the type synthesis of PPMs, the R joint, P joint, and the 4R complicated joint (Pa for short) are selected as the joint unit formed type I-IV limb.

In this paper, it is promised that the limb in PPMs can be indicated by the sequential abbreviated capital letters of joints, for example, PR denotes a limb comprises P joint, R joint and one rod, herein, P joint is the actuated joint.

According to the aforementioned definitions of the type I-IV limb, their topologies can be demonstrated in Table 17.1.

## 17.3 Type Synthesis of PPMs

### 17.3.1 Solutions of PPM Topology

Referring to the composition principle of the PM, the topology of PPMs can be formed by the following four solutions. It is supposed that only one actuator in one actuated limb and the sum of actuated limbs is the same as the DoFs of PPMs.

(1) Solution 1—type I PPM with all DoFs

These PPMs are made up of *f* type I limbs, a fixed base and a moving platform, which can be described as follows.

$$S = S^M \oplus S^C,\ S^M = S_i^M,\ S^C = \Phi,\ f = \lambda^M = \lambda \tag{17.12}$$

(2) Solution 2—type I–II PPM with less DoFs

These PPMs are composed of *f* type I limbs, one type II limb, a fixed base and a moving platform, which can be expressed as

$$S = S^M \oplus S^C,\ S^C = S^C_{i+1},\ S^M = \overline{S^C},\ f = \lambda^M \tag{17.13}$$

(3) Solution 3—type I–III PPM with less DoFs

These PPMs are composed of (*f*-1) type I limbs, one type III limb, a fixed base and a moving platform, which can be described as follows.

$$S = S^M \oplus S^C,\ S^C = S^C_{i+1},\ S^M = \overline{S^C},\ f = \lambda^M \tag{17.14}$$

(4) Solution 4—type IV PPM with less DoFs

These PPMs are made up of *f* type IV limbs, a fixed base and a moving platform, which can be expressed as follows.

$$S = S^M \oplus S^C,\ S^C = \bigcup_{i=1}^{f} S^C_i,\ S^M = \overline{S^C},\ f = \lambda^M \tag{17.15}$$

where the identical limbs can be identified by suffix *i*, and the different limb may by recognized by suffix $(i + 1)$.

## 17.3.2 Type Synthesis of Type I–IV PPMs

To take the PPMs topology for an example, one method of type synthesis is presented in this part, which is described in detail as follows.

Firstly, on the basis of the results of the type synthesis of limbs, different sets made up of diverse topologies of limbs as their elements are formed. Then, the functions are given whose variables is in the corresponding set. Finally, the type syntheses of PPMs are achieved according to the values of the functions.

### 17.3.2.1 Type Synthesis of Type I PPMs

Type I PPMs with all DoFs are made up of three type I limbs, a fixed base and a moving platform, which can process three DoFs motions with two translations and one rotation. The topologies of three type I limbs can (1) identical, (2) different from each other, (3) any two are the same. Without loss of generality, the layout of the three type I limbs was axis-symmetrical, in other words, the planes of the fixed base and the moving platform are the equilateral triangles.

**Definition 1** Set $E^{\mathrm{PI}}$, expressed as

$$E^{\mathrm{PI}} = \{\mathrm{PRR,\ RPR,\ RRP, PRP,\ PRP, RRR, RPR, RRP,\ RRP,\ RPR,\ PRR,\ PRR,\ PRP,}\\ \mathrm{PRPa, PPaR,\ PaPR,\ RPPa,\ PaRP,\ RPaP, RRPa,\ RRPa, RPaR,\ RPaR,\ PaRR,}\\ \mathrm{PaRR,\ PRPa,\ PPaR,\ PaPR, RPPa,\ PaRP, RPaP}\ \}$$

Function 1: $g^{\mathrm{PI}}(x^{\mathrm{PI}})$, $g^{\mathrm{PIA}}(x^{\mathrm{PIA}})$ and $g^{\mathrm{PIB}}(x^{\mathrm{PIB}})$, described as

$$\begin{aligned} g^{\mathrm{PI}}(x^{\mathrm{PI}}) &= x^{\mathrm{PI}}, x^{\mathrm{PI}} \in E^{\mathrm{PI}} \\ g^{\mathrm{PIA}}(x^{\mathrm{PIA}}) &= x^{\mathrm{PIA}}, x^{\mathrm{PIA}} \in E^{\mathrm{PI}} \\ g^{\mathrm{PIB}}(x^{\mathrm{PIB}}) &= x^{\mathrm{PIB}}, x^{\mathrm{PIB}} \in E^{\mathrm{PI}} \end{aligned} \tag{17.16}$$

The results of the type synthesis of type I PPMs with all DoFs can be obtained by the following three conditions.

Condition 1: If $g^{\mathrm{PI}}(x^{\mathrm{PI}}) = g^{\mathrm{PIA}}(x^{\mathrm{PIA}}) = g^{\mathrm{PIB}}(x^{\mathrm{PIB}})$, then the results are 3-$g^{\mathrm{PI}}(x^{\mathrm{PI}})$ or 3-$g^{\mathrm{PIA}}(x^{\mathrm{PIA}})$ or 3-$g^{\mathrm{PIB}}(x^{\mathrm{PIB}})$. The amounts of type I PPMs with all DoFs under condition 1 are 31.

Condition 2: If $g^{\mathrm{PI}}(x^{\mathrm{PI}}) \neq g^{\mathrm{PIA}}(x^{\mathrm{PIA}}) = g^{\mathrm{PIB}}(x^{\mathrm{PIB}})$ or $g^{\mathrm{PI}}(x^{\mathrm{PI}}) = g^{\mathrm{PIA}}(x$-$\mathrm{PIA}) \neq g\mathrm{PIB}(x^{\mathrm{PIB}})$, then the results are 2-$g^{\mathrm{PI}}(x^{\mathrm{PI}})$&1-$g^{\mathrm{PIA}}(x^{\mathrm{PIA}})$ or 2-$g^{\mathrm{PI}}(x^{\mathrm{PI}})$&1-$g^{\mathrm{PIB}}(x^{\mathrm{PIB}})$ or 2-$g^{\mathrm{PIA}}(x^{\mathrm{PIA}})$&1-$g^{\mathrm{PI}}(x^{\mathrm{PI}})$ or 2-$g^{\mathrm{PIB}}(x^{\mathrm{PIB}})$&1-$g^{\mathrm{PI}}(x^{\mathrm{PI}})$. The amounts of type I PPMs with all DoFs under condition 2 are 930.

Condition 3: If $g^{\mathrm{PI}}(x^{\mathrm{PI}}) \neq g^{\mathrm{PIA}}(x^{\mathrm{PIA}}) \neq g^{\mathrm{PIB}}(x^{\mathrm{PIB}})$, then the result is 1-$g^{\mathrm{PI}}(x^{\mathrm{PI}})$ &1-$g^{\mathrm{PIA}}(x^{\mathrm{PIA}})$&1-$g^{\mathrm{PIB}}(x^{\mathrm{PIB}})$. The amounts of type I PPMs with all DoFs under condition 3 are 4492.

where the number before the notation "-" denotes the amounts of the limbs, and the value of the function after the notation "-" represents the topology of the limb.

For example, "2-$g^{\mathrm{PI}}(x^{\mathrm{PI}})$&1-$g^{\mathrm{PIA}}(x^{\mathrm{PIA}})$" stands for a type I PPMs composed of two type I limbs with their topologies are "$g^{\mathrm{PI}}(x^{\mathrm{PI}})$", one type I limb with its topology is "$g^{\mathrm{PIA}}(x^{\mathrm{PIA}})$", a fixed base and a moving platform.

#### 17.3.2.2 Type Synthesis of Type I–II PPMs

Type I–II PPMs with less DoFs are made up of two type I limbs, one type II limb, a fixed base and a moving platform, which can process (a) two DoFs motions with two translations, (b) two DoFs motions with one translations and one rotation, (c) one DoF motion with one translation, (d) one DoF motion with one rotation. The topologies of two type I limbs can (1) identical, (2) different from each other. Without loss of generality, the layout of the two type I limbs was symmetrical relative to the type II limb.

**Definition 2** Set $E^{\mathrm{PII}}$, expressed as

$$E^{\mathrm{PII}} = \{\mathrm{PR,\ RP,\ RR,\ P,\ R,\ PPa,\ RPa,\ PaP,\ PaR,\ Pa}\}$$

Function 2: $g^{\mathrm{PII}}(x^{\mathrm{PII}})$, described as

$$g^{\mathrm{PII}}\left(x^{\mathrm{PII}}\right) = x^{\mathrm{PII}}, x^{\mathrm{PII}} \in E^{\mathrm{PII}} \quad (17.17)$$

The results of the type synthesis of type I–II PPMs with less DoFs can be obtained by the following two conditions.

Condition 1: If $g^{\mathrm{PI}}(x^{\mathrm{PI}}) = g^{\mathrm{PIA}}(x^{\mathrm{PIA}})$, then the results are 2-$g^{\mathrm{PI}}(x^{\mathrm{PI}})$&1-$g^{\mathrm{PII}}(x^{\mathrm{PII}})$ or 2-$g^{\mathrm{PIA}}(x^{\mathrm{PIA}})$&1-$g^{\mathrm{PII}}(x^{\mathrm{PII}})$. The amounts of type I–II PPMs with all DoFs under condition 1 are 310.

Condition 2: If $g^{\mathrm{PI}}(x^{\mathrm{PI}}) \neq g^{\mathrm{PIA}}(x^{\mathrm{PIA}})$, then the results are 1-$g^{\mathrm{PI}}(x^{\mathrm{PI}})$&1-$g^{\mathrm{PIA}}$(x-PIA)&1-gPII$(x^{\mathrm{PII}})$. The amounts of type I–II PPMs with all DoFs under condition 2 are 465.

#### 17.3.2.3 Type Synthesis of Type I–III PPMs

Type I–III PPMs with less DoFs are made up of one type I limbs, one type III limb, a fixed base and a moving platform, which can process (1) two DoFs motions, (2) one DoFs motions. Without loss of generality, the layout of the two type I limbs was symmetrical relative to the line from the center of the fixed base to that of the moving platform.

**Definition 3** Set $E^{\mathrm{PIII}}$, expressed as

$$E^{\mathrm{PIII}} = \{\mathrm{PR},\ \mathrm{RP}, \mathrm{P},\ \mathrm{PR}, \mathrm{RP}, \mathrm{RR},\ \mathrm{RR}, \mathrm{R}, \mathrm{PPa}, \mathrm{RPa},\ \mathrm{PaP},\ \mathrm{PaR}\}$$

Function 3: $g^{\mathrm{PIII}}(x^{\mathrm{PIII}})$, described as

$$g^{\mathrm{PIII}}\left(x^{\mathrm{PIII}}\right) = x^{\mathrm{PIII}}, x^{\mathrm{PIII}} \in E^{\mathrm{PII}}I \quad (17.18)$$

The results of the type synthesis of type I–III PPMs with less DoFs are 1-$g^{\mathrm{PI}}(x^{\mathrm{PI}})$&1-$g^{\mathrm{PIII}}(x^{\mathrm{PIII}})$. The amounts of type I–III PPMs with all DoFs are 372.

#### 17.3.2.4 Type Synthesis of Type IV PPMs

Type IV PPMs with less DoFs are made up of two type IV limbs, a fixed base and a moving platform, which can process general motions in the plane. The topologies of two type IV limbs can (1) identical, (2) different from each other. Without loss of generality, the layout of the two type IV limbs was symmetrical relative to the line from the center of the fixed base to that of the moving platform.

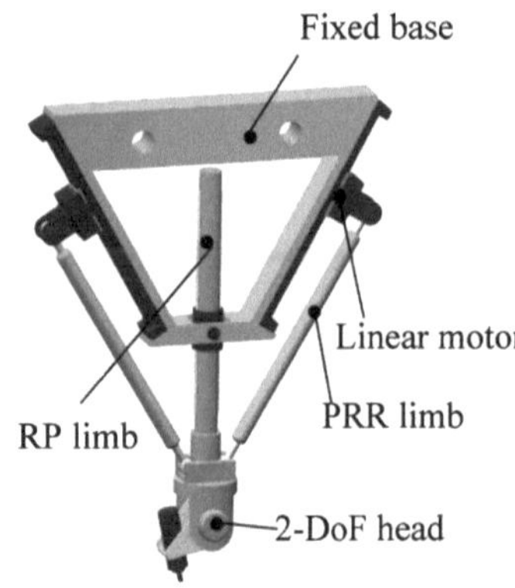


**Fig.17.2** The virtual prototype of Licept PPM

**Definition 4** Set $E^{\mathrm{PIV}}$, expressed as

$$E^{\mathrm{PIV}} = \{\mathrm{PR},\ \mathrm{RP}, \mathrm{RP},\ \mathrm{PR}, \mathrm{PPa}, \mathrm{RPa},\ \mathrm{PaP},\ \mathrm{PaR}\}$$

Function 4: $g^{\mathrm{PIV}}(x^{\mathrm{PIV}})$ and $g^{\mathrm{PIVA}}(x^{\mathrm{PIVA}})$ described as

$$\begin{gathered} g^{\mathrm{PIV}}(x^{\mathrm{PIV}}) = x^{\mathrm{PIV}}, x^{\mathrm{PIV}} \in E^{\mathrm{PIV}} \\ g^{\mathrm{PIVA}}(x^{\mathrm{PIVA}}) = x^{\mathrm{PIVA}}, x^{\mathrm{PIVA}} \in E^{\mathrm{PIV}} \end{gathered} \tag{17.19}$$

The results of the type synthesis of type IV PPMs with less DoFs can be obtained by the following two conditions.

Condition 1: If $g^{\mathrm{PIV}}(x^{\mathrm{PIV}}) = g^{\mathrm{PIVA}}(x^{\mathrm{PIVA}})$, then the results are 2-$g^{\mathrm{PIV}}(x^{\mathrm{PIV}})$ or 2-$g^{\mathrm{PIVA}}(x^{\mathrm{PIVA}})$. The amounts of type IV PPMs with all DoFs under condition 1 are 8.

Condition 2: If $g^{\mathrm{PIV}}(x^{\mathrm{PIV}}) \neq g^{\mathrm{PIVA}}(x^{\mathrm{PIVA}})$, then the results are 1-$g^{\mathrm{PIV}}(x^{\mathrm{PVI}})$&1-$g^{\mathrm{PIVA}}(x^{\mathrm{PIVA}})$. The amounts of type I–II PPMs with all DoFs under condition 2 are 28.

## 17.4 Three Typical PPMs with Linear Motors

Compared to the conventional servo motor, the liner motor has advantages of higher velocity and acceleration, better accuracy, stronger stability, and so on, more and more linear motors are applied in the robots or machines, for example, many machines are manufactured by DMG company.

In this paper, three typical PPMs with linear motors are demonstrated according to the results of the type synthesis of PPMs, which can be seen in Figs. 17.2–17.4.

As shown in Fig. 17.2, the Licept PPM is presented for the first time, which is a hybrid mechanism with four DoFs, its parallel mechanism is a 2-PRR&1-RP PPM, and the serial part is a 2-DoF rotational head. The Licept PPM in Fig. 17.2 can be applied in the drilling, riveting or laser welding of large structural components.

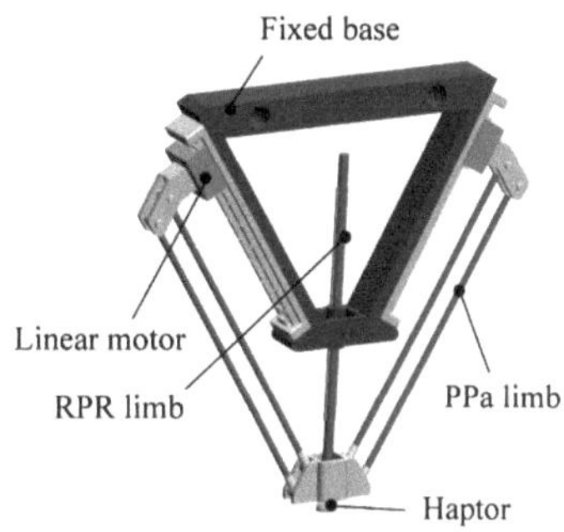


**Fig.17.3** The virtual prototype of Litriangle PPM

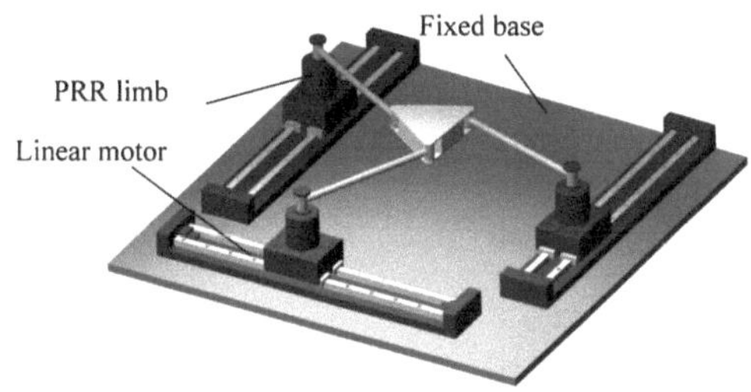


**Fig.17.4** The virtual prototype of 3-PRR PPM

As shown in Fig. 17.3, the Litriangle PPM possesses 2-DoF with two translations is presented for the first time, whose topology is 2-PPa&1-RPR, the Litriangle PPM can be used in the pick-and-place with high velocity in the production line.

As shown in Fig. 17.4, the PPM possesses 3-DoF with two translations and one rotation, its topology is 3-PRR, which can be applied in the motions of high accuracy and large workspace.

## 17.5 Conclusions

This paper concludes with a discussion of type synthesis for the planar parallel mechanism (PPMs) incorporating actuated limbs with zero/one constraint with set conception, whose main conclusions includes

(1) According to contributions to the moving platform of PPMs made by its limbs, these limbs can be classified as four categories: Actuated limb with zero constraint, Fully-constrained limb with zero actuator, Fully-constrained limb with one actuator, and Under-constrained limb with one actuator.

(2) On the basis of the descriptions of actuated space and constrained space of the end for the four type limbs using the screw theory, the type synthesis of the four type limbs are carried out, and the topology amounts of the type I-IV limbs with P joint, R joint, Pa joint are 31, 10, 12, 8 respectively.

(3) The method of type synthesis for PPMs is presented, after defining different sets and functions are defined respectively, whose elements and variables are the topologies of four type limbs. The different function values and the composition principle of PPMs lead to existence of PPMs topologies by four solutions, and the topology-amount of the type I-IV PPMs are 5453, 775, 372, 36, respectively.
(4) To take three typical PPMs with liner motors for examples, their virtual prototypes are obtained with commercial 3-dimensional CAD software, herein, two PPMs are presented for the first time.

**Acknowledgments** The authors of this paper wish to acknowledge the financial support received from the National Natural Science Foundation of China (Grant No. 51075295) and the Tianjin Research Program of Application Foundation and Advanced Technology (Grant No. 11JCZDJC22700).

# References

1. Dasgupta B, Mruthyunjaya TS (2000) The stewart platform manipulator: a review. Mech Mach Theory 35:15–40
2. Wen F, Liang CG (1994) Displacement analysis of the 6-6 Stewart platform mechanisms. Mech Mach Theory 29:547–557
3. Shen HP, Yang TL, Ma LZ (2005) Synthesis and structure analysis of kinematic structure of 6-DOF parallel robotic mechanisms. Mech Mach Theory 40:1164–1180
4. Yao R, Tang XQ, Li TM et al (2010) Error analysis and distribution of 6-SPS and 6-PSS reconfigurable parallel manipulators. Tsinghua Sci Technol 15:547–554
5. Li QC, Hervé JM (2010) 1T2R parallel mechanism without parasitic motion. IEEE T Robotics 26:401–410
6. Gogu G (2007) Structural synthesis of fully-isotropic parallel robots with Schönflies motions via theory of linear transformations and evolutionary morphology. Eur J Mech 26:242–269
7. Huang Z, Li Q-C (2003) Type synthesis of symmetrical lower-mobility parallel mechanisms using the constraint-synthesis method. Int J Robotics Res 22:59–79
8. Kong XW, Gosselin CM, Richard PL (2007) Type synthesis of parallel mechanism with multiple operation modes. J Mech Des 129:595–601
9. Liu XJ, Wang JS, Wu C et al (2009) A new family of spatial 3-DOF parallel manipulators with two translational and one rotational DOFs. Robotica 27:241–247
10. Jin Q, Yang TL (2004) Theory for topology synthesis of parallel manipulators and its application to three-dimension-translation parallel manipulators. J Mech Des 126:625–640
11. Gao F, Zhang Y, Li WM (2005) Type synthesis of 3-DOF reducible translational mechanisms. Robotica 23:239–245
12. Dai JS, Wang D (2007) Geometric analysis and synthesis of the metamorphic robotic hand. J Mech Eng 129:1191–1197

# Chapter 18
# Dynamics Modeling and Analysis of Compaction Force of Crawler on Cone Lamp Pole

**Nan-Jiang Gu and Zhong-Bin Liu**

**Abstract** Static and dynamic model of self-adapting impacted and contractile mechanism of crawling robots on cone lamp pole are built, also contact force and contract action of pressing mechanism are simulated and analyzed in COSMOS-Motion. Research shows that the mechanism compacting force on the pole is evenly distributed when applied at constant force; corresponding relationship of the contracting force and caliber is given, which provides the design basis for the robot crawling with the constant force.

**Keywords** Adaptive institutions · Cone lamp pole · Constant pressure

## 18.1 Introduction

Lamp aspect, as one of the urban appearance, is paid much attention in the modern city. At present, cleaning and detecting of the lamp pole are fulfilled by elevator and maintenance worker, the labor of workers is large, risk and maintenance cost high, traffic blocking caused by cleaning and testing. Application research report of the robot for the lamp pole has not seen at home, while the application research in foreign countries has many shortages which mainly is too large compaction force of robots attached on the light pole surface, thus robot is huge, heavy, which is not appropriate for aerial work, and tends to cause a secondary injury of lamp pole [1–4].

N.-J. Gu (✉) · Z.-B. Liu
Sichuan University of Science & Eng, Zigong, People's Republic of China
e-mail: 28392199@qq.com

Z.-B. Liu
e-mail: lzbxxh@163.com

J. S. Dai et al. (eds.), *Advances in Reconfigurable Mechanisms and Robots I*,
DOI: 10.1007/978-1-4471-4141-9_18, © Springer-Verlag London 2012

According to the above, this paper takes self-adapting impacted and contractile mechanism of crawling robots on cone lamp pole as model, the self-adapting organizations and related parameters are calculated and analyzed, the static and dynamic model of compacting and contracting organizations action of conical pole robot are set up, and simulated with COSMOSMotion, getting the corresponding dynamics index, not only makes stable pressuring lamp pole come true, but also adapts to the change of the conical light pole caliber automatically, provides appropriate adhesion pressure, which builds foundation for the lightweight design of crawling robots on the light pole and improving the adaptive ability.

## 18.2 Self-Adapting Impacted Mechanism

### *18.2.1 Self-Adapting Net Hinged Mechanism Design*

Line constraints of lamp pole are different from those of ground rigid pipes, lamp pole is a vertical linear body with certain disturbance, that moving characteristics are different from those of general cable crawling robots. The main problem of crawling device of robots developed is the excessive pressure against the lamp pole, the primary reason is that compacting force (such as spring tension) loaded from the crawling device makes roller attach tightly to the lamp pole surface. If compaction force is too large, crawling robot tends to secondary injury on the surface.

In addition, the structure characteristics of the conical lamp pole is that lamp pole diameter decreases with height. The crawling process of robot is as follows: At first, the robot is in the low end of the lamp pole whose diameter is large, At this time, the robot attaches itself to the lamp pole with certain preload; Then, the diameter of pole becomes small as robot goes up, general crawling robot mechanism can not adapt to this change. At the same time compacting force decreases, robot will slide on the pole, even fell down.

In order that suitable adhesive force can be produced at the proper time between robot and surface of lamp pole, avoid excessive positive pressure. A new self-adapting pressing mechanism is proposed which is articulated as parallelogram principle (Fig. 18.1). When self-adapting net hinged mechanism is impacted at hinge joint 1 and 2, stretching is produced points to the origin center O.

### *18.2.2 Reaction Analysis from Cylinder to Roller*

Cylinder and roller are supposed to rigid body, and self-adapting net hinged mechanism is supposed to geometrical variant. Three rollers are uniform distributed in the hinged points between 3 and 12 in self-adapting nets mechanism,

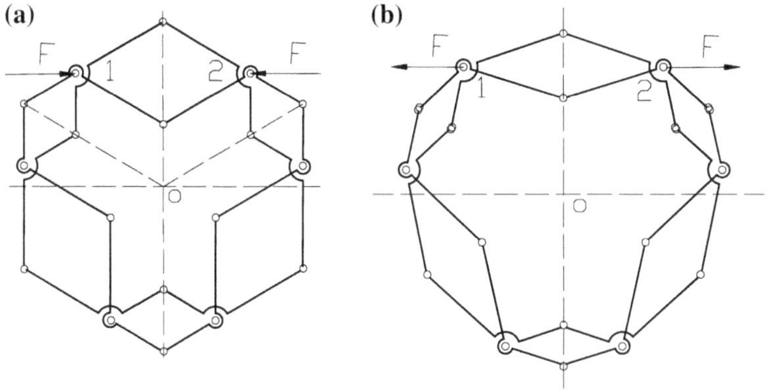


**Fig. 18.1** Self-adapting net hinged mechanism

loaded by F, $N_1$, $N_2$ and $N_3$ are reaction force from cylindrical to three wheels (Fig. 18.2).

From balance equation of load:

$$N_1 = N_2 = N_3 \tag{18.1}$$

Take hinged points 10 and 11 as virtual constraints, and revolute pairs at hinged points 10 and 11 are removed.

$$F_{1X} + F_{2x} + N_1 \cos 30 = F \tag{18.2}$$

$$F_{1y} - F_{2y} + N_1 - N_2 \cos 30 = 0 \tag{18.3}$$

$$F_{1x}(R + r + 2l_2 \cos A) + F_{2y}(R + r) - F(R + r + l_2 \cos A) = 0 \tag{18.4}$$

From (18.2), (18.4)

$$F_{1X} = \frac{F}{2} + \frac{\sqrt{3}N_1(R + r)}{4l_2 \cos A} \tag{18.5}$$

Component composed by hinged points 7, 9, 10 and the component composed by 8, 9, 11 are rods without force, so the force direction at hinged point 5 is as shown in Fig. 18.2c.

$$\begin{aligned} &F_{4y} \cos 30 + F_{4x} \cos 60 - N_1 \cos 30- \\ &F_{65} \cos A \cos 30 - F_{65} \sin A \cos 60 = 0 \end{aligned} \tag{18.6}$$

$$\begin{aligned} &F_{4y} \sin 30 - F_{4x} \sin 60 - N_1 \sin 30- \\ &F_{65} \cos A \sin 30 - F_{65} \sin A \sin 60 = 0 \end{aligned} \tag{18.7}$$

$$F_{4x} L_{04} = F_{65} \times L_{05} \times \sin A \tag{18.8}$$

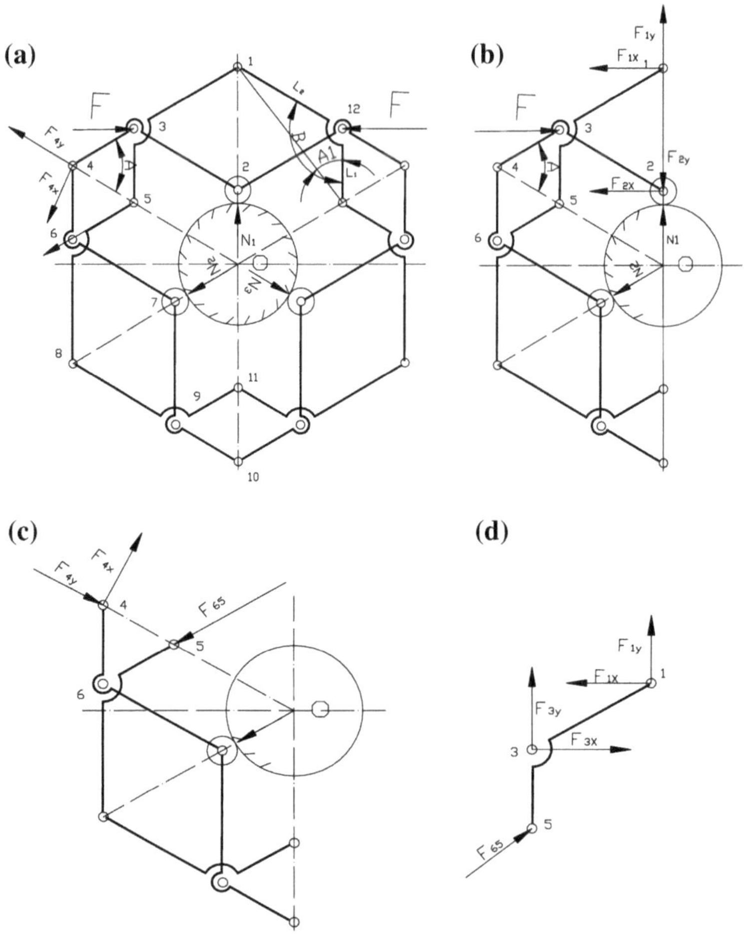


**Fig. 18.2** Plane force analysis of reaction from cylinder to roller

And:

$$L_{04} = l \sin(120 - A + A1)$$

$$l_{05} = l_{04} - 2l_2 \cos A$$

By (18.9), (18.10) and (18.11):

$$F_{65} = -\frac{N_1 l \sin(120 - A + A1)}{2l_1 \sin A \cos A} \tag{18.9}$$

$$\begin{aligned} &F_{1y} l_2 \sin A + F_{1x} l_2 \cos A + \\ &\quad 2F_{65} l_2 \cos A \sin A = 0 \end{aligned} \tag{18.10}$$

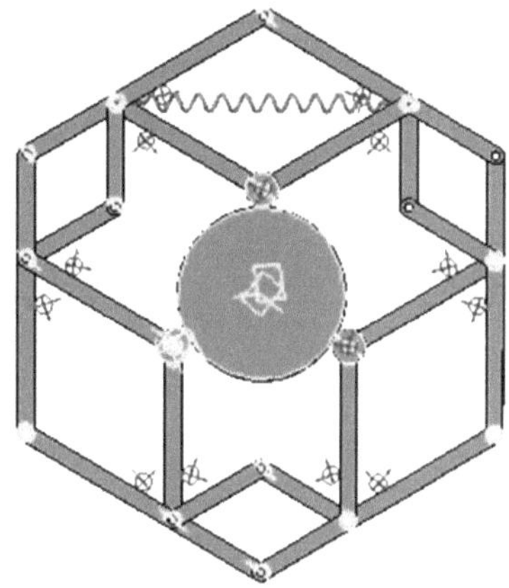

**Fig. 18.3** Model simulation

With symmetric relation of self-adapting net hinged mechanism: $F_{1y} = 0$
And by (18.9), (18.5)

$$N_1 = \frac{Fl_2 \cos A}{2l \sin(120 - A + A1) - \frac{\sqrt{3}(R+r)}{2}} \tag{18.11}$$

## 18.3 Simulation Modeling of Self-Adapting Net Hinged Mechanism

Structure parameters of simulation modeling of self-adapting net hinged mechanism of crawling robot on lamp pole: The weight of components and friction are ignored. as shown in Fig. 18.3, the preloaded force on the two hinged points is 200 N, its parameters are: $L_1 = 60$ mm $L_2 = 100$ mm $\beta$=120 degrees R = 50 mm; r = 10 mm; crashing motion simulation is conducted between roller and cylinder in the center.

From the formula (18.1), (18.11) (Fig. 18.4):

$$N_1 = N_2 = N_3 = \frac{Fl_2 \cos A}{2l \sin(120 - A + A1) - \frac{\sqrt{3}(R+r)}{2}}$$

$$= \frac{200 \times 100 \times \cos 60}{2 \times 140 \sin(120 - 60 + 38.21) - \frac{\sqrt{3}(50+10)}{2}} = 44.411(\mathrm{N})$$

By the formula (18.11) :

$$\mathrm{N}_1 = 44.4111(\mathrm{N})$$

which is in accordance with the simulation results.

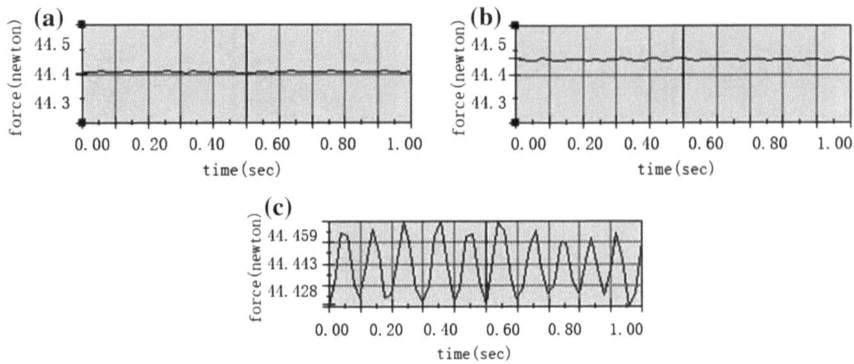


**Fig. 18.4** Simulation result of reaction from cylinder to the three wheels (Y coordinate is the reaction force from cylinder to three rollers)

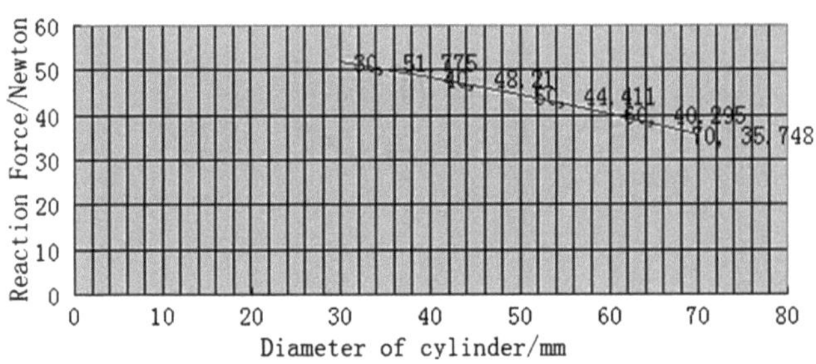


**Fig. 18.5** The relationship between cylinder radius and reaction force

When the component parameters and preloaded force keep constant, while cylindrical radius is changeable: reaction force from cylinder to wheels is shown in Fig. 18.5.

## 18.4 Conclusion

A new type of clamping mechanism is put forward according to the issue that positive pressure of traditional crawling robot loading to the lamp pole is too large. Self-adapting parameters of impacted and contractile mechanism of crawling robot on cone lamp pole are obtained through modeling and analysis, supplies series of elements of allotments to the design of pressing and contracting mechanism.

This mechanism becomes small with diameter of cone pole decreases, big with that increases under certain preloaded force (preloaded force is adjustable), and generate enough positive pressure to make crawling robots attach on the conical pole, avoiding dropping down.

The system dynamics simulation analysis shows that: when constant force is loaded, compacting force on the conical lamp pole is making a uniform distribution. When the caliber of the pole changes, the corresponding modification of contractive force is given, which becomes a reference for robot crawling under constant pressure on the cone pole.

## References

1. Sun H, Chen Z, Ge W (2008) Mechanical principle. Press of Higher Education, Beijing
2. Zhang J, Guo X (2008) Mechanical simulation design of solidworks and COSMOSMotion. Press of Tsinghua University, Beijing
3. Aracil R, Saltarén R, Reinoso O (2003) Parallel robots for autonomous climbing along tubular structures. Robotic Auton Syst 42:125–134
4. Mayhew D, Bachrach B, Rymer WZ, Beer RF (2005) Development of the MACARM—a Novel Cable Robot for Upper Limb Neuro rehabilitation. In: Proceedings of the 2005 IEEE 9th international conference on rehabilitation robotics, Chicago, IL, USA, 2005

# Chapter 19
# Deformation Discipline of the Flexspline in Harmonic Drives with Engagement Output

**Huimin Dong, Zhi Chen, Chunmao Zhang and Delun Wang**

**Abstract** This paper deals with the deformation function of the flexspline under load condition, and its effect on the tooth profile. In order to investigate the deformation discipline under load condition, a kinematic simulation analysis model of ABAQUS for a harmonic drive with engagement output is established. With this model, the nodal displacements of the flexspline neutral layer under different loads are obtained in dynamic transmission process. Fourier function is adopted to fit the deformation function under loads. Based on the deformation function, the tooth profile of the circular spline is modified. The example indicates that with the modified tooth profile under load condition the conjugate region increases, and the deformation function varies little. The work of this paper lays the foundation for optimizing the tooth profile to prolong service life of harmonic drives.

**Keywords** Harmonic drives · ABAQUS · Deformation function · Fourier fitting

## 19.1 Introduction

Harmonic drives (HDs) have been widely used in many industrial fields, such as optical equipment, robots and automobile etc. for its compactness, high precision and high reduction ratio characteristics. The performance of HDs mainly depends on the deformation of the flexspline. At present, however, in the tooth profiles design of HDs, the theoretical deformation function is usually taken into account [1].

H. Dong (✉) · Z. Chen · C. Zhang · D. Wang
School of Mechanical Engineering, Dalian University of Technology, Linggonglu 2# Ganjingzi District, 116024 Dalian People's Republic of China
e-mail: donghm@dlut.edu.cn

J. S. Dai et al. (eds.), *Advances in Reconfigurable Mechanisms and Robots I*,
DOI: 10.1007/978-1-4471-4141-9_19, © Springer-Verlag London 2012

The kinematic model presented in [2, 3] to describe the motion among the components of a HD and meshing equation used to solve the conjugate tooth profile are based on the theoretical deformation function. The deformation of the flexspline is not only related with the shape of the wave generator and type of HDs, but also with load etc. [4, 5]. How the load affects the deformation function and the tooth pair's engagement on HDs with flexspline cup are discussed in [4, 5]. A parameterized finite element model of Abaqus for HDs with flexspline cup is established to analyze the stress distribution on the flexspline cup [6].

In this paper, the deformation function and conjugate tooth profile effects of load in HDs with engagement output are studied. According to the kinematic model and conjugation condition [2], a flexspline tooth profile is given, the conjugate circular spline tooth profile is obtained in term of theoretical deformation of the flexspline. Then, a finite element analysis model of Abaqus for the HD with engagement output is established. Using this model, the deformation function with discrete data can be obtained. The deformation function under different loads is fitted by Fourier function. Comparing the deformation function with the theoretical deformation function, the conjugate involute tooth profile is modified. Finally, the relationships among the load, tooth profile and deformation are discussed in the calculating example.

## 19.2 Kinematic Model of HDs with Engagement Output

A HD with engagement output consists of four components that are a rigid circular spline, an output circular spline, a wave generator and a flexspline, in which the flexspline and the output circular spline have the same number of teeth. The structure form is shown in Fig. 19.1.

### *19.2.1 Theoretical Deformation Function*

According to the kinematic model [2], the theoretical deformation function of the ring flexspline is

$$\begin{cases} w(\theta) = R(\theta) - r_0 \\ v(\theta) = -\displaystyle\int_0^{\theta} w(\theta) d\theta \\ \varphi(\theta) = \cos^{-1}\left(\dfrac{R}{\sqrt{R^2 + \dot{R}^2}}\right) \end{cases} \tag{19.1}$$

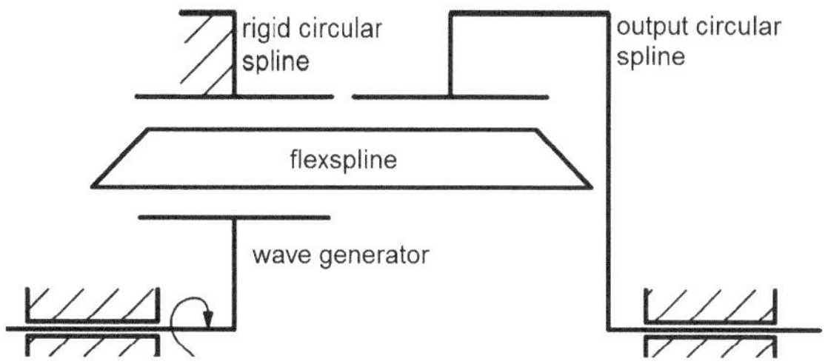


**Fig. 19.1** Structure form of HDS with engagement output

where $R$ is the polar form of the cam profile (the neutral curve), $r_0$ is the radius of the prime circle of the cam, $\theta$ is the rotational angle of the wave generator.

### *19.2.2 Conjugate Condition and Conjugate Tooth Profile*

The basic meshing equation of the HD satisfies the conventional gearing conjugate principle. Assuming the flexspline tooth profile $M(x_f, y_f)$ is known, and the conjugate condition [7] can be written as [2]:

$$\mathrm{V}^{fC} \cdot \mathrm{n}^{f} = R(1 - \omega_C)\sin\psi_f + \dot{w}\cos\psi_f + r_f(\omega_f - \omega_C)\sin(\delta_f - \mu_f) = 0 \quad (19.2)$$

where $\omega_f$, $\omega_C$ are the angular velocity of the flexspline and the circular spline, $\theta_f = \theta + \varphi$, $\psi_f = \delta_f + \theta_f - \theta$, $\delta_f = \tan^{-1}(-\frac{dx_f}{dy_f})$, $r_f = \sqrt{x_f^2 + y_f^2}$, $\mu_f = \tan^{-1}(\frac{y_f}{x_f})$; $\delta_f$, $r_f$ and $\mu_f$ are the tooth profile parameters of the flexspline.

Based on the above equations, the conjugate circular spline tooth profile $M(x_{Ck}, y_{Ck})$ could be calculated [2]:

$$\begin{cases} x_C = R\cos(\theta - \theta_{Ck}) + r_f\cos(\mu_f - \theta_{Ck} + \theta_f) \\ y_C = R\sin(\theta - \theta_{Ck}) + r_f\sin(\mu_f - \theta_{Ck} + \theta_f) \end{cases} \quad k = 1, 2 \qquad (19.3)$$

where $M(x_{C1}, y_{C1})$ represents the rigid circular spline, $M(x_{C2}, y_{C2})$ represents the output circular spline. According to the arc length of the neutral curve is constant, the rotation angle of the rigid circular spline and the output circular spline can be obtained, $\theta_{C1} = \frac{N_F}{R_0 N_{C1}} \int_0^\theta \sqrt{R^2 + \dot{R}^2} d\theta$, $\theta_{C2} = (1 + \frac{2}{N_F})\theta_{C1}$

## 19.3 Deformation Function Under Load Condition

Generally, the deformation of the flexspline is related with the load value, when the load value changes, the radical, circumferential displacement and oscillation angle will change accordingly. The displacements, however, possesses the

**Fig. 19.2** The process flow diagram

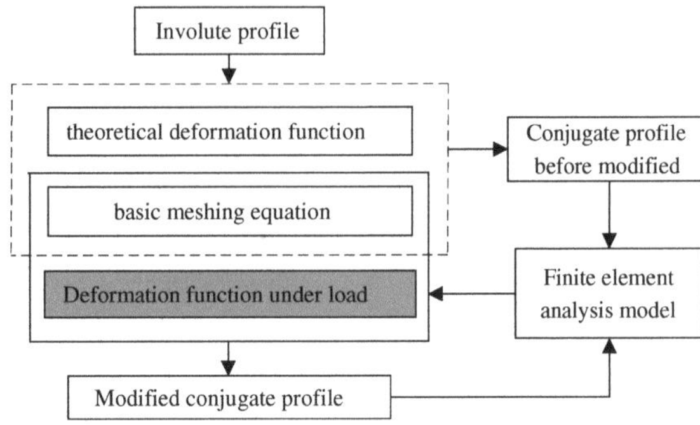


characteristic of periodic changes which could be seen as the superposition of multi harmonic waves.

## *19.3.1 Definition of Deformation Function*

According to the characteristic of the displacement, we choose Fourier function to fit the deformation function under load condition, because Fourier function satisfies the linear combination of multi trigonometric functions. The deformation function can be expressed as:

$$f(\theta) = a_0 + \sum_{k=1}^{n} [a_k \cdot \cos(k \cdot w \cdot \theta) + b_k \cdot \sin(k \cdot w \cdot \theta)] \qquad (19.4)$$

where $n$ is a positive integer related to the fitting precision. $a_0$, $a_k$, $b_k$, $w$are constant, $\theta \in [-180°, 180°]$.

When we fit the deformation function under load with Fourier function, there are several factors to judge the fitting quality including SSE (Sum of squares due to error), RMSE (Root mean squared error), Adjusted R-square (degree-of-freedom adjusted coefficient of determination), R-square (Coefficient of determination), in which if the first two are close to 0, and the latter two approach to 1, indicating the fitting effect is pretty good. On the contrary, the fitting effect is bad.

## *19.3.2 Calculation Procedure of Deformation Function Under Load*

To acquire more reasonable conjugate tooth profile, the calculation procedure of the deformation function should be a multi-iterative process. The process flow diagram is shown in Fig. 19.2:

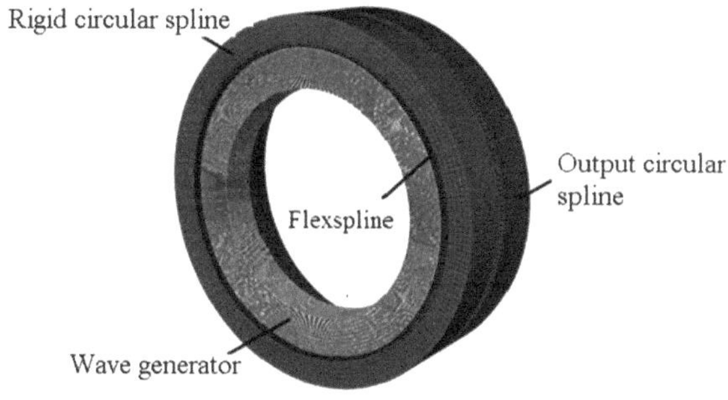


**Fig. 19.3** The finite element analysis model

1. Finite element analysis model: the finite element analysis model of Abaqus is established shown in Fig. 19.3. The material parameters of each component are listed in Table 19.1. The model is modelled with C3D8R element; the solving process is divided into two steps. The first step is simulating the process that the wave generator is installed into the flexspline, forcing the flexspline to produce elastic deformation and prestress. The outer ring of the rigid circular spline is under full constraints, with implicit algorithm to calculate this step; The second step based on the first step add an rated speed $n$ to the wave generator, and torque load $T$ is applied on outer ring of the output circular spline. Then the contact of wave generator and flexspline and the contact of flexspline tooth and circular spline tooth are defined, with explicit algorithm to calculate this step.
2. Deformation function fitting: through the simulation analysis result, the radial, circumferential and oscillation angle displacement are picked up, which need to be fitted by Fourier function. According to Eq. (19.4), the unknown quantities $a_0$, $a_i$, $b_i$ ($i = 1, 2, 3$), and w can be solved.
3. Modification of conjugate profile: Based on the deformation function under load and basic meshing equation, according to Eqs. (19.2), (19.3), the conjugate circular profile can be modified, which can improve performance of meshing.

The above process should be done repeatedly to obtain the reasonable conjugate tooth profile.

## 19.4 An Example

Taking a HD as an example [2], which has number of flexspline tooth $N_F = 240$, number of rigid circular spline $N_{C1} = 242$, number of output circular spline $N_{C2} = 240$, module m = 0.5 mm, radius of the prime circle of the cam $r_0 = 60.31$ mm, radial deformation coefficient $w_0 = 0.425$ mm.

**Table 19.1** Basic constitutive parameters of materials

| Component | Material | Young's modulus E(Pa) | Poisson's ratio $\lambda$ | Density $\rho$ (kg/$mm^3$) |
|---|---|---|---|---|
| Flexspline | 32CrMnSiNiA | 196 | 0.3 | 7.75e-6 |
| Circular spline | 45 | 206 | 0.3 | 7.8e-6 |
| Wave generator | 45 | 206 | 0.3 | 7.8e-6 |

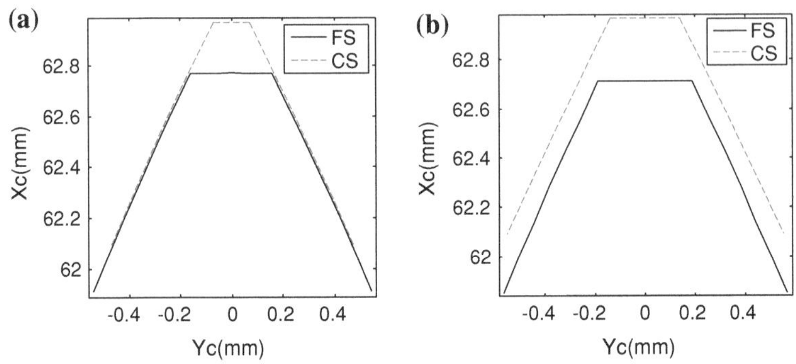


**Fig. 19.4** The flexspline and circular spline tooth profile. **a** The rigid circular spline. **b** The output circular spline

## *19.4.1 Conjugate Tooth Profile*

The given involute tooth profile of a flexspline with pressure angle $\alpha = 20^\circ$ is described [3]:

$$\begin{cases} r_{fi} = \dfrac{r_{fb}}{\cos \alpha_{fi}}, \ \delta_{fi} = \dfrac{\pi}{2} - \alpha_{fi} + \mu_{fi} \\ \mu_{fi} = \dfrac{1}{2N_F}(\pi + 4\xi_f \tan \alpha) + inv\alpha - inv\alpha_{fi} \end{cases}, \quad \alpha_{fi} \in [\alpha_{ff}, \alpha_{fa}] \tag{19.5}$$

where $r_{fb}$ is the base radius of the flexspline, addendum modification $\xi_f = 3.99$, $\alpha_{ff}$ is the dedendum pressure angle and $\alpha_{fa}$ is the addendum pressure angle.

Substituting Eqs. (19.1), (19.5) into Eq. (19.2), the conjugate region $\theta$ can be obtained easily. According to Eq. (19.3), the conjugate circular tooth profile can also be solved. Assuming $\theta \in [-90^\circ, 90^\circ]$, the conjugate region between the flexspline and the rigid circular spline can be solved $\theta_1 \in [2.80^\circ, 5.82^\circ]$. Likewise, the conjugate region between the flexspline and the output circular spline is obtained $\theta_2 \in [-25.77^\circ, -25.47^\circ]$, the corresponding conjugate tooth profile expressed in $S_C$-system is shown in Fig. 19.4.

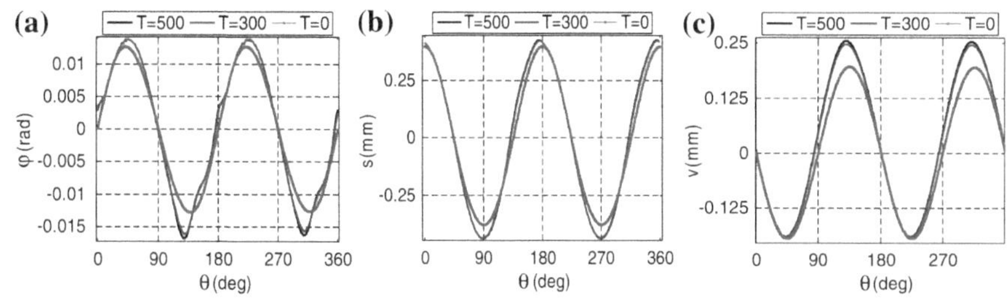


**Fig. 19.5** Comparison of deformation. **a** Radial displacement. **b** Circumferential displacement. **c** Oscillation angle displacement

### 19.4.2 Deformation Discipline by FEAM

Considering the following working conditions (torque $T = 0$, $T = 300$, $T = 500\,\mathrm{N \cdot m}$), the nodal displacements of the neutral layer are shown in Fig. 19.5.

In Fig. 19.5a, it can be seen that the radial displacement amplitude under load compared with the no load increases, the amplitude difference between $T = 0$ and $T = 300$, $500\,\mathrm{N \cdot m}$is 0.07727, 0.08654 mm respectively, in Fig. 19.5b, the circumferential displacement under load moves integrally up than the original, the amplitude difference between $T = 300$, $500\,\mathrm{N \cdot m}$ and $T = 0$ is 0.06074, 0.08654 mm respectively. Figure 19.5c shows that the distortion of the oscillation angle displacement under load occurs, especially appearing in the middle between the long axis and short axis of the wave generator, the amplitude difference is 0.004068, 0.004879 rad respectively.

### 19.4.3 Deformation Function Under Load

From Fig. 19.5, it can be seen when the load value changes, the displacement will change, but it could be seen as the superposition of multi harmonic waves. Adopting Fourier function to fit the deformation function under load is feasible.

Considering that the harmonic gear drive works usually in the rated working condition ($T = 500\,\mathrm{N \cdot m}$), the flexspline deformation under this condition is of critical importance. So in the paper, by the finite element analysis results, the nodal displacement data of the neutral-layer under rated load is picked up and fitted by Fourier function.

The result of Fourier function fitting is shown in Table 19.2 and the fitting evaluation factors are shown in Table 19.3:

From Table 19.3, we can see that SSE and RMSE are very close to 0, Adjusted R-square and R-square approach to 1, which indicates Fourier function fitting curve is very close to the initial data curve.

Table 19.2 Fourier function fitting results

| Fitting results | Radical displacement | Circumferential displacement | Oscillation angle displacement |
|---|---|---|---|
| a0 | 0.0006871 | 0.02999 | 1.538e-6 |
| a1 | 0.4276 | −0.02167 | 0.001268 |
| a2 | −0.01419 | −0.00159 | 0.00065 |
| a3 | −0.003837 | 0.0009123 | −0.000308 |
| b1 | −0.0368 | −0.2188 | 0.01407 |
| b2 | −0.0101 | 0.003781 | −0.0009987 |
| b3 | 0.003055 | 0.0006558 | −0.0003753 |
| w | 0.03485 | 0.03485 | 0.03482 |

Table 19.3 Fitting evaluation factors

| Evaluation factors | Radical displacement | Circumferential displacement | Oscillation angle displacement |
|---|---|---|---|
| SSE | 0.0122 | 0.0004275 | 0.0002956 |
| RMSE | 0.004139 | 0.0007748 | 0.0006443 |
| Adjusted R-square | 0.9998 | 1 | 0.9959 |
| R-square | 0.9998 | 1 | 0.9959 |

In order to give the further explanation that Fourier function can be used to fit the radical, circumferential and oscillation angle displacement. In Fig. 19.6, the comparison between the fitting curve and the original data curve has been done and the maximum error has been given.

From Fig. 19.6a, the maximum error between the radical displacement function fitting and the original data is 0.0144 mm, Likewise, from Fig. 19.6b, the maximum error between the circumferential displacement function fitting and the original data is −0.0052 mm. In Fig. 19.6c, the maximum error between the oscillation angle displacement function fitting and the original data is −0.0018 rad.

Based on the above analysis results, the fitting error is small enough, which is in the acceptable error range, so, Fourier fitting is viable.

### 19.4.4 Modification of Conjugate Tooth Profile

Based on the deformation function under load and basic meshing equation, the modified conjugate profile can be solved. The flexspline and the circular spline tooth profile after modifying are shown in Fig. 19.7, the conjugate region between the flexspline and the modified rigid circular spline is $\theta_1 \in [-3.33°, 12.97°]$, which is 5.4 times than the original conjugate region. Likewise, the conjugate region between the flexspline and the modified output circular spline is

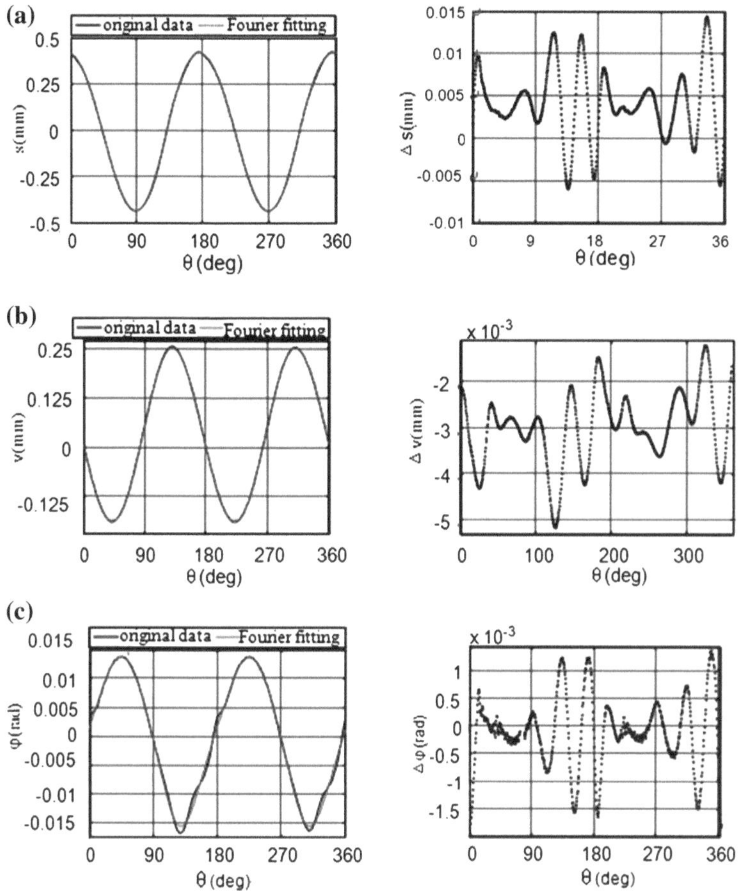


**Fig. 19.6** Fourier function fitting and the fitting error analysis. **a** The radical displacement fitting effort and fitting error analysis. **b** The circumferential displacement fitting effort and fitting error analysis. **c** The oscillation angle displacement fitting effort and fitting error analysis

$\theta_2 \in [-30.08°, -29.26°]$, which is 2.73 times than the original conjugate region. Increasing the conjugate region can help strengthen its transmission steady.

### *19.4.5 Comparison Between the Original and Modified Deformation Function Under Load*

To check that the tooth profile after modifying whether affects the load deformation function or not, the process of finite element analysis has been done again. The analysis results are as follows:

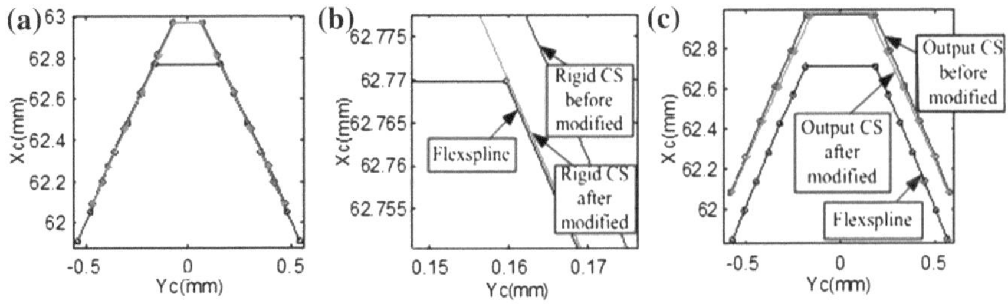


**Fig. 19.7** The comparison between the original and the modified tooth profile. **a** The fixed circular spline. **b** The magnified figure. **c** Output circular spline

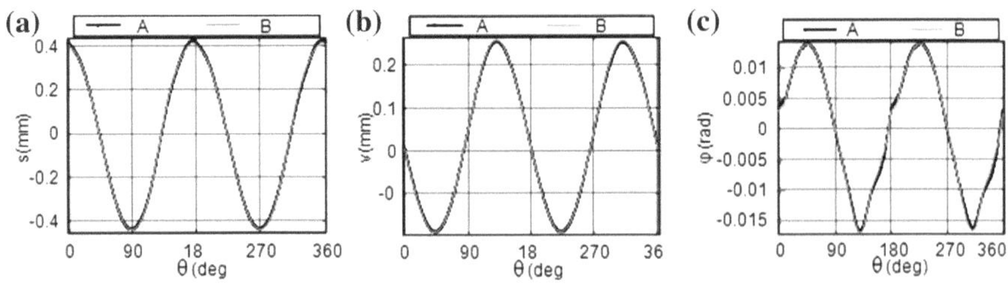


**Fig. 19.8** The comparison between the original and modified load deformation function. **a** Radial displacement. **b** Circumferential displacement. **c** Oscillation angle displacement

In Fig. 19.8, the curve A is the original deformation function under load, the curve B is the modified deformation function, it can be seen that when the tooth profile parameters change, the deformation function change accordingly, but very small.

## 19.5 Summary

(1) The deformation function under load can be regards as the superposition of multi harmonic waves, Fourier function satisfying the linear combination of multi trigonometric functions can be used to fit the displacement function under load, and achieve the very good fitting effect.
(2) Based on the deformation function, the circular spline tooth profile is modified. The conjugate region between the tooth pairs increases.
(3) Comparing the modified and original deformation, it is indicated that the modified tooth profile has little effect on the deformation function.

**Acknowledgments** Authors wish to acknowledge the financial support of the National Natural Science Foundation of P.R. China under Contract No. 50975038.

## References

1. Shen YW, Ye QT (1985) Theory and design of harmonic drive gearing. China Machine Press, Beijing
2. Dong HM, Wang D, Ting K-L (2011) Kinematic fundamentals of planar harmonic drives. ASME J Mech Des 133(1):0110071–0110078
3. Dong HM, Wang D, Ting K-L (2011) Kinematic effect of the compliant cup in harmonic drives. ASME J Mech Des 133(5):0510041–0510047
4. Ivanov MH, Shen YW (1987) The harmonic drive. National Defense Industry Press, Beijing
5. Dong HM (2008) Study of kinematics and meshing characteristic of harmonic gear drives based on the deformation function of the flexspline. Phd E., Thesis, Dalian University of Technology, China
6. Zhou WD (2008) Parameters optimization and dynamic simulation for harmonic gear drive with involute tooth profile. Dalian University of Technology, China
7. Litvin FL, Fuentes A (2004) Gear geometry and applied theory. Cambridge University Press, Cambridge

# Chapter 20
# Development of a New Steering Mechanism for Automobiles

**R. Venkatachalam and A. Padma Rao**

**Abstract** Steering is one of the important systems of an automobile. Ackermann mechanism is popularly used in all the automobiles. Though advances such as power steering have been taking place in providing steering to automobiles, the basic mechanism is Ackermann mechanism only. The Ackermann mechanism does not provide perfect steering conditions. However, it is popular because it involves only revolute pairs in the mechanism. In this paper an attempt is made to propose a new steering mechanism which is simpler than Ackermann mechanism and provides perfect steering always.

**Keywords** Ackermann mechanism · Correct steering · New steering mechanism for automobiles

## 20.1 Introduction

Steering is one of the essential requirements of an automobile. It has been a common practice to provide steering only to front wheels. In fact steering may also be provided only to rear wheels or to all the four wheels. A perfect steering mechanism should ensure that when the axes of rotation of all the four wheels are

R. Venkatachalam
Professor of Mechanical Engineering, National Institute of Technology, 506 004 Warangal India
e-mail: chalamrv@yahoo.com

A. Padma Rao (✉)
Associate Professor of Mechanical Engineering, S.R. Engineering College, 506 371 Warangal India
e-mail: paddu316@yahoo.co.in

J. S. Dai et al. (eds.), *Advances in Reconfigurable Mechanisms and Robots I*,
DOI: 10.1007/978-1-4471-4141-9_20, 

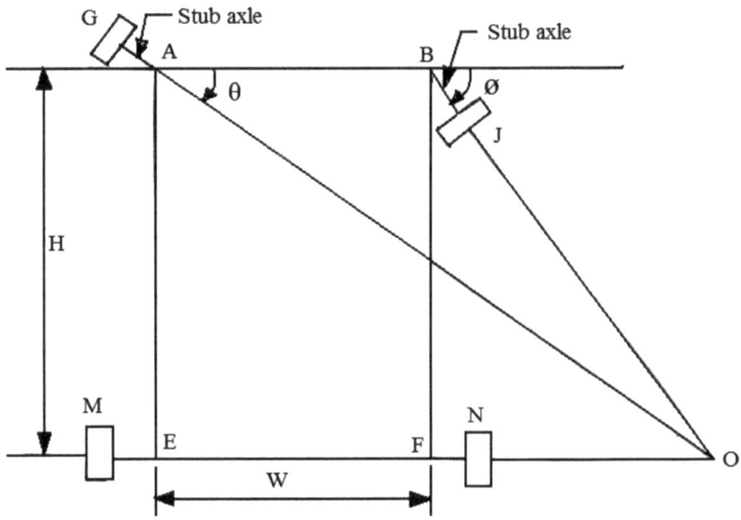


**Fig. 20.1** A car taking *right* turn

extended, they should intersect at one point. With this point as centre, the automobile would take a smooth turn in a circular path. If the steering is given only to front wheels, as is the present practice, the perfect steering condition may be restated as when the axes of the two front wheels are extended they should intersect at a point on the common rear wheel axis. This is shown in Fig. 20.1. Using simple geometric principles the condition for perfect steering may be expressed mathematically [1] as

$$\cot\theta - \cot\phi = W/H \tag{20.1}$$

where, $\theta$ and ø are angles through which the outer and inner stub axles are turned. W and H are called track width and wheel base, respectively. W and H denote the main physical dimensions of the vehicle. Equation (20.1) shows that the angles through which the stub axles are turned are related to physical dimensions of the vehicle. It is also to be observed that the angle ø through which the inner stub axle is turned is more than the angle ø through which the outer stub axle is turned.

Figure 20.2 shows the arrangement of the Ackermann mechanism when the vehicle is moving along a straight path. The Ackermann mechanism is essentially a four bar mechanism ABDC. Links AC and BD are called track arms. Link CD is called track rod. Links AC and BD extended up to G and J, respectively, forming two bell crank levers GAC and JBD. AG and BJ are the stub axles on which the front wheels are free to rotate.

Figure 20.3 shows the Ackermann mechanism providing steering for a right turn. The extensions of the two front wheel stub axles are intersecting at a point O which is not a point on the common rear wheel axis. This clearly shows that the steering provided is not satisfying the condition for perfect steering. The wheels

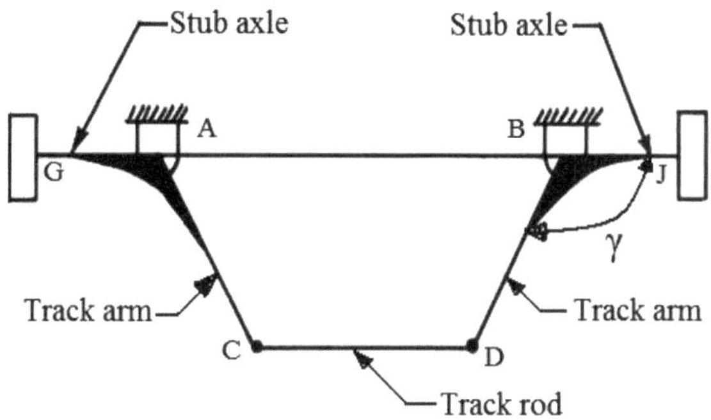


**Fig. 20.2** Ackermann steering mechanism

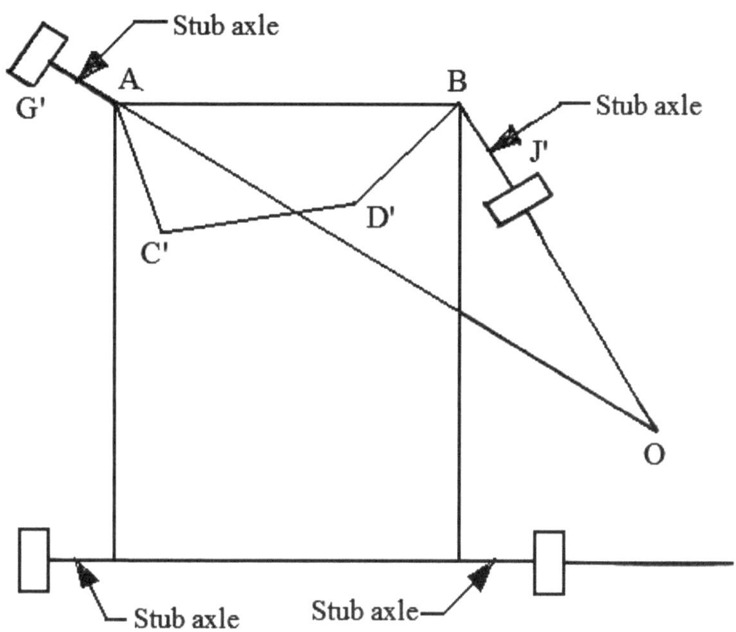


**Fig. 20.3** Steering offered by Ackermann mechanism

may undergo some amount of skidding because of which the tyres undergo more wear.

Many advances have been taking place in steering of an automobile. Attempts have been made to have the steering control through microprocessor [2–6]. When one wheel is turned through an angle, that information is taken by a microprocessor and decides through how much angle should other wheel be turned according to correct steering condition given in Eq. (20.1). The information is then fed to a stepper motor and it turns the outer stub axle through the required angle. Arrangement involving microprocessor, totally eliminates the mechanism. But it suffers from disadvantages such as reliability of operation of microprocessor, slow response of a stepper motor and a non positive drive offered. Mechanical arrangement is foolproof and response is immediate and positive. This is one reason why the present day cars are still depending on Ackermann mechanism. Chicurel [7] presented a concept that allows large steering angles by introducing angle amplifiers between the steering arms and the king pins. The proposed arrangement can be used for both two wheel steering and four wheel steering vehicles. When it is used for four wheel steering, the problem of rear over hang swing reduced drastically.

Hideki [8] proposed a two degree of freedom feedback control for steering rear wheels. The proposed concept was based on a mathematical model constructed in the frequency domain.

Bruce Maclaurin [9] described a model for the steering performance of a tracked vehicle that allows for deformation characteristics of rubber track pads. The model accounted for reduction in friction movement that were observed when radius of turn was increased. Simionescu et al. [10] made a kinematic model of rack and pinion type steering linkage based on which synthesis of mechanism was performed using condition for correct steering. Kamble and Saha [11–13] made extension studies on rack and pinion steering gear. Caserta et al. [14] invented an expandable steering system for vehicles having wheels that move into different positions. In particular, the invention relates to a steering system that has shafts whose length is adjustable to accommodate the movement of vehicle wheels to different lateral positions and elevations relative to the main body of the vehicle. Rahamani Hanzaki et al. [15] made a study on combined kinematic and sensitivity optimization of a rack and pinion steering linkage.

The substitution of electronic system in place of mechanical and hydraulic control is known as *by-wire* technology. In the recent years, a new concept popularly known as *steer-by-wire* (SBW) systems is catching the attention of the researchers. These systems have no mechanical linkage and are expected to improve safety in operation. A typical SBW system consists of a steering wheel angle sensor, resistance torque actuator, steering actuator, front wheel angle sensor, electronic controller and conventional sensor. Feick and Pandit [16] described steer-by-wire system as a mechatronic system. Cedric Wilwert [17] presented a method for evaluating dependability and quality of service of SBW system taking into account both dynamic performance, fault tolerance and static redundancy of the system. Sanket Amberkar [18] suggested design methods to improve system stability. Geoff Rondeau [19] reviewed the steer-by-wire system. However, the basic mechanism is still the Ackermann mechanism only. Though the Ackermann mechanism cannot provide perfect steering, it is widely being used because it is simple, involves only revolute pairs, convenient to operate, foolproof and positive drive.

The authors [20, 21] analyzed the Ackermann mechanism, identified two parameters and evolved a procedure to estimate maximum skidding due to improper steering. It is also observed that the Ackermann mechanism is providing perfect steering only when the vehicle is moving in a straight path. In all other positions, the steering is with errors.

The authors [22, 23] further made attempts to optimize the steering mechanism in order to reduce the errors in the steering. Considerable improvement in the steering has been achieved but the dimensions of the links of the Ackermann mechanism obtained are not very much suitable for space available in the automobile. Hence, suitable dimensions are being adopted sacrificing some amount of steering errors.

The authors also made an attempt [24] to replace the Ackermann mechanism by a belt drive. The method proposed consists of two pulleys to which the stub axles of the front wheels are attached. A belt passing through these pulleys can rotate the stub axles and provide steering. The proposed method is simpler than the Ackermann mechanism and is still providing steering with errors. However, errors are found to be less than those given by Ackermann mechanism.

In this paper an attempt is made to develop a new steering mechanism which may provide perfect steering always. **The work described here is under patent. Patent application no. 704/CHE/2010 dated 17.03.2010**.

## 20.2 Description of the Mechanism

Figure 20.4 shows the schematic arrangement of the proposed mechanism. It consists of two links $L_1$ and $L_2$ which are hinged at A and $B_0$, respectively. The link $L_2$ is attached to gear wheel P which can rotate about $B_0$. Gear wheel P is in mesh with gear wheel Q which can rotate about a point B. The right side stub axle is attached to gear wheel Q while the left side stub axle attached to link $L_1$. If $L_1$ is rotated about A in counter clockwise direction, it pushes $L_2$ through surface to surface contact like a cam and follower mechanism. The link $L_1$ may be considered to have been rotated by the driver through steering wheel. That is, the steering wheel is attached to $L_1$.

In the proposed mechanism $L_1$ and $L_2$ should be in contact always. This may be achieved by using springs which hold these two surfaces together always. It may now be said that the steering wheel may be attached to gear wheel Q instead of $L_1$. Link $L_2$ and so the gear wheel P are made to rotate in clockwise direction. The clockwise rotation of P is converted to counter clockwise rotation of gear wheel Q. Thus the right side stub axle is also rotated in counter clockwise direction. The mechanism is now providing steering for a left turn. The distance AB may be noted as the track width W. The distance $AB_0$ may be represented by $W_1$. The angle made by $AB_0$ with AB is $\beta$. The two parameters $W_1$ and $\beta$ together uniquely locate point $B_0$.

## 20.3 Formulation

The links $L_1$ and $L_2$ transmit the motion like cam and follower mechanism. These two links are forming a higher pair. The motion between the two surfaces involves rolling associated with slipping. What are the profiles to be chosen for $L_1$ and $L_2$

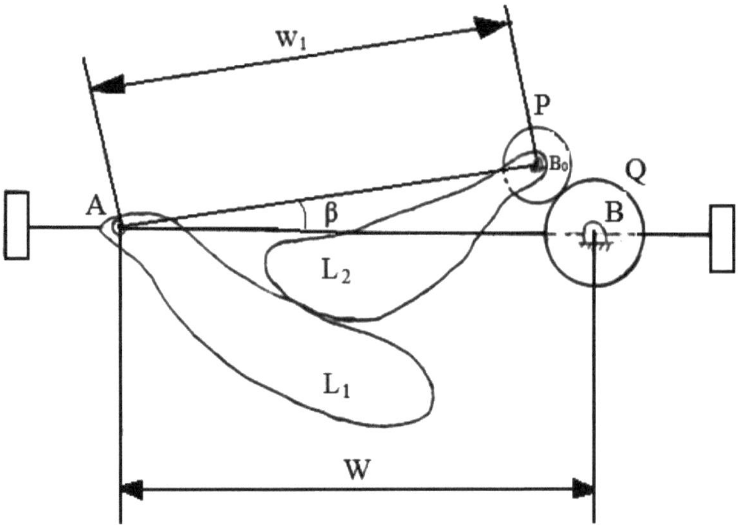


**Fig. 20.4** Schematic arrangement of proposed mechanism

such that the condition for correct steering expressed in Eq. (20.1) is satisfied always? This problem may be attempted as *for a chosen profile for $L_1$ what should the profile of $L_2$ such that the condition for correct steering is satisfied.*

For the purpose of formulating the problem two coordinate systems are chosen. One coordinate system is ($x_0$, $y_0$) having its origin at A and $y_0$ axis along the left side stub axle which is shown in Fig. 20.5 having rotated through an angle $ø_P$. The second coordinate system having its origin at $B_0$ and is being rotated through an angle $ø_P$. The rotation $ø_P$ is converted to rotation ø of the right side stub axle.

$ø_P$ and ø are related through the gear ratio G as

$$G = ø_P/ø \tag{20.2}$$

The profile of $L_1$ may first be expressed conveniently in ($x_0$, $y_0$) coordinate system. The same profile may be expressed in (x, y) coordinate system using the transformation,

$$x_0 = x\cos(\theta + Gø) + y\sin(\theta + Gø) + W_1\sin(\theta + \beta) \tag{20.3}$$

$$y_0 = -x\sin(\theta + Gø) + y\cos(\theta + Gø) + W_1\cos(\theta + \beta) \tag{20.4}$$

Let the profile of $L_1$ be described in ($x_0$, $y_0$) coordinate system as

$$f(x_0, y_0) = 0 \tag{20.5}$$

$x_0$ and $y_0$ may be replaced using Eqs. (20.3) and (20.4). The resulting function describes the same profile $L_1$ in (x, y) coordinate system as

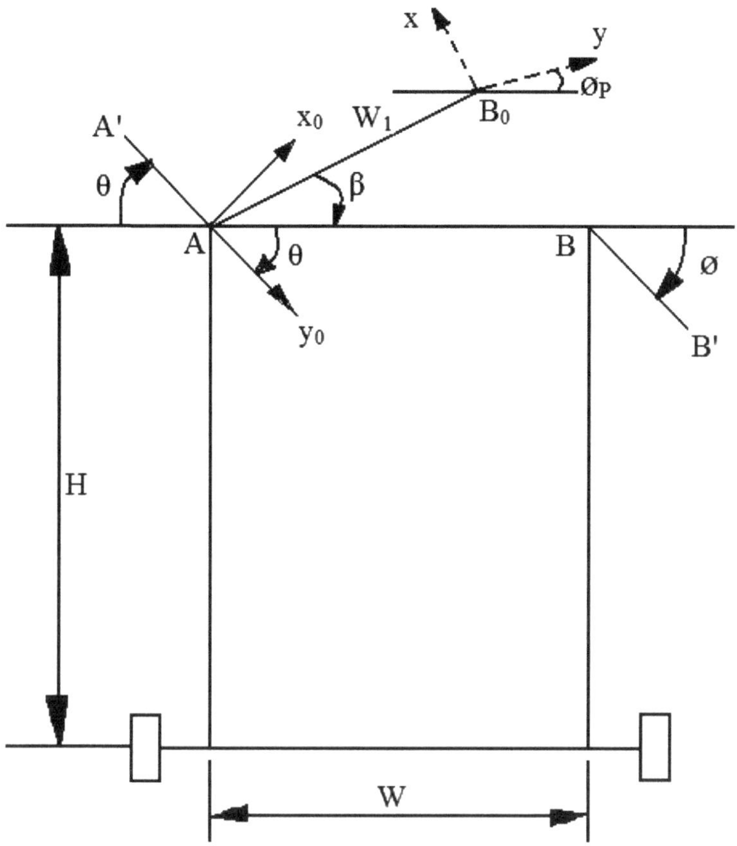


**Fig. 20.5** Defining coordinate axes

$$f(x, y, \theta, \varnothing) = 0 \tag{20.6}$$

However, $\theta$ and ø are related the condition for correct steering. Hence, they are not independent quantities. Either $\theta$ or ø may be retained in Eq. (20.6) and the profile for $L_1$ may be described in (x, y) coordinate system as

$$f(x, y, \theta) = 0 \tag{20.7}$$

$\theta$ is now identified as a parameter in the present context. It denotes the angle of turn of the left side stub axle. By varying the parameter $\theta$ a family of curves each describing a particular position of $L_1$ may be obtained. A curve which goes tangential to the set of family of curves is called an enveloping curve. For obtaining the enveloping curve the following conditions may be applied [25] as

$$f(x\ ,\ y\ ,\theta) = 0 \text{ and } \partial f/\partial\theta = 0 \tag{20.8a, b}$$

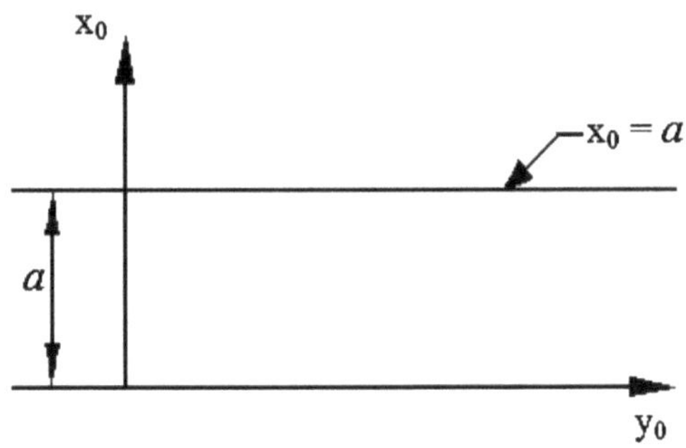


**Fig. 20.6** Straight line profile for $L_2$

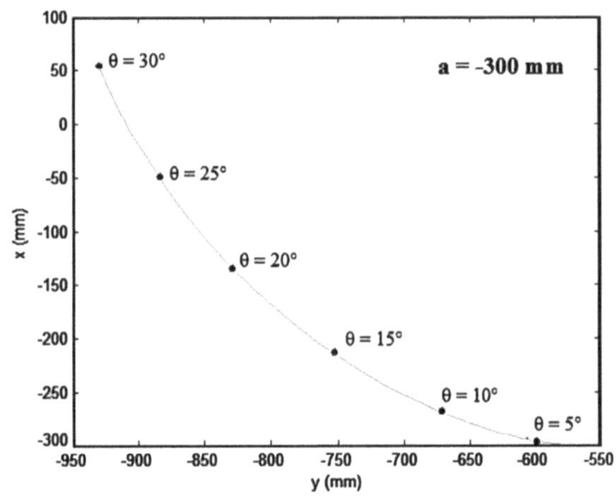


**Fig. 20.7** A typical profile $L_2$ obtained for G = 1, $\beta = 0$, $W_1 = 1120$ mm

Eliminating $\theta$ in the Eq. (20.8a, b), equation for enveloping curve may be obtained. The enveloping curve is nothing but the profile of $L_2$. Thus the pair of surfaces of $L_1$ and $L_2$ can now coordinate the rotations of the stub axles and provide perfect steering always. Eliminating $\theta$ is very much involved because Eq. (20.8a, b) are highly non linear. Hence, it is decided to obtain the enveloping curve numerically.

## 20.4 Algorithm

For a given value of W/H, number of values of $\theta$ may be considered and corresponding values of ø may be computed by solving Eq. (20.1). A particular value G may be taken. The data now available, consists of pairs of values of $\theta$ and ø according to perfect steering condition.

Let a maximum number of values of $\theta$ chosen in working range be $i_{max}$. The step by step procedure may be described as follows.

Step 1: i = 0
Step 2: i = i + 1
Step 3: if i > $i_{max}$ GO TO Step 8
Step 4: Take $\theta_i$ and corresponding $ø_i$ from the memory
Step 5: Solve Eq. (20.8a, b) for the coordinates (x, y)
Step 6: Print i, $\theta$, ø, x, y
Step 7: GO TO Step 2
Step 8: STOP

## 20.5 Results and Discussion

To begin with a straight line profile is chosen for $L_1$. The reasons being it is simple and easy to produce. The equation for $L_1$ is

$$f(x_0, y_0) = x_0 - a = 0 \tag{20.9}$$

The profile looks as in Fig. 20.6. The profile of $L_2$ could be obtained following the procedure described in Sect. 20.4.

Different parameters identified are $W_1$, $\beta$, G and $\boldsymbol{a}$. The authors have tried varieties of cases and studied the performance of mechanism with regard to the parameters identified.

The values of W and H are taken from a Santro car as W = 1360 mm and H = 2380 mm. A typical profile $L_2$ obtained for G = 1, $\beta = 0$, $W_1$ = 1120 mm is shown in the Fig. 20.7. A model study is also performed and the proposed mechanism is fond to give perfect steering throughout.

## 20.6 Conclusions

The significant conclusions based on the present work may be summarized as follows.

i. The requirement of perfect steering condition is described.
ii. A possible substitute of Ackermann mechanism which uses a belt drive arrangement is described.
iii. A mechanism which transmits motion through surface to surface contact like a cam mechanism is proposed.
iv. A steering mechanism using this concept is developed.
v. The method of obtaining the profiles of contact surfaces in order to achieve perfect steering is developed.
vi. A step by step procedure is also described.
vii. Various parameters involved in the mechanism are brought out.

viii. A straight line profile identified by the parameter ***a*** is chosen.
ix. A typical profile of $L_2$ is presented.
x. A model study is also performed and the mechanism is found to give perfect steering throughout.
xi. The proposed mechanism is much simpler than the Ackermann mechanism and offers correct steering always.
xii. A pair of surfaces to offer perfect steering is described. In actual automobile one should have two pairs of surfaces, one for left turn and one for right turn.
xiii. In between the contact surfaces rolling and sliding take place. Smoothness in operation may be achieved by using needle bearings on one of the surfaces, say $L_1$. This will also reduce the wear of the surfaces.
xiv. The attention in this paper is focused towards development of kinematic concept. The concept is also tested through model study. The dynamic effects are at to be investigated.

## References


1. Bevan T (1971) The theory of machines. Longman Group Limited, London, pp 131–135
2. Miller GH (1979) Introduction to microprocessors, SAE Transactions, Report no. 790234, Electrical Engineering, General Motors Institute, Flint, MI
3. Miller GH (1979) Interfacing to microprocessor, SAE Transactions, Report no.790235, Electrical Engineering, General Motors Institute, Flint, MI
4. Paul A, Wilson J (1985) Microprocessor selection and soft wear design, SAE Transactions, Report no. 790236, Wintek Corporation, Lafayetta, IN
5. Basselink BC (2003) Computer controlled steering system for vehicles having two independently driven wheels. Comput Electron Agric 39:209–226
6. Ning J, Wang C, Ma T, Li W (2006) Application of computational steering technique in numerical simulation of explosion problems. Int J Comput Sci Netw Secur 6(10):88–93
7. Chicurel E (1999) A $18\overset{\circ}{0}$ steering interval mechanism. Mech Mach Theory 34:421–436
8. Kuzuya H, Shin S (2000) Development of robust motor servo control for rear steering actuator based on two degree of freedom control system. Mechatronics 10:53–66
9. Maclaurin B (2007) A skid steering model with track pad flexibility. J Terramech 44:95–110
10. Simionescu PA, Smith MR, Tempea I (2000) Synthesis and analysis of the two loop translational input steering mechanism. Mech Mach Theory 35:927–943
11. Kamble N, Saha SK (2005) Evaluation of torque characteristics of rack and pinion steering gear using ADAMS model 2005-01-1064, SAE 2005 Transactions, J Pass Car Mech Syst, June 2005, pp 1217–1222
12. Kamble N, Saha SK (1915) Effect of pinion profile modification on rack and pinion steering gear, 2005-01-1273. In: Proceedings of steering, suspension, tires, wheels, SP 1915, pp 83–90
13. Kamble N, Saha SK (2006) Developing a virtual prototype of a rack and pinion steering system. Int J Veh Syst Model Test 2(1):61–79
14. Caserta AL, Caserta CW (2006) Expandable steering system. United States Patent No. 0 006 019 A1
15. Rahmani Hanzaki A, Rao PVM, Saha SK (2009) Kinematic and sensitivity analysis and optimization of planar rack-and-pinion steering linkages. Mech Mach Theory 44:42–56

16. Feick S, Pandit M (2001) Steer-by-wire as a mechatronics implementation. SAE world congress, 2001-01-0823
17. Wilwert C (2001) Evaluating quality of service and behavioural reliability of steer-by-wire systems. Technical report, Loria-Trio, Germany
18. Amberkar S, Bolourchi F, Demerly J, Millsap S (2004) A control system methodology for steer by wire systems. In: Steering and suspension technology symposium, SAE International paper 2004-01-1106, March 2004
19. Rondeau G (2004) Steer-by-wire systems with integrated torque feedback improve steering performance and reduce cost. Technical report, Thomson industries, Illinois, US
20. Venkatachalam R, Padma Rao A (2010) Analysis of steering offered by Ackermann mechanism. Int J Mech Auto Eng 09(01):37–45
21. Venkatachalam R, Padma Rao A (2010) Analysis of Ackermann mechanism. In: Proceedings of the national conference on advanced in automotive technology (NCAAT 2010), Department of Automobile Engineering, Rajalakshmi Engineering College, Thandalam, Chennai, India, 15–16 July 2010
22. Venkatachalam R, Padma Rao A (2010) Graphical approach for optimum design of steering mechanism. In: Proceedings of the 4th international conference on advances in mechanical engineering, S.V. National Institute of Technology, Surat–395 007, India, 23–25 Sept 2010
23. Venkatachalam R, Padma Rao A (2010) Optimizing the angle of bell crank lever of Ackermann steering mechanism. In: Proceedings of the 10th national conference on industries problems on machines and mechanisms (IPRoMM 2010), MNIT, Jaipur, India, 17–18 Dec 2010
24. Padma Rao A, Analysis, design and development of new steering mechanisms for four wheel road vehicles. PhD thesis submitted to the Department of Mechanical Engineering, National Institute of Technology, Warangal, 2011
25. Shanti Narayan MA (1958) Differential calculus. S. Chand and Company, Delhi, pp 369–383

# Chapter 21
# Constructing Rolling Mechanisms Based on Tetrahedron Units

**Yao-Bin Tian and Yan-An Yao**

**Abstract** In this paper, a *tetrahedron unit* constructed by connecting three links to a node via revolute joints is proposed. A closed loop mechanism is constructed by combining two tetrahedrons units to share a common face with three spherical joints such that the geometric structure of the mechanism is a triangular bipyramid. We show that the structure of the mechanism enables it possess a rolling function when proceeding on the ground. Furthermore, with the similar topology structure, two more closed-loop rolling mechanisms are obtained by combining 4 and 6 units, respectively. Due to the similar rolling principles of the proposed mechanisms, it suffices to analyze the rolling feasibility of the triangular bipyramid mechanism. By analyzing the kinematics, the deformation features of the mechanism are obtained. Upon some proper controls, we show that these features enable the mechanism to roll and switch directions when it proceeds on the ground. These functions are verified by a series of simulations with a 3D model and experiments with a prototype.

**Keywords** Rolling robot · Closed-loop mechanism · Parallel mechanism · Tetrahedron · Triangular bipyramid

---

Y.-B. Tian · Y.-A. Yao (✉)
Department of Mechanical Engineering, Beijing Jiaotong University, Beijing, China
e-mail: yayao@bjtu.edu.cn

Y.-B. Tian
e-mail: tianyaobin@163.com

J. S. Dai et al. (eds.), *Advances in Reconfigurable Mechanisms and Robots I*,
DOI: 10.1007/978-1-4471-4141-9_21, © Springer-Verlag London 2012

## 21.1 Introduction

There are a variety of terrestrial locomotion for mobile robots, such as walking, jumping, crawling, and rolling. In this is paper, we focus on rolling locomotion, which is very effective on the flat ground [1]. Rolling locomotion includes the passive rolling mode and the active rolling mode [2].

The active rolling mode is widely used by spherical robots, which roll with their entire outer surface by controlling the actuators inside. Halme et al. [3] embedded a driving wheel in the interior of the sphere robot, by manipulating the controlling and steering axis connected to the driving wheel, the entire robot can roll and switch direction when it mobiles. Bicchi et al. [4] proposed a spherical robot driven by a small car inside. The car is designed with unicycle kinematics to make the robot move efficiently. Otani et al. [5] designed a new mechanism equipped with a gyro to provide the driving torque for their spherical rolling robot. Rather than using the gyro as the driving wheel, Mukherjee et al. [6, 7] and Javadi et al. [8] proposed the driving mechanisms inside the sphere that distribute weights radially along some spokes fixed to the inner surface of the sphere.

The major merits of the spherical robots are as follows: (1) The touching area with the ground is relatively small; (2) the friction force due to the ground is small; (3) its shape remains unchanged during the rolling procedure. However, the disadvantage is that they require some sophisticated algorithms to control the robot to roll and change directions in order to reach the right target [9, 10].

Different from spherical rolling robots that maintain their appearance during the rolling course, some rolling robots realize the rolling function by deforming their geometry shapes. A common characteristic of these robots is that their centers of mass are changing during the rolling course. Sugiyama and Hirai [11] used soft materials to design a robot that can roll, crawl, and jump by deforming its body. Combining rigid struts with tensional strings, Shibata and Hirai [12] further proposed a kind of deformable structure to realized rolling function.

Without using soft material, some rolling mechanisms are capable of rolling by changing their geometry shape by using different linkage methods. Some reconfigurable robot is able to roll if its topology is a closed loop. It realizes the rolling function by deforming the shape of the loop [13, 14]. Also with modular units, there are a series of modular reconfigurable robots each consisting of multiple loops, the rolling function of these robots is realized by controlling the edge length of some loops [15–17].

In this paper, a *tetrahedron unit* is proposed. Three closed-loop mechanisms are constructed to form some spatial symmetric structures. We show that the mechanisms are capable of rolling and changing directions.

## 21.2 Mechanism Design

### *21.2.1 Mechanism Description*

As shown in Fig. 21.1a, a tetrahedron has 6 edges and 4 vertices. As shown in Fig. 21.1b, each dashed edge in Fig. 21.1a is represented with a physical link, and vertex D is represented with a physical node, each link is connected to the node with a revolute joint, the three links are distributed around the node such that the angle between every two consecutive links is $\pi/3$. For simplicity, such a representation of the tetrahedron is called *a unit.*

As shown in Fig. 21.2a, by connecting two units at their common face, a *triangular bipyramid* can be obtained Similarly, as shown in Fig. 21.2b a closed-loop mechanism with a *triangular bipyramid* shap is composed of two units by connecting the ends of the links via three spherical joints at vertices *A*, *B* and *C*. The triangular bipyramid mechanism consists of 8 links, 6 revolute joints and 3 spherical joints. It is a 3-RSR parallel mechanism. Generally, a 3-RSR parallel mechanism is fixed on other mechanisms or equipments, and used to realize two rotational and one radial motions of the moving platform [18–21]. Different from preexisting conventional applications of the 3-RSR parallel mechanism, we release the fixed platform and use it to realize a rolling function.

Refer to Fig. 21.3a for a tetrahedron *ABCD*. By constructing a tetrahedron on each face of tetrahedron ABCD, the resulting structure is shown in Fig. 21.3b. Further, by representing each newly constructed tetrahedron with a unit, the resulting mechanism is obtained in Fig. 21.3c.

Alternatively, if we add to each face of a triangular bipyramid (Fig. 21.4a) a tetrahedron, then the resulting structure is shown in Fig. 21.4b. Representing each newly added tetrahedron with a unit, the resulting mechanism is obtained in Fig. 21.4c.

By controlling the units in any mechanism described above, the mechanism can be deformed into various shapes with alternative links lying on the ground, therefore realizing the rolling motion. In the subsequent section, taking triangular bipyramid mechanism as an example, we show that the spatial mechanism can be controlled properly by transforming it into a four bar mechanism such that it can roll and switch directions when proceeding on the ground.

### *21.2.2 Kinematic Analysis*

Due to the limit of space, in this section, we only analyze the kinematics of the triangular bipyramid mechanism to illustrate the idea of deforming the shape of the mechanism to achieve the rolling function. The deformation of the shape of the mechanism varies its center of mass, and once the center of mass is out of the

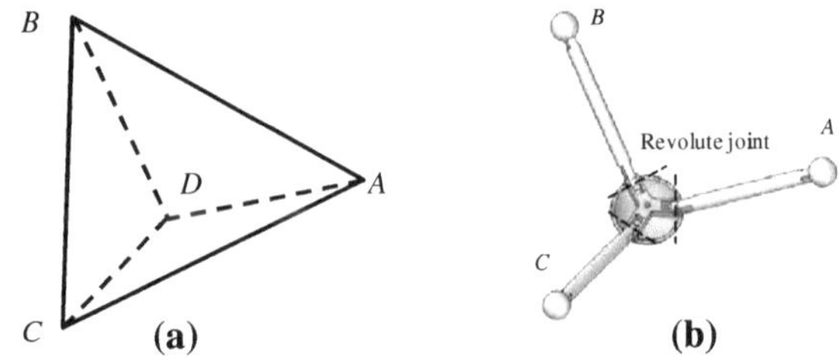


**Fig. 21.1** The sketch of unit: **a** the geometric tetrahedron, **b** the mechanism representation of a unit

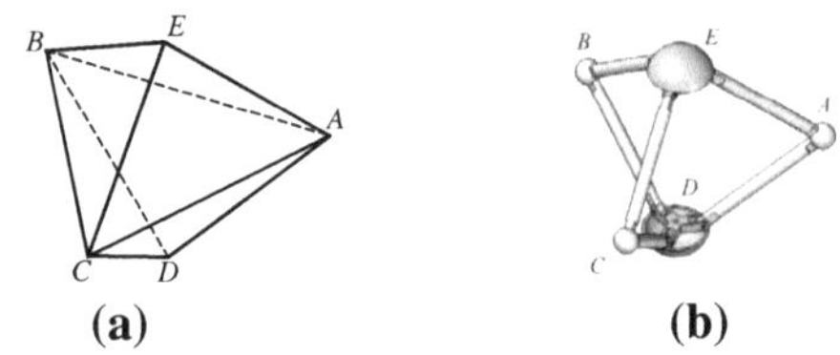


**Fig. 21.2** The *triangular bipyramid*: **a** the geometric of *triangular bipyramid*, **b** the *triangular bipyramid* mechanism

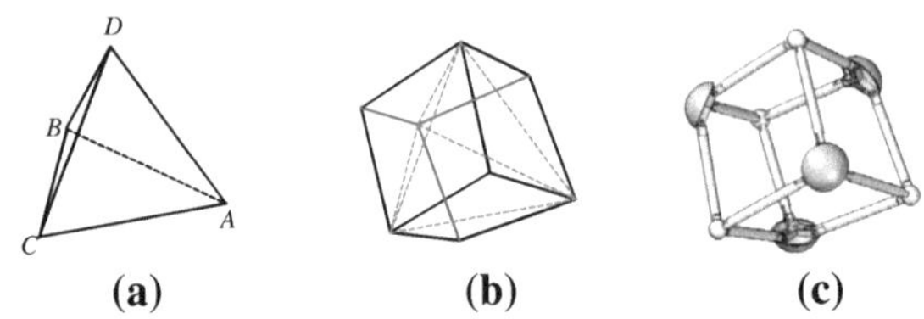


**Fig. 21.3** The construction procedure of the mechanism with four tetrahedron units: **a** a tetrahedron *ABCD*, **b** add a units to each face of the tetrahedron *ABCD*, **c** the 3D mechanism model

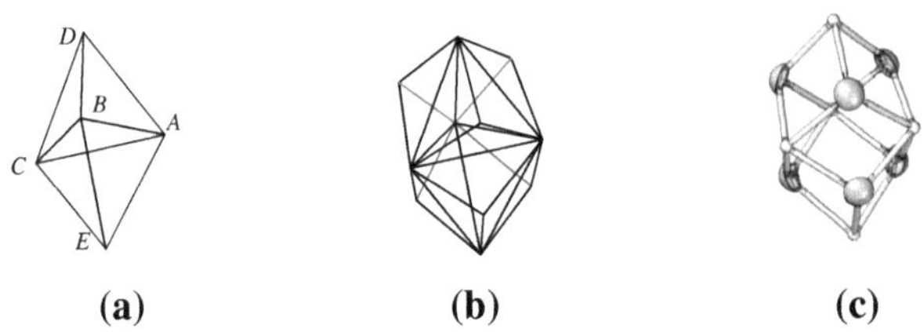


**Fig. 21.4** The construction procedure of the mechanism with four tetrahedron units: **a** *triangular bipyramid ABCDE*, **b** add a unit to each face of the triangular bipyramid ABCDE, **c** the 3D mechanism model

supporting area of the mechanism, the mechanism will become unbalanced and roll on the ground.

A simplified model of the triangular bipyramid mechanism is shown in Fig. 21.5. To analyze the position of *E*, fix node *D* of the mechanism on the ground initially. For the ease of discussion, a coordinate system is built up. The origin of the system is *D*, the *z*-axis is vertical to the ground, the *x*-axis is orthogonal to the link *BD*, and the *y*-axis is orthogonal to the plane of *xDz*. As link *DA*, *DB* and *DC* rotate about node *D* for the angle of $\theta_1$, $\theta_2$ and $\theta_3$.

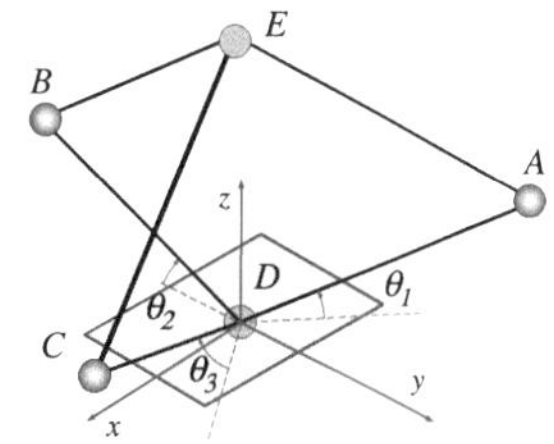


**Fig. 21.5** The mathematic model of *triangular bipyramid* mechanism

Let $L$ be the length of each link. According to the mechanism, the positions of $A$, $B$ and $C$ are determined by Eq. (21.1).

$$\begin{bmatrix} x_A & x_B & x_C \\ y_A & y_B & y_C \\ z_A & z_B & z_C \end{bmatrix} = \begin{bmatrix} -\sqrt{3}L\cos\theta_1/2 & 0 & \sqrt{3}L\cos\theta_3/2 \\ L\cos\theta_1/2 & -L\cos\theta_2 & L\cos\theta_3/2 \\ L\sin\theta_1 & L\sin\theta_2 & L\sin\theta_3 \end{bmatrix} \tag{21.1}$$

Due to the length of link $EA$, $EB$ and $EC$, given the positions of $A$, $B$ and $C$, the position of $E$ can be determined by Eq. (21.2).

$$\begin{cases} (x_E - x_A)^2 + (y_E - y_A)^2 + (z_E - z_A)^2 = L^2 \\ (x_E - x_B)^2 + (y_E - y_B)^2 + (z_E - z_B)^2 = L^2 \\ (x_E - x_C)^2 + (y_E - y_C)^2 + (z_E - z_C)^2 = L^2 \end{cases} \tag{21.2}$$

According to Eq. (21.2), we can express $x_E$ and $y_E$ with $z_E$ respectively as follow:

$$\begin{cases} x_E = z_E[k_{11}(k_{22} - k_{12}) + k_{12}(k_{11} - k_{21})]/(k_{11} - k_{21}) \\ y_E = z_E(k_{22} - k_{12})/k_{11} - k_{21} \end{cases} \tag{21.3}$$

where $k_{11} = \dfrac{y_A - y_B}{x_B - x_A}$ $k_{12} = \dfrac{z_A - z_B}{x_B - x_A}$ $k_{21} = \dfrac{y_A - y_C}{x_C - x_A}$ $k_{22} = \dfrac{z_A - z_C}{x_C - x_A}$

To express the coordinates of $E$ in a succinct manner, define $t_{11}$ and $t_{12}$ in Eq. (21.4).

$$\begin{bmatrix} t_{11} & t_{12} \end{bmatrix} = \begin{bmatrix} \dfrac{k_{11}(k_{22} - k_{12}) + k_{12}(k_{11} - k_{21})}{k_{11} - k_{21}} & \dfrac{k_{22} - k_{12}}{k_{11} - k_{21}} \end{bmatrix} \tag{21.4}$$

Based on Eq. (21.2) and (21.3), the position of $E$ can be expressed in Eq. (21.5).

$$\begin{bmatrix} x_E & y_E & z_E \end{bmatrix}^T = \begin{bmatrix} 2t_{11}(t_{11}x_A + t_{12}y_A + z_A)/(t_{11}^2 + t_{12}^2 + 1) \\ 2t_{12}(t_{11}x_A + t_{12}y_A + z_A)/(t_{11}^2 + t_{12}^2 + 1) \\ 2(t_{11}x_A + t_{12}y_A + z_A)/(t_{11}^2 + t_{12}^2 + 1) \end{bmatrix} \tag{21.5}$$

Particularly, let $\theta_1 = \theta_3$, then $y_A = y_C$, $z_A = z_C$. That leads to $t_{11} = 0$. According to Eq. (21.5), the position $x_E$ is equal to zero. This means that $E$ is

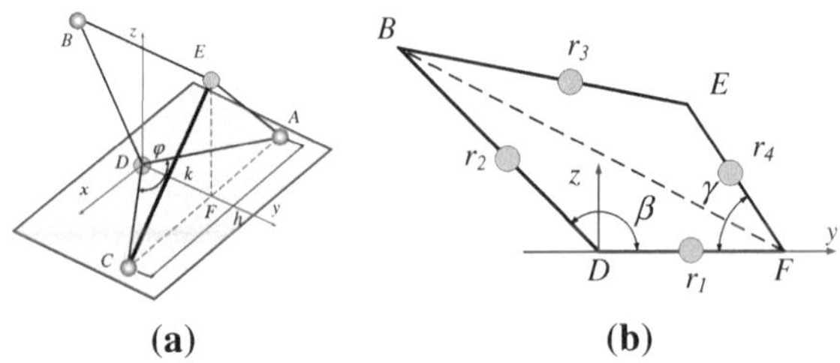


**Fig. 21.6** Deforming into a planar four-bar mechanism

moving in the $yDz$ plane. As shown in Fig. 21.6a, let the link $DA$ and link $DC$ be lying on the ground to form an entity. In the view of metamorphic, when the link DA and DC are locked by the actuators, they can be treated as a single link. By the symmetry of the mechanism, link $CE$ and link $AE$ will become a single link. These two new links obtained can rotate about the axis passing through $A$ and $C$. In this manner, as shown in Fig. 21.6b, the original spatial mechanism can be treated as a planar four-bar mechanism.

As shown in Fig. 21.6a, the origin of the system is located at $D$. The $x$-axis is orthogonal to link $DB$. Link $DA$ and link $DC$ are lying on the ground to form an angle of $\varphi$. According to Eq. (21.1), the coordinates of $A$ and $C$ are obtained, with which we can represent $DA$ and $DC$ each with a three dimensional vector. Therefore, $\varphi$ can be determined by *vectorial angle cosine* as shown in Eq. (21.6). Let $F$ be the intersection point of $AC$ and the $y$-axis. Let $k$ be the $y$-coordinate of $F$, then $k$ is a step length of the rolling process. Additionally, let $h$ be the width of the robot.

$$\varphi = ar\cos(1 - 1.5\cos^2\theta_1) \tag{21.6}$$

As shown in Fig. 21.6b, let the angle between link $DB$ and link $DF$ be $\beta$, link $DF$ and link $EF$ have the same length of $k$. Therefore, the positions of $B$, $E$ and $F$ can be expressed in Eq. (21.7).

$$\begin{bmatrix} y_B & y_E & y_F \\ z_B & z_E & z_F \end{bmatrix} = \begin{bmatrix} L\cos\beta & k(1-\cos\gamma) & k \\ L\sin\beta & k\sin\gamma & 0 \end{bmatrix} \tag{21.7}$$

Let $r_i$ be the center of mass of the $i$th link. Note that each $r_i$ can be determined by the coordinates of $B$, $E$ and $F$. Let $m$ denote the weight of link $DB$ ($BE$), then the weight of $DF$ ($EF$) is 2 m. Therefore, the center of mass of the mechanism in the $yDz$ plane can be expressed in Eq. (21.8).

$$r_y = \frac{2L\cos\beta - 3k\cos\gamma + 5k}{12} \tag{21.8}$$

where $\gamma = 2\arcsin(L\sin\beta/\sqrt{L^2 + k^2 - 2kL\cos\beta})$.

As shown in Fig. 21.6b, the mechanism can only roll along the $y$-axis. The supporting area along the $y$-axis is $[0, k]$. If $r_y$ is out of the supporting area, then the

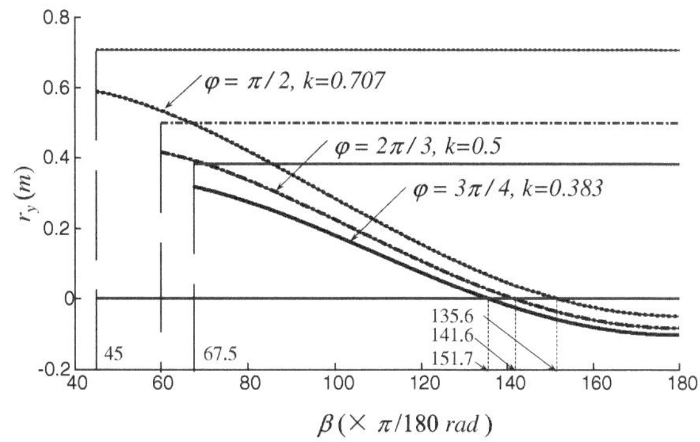


**Fig. 21.7** The trend of $r_y$ with respect to $\beta$ and $\varphi$

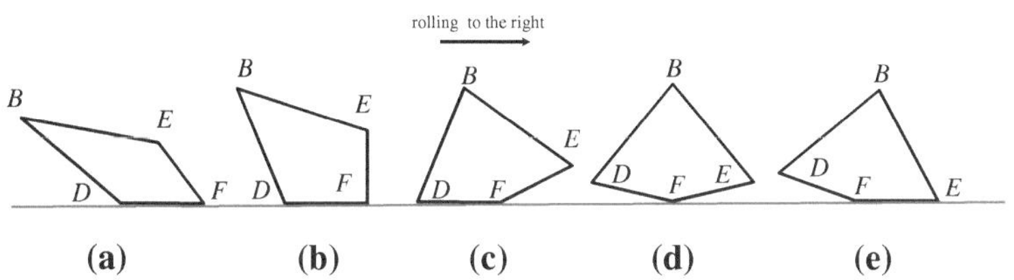


**Fig. 21.8** The procedure of rolling to the *right*

mechanism will roll along the *y*-axis. To illustrate the relation of $r_y$ with $\beta$ and $\varphi$, three values of $\varphi$ are studied. For each value of $\varphi$ chosen from the set $\{\pi/2, 2\pi/3, 3\pi/4\}$, set $m = 0.5$ kg, $L = 1$ m, then the trend of $r_y$ is plotted in Fig. 21.5.

As shown in Fig. 21.7, given any fixed $\varphi$, $r_y$ is monotonically decreasing with the increment of $\beta$; given any fixed $\beta$, $r_y$ is monotonically decreasing with the increment of $\varphi$. By Eq. (21.7), *k* (the upper limit of the supporting area) decreases with the increment of $\varphi$. Further notice that $r_y$ cannot be large than *k* until the motor provides a proper inertial force, and that $r_y$ will surpass the lower limit when $\beta$ is large enough.

As $\beta$ decreases $r_y$ is moving to the right. A rolling procedure of the mechanism is shown in Fig. 21.8. During this procedure, the rolling direction is fixed while the length of a rolling step, i.e., *k*, is controllable by changing $\varphi$. When the rolling step is accomplished, as shown in Fig. 21.9, the mechanism will have link *EA* and link *EC* lying on the ground, resulting in the state that *FE* becomes the new supporting area.

Note that, if $\varphi$ is decreasing, the step length *k* is increased and the width *h* is reduced. This characteristic is useful when the mechanism is rolling along a narrow passage.

Similarly, as $\beta$ increases $r_y$ is moving to the left. A rolling procedure of the mechanism is shown in Fig. 21.10. During this procedure, the rolling direction is fixed while the length of a rolling step, i.e., *k*, is controllable by changing $\varphi$. When the rolling step is accomplished, the mechanism will have link *EA* and link *EC*

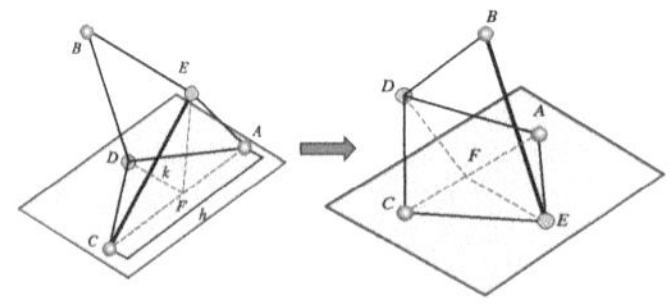


**Fig. 21.9** The state of the mechanism with link *EA* and link *EC* lying on the ground

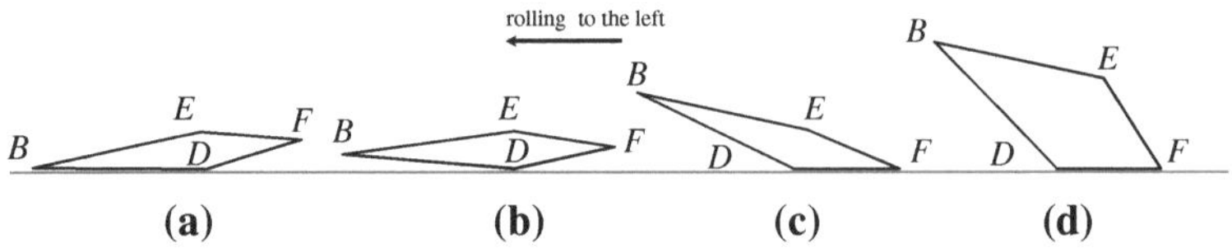


**Fig. 21.10** The procedure of rolling to the *left*

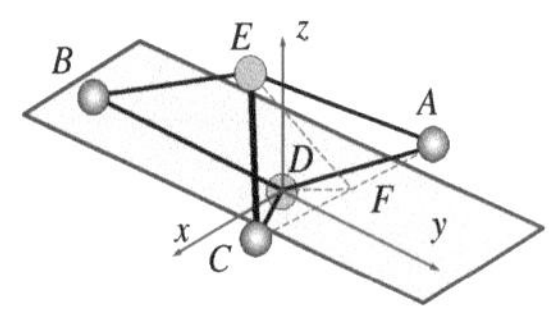


**Fig. 21.11** The state of the mechanism with link *DB* lying on the ground

lying on the ground (refer to Fig. 21.11), leading to the state with *FE* becomes the new supporting area. In this state, the mechanism is symmetrical with respect to the *yDz* plane.

If link *DA* is rotated to increase the height of *A*, then $r_x$ becomes positive. Thereafter, the mechanism rolls about link *DB* to the positive *x*-axis. As a result, link *DB* and link *DC* are both lying on the ground.

Alternatively, if link *DC* is rotated to increase the height of *C*, then $r_x$ becomes negative. Thereafter, the mechanism rolls about link *DB* to the negative *x*-axis. Link *DB* and link *DA* are both lying on the ground.

To summarize, the mechanism is capable of altering its rolling direction when it is in the state with link *DB* lying on the ground only.

## 21.3 Simulation

To verify the rolling states of the mechanism, a virtual mechanism is constructed in the Solid Edge software. Based on this model, a series of simulations are performed in the ADAMS software. Figure 21.12 shows a rolling procedure for the mechanism. The procedure starts with a state of the mechanism with two links lying on the ground and ends with only one link lying on the ground. Some arrows are artificially added to indicate the driven links.

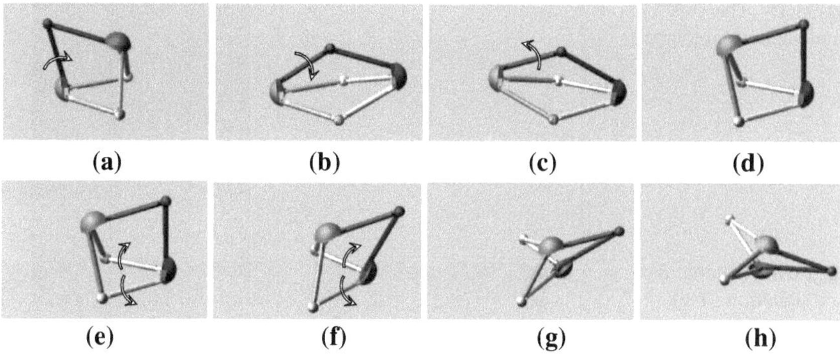


**Fig. 21.12** A rolling procedure (**a–h**)

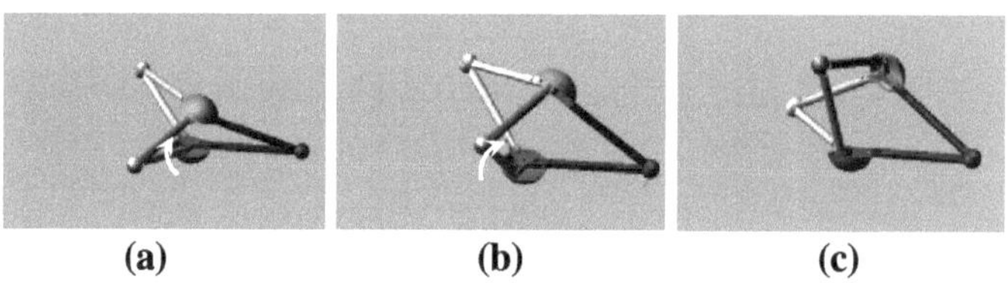


**Fig. 21.13** Rolling to one side of the mechanism

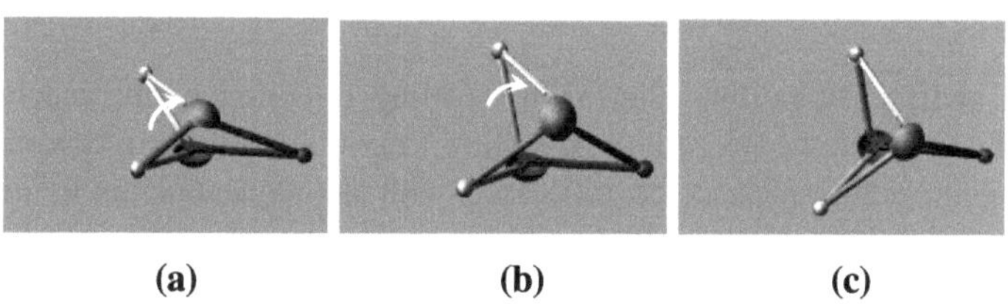


**Fig. 21.14** Rolling to the other side of the mechanism

As shown in Fig. 21.13h, when the mechanism is in the state with only one link lying on the ground, it can choose one more link to lie on the ground to maintain its balance in practice. In doing so, it alters its rolling direction in reality. As shown in Figs. 21.13 and 21.14, the mechanism can alter its rolling directions in two manners.

## 21.4 Prototype

To test the feasibility and efficiency of the rolling mechanism in practice, a prototype is manufactured (Fig. 21.15). Based on the prototype, a rolling experiment is performed. As shown in Fig. 21.16, the mechanism can roll along a fixed

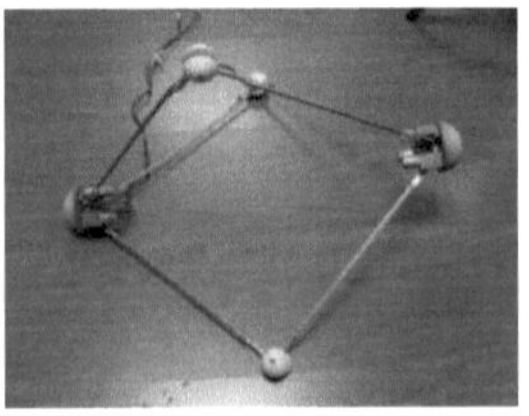

**Fig. 21.15** The manufactured prototype

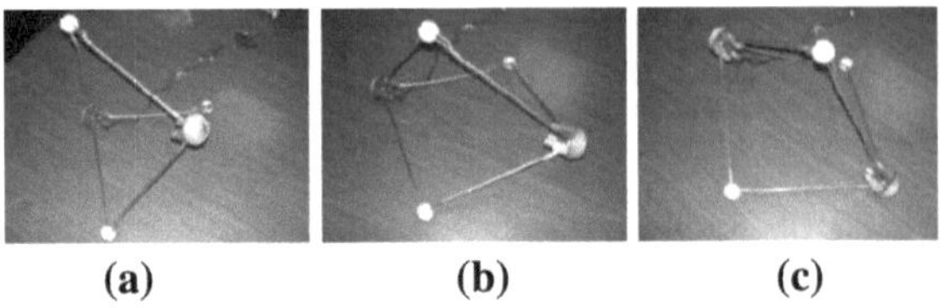

**Fig. 21.16** Rolling along a fixed direction (**a**–**c**)

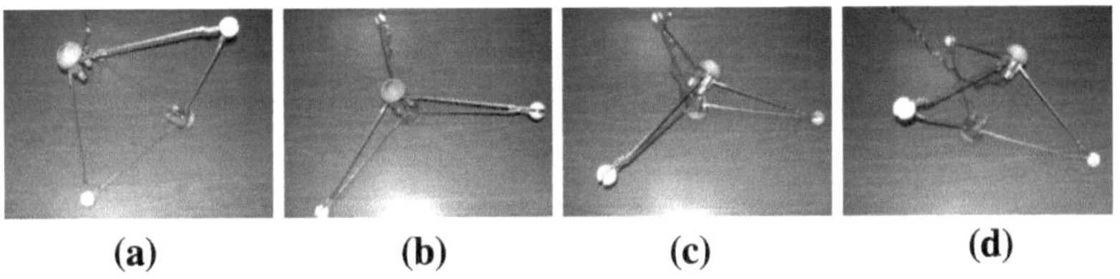

**Fig. 21.17** A procedure of altering the rolling direction (**a**–**d**)

direction swiftly. Regarding altering the rolling direction of the mechanism, it suffices to consider the procedure of rolling to one side of the mechanism since rolling to the other side can be processed with the same operations due to the symmetry of the mechanism. Figure 21.17 shows the procedure of altering the rolling direction of the mechanism to one side.

## 21.5 Discussions and Conclusions

In this paper, a mechanism representing the shape of a tetrahedron is proposed. It is constructed by connecting three links to a node via revolute joints. Each such mechanism is a unit, or a brick, used to build three rolling mechanisms. The rolling principles of the mechanisms are analyzed by studying the kinematics of the simplest model, i.e., the triangular bipyramid mechanism. Based on the kinematics analysis, the deformation characteristics of the mechanism are revealed. With these characteristics, the rolling and direction switching functions are shown. These functions are verified by a series of simulations with 3D model mechanism in ADAMS and practical experiments with a real prototype.

In addition, our proposed triangular bipyramid mechanism can change the length of its rolling step and the width of the body during the rolling process. With this feature, the mechanism is capable to rolling across narrow passages. This can be very useful in exploring the unstructured environments.

In the future research, we would like to combine even more tetrahedron units to generate different mechanisms with various functionalities.

**Acknowledgments** This work has been supported by National Natural Science Foundation of China (51175030), Science Foundation of Beijing Jiaotong University (M11JB00290) and Science Foundation of Beijing Jiaotong University (2011JBM277).

## References

1. Ylikorpi T, Suomela J (2006) Ball shaped robots: an historical overview and recent development at TKK. Field Serv Robotics 25(6):343–354
2. Armour RH, Vincent JFV (2006) Rolling in nature and robotics: a review. J Bionic Eng 3(4):195–208
3. Halme A, Schönberg T, Wang Y (1996) Motion control of a spherical mobile robot. In: Proceedings of IEEE international workshop on advanced motion control, IEEE Press, Mie, pp 259–264
4. Bicchi A, Balluchi A, Prattichizzo D et al (1997) Introducing the sphericle: an experimental testbed for research and teaching in non-holonomy. In: Proceedings of IEEE international conference on robototics and automation, IEEE Press, New Mexico, pp 2620–2625
5. Otani T, Urakuboi T, Maekawa S et al (2006) Position and attitude control of a spherical rolling robot equipped with a gyro. In: 9th IEEE international workshop on advanced motion control, IEEE Press, Istanbul, pp 416–421
6. Mukherjee R, Minor MA, Pukrushpan JT (1999) Simple motion planning strategies for spherobot: a spherical mobile robot. In: Proceedings of IEEE international conference on decision and control, IEEE Press, Arizona, pp 2132–2137
7. Mukherjee R, Minor MA, Pukrushpan JT (2002) Motion planning for a spherical mobile robot: revisiting the classical ball-plate problem. ASME J Dyn Syst T 124:502–511
8. Javadi AH, Mojabi P (2004) Introducing glory: a novel strategy for an omnidirectional spherical rolling robot. ASME J Dyn Syst T 126:678–683
9. Joshi VA, Banavar RN (2009) Motion analysis of a spherical mobile robot. Robotica 27:343–353
10. Joshi VA, Banavar RN, Hippalgaonkar R (2010) Design and analysis of a spherical mobile robot. Mech Mach Theory 45:130–136
11. Sugiyama Y, Hiral S (2006) Crawling and jumping by a deform robot. Int J Robot Res 25(5–6):603–620
12. Shibata M, Hiral S (2009) Rolling locomotion of deformable tensegrity structure. In: 12th International conference on climbing and walking robots and the support technologies for mobile machines, Istanbul, pp 479–486
13. Sastra J, Chitta S, Yim M (2009) Dynamic rolling for a modular loop robot. Int J Robot Res 28(6):758–773
14. Yim M, Duff DG, Roufas KD (2000) PolyBot: a modular reconfigurable robot. In: Proceeding of the 2000 IEEE international conference on robotics and automation, IEEE Press, San Francisco, pp 514–520
15. Lee WH, Sanderson AC (2002) Dynamic rolling locomotion and control of modular robots. IEEE T Robotic Autom 18(1):32–41

16. Clark PE, Rilee ML, Curtis SA et al (2004) BEES for ANTS: space mission applications for the autonomous nanotechnology swarm. In: Proceedings of first AIAA intelligent systems technical conference, Session 29-IS-13. Abrahantes
17. Lyder A, Franco R, Garcia M et al (2008) Mechanism design of odin, an extendable heterogeneous deformable modular robot. In: Proceeding of 2008 IEEE/RSJ international conference on intelligent robots and systems acropolis convention center, IEEE Press. Nice, pp 883–888
18. Hui R, Ouellet A, Wang A et al (1995) Mechanisms for haptic feedback. IEEE international conference on robotics and automation, IEEE Press, Nagoya, pp 2138–2143
19. Dunlop DR, Jones TP (1997) Position analysis of a 3-DOF parallel manipulator. Mech Mach Theory 38(8):903–920
20. Dunlop DR, Jones TP (1999) Position analysis of a two DOF parallel mechanism—the canterbury tracker. Mech Mach Theory 34:599–614
21. Gregorio RD (2004) Kinematics of the 3-RSR wrist. IEEE T Robotic Autom 20(4):750–754

# Chapter 22
# A Compliant 5-bar Tristable Mechanism Utilizing Metamorphic Transformation

**Guimin Chen, Yi Liu and Yanjie Gou**

**Abstract** Compliant multistable mechanisms, which are capable of steadily staying at multiple distinct positions without power input, have many potential applications in switches, valves, closures, relays, statically-balanced mechanisms, reconfigurable robots, and large-displacement micro actuators. In this paper, we propose a new idea of utilizing metamorphic transformations to develop compliant multistable mechanisms. By distributing the prescribed stable equilibrium positions into different metamorphic working phases, the design of a compliant multistable mechanism can be greatly simplified. The idea is demonstrated by a tristable mechanism that can metamorphically transform from a compliant 5-bar mechanism into a compliant 4-bar mechanism in a certain range of motion. The kinetostatic solution of this tristable mechanism is formulated and the kinetostatic results confirm that the mechanism has two deflected stable equilibrium positions besides its initial assembly position, with one occuring in the 4-bar working phase and the other in the 5-bar working phase. Although the discussion is limited to a planar 5-bar mechanism, the idea of utilizing metamorphic transformations to achieve multistable behivors can surely be extended to other types of linkages.

**Keywords** Compliant 5-bar mechanism · Multistable mechanism · Tristable mechanism · Metamorphic mechanism

G. Chen (✉) · Y. Liu · Y. Gou
School of Mechatronics, Xidian University, 710071 Xi'an, China
e-mail: guimin.chen@gmail.com

J. S. Dai et al. (eds.), *Advances in Reconfigurable Mechanisms and Robots I*,
DOI: 10.1007/978-1-4471-4141-9_22, © Springer-Verlag London 2012

## 22.1 Introduction

Compliant multistable mechanisms are often considered as a class of special-purpose compliant mechanisms, which are capable of steadily staying at multiple distinct positions without power input [1]. Because compliant mechanisms store and release energy in their flexible members during motion, these stable positions correspond to the local minima of the strain energy stored in the mechanisms. Multistable mechanisms have been widely used in various switches, valves, closures, relays, and clasps. They can serve as negative stiffness building blocks for achieving static balancing in compliant mechanisms [2, 3]. They also find possible applications in reconfigurable robots [4], landing gear mechanisms [5], and large-displacement micro actuators [6].

Compliant multistable mechanisms can be roughly divided into two categories depending on the number of stable positions: compliant bistable mechanisms [1] and compliant mechanisms with three or even more stable positions [7]. In the rest of this paper, the term "multistable" will be used to exclusively represent "having more than two stable positions," if not specifically state. There has been considerable work devoted to various configurations of bistable compliant mechanisms [8–16]. In contrast, fewer multistable configurations [17–20] have been found, and combining multiple bistable elements has been adopted as a traditional way to develop multistable mechanisms [6, 7, 21–23]. Therefore, identifying new compliant configurations that may exhibit multistability is still of the researchers' interest in the community of compliant mechanism.

In this paper, we propose a unique idea of utilizing metamorphic transformation techniques [24] to develop compliant multistable mechanisms. By incorporating metamorphic transformations into a compliant mechanism, it can change mobility and possess multiple working phases within its range of motion. Therefore, we think the design of a compliant multistable mechanism can be greatly simplified by distributing the prescribed stable equilibrium positions into different working phases. In the following, we employ a compliant 5-bar tristable mechanism (having three stable equilibrium positions) which can metamorphically transforms into a compliant 4-bar mechanism in a certain range of motion to demonstrate the idea.

Metamorphic mechanisms, originated from the research of decorative carton folds and artifacts [24], can change their mobility and topology due to the change of the effective number of links and joints during motion. A lot of work has been done on representation [25, 26], synthesis [27] and analysis [28] of metamorphic mechanisms. There has also been some work on exploring the integration of metamorphic mechanisms with compliant mechanisms, for example, Parise et al. [29, 30] presented compliant ortho-planar metamorphic mechanisms, and Ding et al. [31] and Li et al. [32] studied a few metamorphic mechanisms employing compliant segments that can transform from one degree-of-freedom mechanisms into structures.

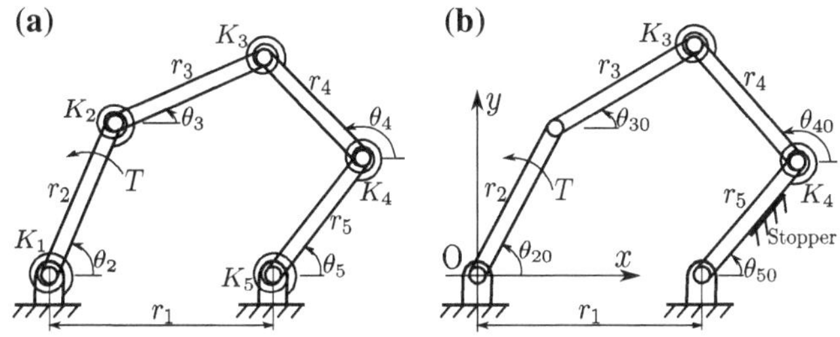


**Fig. 22.1** Pseudo-rigid-body models of compliant 5-bar mechanisms: **a** a general compliant 5-bar mechanism, and **b** a compliant 5-bar mechanism utilizing metamorphic transformation

The rest of the paper is organized as follows: A metamorphic compliant 5-bar mechanism based on pseudo-rigid-body model is introduced in Sect. 22.2. Section 22.3 formulates the kinetostatics of the compliant 5-bar mechanism. A metamorphic compliant 5-bar mechanism exhibiting tristability is designed in Sect. 22.4. Section 22.5 gives some concluding remarks.

## 22.2 Compliant 5-bar Mechanism Utilizing Metamorphic Transformation

We limit our discussion to a compliant mechanism that can be modeled as a 5-bar model using the pseudo-rigid-body (PRB) method [1], as shown in Fig. 22.1a. The PRB method, which simplifies the nonlinear deflections of flexible segments as the motion of rigid links connected by torsional springs, enables us to apply the knowledge of rigid-body mechanisms to the synthesis and analysis of compliant mechanisms. Once a PRB model exhibiting desired performance is located, the rigid-body replacement synthesis method [5] may be employed to convert the model into a compliant mechanism design.

The 5-bar model shown in Fig. 22.1a contains five links, with link lengths $r_1, r_2, r_3, r_4$, and $r_5$. Link 1 is the ground link located along the $x$-axis. The spring constants of the five connecting joints are $K_1, K_2, K_3, K_4$ and $K_5$ ($K_i = 0$ for articulated joints, $K_i = +\infty$ for rigid connections and $0 < K_i < +\infty$ for torsional springs, $i = 1, 2, \ldots, 5$). The angle of each link with respect to the horizontal is represented by $\theta_2, \theta_3, \theta_4$, and $\theta_5$. It is assumed that the 5-bar model is subject to a single input at $r_2$. Knowing that a planar 5-bar mechanism has two degrees of freedom (DOF), this model is underactuated with one DOF unspecified [33]. Nonetheless, it may still have a local equilibrium position for a given input because the mechanism will tend to the position where the overall potential energy stored in its flexible elements is at a local minimum (according to the principle of minimum potential energy) [33]. In addition, this underactuated model contains at least two torsional springs among $K_2, K_3, K_4$ and $K_5$ (note that $K_1$ is excluded because link 2 is the driving link) to prevent pure rigid-body motion of the model in its mobility range [34].

Our experience indicates that to realize a tristable mechanism with the compliant 5-bar model is quite difficult. However, this can be greatly simplified by introducing a metamorphic transformation into the model, as will be illustrated in the Sect. 22.4. We add a stopper to the model for metamorphic purpose, as shown in Fig. 22.1b. When link 5 is attached to the stopper, its clockwise rotation is blocked. If link 5 tends to rotate clockwise from this position, it becomes a part of the ground link and the PRB model changes its topological configuration to a 4-bar model (the number of effective links is decreased to 4), that is to say, the metamorphism of the model occurs. In the PRB model, this metamorphic process can be represented by changing $K_5$ to $+\infty$ (i.e., a rigid connection).

## 22.3 Kinetostatic Solution for Compliant 5-bar Mechanism

The loop closure equations of the model shown in Fig. 22.1a are given as

$$\begin{cases} r_2\cos\theta_2 + r_3\cos\theta_3 = r_4\cos\theta_4 + r_5\cos\theta_5 + r_1 \\ r_2\sin\theta_2 + r_3\sin\theta_3 = r_4\sin\theta_4 + r_5\sin\theta_5 \end{cases} \tag{22.1}$$

and the total potential energy stored in the model as

$$V = \sum_{i=1}^{5} \frac{1}{2} K_i \psi_i^2 \tag{22.2}$$

where

$$\begin{cases} \psi_1 = \theta_2 - \theta_{20} \\ \psi_2 = \theta_2 - \theta_{20} - \theta_3 + \theta_{30} \\ \psi_3 = \theta_4 - \theta_{40} - \theta_3 + \theta_{30} \\ \psi_4 = \theta_5 - \theta_{50} - \theta_4 + \theta_{40} \\ \psi_5 = \theta_5 - \theta_{50} \end{cases} \tag{22.3}$$

and $\theta_{20}, \theta_{30}, \theta_{40}$, and $\theta_{50}$ are the initial angles of the links (torsional springs are undeflected, as-fabricated position). The torque applied on $r_2$ can be expressed as (with respect to $\theta_2$)

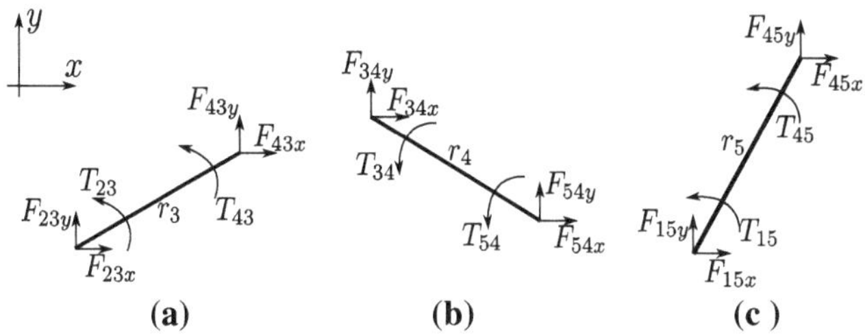


**Fig. 22.2** Free-body diagrams of links: **a** $r_3$, **b** $r_4$, and **c** $r_5$

$$T = \frac{\partial V}{\partial \theta_2} = \sum_{i=1}^{5} K_i \psi_i \frac{\partial \psi_i}{\partial \theta_2} \tag{22.4}$$

where

$$\begin{cases} \dfrac{\partial \psi_1}{\partial \theta_2} = 1 \\ \dfrac{\partial \psi_2}{\partial \theta_2} = 1 - h_{32} \\ \dfrac{\partial \psi_3}{\partial \theta_2} = h_{42} - h_{32} \\ \dfrac{\partial \psi_4}{\partial \theta_2} = h_{52} - h_{42} \\ \dfrac{\partial \psi_5}{\partial \theta_2} = h_{52} \end{cases} \tag{22.5}$$

and $h_{32}, h_{42}$, and $h_{52}$ are the kinematic coefficients with respect to $\theta_2$.

Figure 22.2 shows the free-body diagrams for $r_3, r_4$ and $r_5$. The equations of static equilibrium for $r_3$ are given as

$$\begin{cases} F_{23x} + F_{43x} = 0 \\ F_{23y} + F_{43y} = 0 \\ T_{23} + T_{43} + r_3 F_{43y} \cos\theta_3 - r_3 F_{43x} \sin\theta_3 = 0 \end{cases} \tag{22.6}$$

The equations for $r_4$ are

$$\begin{cases} F_{34x} + F_{54x} = 0 \\ F_{34y} + F_{54y} = 0 \\ T_{34} + T_{54} + r_4 F_{54y} \cos\theta_4 - r_4 F_{54x} \sin\theta_4 = 0 \end{cases} \tag{22.7}$$

and the equations associated with $r_5$ are

**Table 22.1** The design parameters of the tristable metamorphic 5-bar mechanism

| $r_1$ | $r_2$ | $r_3$ | $r_4$ | $r_5$ | $K_3$ | $K_4$ | $\theta_{20}$ | $\theta_{30}$ |
|---|---|---|---|---|---|---|---|---|
| 0.186 m | 0.113 m | 0.132 m | 0.097 m | 0.152 m | 0.2216 N m/rad | 0.4966 N m/rad | 95° | 37.8° |

$$\begin{cases} F_{45x} + F_{15x} = 0 \\ F_{45y} + F_{15y} = 0 \\ T_{45} + T_{15} + r_5 F_{45y} \cos\theta_5 - r_5 F_{45x} \sin\theta_5 = 0 \end{cases} \tag{22.8}$$

where $T_{ij}$ is determined in the following formulas ($F_{ijx} = -F_{jix}$):

$$\begin{cases} T_{12} = -K_1(\theta_2 - \theta_{20}) \\ T_{32} = -T_{23} = -K_2(\theta_2 - \theta_{20} - \theta_3 + \theta_{30}) \\ T_{43} = -T_{34} = -K_3(\theta_4 - \theta_{40} - \theta_3 + \theta_{30}) \\ T_{54} = -T_{45} = -K_4(\theta_5 - \theta_{50} - \theta_4 + \theta_{40}) \\ T_{15} = -K_5(\theta_5 - \theta_{50}) \end{cases} \tag{22.9}$$

Combining these static equilibrium equations results in the following equation:

$$\begin{aligned} & K_2 r_4 r_5(\theta_2 - \theta_{20} - \theta_3 + \theta_{30}) \sin(\theta_4 - \theta_5) \\ & + K_3 r_5(\theta_3 - \theta_{30} - \theta_4 + \theta_{40})[r_4 \sin(\theta_4 - \theta_5) - r_3 \sin(\theta_3 - \theta_5)] \\ & + K_4 r_3(\theta_4 - \theta_5 - \theta_{40} + \theta_{50})[r_4 \sin(\theta_3 - \theta_4) + r_5 \sin(\theta_3 - \theta_5)] \\ & + K_5 r_3 r_4(\theta_5 - \theta_{50}) \sin(\theta_3 - \theta_4) = 0 \end{aligned} \tag{22.10}$$

After being differentiated with respect to $\theta_2$, Eqs. (22.9) and (22.1) are used to solve for the kinematic coefficients, i.e., $h_{32}, h_{42}$ and $h_{52}$. Then, the force-deflection characteristics of the mechanism can be obtained using Eq. (22.2).

By setting the stiffness of one of the torsional springs to $+\infty$, these equations can be directly employed to analyze compliant 4-bar mechanisms.

## 22.4 Example and Results

This section presents a design of the metamorphic compliant 5-bar mechanism that exhibits tristablilty. We assume that the stopper is located at the place where the clockwise rotation of link 5 is blocked when the mechanism is at its the initial assembly position (or first stable equilibrium position where all the torsional springs are undeflected, i.e., $\theta_i = \theta_{i0}, i = 2, 3, 4, 5$.) If link 5 tends to rotate clockwise when $\theta_5 = \theta_{50}, K_5$ is set to $+\infty$ in the model to block the motion.

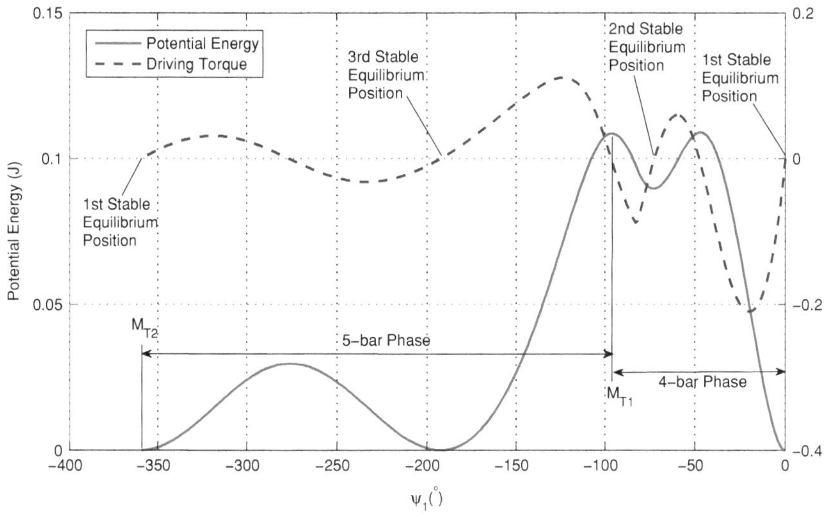


**Fig. 22.3** The curves of energy and driving torque versus $\psi_1$

The design parameters of a tristable metamorphic mechanism are list in Table 22.1. The other two parameters at the initial position, $\theta_{40}$ and $\theta_{50}$, can be calculated from Eq. (22.1): $\theta_{40} = 154.7^\circ$ and $\theta_{50} = 91.5^\circ$. The curves of the potential energy and the driving torque obtained from the kinetostatic solution are plotted as functions of $\psi_1$ in Fig. 22.3. From Fig. 22.3 we see that when moving link 2 clockwise (with decreasing $\psi_1$), the mechanism first works as a compliant 4-bar mechanism until $\psi_1 = -96^\circ$ (denoted as $M_{T1}$ in the figure) where the metamorphic transformation takes place, and then works as a compliant 5-bar mechanism. Knowing that the minima on the potential energy curve correspond to the stable equilibrium positions, it can be seen from Fig. 22.3 that the second stable equilibrium position occurs at $\psi_1 = -74^\circ$ ($\theta_2 = 21^\circ$,) while the third stable equilibrium position at $\psi_1 = -188^\circ(\theta_2 = -93^\circ)$. When $\psi_1$ reaches 360°, the mechanism returns to its first stable equilibrium position. This position, denoted as $M_{T2}$ in Fig. 22.3, can be considered as another metamorphic position because the mechanism works as a compliant 5-bar mechanism if link 2 moves counter-clockwise, while the mechanism works as a compliant 4-bar mechanism if link 2 moves clockwise. The two deflected stable configurations are shown in Fig. 22.4, with one occuring in the 4-bar working phase and the other in the 5-bar working phase. We are now prototyping this tristable mechanism which will be presented in future version of this paper.

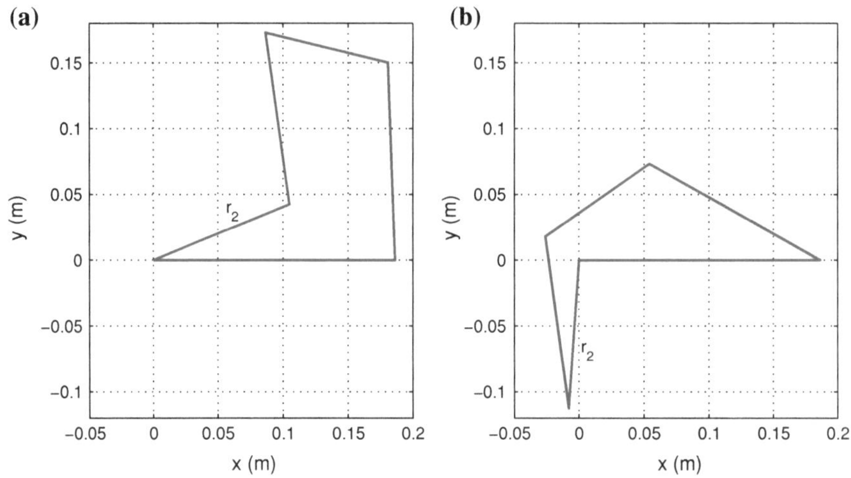


**Fig. 22.4** The two deflected stable equilibrium configurations: **a** the second stable equilibrium position, and **b** the third stable equilibrium position

## 22.5 Conclusions

This paper presented a method of utilizing metamorphic transformations to develop compliant multistable mechanisms. The feasibility of the idea was demonstrated by a tristable mechanism that can metamorphically transform from a 5-bar mechanism into a 4-bar mechanism in a certain range of motion.

Although the discussion is limited to a planar 5-bar model, the idea of utilizing metamorphic transformations to achieve multistable behivors can surely be extended to other types of linkages, for example, planar 6-bar linkages and five-bar spherical linkages [26]. It is also interesting to explore other features of compliant mechanisms employing metamorphic transformation techniques besides multistabilities.

**Acknowledgements** The authors gratefully acknowledge the financial support from the National Natural Science Foundation of China under Grant No. 50805110 and 51175396, and the program for new century excellent talents in university under Grant No. NCET-11-0689.

## References

1. Howell LL (2001) Compliant mechanisms. Wiley, New York
2. Tolou M, Estevez P, Herder JL (2011) Collinear-type statically balanced compliant micro mechanism (SB-CMM): experimental comparison between pre-curved and straight beams. Proceedings of the ASME design engineering technical conferences & computers and information in engineering conference, 28–31 August 2011, Washington, DC, USA, DETC2011-47678

3. Chen G, Zhang S (2011) Fully-compliant statically-balanced mechanisms without prestressing assembly: concepts and case studies. Mech Sci 2(2):169–174
4. Hafez M, Lichter MD, Dubowsky S (2003) Optimized binary modular reconfigurable robotic devices. IEEE-ASME T Mech 8(1):304–326
5. Pucheta MA, Cardona A (2010) Design of bistable compliant mechanisms using precision-position and rigid-body replacement methods. Mech Mach Theory 45(2):304–326
6. Gerson Y, Krylov S, Ilic B, Schreiber D (2012) Design considerations of a large-displacement multistable micro actuator with serially connected bistable elements. Finite Element Anal Des 49(1):58–69
7. Oh YS, Kota S (2009) Synthesis of multistable equilibrium compliant mechanisms using combinations of bistable mechanisms. J Mech Design, 131:021002
8. Jensen BD, Howell LL, Salmon LG (1999) Design of two-link, in-plane, bistable compliant micro-mechanisms. J Mech Des 121(3):416–423
9. Jensen BD, Parkinson MB, Kurabayashi K, Howell LL, Baker MS (2001) Design optimization of a fully-compliant bistable micro-mechanism. Proceedings of ASME IMECE, vol 2, pp 2931–2937, 11–16 Novemb, New York, USA
10. Masters ND, Howell LL (2003) A self-retracting fully compliant bistable micromechanism. J Microelectromech Syst 12:273–280
11. Jensen BD, Howell LL (2003) Identification of compliant pseudo-rigid-body mechanism configurations resulting in bistable behavior. J Mech Des 125:701–708
12. Jensen BD, Howell LL (2004) Bistable configurations of compliant mechanisms modeled using four links and translational joints. J Mech Des 126:657–666
13. Qiu J, Lang JH, Slocum AH (2004) A curved-beam bistable mechanism. J Microelectromech Syst 13(2):137–146
14. Wilcox DL, Howell LL (2005) Fully compliant tensural bistable micromechanisms (FTBM). J Microelectromech Syst 14(6):1223–1235
15. Sönmez Ü, Tutum CC (2008) A compliant bistable mechanism design incorporating elastica buckling beam theory and pseudo-rigid-body model. J Mech Des 130(4):042304
16. Smith CL, Lusk CP (2011) Modeling and parameter study of bistable spherical compliant mechanisms. Proceedings of ASME IDETC/CIE 2011, 28–31 August, Washington, DC, USA, DETC2011-47397
17. Pendleton TM, Jensen BD (2007) Development of a tristable compliant mechanism. Proceedings of 12TH IFToMM world congress, A835, 18–21 June, Besancon, France
18. Chen G, Aten QT, Zirbel S, Jensen BD, Howell LL (2010) A tristable mechanism configuration employing orthogonal compliant mechanisms. J Mech Robot-T ASME 2(1):014501
19. Halverson PA, Howell LL, Magleby SP (2010) Tension-based multi-stable compliant rolling-contact elements. Mech Mach Theory 45(2):147–156
20. Chen G, Gou Y, Zhang A (2011) Synthesis of compliant multistable mechanisms through use of a single bistable compliant mechanism. J Mech Des 133(8):081007
21. Han JS, Muller C, Wallrabe U, Korvink JG (2007) Design, simulation, and fabrication of a quadstable monolithic mechanism with X- and Y-directional bistable curved beams. J Mech Des 129(11):1198–1203
22. Chen G, Wilcox DL, Howell LL (2009) Fully compliant double tensural tristable micromechanisms (DTTM). J Micromech Microeng 19(2):025011
23. Pham H-T, Wang D-A (2011) A quadristable compliant mechanism with a bistable structure embedded in a surrounding beam structure. Sensor Actuat A-Phys 167(2):438–448
24. Dai JS, Jones JR (1999) Mobility in metamorphic mechanisms of foldable/erectable kinds. J Mech Des 121(3):375–382
25. Dai JS, Jones JR (2005) Matrix representation of topological changes in metamorphic mechanisms. J Mech Des 127(7):837–840
26. Lan ZH, Du R (2008) Representation of topological changes in metamorphic mechanisms with matrices of the same fimension. J Mech Des 130(7):074501

27. Zhang L, Wang D, Dai JS (2008) Biological modeling and evolution based synthesis of metamorphic mechanisms. J Mech Des 130(7):072303
28. Zhang K, Dai JS, Fang Y (2010) Topology and constraint analysis of phase change in the metamorphic chain and its evolved mechanism. J Mech Des 132(12):121001
29. Parise JJ, Howell LL, Magleby SP (2000) Ortho-planar mechanisms. Proceedings of the 26th biennial mechanisms and robotics conference, Baltimore, MD, paper no. DETC2000/MECH-14193
30. Carroll DW, Magleby SP, Howell LL, Todd RH, Lusk CP (2005) Simplified manufacturing through a metamorphic process for compliant ortho-planar mechanisms. Proceedings of ASME IMECE, Oriando, FL, USA, 5–11, Novemb, IMECE2005-82093
31. Yang Y, Ding X, Dai J (2007) Design and analysis of an umbrella foldable complant metamorphic mechanism. Acta Aeronaut Astronaut Sinica 28(4):1014–1017 (in Chinese)
32. Li D, Zhang Z, Chen G (2011) Structure synthesis of compliant metamorphic mechanisms based on adjacency matrix operations. Chin J Mech Eng-En 24(4):522–528
33. Aten QT, Zirbel S, Jensen BD, Howell LL (2011) A numerical method for position analysis of compliant mechanisms with more degrees of freedom than inputs. J Mech Des 133(6):061009
34. Her I, Midha A (1987) A compliance number concept for compliant mechanisms, and type synthesis. J Mech Transm-T ASME 109(3):348–357

# Part III
# Reconfigurable Parallel Mechanisms

# Chapter 23
# Type Synthesis of Partially Decoupled 2-DOF Parallel Mechanisms with Two 1T1R Operational Modes

**Xianwen Kong**

**Abstract** This paper deals with the type synthesis of partially decoupled 2-DOF parallel mechanisms with two 1T1R (one translation and one rotation) operational modes—a novel class of reconfigurable parallel mechanisms which can switch operational modes without disassembly. In the two 1T1R operational modes, the axes of rotation of the moving platform are different. A bi-mode revolute joint is defined to simplify the representation of motion patterns of the moving platform. Using the vitual-chain approach, the type synthesis of partially decoupled 2-DOF 1T1R parallel mechanisms and partially decoupled 2-DOF parallel mechanisms with two 1T1R operational modes is carried out. The proposed mechanisms need one less actuated joint than the existing parallel mechanisms with two 1T1R operational modes. This work provides a solid foundation for developing energy-efficient parallel mechanisms with two 1T1R operational modes.

**Keywords** Type synthesis · Parallel mechanism · Screw theory · Virtual-chain approach · Parallel mechanism with multiple operation modes

## 23.1 Introduction

Several disassembly-free reconfigurable parallel mechanisms (PMs), which can switch operational modes without disassembly, have been proposed in recent years. Disassembly-free reconfigurable PMs include PMs with multiple operational modes (PMwmoms) [1–10], which are not kinematically redundant, and kinematically redundant PMs in which several joints are used for reconfiguration

X. Kong (✉)
Heriot-Watt University, Edinburgh, UK
e-mail: X.Kong@hw.ac.uk

J. S. Dai et al. (eds.), *Advances in Reconfigurable Mechanisms and Robots I*,
DOI: 10.1007/978-1-4471-4141-9_23, © Springer-Verlag London 2012 

[11, 12]. This paper is about PMwmoms. In addition to the analysis of PMwmoms [1–5], efforts have been made to develop a systematic approach to the type synthesis of PMwmoms [6–10], which refers to finding all the types of PMs that can generate two or more specified motion patterns. Meanwhile, some PMwmoms have been put into use in several applications [1, 6]. For example, a 2-DOF PM with two 1T1R (one translation and one rotation) operational modes [6] has been used as a novel swivel head for CNC centers. Despite the above advances, the type synthesis of PMwmoms, which is the first step in the research on PMwmoms, is still an open issue.

In a PM with two 1T1R operational modes proposed in the literature [6, 13], three actuated joints are needed to control the PM in each operational mode. Since a PM with two 1T1R operational modes has two DOF in each operational mode, it deserves to identify PMs with two 1T1R operational modes in which two actuated joints are needed to control the PMs. This may help develop energy-efficient PMs with two 1T1R operational modes.

This paper deals with the type synthesis of partially decoupled 2-DOF PMs with two 1T1R operational modes in which two actuated joints are needed to control the PMs. In one operational mode, the moving platform of the PM can translate along the Z-axis and rotate about a line parallel to the X-axis. In another operational mode, the moving platform can translate along the Z-axis and rotate about a line parallel to the Y-axis. To the author's knowledge, no work has been published on this class of PMs with two 1T1R operational modes.

The virtual-chain approach to the type synthesis of PMs [14] is briefly recalled in Sect. 22.2. In Sect. 23.3, the type synthesis of partially decoupled 2-DOF 1T1R PMs is presented. In Sect. Sect. 23.3, the type synthesis of partially decoupled 2-DOF PMs with two 1T1R operational modes is carried out. Finally, conclusions are drawn.

Due to space limitation, we make the assumption that the only P (prismatic) joints in a PM with two 1T1R operational modes are located on the base and along the same direction.

## 23.2 Virtual-Chain Approach to the Type Synthesis of Parallel Mechanisms

In this section, the virtual-chain approach to the type synthesis of PMs [14][1] will be briefly recalled. Using this approach, motion pattens are represented using virtual chains, and the type synthesis of PMs is reduced to the type synthesis of

[1] Different methods for the type synthesis of PMs with a single operational mode have also been proposed a number of researchers, such as Profs. R.I. Alizade, O. Altuzarra, J.S. Dai, R. Di Gregorio, Y.F. Fang, F. Gao, G. Gogu, J. M. Hervé, Z. Huang, C.-C. Lee, Q.-C. Li, Z. Li, J.M. Rico, L.-W. Tsai, and T.-L. Yang (in alphabet order). Their publications are not listed here due to space limitation. Inactive joints, which are very useful in the type synthesis of PMwmoms, are not used in the PMs obtained using these methods.

single-loop kinematic chains. PMs with full-cycle mobility (also finite motion) can be constructed using several classes of compositional units (CUs). The instantaneous constraint analysis of CUs and PMs based on the fundamental screw theory, which is essential in the type synthesis, will not be discussed in this paper since it has been well documented in the literature (see [14] for example).

## *23.2.1 Virtual Chain*

A virtual chain is defined as a serial or parallel kinematic chain whose moving platform has a given motion pattern [14]. For example, the virtual chain associated with 3-DOF translational PMs is a PPP virtual chain, which is a serial chain composed of three P joints. With the introduction of the virtual chain, a translational PM is called a PPP= (PPP equivalent) PM [14].

Using the conventional kinematic joints, we need to use a virtual chain to represent the motion pattern of a PMwmom in each operational mode [7, 10]. With the concept of variable kinematic joint [15], we may represent the motion patterns of a PMwmom using one virtual chain.

### 23.2.1.1 Bi-Mode Revolute Joint: $R_+$

To define the virtual chain of a 2-DOF PM with two 1T1R operational modes, let us first define a bi-mode R (revolute) joint (Figs. 23.1 and 23.2), denoted by $R_+$, as a kinematic joint which functions as an R joint along two different axes in its two operational modes (Figs. 23.1b, c and 23.2b, c). An $R_+$ can take different forms, such as a kinematotropic 4R linkage (Fig. 23.1a) or a U (universal joint, Fig. 23.2a). The bi-mode is realized by the constraints within the $R_+$ in the former case (Fig. 23.1) or by the constraints imposed on the $R_+$ joint by the other part of a mechanism in the latter case (Fig. 23.2). It is noted that an $R_+$ joint has two instantaneous DOF in its transition configuration (Figs. 23.1a and 23.2a) through which it can transfer from one operational mode (Figs.23.1b and 23.2b) to another (Figs. 23.1c and 23.2c).

### 23.2.1.2 $PR_+$ Virtual Chain

With the aid of the $R_+$ joint, the motion pattern of the moving platform of a 2-DOF PM with two 1T1R operational modes can be described using a $PR_+$ virtual chain (Fig. 23.3), which is composed of a P joint along Z-axis and an $R_+$ joint. Kinematically, the $PR_+$ virtual chain is equivalent to a $PR_X$ virtual chain (Fig. 23.4a) in the $PR_{+X}$ operational mode (Fig. 23.3b) or the $PR_Y$ virtual chain (Fig. 23.4b) in the $PR_{+Y}$ operational mode (Fig. 23.3c). Here, $R_X$ and $R_Y$ denote R joints whose axes are parallel to *X*- and *Y*-axes respectively.

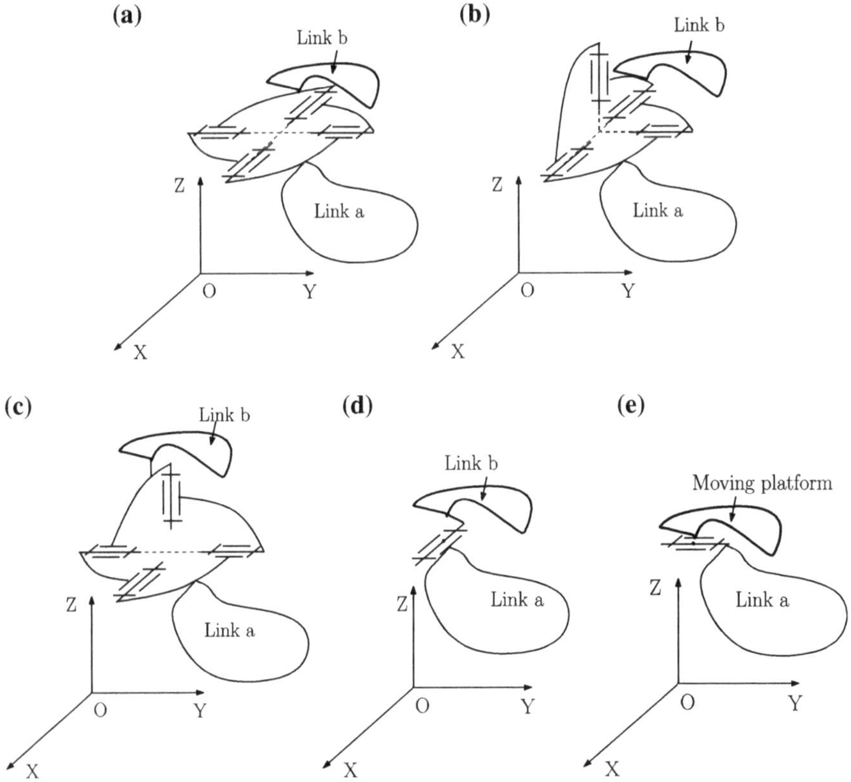


**Fig. 23.1** Bi-mode R joint $R_+$ : case 1 **a** Transition configuration; **b** $R_{+X}$ mode; **c** $R_{+Y}$ mode; **d** Kinematic representation of the $R_X$ mode; **e** Kinematic representation of the $R_Y$ mode

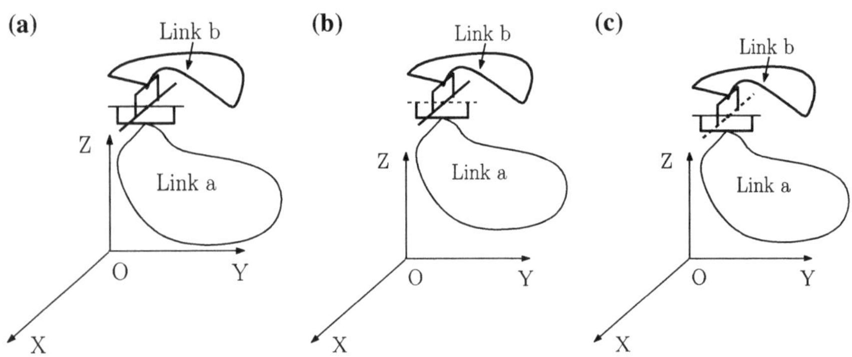


**Fig. 23.2** Bi-mode R joint $R_+$ : case 2 **a** Transition configuration; **b** $R_{+X}$ mode; **c** $R_{+Y}$ mode

Throughout this paper, the joints whose axes or directions are parallel to a coordinated axis, have common points or are collinear are drawn in such a way that these features can be easily observed from the figures. The R joints with parallel

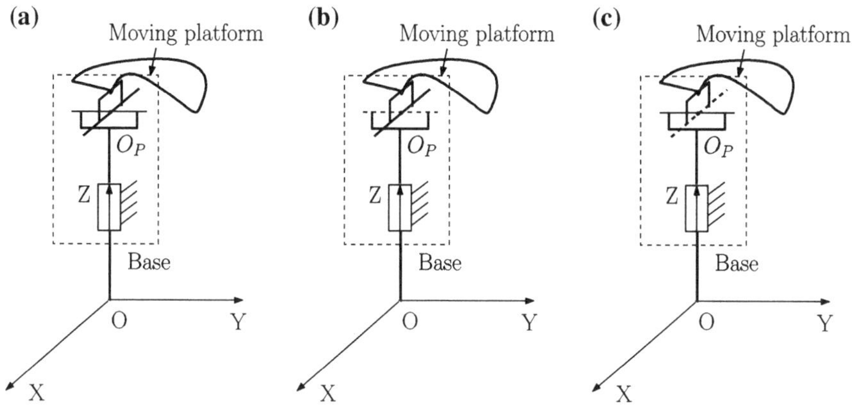


**Fig. 23.3** $PR_+$ virtual chain **a** Transition configuration; **b** $PR_{+X}$ mode; **c** $PR_{+Y}$ mode

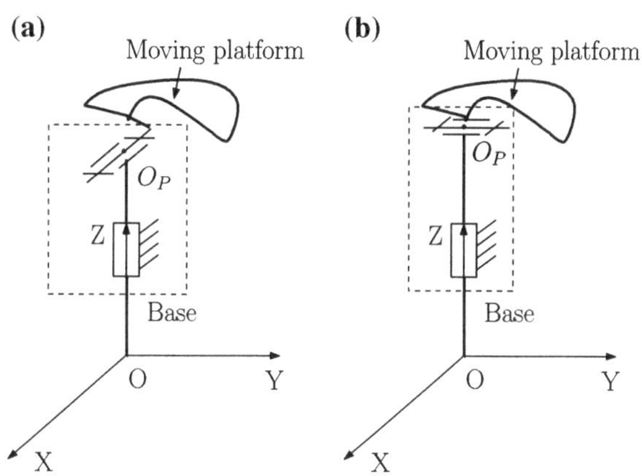


**Fig. 23.4** PR virtual chain **a** $PR_X$; **b** $PR_Y$

axes, which are not necessarily parallel to the coordinate axes, are indicated by $A_i$. The axis of an inactive R joint within the $R_+$ joint is indicated by a dashed line and the other inactive joints are indicated by $I_i$. A PMwmom is shown in its transition configuration where it can switch from one operational mode to another.

As in [14], 2-DOF 1T1R PMs and 2-DOF PMs with two 1T1R operational modes are called 2-DOF PR= (PR equivalent) PMs and 2-DOF $PR_+$ = ($PR_+$equivalent) PMs respectively. In the following, the $R_+$ joints will take the form of a U joint since the type synthesis of $PR_+$= PMs is general in this case and the results can be readily applied to the cases where the $R_+$ joints take the other forms.

## 23.2.2 Compositional Units

A CU is a serial kinematic chain composed of at least two kinematic joints which satisfy certain invariant kinematic and constraint characteristics [14]. The CUs

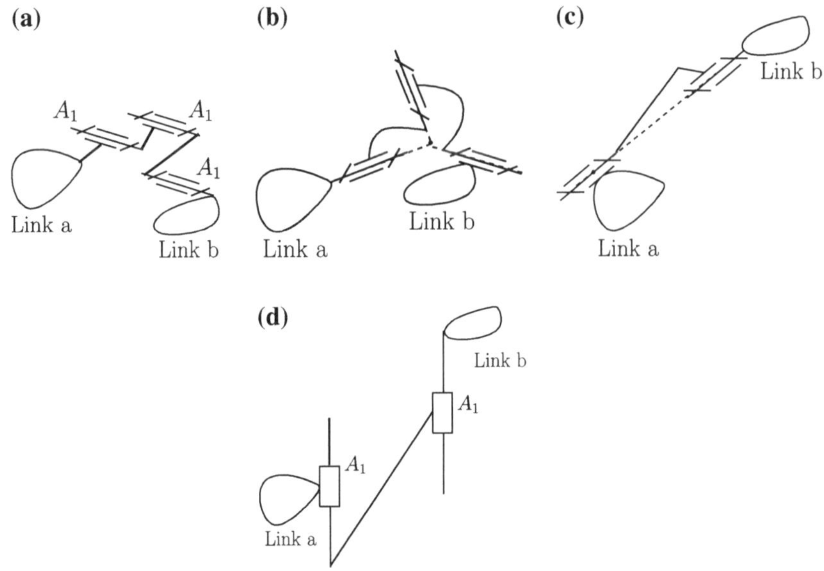


**Fig. 23.5** Compositional units **a** Planar CU; **b** spherical CU; **c** Codirectional CU; **d** Coaxial CU

(Fig. 23.5) that can be used to construct 2-DOF PR= PMs include:[2] planar CUs (Fig. 23.5a), spherical CUs (Fig. 23.5b), codirectional CUs (Fig. 23.5c), and coaxial CUs (Fig. 23.5d). In a planar CU, all the links move along parallel planes. In a spherical CU, the axes of all the R joints pass through a common point. In a codirectional CU, all the P joints are along the same direction. In a coaxial CU, the axes of all the R joints are coaxial.

## 23.3 Type Synthesis of Partially Decoupled 2-DOF PR= Parallel Mechanisms

In this section, the type synthesis of partially decoupled 2-DOF PR= PMs will be dealt with using the virtual-chain approach [14]. In a partially decoupled 2-DOF PR= PM, one output, the translation of the moving platform, is controlled by one actuated joint, while the remaining output, the rotation of the moving platform, is further controlled by another actuated joint.

In a partially decoupled 2-DOF PR= PM, the PR virtual chain is in fact a physical leg of the PM. Therefore, the type synthesis of 2-legged partially

[2] The parallelaxis CUs, spatial translation CUs, and planar translation CUs can also be used to construct 2-DOF PR= PMs. They are omitted here since the PMs obtained using these CUs do not satisfy the assumptions that we made in Sect. 1.

decoupled 2-DOF PR= PMs is reduced to the type synthesis of 2-DOF single-loop mechanisms involving a PR virtual chain. Since the type synthesis of non-overconstrained single-loop mechanisms is trivial, we will focus on overconstrained 2-legged partially decoupled 2-DOF PR= PMs.

From [14], 2-DOF single-loop mechanisms involving a PR virtual chain (Fig. 23.6) can be constructed by connecting one or two CUs listed above. For example, the 2-DOF single-loop mechanism in Fig. 23.6a is composed of a planar CU. The 2-DOF single-loop mechanisms in Fig. 23.6b, c, d are respectively composed of (a) two planar CUs, (b) a planar CU and a spherical CU, and (c) a codirectional CU and a coaxial CU.

From the PMs obtained using the above procedure, additional partially decoupled 2-DOF PR= PMs can also be obtained by inserting $n(n \leq c_r)$ inactive joints [14] to the single-loop mechanisms. Here, $c_r$ denotes the number of redundant constraints in a single-loop mechanism. For example, the partially decoupled 2-DOF PR= PM in Fig. 23.6e is obtained by inserting an inactive R joint, $I_1$, to the partially decoupled 2-DOF PR= PM in Fig. 23.6a, for which we have $c_r = 3$.

It is noted that some PMs, such as the one shown in Figs. 23.6d, may be regarded as degenerated cases in the type synthesis of PMs with a single operational mode. However, such PMs are useful to the type synthesis of PMwmoms.

## 23.4 Type Synthesis of Partially Decoupled 2-DOF $PR_+ =$ Parallel Mechanisms

From the types of partially decoupled 2-DOF PR= PMs obtained in Sect. 23.3, the partially decoupled 2-DOF $PR_+ =$ PMs can be obtained using the following 3-step procedure.

### 23.4.1 *Step 1: Type Synthesis of Partially Decoupled 2-DOF* $PR_{+i}=$ Parallel Mechanisms

Types of partially decoupled 2-DOF $PR_{+i}=$ PMs ($i = X$ or $Y$) can be obtained by inserting an inactive joint to partially decoupled 2-DOF PR= PMs with $c_r \geq 1$. For example, the partially decoupled 2-DOF $PR_{+X}=$ PM in Fig. 23.7a is obtained by inserting an inactive R joint to the PM in Fig. 23.6e. This inactive R joint and the R joint within the PR virtual chain in Fig. 23.6e form the $R_{+X}$ joint in Fig. 23.7a. Figure 23.7b shows a partially decoupled 2-DOF $PR_{+Y} =$ PM.

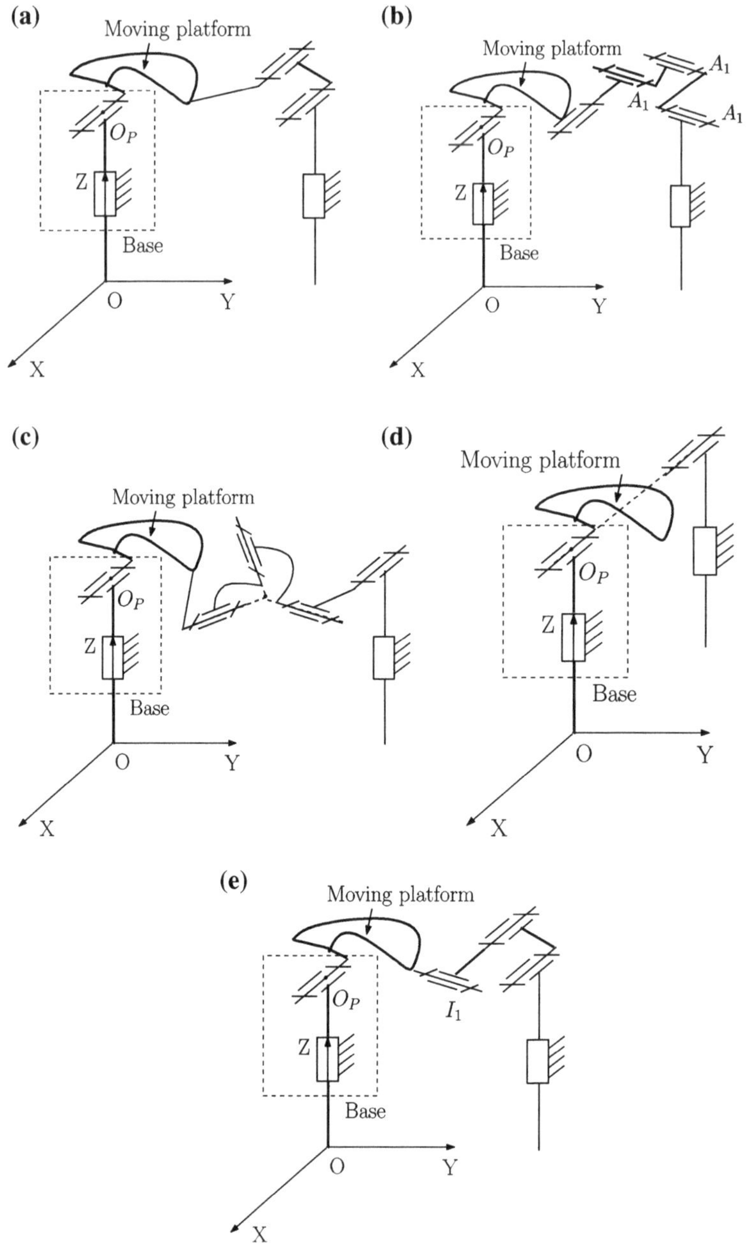


**Fig. 23.6** Partially decoupled 2-DOF PR= parallel mechanisms **a** Case 1; **b** Case 2; **c** Case 3; **d** Case 4; **e** Case 5

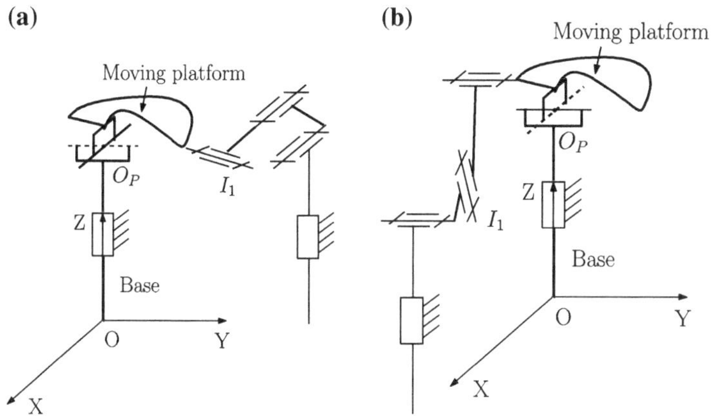


**Fig. 23.7** Partially decoupled 2-DOF $PR_{+i}=$ parallel mechanisms **a** $PR_{+X}=$ PM; **b** $PR_{+Y}=$ PM

### 23.4.2 *Step 2: Type Synthesis of Potential Partially Decoupled 2-DOF $PR_+=$* **Parallel Mechanisms**

Potential partially decoupled 2-DOF $PR_+=$ PMs can be obtained by integrating a partially decoupled 2-DOF $PR_{+X}=$ PM and a partially decoupled $PR_{+Y}=$ PM with a common $PR_{+X}=$ virtual joint. Figure 23.8 shows a 3-legged potential partially decoupled 2-DOF $PR_+=$ PM obtained by assembling a partially decoupled 2-DOF $PR_{+X}=$ PM in Fig. 23.7a and a partially decoupled 2-DOF $PR_{+Y}=$ PM in Fig. 23.7b with a common $PR_{+X}=$ virtual chain. However, a potential partially decoupled 2-DOF $PR_+=$ PM may not be a partially decoupled 2-DOF $PR_+=$ PM. In the PM shown in Fig. 23.8, both the R joints within the $R_+$ joint are inactive due to the constraints imposed by leg 3 and leg 2. Therefore, the moving platform of this PM has 1-DOF and can only translate along the Z axis.

### 23.4.3 *Step 3: Type Synthesis of Partially Decoupled 2-DOF* **$PR_+=$ Parallel Mechanisms**

From a potential partially decoupled 2-DOF $PR_+=$ PM obtained in Step 2, we may identify the conditions of the transition configuration of the partially decoupled 2-DOF $PR_+=$ PMs by making the legs of partially decoupled 2-DOF $PR_{+X}=$ (or $PR_{+Y}=$) PMs in the potential PM satisfy the conditions for legs of partially decoupled 2-DOF $PR_{+Y}=$ (or $PR_{+X}=$) PMs.

For example, in order for the potential partially decoupled 2-DOF $PR_+=$ PM in Fig. 23.8 to be a partially decoupled 2-DOF $PR_+=$ PM, the inactive R joint, $I_2$,

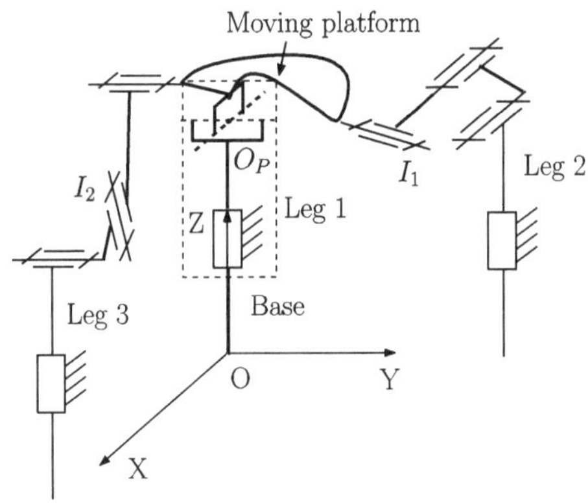


**Fig. 23.8** Type synthesis of potential partially decoupled 2-DOF $PR_+ =$ parallel mechanisms

in Leg 3 should be collinear with the axis of the $R_{+X}$ joint (see Fig. 23.6d) and the inactive R joint, $I_1$, in Leg 2 should be collinear with the axis of the $R_{+Y}$ joint (Fig. 23.6d). This leads to the partially decoupled 2-DOF $PR_+ =$ PM shown in Fig. 23.9a, the simplified version of which is shown in Fig. 23.9b.

Following the above procedure, a large number of partially decoupled 2-DOF $PR_+ =$ PMs have been obtained. Figures 23.9b, 23.10a, b show three $PR_+ =$ PMs that may have potential applications.

From the partially decoupled 2-DOF $PR_+ =$ PMs obtained using the above procedure, one may get their variations by releasing some or all of their redundant constraints. For example, we may release one redundant constraint in the partially decoupled 2-DOF $PR_+$-2-PRS $PR_+ =$ PM shown in Fig. 23.10b by replacing the $R_+$ joint with an S (spherical) joint whose center coincides with the center of the $R_+$ joint. Then we obtain a non-overconstrained partially decoupled 2-DOF PS-2-PRS $PR_+ =$ PM (Fig. 23.10c).

In a partially decoupled 2-DOF $PR_+ =$ PM, the two actuated joints are the P joints on the base except the one associated with the $PR_+$ virtual chain. Unlike the existing 2-DOF parallel mechanisms with two 1T1R operational modes in which three actuated joints are needed [6, 13], two actuated joints are used to control the motion of the moving platform in the PMs proposed in this paper.

Figure 23.11 shows the transition of a partially decoupled 2-DOF PS-2-PRS $PR_+ =$ PM between the $PR_{+X}$ operational mode (Fig. 23.11a) and $PR_{+Y}$ operational mode (Fig. 23.11e) through the transition configuration (Fig. 23.11c). In this PM, the two actuated joints are the P joints in the PRS legs. In the $PR_{+X}$ operational mode (Fig. 23.11a), the moving platform can translate along the Z-axis and rotate about a line that passes $O_p$ and is parallel to the X-axis. In the $PR_{+Y}$ operational mode (Fig. 23.11a), the moving platform can translate along the Z-axis and rotate about a line that passes $O_p$ and is parallel to the Y-axis. In a transition configuration, the PM is in a constraint singular configuration [2], and the moving platform has three instantaneous DOF, which can be represented by the translation along the Z-axis and two rotations permitted by both operational modes. To make

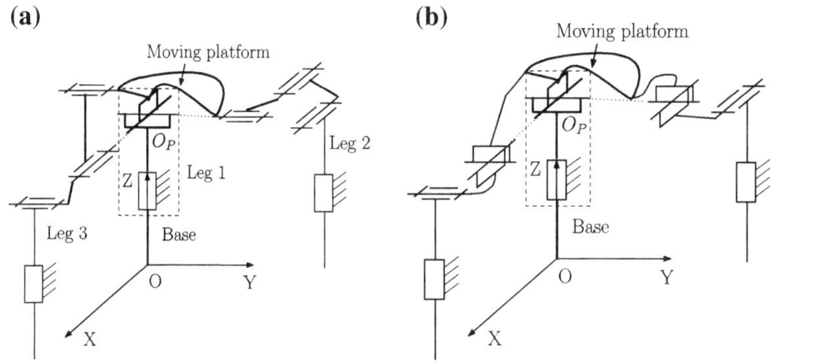


**Fig. 23.9** Type synthesis of partially decoupled 2-DOF $PR_+$ = parallel mechanisms (in transition configuration) **a** A $PR_+$=PM; **b** -2-PRU (Case 1)

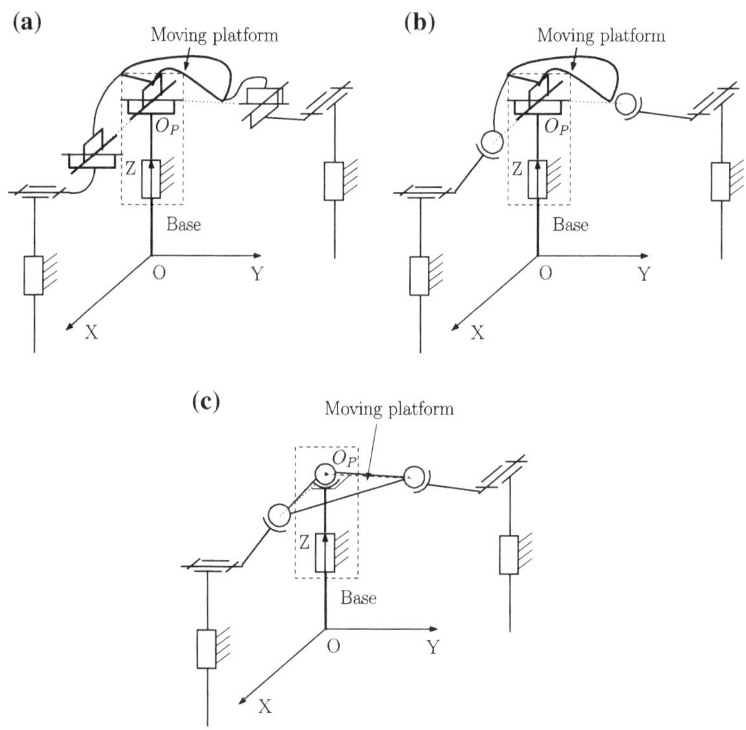


**Fig. 23.10** Types of partially decoupled 2-DOF $PR_+$= parallel mechanisms (in transition configuration) **a** $PR_+$-2-PRU; (Case 2) **b** $PR_+$-2-PRS; **c** PS-2-PRS

the PM move between a transition configuration and an operational mode, one needs to lock/release one R joint as indicated in (Figs. 23.11b, c) in the transition configuration.

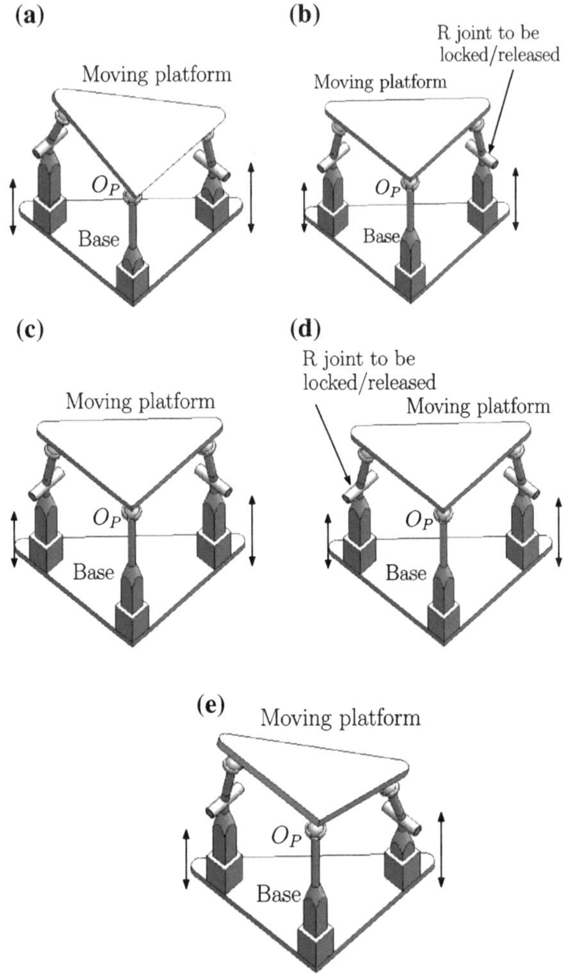


**Fig. 23.11** Reconfiguration of a partially decoupled 2-DOF PS-2-PRS $PR_+$ = parallel mechanism **a** $PR_{+X}$ operational mode; **b** $PR_{+X}$ operational mode ⇔ transition configuration; **c** Transition configuration; **d** Transition configuration ⇔ $PR_{+Y}$ operational mode; **e** $PR_{+Y}$ operational mode

## 23.5 Conclusions

An approach to the type synthesis of partially decoupled 2-DOF PMs with two 1T1R operational modes—partially decoupled 2-DOF $PR_+$ = PMs—has been proposed. Types of partially decoupled 2-DOF $PR_+$ = PMs have been obtained. Unlike the existing 2-DOF $PR_+$ = PMs in which three actuated joints are needed in each operational mode, two actuated joints are needed for a partially decoupled 2-DOF $PR_+$ = PM. The bi-mode revolute ($R_+$) joint defined in this paper and other variable kinematic joints [15] may provide an effective representation of motion patterns of PMwmoms.

This work provides a solid starting point for designing energy-efficient 2-DOF $PR_+ = PMs$. The approach proposed in this paper can also be readily used to generate coupled 2-DOF $PR_+ = PMs$, including all the types proposed in the literature.

**Acknowledgements** The author would like to thank the Engineering and Physical Science Research Council (EPSRC), United Kingdom, for the support under grant No. EP/I016333/1 and the Royal Society, United Kingdom, through an International Joint Project No. JP100715. Mr Andrew Johnson is acknowledged for creating the CAD model of Fig. 23.11.

## References

1. Hunt KH (1973) Constant-velocity shaft couplings: a general theory. ASME J Eng Ind 95:455–464
2. Zlatanov D, Bonev IA, Gosselin CM (2002) Constraint singularities as C-space singularities. Advances in robot kinematics—theory and applications, Kluwer Academic Publishers, The Netherlands, pp 183–192
3. Fanghella P, Galleti C, Gianotti E (2006) Parallel robots that change their group of motion. Advances in Robot Kinematics. Springer, New York, pp 49–56
4. Hernandez A, Altuzarra O, Pinto C, Amezua E (2008) Transitions in the velocity pattern of lower mobility parallel manipulators. Mech Mach Theory 43(6):738–753
5. Walter DR, Husty ML, Pfurner M (2008) The SNU 3-UPU parallel robot from a theoretical viewpoint. Proceedings of the second international workshop on fundamental issues and future research directions for parallel mechanisms and manipulators, Montpellier, France, pp151–158, 21–22 Sept 2008
6. Refaat S, Hervé JM, Nahavandi S, Trinh H (2007) Two-mode overconstrained three-DOFs rotational-translational linear-motor-based parallel kinematics mechanism for machine tool applications. Robotica 25:461–466
7. Kong X, Gosselin CM, Richard P-L (2007) Type synthesis of parallel mechanisms with multiple operation modes. ASME J Mech Des 129(7):595–601
8. Li Q, Herve JM (2009) Parallel mechanisms with bifurcation of Schoenflies motion IEEE Trans. Robot. 25(1):158–164
9. Gogu G (2011) Maximally regular T2R1-type parallel manipulators with bifurcated spatial motion. ASME J Mech Robot 3(1):011010
10. Kong X (2011) Type synthesis of 3-DOF parallel manipulators with both planar and translational operation modes. Proceedings of the ASME 2011 international design engineering technical conference & computers and information in engineering conference, DETC2011-48510, Washington, USA
11. Gan DM, Dai JS, Liao QZ (2009) Mobility analysis of two types of metamorphic parallel mechanisms. ASME J Mech Robot 1:041007-1-9
12. Borras J, Thomas F, Ottaviano E, Ceccarelli M (2009) A reconfigurable 5-DoF 5-SPU parallel platform. Proceedings of the 2009 ASME/IFToMM international conference on reconfigurable mechanisms and robots, London, UK, pp 617–623
13. Li Q, Hervé JM (2010) 1T2R parallel mechanisms without parasitic motion. IEEE Trans Robot 26(3):401–410
14. Kong X, Gosselin C (2007) Type Synthesis of parallel mechanisms. Springer, The Netherlands
15. Yan HS, Kuo CH (2006) Topological representations and characteristics of variable kinematic joints. ASME J Mech Des 128(2):384–391

# Chapter 24
# Reconfiguration and Actuation Scheme of 3rT*PS* Metamorphic Parallel Mechanisms with Parallel Constraint Screws

**Dongming Gan, Jian S. Dai, Jorge Dias and Lakmal Seneviratne**

**Abstract** This paper presents reconfiguration with mobility change of a class of metamorphic parallel mechanisms consisting of three reconfigurable rT*PS* limbs with parallel bracket rotation axes. Stemming from the reconfiguration of the reconfigurable Hooke (rT) joint, the rT*PS* limb has two phases with one having full mobility and the other giving a constraint force to the platform based on the constraint screws. The platform constraint screw systems show that the new metamorphic parallel mechanisms have four topologies by altering the limb phases with mobility change among 1R2T (one rotation with two translations), 2R2T, 3R2T and mobility 6. Special topologies are identified considering the limb arrangement. Following these, actuation scheme is discussed by covering all the topologies of the metamorphic parallel mechanisms based on constraint screws.

D. Gan (✉) · J. Dias · L. Seneviratne
Khalifa University of Science, Technology and Research,
127788 Abu Dhabi, UAE
e-mail: dongming.gan@kustar.ac.ae

J. Dias
e-mail: jorge.dias@kustar.ac.ae

L. Seneviratne
e-mail: lakmal.seneviratne@kcl.ac.uk

J. S. Dai · L. Seneviratne
King's College London, University of London, London,
WC2R2LS, UK
e-mail: jian.dai@kcl.ac.uk

J. Dias
University of Coimbra, Faculty of Science and Technology,
Coimbra, Portugal

J. S. Dai et al. (eds.), *Advances in Reconfigurable Mechanisms and Robots I*,
DOI: 10.1007/978-1-4471-4141-9_24, © Springer-Verlag London 2012

**Keywords** Reconfiguration · Mobility change · Constraint screw · Singularity · Parallel mechanism

## 24.1 Introduction

Metamorphic parallel mechanisms (MPMs) [1] are a class of mechanisms that possess adaptability and reconfigurability to change permanent finite mobility based on topological structure change. Metamorphic parallel mechanisms keep the advantages of traditional parallel mechanisms in terms of high load-carrying capacity, good positioning accuracy and low inertia but with ability of reconfiguration [2] and mobility change. This meets the requirements of rapid customization and diverse environment changing in parallel mechanism applications.

The metamorphic parallel mechanisms synthesized in this paper are based on a reconfigurable r*TPS* limb with two topology phases stemming from a reconfigurable Hooke (rT) joint [3]. Based on this rT joint, various metamorphic parallel mechanisms [1, 4] have been presented with a general construction method introduced using screw theory [5]. Using geometric constraints of the joints to change the constraint system of the mechanisms shows one way to obtain metamorphic mechanisms [6–9] while the other way is to change the number of links by link coincidence and geometric constraint [10–12].

Many methods have been proposed for mobility analysis of parallel mechanisms, such as screw theory [13–15] and linear transformation [16, 17]. As screws naturally present the relations of geometric constraints among the joints and limbs in parallel mechanisms, they show great advantage in describing the topology change and actuation analysis of metamorphic parallel mechanisms as illustrated in this paper.

## 24.2 Two Phases of the Reconfigurable r*TPS* Limb

The reconfigurable r*TPS* limb consists of a reconfigurable Hooke (rT) joint, a prismatic joint and a spherical joint. The reconfigurability of this limb stems from the configuration change of the rT joint which has two rotational degrees of freedom (DOFs) about two perpendicularly intersecting rotational axes (radial axis and bracket axis) as in Fig. 24.1. A grooved ring is used to house the radial axis and make it have the ability of altering its direction by rotating and fixing freely along the groove. This allows the radial rotation axis change with respect to the limb, resulting in two typical phases of the r*TPS* limb as in Fig. 24.1. While in Fig. 24.1a, the radial axis is perpendicular to the limb (prismatic joint) which is denoted as $(\mathrm{rT})_1 PS$, it is collinear with the limb (prismatic joint) passing through the spherical joint center in Fig. 24.1b and the limb phase is symbolized as $(\mathrm{rT})_2 PS$.

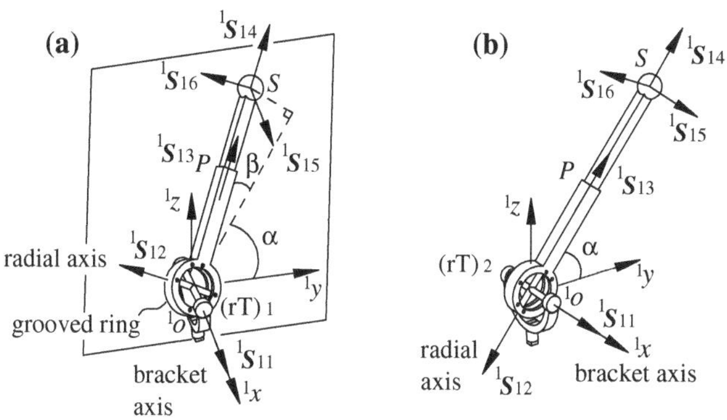


**Fig. 24.1** Two phases of the rT*PS* limb. **a** $(\mathrm{rT})_1PS$, **b** $(\mathrm{rT})_2PS$

Set a limb coordinate system $^1o^1x^1y^1z$ at the rT joint center with $^1x$ axis collinear with the bracket axis and $^1y$ axis perpendicular to the bracket surface as in Fig. 24.1a, the twist system of the $(\mathrm{rT})_1PS$ limb is given as:

$$\{^1\mathbf{S}_1\} = \left\{ \begin{array}{l} ^1\mathbf{S}_{11} = [1 \quad 0 \quad 0 \quad 0 \quad 0 \quad 0] \\ ^1\mathbf{S}_{12} = [0 \quad -s\alpha \quad c\alpha \quad 0 \quad 0 \quad 0] \\ ^1\mathbf{S}_{13} = [0 \quad 0 \quad 0 \quad -s\beta \quad c\beta c\alpha \quad c\beta s\alpha] \\ ^1\mathbf{S}_{14} = [0 \quad c\alpha \quad s\alpha \quad 0 \quad ls\beta s\alpha \quad -ls\beta c\alpha] \\ ^1\mathbf{S}_{15} = [1 \quad 0 \quad 0 \quad 0 \quad lc\beta s\alpha \quad -lc\beta c\alpha] \\ ^1\mathbf{S}_{16} = [0 \quad -s\alpha \quad c\alpha \quad lc\beta \quad ls\beta c\alpha \quad ls\beta s\alpha] \end{array} \right\} \tag{24.1}$$

where the first two twists are generated from the rT joint, the third is from the prismatic joint, and the last three are for the spherical joint. $\beta$ is the angle between the limb and its projection on plane $^1y^1o^1z$ passing through rT joint center and perpendicular to the bracket axis ($^1x$), $\alpha$ is the angle between axis $^1y$ and the limb projection on plane $^1y^1o^1z$. $l$ is the distance between the rT joint centre and the spherical joint centre. Labels $c\alpha$ and $s\alpha$ denote Cosin($\alpha$) and Sin($\alpha$) respectively. In the twist notation $^1\boldsymbol{S}_{ij}$, the first subscript $i$ denotes the limb number, the second subscript $j$ denotes the joint number within the limb and the leading superscript indicates the local frame.

The six screws in (24.1) form a six-system, showing that the $(\mathrm{rT})_1PS$ limb has mobility six and does not supply constraint to the platform connected to it.

For the $(\mathrm{rT})_2PS$ limb as in Fig. 24.1b, radial axis of the rT joint is collinear with the prismatic joint passing through the spherical joint center. Thus, the spherical joint center $A$ cannot move out of the plane $^1y^1o^1z$. Based on the coordinate system in Fig. 24.1b, the twist system of the $(\mathrm{rT})_2PS$ limb can be given as

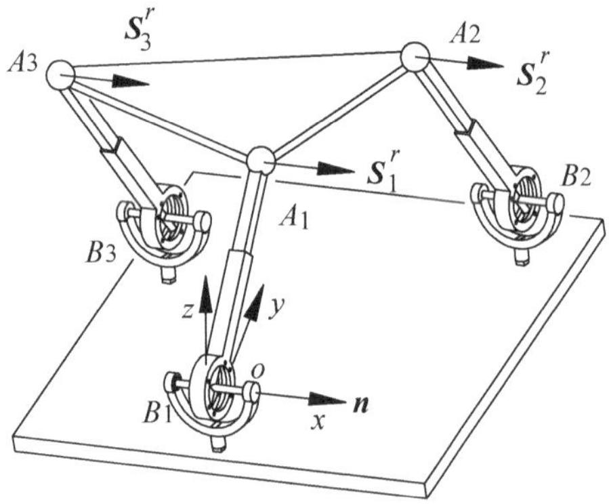


**Fig. 24.2** The 3(rT)$_2$*PS* with parallel constraint screws

$$
\{{}^1\mathbf{S}_1\} = \left\{ \begin{array}{l}
{}^1\mathbf{S}_{11} = [1 \quad 0 \quad 0 \quad 0 \quad 0 \quad 0] \\
{}^1\mathbf{S}_{12} = [0 \quad -c\alpha \quad -s\alpha \quad 0 \quad 0 \quad 0] \\
{}^1\mathbf{S}_{13} = [0 \quad 0 \quad 0 \quad 0 \quad c\alpha \quad s\alpha] \\
{}^1\mathbf{S}_{14} = [0 \quad c\alpha \quad s\alpha \quad 0 \quad 0 \quad 0] \\
{}^1\mathbf{S}_{15} = [1 \quad 0 \quad 0 \quad 0 \quad ls\alpha \quad -lc\alpha] \\
{}^1\mathbf{S}_{16} = [0 \quad -s\alpha \quad c\alpha \quad l \quad 0 \quad 0]
\end{array} \right\} \tag{24.2}
$$

It is clear that twists ${}^1\boldsymbol{S}_{12}$ and ${}^1\boldsymbol{S}_{14}$ are the same and the six twists form a five-system. Thus, there is one reciprocal screw in the limb constraint system as

$$
\{{}^1\mathbf{S}_1^r\} = {}^1\mathbf{S}_1^r = [1 \quad 0 \quad 0 \quad 0 \quad ls\alpha \quad -lc\alpha] \tag{24.3}
$$

This gives a constraint force acting along a line passing through the spherical joint centre with a direction parallel to the bracket axis of the rT joint. Thus, the (rT)$_2$*PS* limb has five degrees of freedom (DOFs), one less than the (rT)$_1$*PS* phase.

When constructing parallel mechanisms with the rT*PS* limbs, the mechanisms will have ability of mobility change by altering the rT*PS* limbs into these two phases. In the following, the (rT)$_2$*PS* limb phases will be used by connecting the rT joints to the base with parallel bracket axis and the spherical joints to the moving platform. Then altering the limbs into (rT)$_1$*PS* phases will generate new mechanism topologies, which gives a class of metamorphic parallel mechanisms (MPMs) with parallel constraint screws.

## 24.3 3rT*PS* MPMs with Parallel Constraint Screws

### 24.3.1 Mobility Analysis

From (24.3), it can be seen that the $(\mathrm{rT})_2PS$ limb gives one constraint force to the platform with the direction parallel to its bracket axis. Thus, by arranging the direction of the bracket axis of the rT joint, the constraint to the platform can be represented. In this paper, the rT joints are connected to the base with parallel bracket axes, leading to parallel constraint forces expressed by screws on the platform as in Fig. 24.2.

Let points $A_i$ and $B_i$ denote the spherical joint center and the rT joint center in limb $i$ ($i = 1, 2, 3$) respectively. Locate a global coordinate system $oxyz$ at the center ($A_1$) of the rT joint in limb 1 with $x$ axis collinear with the bracket axis and $y$ axis perpendicular to the bracket plane. Let $\boldsymbol{a}_i$ and $\boldsymbol{b}_i$ denote the vectors of points $A_i$ and $B_i$ in the coordinate system $oxyz$, $l_i$ be the distance between the spherical joint center $A_i$ and the rT joint center $B_i$.

The constraint system of the $3(\mathrm{rT})_2PS$ parallel mechanism in Fig. 24.2 can be given as:

$$\{\mathbf{S}^r\} = \left\{ \begin{array}{l} \mathbf{S}_1^r = [\mathbf{n} \quad \mathbf{a}_1 \times \mathbf{n}] \\ \mathbf{S}_2^r = [\mathbf{n} \quad \mathbf{a}_2 \times \mathbf{n}] \\ \mathbf{S}_3^r = [\mathbf{n} \quad \mathbf{a}_3 \times \mathbf{n}] \end{array} \right\} = \left\{ \begin{array}{llllll} [1 & 0 & 0 & 0 & l_1 s\alpha_1 & -l_1 c\alpha_1], \\ [1 & 0 & 0 & 0 & b_{2z} + l_2 s\alpha_2 & -b_{2y} - l_2 s\alpha_2], \\ [1 & 0 & 0 & 0 & b_{3z} + l_3 s\alpha_3 & -b_{3y} - l_3 s\alpha_3] \end{array} \right\} \tag{24.4}$$

where $\boldsymbol{n} = (1,0,0)$ is the direction of the bracket axes, $\boldsymbol{a}_i = \boldsymbol{b}_i + l_i(0, c\alpha_i, s\alpha_i) = (b_{ix}, b_{iy}, b_{iz}) + l_i(0, c\alpha_i, s\alpha_i)$, $\alpha_i$ is the angle between limb $i$ and the line passing through rT joint center $B_i$ with direction parallel to $y$ axis.

Equation (24.4) gives three parallel constraint forces which constrain the translation along $x$-axis and rotations about $y$ and $z$ axes. Thus, the $3(\mathrm{rT})_2PS$ parallel mechanism with parallel constraint screws has three DOFs with one rotation about $x$-axis and two translations along $y$ and $z$ axes.

### 24.3.2 Special Topologies

When considering particular arrangements of the base and platform joints, the $3(\mathrm{rT})_2PS$ parallel mechanism will have different topologies and some special cases are identified below.

A special one is that the three spherical joint centers ($A_1$, $A_2$ and $A_3$) on the platform are in line as in Fig. 24.3a. The last two constraint screws in (24.4)

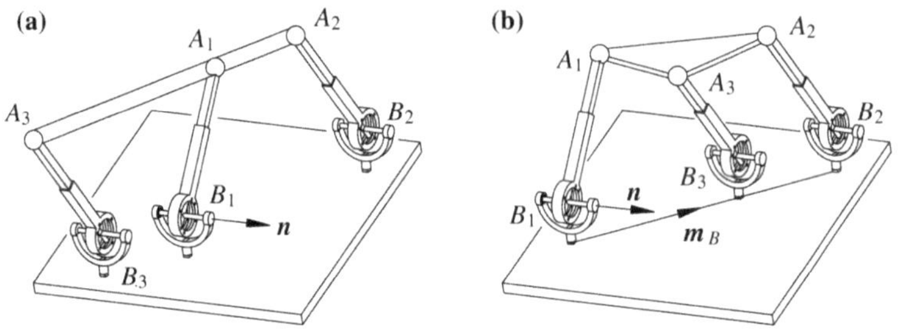


**Fig. 24.3** Two special topologies. **a** 3(rT)$_2$*PS*-a with $A_1$, $A_2$, $A_3$ inline, **b** 3(rT)$_2$*PS*-b with $B_1$, $B_2$, $B_3$ inline

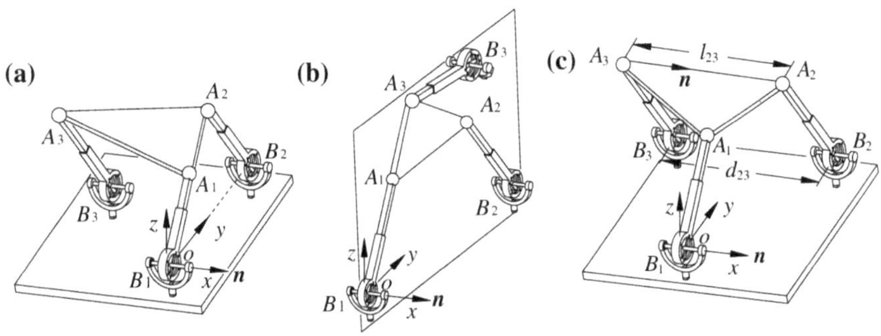


**Fig. 24.4** Three more special topologies. **a** 3(rT)$_2$*PS*-c, **b** 3(rT)$_2$*PS*-d, **c** 3(rT)$_2$*PS*-e

constrain the same rotation between line $A_1A_2$ and norm $\boldsymbol{n}$. One of limb 2 and limb 3 becomes redundant, resulting in a local rotational DOF about line $A_1A_2$ and the mechanism has two translations and two rotations.

Another special topology is that the three rT joint centers ($B_1$, $B_2$ and $B_3$) on the base are in line with unit vector $\boldsymbol{m}_B$ as in Fig. 24.3b, the platform still has two translations and one rotation. However, when $\boldsymbol{m}_B = \boldsymbol{n}$, the three prismatic joints cannot be chosen as actuations at the same time.

Two other special topologies of the 3(rT)$_2$*PS* with parallel constraint screws can be obtained when locating the rT joint center $B_2$ on *yoz* plane with $b_{2x} = 0$ as in Fig. 24.4a and setting both rT joint centers ($B_2$ and $B_3$) on *yoz* plane with $b_{2x} = b_{3x} = 0$ as in Fig. 24.4b. Both two topologies have two translations and one rotation and the topology in Fig. 24.4b is a planar parallel mechanism.

One more special topology can be obtained by setting line $B_iB_j$ parallel to $\boldsymbol{n}$ with distance $l_{ij}$ between two spherical joint centers $A_iA_j$ equal to the distance $d_{ij}$ between two rT joint centers $B_iB_j$ as in Fig. 24.4c in which $l_{23} = d_{23}$. In this case, the two constraint screws in limb 2 and limb 3 are collinear and dependent. The platform will have two independent constraint screws with four reciprocal motion screws including two rotations and two translations. However, the geometric constraint $l_{23} = d_{23}$ makes line $A_2A_3$ constantly parallel with $x$ axis and the

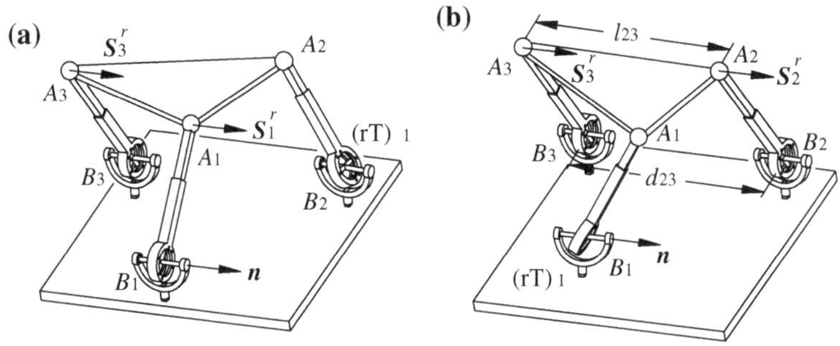


**Fig. 24.5** The $2(\mathrm{rT})_2PS$-$1(\mathrm{rT})_1PS$ topologies. **a** general case (4DOFs-2R2T), **b** $d_{23} = l_{23}$ (3DOFs-1R2T)

platform can only rotate about line $A_2A_3$. Thus, the mechanism still has one rotation and two translation DOFs, which is a structure singularity case.

## 24.4 Topology Reconfiguration

Altering the $(\mathrm{rT})_2PS$ limbs in the previous $3(\mathrm{rT})_2PS$ parallel mechanisms into the phase $(\mathrm{rT})_1PS$ will result in various new mechanism topologies with increased mobility. After changing the phase of one limb, all the $3(\mathrm{rT})_2PS$ parallel mechanisms become the topology $2(\mathrm{rT})_2PS$-$1(\mathrm{rT})_1PS$ in Fig. 24.5a that has two parallel constraint screws following the first two in (24.4). One constraint less makes the $2(\mathrm{rT})_2PS$-$1(\mathrm{rT})_1PS$ one more DOF than the $3(\mathrm{rT})_2PS$. Based on the constraint screw analysis, the $2(\mathrm{rT})_2PS$-$1(\mathrm{rT})_1PS$ parallel mechanism has four DOFs with two translations and two rotations (2R2T).

A special case is the $3(\mathrm{rT})_2PS$-e in Fig. 24.4c. When changing limb 2 or limb 3 from phase $(\mathrm{rT})_2PS$ to $(\mathrm{rT})_1PS$ in Fig. 24.4c, the mechanism changes to the case in Fig. 24.5a. However, when changing the phase of limb 1, the mechanism becomes topology $2(\mathrm{rT})_2PS$-$1(\mathrm{rT})_1PS$ in Fig. 24.5b which has mobility three with one rotation and two translations (1R2T). Thus, the mechanism does not change mobility due to the structure singularity.

When further changing one more limb phases, both the topologies in Fig. 24.5 change to the same topology $1(\mathrm{rT})_2PS$-$2(\mathrm{rT})_1PS$ as in Fig. 24.6a, which has one constraint screw that limited the translation along $\boldsymbol{n}$. Thus, this mechanism has five DOFs with three rotations and two translations (3R2T). It can be noticed that the $2(\mathrm{rT})_2PS$-$1(\mathrm{rT})_1PS$ in Fig. 24.5b changes its mobility from 3 to 5 in this reconfiguration.

When changing the third limb to phase $(\mathrm{rT})_1PS$, the mechanism becomes another topology $3(\mathrm{rT})_1PS$ as in Fig. 24.6b that does not have any constraint screw and has full mobility 6.

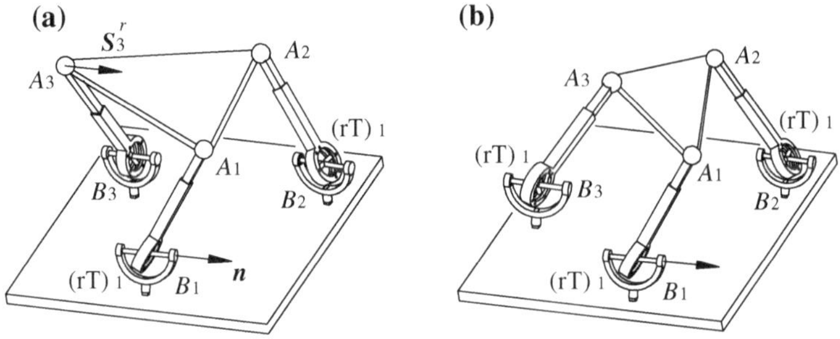


**Fig. 24.6** Topologies $1(\mathrm{rT})_2PS$-$2(\mathrm{rT})_1PS$ and $3(\mathrm{rT})_1PS$. **a** $1(\mathrm{rT})_2PS$-$2(\mathrm{rT})_1PS$ (5DOFs-3R2T), **b** $3(\mathrm{rT})_1PS$ (6DOFs)

One topology that should be mentioned is the $3(\mathrm{rT})_2PS$-d in Fig. 24.4b, in which all three limbs are constrained in one plane and it is a planar parallel mechanism. By changing the limb phases one by one, the topology becomes that in Fig. 24.5a and the two in Fig. 24.6. Thus, a planar parallel mechanism becomes a spatial parallel mechanism while the mobility changes from 3 to 6.

## 24.5 Actuation Scheme for the Reconfigurable Topologies

The actuation inputs of a parallel mechanism work with limb constraints to determine the platform position and orientation. In the $(\mathrm{rT})_2PS$ limb, the rotation about the bracket axis or the translation along the prismatic joint can be used as inputs. Based on Fig. 24.1b and the limb motion screws in (24.1), these two inputs can result in constraint screws as:

$$\begin{cases} {}^1\mathbf{S}^r_{rTba} = \begin{bmatrix} 0 & -s\alpha & c\alpha & l & 0 & 0 \end{bmatrix} \\ {}^1\mathbf{S}^r_{Pa} = \begin{bmatrix} 0 & c\alpha & s\alpha & 0 & 0 & 0 \end{bmatrix} \end{cases} \tag{24.5}$$

where ${}^1\mathbf{S}^r_{rTba}$, a constraint force passing through the spherical joint center and perpendicular to both the bracket axis and the limb, is the input constraint from the bracket axis rotation. ${}^1\mathbf{S}^r_{Pa}$, a constraint force along the limb, is the input constraint from the prismatic joint.

The $3(\mathrm{rT})_2PS$ with parallel constraint screws has three DOFs and needs three actuation inputs. Based on (24.5), each limb gives two possible actuations and there are six for the three limbs. Taking three of the six actuations should constrain the three DOFs of the mechanism. For the $3(\mathrm{rT})_2PS$ parallel mechanism in Fig. 24.2, when choosing two prismatic joint inputs in limb 1 and limb 2 with a bracket rotation input in limb 3, the input constraints for the platform are:

$$\{\mathbf{S}_a^r\} = \left\{ \begin{array}{l} \mathbf{S}_{1Pa}^r = [\mathbf{u}_1 \quad 0] \\ \mathbf{S}_{2Pa}^r = [\mathbf{u}_2 \quad \mathbf{b}_2 \times \mathbf{u}_2] \\ \mathbf{S}_{3rTba}^r = [\mathbf{n} \times \mathbf{u}_3 \quad \mathbf{a}_3 \times (\mathbf{n} \times \mathbf{u}_3)] \end{array} \right\} \tag{24.6}$$

where $\boldsymbol{u}_i$ is the unit vector along limb $i$.

The three input forces with the three geometric constraint forces in (24.4) form a six-system, showing that the selected actuation inputs can fix the required position and orientation of the platform. Following this way, actuations for the 3(rT)$_2$*PS* parallel mechanisms including the special topologies can be concluded as:

(a) Any three of the six inputs (the prismatic translation and bracket rotation in each limb) can be actuations for the general 3(rT)$_2$*PS* in Fig. 24.2, the 3(rT)$_2$*PS*-a, 3(rT)$_2$*PS*-b, 3(rT)$_2$*PS*-c and 3(rT)$_2$*PS*-d except the case that the three rT joint centers are on $x$ axis of the 3(rT)$_2$*PS*-b which cannot choose three prismatic joints due to the constraint screw redundancy with a rotation about $x$-axis left;
(b) The 3(rT)$_2$*PS*-a needs three from the six inputs with an extra for the local rotation of the platform;
(c) At least one of the three inputs should be from limb 2 for actuations of the 3(rT)$_2$*PS*-e parallel mechanism.

When altering the (rT)$_2$*PS* limb to phase (rT)$_1$*PS*, one more actuation choice appears from the rotation about the radial axis with an input constraint force passing through the spherical joint center and parallel to the bracket axis as described by (24.3). Based on this and the above analysis, the actuation scheme rule is:

(a) At least one of the actuations should be from the (rT)$_1$*PS* limb in the reconfigured topologies 2(rT)$_2$*PS*-1(rT)$_1$*PS* in Fig. 24.5a, 1(rT)$_2$*PS*-2(rT)$_1$*PS* and 3(rT)$_1$*PS* in Fig. 24.6;
(b) At least two should be from the (rT)$_1$*PS* limb in the reconfigured topology 2(rT)$_2$*PS*-1(rT)$_1$*PS* with structure singularity in Fig. 24.5b.

## 24.6 Conclusion

This paper introduced a class of metamorphic parallel mechanisms consisting of three reconfigurable rT*PS* limbs with parallel bracket axes. Based on screw theory, mobility of the rT*PS* limb in its two phases was analyzed while in one phase it had mobility six and in the other it had one DOF less, providing a constraint force to the platform. Using this property, the 3rT*PS* metamorphic parallel mechanisms demonstrated reconfiguration and mobility change by altering the phases of the r*TPS* limbs, which was illustrated in constraint screws. Some special topologies were investigated, in which one topology showed mobility less than normal case due to structure singularity and another one could change the platform motion from planar to spatial. Following this, the actuation constraint screws were calculated to select possible actuations for the various topologies of the parallel mechanisms. The results showed that the rotations about the bracket axis and the radial axis, the translation of the prismatic joint can be the actuation inputs while one should be chosen from the $(rT)_1PS$ limbs.

## References


1. Gan DM, Dai JS, Liao QZ (2009) Mobility analysis of two types of metamorphic parallel mechanisms. ASME J Mech Rob 1:041007_1–9
2. Fanghella P, Galletti C, Giannotti E (2006) Parallel robots that change their group of motion. In: Lenarčič J, Parenti-Castelli V (eds) Advances in robot kinematics: mechanisms and motion. Springer, Dordrecht, pp 49–56
3. Gan DM, Dai JS (2011) China Patent, Application Number 201110054834.7
4. Gan DM, Dai JS, Liao QZ (2010) Constraint analysis on mobility change of a novel metamorphic parallel mechanism. Mech Mach Theory 45(12):1864–1876
5. Gan DM, Dai JS, Caldwell DG (2011) Constraint-based limb synthesis and mobility-change aimed mechanism construction. ASME J Mech Des 133(5):051001_ 1–9
6. Kuo CH, Yan HS (2007) On the mobility and configuration singularity of mechanisms with variable topologies. ASME J Mech Des 129:617–624
7. Zhang KT, Dai JS, Fang YF (2010) Topology and constraint analysis of phase change in the metamorphic chain and its evolved mechanism. ASME J Mech Des 132(12):121001(1–11)
8. Grosch P, Di Gregorio R, López J, Thomas F (2010) Motion planning for a novel reconfigurable parallel manipulator with lockable revolute joints. In: Proceedings of the 2010 IEEE international conference on robotics and automation, IEEE Press, Anchorage, Alaska, 3–8 May 2010
9. Grosch P, Di Gregorio J, Thomas F (2010) Generation of under-actuated manipulators with non-holonomic joints from ordinary manipulators. ASME J Mech Robotics 2(1):11005–11012
10. Dai JS, Rees JJ (1999) Mobility in metamorphic mechanisms of foldable/erectable kinds. ASME J Mech Des 121(3):375–382
11. Leonesio M, Bianchi G, Manara P (2007) A general approach for self-locking analysis in closed kinematic chains. In: Proceedings of the 12th world congress in mechanism and machine theory, Besancon, France, June 2007, pp 141–147
12. Dai JS, Wang D (2007) Geometric analysis and synthesis of the metamorphic robotic hand. ASME J Mech Des 129(11):1191–1197
13. Huang Z, Li QC (2002) General methodology for the type synthesis of lower-mobility symmetrical parallel manipulators and several novel manipulators. Int J Robot Res 21(2):131–145
14. Kong XW, Gosselin CM (2007) Type synthesis of parallel mechanisms. Springer, Germany
15. Dai JS, Huang Z, Lipkin H (2006) Mobility of overconstrained parallel mechanisms. ASME J Mech Des 128:220–229
16. Gogu G (2008) Structural synthesis of parallel robots, Part 1: methodology. Springer, Dordrecht
17. Gogu G (2012) T2R1-type parallel manipulators with bifurcated planar-spatial motion. Eur J Mech/A-Solids 33:242–269

# Chapter 25
# Topological Structure Synthesis of 3-Translation Parallel Mechanisms

**Huiping Shen, Tongzhu Yu, Jiaming Deng, Zheng Wang and Tingli Yang**

**Abstract** Using topological structure synthesis method based on the Position and Orientation Characteristic (POC) set, 26 kinds of 3-translation parallel mechanisms are synthesized, and 19 of which are presented for the first time. This method may be applicable for both topological structure synthesis of non-over-constrained and generally over-constrained mechanisms.The geometric and physical meaning of this method is clear and easy to understand for designers.

**Keywords** Parallel mechanisms · 3-translation · POC set · Topological structure

## 25.1 Introduction

In recent years, mechanism researchers were focusing on topological structure synthesis of parallel mechanisms which is an open challenging problem. There are three kinds of main existing methods, i.e. (1) the screw theory based method [1–3]; (2) the displacement subgroup based method [4, 5]; (3) the POC set based method [6–9], includes following characteristics: (1) the synthesized mechanisms are non-instantaneous one; and (2) simple mathematical tools have been used. Its geometric meaning and physical meaning are easy for designers to understand and use.

3-translation robots has a broad application in modern manufacturing such as rapidly assembling, gelatinizing, assembling, conveying, detecting, packaging and many other fields, of which Delta robot is the most successful application. From the point of view of controlling, decoupling is essential for the input and output.

H. Shen (✉) · T. Yu · J. Deng · Z. Wang · T. Yang
Institute for Robot, Changzhou University, Changzhou, Jiangsu, China
e-mail: shp65@126.com

J. S. Dai et al. (eds.), *Advances in Reconfigurable Mechanisms and Robots I*,
DOI: 10.1007/978-1-4471-4141-9_25, © Springer-Verlag London 2012

So it is of vital academic significance and practical value to design 3-translation parallel mechanisms with decoupled input and output.

In this paper, 26 kinds of 3-translation parallel mechanisms are synthesized by the POC set based method, including seven kinds of previously designed mechanisms and nineteen kinds of new mechanisms. The POC set method is based on the units of Single-Opened-Chain (SOC) method [6, 8], which will become more systemic and integral.

## 25.2 General Process for the Topological Structure Synthesis of Parallel Mechanism

Theory and method for the structure synthesis based on POC set are described in the document [6]. Here, there only gives the main steps as follows.

*Step 1*. Determine the POC set of parallel mechanism

Determine the expected POC set of parallel mechanisms.

*Step 2*. Synthesis of the branch structures

POC equation for serial mechanisms:

$$M_s = \bigcup_{i=1}^{m} M_{J_i} \tag{25.1}$$

where, $M_s$ — POC set of the end link relative to the frame link; $M_{J_i}$ — POC set of the $i$th kinematic pair.

Synthesize the branch structure according to Eq. (25.1).

*Step 3*. Determine the scheme of branch combination

Here lists all possible combination types of branch which satisfies POC equation.

*Step 4*. Determine the geometric condition of branch assembled between the two platforms

POC equation for parallel mechanisms:

$$M_{Pa} = \bigcap_{j=1}^{v+1} M_{b_j} \tag{25.2}$$

where, $M_{Pa}$ — POC set of parallel mechanisms; $M_{b_j}$ — POC set of the $j$th branch.

This determines the geometric condition of branches assembled between the two platforms.

*Step 5*. Examine DOF

According to Eqs. (25.3) and (25.4), examine DOF of mechanism:

$$F = \sum_{i=1}^{m} f_i - \sum_{j=1}^{v} \xi_{L_j} \tag{25.3}$$

$$\xi_{L_j} = \dim \cdot \left\{ \left( \bigcap_{i=1}^{j} M_{b_i} \right) \bigcup M_{b_{(j+1)}} \right\} \tag{25.4}$$

Where, $F$ — DOF of parallel mechanisms; $f_i$ — DOF of the $i$th kinematic pair; $m$ — Number of kinematic pair; $v$ — Number of independent loop; $M_{bi}$ — POC set of the $i$th branch; $M_{b_{(j+1)}}$ — POC set of the ($j$ + $1$)th branch. $\xi_{L_j}$ — Number of independent finite elements of the $j$th loop;

*Step 6*. Determine inactive pairs of parallel mechanisms

Making sure if there is an inactive pair according to the inactive pair of decision rules. If exists, it should be deleted. Then examines the POC set of the movable platform of parallel mechanisms again.

*Step 7*. Chose input pairs of parallel mechanisms

Chose input pairs of parallel mechanisms based on input pair of decision rules.

*Step 8*. Determine parallel mechanisms topological structure and draw a sketch

According to the steps above, determines parallel mechanisms topological structure, including: (a) topological structure and combination types of the branch; (b) topological structure of the movable platform and the base; (c) the input pairs.

*Step 9*. Characteristics analysis of parallel mechanisms

(1) BKC type and coupling-degree: Determines BKC type and the coupled-degree value of mechanism according to the judgment method of BKC.
(2) Decoupling: Determines the input and output decoupling of mechanisms based on decoupling principle of topological structure.
(3) Topological structure characteristic: Determines the characteristic parameters of topological structure, which provides the basis for the optimization of structure.

## 25.3 Topological Structure Synthesis of 3-Translation Parallel Mechanisms

*Step 1*. Determine the expected POC equation for parallel mechanisms: $M_{Pa}$

The expected POC equation for 3-translation parallel mechanisms can be expressed as

$$M_{Pa} = \begin{bmatrix} t^3 \\ r^0 \end{bmatrix} \tag{25.5}$$

where, $t^3$ — 3 independent finite translation characteristics; $r^0$ — 0 independent finite rotation characteristics.

*Step 2*. Synthesis of branch structure

(1) POC set of branch According to the POC method $M_{b_i} \supseteq M_{Pa}$, POC set which satisfies the above POC equation for parallel mechanisms can be selected as below:

**Table 25.1** Structures type with simple branch (*SOC*) and complex branch (*HSOC*)

| $M_{bi}$ | SOC | HSOC |
|---|---|---|
| $\begin{bmatrix} t^3 \\ r^0 \end{bmatrix}$ | 1 $SOC\{-P-P-P-\}$ | 9 $HSOC\{-P//R//\diamondsuit P^{(4R)}//R-\}$ |
| $\begin{bmatrix} t^3 \\ r^1 \end{bmatrix}$ | 2 $SOC\{-R//R//R-P-\}$ | 10 $HSOC\{-P//\diamondsuit P^{(4U)}//R^{(4U)}//R-\}$ |
| | 3 $SOC\{-R//R-P-P-\}$ | |
| | 4 $SOC\{-R-P-P-P-\}$ | 11 $HSOC\{-P-\diamondsuit P^{(4S)}-\diamondsuit P^{(4S)}-R^{(4S)}-R^{(4S)}//R-\}$ |
| $\begin{bmatrix} t^3 \\ r^2 \end{bmatrix}$ | 5 $SOC\{-R//R//R-R//R-\}$ | 12 $HSOC\{-\diamondsuit(P^{(5R)}-P^{(5R)})\perp R\perp R//R-\}$ |
| | 6 $SOC\{-R//R-P-R//R-\}$ | |
| | 7 $SOC\{-R-P-P-R//R-\}$ | 13 $HSOC\{-\diamondsuit(P^{(6R)}-P^{(6R)})\perp R^{(6R)}\perp R//R-\}$ |
| $\begin{bmatrix} t^3 \\ r^3 \end{bmatrix}$ | 8 $SOC\{-S-P-S-\}$ | |

$$M_{b_i} = \begin{bmatrix} t^3 \\ r^0 \end{bmatrix}, \begin{bmatrix} t^3 \\ r^1 \end{bmatrix}, \begin{bmatrix} t^3 \\ r^2 \end{bmatrix}, \begin{bmatrix} t^3 \\ r^3 \end{bmatrix},$$

(2) Structure type of branch
Satisfy POC sets of four kinds of branches above, the structure type of the corresponding with simple branch (*SOC*) and complex branch (*HSOC*) as shown in Table 25.1.

*Step 3*. The scheme of branch combination

We can synthesize the scheme of combination as shown in Table 25.2 according to the simple branch (*SOC*) and complex branch (*HSOC*) in Table 25.1.

For example, selecting the first scheme, A1, in the Table 25.2, as the three branches have identical topological structure, i.e. $3-SOC\{-R_{i1}//R_{i2}//R_{i3}-P_i-\}(i=1,2,3)$.

*Step 4*. Determine the geometric condition of branch assembled between two platforms

According to the expected POC equation for parallel mechanisms, its DOF is 3.

(1) Determine the POC equation for each branch
According to the POC equation for serial mechanisms [6], POC equations for the three branches are independent:

$$M_{b_i} = \begin{bmatrix} t^3 \\ r^1 \end{bmatrix} \quad (i = 1, 2, 3)$$

(2) Establish the POC equation for parallel mechanisms
Substitute POC equation for each branch's end link into the POC equation for parallel mechanism:

$$M_{Pa} = \begin{bmatrix} t^3 \\ r^0 \end{bmatrix} \Leftarrow \begin{bmatrix} t^3 \\ r^1 \end{bmatrix} \cap \begin{bmatrix} t^3 \\ r^1 \end{bmatrix} \cap \begin{bmatrix} t^3 \\ r^1 \end{bmatrix}$$

**Table 25.2** All possible branch combination forming parallel mechanisms

| A type: all branches of structure are identical | | B type: partial branches of structure are identical | |
|---|---|---|---|
| A1 | $3-SOC\{-R//R//R-P-\}$ | B1 | $2-SOC\{-R//R//R-P-\}$<br>$1-SOC\{-R//R//R-R//R-\}$ |
| A2 | $3-SOC\{-R//R//R-R//R-\}$ | B2 | $2-SOC\{-R//R//R-P-\}$<br>$1-SOC\{-R//R-P-R//R-\}$ |
| A3 | $3-SOC\{-R//R-P-R//R-\}$ | B3 | $2-SOC\{-R//R//R-R//R-\}$<br>$1-SOC\{-R//R-P-R//R-\}$ |
| A4 | $3-HSOC_1\{-P//R-\Diamond P^{(4R)}//R-\}$ | B4 | $2-SOC\{-R//R-P-R//R-\}$<br>$1-SOC\{-R//R//R-P-\}$ |
| A5 | $3-HSOC_2\{-P//\Diamond P^{(4U)}//\Diamond P^{(4U)}//R-\}$ | B5 | $2-SOC\{-R//R//R-P-\}$<br>$1-SOC\{-S-P-S-\}$ |
| A6 | $3-HSOC_3\{-P//\Diamond P^{(4S)}//\Diamond P^{(4S)}//\Diamond R^{(4S)}//\Diamond R^{(4S)}//R-\}.$ | B6 | $2-SOC\{-R//R-P-R//R-\}$<br>$1-SOC\{-R//R//R-R//R-\}$ |
| C type: every branches of structure are different | | B7 | $2-SOC\{-R//R//R-P-\}$<br>$1-HSOC\{-P//R-\Diamond P^{(4R)}//R-\}$ |
| C1 | $1-SOC\{-R//R//R-P-\}$<br>$1-HSOC\{-\Diamond(P^{(5R)}-P^{(5R)})\perp R\perp R//R-\}$ | B8 | $2-SOC\{-R//R//R-R//R-\}$<br>$1-HSOC\{-P//R-\Diamond P^{(4R)}//R-\}$ |
| C2 | $1-SOC\{-R//R//R-R//R-\}$<br>$1-HSOC\{-\Diamond(P^{(5R)}-P^{(5R)})\perp R\perp R//R-\}$ | B9 | $2-SOC\{-R//R-P-R//R-\}$<br>$1-HSOC\{-P//R-\Diamond P^{(4R)}//R-\}$ |
| C3 | $1-SOC\{-R//R-P-R//R-\}$<br>$1-HSOC\{-\Diamond(P^{(5R)}-P^{(5R)})\perp R\perp R//R-\}$ | B10 | $2-SOC\{-R//R//R-P-\}$<br>$1-HSOC\{-P//\Diamond P^{(4U)}//\Diamond P^{(4U)}//R-\}$ |
| C4 | $1-SOC\{-S-P-S-\}$<br>$1-HSOC\{-\Diamond(P^{(5R)}-P^{(5R)})\perp R\perp R//R-\}$ | B11 | $2-SOC\{-R//R//R-R//R-\}$<br>$1-HSOC\{-P//\Diamond P^{(4U)}//\Diamond P^{(4U)}//R-\}$ |
| C5 | $1-SOC\{-R//R//R-P-\}$<br>$1-HSOC\{-\Diamond(P^{(6R)}-P^{(6R)})\perp R^{(6R)}\perp R//R-\}$ | B12 | $2-SOC\{-R//R-P-R//R-\}$<br>$1-HSOC_2\{-P//\Diamond P^{(4U)}//\Diamond P^{(4U)}//R-\}$ |
| C6 | $1-SOC\{-R//R//R-R//R-\}$<br>$1-HSOC\{-\Diamond(P^{(6R)}-P^{(6R)})\perp R^{(6R)}\perp R//R-\}$ | | |
| C7 | $1-SOC\{-R//R-P-R//R-\}$<br>$1-HSOC\{-\Diamond(P^{(6R)}-P^{(6R)})\perp R^{(6R)}\perp R//R-\}$ | | |
| C8 | $1-SOC\{-S-P-S-\}$<br>$1-HSOC\{-\Diamond(P^{(6R)}-P^{(6R)})\perp R^{(6R)}\perp R//R-\}$ | | |

The symbol "$\Leftarrow$" in the equation means that the POC set of the 3-translation parallel mechanism is the expected goal of the intersection operation of the POC set of branches.

(3) Determine the geometric condition of branches assembled on the movable platform

According to the inverse operation of the intersection operation of the POC set, in order to achieve three translation outputs of the movable platform, the geometric condition of branch assembled on the movable platform is only that pair axes of $P_{14}$, $P_{24}$ and $P_{34}$ are allocated arbitrarily in space, then the equation above becomes

$$M_{Pa} = \begin{bmatrix} t^3 \\ r^0 \end{bmatrix} \Leftarrow \begin{bmatrix} t^3 \cap t^3 \cap t^3 \\ r^1(R_{11}) \cap r^1(R_{21}) \cap r^1(R_{31}) \end{bmatrix}$$

(4) Determine the geometric condition of branches assembled on the base
According to the inverse operation of the intersection operation of the POC set, in order for the intersection operation of the POC set between the first branch and the second branch can reduce a rotation element. So the geometric condition of two branches assembled on the base should be that pair axes of R on the base are not parallel, (that is $R_{11} \not\parallel R_{21}$), then the equation above becomes

$$M_{Pa} = \begin{bmatrix} t^3 \\ r^0 \end{bmatrix} \Leftarrow \begin{bmatrix} t^3 \\ r^0 \cap r^1(//R_{31}) \end{bmatrix}$$

According to the inverse operation of the intersection operation of the POC set, assembling with the third branch to reduce a rotation element of the equation above. So the geometric condition of the third branch assembled on the base is that the rotation direction of $r^1(//R_{31})$ is not parallel to the plane composed by the first two rotation pairs of $R_{11}$ and $R_{21}$, then the equation above becomes

$$M_{Pa} = \begin{bmatrix} t^3 \\ r^0 \end{bmatrix} \Leftarrow \begin{bmatrix} t^3 \\ r^0 \end{bmatrix}$$

Therefore, the geometric condition of branches assembled on the base can be summarized as: axes of $R_{11}$, $R_{21}$ and $R_{31}$ are allocated arbitrarily in space.

(5) Draw a sketch of parallel mechanism
The geometric condition of three branches assembled on the two platforms can be described as follows:

(a) Topological structure of the movable platform: axes of $P_{14}$, $P_{24}$ and $P_{34}$ can be allocated arbitrarily in space.
(b) Topological structure of the base: axes of $R_{11}$, $R_{21}$ and $R_{31}$ can be allocated arbitrarily in space.

In conclusion, a sketch of parallel mechanism is drawn as shown in Fig. 25.1a.

But the intersection operation of the POC set also depends on DOF of mechanism, and it need to examine whether the DOF of mechanism satisfy the requirement. i.e. $DOF = 3$.

*Step 5*. Examine DOF of mechanism

According to Eq. (25.3), examining DOF of the mechanism as shown in Fig. 25.1a.

(1) Determine the number $\xi_{L_1}$ of independent finite elements of the first loop

(a) According to Eq. (25.4) and considering $R_{11} \not\parallel R_{21}$, $\xi_{L_1}$ of the first loop composed of the first branch and the second branch is

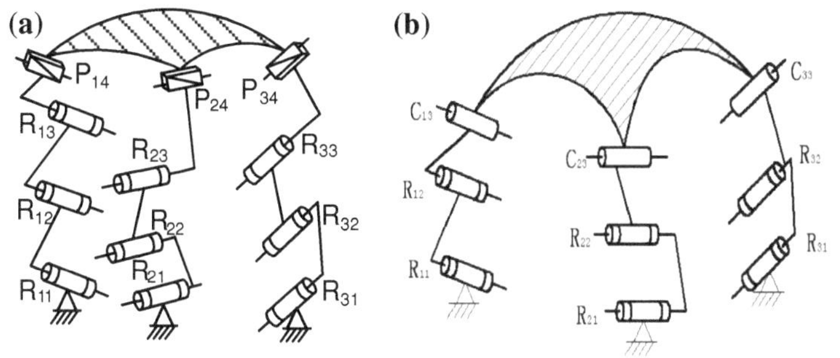


**Fig. 25.1** 3-translation parallel mechanisms. **a** $3-SOC\{-R//R//R-P-\}$, **b** $3-SOC -R//R//C-\}$

$$\xi_{L_1} = \dim\cdot(M_{b_1}\cup M_{b_2}) = \dim\cdot\left(\begin{bmatrix} t^3 \\ r^1(//R_{11})\end{bmatrix}\cup\begin{bmatrix} t^3 \\ r^1(//R_{21})\end{bmatrix}\right)$$
$$= \dim.\left(\begin{bmatrix} t^3 \\ r^2(//\diamondsuit(R_{11},R_{21}))\end{bmatrix}\right) = 5$$

(b) According to Eq. (25.3), DOF of sub-parallel mechanism composed of the first branch and the second branch is

$$F_{(1-2)} = \sum_{i=1}^{m} f_i - \sum_{j=1}^{v}\xi_{L_j} = 8-5 = 3$$

(c) According to Eq. (25.2) and $F_{(1-2)}=3$, the POC set of the movable platform of sub-parallel mechanisms composed of the first branch and the second branch is

$$M_{Pa(1-2)} = M_{b_1}\cap M_{b_2} = \begin{bmatrix} t^3 \\ r^1(//R_{11})\end{bmatrix}\bigcap\begin{bmatrix} t^3 \\ r^1(//R_{21})\end{bmatrix} = \begin{bmatrix} t^3 \\ r^0\end{bmatrix}$$

(2) Determine the number $\xi_{L_2}$ of independent finite elements of the second loop According to Eq. (25.4) and considering that pair axes of $R_{11}, R_{21}$ and $R_{31}$ are allocated arbitrarily in space, $\xi_{L_2}$ is

$$\xi_{L_2} = \dim\cdot\left(M_{Pa(1-2)}\cup M_{b_3}\right) = \dim\cdot\left(\begin{bmatrix} t^3 \\ r^0\end{bmatrix}\cup\begin{bmatrix} t^3 \\ r^1(||R_{31})\end{bmatrix}\right)$$
$$= \dim\cdot\left(\begin{bmatrix} t^3 \\ r^1(//R_{31})\end{bmatrix}\right) = 4$$

(3) Determine DOF of parallel mechanisms According to Eq. (25.3), DOF of the parallel mechanism is

$$F = \sum_{i=1}^{m} f_i - \sum_{j=1}^{v} \xi_{L_j} = 12 - (5 + 4) = 3$$

Therefore, $DOF = 3$ satisfies design requirements.

The intersection operation of the POC set is restrained by inactive pairs of mechanism. Therefore, there is need to examine whether or not the mechanism has inactive pairs. If exists, we need to delete them.

*Step 6*. Determine inactive pairs of the mechanism

(1) Determine inactive pairs of mechanism According to the topological structure and DOF of the mechanism, examine whether or not the mechanism has inactive pairs based on decision rules of inactive pair.

(a) Supposing that the $R_{31}$ pair is fixed and get a new mechanism, topological structure of the third branch becomes $SOC\{-R_{32}//R_{33} - P_{34}-\}$. According to Eq. (25.1), the POC set of the branch is

$$M_{b_3} = \begin{bmatrix} t^1(\perp R_{33}) \\ r^1(//R_{33}) \end{bmatrix} \cup \begin{bmatrix} t^1(//P_{34}) \\ r^0 \end{bmatrix} = \begin{bmatrix} t^2(//\diamond(R_{33}, P_{34})) \\ r^1(//R_{33}) \end{bmatrix}$$

Since the DOF of the third branch is three, $M_{b_3}$ has three independent elements.

(b) Determining the number $\xi_{L_1}$ of independent finite elements of the first loop. Since the first branch and second branch are not changed, $\xi_{L_1}$ is equal to the result of Step 5, i.e. $\xi_{L_1} = 5$.

(c) Determining the $\xi_{L_2}$ value of independent finite elements of the second loop. According to Step 5, we only need to replace the POC set of the third branch in Step 5. That is

$$\xi_{L_2} = \dim\cdot\left(M_{Pa(1-2)} \cup M_{b_3}\right) = \dim\cdot\left(\begin{bmatrix} t^3 \\ r^0 \end{bmatrix} \cup \begin{bmatrix} t^2(//\diamond(R_{33}, P_{34})) \\ r^1(//R_{33}) \end{bmatrix}\right)$$
$$= \dim\cdot\left(\begin{bmatrix} t^3 \\ r^1 \end{bmatrix}\right) = 4$$

(d) Determining the DOF of the new mechanism According to Eq. (25.3), DOF of the new mechanism is

$$F^* = \sum_{i=1}^{m} f_i - \sum_{j=1}^{v} \xi_{L_j} = 11 - (5 + 4) = 2$$

Compared to the previous mechanism, since DOF of the new mechanism has reduced one, therefore, $R_{31}$ pair is not an inactive pair. Similarly, we can prove all pairs are not inactive pairs in this way.

(2) Examine the POC equation for the parallel mechanism Since the mechanism as shown in Fig. 25.1a does not contain inactive pairs, we need not examine the POC equation for the parallel mechanism.

*Step 7*. Select the input pair of parallel mechanisms

Select input pair of parallel mechanisms based on decision rules. Determine whether $R_{11}$, $R_{21}$ and $R_{31}$ pairs of the mechanism as shown in Fig. 277.1a are input pairs simultaneously.

(1) Supposing that $R_{11}$, $R_{21}$ and $R_{31}$ pairs are fixed and get a new mechanism, topological structure of its branches become $SOC\{-R_{i2}//R_{i3}-P_{i4}-\}$ $(i=1,2,3)$. According to Eq. (25.1), the POC set of branch end link is

$$M_{b_i} = \begin{bmatrix} t^1(\perp R_{i3}) \cup t^1(//R_{i4}) \\ r^1(//(R_{i3}) \cup r^0) \end{bmatrix} = \begin{bmatrix} t^2 \\ r^1(//R_{i3}) \end{bmatrix}$$

Since the DOF of the branch is three, $M_{b_i}$ has three independent elements.

(2) Determining the number $\xi_{L_1}$ of independent finite elements of the first loop.

(a) According to Eq. (25.4), and considering $R_{12}//R_{22}$, $\xi_{L_1}$ of the first loop composed of the first branch and the second branch is

$$\begin{aligned}\xi_{L_1} &= \dim\cdot(M_{b_1}\cup M_{b_2}) = \dim\cdot\left(\begin{bmatrix} t^2 \\ r^1(//R_{12}) \end{bmatrix} \cup \begin{bmatrix} t^2 \\ r^1(//R_{22}) \end{bmatrix}\right) \\ &= \dim\cdot\left(\begin{bmatrix} t^3 \\ r^2(//\diamond(R_{12}//R_{22})) \end{bmatrix}\right) = \dim\cdot\left(\begin{bmatrix} t^3 \\ r^2 \end{bmatrix}\right) = 5\end{aligned}$$

(b) According to Eq. (25.3), DOF of the sub-parallel mechanism composed of the first branch and the second branch is

$$F_{(1-2)} = \sum_{i=1}^{m} f_i - \sum_{j=1}^{v} \xi_{L_j} = 6-5 = 1$$

(c) According to Eq. (25.2) and $F_{(1-2)}=1$, the POC set of the movable platform of sub-parallel mechanism composed of the first branch and the second branch is

$$M_{Pa(1-2)} = M_{b_1}\cap M_{b_2} = \begin{bmatrix} t^2 \\ r^1(//R_{12}) \end{bmatrix} \bigcap \begin{bmatrix} t^2 \\ r^1(//R_{22}) \end{bmatrix} = \begin{bmatrix} t^1 \\ r^0 \end{bmatrix}$$

Since the DOF of the branch is 1, $M_{Pa(1-2)}$ only has one independent element.

(3) Determining the number $\xi_{L_2}$ in independent finite elements of the second loop. According to Eq. (25.5) and considering that pair axes of $R_{11}$, $R_{21}$ and $R_{31}$ are allocated arbitrarily in space, $\xi_{L_2}$ is

$$\xi_{L_2} = \dim\cdot\left(M_{Pa(1-2)} \cup M_{b_3}\right) = \dim\cdot\left(\begin{bmatrix} t^1 \\ r^0 \end{bmatrix} \cup \begin{bmatrix} t^2 \\ r^1(//R_{32}) \end{bmatrix}\right) = \dim\cdot\left(\begin{bmatrix} t^3 \\ r^1 \end{bmatrix}\right)$$
$$= 4$$

(4) Determining DOF of the new parallel mechanism According to Eq. (25.3), DOF of the new parallel mechanism is

$$F = \sum_{i=1}^{m} f_i - \sum_{j=1}^{v} \xi_{L_j} = 9 - (5 + 4) = 0$$

Since DOF of the new parallel mechanism is 0, $R_{11}$, $R_{21}$ and $R_{31}$ pairs on identical platform of mechanism as shown in Fig. 25.1a can be input pairs simultaneously.

*Step 8*. Determine parallel mechanisms topological structure

The parallel mechanism topological structure as shown in Fig. 25.1a can be described as follows:

(1) Topological structures of three branches are identical: $SOC\{-R_{i1}//R_{i2}//R_{i3} - P_{i4}-\}(i = 1, 2, 3)$.
(2) Topological structure of the movable platform: pair axes of $P_{14}$, $P_{24}$ and $P_{34}$ are allocated arbitrarily in space.
(3) Topological structure of the base: pair axes of $R_{11}$, $R_{21}$ and $R_{31}$ are allocated arbitrarily in space. It should be noted that $P_{14}$, $P_{24}$ and $P_{34}$ pairs on the movable platform and $R_{11}$, $R_{21}$ and $R_{31}$ pairs on the base are allocated arbitrarily in space. The direction and position of three axes can be determined in accordance with design requirement (e.g. symmetry). It is clear that three axes should be not parallel to identical plane.
(4) $R_{11}$, $R_{21}$ and $R_{31}$ pairs on the base are input pairs.

*Step 9*. Characteristic analysis of parallel mechanisms

(1) BKC type and coupling degree

(a) Determining $SOC_1$: Since the mechanism has three identical branches, we choose any loop as the first Single-Opened-Chain. i.e. $SOC_1$ is:

$$SOC\{-R_{11}//R_{12}//R_{13} - P_{14} - P_{24} - R_{23}//R_{22}//R_{21}-\}$$

(b) Determining the constraint degree $\Delta_1$ of $SOC_1$: According to Sect. 3.5, the number $\xi_{L_1}$ of independent finite elements of the first loop is $\xi_{L_1} = 5$. The constraint degree $\Delta_1$ of $SOC_1$ is:

**Table 25.3** All branches of the parallel mechanisms are identical

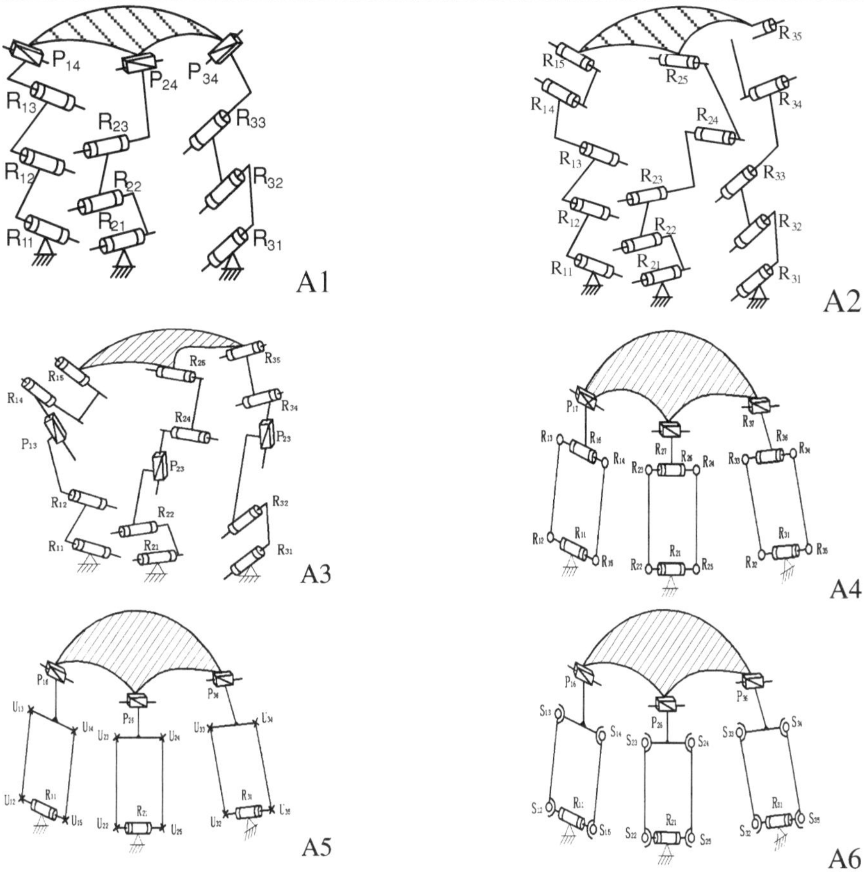


$$\Delta_1 = \sum\nolimits_{i=1}^{m_1} f_i - I_1 - \xi_{L_1} = 8 - 2 - 5 = +1$$

(c) Determining $SOC_2$: obviously, the second Single-Opened-Chain, i.e. $SOC_2$ is:

$$SOC\{-R_{31}//R_{32}//R_{33} - P_{34}-\}$$

(d) Determining the constraint degree $\Delta_2$ of $SOC_2$: According to Sect. 3.5, the number $\xi_{L_2}$ of independent finite elements of the second loop is $\xi_{L_2} = 5$. The constraint degree $\Delta_2$ of $SOC_2$ is:

$$\Delta_2 = \sum\nolimits_{i=2}^{m_2} f_i - I_2 - \xi_{L_2} = 4 - 1 - 4 = -1$$

**Table 25.4** Two branches of the parallel mechanisms are identical

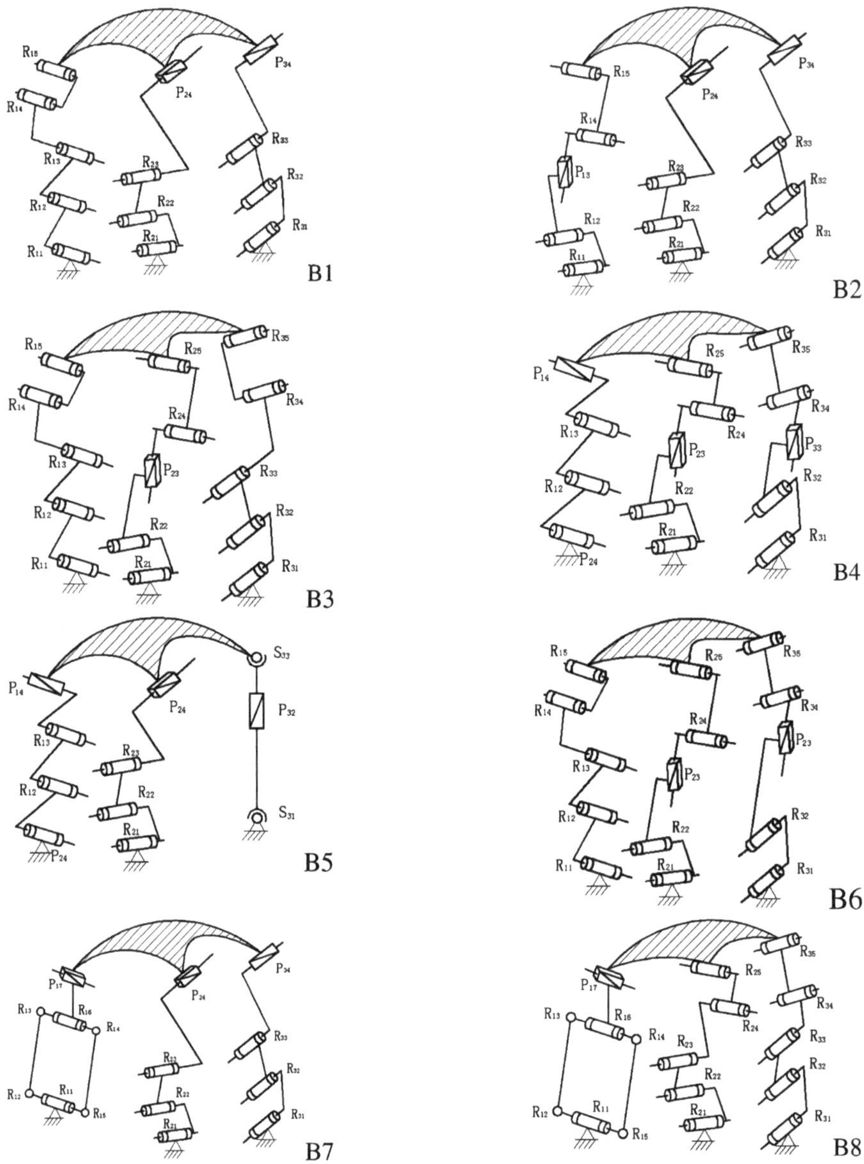


(continued)

**Table 25.4** (continued)

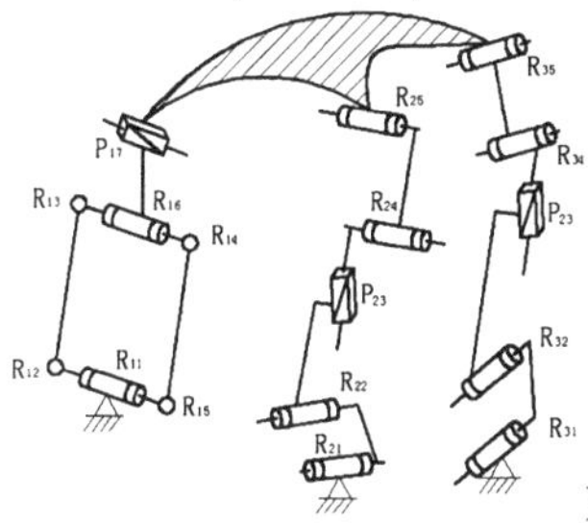

B9

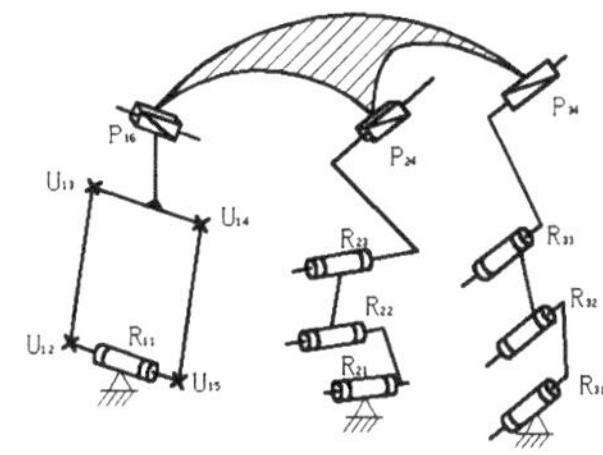

B10

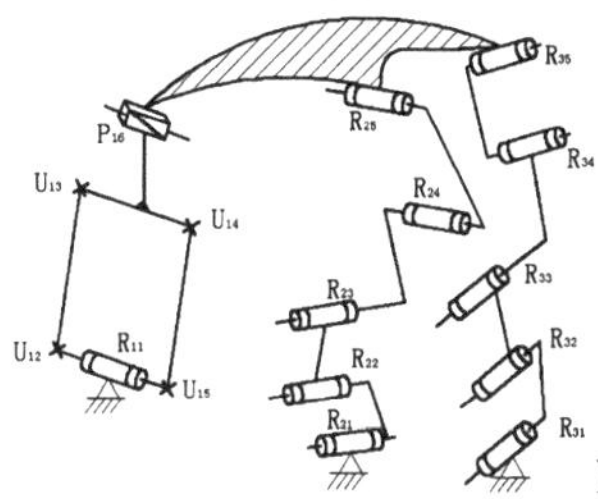

B11

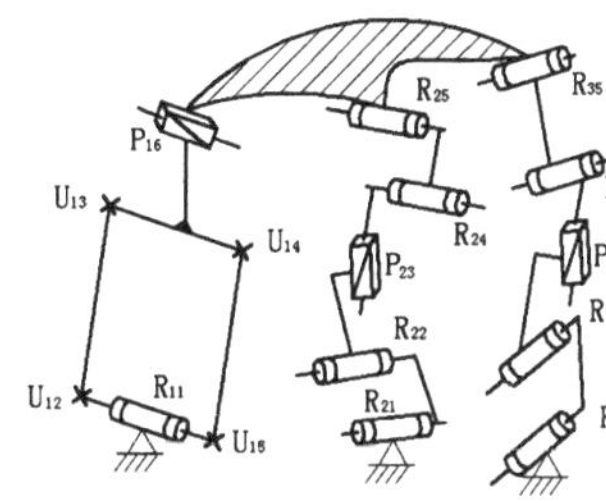

B12

(e) Determining BKC type and coupling degree $k$ of the mechanism: According to the judgment method of BKC, this mechanism only has one BKC type and its coupling degree is:

$$k = \frac{1}{2}\sum_{j=1}^{v} |\Delta_j| = \frac{1}{2}(|+1| + |-1|) = 1$$

(2) DOF Type

(3) The input and output decoupling Depending on decoupling principle of topological structure and the mechanism has complete degree of freedom, so it does not have an input and output decoupling.

(4) Topological structure characteristic of parallel mechanisms The characteristic parameters of parallel mechanisms topological structure, which show the whole performance of mechanism, could be used to compare performance of mechanism of different topological structures and provide the basis for the optimization of structure types.

For the mechanisms as shown in Fig. 25.1a, as prismatic pair $P_{i4}$ and rotation pair $R_{i3}$ can be replace by cylinder pair $C_{i3}(i = 1 - 3)$, the mechanism topological structure becomes $3 - SOC\{-R_{i1}//P_{i2}//C_{i3}-\}(i = 1 - 3)$. In the same way, we synthesize a parallel mechanism reported in document [8] as shown in Fig. 25.1b.

**Table 25.5** All branches of the parallel mechanisms are different

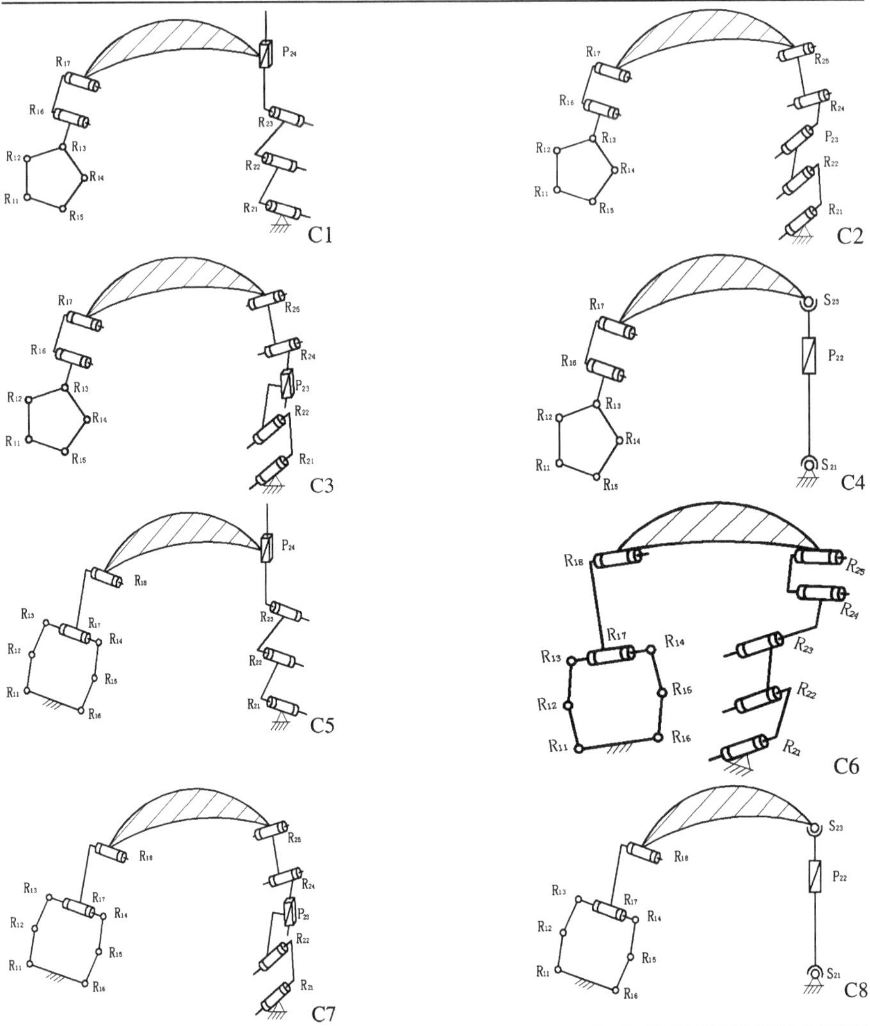


## 25.4 Classification of Structure Type of Parallel Mechanisms

According to the structure synthesis method based on POC set, we get 26 three-translation parallel mechanisms as shown in Tables 25.3, 25.4 and 25.5.

(1) Check if parallel mechanism has identical branches: ① All branches of parallel mechanisms are identical: A1–A6 in Table 25.3. ② Two branches of the parallel mechanisms are identical: B1–B12 in Table 25.4. ③ All branches of parallel mechanisms are different: C1–C8 in Table 25.5.
(2) Parallel mechanism has complex branch: C1–C8 in Table 25.5.

(3) Parallel mechanism has partial decoupling: C1–C8 in Table 25.5. Generally, these mechanisms always have complex branches.
(4) Coupling degree analysis: For example, parallel mechanisms with coupling degree $\kappa = 2$, i.e. A2 and A3 in Table 25.3; $\kappa = 1$, i.e. B1 and B5 in Table 25.4; $\kappa = 0$, i.e. C1 in Table 25.5 and so on. Obviously, the less the coupling degree is, the simpler kinematics analysis will be.

## 25.5 Conclusions

Using the method based on POC set, 26 kinds of 3-translation parallel mechanisms are synthesized in this article, and four kinds of parallel mechanism, i.e. A1, A2, A3, A4 in Table 25.3, have been reported in the document [8]; the other three kinds of parallel mechanisms, i.e. B12 in Table 25.4 and C1, C5 in Table 25.5 have also been reported in the document [10]; the remaining nineteen kinds of parallel mechanisms are presented firstly in this paper. Further more, parallel mechanisms shown in Tables 25.3, 25.4 and 25.5 are classified by branch types, which can satisfy the engineering requirement.

This method is not only applicable for the topological structure synthesis of non-over-constrained and generally over-constrained mechanisms, but also easy for mathematical operation. The geometrical and physical meaning of this method is clear, which is easy to understand and apply for designers.

**Acknowledgments** The financial support of National Science Foundation of China (NSFC-51075045, 50875261), Jiang Su Key Technology R&D and Independent Innovation Project (BE2010074).

## References

1. Hunt KH (1978) Kinematic geometry of mechanisms. Claredon Press, Oxford
2. Frisoli A, Checcacci F et al (2000) Synthesis by screw algebra of translating in-parallel actuated mechanisms. Adv Robot Kinemat 433–440
3. Tsai LW (1999) The enumeration of a class of three-DOF parallel manipulators. In: Proceedings of the 10th world congress of the theory of machine and mechanisms, Finland, pp.1121–1126
4. Hervé JM (1978) Analyse Structure des mécanismes par group des déplacements. Mech Mach Theory 13:437–450
5. Hervé JM, Sparacino F (1991) Structural synthesis of parallel robots generating spatial translation. In: 5th IEEE international conference on, advanced robotics, pp.808–813
6. Yang TL (2012) Topological structure design theory for robot mechanisms. China Science Press, Beijing

7. Jin Q, Yang TL (2001) Structure Synthesis of a Class of Five-DOF (three translations and two rotations) parallel robot mechanisms based on single-opened-chain units. In: Proceedings ASME design engineering technical conferences, DETC/DAC-21153
8. Yang TL, Jin Q, LIU AX (2002) Structural synthesis and classification of the 3DOF translational parallel robot mechanisms based on the units of single-opened-chain. J Mech Engin 38(8):31–36
9. Jin Q, Yang TL (2004) Synthesis and analysis of a group of 3-DOF-of-freedom decoupling parallel mechanisms. J Mech Design 126(2):301–306
10. Shen HP, Yang TL, MA LZ (2005) Kinematics structural synthesis of 3-translational weakly-coupled parallel mechanisms based on hybrid chains. J Mech Eng 41(4):22–27

# Chapter 26
# Development of Reconfigurable Spherical Motion Generator

**Shiu Hang Ip, Chao Chen, Richard P. H. Chen and Denny Oetomo**

**Abstract** In this paper, a novel Reconfigurable Spherical Motion Generator is proposed. The aim of the proposed redundant parallel manipulator is to provide unlimited spherical motion. The conceptual design is first presented and followed by three different prototypes. The motivations of each prototype are explained and the structures of each prototype are described in detail. During the development of RSMG, different problems have been addressed and solved by the following up prototypes.

**Keywords** Parallel manipulator · Redundant actuator · Reconfigurability

## 26.1 Introduction

Researchers have attempted to generate spherical motion for different applications, such as flight simulators and robotic eyes. There are motor-type spherical motion generators by electromagnetic drives or piezoelectric drives. Electromagnetic

S. H. Ip (✉) · C. Chen · R. P. H. Chen
Department of Mechanical and Aerospace Engineering, Monash University, Melbourne, VIC, Australia
e-mail: shiu.ip@monash.edu

C. Chen
e-mail: chao.chen@monash.edu

R. P. H. Chen
e-mail: richard.chen@monash.edu

D. Oetomo
Department of Mechanical Engineering, The University of Melbourne, Parkville, VIC, Australia
e-mail: doetomo@unimelb.edu.au

J. S. Dai et al. (eds.), *Advances in Reconfigurable Mechanisms and Robots I*,
DOI: 10.1007/978-1-4471-4141-9_26, © Springer-Verlag London 2012

drives provide spherical motion by changing the electromagnetic polarity of the stator for a desired spherical motion [1] and the real time control is further improved by analyzing the required torque of the spherical motor in [2]. Piezoelectric drives are similar to electromagnetic drives, they change the piezoelectricity polarity instead of electromagnetic polarity. The concept of piezoelectric motor was introduced in [3]. A three degree-of-freedom piezoelectric motor was introduced in [4]. Spherical robots have been proposed recently. There are six types of spherical robots classified based on the corresponding actuation [5] such as actuation by changing the internal mass distribution of the robot in [6, 7]. In [8], these six types of spherical robots are further summarised down to two kinds based on their motion principles. Atlas spherical motion platform is a design which allows unlimited angular displacement rotating about any axis passing thought the geometric centre of the sphere [9]. Three-dof parallel manipulators are proposed such as the Agile Eye [10], spherical star triangle manipulator [11], spherical wrists [12, 13], and shoulder joint in humanoid "Kojiro" [14]. However, the existence of internal and boundary singularities [15, 16], and joint limits, prevents these devices from achieving an unlimited spherical workspace.

The workspace-boundary singularities and workspace-interior singularities of parallel manipulators are identified in [17]. Reconfigurablility may provide the manipulators the capability to avoid singularities. Reconfigurable parallel manipulators have been studied. Cheope [18] was proposed as a reconfigurable redundant manipulator. In [19], a reconfigurable planar parallel manipulator was introduced. The reconfigurability under this investigation is to remove any singular configuration of a type of manipulator by changing its structure. In [20], kinematotropic mechanisms were studied. Such manipulators can reconfigure into different kinds of manipulators which allow them to behave differently.

Reconfigurable spherical motion generator (RSMG) is a spherical motion generator that aims to provide unlimited spherical motion. By employing parallel robotic arms, it is able to generate rigid and robust motion. The workspace limit of these parallel robotic arms are removed by its reconfigurability.

In this paper, the conceptual design of RSMG is presented in Sect. 26.2. The development of RSMG is presented in Sect. 26.3. The structure of three different prototypes are described in detail which includes the methodology of conception, discovery of problems and development based on the previous prototypes

## 26.2 Conceptual Design

The conceptual design of the RSMG is shown in Figs. 26.1 and 26.2. The proposed manipulator consists of four motor-actuated robotic arms and one spherical rotor. Each robotic arm is a two-link serial manipulator, with the motion of the upper link actuated by the motor, while the lower is passive. The encoders are mounted to measure the relative angular displacements of upper and lower links. The end effector of each two-link arm is fitted with a gripper that attaches to the rotor. Due

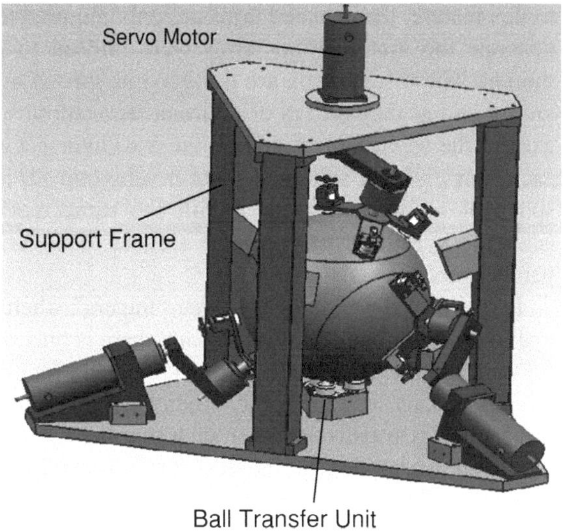


**Fig. 26.1** CAD model of the conceptual design of RSMG

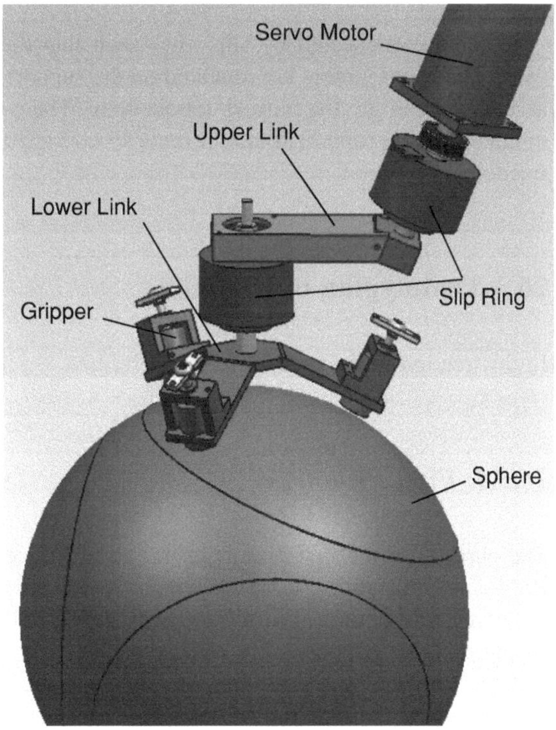


**Fig. 26.2** The conceptual design of RSMG with one arm

to this feature, the actuated arms are only required to overcome the rotor inertia to generate the motion. Four arms were utilised to produce the 3-DOF spherical motion. These four arms are designed in special arrangements, the servo motors are located at the vertices of a virtual tetrahedron which is form by the rotational axes of the servo motors and all the axes intersect at a single point. Furthermore, each arm yields a spherical motion subgroup [21]. The intersection of the four spherical motion subgroup is still the same motion subgroup. The redundant actuation is utilised to overcome singular configurations of the overall resulting parallel mechanism.

In the lower links, there are three "fingers" where one gripper is attached on the end of each "gripper". In any instant, there is only one "finger" grabbing the rotor for each robotic arm, in other words, there are four contact points from the robotic arms in contact with the rotor. When the finger actuates down towards the sphere, the gripper grip onto the rotor such that the whole system is set. To achieve the unlimited workspace of the RSMG, it is necessary to establish an intelligent switching algorithm to appropriately reconfigure the mechanism when approaching singularity. Singularity detection is done by evaluating the explicit expression for the singular configurations and determining the moment when the RSMG has entered a region close to singularities. When this occurs, the mechanism is reconfigured.

Wires are connected by slip rings such that the wires on the linkages never twist. The servo motors are mounted on the support frame so that they are located at the vertices of the virtual tetrahedron. There are three ball transfer units mounted on the support frame in order to constrain the rotor in a three degree of freedom orientation.

## 26.3 Prototyping of RSMG

In this section, three prototypes are described.

### *26.3.1 First Prototype*

The purpose of the first prototype was to test the concept of reconfigurability of RSMG and examined the engagement of the grippers and the rotor. The first prototype was an on-armed prototype as shown in Fig. 26.3. The rotor was a steel sphere and the grippers were electromagnets. The attachment between the rotor and the grippers was done by activating the electromagnets and they generate magnetic flux to attach the rotor to the electromagnets. It was a pull-type RSMG because the grippers were pulling the rotor to the robotic arms. The slip rings allowed the wires to be attached out of the shaft and prevent twisting. The virtual tetrahedron was pointed such that a vertex points upwards, as shown in Fig. 26.3.

**Fig. 26.3** The first prototype of RSMG

Due to this setting, there was one servo motor seating above the rotor and other three laying on the basement of the whole supporting frame.

It was found that the concept was working as the electromagnets were able to transfer motion to the rotor. However, there were issues necessary to be improved as follows:

(1) The robotic arms were heavy due to the linkage and the slip rings.
(2) Sliding occurs due to insufficient magnetic force.
(3) The positioning of the ball transfer units were not able to fully constraint the position of the rotor. When the robotic arm was actuated with an activated electromagnet, the spherical motion provided form the arm would drive the rotor to spin away from the desired constrained rotor position. It also meant that when four arms were placed at the corresponding positions and actuate, this redundancy would damage the robotic arms as the supporting frame itself was not able to absorb the torques and forces from the robotic arms.
(4) The gripper was not able to attach onto the robot continuously.
(5) The direct drive could not handle the heavy robotic arm in certain configurations. It was necessary to gear up the actuator torques for a robust and accurate motion.

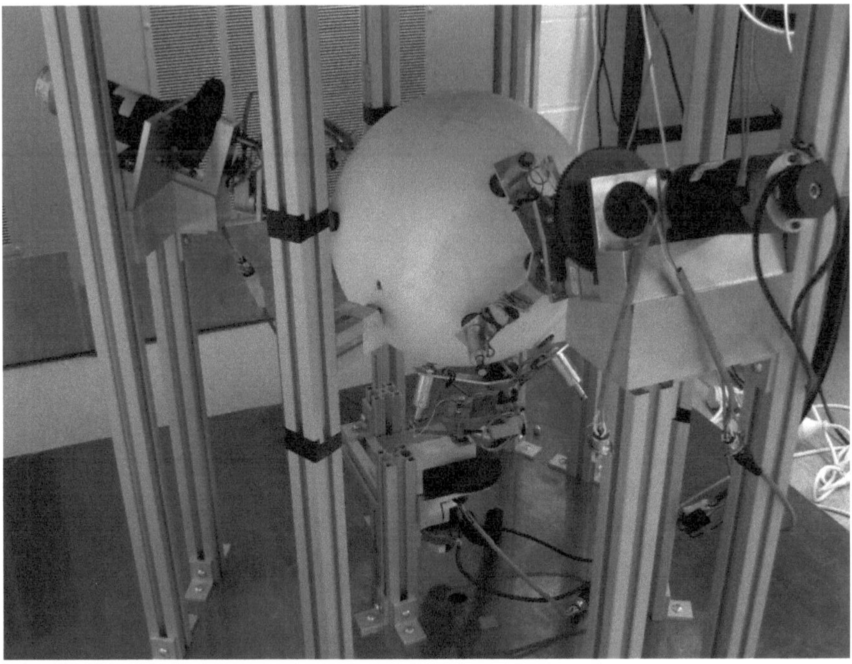

**Fig. 26.4** The second prototype of RSMG

### *26.3.2 Second Prototype*

In order to solve the above issues in the first prototype, a second prototype was proposed. The main tasks of the second prototype were to validate this push-type concept and to test the control schemes. The second prototype was a full-armed-prototype which meant it obtained four robotic arms as shown in Fig. 26.4. The robotic arms, the linkages were reduced in weight but still strong enough to transfer the torque during operation. The rotor was a plastic sphere and the grippers are solenoids. The attachment between the rotor and the grippers was done by activating the push-type solenoids and they punched out the solenoid pins to attach the rotor. Hence, it was called push-type RSMG. The slip rings employed in the second prototype were much smaller and they allowed the wires to be hidden inside the shafts. Gears were installed to gear up the torque such that the different configurations could be held robustly. The positioning of the virtual tetrahedron was rearranged such that one of the vertex was pointing downwards as shown in Fig. 26.4. Due to this setting, there was one servo motor locating below the rotor and the other three hanging above the rotor, supported by the supporting frame. Since the virtual tetrahedron was now pointing downwards, the three ball transfer units were able to be further separated away from each other for a better support. One major improvement in this prototype was its ability to be tuned for better performance.

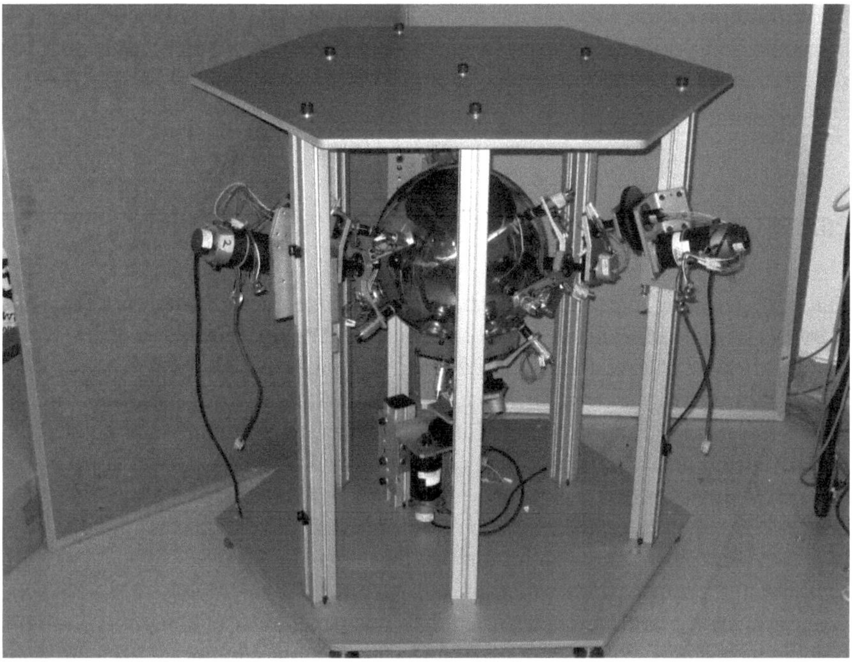

Fig. 26.5 The third prototype of RSMG

The positioning of the rotor was supported by three ball transfer unit which were further separated from each other. However, this prototype was still sensitive to tolerance issues. The performance of the RSMG was highly dependent on the reliability of the point contact forces between the robotic arms and the sphere. If the contact force was not optimal, the sphere would not track accurately. The magnetic force applied by a solenoid was difficult to predict if the stroke length of the solenoid was not within a reasonable range. This was mainly due to fabrication tolerances.

### 26.3.3 Third Prototype

The third prototype was built to eliminate the tolerance issues in the second prototype and achieve the unlimited spherical motion. The third prototype was a full-armed-prototype which meant it obtained four robotic arms and it is a push-type RSMG as shown in Fig. 26.5. The actuation methodology was the same as the second prototype. However, in order to eliminate the fabrication tolerances which appeared in the second prototype, in the design of the third prototype, it has been considered carefully. The improved fine tuning mechanisms on the third prototype permit accurate position adjustments after assembly. For example, the robotic arms of the third prototype were placed further away from the rotor. Couplings between solenoids and grippers are used to maintain the strokes of solenoids within reasonable ranges. The third prototype of RSMG is currently under test with different control schemes.

## 26.4 Future Work

The next step is to achieve a complete reconfigurable control which has two parts, a lower level synchronized control scheme and a upper level reconfiguration scheme. RSMG is a redundant parallel manipulator, it is necessary to synchronize the motion of the actuators to reduce the internal conflicting forces and torque. About the reconfiguration scheme, it is necessary to first study the behavior of RSMG during reconfiguration. Reconfiguration of RSMG introduces errors which can be classified in two types, individual error and co-operation error. Individual error occurs when the current solenoid is inactivated and the next solenoid is going to be activated, there exists an instant that the contact point is not fixed on the rotor such that the gripper slides which introduces error. Co-operation error may occur when two robotic arms are inactivated their own solenoids such that there are only two contact points on the rotor. Hence, a fast reconfiguration scheme must be fulfilled.

## 26.5 Conclusion

The development of a reconfigurable spherical motion generator (RSMG) was discussed. The RSMG utilizes four robotic arms to actuate a spherical rotor. The arms can be reconfigured by alternating the engaging solenoids, to avoid the boundary of the workspace and achieve unlimited spherical motion. To achieve fast motion, light robotic arms were designed. It was found that the push-type RSMG worked better than the pull-type RSMG. Accurate assembly was achieved by employing fine tuning mechanisms.

**Acknowledgments** The support from ARC Discovery DP0985120 is acknowledged.

## References

1. Chirikjian GS, Stein D (1999) Kinematic design and commutation of a spherical stepper motor. IEEE/ASME Trans Mechatron 4(4):342–35
2. Yan L, Chen I, Yang G, Lee K (2006) Analytical and experimental investigation on the magnetic field and torque of a permanent magnet spherical actuator. IEEE/ASME Trans Mechatron 11(4):409–418

3. Koc B, Bouchilloux P, Uchino K (2000) Piezoelectric micromotor using a metal-ceramic composite structure. IEEE Trans Ultrason Ferroelectr Freq Control 47(4):836–843
4. Khoo T, Dang D, Friend J, Oetomo D, Yeo L (2009) Triple degree-of-freedom piezoelectric ultrasonic micromotor via flexural-axial coupled vibration. IEEE Trans Ultrason Ferroelectr Freq Control 56(8):1716–1724
5. Armour RH, Vincent JFV (2006) Rolling in nature and robotics: a review. J Bionic Eng 3(4):195–208
6. Alizadeh HV, Mahjoob MJ (2009) Effect of incremental driving motion on a vision-based path planning of a spherical robot. In: Second international conference on computer and electrical engineering, 2009, pp 299–303
7. Ghanbari A, Mahboubi, S, Fakhrabadi MMS (2010) Design, dynamic modeling and simulation of a spherical mobile robot with a novel motion mechanism,. In: IEEE/ASME international conference on mechatronic and embedded systems and applications, MESA 2010, pp 434–439
8. Zhao B, Li M, Yu H, Hu H, Sun L (2010) Dynamics and motion control of a two pendulums driven spherical robot. IEEE/RSJ international conference on intelligent robots and systems, pp 147–153
9. Hayes MJD, Langlois RG (2005) A novel kinematic architecture for six dof motion platforms. CSME Trans 29(4):701–709
10. St. Pierre E, Gosselin C, Gagne M (1996) On the development of the Agile Eye. IEEE Robotics Autom Mag 3(4):29–37
11. Enferadi J, Tootoonchi AA (2009) A novel spherical parallel manipulator: forward position problem, singularity analysis, and isotropy design. Robotica 27:663–676
12. Gregorio RD (2001) Kinematics of a new spherical parallel manipulator with three equal legs: the 3-urc wrist. J Robotic Syst 18(5):213–219
13. Innocenti C, Parenti-Castelli V (1993) Echelon form solution of direct kinematics for the general fully-parallel spherical wrist. Mech Mach Theory 4(4):553–561
14. Sodeyama Y, Nishino T, Namiki Y, Nakanishi Y, Mizuuchi I, Inaba M (2008) The designs and motions of a shoulder structure with a spherical thorax, scapulas and collarbones for humanoid "kojiro." In: 2008 IEEE/RSJ international conference on intelligent robotics and systems, sept 2008, pp 1465–1470
15. Oetomo D, Ang M Jr (2009) Singularity robust algorithm in serial manipulators. Robotics Comput-Integr Manuf 25(1):122–134
16. Gosselin C, Wang J (2002) Singularity loci of a special class of spherical three-degree-of-freedom parallel mechanisms with revolute actuators. Int J Robotics Res 21(7):649–659
17. Gosselin C, Angeles J (1990) Singularity analysis of closed-loop kinematic chains. IEEE Trans Robotics Autom 6(3):281–290
18. Tosi D, Legnani G, Pedrocchi N, Righettini P, Giberti H (2010) Cheope: a new reconfigurable redundant manipulator. Mech Mach Theory 45:611–626
19. Fisher R, Podhorodeski RP, Nokleby SB (2004) Design of a reconfigurable planar parallel manipulator. J Robotic Syst 21(12):665–675
20. Galletti C, Fanghella P (2001) Single-loop kinematotropic mechanisms. Mech Mach Theory 36(6):743–761
21. Herve J (1999) The lie group of rigid body displacements, a fundamental tool for mechanism design. Mech Mach Theory 34(5):719–730

# Chapter 27
# Reconfiguration Analysis of a Fully Reconfigurable Parallel Robot

**Allan Daniel Finistauri and Fengfeng Xi**

**Abstract** This paper presents a new method for the topological reconfiguration of a parallel robot. Using the existing structure of a full six degree-of-freedom parallel robot, limited mobility modes can be realized easily without the need to remove branch modules from the robot structure. Instead, branch modules are reconfigured from an unconstrained-active to a constrained-passive state by means of hybrid active/passive motors and reconfigurable universal-to-revolute joints. In doing so, the robot is capable of assuming a configuration that uses only the degrees-of-freedom required to complete a given task. However, due to the system setup, there are multiple candidate configurations available, each with its own workspace and reach capabilities, thus guidance is needed in selecting the appropriate configuration. An isomorphic and workspace analysis are performed to identify the capabilities of each configuration. To accomplish this, a branch-based mobility analysis, and a parametric kinematic constraint equation are formulated. It is shown that limited mobility modes with different isomorphic configurations can be synthesized automatically with this method.

**Keywords** Reconfigurable parallel robots · Kinematics · Mobility

A. D. Finistauri · F. Xi (✉)
Department of Aerospace Engineering, Ryerson University,
350 Victoria Street, Toronto, ON M5B 2K3, Canada
e-mail: fengxi@ryerson.ca

A. D. Finistauri
e-mail: dfinista@ryerson.ca

J. S. Dai et al. (eds.), *Advances in Reconfigurable Mechanisms and Robots I*,
DOI: 10.1007/978-1-4471-4141-9_27, © Springer-Verlag London 2012 

## 27.1 Introduction

The need to expand the role of traditional mechanisms has within the last few decades led to the development of ones that are modular and reconfigurable. Modularity keeps cost low by reusing common components, while reconfiguration expands the capabilities of what is achievable from a single-purpose mechanism. The combination of these two design concepts has led to the development of modular, reconfigurable robots (MRRs), particularly serial-style robots, however, recently modular and reconfigurable parallel robots have been developed.

Parallel robots are attractive candidates for reconfiguration due to the modular nature of their branch systems where reconfiguration can be classified as being geometric, topological or a combination of the two (classified as group reconfiguration) [1]. Geometric reconfiguration is the process of adjusting the size or orientation of branch components without rearranging their sequence. Two recent examples include the reconfigurable tricept machine tool in which standard components are used to create different sized tricepts [2], and the parallel robot with reconfigurable Hooke joints where one axis of the Hooke joint is reconfigurable [3]. Redundancy within the robot structure allows for geometric reconfiguration as seen with the variable geometry planar parallel robot which is capable of singularity avoidance [4]. The tricept machine tools with adjustable platforms allow for optimal configurations for given tasks [5, 6]. Topological reconfiguration is more complex and is associated with the re-sequencing of joints within the branch structure, usually to achieve a different motion profile for the platform. For instance, the reconfigurable parallel robot with multiple actuation modules and links lead to a multitude of parallel robot configurations [7]. Xi et al. [8] presented a reconfigurable parallel-robot with varying platform degrees-of-freedom and detachable serial arms. Redundancy can also be used as seen in the topological reconfigurable four-branch parallel robot [9]. Group morphing as mentioned is a combination of geometric and topological reconfiguration and is seen in advanced systems such as variable geometry truss mechanisms (VGTM). One such example is the modular wing truss capable of multiple level of reconfiguration to achieve unique wing shapes [10].

This paper presents a new type of reconfigurable parallel robot capable of topological reconfiguration. By using the existing structure of a six degree-of-freedom (DOF) parallel robot, limited mobility modes (configurations with fewer than six degrees-of-freedom) can be synthesized easily without the need to remove branch modules from the robot structure. Instead, branch modules are reconfigured from an unconstrained-active to a constrained-passive state by means of hybrid active/passive motors and reconfigurable universal-to-revolute joints. In doing so, the robot is capable of assuming a configuration that uses only the degrees-of-freedom required to complete a given task. However, due to the system setup, reconfiguration results in multiple candidate configurations when reconfiguring from one mobility mode to another. Also, each configuration has its own associated workspace and reach capabilities, thus guidance is needed in selecting the appropriate configuration. An isomorphic and workspace analysis are performed to identify the capabilities of each configuration. To accomplish this task, a branch-based mobility analysis, and a parametric kinematic constraint equation are formulated. It is shown that limited degree-of-freedom parallel robots with different isomorphic configurations can be synthesized automatically with this method.

The paper is organized as follows. The system is introduced, including a mobility study and identification of isomorphic and degenerate configurations. A parametric constraint kinematics system is formed and used in evaluating the workspace of non-degenerate isomorphic configurations. Finally, conclusions are given.

## 27.2 System Description

The system being developed is a unique parallel robot system in which reconfiguration occurs within the structure itself and does not require outside assistance for reconfiguration. This allows the robot to reconfigure itself to have only the number of controllable degrees-of-freedom required to complete a specific task. To realize this, the parallel robot is fit with advanced components that allow for reconfiguration to occur while maintaining the mobility and connectivity requirements of the system. As shown in Fig. 27.1, the parallel robot has six branches in which each is of universal-prismatic-spherical (UPS) architecture where the prismatic joint is active. Here, the specialized universal joints allows the second rotation axis to lock, similar to those shown in [8], and [11]. The prismatic joints are hybrid active/passive motors which allow the clutch to disengage and the joint to react passively according to the resultant motion of the robot, similar to other systems [12]. These reconfigurations maintain mobility and connectivity requirements as detailed in Sect. 27.2.1.

### *27.2.1 Branch Module-Based Mobility Analysis*

In order to understand the capabilities of the robot after reconfiguration, a mobility analysis must be performed. The mobility $M$, of a mechanical system is defined as

$$M = \lambda(n_l - n_j - 1) + \sum_{j=1}^{n_J} f_j \tag{27.1}$$

where $\lambda$ is the system order, respectively, $n_l$ is the number of links, $n_j$ is the number of joints, and $f_j$ is the degree-of-freedom number for the $jth$ joint. If we consider the robot to be a modular system consisting of separate base, platform and branch modules, then Eq. (27.1) can be rewritten in terms these branch modules

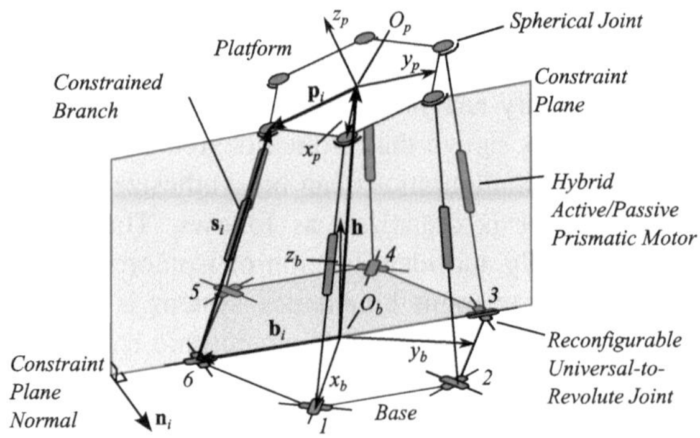


**Fig. 27.1** Fully reconfigurable parallel robot showing reconfigurable universal-to-revolute joints, and hybrid active/passive prismatic motor and vector loop describing constraint kinematic motion plane

$$M = \sum_{i=1}^{n_b} \left[ \lambda \left( n_{l,i} - n_{j,i} \right) + C_i \right] + \lambda \tag{27.2}$$

where $n_b$ is the number of branches, $n_{l,i}$ and $n_{j,i}$ are the number of links and joints in the *ith* branch, respectively, and $C_i = \sum_{j=1}^{n_{j,i}} f_j$ is the connectivity of the *ith* branch. Note that the two links for the base and platform have been added to the equation such that $n_l = \sum_{i=1}^{n_b} n_{l,i} + 2$. Equation (27.2) is generic for a parallel robot in that each branch can be defined individually for its number of links, joints and connectivity.

Tsai [13] has shown using Euler's equation that for each branch, the following relationship between the mobility, connectivity and system order must hold

$$M \leq C_i \leq \lambda \tag{27.3}$$

In Eq. (27.3), we see that the connectivity of the *ith* branch must be no less than the mobility of the robot, and no greater than the system order, establishing the branch mobility and connectivity requirements. From Eq. (27.3) and assuming that only one actuator is situated on each branch, then the connectivity configurations of traditional, spatial ($\lambda = 6$) parallel robots are shown in Table 27.1. We see that in order to satisfy the mobility and connectivity requirements, for each degree-of-freedom removed from the system, one branch of $C_i = 6$ is removed, and a degree-of-freedom is removed from one of the remaining branches, traditionally from one of the remaining $C_i = 6$ branches. With a mobility of less than three, this is impossible due to the limited number of branches.

The addition of constraints by subtraction of branches and degrees-of-freedom as shown above is only one way of modifying the parallel robot structure. Tricept-style robots add constraints by use of a passive branch with specific connectivity in the centre of the robot. Using this notion, constraints can be added to a parallel robot while maintaining the imposed mobility and connectivity requirements by means of simply removing a degree-of-freedom from an existing $C_i = 6$ branch while simultaneously removing the controllability of one of the remaining actuators. These two reconfigurations need not occur on the same branch and can be

Table 27.1 Mobility and connectivity of traditional and extended parallel robots

| | | Traditional parallel robot | | Extended parallel robot | |
|---|---|---|---|---|---|
| Mobility | Constraints | Branches | Connectivity | Branches | Connectivity |
| 6 | 0 | 6 | 6,6,6,6,6,6 | 6 | 6,6,6,6,6,6 |
| 5 | 1 | 5 | 6,6,6,6,5 | 6 | 6,6,6,6,6,5 |
| 4 | 2 | 4 | 6,6,5,5 | 6 | 6,6,6,6,5,5 |
| 3 | 3 | 3 | 5,5,5 | 6 | 6,6,6,5,5,5 |
| 2 | 4 | 2 | 4,4 | 6 | 6,6,5,5,5,5 |
| 1 | 5 | 1 | 1 | 6 | 6,5,5,5,5,5 |

applied recursively to form the extended parallel robot configurations listed in Table 27.1. Assuming that these reconfiguration do occur on the same branch, we then have a parallel robot system with a series of $n_{b,u}$ unconstrained-active branches, and $n_{b,c}$ constrained-passive branches. Thus Eq. (27.2) can be rewritten as

$$M = n_{b,u}\left[\lambda\left(n_{l,u} - n_{j,u}\right) + C_u\right] + n_{b,c}\left[\lambda\left(n_{l,c} - n_{j,c}\right) + C_c\right] + \lambda \tag{27.4}$$

where subscripts 'u' and 'c' stand for unconstrained and constrained, respectively.

When reconfiguration to a limited mobility mode occurs, an unconstrained-active UPS branch reconfigures to a constrained-passive branch of revolute-prismatic-spherical (RPS) joint architecture. Note that the branch reconfiguration satisfies the mobility and connectivity requirements imposed by Eq. (27.3). Other reconfigurable branch configurations can be used [11] however we see that any unconstrained branch has $C_u = 6$, $n_{l,u} = 2$, and $n_{j,u} = 3$ and any constrained branch has $C_c = 5$, $n_{l,c} = 2$, and $n_{j,c} = 2$. Using these values and $\lambda = 6$, Eq. (27.4) reduces to

$$M = 6 - n_{b,c} \tag{27.5}$$

This confirms that for the configurations listed in Table 27.1 for the extended parallel robots, the mobility of the system is dependent on the number of constrained branches. With this system, up to five unconstrained branches in any combination can be reconfigured to a constrained branch, thus allowing the reconfigurable parallel robot to assume any of the five different limited mobility modes without the need to remove branches from the system. For the remainder of the paper, we will focus our analysis of the 5-, 4- and 3-DOF mobility modes.

### 27.2.2 Isomorphic and Degenerate Configuration Identification

The ability to select any combination of branches to be constrained leads to limited mobility cases with isomorphic configurations. Understanding which configurations are isomorphic is important in that any analysis of one configuration can be

**Table 27.2** Isomorphic and degenerate configurations of the reconfigurable parallel robot

| *M* | Type | Passive branch numbers | Branch constraint type |
|---|---|---|---|
| 5 | 1 | 1, 2, 3, 4, 5, 6 | Single branch constraint |
| 4 | 1–2 | 1–2, 2–3, 3–4, 4–5, 5–6, 6–1 | Adjacent branches constrained |
| | 1–3 | 1–3, 2–4, 3–5, 4–6, 5–1, 6–2 | Single spaced-branch constraint |
| | 1–4 | 1–4, 2–5, 3–6 | Degenerate configurations |
| 3 | 1–2–3 | 1–2–3, 2–3–4, 3–4–5, 4–5–6, 5–6–1, 6–1–2 | Adjacent branches constrained |
| | 1–2–4 | 1–2–4, 2–3–5, 3–4–6, 4–5–1, 5–6–2, 6–1–3 | Degenerate configurations |
| | 1–2–5 | 1–2–5, 2–3–6, 3–4–1, 4–5–2, 5–6–3, 6–1–4 | Degenerate configurations |
| | 1–3–5 | 1–3–5, 2–4–6 | Single spaced-branch constraints |

applied directly to the associative isomorphisms. Here, isomorphic configurations are present due to the fact that the branches are equally spaced at 60° intervals around the base and platform, thus isomorphic configurations are periodically spaced around the robot. For instances, for a robot with 5-DOF, any branch selected as constrained is an isomorphic configuration of the others. The isomorphic configurations for the other robot mobility modes are listed in Table 27.2. The robot type indicates the sequence of constrained branches. For instance, the 1–3 type is a 4-DOF parallel robot with branches 1, and 3 designated as constrained and passive, with isomorphic configurations consisting of any two passive branches separated by one active branch.

Identifying degenerate configurations is also essential for reconfiguration. For this robot, degenerate configurations are those in which the constraints imposed by reconfiguration result in reachable workspace collapsing from bound volumes to planar areas. These configurations are thus degenerate due to their inability to provide required spatial motion and are identified in Table 27.2. Any configuration in which two constrained branches are separated by two unconstrained branches (180° separation) are deemed degenerate, i.e. any 1–4, 2–5 or 3–6 combination of passive branches. This is due to the orientation of the constraint planes, which will be discussed in Sect. 27.3. Any non-degenerate type is also identified by the type of branch constraint.

## 27.3 Parametric Constraint Kinematics Formulation

### *27.3.1 Parametric Constraint Equation*

The kinematics of a parallel robot are solved by forming vector loop equations for each branch as seen in Fig. 27.1. For the *ith* branch, the following loop equation holds

$$\mathbf{b}_i + \mathbf{s}_i = \mathbf{h} + \mathbf{p}_i \tag{27.6}$$

where for the *ith* branch, $\mathbf{b}_i$, $\mathbf{s}_i$, and $\mathbf{p}_i$ are the base, branch and platform vectors, respectively, and $\mathbf{h}$ is the height of the platform about the base. All of the vectors are expressed in the base-fixed coordinate frame $O_b$.

For a 6-DOF parallel robot, $\mathbf{h} = \{h_x, h_y, h_z\}^T$ and $\mathbf{p}_i = \mathbf{R}_x(\theta_x)\mathbf{R}_y(\theta_y)\mathbf{R}_z(\theta_z)\mathbf{p}'_i$ contain the six controllable degrees-of-freedom. Here $\mathbf{p}'_i = \left\{p'_{i,x}, p'_{i,y}, p'_{i,z}\right\}^T$ is the local representation of the platform vector and is expressed in the platform-fixed coordinate frame $O_p$. $\mathbf{R}_p = \mathbf{R}_x(\theta_x)\mathbf{R}_y(\theta_y)\mathbf{R}_z(\theta_z)$ is the rotation matrix that expresses $\mathbf{p}'_i$ within the base-fixed coordinate frame $O_b$ where

$$\mathbf{R}_x = \begin{bmatrix} 1 & 0 & 0 \\ 0 & \cos\theta_x & -\sin\theta_x \\ 0 & \sin\theta_x & \cos\theta_x \end{bmatrix} \quad \mathbf{R}_y = \begin{bmatrix} \cos\theta_y & 0 & \sin\theta_y \\ 0 & 1 & 0 \\ -\sin\theta_y & 0 & \cos\theta_y \end{bmatrix} \quad \mathbf{R}_z = \begin{bmatrix} \cos\theta_z & -\sin\theta_z & 0 \\ \sin\theta_z & \cos\theta_z & 0 \\ 0 & 0 & 1 \end{bmatrix}$$

For simplicity, $\mathbf{p}_i$ will be used in lieu of $\mathbf{R}_p\mathbf{p}'_i$. Equation (27.6) can then be rearranged to solve for the only unknown, the branch vector length

$$s_i = |\mathbf{h} + \mathbf{p}_i - \mathbf{b}_i| \tag{27.7}$$

As seen in Eq. (27.5) the reconfigurable parallel robot can have anywhere from zero to five constrained branches. Each constrained branch results in an additional dependent motion to be solved as a function of the remaining independent variables $\mathbf{q} = \{h_x, \ h_y, \ h_z, \ \theta_x, \ \theta_y, \ \theta_z\}^T$. The constraint equation(s) are in the form of $\vec{q}_D = f(\vec{q}_I)$ where $\vec{q}_D$ and $\vec{q}_I$ are the dependent and independent variables, respectively. For a constrained branch, the spherical joint attached to the platform is constrained to move in a plane as shown in Fig. 27.1. From this we can define the following constraint equation

$$(\mathbf{h} + \mathbf{p}_i) \cdot \mathbf{n}_i = 0, \forall i \in \{n_{b,c}\} \tag{27.8}$$

where $\mathbf{n}_i$ is the constraint plane normal for branch $i$ within the set of constrained branches. Since the constrained branch is of RPS joint architecture, $\mathbf{n}_i$ is parallel to the rotation axis of the revolute joint at the base. Often, but not required, the constraint plane passes through the origin of the base coordinate frame $O_b$ and is perpendicular to the base of the robot. This orientation of the constraint planes is used here and is the reason why degenerate configurations are possible (see Table 27.2).

### *27.3.2 Constraint Equation Solving*

Using the reconfigurable robot setup, Eq. (27.8) can be expanded, and for each constrained branch

$$
\begin{aligned}
& n_{i,x}\left(h_x + p'_{i,x}\cos\theta_y\cos\theta_z - p'_{i,y}\cos\theta_y\sin\theta_z\right)+ \\
& n_{i,y}\Big[h_y + p'_{i,x}\left(\sin\theta_x\sin\theta_y\cos\theta_z + \cos\theta_x\sin\theta_z\right)- \\
& \qquad p'_{i,y}\left(\sin\theta_x\sin\theta_y\sin\theta_z + \cos\theta_x\sin\theta_z\right)\Big] = 0
\end{aligned} \tag{27.9}
$$

Equation (27.9) forms the equation for each $n_{b,c}$ constrained branch. There is no specification on which platform motions are independent and which are dependent, thus offering a high level of flexibility when deciding which configuration is best suited for a particular task. However we do note that $h_z$ is not present in Eq. (27.9), due to fact that $n_{i,z} = 0$. This is particular to this reconfigurable parallel robot, as physically, there is no restraint on the orientation of the constraint plane. Here, the constraint plane(s) are always perpendicular to the base of the robot, hence $n_{i,z} = 0$. This requires that $h_z$ is chosen as an independent motion for all tasks for this particular robot setup.

Depending though on the given task and the associative constraint(s), the solution to Eq. (27.8) can either be linear or nonlinear. It is beyond the scope of this work to delve into the solution process for each individual constraint combination, however some discussion is given here. For the 5-DOF limited mobility mode, all constraints are linear and can be solved directly. For the 4-DOF limited mobility modes, a linear solution can be obtained for most constraint combinations. In the case where two angular constraints are required, then a multiple solutions to the trigonometric equations are possible, so evaluation of each in terms of satisfying the constraints must be checked. Alternatively, this case can be solved numerically. Finally, for the 3-DOF limited mobility modes, there will always be at least one angular platform constraint. Equation (27.9) can be rearranged accordingly for each branch and for most cases, a linear solution to the constrained motions can be found. For the specific case in which all three angular constraints are required, a numerical solution is required. Specific cases are listed in [8] and nonlinear solutions are attainable through suitable numerical methods such as Newton-Raphson.

## 27.4 Workspace Evaluation

The size and shape of a workspace is partially influenced by the orientation of the constraint plane(s), $\mathbf{n}_i$. However, due to the isomorphic nature of the robot, the orientation of the branches with respect to the global coordinate frame, and the selection of constrained branches can be selected such that a direct comparison between two configurations with the same mobility can be achieved directly. For each case, the robot is aligned, and passive branches are selected such that symmetry in the workspace is seen in the the $xz$-plane. In doing so, a direct comparison between minimal and maximal platform variables is possible. The base and

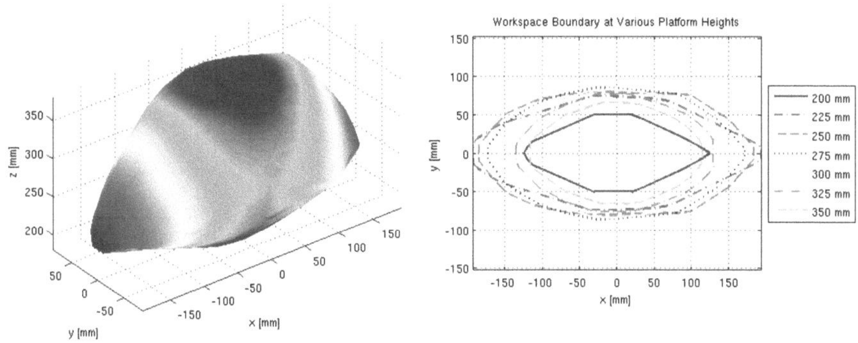


**Fig. 27.2** Workspace evaluation of the reconfigurable robot showing *angle-view* workspace volume and *top-view* workspace boundaries for non-degenerative 5-DOF limited mobility modes

platform vector lengths are 250 and 150 mm, respectively. The stroke limits of the hybrid active/passive prismatic motors are defined as $200\,\text{mm} \le s_i \le 400\,\text{mm}$. The operational limit of the spherical joint is restricted to be within 30° of the home configuration. Here, we define the home configuration of the parallel robot to be $\text{h}_\text{h} = \{0, 0, 300^\text{T}$, mm, and no platform rotation. At this configuration, the direction of each branch is simply $\hat{\mathbf{s}}_{i,\text{h}} = \left|\mathbf{h}_\text{h} + \mathbf{p}'_i - \mathbf{b}_i\right| / \left|\mathbf{s}_{i,\text{h}}\right|$. Thus we can see that the operational limit of each spherical joint is $\cos^{-1}\left(\hat{\mathbf{s}}_{i,\text{h}} \cdot \hat{\mathbf{s}}_i\right) \le \pi/6$.

The workspaces for the non-degenerate limited mobility modes are show in Figs. 27.2, 27.3, and 27.4 with a summary given in Table 27.3. For each case, the workspace is first decomposed into 10 mm intervals in the global $z$-direction in order to find the boundary of the workspace at that particular level. The individual boundaries are then unified to create a bound volume as shown in the figures below. In all cases, the minimum and maximum bounds for $h_z$ is 180 and 380 mm, respectively, yielding a 200 mm height differential in terms of reachable space. Also shown are workspace boundaries at 25 mm intervals ranging from 200 to 350 mm, highlighting the bulk of the workspace in terms of reachable space.

For the 5-DOF limited mobility mode, there is only one isomorphic type. Here, $\mathbf{q}_D = \theta_z$, as most 5-DOF tasks do not require independent rotation about the platform $z$-axis. For instance, machining operations do not require independent $\theta_z$ rotational control. From Fig. 27.2, the workspace is symmetric about the constraint plane and elongated along this plane due to the branch constraint. The workspace bulges out slightly on the end opposite to the constrained branch, but allows for significant reach in all directions. For a given task, the passive branch can be selected accordingly to ensure that the task remains encapsulated within the workspace.

The two non-degenerate 4-DOF limited mobility isomorphic types show symmetry about the plane that is angularly equidistant between the two constraints planes (Figs. 27.3a and b). Here, the dependent variables are usually $\mathbf{q}_D = \{\theta_x, \theta_y\}^\text{T}$, or $\mathbf{q}_D = \{h_x, h_y\}^\text{T}$, depending if Cartesian with $\theta_z$ motion, or

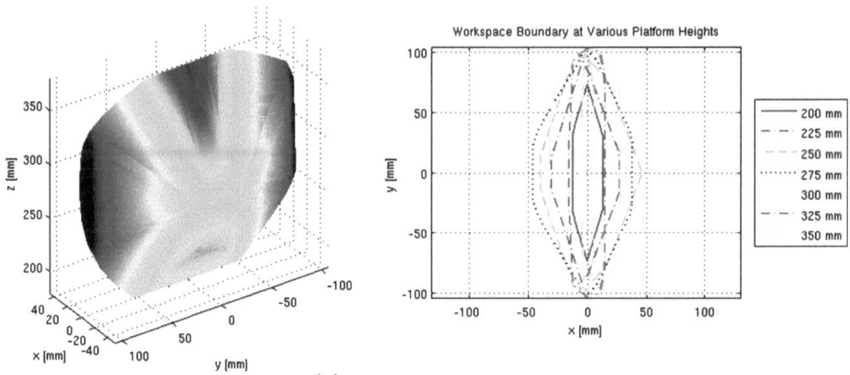


**(a)** 1-2 Isomorphic Type

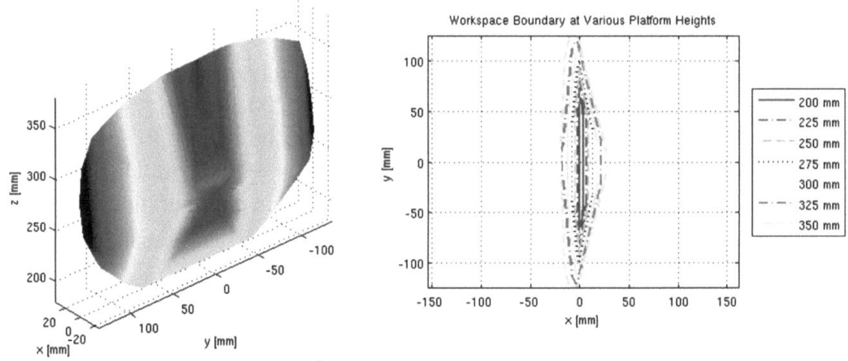


**(b)** 1-3 Isomorphic Type

**Fig. 27.3** Workspace evaluation of the reconfigurable robot showing *angle-view* workspace volume and *top-view* workspace boundaries for non-degenerative 4-DOF limited mobility modes

spherical with $h_z$ motion is required. Once again, the passive branch selection is independent of dependent variable selection. From Fig. 27.3, we see that as the angle between the two constrained branches is increased, the resultant workspace volume decreases, whereas the maximum reach increases. The 1–2 isomorphic type as compared to the 1–3 isomorphic type, has a shorter reach in the $y$-direction, but has a longer reach in the $x$-direction, resulting in a larger workspace volume. If then a Cartesian plus $\theta_z$ motion is desired, a trade-off between workspace volume and maximum reach becomes a dominant factor in selecting which isomorphic type is better for a given task.

In the case of the 3-DOF limited mobility mode, the 1–2–3 isomorphic type displays symmetry about the middle constraint plane, $\mathbf{n}_2$ (Fig. 27.4), whereas for the 1–3–5 isomorphic type, the workspace is periodically symmetric about each constraint plane (Fig. 27.4). Here, the dependent variables are defined as $\mathbf{q}_D = \{h_x, h_y, \theta_z\}^{\mathrm{T}}$ which is common for machining tasks. Other dependent motion combinations can be Cartesian where $\mathbf{q}_D = \{\theta_x, \theta_y, \theta_z\}^{\mathrm{T}}$, or spherical where $\mathbf{q}_D = \{h_x, h_y, h_z\}^{\mathrm{T}}$. Once again, we note that if spherical motion is required, then at least

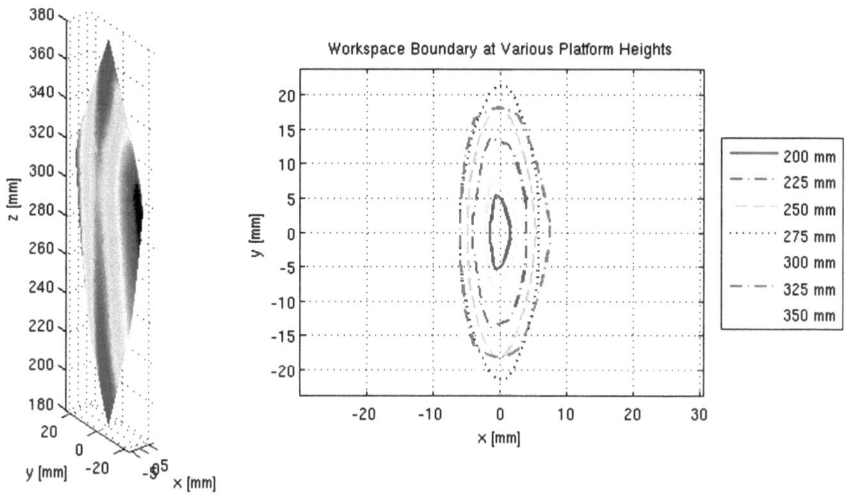


**(a)** 1-2-3 Isomorphic Type

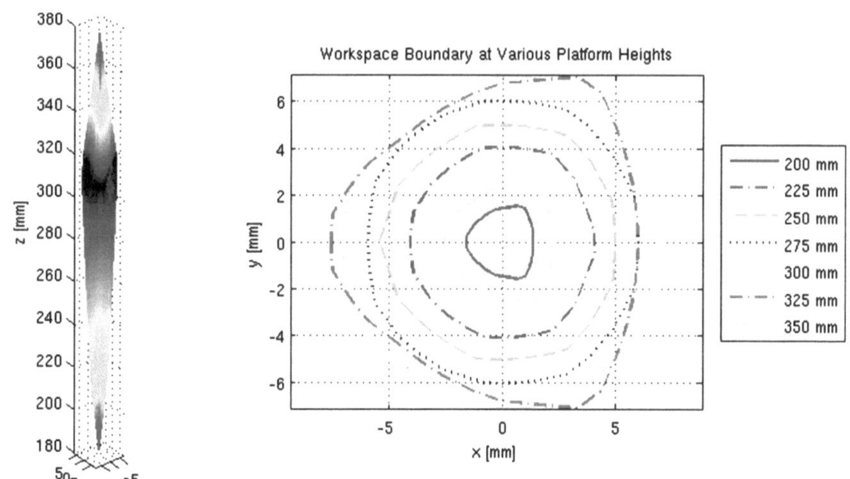


**(b)** 1-2-3 Isomorphic Type

**Fig. 27.4** Workspace evaluation of the reconfigurable robot showing *angle-view* workspace volume and *top-view* workspace boundaries for non-degenerative 3-DOF limited mobility modes

one constraint plane must have a nonzero $n_{i,z}$ component. A direct comparison the workspaces for this limited mobility mode is difficult due to the difference in workspace shape. For the 1–2–3 isomorphic type, the workspace is elongated, allowing for an extended reach. We can see that the workspace of the 1–3–5 isomorphic type could possibly fit within that of the 1–2–3 isomorphic type. However, we not that the distribution of branches in the 1–3–5 isomorphic type is

**Table 27.3** Workspace summary of the reconfigurable parallel robot

| | $M$ | 5-DOF | 4-DOF | | 3-DOF | |
|---|---|---|---|---|---|---|
| | Type | 1 | 1–2 | 1–3 | 1–2–3 | 1–3–5 |
| $h_x$(mm) | Minimum | −191.0 | −50.1 | −24.8 | −7.0 | −7.6 |
| | Maximun | 191.0 | 45.5 | 33.1 | 7.6 | 7.0 |
| $h_y$(mm) | Minimum | −86.0 | −106.8 | −136.4 | −23.8 | −7.4 |
| | Maximum | 86.0 | 106.8 | 136.4 | 23.8 | 7.4 |
| $\theta_x$(°) | Minimum | −26.0 | −52.0 | −56.0 | −26.0 | −26.0 |
| | Maximum | 26.0 | 52.0 | 56.0 | 26.0 | 26.0 |
| $\theta_y$(°) | Minimum | −32.0 | −44.0 | −48.0 | −25.0 | −24.0 |
| | Maximum | 32.0 | 52.0 | 48.0 | 24.0 | 25.0 |
| $\theta_z$(°) | Minimum | −36.8 | −44.0 | −30.0 | −3.3 | −3.0 |
| | Maximum | 36.8 | 44.0 | 30.0 | 3.3 | −3.0 |
| Maximum reach (mm) | | 191.3 | 106.8 | 136.4 | 23.9 | 7.8 |
| Volume (mm$^3$) | | 6,331,180 | 1,238,026 | 946,572 | 45,718 | 14,301 |

more common of traditional 3-DOF robots and might provide more consistent stiffness properties throughout the workspace as compared to the 1–2–3 isomorphic type which has all active branches on one side of the robot. From a pure reach and workspace volume perspective though, the 1–2–3 isomorphic type outperforms the 1–3–5 isomorphic type.

The resultant workspaces are an indication of the capabilities of the robot for each limited mobility state. The resultant volume, maximum reach and other kinematic values listed in Table 27.3 provides guidance for reconfiguration. Depending on the task to be accomplished, the appropriate isomorphic configuration can be selected to make sure that the task is encapsulated within the workspace. If this is not the case, then the reconfigurable robot can intelligently switch between appropriate isomorphic configurations to further enhance its capabilities by superpositioning each of the workspace manifolds. Finally, though not addressed here, repositioning the orientation of the constraint planes can further expand the robot's reach and is currently being investigated.

## 27.5 Conclusions

In this paper we provide guidance on the topological reconfiguration of a fully reconfigurable parallel robot which has not previously been studied. By using the existing structure of a full six degree-of-freedom parallel robot, reconfiguration to limited mobility configurations is realized by means of hybrid active/passive prismatic motors and reconfigurable universal-to-revolute joints. The robot is able to assuming a configuration with only the degrees-of-freedom required to complete a particular task, buy reconfiguring to different mobility modes. Branch-based mobility was established, highlighting the mobility and connectivity requirements in reconfiguring an unconstrained-active branch to a constrained-passive branch. By analyzing the isomorphisms of non-degenerate configurations, their kinematic constraints and resultant workspace volume and reach, highlight the capabilities of each configuration. This providing the necessary guidance as to which configuration is best suited for a given task. Further analysis into the orientation of the passive constraint planes, as well as combining geometric and topological reconfiguration within the system is ongoing and will produce reconfigurable robots with greater reconfiguration capabilities.

## References


1. Xi F, Li Y, Wang H (2011) Module-based method for design and analysis of reconfigurable parallel robots. Front Mech Eng 6(2):151–159
2. Bi ZM, Wang L (2009) Optimal design of reconfigurable parallel machining systems. Robot Com-Int Manuf 25:951–961
3. Gan D, Dai JS, Liao Q (2010) Constraint analysis on mobility change of a novel metamorphic parallel mechanism. Mech Mach Theory 45:1864–1876
4. Kotlarski J, Abdellatif H, Ortmaier T, Heimann B (2009) Enlarging the useable workspace of planar parallel robots using mechanisms of variable geometry. In: Proceedings of international conference on reconfigurable mechanisms and robots, 2009, pp 94–103
5. Bi ZM, Kang B (2010) Enhancement of adaptability of parallel kinematic machines with an adjustable platform. J Manuf Sci Eng 132:061016–10610169
6. Sun T, Song Y, Li Y, Zhang J (2010) Workspace decomposition based dimensional synthesis of a novel hybrid reconfigurable robot. J Mech Robot 2:031009–10310098
7. Dash AK, Chen I-M, Yeo SH, Yang G (2005) Task-oriented configuration design for reconfigurable parallel manipulator systems. Int J Comp Integ M 18(7):615–634
8. Xi F, Xu Y, Xiong G (2006) Design and analysis of a re-configurable parallel robot. Mech Mach Theory 41:191–211
9. Vertuan A, Legnani G, Adamini R et al (2009) Performance analysis of a reconfigurable redundant parallel manipulator. In: Proceedings of international conference on reconfigurable mechanisms and robots, pp 593–601
10. Finistauri AD, Xi F (2009) Type synthesis and kinematics of a modular variable geometry truss mechanism for aircraft wing morphing. In: Proceedings of international conference on reconfigurable mechanisms and robots, 2009, pp 470–477
11. Finistauri AD, Xi F, Petz B (2008) Architecture design and optimization of an on-the-fly reconfigurable parallel robot. In: H Wu (ed) Parallel manipulators, towards new applications. I-Tech Education and Publishing, Vienna, pp 379–404
12. Liu G, He X, Yuan J et al (2008) Development of modular and reconfigurable robot with multiple working modes. In: Proceedings IEEE international conference on robotics and automation. doi:10.1109/ROBOT.2008.4543746
13. Tsai L-W (1998) Systematic enumeration of parallel manipulators, vol 98-33. In: . Technical report, Department of Mechanical Engineering, Institute for System Research, University of Maryland, pp 1–11

# Chapter 28
# Type-Changeable Kinematic Pair Evolved Reconfigurable Parallel Mechanisms

**Ketao Zhang, Evangelos Emmanouil, Yuefa Fang and Jian S. Dai**

**Abstract** This paper presents a type-changeable kinematic pair with variable topology. The geometry properties and topological phases of the type-changeable kinematic joint are revealed. A novel reconfigurable parallel mechanism is evolved from the type-changeable kinematic pair according to geometric conditions of assemblages for parallel mechanisms. The platform of the reconfigurable parallel mechanism is capable of implementing various functions accompanying to the phase change of the integrated type-changeable kinematic pair. The platform has 6 DOFs in the source phase and can change its mobility to 5, 4 and 3. Experiments have been carried out for identification of a $3R_ePS_e$ subphase controller.

**Keywords** Type-changeable kinematic pair · Reconfigurable parallel mechanism · Screw theory

K. Zhang (✉) · E. Emmanouil · J. S. Dai
Centre for Robotics Research, King's College London, University of London, Strand, London, WC2R 2LS, UK
e-mail: ketao.zhang@kcl.ac.uk

E. Emmanouil
e-mail: evangelos.emmanouil@kcl.ac.uk

J. S. Dai
e-mail: jian.dai@kcl.ac.uk

K. Zhang · Y. Fang
Department of Mechanical Engineering, Beijing Jiaotong University, 100044 Beijing, People's Republic of China
e-mail: yffang@bjtu.edu.cn

J. S. Dai et al. (eds.), *Advances in Reconfigurable Mechanisms and Robots I*,
DOI: 10.1007/978-1-4471-4141-9_28, © Springer-Verlag London 2012 

## 28.1 Introduction

Parallel mechanisms [1, 2] with invariant topological structure which leads to specific type of motion in a general configuration are used in the fields such as manufacturing, space exploration and medical robotics [3]. The extensive applications of parallel mechanisms and robot manipulators raise numerous theoretical and experimental challenges. In particular, a parallel mechanism usually has fixed mobility and does not have the ability to change it, resulting in limited flexibility to accommodate changes required in the industrial environments [4].

New mechanisms including metamorphic mechanisms and kinematotropic mechanisms and other developments in reconfigurable mechanisms can be developed into reconfigurable robot manipulators, evolutionary robot limbs and devices to adapt to changed conditions and deal with complicated tasks. The study of this new class of mechanisms started in the mid-1990s when Dai [5] addressed the reconfigurable assembly and packaging system for various artifacts and decorative gifts and boxes, and Wohlhart [6] presented the kinematotropic linkage together with the position parameter variation. In 1998, the metamorphic mechanism [7] was proposed following the concept of metamorphosis in the sense of evolution. This new kind of mechanisms is capable of changing its structure, topological configuration and subsequently mobility via either link annex or joint property change. The kinematotropic linkages are coined as one kind of linkages that change their full-cycle mobility at certain configuration which was called branch singularity [8]. The mobility change of this kind of mechanisms is induced by a certain geometric constraint to limit the motion and lead to different motion branches. Calletti and Fanghella [9] presented various single and multi-loop kinematotropic mechanisms. Fanghella et al. [10] addressed the parallel robots whose platforms can change their subgroups of displacement. The mobility change derived from varying topology is the characteristic of reconfigurable mechanisms with variable topologies. Yan and Kuo [11] emphasized the meaning of topology variation of the variable kinematic joints based on descriptive joint representation, and carried out the configuration synthesis of mechanisms with variable topologies [12].

Though the study of the new mechanisms with variable topologies has been made, literatures focused on typical parallel mechanisms with ability either to alter motion pattern or to perform mobility change are very few. Zlatanov et al. [13] presented the DYMO, a multi-operational parallel mechanism and examined the different operation modes separated by constraint singularity of the parallel mechanism. Kong et al. [14] synthesized the 3-DOF parallel mechanisms with both spherical and translational modes. Recently, Zhang et al. [4, 15, 16] presented a type-changeable kinematic pair extracted from origami folds and investigated the topological reconfiguration and mobility change of the evolved parallel mechanism. Gan et al. [17] proposed the design of a new rT joint by integrating an extra rotational degree of freedom in the well-known Hooke's joint and investigated the

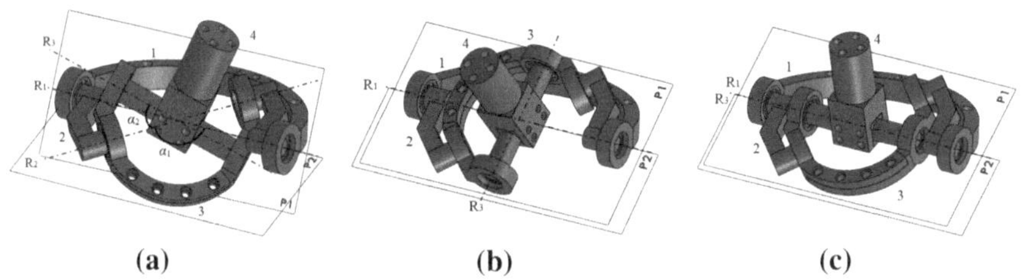

Fig. 28.1 The type-changeable kinematic pair. **a** Source phase. **b** First subphase. **c** Second subphase

mobility change of two types of metamorphic parallel mechanisms which employ the rT joint.

This paper presents a novel reconfigurable parallel mechanism evolved from the type-changeable kinematic pair. The reconfigurable parallel mechanism can change its configuration from fully 6-DOF to 3-DOF. The constraints change induced mobility variations of the platform are analyzed based on reciprocity of screws [18–20]. Experiments have been carried out for identification of a $3R_ePS_e$ subphase controller.

## 28.2 Geometry of the Type-Changeable Kinematic Pair

The type-changeable kinematic pair in Fig. 28.1 is inspired and extracted from an artifact [15]. Metamorphosis of the type-changeable kinematic pair is induced by its embodied geometry which distinguishes this equivalent kinematic pair from a general 3R spherical kinematic chain. The geometric characteristic of the type-changeable kinematic pair is determined by the angle length between adjacent joint axes $\alpha_1$ and $\alpha_2$, where $\alpha_1 = 135°$ and $\alpha_2 = 45°$.

Consider the geometry of nonadjacent joint axes $R_1$ and $R_3$, the type-changeable kinematic pair is able to change its topological configuration and implement different working functions in the working cycle resorting to link annex. The type-changeable kinematic pair in its original configuration in Fig. 28.1a generates equivalent spherical motion and acts as a metamorphic source generator that contains subphase configuration and completes the transition between subphase configurations. Applying the geometric constraint in the process of motion of the type-changeable kinematic pair, two subphase configurations can be generated.

The type-changeable kinematic pair reduces its link number by annexing link 3 to link 2 at the configuration in Fig. 28.1b. In this stable topological configuration, the link 4 has only two independent rotational DOFs with respect to the base link 1 and can be taken as an equivalent Hooke's pair. The type-changeable kinematic pair changes into the second stable subphase in Fig. 28.1c when the link 3 and link 2 are annexed in the typical configuration with collinear joint axes $R_1$ and $R_3$. In

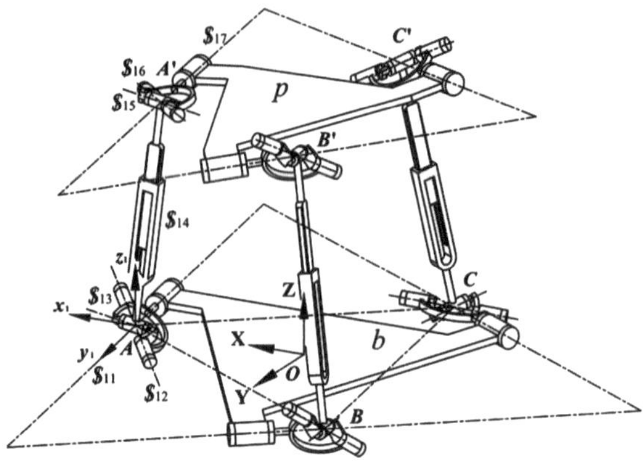


**Fig. 28.2** Kinematic model of the reconfigurable parallel mechanism

this configuration, the kinematic pair acts as an equivalent compound rotating pair and the output link 4 has one rotational DOF with respect to the base link 1. The source configuration, first subphase and second subphase of the kinematic pair are labeled $S_e$, $U_e$ and $R_e$ respectively, where the subscript 'e' denotes the equivalence of the motion to a commonly used kinematic pair.

## 28.3 Mobility Change of the Reconfigurable Parallel Mechanism

A reconfigurable parallel mechanism in Fig. 28.2 is symmetrically constructed with three limbs by integrating the type-changeable kinematic pair at both end of the prismatic joint in each limb.

As illustrated in Fig. 28.3, the common points of the three lower $S_e$ pairs located in the base plane form an equilateral triangle $\triangle ABC$. The axes of revolute joints connecting to the base are coplanar. At the other end, the common points of the three upper $S_e$ pairs located in the platform form an equilateral triangle $\triangle A'B'C'$ and the axes of revolute joints connecting to the platform are coplanar. The coordinate frame O-XYZ attached at the center of the base is set as the global frame. The X-axis is perpendicular to $BC$, Y-axis is parallel to $BC$, and Z-axis is normal to the base following the right-handed rule.

### *28.3.1 Working Phase with 6 DOFs*

The reconfigurable parallel mechanism in Fig. 28.2 is in the source configuration as all the integrated type-changeable kinematic pairs are in their source phase. It implies that all the joints are active without link annex. The origin of the local

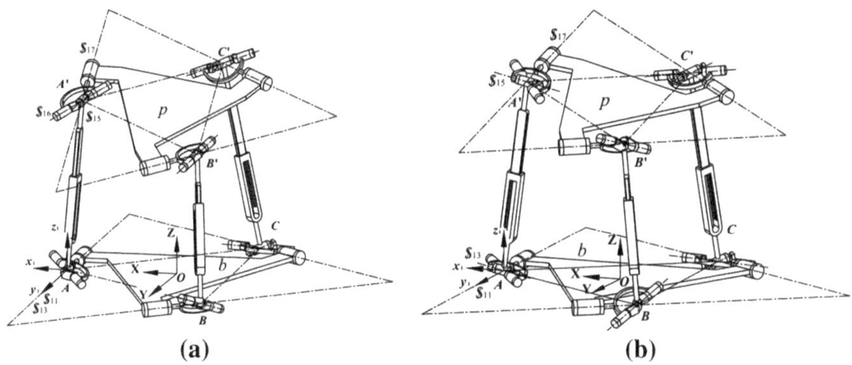


**Fig. 28.3** Subphases $R_ePS_e$-$2S_ePS_e$ and $U_ePU_e$-$2S_ePS_e$

reference frames $o_i$-$x_iy_iz_i$ is located at the common points $A$, $B$ and $C$ of lower $S_e$ pairs in the three limbs. The $y_i$-axis is collinear with the axis of joint $\boldsymbol{S}_{i1}$, $z_i$-axis is normal to the base plane and pointing upward. The $x_i$-axis is perpendicular to $\boldsymbol{S}_{i1}$ and completes a right-handed coordinate frame. In this source configuration, the screw system of limb $AA'$ expressed in the local frame $o_1$-$x_1y_1z_1$ is given by

$$\mathbb{S}_1 = \begin{cases} \boldsymbol{S}_{11} = [0 \quad 1 \quad 0 \quad 0 \quad 0 \quad 0]^{\mathrm{T}} \\ \boldsymbol{S}_{12} = [l_2 \quad \sqrt{(l_2)^2 + (n_2)^2} \quad n_2 \quad 0 \quad 0 \quad 0]^{\mathrm{T}} \\ \boldsymbol{S}_{13} = [l_3 \quad m_3 \quad n_3 \quad 0 \quad 0 \quad 0]^{\mathrm{T}} \\ \boldsymbol{S}_{14} = [0 \quad 0 \quad 0 \quad a \quad b \quad c]^{\mathrm{T}} \\ \boldsymbol{S}_{15} = [l_3 \quad m_3 \quad n_3 \quad bn_3 - cm_3 \quad cl_3 - an_3 \quad am_3 - bl_3]^{\mathrm{T}} \\ \boldsymbol{S}_{16} = [l_6 \quad m_6 \quad n_6 \quad bn_6 - cm_6 \quad cl_6 - an_6 \quad am_6 - bl_6]^{\mathrm{T}} \\ \boldsymbol{S}_{17} = [l_7 \quad m_7 \quad n_7 \quad bn_7 - cm_7 \quad cl_7 - an_7 \quad am_7 - bl_7]^{\mathrm{T}} \end{cases} \tag{28.1}$$

where $(l_2, \sqrt{(l_2)^2+(n_2)^2}, n_2), (l_3, m_3, n_3), (l_6, m_6, n_6)$ and $(l_7, m_7, n_7)$ are the unit vectors pointing in the direction of screws $\boldsymbol{S}_{12}$, $\boldsymbol{S}_{13}$, $\boldsymbol{S}_{16}$ and $\boldsymbol{S}_{17}$, $(a, b, c)$ is the coordinates of point $A'$ in the local frame.

The twist screw system in Eq. (28.1) is a 6-system and there is no reciprocal screw. Under such conditions, the three limbs do not impose any constraints on the platform and the mobility of the parallel mechanism is 6.

When the three type-changeable kinematic pairs connected to the base change their configuration to the $U_e$ subphase, the parallel mechanism is working in the $3U_ePS_e$ subphase. In this topological subphase, the joints with screw $\boldsymbol{S}_{i2}$ are inactive due to link annex of lower $S_e$ pair in each limb. The remaining screws except $\boldsymbol{S}_{i2}$ of the twist screw system in Eq. (28.1) are independent and form a 6-system. It shows that the three limbs do not contribute any constraint to the platform. Hence, the platform of the parallel mechanism in this subphase is free of any constraints and has 6 DOFs.

### 28.3.2 Working Phase with 5 DOFs

When the lower $S_e$ pair of one limb changes its configuration from the source phase to the $R_e$ subphase, the parallel mechanism will changes to $R_ePS_e$-$2S_ePS_e$ structure in Fig. 28.3a. For instance, the $S_e$ pair with common point $A$ changes to $R_e$, the twist system of limb $AA'$ becomes

$$\mathbb{S}_1 = \begin{cases} \mathbf{S}_{11} = \begin{bmatrix} 0 & 1 & 0 & 0 & 0 & 0 \end{bmatrix}^{\mathrm{T}} \\ \mathbf{S}_{13} = \begin{bmatrix} 0 & 1 & 0 & 0 & 0 & 0 \end{bmatrix}^{\mathrm{T}} \\ \mathbf{S}_{14} = \begin{bmatrix} 0 & 0 & 0 & a & 0 & c \end{bmatrix}^{\mathrm{T}} \\ \mathbf{S}_{15} = \begin{bmatrix} 0 & 1 & 0 & -c & 0 & a \end{bmatrix}^{\mathrm{T}} \\ \mathbf{S}_{16} = \begin{bmatrix} l_6 & m_6 & n_6 & -cm_6 & cl_6 - an_6 & am_6 \end{bmatrix}^{\mathrm{T}} \\ \mathbf{S}_{17} = \begin{bmatrix} l_7 & m_7 & n_7 & -cm_7 & cl_7 - an_7 & am_7 \end{bmatrix}^{\mathrm{T}} \end{cases} \tag{28.2}$$

The constraint-screw system which is reciprocal to the twist system in Eq. (28.2) can be yielded as

$$\mathbb{S}_1^{\mathrm{r}} = \left\{ \mathbf{S}_{11}^{\mathrm{r}} = \begin{bmatrix} 0 & 1 & 0 & -c & 0 & a \end{bmatrix}^{\mathrm{T}} \right\} \tag{28.3}$$

This shows that the limb $AA'$ exerts one constraint force on the platform. The constraint force is parallel to $\boldsymbol{S}_{i1}$ and passing through the upper common point $A'$. Thus the translational motion parallel to the $Y$-axis of the platform is restricted by the constraint force and the platform has 5 DOFs.

When both the lower and upper $S_e$ pairs of one limb change their configuration from the source phase to the $U_e$ subphase, the parallel mechanism changes to $U_ePU_e$-$2S_ePS_e$ structure. For instance, the $S_e$ pairs of limb $AA'$ change to $U_e$ subphase in Fig. 28.3b. The local reference frame of limb $AA'$ is set up according to the geometry. The $y_1$-axis is collinear with the axis of joint $\boldsymbol{S}_{11}$ and $x_1$-axis is collinear with $\boldsymbol{S}_{13}$. The $z_1$-axis is pointing upward and completes a right-handed coordinate frame. Hence, the twist system of limb $AA'$ with $U_ePU_e$ structure is given by

$$\mathbb{S}_1 = \begin{cases} \mathbf{S}_{11} = \begin{bmatrix} 0 & 1 & 0 & 0 & 0 & 0 \end{bmatrix}^{\mathrm{T}} \\ \mathbf{S}_{13} = \begin{bmatrix} 1 & 0 & 0 & 0 & 0 & 0 \end{bmatrix}^{\mathrm{T}} \\ \mathbf{S}_{14} = \begin{bmatrix} 0 & 0 & 0 & 0 & b & c \end{bmatrix}^{\mathrm{T}} \\ \mathbf{S}_{15} = \begin{bmatrix} 1 & 0 & 0 & 0 & c & -b \end{bmatrix}^{\mathrm{T}} \\ \mathbf{S}_{17} = \begin{bmatrix} 0 & \cos\theta & \sin\theta & b\sin\theta - c\cos\theta & 0 & 0 \end{bmatrix}^{\mathrm{T}} \end{cases} \tag{28.4}$$

where $\theta$ is the angles measured from joint with screw $\boldsymbol{S}_{17}$ to the $y_1$-axis.

The constraint-screw system is reciprocal to the twist system in Eq. (28.4) and can be calculated as

$$\mathbb{S}_1^{\mathrm{r}} = \left\{ \mathbf{S}_{11}^{\mathrm{r}} = \begin{bmatrix} 1 & 0 & 0 & 0 & 0 & c\cot\theta - b \end{bmatrix}^{\mathrm{T}} \right\} \tag{28.5}$$

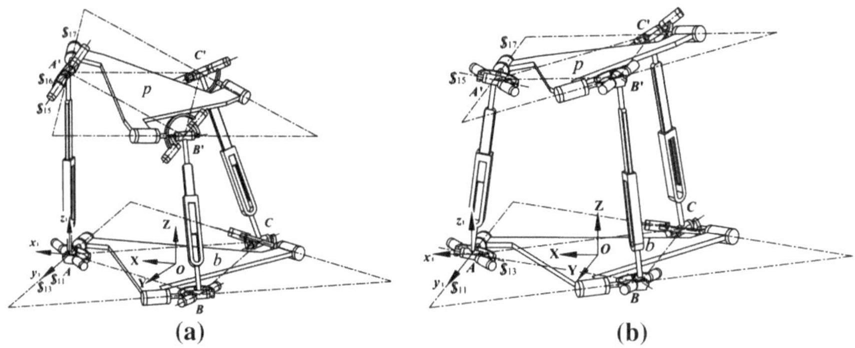


**Fig. 28.4** Subphases $2R_ePS_e$-$S_ePS_e$ and $2U_ePU_e$-$S_ePS_e$

The above screw is a constraint force which passes through the intersection of $\boldsymbol{S}_{11}$ and $\boldsymbol{S}_{17}$ and is parallel to $\boldsymbol{S}_{13}$. Under such a condition, the platform has 5 DOFs with one translational motion restricted.

### 28.3.3 Working Phase with 4 DOFs

When the lower $S_e$ pairs in two limbs change their configuration from source phase to the $R_e$ subphase, the parallel mechanism changes its structure to $2R_ePS_e$-$S_ePS_e$ in Fig. 28.4a.

In this subphase, limb $AA'$ exerts one constraint force passing through common point $A'$ and limb $BB'$ exerts one constraint force passing through common point $B'$. As a result, two translations of the platform are restricted and the platform has 4 DOFs with one translation and three rotations.

When both the lower and upper $S_e$ pairs in two limbs change their configuration from the source phase to the $U_e$ subphase, the parallel mechanism changes its structure to $2U_ePU_e$-$S_ePS_e$ in Fig. 28.4b. In this topological subphase, the two limbs $AA'$ and $BB'$ exert two constraint forces to the platform. Hence, the platform has 4 DOFs with one translation and three rotations.

### 28.3.4 Working Phase with 3 DOFs

When the lower Se pairs of all the three limbs change their configuration from the source phase to the Re subphase, the parallel mechanism changes its structure to $3R_ePS_e$ [16]. In this subphase, the three limbs exert three constraint forces which are parallel to the base without common point. Thus the platform has 3 DOFs with one translation and two rotations.

When all the six $S_e$ pairs in the three limbs change their configuration from the source phase to the $U_e$ subphase, the parallel mechanism changes to $3U_ePU_e$ structure [16]. In this subphase, the joints axes characterized with screw $\boldsymbol{S}_{i1}$ and $\boldsymbol{S}_{i7}$ in each limb are parallel. The local reference frame is set up as that in Fig. 28.3b and the twist system of the limb $AA'$ is

$$\mathbb{S}_1 = \begin{cases} \mathbf{S}_{11} = [0 \quad 1 \quad 0 \quad 0 \quad 0 \quad 0]^{\mathrm{T}} \\ \mathbf{S}_{13} = [1 \quad 0 \quad 0 \quad 0 \quad 0 \quad 0]^{\mathrm{T}} \\ \mathbf{S}_{14} = [0 \quad 0 \quad 0 \quad a \quad b \quad c]^{\mathrm{T}} \\ \mathbf{S}_{15} = [1 \quad 0 \quad 0 \quad 0 \quad c \quad -b]^{\mathrm{T}} \\ \mathbf{S}_{17} = [0 \quad 1 \quad 0 \quad -c \quad 0 \quad a]^{\mathrm{T}} \end{cases} \tag{28.6}$$

The wrench system of the platform reciprocal to the motion screws can be yielded as

$$\mathbb{S}_1^{\mathrm{r}} = \left\{ \mathbf{S}_{11}^{\mathrm{r}} = [0 \quad 0 \quad 0 \quad 0 \quad 0 \quad 1]^{\mathrm{T}} \right\} \tag{28.7}$$

The screw in Eq. (28.7) presents a couple which is perpendicular to the plane formed by joint axes $\boldsymbol{S}_{11}$ and $\boldsymbol{S}_{13}$. Hence, the three limbs exert three constraint couples on the platform and the platform has three pure translations.

## 28.4 Inverse Kinematics for the $3R_ePS_e$ Configuration Controller and Experimental Data

The proposed reconfigurable parallel mechanism has been fabricated and the physical device actuated with DC motors is illustrated in Fig. 28.5.

According to the geometry of the $3R_ePS_e$ configuration, the common points $A'$, $B'$ and $C'$ can be expressed in the global coordinate frame O-XYZ as

$$\mathbf{A}' = [-r_a W_{12} \quad -r_a W_{22} \quad Z - r_a W_{32}]^{\mathrm{T}} \tag{28.8}$$

$$\mathbf{B}' = \left[\tfrac{1}{2} r_a(\sqrt{3}W_{11} + W_{12}) \quad \tfrac{1}{2} r_a(\sqrt{3}W_{21} + W_{22}) \quad Z + \tfrac{1}{2} r_a(-\sqrt{3}W_{31} + W_{32})\right]^{\mathrm{T}} \tag{28.9}$$

$$\mathbf{C}' = \left[\tfrac{1}{2} r_a(-\sqrt{3}W_{11} + W_{12}) \quad \tfrac{1}{2} r_a(-\sqrt{3}W_{21} + W_{22}) \quad Z + \tfrac{1}{2} r_a(\sqrt{3}W_{31} + W_{32})\right]^{\mathrm{T}} \tag{28.10}$$

in which $r_a = 17$ mm is the radius of the circle defined by $A'$, $B'$ and $C'$ and $W_{ij}$ ($i = 1, 2, 3$ and $j = 1, 2, 3$) are the coefficients derived using the D-H method on the platform and expressed in terms of $\theta_i$, the orientation of the platform about the Z, X and Y axis.

The common point A B and C in the global frame are given by

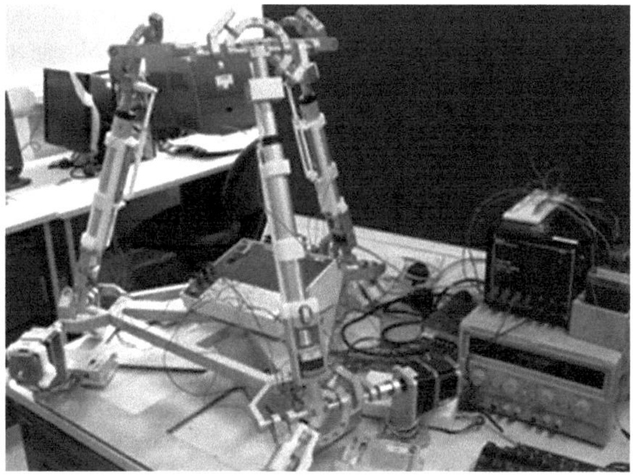

**Fig. 28.5** Subphase $3R_ePS_e$ of the reconfigurable parallel mechanism

$$\mathbf{A} = \begin{bmatrix} r_b\cos\phi_1 \\ r_b\sin\phi_1 \\ 0 \end{bmatrix}, \quad \mathbf{B} = \begin{bmatrix} r_b\cos\phi_2 \\ r_b\sin\phi_2 \\ 0 \end{bmatrix} \text{ and } \mathbf{C} = \begin{bmatrix} r_b\cos\phi_3 \\ r_b\sin\phi_3 \\ 0 \end{bmatrix} \tag{28.11}$$

where $r_b = 340$ mm is the radius of the circle defined by $A$, $B$ and $C$ and $\phi_1 = 0$, $\phi_2 = 2\pi/3$ and $\phi_3 = 4\pi/3$.

Hence, the length $L_i$ of each leg can be obtained and the desired displacements $D_i$ of the prismatic joints are

$$D_i = L_i - l_i \tag{28.12}$$

where $l_i$ is the certain minimum length of the leg, fixed during manufacturing.

The height of the centre of the platform is set to 525 mm and measurements for different angle combinations are taken by using a hanging weight and a protractor.

The following error means ($\mu$) and standard deviations ($\sigma$) where observed (Table 28.1). Height: $\mu = 0.21$ cm, $\sigma = 0.054$ cm. $\theta_2$: $\mu = 2.06°$, $\sigma = 0.165°$. $\theta_3$: $\mu = -1.07°$, $\sigma = 0.457°$. These error values are mainly due to imperfections in the manufacturing, attributed to the complexity of the metamorphic joints, as well as the base of the mechanism not being perfectly level.

The following measurements are taken by setting one of the angles to zero and changing the other angle from 0 up to the maximum achievable value (Table 28.2).

## 28.5 Conclusions

This paper presented a type-changeable kinematic pair and its evolved reconfigurable parallel mechanism. The reconfigurable parallel mechanism is able to change its topological configuration resorting to phase change of the integrated type-changeable kinematic pair. The constraints corresponding to the source phase

**Table 28.1** Position accuracy measurements for different orientations

| Set | | | Measured | | |
|---|---|---|---|---|---|
| Height (cm) | $\theta_2$ (deg) | $\theta_3$ (deg) | Height (cm) | $\theta_2$ (deg) | $\theta_3$ (deg) |
| | 10 | 10 | 50.5 | 12 | 8.5 |
| | 10 | 7 | 50.5 | 12 | 6.5 |
| | 10 | 3 | 50.5 | 12 | 3 |
| | 7 | 10 | 50.6 | 9 | 9 |
| 50.3 | 7 | 7 | 50.4 | 9 | 6 |
| | 7 | 3 | 50.5 | 9 | 2.5 |
| | 3 | 10 | 50.6 | 4 | 8 |
| | 3 | 7 | 50.5 | 4 | 6 |
| | 3 | 3 | 50.5 | 4 | 3 |
| | 0 | 0 | 50.5 | −0.5 | 2 |

**Table 28.2** Orientation accuracy measurements

| $\theta_2$ for $\theta_3 = 0°$ | | $\theta_3$ for $\theta_2 = 0°$ | |
|---|---|---|---|
| Set (deg) | Measured (deg) | Set (deg) | Measured (deg) |
| 0 | 2 | 0 | −0.5 |
| 1 | 3 | 1 | 1 |
| 2 | 4 | 2 | 1 |
| 4 | 6.5 | 3 | 2 |
| 5 | 7 | 4 | 2.5 |
| 6 | 8 | 5 | 4 |
| 8 | 10 | 6 | 5 |
| 10 | 12 | 7 | 6 |
| | – | 8 | 7 |
| | – | 9 | 8 |
| | – | 10 | 8.5 |
| | – | 12 | 11 |
| | – | 15 | 13 |
| | – | 18 | 16.5 |

and subphases were analyzed and the motion characteristics of the platform are revealed based on screw theory. The analysis reveals that the platform of the mechanism is capable of varying from the source phase with 6 DOFs to subphases with 5, 4 and 3 DOFs. Experiments for identification of a $3R_ePS_e$ subphase controller have been carried out with investigation of inverse kinematics.

**Acknowledgments** The authors thank the support of European Commission—Framework 7 Programme under grant number 270436 and the support of the National Natural Science Foundation of China under grant number 51075025.

## References


1. Dasgupta B, Mruthyunjaya TS (2000) The Stewart platform manipulator: a review. Mech Mach Theory 35:15–40
2. Merlet J-P (1999) Parallel robots. Kluwer Academic Publishers, New York
3. Pennock GR, Raghavan M (2006) Spatial mechanisms and robot manipulators. ASME J Mech Des 128:149–150
4. Zhang KT, Dai JS, Fang YF (2010) Topology and constraint analysis of phase change in the metamorphic chain and its evolved mechanism. ASME J Mech Des 132:121001
5. Dai JS (1996) Conceptual study of the dexterous reconfigurable assembly and packaging system. Science and Technology Report, Unilever Research
6. Wohlhart K (1996) Kinematotropic linkages. Recent advances in robot kinematics. In: Lenarcic J, Parenti Castelli V (eds). Kluwer, Dordrecht
7. Dai JS, Rees Jones J (1998) Mobility in metamorphic mechanisms of Folda-ble/Erectable kinds. In: Proceedings of the 25th ASME Biennial mechanisms and robotics conference, DETC-5902, Atlanta, USA, 1998
8. Gogu G (2009) Branching singularities in kinematotropic parallel mechanisms. In: Proceedings of the 5th international workshop on computational kinematics, Duisburg, Germany, 2009, pp. 341–348
9. Calletti C, Fanghella P (2001) Single-loop kinematotropic mechanism. Mech Mach Theory 36:743–761
10. Fanghella P, Galletti C, Giannotti E (2006) Parallel robots that change their group of motion. Advances in robot kinematics, In: Jadran L, Roth B (eds). The Netherlands, pp. 49–56
11. Yan H-S, Kuo C-H (2006) Topological representations and characteristics of variable kinematic joints. ASME J Mech Des 128:384–391
12. Yan H-S, Kang C-H (2009) Configuration synthesis of mechanisms with variable topologies. Mech Mach Theory 44:896–911
13. Zlatanov D, Bonev IA, Gosselin CM (2002) Constraint singularities as C-Space singularities. Advances in robot kinematics—theory and applications, pp. 183–192
14. Kong X, Gosselin CM, Richard PL (2007) Type synthesis of parallel mechanisms with multiple operation modes. ASME J Mech Des 129:595–601
15. Zhang KT, Dai JS, Fang YF (2009) A new metamorphic mechanism with ability for platform orientation switch and mobility change. In: ASME/IFToMM international conference on reconfigurable mechanisms and robots, London, UK, 2009, pp. 626–632
16. Zhang KT, Dai JS, Fang YF (2011) Geometry and constraint based design of metamor-phic parallel mechanisms. In: Proceedings of the ASME 2011 international design engineering technical conferences, DETC2011—47703, Washington DC, USA, 2011
17. Gan D, Dai JS, Liao Q (2009) Mobility change in two types of metamorphic parallel mechanisms. ASME J Mech Robotics 1:041007
18. Hunt KH (1990) Kinematic geometry of mechanisms. Oxford University Press, New York
19. McCarthy JM (2000) Geometric design of linkages. Springer-Verlag, New York
20. Dai JS, Rees JJ (2001) Interrelationship between screw systems and corresponding reciprocal systems and applications. Mech Mach Theory 36:633–651

# Chapter 29
# Developing a New Concept of Self Reconfigurable Intelligent Swarm Fixtures

**Luis de Leonardo, Matteo Zoppi, Li Xiong, Serena Gagliardi and Rezia Molfino**

**Abstract** As manufacture trends move towards life-cycle design, sustainable production, geometrical complexity, short time-to-market, small and variable batch production and mass customization; manufacture equipment struggles to keep pace and provide the required flexibility, adaptability and automation. This paper describes the new concept of self-reconfigurable intelligent swarm fixtures. Following the conception of this new fixturing system, it arises under European commission 7th framework programme the SwarmItFIX project that gives an application and an insight towards the new fixture technology. The paper gives a description of the solution developed in SwarmItFIX focusing on the physical prototype demonstrator constructed to validate the concept as an intended new fixture benchmark for manufacturing sheet metal panels.

**Keywords** Aeronautics · Fixture technology · Robotics · Manufacturing

---

L. de Leonardo (✉) · M. Zoppi · L. Xiong · S. Gagliardi · R. Molfino
DIMEC-Universita' di Genova, Via all'Opera Pia 15A, Genoa, Italy
e-mail: deleonardo.luis@gmail.com

M. Zoppi
e-mail: zoppi@dimec.unige.it

L. Xiong
e-mail: lixiong@dimec.unige.it

S. Gagliardi
e-mail: serenagagliardi85@alice.it

R. Molfino
e-mail: molfino@dimec.unige.it

J. S. Dai et al. (eds.), *Advances in Reconfigurable Mechanisms and Robots I*,
DOI: 10.1007/978-1-4471-4141-9_29, © Springer-Verlag London 2012

## 29.1 Introduction

Thin metal sheets and composite components with 3D geometries are being increasingly used in the aerospace, automotive and other industries keeping with trends such as life cycle design, sustainable production and geometric complexity [9]. The effects of these trends make manufacture equipment and fixtures to evolve in flexibility, reconfigurability and autonomy. The demand for flexibility is pushed by short time-to-market, small and variable batch production and mass customization [5].

Today smartest adaptable flexible fixture systems (FFS), can be classified in modular flexible fixture systems (MFFS), single structure flexible fixture systems (SSFFS) and robotic fixtureless assemblies (RFAs), a brief description and evaluation regarding their use in flexible automated manufacturing of thin walled parts is given in [12]. Extensive research has been done in the field proving the interest of both researches and industrialists, it is reported in [6] the development of a pin-bed layout combining the use of phase-change materials. In [1] a new device capable to support thin and compliant workpieces securely using a low temperature melting alloy is introduced. In [15], a modular reconfigurable fixturing system of thin-walled flexible objects subjected to a discrete number of point forces is presented. In [16] a reconfigurable fixturing technique for robotic setup is described. [2] Proposes a reconfigurable fixture for robotic assembly, with suction cups as holding principle. An in depth analysis of the pin-type tool is reported in [13]. The methodology to carry out the reconfiguration autonomously is of major interest; [8] presents different methods used to reconfigure FFS using a parallel kinematic manipulator (PKM) device, sorting the reconfiguration techniques according to their key features.

Self-reconfigurable intelligent swarm fixture (SwarmItFIX) project, funded by the European Commission, presents a novel concept of a fixture system combining flexibility, self-reconfigurability, automation and swarm collaboration. A self-reconfigurable intelligent swarm fixture system is developed; composed of mobile agents capable to freely move on a bench and reposition below the supported panel while behaving as a swarm [7], with the focus on machining large metal sheets, especially for the aerospace industry, such as fuselage sections, aerofoils and other panels; but applicable as well to automotive, train industries and others. Each agent is composed of:

- Mobile base.
- Parallel manipulator (PM).
- End effector capable of supporting the part adapting to the local geometry [12].

Once in the desired location, the agent docks to the bench and adjusts the position of the end effector. Thanks to the architecture of the agent, the adjustment can be carried out in two levels:

- Fine tuning of the end effector's position due to the PM increasing the resolution provided by the mobile base [21].
- The end effector capability to adapt exactly to the local geometry of the part being supported.

## 29.2 The Concept of Self Reconfigurable Intelligent Swarm Fixtures

The new fixture concept arises from the necessity of today's manufacture industry of more flexibility, adaptability and autonomy in the employed fixture systems. The envisioned concept is a fixture, flexible in many senses, such as the number of dynamic supports (agents) used, free movement under the workpiece to reach the operative area in which the support is locally needed by moving either the mobile base, the PM or the end effector, to the most suitable pose providing the required stiff support with outmost resolution giving outstanding flexibility and adaptability.

The complex repositioning of the agents is carried out under the following constraints:

- Respect of the geometry of the workpiece.
- Coordination of the cooperative support action (path planning).
- Fast reposition of the supports [11].

By achieving the aforementioned goals, large thin sheets can be manufactured without the need of neither replacing fixtures nor dismounting the workpiece from the fixturing system increasing then, the production rate and the quality of the workpiece; and decreasing overall costs and manufacturing time. The dynamic support is given by concentration of the swarm fixture agents, constantly and actively supporting the workpiece, near the area were the tool center point (TCP) is; while the further regions are supported by a minimum number of agents. The swarm behavior control and path planning intelligence replaces the need for pin array fixtures. The cooperative swarm behavior of the agents' controller is fully described in [18]; the fixture reconfiguration is driven by the nominal geometry and manufacturing cycle given by the CAD/CAM software. The planner uses this information integrating reference supporting strategies, studied and implemented for the specific environment but embedding general logics.

The concept gives a new insight in fixture technology not only by increasing the adaptability and flexibility of the fixture but doing so by adding intrinsic modularity and scalability. Given the geometry of a family of workpieces, the user can select the size and geometry of the bench, type and number of agents with different end effectors varying in shape, size, adaptability and adhesion principles, that best suit the specific needs.

### 29.2.1 *Profitability Analysis of Swarm Fixtures*

These intelligent devices solve a problem of growing importance for aircraft and automotive industries where the amount of material is reduced for costs and life cycle eco-consistency reasons. These sectors represent in Europe, according to

ACEA 2001 and EADS 2007 data, more than 450 billion €. Near future applications are also foreseen in the railway, metal construction and furniture, white industry and ship building sectors. They will allow users to produce at shorter time to market, improved quality, lower cost and reduced resources consumption. It is known that 10–20 % of the total cost of manufacture comes from the fixturing system [8]. With this new system the fixture storage and management can be reduced of more than 400 %. Compared to traditional methods adopted for sheet manufacturing, the use of the new highly reconfigurable fixtures, whose set-up time is in the range of minutes and may be performed in hidden time, will allow to cut storage size and re-tooling times, which, for complex and delicate thin sheets, are in the range of hours and several thousands euros.

Product innovation, technological development, sustainability and efficiency of the manufacturing system strongly depend on innovations and developments on fixtures and tools. In particular the new fixtures and associated manufacturing technology will allow indirect benefits through the reduction of consumption of material and energy resources. At the same time re-fixturing operations require an attentive and very accurate manual re-positioning of the thin sheet on the fixture for next operation; the re-positioning errors are the main cause of defects. The use of the new fixture will avoid many manual re-positioning steps because it is the fixture itself that is programmed to adapt the supporting points to the different sequential operations and the re-positioning program, before the actuation, can be checked through simulation.

This new concept of fixtures are expected to produce an impact chain by representing instrumental goods that first will impact the fixtures market offering a manufacturing technology of thin sheets that saves production time and costs mainly in mass customized markets. This technology will extend the use of light high strength materials in a wide series of final products that will positively influence the users of these products and the environmental issues.

In Table 29.1, an estimate of the probable savings one SwarmItFIX installation can provide to the aerospace and automotive sectors is presented. The data have been gathered directly from end-users for the aerospace sector. In the case of the automotive sector, the savings are lower but yet present, although sheet sizes are much smaller than in aerospace.

## 29.3 SwarmItFIX Physical Demonstrator

Within SwarmItFIX the first embodiment of the concept of swarm fixtures into a complete system is created. The prototype, following the main idea consists of: a workbench with embedded air and power supply, two agents with mobile base, Exechon PKM and two possible end effectors. The physical prototype with two swarm agents was installed in Piaggio Aero Industries premises in Finale Ligure, Italy for trials, see Fig. 29.1; and is capable of supporting the workpiece with only two agents and the help of four passive supports.

**Table 29.1** Estimate of main savings offered by one SwarmItFIX installation in the aerospace manufacturing sector

| Aerospace sector | | Automotive sector | |
|---|---|---|---|
| Savings | | Savings | |
| Conventional fixturing | 2,000,000 € | Conventional fixturing | 840,000 € |
| Fixture storage and management | 100,000 €/year × 5 years | Fixture storage and management | 115,000 € /year × 5 years |
| 95% Reduction of set-up times | 720,000 €/year × 5 years | 75% Reduction of set-up times | 600,000 €/year × 5 years |
| Lower time cycle, considered 975 h/year saved | 78,000 €/year × 5 years | Less re-positioning of part, less manufacturing operations to adapt the support device to the part local shape, considered 1200 h/year saved | 98,500 €/year × 5 years |
| Total saving during the first 5 years | 6,490,000 € | Total saving during the first 5 years | 4,907,500 € |

**Fig. 29.1** Final installation of the SwarmItFIX fixture system prototype

## *29.3.1 SwarmItFIX Workbench*

The workbench provides a suitable surface in which to move and securely maintain the fixture in the desired location, air and electrical supply for the whole system is embedded in it. The bench is a planar design with 52 pin modules arranged in interspersed columns in which the agent can dock. To be able to sustain the harsh manufacturing environment filled with swarf and cooling fluids, the workbench implements a cleaning system in which from the locking pin modules the chips are blown, allowing a non-stop operation for longer time periods and without human intervention. The exploded view of the pin module is shown in Fig. 29.2a. The installed bench can be appreciated in Fig. 29.1. The bench and hence the whole system can be mounted in different positions;

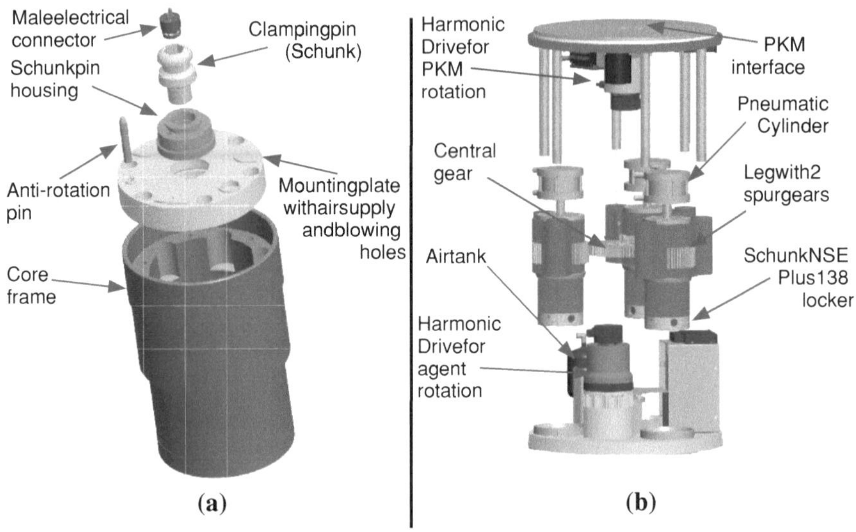


**Fig. 29.2** SwarmItFIX mobile base and Bench pin module

horizontally (floor), vertically (walls) or even upside-down (ceiling), increasing the modularity and customization of the fixture system.

## *29.3.2 Mobile Base Architecture and Locomotion Principle*

The mobile base is the support of the PM, hosting all the electrical components necessary for the communication and control. It gives the first position tuning of the system, using the swing locomotion method [20]; and a 3 legged design, with modified commercial Schunk lockers for the docking mechanism, described in depth in [10], the exploded view is shown in Fig. 29.2, letting appreciate all the components, for the physical prototype see Fig. 29.1.

The three legs are inserted alternatively in the bench taking electrical power and air supply from it and transmitting it to the whole agent. The base rotates about one locked leg while the other two are inside the base; it rotates at an angle and stops with the unused legs above free bench pins. Then, the latter legs slides down one at a time, two legs are again pulled in leaving one clamped. The presence of one leg always inserted in the bench guarantees the electrical and air supply; and the special gear transmission provides the necessary torque for the agent to rotate orienting at the same time the retracted legs. Since the rotation is about a single leg, the locking force has to be greater than 10 kN. This force is provided by the Schunk components that together provide:

- holding force of 75 kN
- draw in force of 18 kN
- repeated accuracy $<0.005$ mm.

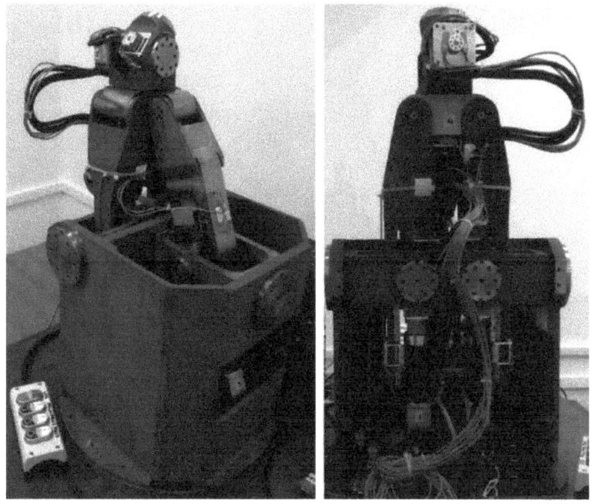

**Fig. 29.3** Exechon 150X PKM prototype

To actuate the legs assuring complete attachment/detachment from the electrical connectors, pneumatic cylinders are used with maximum stroke of 45 mm and minimum operation time manually regulable of less than 0.5 s in either direction.

### *29.3.3 Parallel Kinematic Machine*

The PM provides the necessary workspace, stiffness, degrees of freedom and support of the end effector, it has to fine tune the position at the optimal point of the workpiece required for support, taking into account the nominal geometry, machining operation, number and locations of cooperative agents in the fixture system. A 6 DoF hybrid Exechon architecture detailed in [4, 21] is used. The Exechon PKM 150X, see Fig. 29.3a, is a variant of the Exechon's tripod [17], smaller in dimension and with a 3-DoF spherical wrist. As stated in [14], the use of as few as two agents with mounted Exechon PKM to machine a complete panel will not only reduce the fixturing cost and the need of dedicated fixtures, but also will increase the up-time in production.The kinematics of the PKM used in SwarmItFIX are reported in [21].

### *29.3.4 End Effector*

The end effector (or head) directly contacts the workpiece, it is capable of acquiring the local shape of the machined part and adhering itself to it, to give the necessary stiffness for the machining operations. Two head designs were developed in SwarmItFIX, magneto-rheological fluid head (MRF) and vacuum-clustered incompressible grains head (sand) [3, 19]. The main common requirements are:

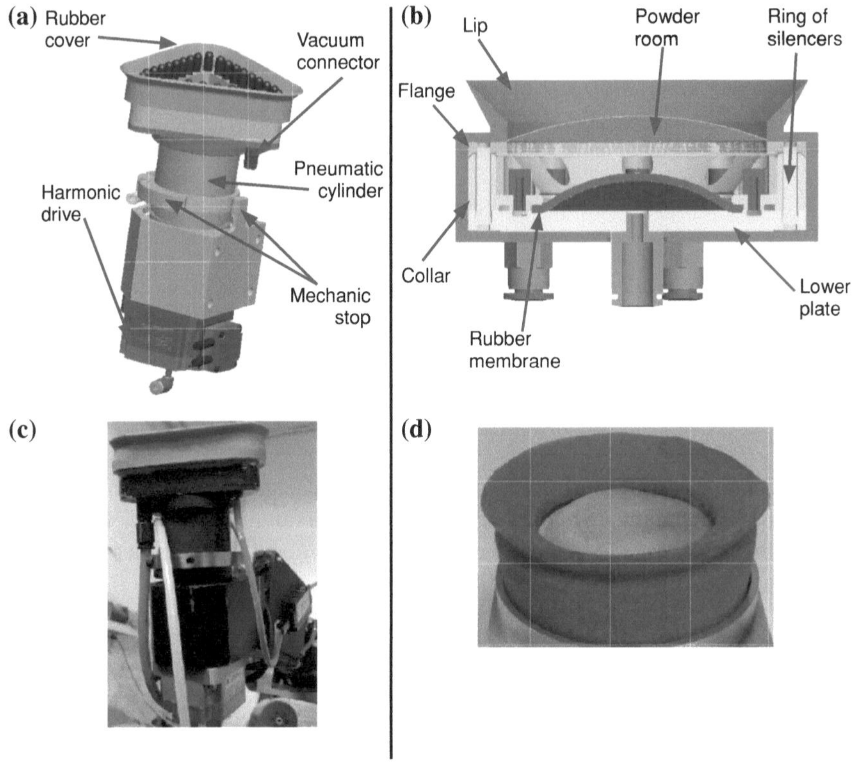


**Fig. 29.4** SwarmItFIX developed end effectors

- adhesion at one side of the workpiece only (no geometric or closure grasp possible)
- non magnetic adhesion possible (non-ferromagnetic workpiece)
- Glues and sticky tapes not feasible (smooth workpiece surface with delicate protective layers not to be damaged)
- high adhesion force per unit area [11].

The MRF head, Fig. 29.4a and c, has a triangular shape to minimize the distance between its perimeter and the tool. A crown of small pistons communicating through a network of channels are distributed along the perimeter of the head. The pistons are semi drowned in a room full of magneto-rheological fluid and are constrained to shift along the head axis. When the magneto-rheological fluid is not magnetized its viscosity is low enough to let the pistons translate copying the workpiece shape along the head perimeter, once the shape is copied, a pneumatic cylinder drives a permanent magnet close to the network of channels containing the MR fluid increasing its viscosity nearly to solid state blocking all pistons from further translation. Vacuum is applied inside the perimeter of the head reproducing the function of a suction cup. The orientation of the head triangle is commanded by a dedicated degree of freedom.

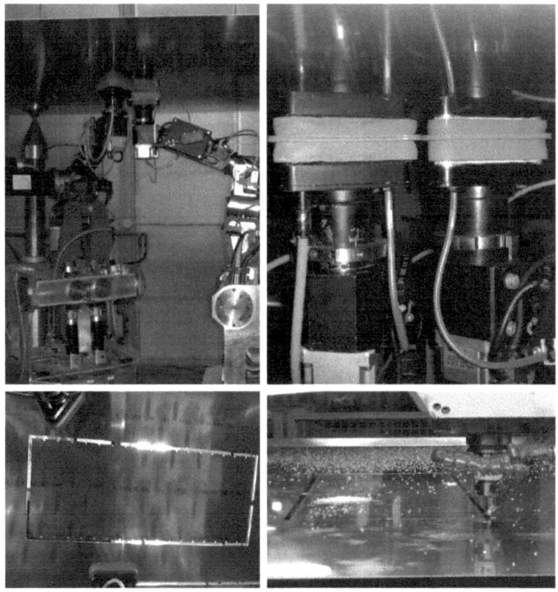

**Fig. 29.5** SwarmItFIX prototype during manufacturing operations

The sand head, Fig. 29.4b and d, is of circular shape not needing an additional rotation to orient it with respect to the workpiece. The adaptation is achieved putting granular material in an compliant envelope and the adhesion used is vacuum. The head is covered by an outer lip that gives the necessary sealing for the vacuum to be produced. Initially the sand inside the head's body can take any shape just by pushing it against the workpiece; the vacuum is then generated compacting the grains in the envelope and generating head-workpiece adhesion. To recover the original compliance, a rubber membrane is located at the bottom of the sand chamber, introducing compressed air expands the rubber membrane relocating the powder and filling the whole space of the chamber.

## *29.3.5 Results*

The prototype was installed in Piaggio Aero premises in Finale Ligure, Italy; under a 5 axis vertical machining center from Jobs S.p.a., Jomach 032; and manufacturing operation trials were realized utilizing thin aluminum sheets of 1.2 mm thickness, the operations were mainly those of drilling 5 mm wholes and milling of a rectangular shape with an 8 mm mill tip. The different locations of the heads are selected by the planner in order to assure correct support when the CNC machine is in the vicinity of the heads, or better said, when the heads have achieved their final position and exert the prescribed suction force at a correct position and orientation to assure the support of the metal sheet. During the trials was confirmed that the setup time of the system when turned on is of less than 10 minutes, this is only for the need of synchronization of the agents, after that it is only needed to locate the workpiece in the peripheral passive supports and run the software. If the system is never turned off only few minutes are needed to mount the workpiece and start the machining operations, hence reducing the machining time. Given the nature of the trials to be the first ones ever realized, the velocities were constrained, however, the mobile base is capable of repositioning the agent in under 4 s and a total relocation of support point can be done in less than 10 s. In Fig. 29.5 the SwarmItFIX functioning prototype is shown. Videos of the working prototype are available in: http://www.youtube.com/user/LMdeLeonardo?feature=mhee

## 29.4 Conclusions

The paper presents the concept of reconfigurable swarm fixtures as the new intended benchmark for future fixture systems and the first prototype ever created in the first complete demonstrator of this type of fixture under the SwarmItFIX European project. The focus is on the final solutions implemented for all the components that constitute the system. The final prototype constructed along with the final installation is presented. A profitability analysis is given providing the first tools for an economical decision to be taken regarding whether the system investment is substantiable to a specific application or not. The final prototype, was subjected to an initial experimental phase in the premises of Piaggio Aero Industries in Finale Ligure, Italy to verify its functionality and demonstrate applicability and advantages. Further work will be done to continue improving the prototype, rigourously experimental phases will be done in the near future with the clear intention of bringing the concept and system developed to a commercial product.

**Acknowledgments** The research is developed within the SwarmItFIX project funded under the Seventh Framework Programme (Collaborative Project 214678). We acknowledge the assistance of the European Commission and the other partners of the project.

## References

1. Aoyama T, Kakinuma Y (2005) Development of fixture devices for thin and compliant workpieces. CIRP Ann Manuf Technol 54(1):325–328
2. Arzanpour S, Fung J, Mills JK, Cleghorn WL (2006) Flexible fixture design with applications to assembly of sheet metal automotive body parts. Assembly Autom 26:143–153
3. Avvenente R, Khan A, Li X, Zoppi M, Zlatanov D, Molfino R (2010) Development and analysis of a shape-conformable supporting head for a self-reconfigurable intelligent swarm fixture system. In: Robotics (ISR), 2010 41st international symposium on and 2010 6th German conference on robotics (ROBOTIK), pp 1–8

4. Bi ZM, Jin Y (2011) Kinematic modeling of exechon parallel kinematic machine. Robot Comput Integr Manuf 27:186–193
5. Blecker T, Abdelkafi N (2006) Complexity and variety in mass customization systems: analysis and recommendations. Manag Decis 44:908–929
6. Grippo PM, Thompson BS, Gandhi MV (1988) A review of flexible fixture systems for computer-integrated manufacturing. Int J Comput Integr Manuf 1(2):124–135
7. Patent IT2008GE00003 (2008) Dispositivo riconfigurabile multi-agente per il sostegno di corpi a geometria complessa, January 2008. Applicants: R. Molfino, M. Zoppi
8. Jonsson M, Ossbahr G (2010) Aspects of reconfigurable and flexible fixtures. Prod Eng 4:333–339
9. Kurk F, Eagan P (2007) The value of adding design-for-the-environment to pollution prevention assistance options. J Clean Prod 16:722–726
10. Li X, Zoppi M, De Leonardo L, Molfino R (2011) Design of a mobile base for a self-reconfigurable intelligent swarm fixture system. In: International conference on climbing and walking robots and the support technologies for mobile machines (CLAWAR 2011), Paris, 6–8 Sept
11. Molfino R, Zoppi M (2011) The robotic swarm concept in fixtures for transport industry. In: Proceedings of the ASME 2010 international design engineering technical conferences and computers and information in engineering conference. Washington, DC, 8–31 August
12. Molfino R, Zoppi M, Zlatanov D (2009) Reconfigurable swarm fixtures. In: ASME/IFToMM international conference on reconfigurable mechanisms and robots, 2009 (ReMAR 2009), pp 730 –735, 22–24 June
13. Munro C, Walczyk D (2007) Reconfigurable pin-type tooling: a survey of prior art and reduction to practice. J Manuf Sci Eng 129:551–565
14. Neumann K-E (2008) IEEE/IFR Invention and Entrepreneurship Award 2011
15. Sela MN, Gaudry O, Dombre E, Benhabib B (1997) A reconfigurable modular fixturing system for thin-walled flexible objects. Int J Adv Manuf Technol 13:611–617
16. Shen Y, Shirinzadeh B (2001) Dynamic analysis of reconfigurable fixture construction by a manipulator. Robot Comput Integr Manuf 17(5):367–377
17. Patent WO2006062466 (2006) Parallel kinematic machine with an active measuring system, 2006. Applicant: Exechon AB, Neumann K-E, Sweden
18. Zielinski C, Kornuta T, Trojanek P, Winiarski T, Walecki M (2011) Specification of a robot-based reconfigurable fixture control system. In: Eighth international workshop on robot motion and control (ROMOCO 2011), 15–17 June
19. Zoppi M, Gagliardi S, Molfino R (2011) Multi-agent fixtures for automotive and aeronautical industries: design and prototyping of agent adaptable heads. In: The 5th international conference on integrated modeling and analysis in applied control and automation (IMAACA 2011), 12–14 Sept
20. Zoppi M, Xion L, Zlatanov D, Molfino R (2010) The development of the locomotion subsystem of a self-reconfigurable mobile fixture. In: International conference on mechanisms and machine science, Shanghai, 21–25 July
21. Zoppi M, Zlatanov D, Molfino R (2010) Kinematics analysis of the exechon tripod. In: Proceedings of the ASME DETC, 34th annual mechanisms and robotics conference (MR), Montreal

# Chapter 30
# Configuration Change and Mobility Analysis of a New Metamorphic Parallel Mechanism Used for Bionic Joint

**Guoguang Jin and Boyan Chang**

**Abstract** Traditional bionic joints including wrist, waist, ankle and shoulder have three revolute degrees of freedom, which can be treated as spherical joints, and its structure is prone to be damaged with impact force through the center of spherical joint. In this paper, a novel metamorphic parallel mechanism with two configurations used for bionic joint design is presented. The origin kinematic chain for this metamorphic mechanism, which is also the first configuration mechanism, consists of a moving platform, a base plate and four connecting legs between moving platform and base plate. In contrast with traditional ankle joint, a constrained translational degree of freedom can be activated in origin kinematic chain to avoid impact damage. The origin kinematic chain can transform to the second configuration mechanism which is in possession of three revolute degrees of freedom and equivalent to normal spherical joint. The transforming process is represented by new incidence matrix and the motion characteristics of this metamorphic mechanism in different configuration corresponding to each work-stage are analysed using screw theory.

**Keywords** Metamorphic mechanisms · Bionic joint · Incidence matrix · Screw theory · Mobility

## 30.1 Introduction

Bionic robot is the best combination of advanced bionics technology and robotics in a variety of application [1, 2]. A robot like human being or other animals is an imagination for the common people and it is depicted vividly in science fiction.

G. Jin (✉) · B. Chang
School of Mechanical Engineering, Tianjin Polytechnic University, Tianjin, China
e-mail: jinguoguang@tjpu.edu.cn

J. S. Dai et al. (eds.), *Advances in Reconfigurable Mechanisms and Robots I*,
DOI: 10.1007/978-1-4471-4141-9_30, © Springer-Verlag London 2012

However, it is almost impossible to use machinery to completely imitate the organism. Unless doing the full research on the characteristics of biological structure and motion [3] and establishing the reasonable simplified model, generalized bionic joints and robots consist of these joints can be designed and developed. Ankle [4], waist [5, 6], wrist and shoulder joints [7], the most common bionic joints used in robot, can be simplified in the form of spherical joint and achieved by 3-DOF spherical parallel mechanism which has been widely used. Gosselin proposed spherical three-degree-of-freedom parallel manipulators [8] and designed a novel device named agile eye [9]. Liu proposed a serial-parallel 7-DOF redundant anthropomorphic arm with spherical 3-DOF spherical parallel mechanism acting as the shoulder and wrist [7]. Jin analyzed the workspace of 3-RRR spherical parallel manipulator and proposed a waist joint [5].

But in certain extreme conditions such as colliding with hard things, spherical joint would suffer from huge reaction force through the center of spherical parallel mechanism and its structure is prone to be damaged because it has no translational mobility in the direction of radius to lessen and absorb the energy of impact. Take ankle joint for example, Dai analyzed the human ankle complex and indicated its structures are most commonly sprained mainly due to the greater range of inversion movement at times of high and rapid loading of the foot, such as when the foot hits the ground during landing from a jump or fall [10]. Therefore, it's expected the structure can make a change and an extra translational degree of freedom can be introduced.

During the past decade, mechanisms with capability that can change their structure and mobility in extreme conditions and achieve different tasks have attracted great attention especially since metamorphic mechanism was proposed according to the study of foldable and erectable artifacts cartons in 1998 [11, 12]. The potential use of this kind of mechanism is substantially large in terms of freedom and geometric size required to change [13]. A novel robotic hand with a metamorphic palm was proposed in [14] based on the metamorphosis principle. Zhang and Dai proposed a new metamorphic mechanism with ability for platform orientation switch [15]. Ding investigated the topology and configuration of an assembly-circles artifact and provided some fundamental ideas for design of metamorphic mechanism [16]. Dai designed a gecko-like robot and made a bio-mimetic study of this discontinuous-constraint metamorphic mechanism [4]. In order to describe and analyze the topological changes in metamorphic mechanism, Professor Dai proposed a new way, in Ref. [17], for modeling configuration changes of metamorphic mechanisms by introducing adjacency matrix and its mathematical operation. Professor Liu discussed three basic metamorphic ways [18], and Professor Yan made scientific researches on topological representations and characteristic analysis of variable kinematic joints with matrices in Ref. [19]. Professor Jin established the model of topological changes with Huston lower body arrays and adjacency matrix in Ref. [20, 21], and then introduced incidence matrix with type of kinematic joints and its logical operation to describe the variable topology [22]. The incidence matrix method can describe all types of variable topological characteristics intuitively and it's easy to construct the metamorphic matrix for structure analysis and synthesis of metamorphic mechanism. By contrast with adjacency matrix, this method is more conducive to describe changes in kinematic joints because of the structure of incidence matrix (its rows and columns represent component's and joint's number respectively) as well as the convenience of Boolean calculation.

In this paper, a spatial metamorphic parallel mechanism with two configurations used for bionic joint design is presented. The origin kinematic chain for this metamorphic mechanism, which is also the first configuration mechanism, consists of a moving platform, a base plate and four connecting legs between moving platform and base plate. In contrast with traditional ankle joint, a redundant translational degree of freedom is introduced in origin kinematic chain to avoid impact damage when extreme conditions occur. The origin kinematic chain can transform to the second configuration mechanism which is in possession of three revolute degrees of freedom and equivalent to normal spherical joint. The transforming process is represented by new incidence matrix and the motion characteristics of this metamorphic mechanism in different configuration corresponding to each work-stage are analyzed using screw theory.

## 30.2 Structure of the Metamorphic Mechanism and its Configuration Change

The new metamorphic mechanism, shown in Fig. 30.1, contains a middle leg (UC chain or $^u$R$^v$R$^w$C chain shown in Fig. 30.2) numbered leg-0 and three identical legs ($^i$P$^j$R$^k$R$^k$P$^l$R chain shown in Fig. 30.3) numbered leg-$n$ ($n = 1, 2, 3$), where R, P, U and C represent Revolute joint, Prismatic joint, Cylindrical joint and Universal joint respectively. Its structure and link connectivity can be represented in incidence matrix form [22]. In the matrix each link of the mechanism in its current configuration is identified by a number from 1 to $n$. The rows of the matrix from 1 to $n$ take these numbers in sequence. Kinematic joints between the $i$th link and $j$th link is given in the zeroth row. When two links are connected, the entry of the corresponding row and column is given as 1. When two links are disconnected, the entry of the corresponding row and column is given as 0. The incidence matrix of the metamorphic mechanism in Fig. 30.1 is

$$\boldsymbol{A}_0 = \begin{bmatrix} {}^i\mathrm{P} & {}^j\mathrm{R} & {}^k\mathrm{R} & {}^k\mathrm{P} & {}^l\mathrm{R} & \mathrm{C} & \mathrm{U} \\ 1^{1\times3} & 0^{1\times3} & 0^{1\times3} & 0^{1\times3} & 0^{1\times3} & 0 & 1 \\ E^{3\times3} & E^{3\times3} & 0^{3\times3} & 0^{3\times3} & 0^{3\times3} & 0 & 0 \\ 0^{3\times3} & E^{3\times3} & E^{3\times3} & 0^{3\times3} & 0^{3\times3} & 0 & 0 \\ 0^{3\times3} & 0^{3\times3} & E^{3\times3} & E^{3\times3} & 0^{3\times3} & 0 & 0 \\ 0^{3\times3} & 0^{3\times3} & 0^{3\times3} & E^{3\times3} & E^{3\times3} & 0 & 0 \\ 0^{1\times3} & 0^{1\times3} & 0^{1\times3} & 0^{1\times3} & 1^{1\times3} & 1 & 0 \\ 0 & 0 & 0 & 0 & 0 & 1 & 1 \end{bmatrix} \quad (30.1)$$

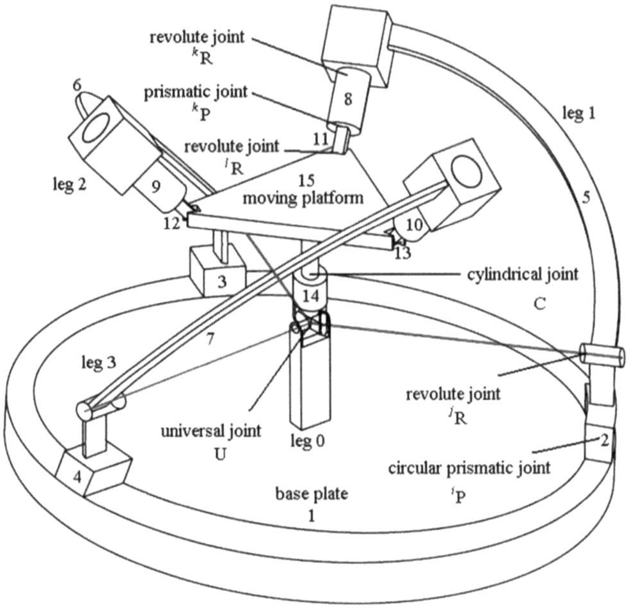


**Fig. 30.1** UC-3($^i$P$^j$R$^k$R$^k$P$^l$R) metamorphic parallel mechanism

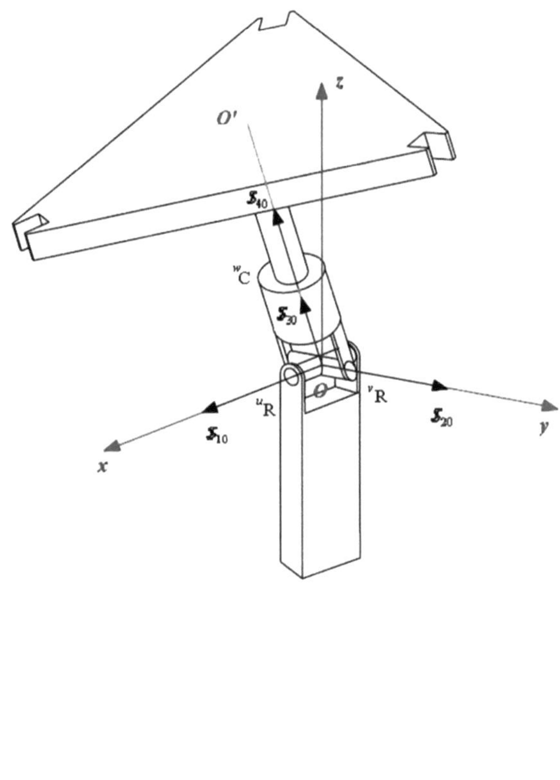


**Fig. 30.2** $^u$R$^v$R$^w$C kinematic chain

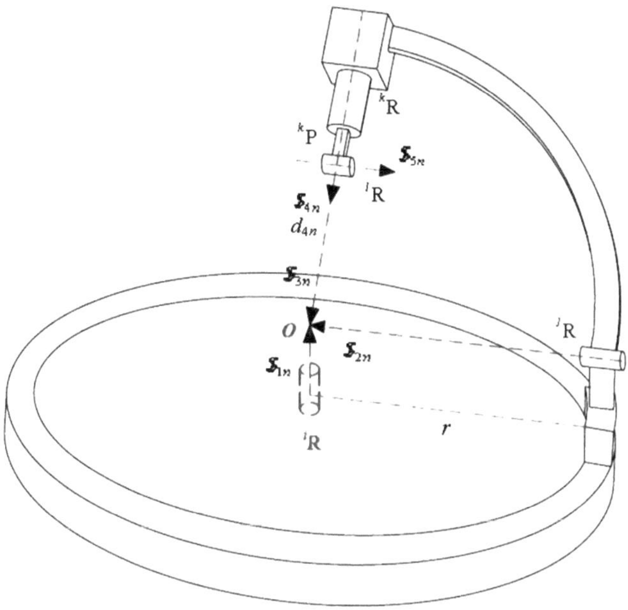


**Fig. 30.3** $^i$P$^j$R$^k$R$^k$P$^l$R kinematic chain

This incidence matrix defines a topological configuration of a mechanism during motion and its configuration change can be expressed by a series of logical operations

$$A^{p+1} = A^{p}_{(n_p+1)\times m_p} \oplus B^{(p,p+1)} \tag{30.2}$$

where $A^p$ is the matrix of configuration $p$, and $B^{(p,p+1)}$ is metamorphic matrix with different form according to the type of topological changes (four types of topological changes involving change in the number of components, change in the number of kinematic joints, change in the type of kinematic joints and change of link connectivity [22]) from configuration $p$ to configuration $p+1$, $\oplus$ is the symbol of XOR operation.

In the second configuration, cylindrical joint in middle leg is a variable kinematic joint [19] and can be transformed to revolute joint by constraining the translational degree of freedom. Metamorphic matrix of changing kinematic joint $r_j$ from $C_p$ to $C_{p+1}$ is

$$B^{r_j}_{1\times m_p}(C_{p+1}, C_p) = \begin{bmatrix} 0_{1\times(j-1)} & \Delta r_j & 0_{1\times(m_p-j)} \end{bmatrix}_{1\times m_p} \tag{30.3}$$

$$\Delta r_j = C_{p+1} \oplus C_p \tag{30.4}$$

**Table 30.1** Joint codes of commonly used kinematic pairs

| Pair type | Joint code |
|---|---|
| Revolute joint | 0001 |
| Prismatic joint | 0010 |
| Helical joint | 0011 |
| Gear joint | 0100 |
| Cylindrical joint | 0101 |
| Universal joint | 0110 |
| Spherical joint | 0111 |
| Flat joint | 1000 |

$C_p$, shown in Table 30.1, represents the joint codes of kinematic pairs used in configuration $p$.

The metamorphic matrix $\boldsymbol{B}^{(p,p+1)}$ can be listed as follows

$$\boldsymbol{B}^{r_j}_{1\times m_p}(\mathrm{R},\mathrm{C}) = \begin{bmatrix} 0_{1\times(j-1)} & 0110 & 0_{1\times(m_p-j)} \end{bmatrix}_{1\times m_p} \tag{30.5}$$

where $j = 16$, $m_p = 17$

$$\boldsymbol{B}^{(0,1)} = \begin{bmatrix} 0_{1\times 15} & 0110 & 0 \end{bmatrix}$$

$$\boldsymbol{A}^1 = \boldsymbol{A}^0 \oplus \boldsymbol{B}^{(0,1)} = \begin{bmatrix} {}^i\mathrm{P} & {}^j\mathrm{R} & {}^k\mathrm{R} & {}^k\mathrm{P} & {}^l\mathrm{R} & \mathrm{R} & \mathrm{U} \\ 1^{1\times 3} & 0^{1\times 3} & 0^{1\times 3} & 0^{1\times 3} & 0^{1\times 3} & 0 & 1 \\ E^{3\times 3} & E^{3\times 3} & 0^{3\times 3} & 0^{3\times 3} & 0^{3\times 3} & 0 & 0 \\ 0^{3\times 3} & E^{3\times 3} & E^{3\times 3} & 0^{3\times 3} & 0^{3\times 3} & 0 & 0 \\ 0^{3\times 3} & 0^{3\times 3} & E^{3\times 3} & E^{3\times 3} & 0^{3\times 3} & 0 & 0 \\ 0^{3\times 3} & 0^{3\times 3} & 0^{3\times 3} & E^{3\times 3} & E^{3\times 3} & 0 & 0 \\ 0^{1\times 3} & 0^{1\times 3} & 0^{1\times 3} & 0^{1\times 3} & 1^{1\times 3} & 1 & 0 \\ 0 & 0 & 0 & 0 & 0 & 1 & 1 \end{bmatrix} \tag{30.6}$$

## 30.3 Mobility Analysis of the Metamorphic Mechanism

### *30.3.1 Modified Grubler-Kutzbach Criterion*

In order to calculate the mobility of a mechanism, Huang Z proposed the Modified Grubler-Kutzbach Criterion [23, 24].

$$M = d(n - g - 1) + \sum_{i=1}^{g} f_i + v - \xi \tag{30.7}$$

where, $M$ denotes the mobility of a mechanism, $d$ is the order of a mechanism, $n$ is the number of links including the frame, $g$ is the number of kinematic joints, $f_i$ is the number of freedoms of the $i$th joint, $v$, named $v$-factor, is the number of redundant constraints except the number having been accounted in common constraints. $\xi$ is the number of local freedoms. The order of a mechanism $d$ is given by

$$d = 6 - \lambda \tag{30.8}$$

where, $\lambda$ is the common constraint of the mechanism.

The core of the problem of mobility analysis is how to determine the redundant constraints including the common constraint $\lambda$ and the number of redundant constraints $v$.

### 30.3.2 The Mobility Analysis of Each Work-Stage Based on Screw Theory

As is shown in Fig. 30.1, the middle leg connects the center of moving platform $O'$ and the center of base plate $O$. The axes of ${}^u$R and ${}^v$R, ${}^v$R and ${}^w$C are perpendicular and intersect at $O$. In addition, the axis of ${}^w$C is perpendicular to moving platform and intersects at $O'$. In the three identical legs, the circular prismatic joint ${}^i$P is equal to ${}^i$R shown in Fig. 30.2 [25] and the axes of ${}^i$R, ${}^j$R, ${}^k$R and ${}^k$P intersect at $O$. The axis of ${}^l$R is perpendicular to ${}^k$P and coplanar with moving platform.

The local coordinate system is shown in Fig. 30.3. Its $x$-axis is along the axis of ${}^u$R fixed to the base and $y$-axis along the axis of ${}^v$R.

The screw system of the three identical legs can be listed as

$$\begin{cases} \$_{1n} = (\boldsymbol{S}_{1n}; 0) \\ \$_{2n} = (\boldsymbol{S}_{2n}; 0) \\ \$_{3n} = (\boldsymbol{S}_{3n}; 0) \\ \$_{4n} = (0; \boldsymbol{S}_{3n}) \\ \$_{5n} = (\boldsymbol{S}_{5n}; (r - d_{4n})\boldsymbol{S}_{3n} \times \boldsymbol{S}_{5n}) \end{cases} \tag{30.9}$$

Its reciprocal screw system is obtained by observation as follows

$$\$^r_{1n} = (\boldsymbol{S}_{5n}; 0) \tag{30.10}$$

The screw system of the middle leg can be listed as

$$\begin{cases} \$_{10} = (\boldsymbol{S}_{10}; 0) \\ \$_{20} = (\boldsymbol{S}_{20}; 0) \\ \$_{30} = (\boldsymbol{S}_{30}; 0) \\ \$_{40} = (0; \boldsymbol{S}_{30}) \end{cases} \tag{30.11}$$

Its reciprocal screw system is obtained by observation as follows

$$\begin{cases} \$^r_{10} = (\boldsymbol{S}_{10} \times \boldsymbol{S}_{30}; 0) \\ \$^r_{20} = (\boldsymbol{S}_{20} \times \boldsymbol{S}_{30}; 0) \end{cases} \tag{30.12}$$

From the constraint screw system Eqs. 30.10 and 30.12, The four legs exert five constraint forces on the moving platform. The five constraint forces are coplanar and concurrent and this plane determined by constraint forces is always parallel with moving platform. Therefore, only two of the constraint forces are linearly independent and corresponding two translational degrees of freedom along moving platform are constrained, i.e. $v = 3$. The constraint forces provided by each leg constraint system are not coaxial, i.e. $\lambda = 0$.

Based on the Modified Grubler-Kutzbach Criterion, the mobility of first configuration mechanism is

$$\begin{aligned} M_1 &= d(n - g - 1) + \sum_{i=1}^{g} f_i + v - \xi \\ &= 6 \times (15 - 17 - 1) + 19 + 3 - 0 \\ &= -18 + 19 + 3 = 4 \end{aligned} \tag{30.13}$$

In this configuration, the moving platform has four degrees of freedom including three rotational freedoms and one translational freedom in the direction of radius.

As for the second configuration, the translational freedom in the direction of radius is constrained by using variable kinematic joint [19] transforming cylindrical joint to revolute joint. In this configuration, the screw system of the middle leg makes a change and can be listed as

$$\begin{cases} \$_{10} = (\boldsymbol{S}_{10}; 0) \\ \$_{20} = (\boldsymbol{S}_{20}; 0) \\ \$_{30} = (\boldsymbol{S}_{30}; 0) \end{cases} \tag{30.14}$$

Its reciprocal screw system is obtained by observation as follows

$$\begin{cases} \$^r_{10} = (\boldsymbol{S}_{10} \times \boldsymbol{S}_{30}; 0) \\ \$^r_{20} = (\boldsymbol{S}_{20} \times \boldsymbol{S}_{30}; 0) \\ \$^r_{30} = (\boldsymbol{S}_{30}; 0) \end{cases} \tag{30.15}$$

From the constraint screw system Eqs. 30.10 and 30.15, The four legs exert six constraint forces on the moving platform and only three of the constraint forces are linearly independent and corresponding all three translational degrees of freedom are constrained, i.e. $v = 3$. The constraint forces provided by each leg constraint system are not coaxial, i.e. $\lambda = 0$.

Based on the Modified Grubler-Kutzbach Criterion, the mobility of first configuration mechanism is

$$\begin{aligned} M_2 &= d(n-g-1) + \sum_{i=1}^{g} f_i + v - \xi \\ &= 6 \times (15 - 17 - 1) + 18 + 3 - 0 \\ &= -18 + 18 + 3 = 3 \end{aligned} \tag{30.16}$$

In this configuration, the moving platform has three rotational freedoms.

## 30.4 Conclusion

A novel metamorphic parallel mechanism with two configurations used for bionic joint design is presented. In contrast with traditional ankle joint, a constrained translational degree of freedom is introduced and it can be activated to avoid impact damage. The origin kinematic chain has three rotational freedoms and one translational freedom in the direction of radius and can transform to the second configuration mechanism which is in possession of three revolute freedoms and equivalent to normal spherical joint. The transforming process is represented by new incidence matrix and the motion characteristics of this metamorphic mechanism in different configuration are calculated based on screw theory.

**Acknowledgments** This work was partially supported by the National Natural Science Foundation of China under grant No. 50675154 and Tianjin Research Program of Application Foundation and Advanced Technology under grant No. 09JCYBJC04600.

## References

1. Altendorfer R et al (2001) RHex: a biologically inspired hexapod runner. Auton Robots 11:207–213
2. Collins S et al (2005) Efficient bipedal robots based on passive-dynamic walkers. Science 307(18):1082–1085
3. Dickinson MH et al (2000) How animals move: an integrative view. Science 288(7):100–106
4. Dai ZD, Li HK (2007) A biomimetic study of discontinuous-constraint metamorphic mechanism for gecko-like robot. J Bionic Eng 4(2):91–95
5. Jin ZL, Rong Y (2007) Design of a waist joint based on three branches unequal spaced distribution spherical parallel manipulator. China Mech Eng 18(22):2697–2699
6. Wu WK, Jin ZL, Gao F (2008) The kinematic analyse of a new waist's joint based on a four degree parallel manipulator. Mach Des Res 24(1):42–45
7. Liu XJ, Wang JS, Gao F, Jin ZL (2002) Design of a serial-parallel 7-DOF redundant anthropomorphic arm. China Mech Eng 13(2):101–105
8. Gosselin CM, Sefrioui J, Richard MJ (1990) On the direct kinematics of spherical three-degree-of-freedom parallel manipulators. Tans ASME, J Mech Des 116(2):594–598
9. Gosselin CM, Jean-Franqois H (1994) The agile eye: a high-performance three-degree-of freedom camera-orienting device. IEEE 1994:781–786
10. Dai JS, Zhao TS (2004) Sprained ankle physiotherapy based mechanism synthesis and stiffness analysis of robotic rehabilitation device. Auton Robots 16:207–218

11. Dai JS, Jones JR (1998) Mobility in metamorphic mechanisms of foldable/erectable kinds. In: Proceedings of ASME design engineering technical conference, Atlanta, Georgia, pp 13–16
12. Dai JS, Jones JR (1999) Mobility in metamorphic mechanisms of foldable/erectable kinds. ASME Trans, ASME J Mech Des 121(3):375–382
13. Zhang WX, Ding XL (2009) A method for designing metamorphic mechanisms and its application. ASME/IFToMM international conference on reconfigurable mechanism and robots, London, United Kingdom, pp 171–174
14. Dai JS (2004) Robotic hand with palm section comprising several parts able to move relative to each other. U.K. patent No.GB20040009548, 29 April 2004
15. Zhang KT, Dai JS, Fang YF (2009) A new metamorphic mechanism with ability for platform orientation switch and mobility change. ASME/IFToMM international conference on reconfigurable mechanism and robots, London, United Kingdom, pp 626–632
16. Ding XL, Yang Y (2010) Reconfiguration theory of mechanism from a traditional artifact. ASME Trans, ASME J Mech Des 132:114501–114509
17. Dai JS, Jones JR (2005) Matrix representation of topological changes in metamorphic mechanisms. ASME Trans, ASME J Mech Des 127:837–840
18. Liu CH, Yang TL (2004) Essence and characteristics of metamorphic mechanism and their metamorphic ways. In: Proceedings of the 11th world congress in mechanism and machine science, Tianjin, China, pp 1285–1288
19. Yan HS, Kuo CH (2006) Topological representations and characteristics of variable kinematic joints. ASME Trans, ASME J Mech Des 128:384–391
20. Jin GG, Ding XL, Zhang QX (2004) Research on configuration-complete dynamics modeling and numerical simulation of metamorphic mechanism. Chin J Aeronaut 25(4):401–405
21. Wu YR, Jin GG, Li DF, Yang SM, Dai JS (2007) Adjacent matrix method describing the structure changing of metamorphic mechanisms. Chin J Mech Eng 43(7):23–26
22. Chang BY, Jin GG, Dai JS, Yang SM, Feng ZY (2011) Incidence matrix representation of configuration changes in metamorphic mechanisms. Mech Sci Technol Aerospace Eng 30(3):507–511.516
23. Huang Z, Liu JF, Zeng DX (2009) A general methodology for mobility analysis of mechanisms based on constraint screw theory. Sci China Ser E-Tech Sci 52(5):1337–1347
24. Huang Z, Liu JF, Li QC (2008) A unified methodology for mobility analysis based on screw theory. In: Wang L, Xi J (eds) Smart device and machines for advanced manufacturing. Springer-Verlag, London, pp 49–78
25. Li QC, Wu CY, Shen WP (2006) Novel 3-$P_C R_N S$ Spherical 3-DOF Parallel Mechanism. Chin J Mech Eng 42(11):44–48

# Chapter 31
# Optimal Design of a New Parallel Kinematic Machine for Large Volume Machining

**Yan Jin, Zhuming Bi, Colm Higgins, Mark Price, Weihai Chen and Tian Huang**

**Abstract** Although numerous PKM topologies have been invented recently, few of them have been successfully put into production. A good topology can only provide good performance unless its geometrical parameters are optimized. This paper studies the dimensional synthesis of a new PKM which has shown great potential for large volume high performance manufacturing. A new optimization approach is proposed for design optimization, with a new performance index composed of weight factors of both Global Conditioning Index (GCI) and actuator stroke. Maximizing GCI will ensure the effectiveness of the workspace, while minimizing actuator stroke leads to reduced machine cost and increased efficiency. Results show that the proposed optimization method is valid and effective. The PKM with optimized dimensions has a large workspace to footprint ratio and a large well-conditioned workspace, which ensures its suitability for large volume machining.

Y. Jin (✉) · C. Higgins · M. Price
Queen's University, Belfast BT9 5AH, UK
e-mail: y.jin@qub.ac.uk

C. Higgins
e-mail: c.higgins@qub.ac.uk

M. Price
e-mail: m.price@qub.ac.uk

Z. Bi
Indiana University Purdue University, Fort Wayne, IN 46805, USA
e-mail: biz@ipfw.edu

W. Chen
Beihang University, Beijing 100191, China
e-mail: w.chen@buaa.edu.cn

T. Huang
Tianjin University, Tianjin 300072, China
e-mail: htiantju@public.tpt.tj.cn

J. S. Dai et al. (eds.), *Advances in Reconfigurable Mechanisms and Robots I*,
DOI: 10.1007/978-1-4471-4141-9_31, 

**Keywords** Parallel kinematic machine · Dimension optimization · High performance manufacturing

## 31.1 Introduction

Large volume manufacturing companies, e.g., aerospace manufacturers, are looking for cost effective flexible solutions to meet the ever increasing customer demands towards high speed and high quality [1]. Parallel kinematic machines (PKMs) shows the huge potential to meet these requirements [2], and they have attracted a lot of attention from universities to industries over the last three decades. Numerous types of PKMs have been proposed in literature, but few of them have been successfully commercialized and utilized in production [3]. The major reason for that is the small workspace and the limited flexibility. To overcome these drawbacks while maintaining the merits of high stiffness, speed and accuracy, the current trend in large volume manufacturing is to utilize hybrid parallel kinematic machines (HPKM) for five-axis machining [4–9]. Research has shown that HPKMs can offer competitive advantages comparing to conventional CNC machines [6]. However, there is still a long way to go before putting them into productive work, such as in large volume high precision manufacturing. The Tricept machine is by far the most successful machine for high stiffness manufacturing, with more than 300 currently in production. But it is not stiff enough for some high precision manufacturing tasks, such as aircraft assembly [6]. To overcome the limitations of Tricept, Neumann [6] recently patented a novel HPKM called Exechon machine tool, which showed a better performance comparing to Tricept. As shown in Fig. 31.1a, an integrated HPKM Exechon system has been designed and implemented, and it is constructed by a parallel kinematic architecture connected by a two-DOF (degree of freedom) wrist at the end. Figure 31.1b shows a zoomed view of the actual physical model with some joints depicted. This paper will deal with dimension synthesis of the Exechon PKM (without accounting the 2-DOF wrist), which has one translation DOF and two rotational DOF [10].

PKM design basically involves two steps, i.e., topology synthesis and dimension synthesis (optimization), both of which are very important for design of a suitable PKM to achieve specified performance. A good topology can only provide good performance unless its geometrical parameters are optimized. The existing methods of dimension synthesis can be classified into two categories, i.e., objective-function based optimal design and performance-chart based design [11–20]. Defining suitable performance indices and reducing the number of design variables, as well as employing proper optimization algorithms, are still challenging issues and have caught a lot of attention recently [21–26].

Most literature focus on design of PKMs with symmetrical architectures associated with pure translation or pure rotational DOF, which makes ease of analysis and dimensional optimization because the Jacobian matrix is homogeneous

**Table 31.1** Optimization results of the PKM

| $w_1$ | $w_2$ | $\lambda_1$ | $\lambda_2$ | $\lambda_3$ | GCI | Stroke |
|---|---|---|---|---|---|---|
| 0.5 | 0.5 | 0.3701 | 0.5926 | 1.5335 | 0.3899 | 0.7335 |
| 1.0 | 0 | 0.3965 | 0.7557 | 1.5692 | 0.4066 | 0.7692 |
| 0 | 1.0 | 0.4277 | 0.5001 | 1.5062 | 0.3598 | 0.7062 |

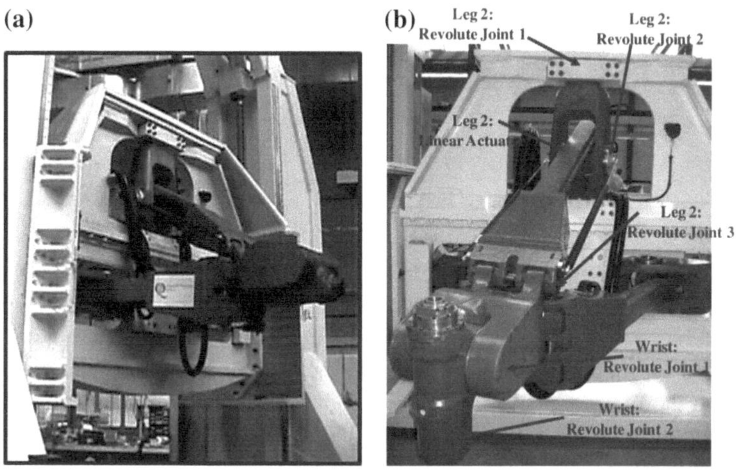


**Fig. 31.1** Physical model of a 5-DOF hybrid Exechon machine

provided that all actuators are of the same type. This paper deals with a PKM with asymmetrical architecture associated with mixed translation and rotational DOF. A new performance function, which combines the workspace conditioning measure and actuator efficiency, is proposed for dimension optimization. Results show that the newly developed method is very effective and very useful for design a series of these PKMs with disparate scales. The results also show that the Exechon PKM has a large workspace to footprint ratio, and very good conditioning over the whole workspace. This makes it the suitable machine tool for large volume machining, e.g., milling in aircraft assembly.

This paper is organized as follows. Section 31.2 describes the architecture of the Exechon PKM. Section 31.3 presents the kinematics of the PKM. Section 31.4 formulates the generalize Jacobian of the PKM. Sections 31.5 and 31.6 introduce the workspace analysis and dimension optimization respectively. Section 31.7 concludes this paper.

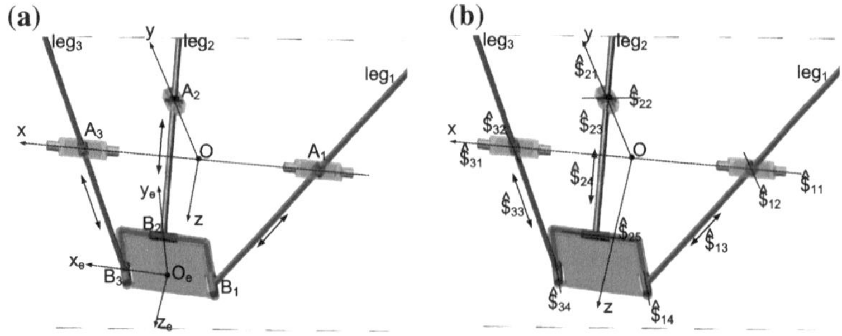


Fig. 31.2 Schematic diagram of Exechon PKM. a Coordinates and points. b Twists description

## 31.2 Description of the PKM

Figure 31.2 shows the parallel architecture of the Exechon in which the end-effecter platform is supported by three legs denoted as $leg_1$, $leg_2$, and $leg_3$, respectively. $Leg_1$ and $leg_3$ have an identical architecture in which a universal joint is mounted on the base, followed by a linear actuator and an revolute joint connecting to the moving platform. $Leg_2$ is slightly different from $leg_1$ and $leg_3$ as it has one more rotary DOF about the actuator axis. Kinematically, $leg_2$ can be regarded as being constructed by an spherical joint on the base, followed by the linear actuator and a revolute joint connecting to the moving platform. At home position, $leg_1$ and $leg_3$ are symmetrical with respect to $leg_2$. Assume points $A_i$ and $B_i\,(i = 1, 2, 3)$ denote the attachment points at the corresponding joint centers to the base and the moving platform respectively. The base coordinate system $O - xyz$ is defined as shown in Fig. 31.2a. Point $O$ is the central point of $A_1A_3$, $x$-axis is pointing from $A_1$ to $A_3$. $y$-axis is perpendicular to $A_1A_3$ and towards $A_2$, $z$-axis is obtained by right-hand rule. Similarly, the platform coordinate system $O_e - x_ey_ez_e$ is also defined as shown in Fig. 31.2b. Point $O_e$ is the central point of $B_1B_3$, $x_e$-axis is pointing from $B_1$ to $B_3$. $y_e$-axis is perpendicular to $A_1A_3$ and towards $B_2$, $z_e$-axis can then be obtained by right-hand rule. Let $\hat{\$}_{ij}\,(s_{ij})$ represent a unit screw (vector) along the $jth$ joint of the $ith$ leg, the geometrical constraints of the architecture can be described as follows.

- $s_{12}$, $s_{14}$, $s_{34}$ and $s_{32}$ are parallel to each other;
- $s_{11}$ and $s_{31}$ are coincident;
- Actuator axes $s_{i3}\,(i = 1, 3)$ are perpendicular to both $s_{i4}$ and $s_{i2}$;
- Actuator axis $s_{24}$ is perpendicular to both $s_{25}$ and $s_{22}$;

Mobility analysis [10] showed that the moving platform has three DOF including one translation along $z$ direction and two rotations about $x$- and $y$-axes, respectively.

## 31.3 Kinematics

Both the inverse and forward kinematics of the PKM was studied by Bi and Jin [7, 10]. The inverse kinematics of the Exechon hybrid parallel-serial architecture was also recently studied by Zoppi et al. [27]. It is found that only one unique solution exists for the inverse kinematics of the Exechon PKM. This makes ease of control and is regarded as one advantage of this PKM architecture as most PKMs have multiple inverse kinematic solutions.

## 31.4 Generalized Jacobian

The conventional Jacobian matrix of the Exechon PKM was formulated as a $3 \times 6$ matrix by Jin et al. [10]. The analysis shows that no singular configuration exists by considering the physical constraints of the PKM. As the Jacobian matrix is used to transform both the velocity and the force systems from the actuator input to moving platform output, its condition is of great interest in design [21]. The condition index $C_I$ of a Jacobian matrix is often used to evaluate the kinetostatic performance of a manipulator. However, care must be taken for applying the condition index, because the elements of Jacobin matrix have inhomogeneous units when the moving platform motion includes both rotation and translation, i.e., no physical unit for orientation but physical unit for position such as meter [28–31]. As the Exechon PKM has two rotational DOF about $x$- and $y$-axes respectively, and one translational DOF along z-axis, the conventional Jacobian cannot be used directly to evaluate its conditioning. Therefore, the dimensionless homogenous Jacobian [30, 31] is formulated in this section and its condition index is utilized as a performance measure for the design optimization later on. As a result, the instantaneous kinematics between the twist of the moving platform $\$_t$ and the joint (both active and passive) velocity $\delta\rho$, is as follows.

$$J\$_t = \delta\rho, \tag{31.1}$$

where

$$J = \begin{bmatrix} J_a \\ J_c \end{bmatrix}, \quad J_a = \begin{bmatrix} [OA_1 \times u_1]^T & u_1^T \\ [OA_2 \times u_2]^T & u_2^T \\ [OA_3 \times u_3]^T & u_3^T \end{bmatrix}, \quad J_c = \begin{bmatrix} [OA_1 \times s_{12}]^T/r_{44} & s_{12}^T/r_{44} \\ [OA_2 \times s_{25}]^T/r_{55} & s_{25}^T/r_{55} \\ [OA_3 \times s_{32}]^T/r_{66} & s_{32}^T/r_{66} \end{bmatrix},$$

$r_{44} = s_{12} \cdot (A_1B_1 \times s_{11})$, $r_{55} = OA_2 \times s_{25} \cdot (s_{24} \times s_{25}) + OB_2 \times (s_{24} \times s_{25}) \cdot s_{25}$, $r_{66} = s_{32} \cdot (A_3B_3 \times s_{31})$. $J$ is known as the generalized Jacobian.

With the generalized Jacobian, the dimensionless Jacobian $Jpa$ can then be calculated by

$$Jpa = \left[ J_{pa0} Ad^T_{g^{R'}_{R_0}} (J^T J)^{-1} J_a^T \right]^{-1}, \tag{31.2}$$

where $J_{pa0}$ is a $3 \times 6$ matrix where each row represents a twist of one feature point on the moving platform. $Ad_{g^{R'}_{R_0}} = \begin{bmatrix} R & 0 \\ 0 & R \end{bmatrix}$ where R is the orientation matrix of platform frame relative to the base frame. The readers are referred to [30] for more details.

## 31.5 Workspace Analysis

As the mechanism holds a three leg architecture, the reachable workspace of the moving platform is formed by the intersections of the three reachable workspaces of the three legs. An analytical method of calculating the reachable workspace of the Exechon-PKM is introduced in [10].

## 31.6 Dimensional Optimization

This section deals with dimension optimization of the PKM for achieving maximum effective workspace as well as maximizing actuators efficiency. Effective workspace is defined as the workspace which have good conditioning and is suitable for machining tasks [32]. Conditioning index will be employed in this paper as one measure to evaluate the effectiveness of the workspace of the PKM in terms of velocity, statics and accuracy, as most literatures did [11–16, 20]. Little research has considered actuator efficiency in the past. Bi and Jin [7] identified that the active joint workspace are far lack of utilization if actuator's displacements are neglected in dimensional optimization. Thus the actuator efficiency has to be taken into account.

### *31.6.1 Design Parameters*

To determine the geometry of the PKM, following parameters need be considered.

- $l_1$: the distance between points $A_1$ and $A_3$ on the base;
- $l_2$: the distance between the center point of $A_1A_3$ to point $A_2$;
- $l_3$: the distance between points $B_1$ and $B_3$ on the moving platform;
- $l_4$: the distance between the center point of $B_1B_3$ to point $B_2$;
- $[L_{min}, L_{max}]$: the motion range of actuators, i.e., the magnitude of $A_iB_i$, $(i = 1, 2, 3)$.

Without loss of generality, it is assumed that the ratio of $O_eB_2$ and $B_1B_3$ is $l_4/l_3 = 0.75$ for making the three attachment points $B_i$ form a relatively symmetrical triangle. Also assume $L_{min} = 0.8$ so that the resulted workspace volume is close to one, which will be explained in the objective function. All other variables are normalized by $l_1$. Thus the three variables to be optimized are defined as:

$$\lambda_1 = l_3/l_1, \tag{31.3}$$

$$\lambda_2 = l_2/l_1, \tag{31.4}$$

$$\lambda_3 = L_{max}/l_1. \tag{31.5}$$

The constraints of the three variables are as follows. $0.25 \leq \lambda_1 \leq 0.5$, $0.5 \leq \lambda_2 \leq 1.0$, $1.0 \leq \lambda_3 \leq 2.0$. For these passive revolute joint axis $s_{i2}$, their physical constraints are within $[-70°, +70°]$. Mathematically, they can be expressed as:

$$-70° \leq acos(s_{i1} \cdot u_i) \leq +70°. \tag{31.6}$$

### 31.6.2 Objective Function

For large volume machining, it is crucial to have a large workspace to footprint ratio as well as good conditioning in the entire workspace. The design objective in this paper is for minimizing actuator strokes while achieving a specified sized (normalized) effective workspace, i.e., 1 in this case, which is defined as:

$$MIN : F(\lambda_1, \lambda_2, \lambda_3) = w_1/GCI + w_2 * \lambda_3, \tag{31.7}$$

subject to

$$0.25 \leq \lambda_1 \leq 0.5, \tag{31.8}$$

$$0.5 \leq \lambda_2 \leq 1.0, \tag{31.9}$$

$$1.0 \leq \lambda_3 \leq 2.0, \tag{31.10}$$

$$V = 1\,\mathrm{m}^3, \tag{31.11}$$

where $w_1$ and $w_2$ are weight factors of the two parts of the objective function respectively. $V$ denotes the reachable workspace and $GCI$ refers to the global condition index [11], which is defined as

$$GCI = \frac{\int_V C_I d_V}{\int_V d_V}, \tag{31.12}$$

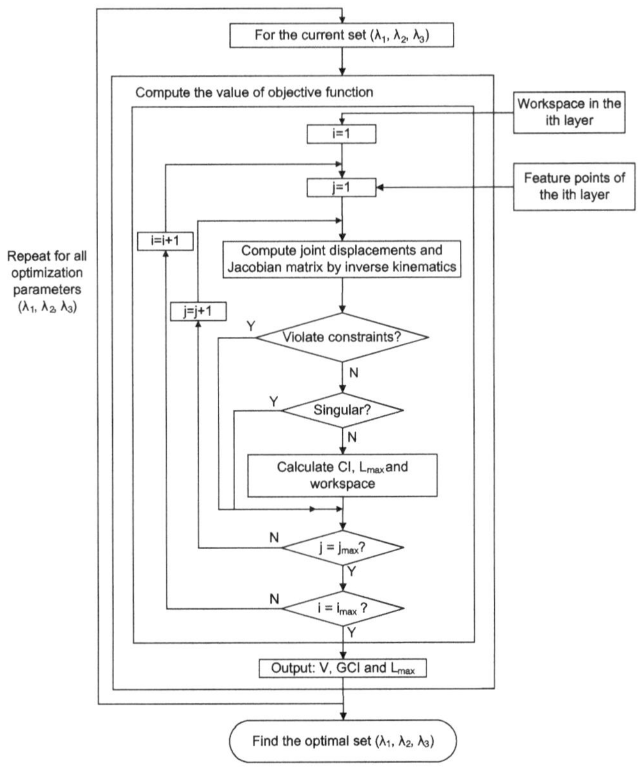


**Fig. 31.3** Optimization procedure of Exechon PKM

where $C_I$ denotes the local condition index, and $L_{max}$ denotes the maximum leg length of $A_iB_i$, and it should be minimized so as to minimize the actuator stroke. Without loss of generality, it is assumed that $L_{max} = L_{min} + stroke$. With a given value of $L_{min}$, the motion range of each leg will be $[L_{min}, L_{max}]$. $L_{max}$ is normalized by $l_1$. So the design objective is formulated as maximize the *GCI* while minimize the actuator stroke in a specific sized workspace. The target workspace volume is one, which means the ratio between workspace volume and the footprint area is at least greater than two. This signifies that the space has been well utilized and it will make ease of the installation of a number of such PKM modules for large production systems.

### *31.6.3 Optimization Algorithm and Results*

The complex optimization [33] is employed to search for the solution in MATLAB environment. The optimization procedure is shown in Fig. 31.3. For each set of $(\lambda_1, \lambda_2, \lambda_3)$, the objective function is evaluated for obtaining the effective

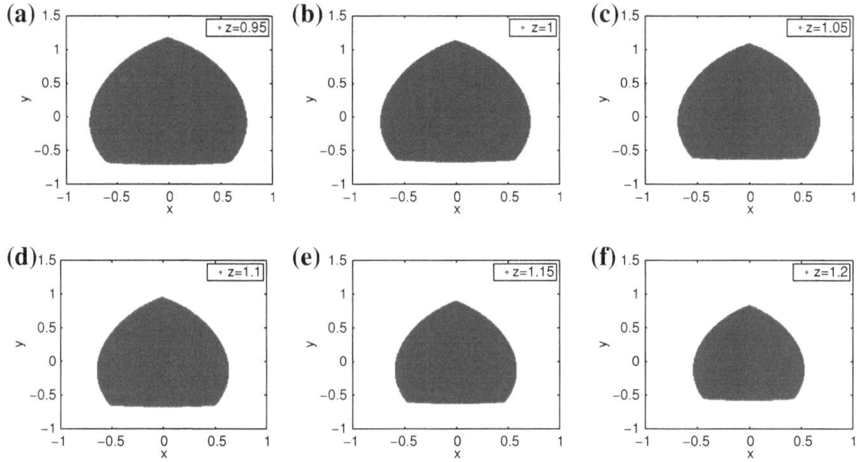


**Fig. 31.4** Workspace of the Exechon PKM

workspace with minimized actuator stroke. The 3-D reachable workspace is divided into a number of layers along $z$-axis with a resolution of $\Delta z = 0.05$, and a number of grids are then generated in each layer with a resolution of $\Delta x = \Delta y = 0.05$. The center point of each grid is then taken as the feature point of the workspace. During optimization process, all feature points within the workspace resulted from the analytical method will be generated first as global known data. In computing the objective function, each of these points is assessed by using the inverse kinematic model of the PKM. If any constraint is violated or the Jacobian matrix is near singular $(C_I < 0.05)$, the point will be excluded out of the workspace. In this way, the effective workspace as well as the GCI can be calculated. After a number of repetitions, the optimal set of $(\lambda_1, \lambda_2, \lambda_3)$ can be obtained. The convergence tolerance is set at $1e-4$ for all design variables and the object function.

Table 31.1 shows the optimization results associated with various sets of weights in the objective function. It is observed that there is a trade-off between the *GCI* and the actuator stroke for achieving a certain sized workspace. When the optimization is conducted for maximum *GCI* only, the largest actuator stroke 0.7692 will be required. When the optimization is conducted for minimizing actuator stroke only, the lowest *GCI* 0.3598 is resulted, and the actuator stroke is 0.7062. When both objectives are weighted equally, i.e., $w_1 = w_2 = 0.5$, both *GCI* and actuator stroke are resulted into intermediate values (GCI = 0.3899, stroke = 0.7335). Therefore, one conclusion is drawn that the cost of increment of *GCI* is a longer actuator stroke. Depending on the specific application and design constraints, a certain set of optimal values can be selected. The corresponding workspace shape and global conditioning distribution can be easily obtained. Taking the optimal set of parameter values of $w_1 = w_2 = 0.5$ as an example, its workspace and conditioning atlas in base frame are shown in Figs. 31.4 and 31.5 respectively. It can be observed that the workspace is in a wedge-like shape with a

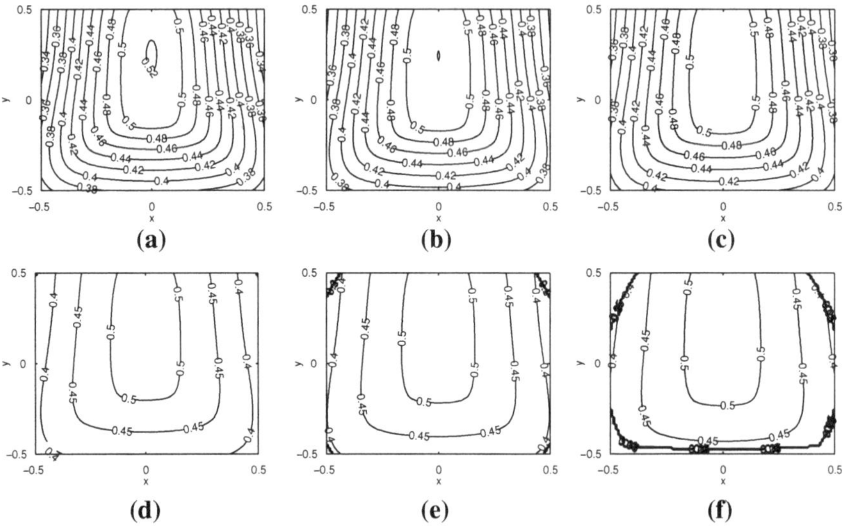


**Fig. 31.5** Conditioning distribution of workspace **a** z = 0:95, **b** z = 1:0, **c** z = 1:05, **d** z = 1:1, **e** z = 1:15, **f** z = 1:2

peach like section area. The ratio of workspace volume to footprint area is about six, which denotes a rather good space utilization. The distribution of condition index is symmetrical about $y$-axis, and a large part of the central area in the workspace ($x \in [-0.4, 0.4], y \in [-0.5, 0.5]$) has a rather good uniform conditioning, i.e., $C_I > 0.4$.

## 31.7 Conclusion

This paper is the first study of the dimensional optimization of a PKM of non-symmetrical architecture with mixed rotational and translational DOF. For the first time, minimizing actuators' stoke is utilized as the design objective. A new optimization approach is proposed for dimension synthesis of PKMs with an objective to maximize the workspace conditioning while minimizing actuators' stoke for achieving specified workspace volume. Optimization results show that the optimization algorithm is effective in design with various sets of weights in the performance function. Results also show that the PKM has a large workspace to footprint ratio (about six). The workspace has a rather good uniform conditioning which is suitable for large volume machining tasks. Kinematic analysis shows that the new PKM has a simple unique solution for inverse kinematics, and no singularity in workspace. The presented research showed that this PKM has great potential to meet the industrial demands today and future.

**Acknowledgments** The authors gratefully acknowledge the help from the team members and industrial partners of the PKAAA project. Funding support from Investment Northern Ireland is acknowledged. Funding support from Royal Academy of Engineering Research Exchange is also acknowledged.

## References

1. Summers M (2005) Robot capability test and development of industrial robot positioning system for the aerospace industry. In: SAE aerospace manufacturing and automated fastening conference, Dallas, pp 2005–01–3336, Oct 2005
2. Weck M, Staimer D (2002) Parallel kinematic machine tools—current state and future potentials. CIRP Ann Manuf Technol 51(2):671–683
3. Rehsteiner F, Neugebauer R, Spiewak S, Wieland F (1999) Putting parallel kinematics machines (pkm) to productive work. CIRP Ann Manuf Technol 48(1):345–350
4. Neumann K (1988) Robot. US patent, US4732525
5. Tonshoff H (1998) A systematic comparison of parallel kinematics. In: First European–American forum on parallel kinematic machines, Milan, Italy
6. Neumann K (2006) The key to aerospace automation. In: SAE aerospace manufacturing and automated fastening conference and exhibition, Detroit, pp 2006–01–3144
7. Bi Z, Jin Y (2011) Kinematic modeling of exechon parallel kinematic machine. Robotics Comput Integr Manuf 27:186–193
8. Eastwood S (2004) Error mapping and analysis for hybrid parallel kinematic machines. PhD thesis, Unversity of Nottingham, Nottingham
9. Liu H, Huang T, Zhao X, Mei J, Chetwynd D (2007) Optimal design of the trivariant robot to achieve a nearly axial symmetry of kinematic performance. Mech Mach Theory 42(12): 1643–1652
10. Jin Y, Bi Z, Gibson R, McToal P, Morgan M, McClory C, Higgins C (2011) Kinematic analysis of a new over-constrained parallel kinematic machine. In: Proceedings of the 13th world congress in mechanism and machine science, Guanajuato, Mexico, pp A7–282, 19–25 June
11. Gosselin C, Angeles J (1991) A global performance index for the kinematic optimization of robotic manipulators. ASME J Mech Des 113:220–226
12. Liu X, Wang J (2007) A new methodology for optimal kinematic design of parallel mechanisms. Mech Mach Theory 42:1210–1224
13. Monsarrat B, Gosselin C (2003) Workspace analysis and optimal design of a 3-leg 6-dof parallel platform mechanism. IEEE Trans Robotics Autom 19(6):954–966
14. Castelli G, Ottaviano E, Ceccarelli M (2008) A fairly general algorithm to evaluate workspace characteristics of serial and parallel mnipulators. Mech Based Des Struct Mach 36:14–33
15. Li Y, Xu Q (2006) A new approach to the architecture optimization of a general 3-puu translational parallel manipulator. J Intell Robotic Syst 46(1):59–72
16. Chablat D, Wenger P (2003) Architecture optimization of a 3-dof translational parallel mechanism for machining applications, the orthoglide. IEEE Trans Robotics Autom 19(3):403–410
17. Jin Y, Chen I, Yang G (2006) Kinematic design of a 6-dof parallel manipulator with decoupled translation and rotation. IEEE Trans Robotics 22(3):545–551
18. Carbone G, Ottaviano E, Ceccarelli M (2007) An optimum design procedure for both serial and parallel manipulators. Proc IMechE Part C J Mech Eng Sci 221:829–843
19. Lou Y, Liu G, Li Z (2008) Randomized optimal design of parallel manipulators. IEEE Trans Autom Sci Eng 5(2):223–233

20. Pierrot F, Nabat V, Krut S, Poignet P (2009) Optimal design of a 4-dof parallel manipulator: from academia to industry. IEEE Trans Robotics 25(2):213–224
21. Merlet J (2006) Jacobian, manipulability, condition number, and accuracy of parallel robots. ASME J Mech Des 128(1):199–206
22. Cardou P, Bouchard S, Gosselin C (2010) Kinematic-sensitivity indices for dimensionally nonhomogeneous jacobian matrices. IEEE Trans Robotics 26(1):166–173
23. Liu X, Gao F (2000) Optimum design of 3-dof spherical parallel manipulators with respect to the conditioning and stiffness indices. Mech Mach Theory 35:1257–1267
24. Huang T, Li M, Zhao X, Mei J, Whitehouse D, Hu S (2005) Conceptual design and dimensional synthesis for a 3-dof module of the trivariant-a novel 5-dof reconfigurable hybrid robot. IEEE Trans Robotics Autom 21(3):449–456
25. Ottaviano E, Ceccarelli M (2002) Optimal design of capaman (cassino parallel manipulator) with a spefied orientation workspace. Robotica 20:159–166
26. Altuzarra O, Pinto C, Sandru B, Hernandez A (2011) Optimal dimensioning for parallel manipulators: workspace, dexterity, and energy. ASME J Mech Des 133:041007–041017
27. Zoppi M, Zlatanov D, Molfino R (2010) Kinematic analysis of the exechon tripod. In: ASME 2010 international design engineering technical conference & computers and information in engineering conference, Montreal, Quebec, Canada, pp 1–8
28. Tandirci M, Angeles J, Ranjbaran F (1992) The characteristic point and the characteristic length of robotic manipulators. In: Proceedings of ASME 22nd biennial conference on robotics, spatial mechanisms, and mechanical systems, Scottsdale, pp 203–208, 13–16 Sept
29. Stocco L, Sacudean S, Sassani F (1999) On the use of scaling matrices for task-specific robot design. IEEE Trans Robotics Autom 15(5):958–965
30. Liu H, Huang T, Chetwynd D (2011) A method to formulate a dimensionally homogeneous jacobian of parallel manipulators. IEEE Trans Robotics 27(1):150–156
31. Huang T, Liu H, Chetwynd D (2011) Generalized jacobian analysis of lower mobility manipulators. Mech Mach Theory 46(6):831–844
32. Chanal H, Duc E, Ray P (2006) A study of the impact of machine tool structure on machining processes. Int J Mach Tools Manuf 46(2):98–106
33. Rao S (1996) Engineering optimization: theory and practice, 3rd edn. Wiley, New York

# Chapter 32
# A Study of the Instantaneous Kinematics of the 5-RSP Parallel Mechanism Using Screw Theory

**Ernesto Rodriguez-Leal, Jian S. Dai and Gordon R. Pennock**

**Abstract** This paper presents a detailed study of the instantaneous kinematics of the 5-*R̲SP* parallel mechanism with centralized motion. The study uses screw theory to investigate the mobility and the singular configurations of the mechanism. The constraint-screw set of the platform is obtained from an analysis of the motion-screw sets comprised by each kinematic chain. The analysis shows that the platform has a screw motion, that is, a one degree-of-freedom motion consisting of a rotation and a translation about an invariant axis. The motion-screw sets are also used to obtain the Jacobian matrix of the mechanism which provides closed-form solutions for the inverse and forward instantaneous kinematic problems. This matrix also provides insight into the singular configurations by investigating the constraint-screws and the motion-screws of the platform in these configurations. Finally, two numerical examples and a motion simulation of the mechanism are presented to illustrate the significance of the analytical results.

**Keywords** Parallel mechanisms · Screw theory · Mobility analysis · Jacobian matrix · Singularity analysis

E. Rodriguez-Leal (✉)
Tecnológico de Monterrey, 2501, Garza Sada, Mexico
e-mail: ernesto.rodriguez@itesm.mx

J. S. Dai
King's College London, Strand, London, WC2R2LS, UK
e-mail: jian.dai@kcl.ac.uk

G. R. Pennock
Purdue University, West Lafayette, IN 47907, USA
e-mail: pennock@ecn.purdue.edu

J. S. Dai et al. (eds.), *Advances in Reconfigurable Mechanisms and Robots I*,
DOI: 10.1007/978-1-4471-4141-9_32, © Springer-Verlag London 2012

## 32.1 Introduction

A comprehensive treatise on screw theory was published by Ball in 1900 [1]. The theory lay dormant, however, for many years until several researchers including von Mises [2], Brand [3], and Dimentberg [4], demonstrated the important role of the theory in applied mechanics. The first application of the theory to the kinematic analysis of closed kinematic chains and spatial mechanisms is believed to be the work of Hunt [5]. Since that time numerous researchers have used this technique to provide geometrical insight into the instantaneous kinematics of multi-degree-of-freedom manipulators [6–8]. More recently, the theory was used to reveal insight into the duality between kinematics and statics [9, 10] and this opened the door into the design of novel parallel robot manipulators [11]. One of the current applications of screw theory is to investigate the mobility of spatial parallel mechanisms [12]. For example, Dai et al. [13] studied the screw systems of parallel mechanisms and their interrelationships, relating the screw systems to motion and constraints and identifying the constraints as (i) platform constraints, (ii) mechanism constraints, or (iii) redundant constraints. Their work resulted in a new approach to the mobility problem based on the decomposition of the motion-screw system and the constraint-screw systems.

Another application of screw theory in kinematics is the determination of the Jacobian matrix [7], which is fundamental to the velocity and trajectory control planning of the mechanism. Typically, this matrix is obtained by multiplication of the platform twist by the orthogonal product of the reciprocal screws [14]. The identification of singular configurations is important for the design and control of parallel mechanisms. Gosselin and Angeles [15] defined three types of kinematic singularities in relation with the rank deficiency of the Jacobian matrices for the inverse and direct kinematics problems, namely: (i) the inverse kinematics singularities, the (ii) direct kinematics singularities, and (iii) when the inverse and direct kinematics Jacobian matrices are both rank deficient. An interesting method for analyzing singularities of parallel mechanisms was proposed by Zhao et al. [16] where the terminal constraints of the mechanism are investigated using screw theory. This method allows the identification of singularities without the need of Jacobian matrices and provides a better physical insight into the constraints imposed on the platform.

The authors have proposed a new $5\text{-}\underline{R}SP$ parallel mechanism (see Fig. 32.1) with centralized motion in [17], i.e. that the centroid of the platform is intersected by an invariant axis for all positions of the mechanism. This mechanism consists of a platform connected to a base by five identical *RSP* kinematic chains (where *R* denotes a revolute joint, *S* denotes a spheric joint, *P* denotes a prismatic joint, and the underline denotes the active joint). The inverse and forward kinematics were solved in [17], resulting in closed form equations. Furthermore, the paper approached the instantaneous kinematics by differentiating the closed loop vector equation. The mobility of the mechanism was explained by geometrical insight supported by the position analysis and validated by numerical examples and

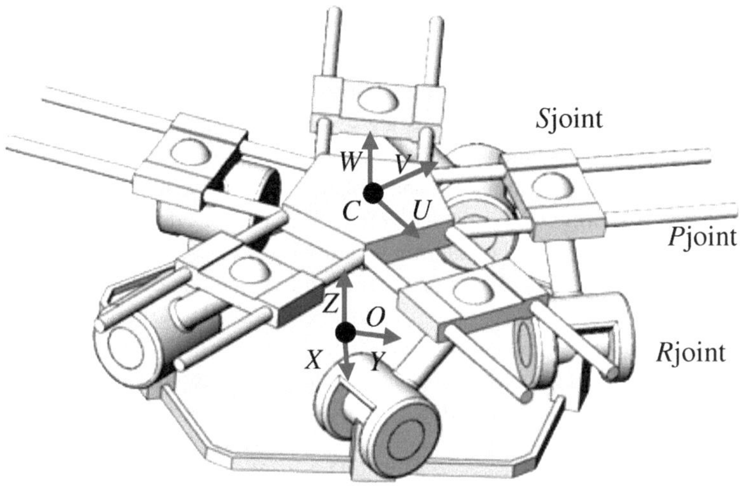


**Fig. 32.1** The 5-*R̲SP* parallel mechanism

motion simulations, showing that the platform of this mechanism exhibits a screw motion, i.e. a translation along and a rotation about the *Z*-axis.

This paper uses screw theory for obtaining the mobility of the 5-*R̲SP* parallel mechanism previously studied in [17], by applying the method proposed by Dai et al. [13], revealing the type and geometry of the constraints exerted by each kinematic chain on the platform. Furthermore, the paper presents solutions to the forward and inverse instantaneous kinematic problems following the methods proposed by Mohamed and Duffy [6], and Joshi and Tsai [7] for obtaining a screw-based Jacobian matrix. The results in this paper are compared to the velocity analysis conducted in [17], showing the conciliation of the screw-based and the differentiation-based Jacobian matrices. In addition, the paper uses the Jacobian matrix to identify the singular configurations of the parallel mechanism. Finally, the platform constraint-screw sets and the motion-screw sets are evaluated in the singular configurations in order to understand the change of mobility.

The paper is arranged as follows. Section 32.2 is a description of the geometry of the parallel mechanism and the notation that is adopted here. Section 32.3 is a study of the mobility of the mechanism. Section 32.4 uses screw theory to obtain closed-form solutions to the instantaneous kinematic problems. Section 32.5 presents the singularity analysis of the platform. Section 32.6 includes a numerical example, where the mobility and the solutions to the kinematic problems are compared with a motion simulation of the parallel mechanism. Finally, Sect. 32.7 presents some important conclusions and suggestions for future research on this new class of parallel mechanism.

## 32.2 Notation and Description of an Arbitrary Leg

Figure 32.2 is a schematic representation of leg $i$ ($i = 1{-}5$) which is an *RSP* kinematic chain. To describe the geometry of this leg several Cartesian reference fames (henceforth referred to simply as fames) are proposed. A global frame, denoted as G ($X$, $Y$, $Z$), is attached to the base with origin at $O$; and a platform frame, denoted as H ($U$, $V$, $W$), is attached to the platform with origin at $C$ (see Fig. 32.1). Consider that $p_x$, $p_y$, and $p_z$ are the Cartesian coordinates of the platform centroid $C$; and $\alpha$, $\beta$, and $\gamma$ are the projection angles about the $X$-, $Y$-, and $Z$-axes that express the orientation of the platform. Seven local frames are also proposed ($k = I, \ldots, VII$) where the $k$th local frame is denoted as ${}^{k}$G (${}^{k}X$, ${}^{k}Y$, ${}^{k}Z$). The first local frame is originally coincident with G, then rotated by the angle $\psi_i$ about the $Z$-axis and translated a distance $a$ along the current ${}^{I}X$-axis. Note that this local frame is positioned at the location of joint $\mathrm{j}_{i1}$ (point $A_i$) and the angles $\psi_i$ are chosen as $\psi_1 = 0°$, $\psi_2 = 72°$, $\psi_3 = 144°$, $\psi_4 = 216°$, and $\psi_5 = 288°$ (i.e., the joints $\mathrm{j}_{i1}$ are placed symmetrically on the base). The second local frame is obtained by rotating ${}^{I}$G by the angle $\varphi$ around the $Z = {}^{I}Z$-axis. Note that the unit vector $\boldsymbol{s}_{i1}$, which represents the axis of joint $\mathrm{j}_{i1}$, is parallel to the ${}^{II}X$-axis. Also note that the angle $\varphi$, which is considered a constant for all five legs, has a significant influence on the behavior of the screw motion of the platform. The third local frame is centered at joint $\mathrm{j}_{i1}$ but is rotated by the angle $\theta_{i1}$ around the ${}^{II}X = {}^{III}X$-axis which allows the action of the joint. The fourth local frame is translated a distance $d$ along the current ${}^{III}Y$-axis and placed coincident with $B_i$. The origin of the fifth, sixth, and seventh local frames are coincident with $B_i$ but are rotated by the angle $\theta_{i2}$ about the ${}^{II}X = {}^{III}X = {}^{IV}X = {}^{V}X$-axis, by the angle $\theta_{i3}$ about the ${}^{V}Y = {}^{VI}Y$-axis, and by the angle $\theta_{i4}$ about the ${}^{VI}Z = {}^{VII}Z$-axis, respectively. Note that $\theta_{i2}$, $\theta_{i3}$, and $\theta_{i4}$ denote the angles of rotation of joints $\mathrm{j}_{i2}$, $\mathrm{j}_{i3}$, and $\mathrm{j}_{i4}$, which represent the $S$ joint in the kinematic chain. Also note that the joint axes $\boldsymbol{s}_{i3}$, $\boldsymbol{s}_{i4}$, and $\boldsymbol{s}_{i5}$ represent a $S$ joint, considering that the axes intersect at $B_i$ and are orthogonal to each other. When the frame ${}^{VII}$G is translated a distance -$b_i$ along the ${}^{VII}X$-axis and then rotated by the angle -$\chi_i$ (where $\chi_1 = 0°$, $\chi_2 = 72°$, $\chi_3 = 144°$, $\chi_4 = 216°$, and $\chi_5 = 288°$) about the ${}^{VII}Z$-axis, then the frame would be coincident with frame H (Note that the angles $\chi_i$ are chosen here to be the same as the angles $\psi_i$). An additional rotation of frame H by the angle -$\gamma$ about the $W$-axis would make the frame G be parallel to the frame H.

The following section presents a mobility analysis of the 5-$\underline{R}SP$ parallel mechanism using screw theory. The individual motion of each leg will be treated as an independent twist applying a wrench to the platform. The section shows that the mobility of the platform can then be determined by the effect of the linear combination of the wrench systems for all the legs.

## 32.3 Mobility Analysis

The leg $i$ of the mechanism, see Fig. 32.2, contains a system of five joint screws where joint $\mathrm{j}_{i1}$ is the only actuated joint; i.e., the remaining joints in the leg are passive. For convenience, the joint screws will be expressed in the global frame G and can be written as

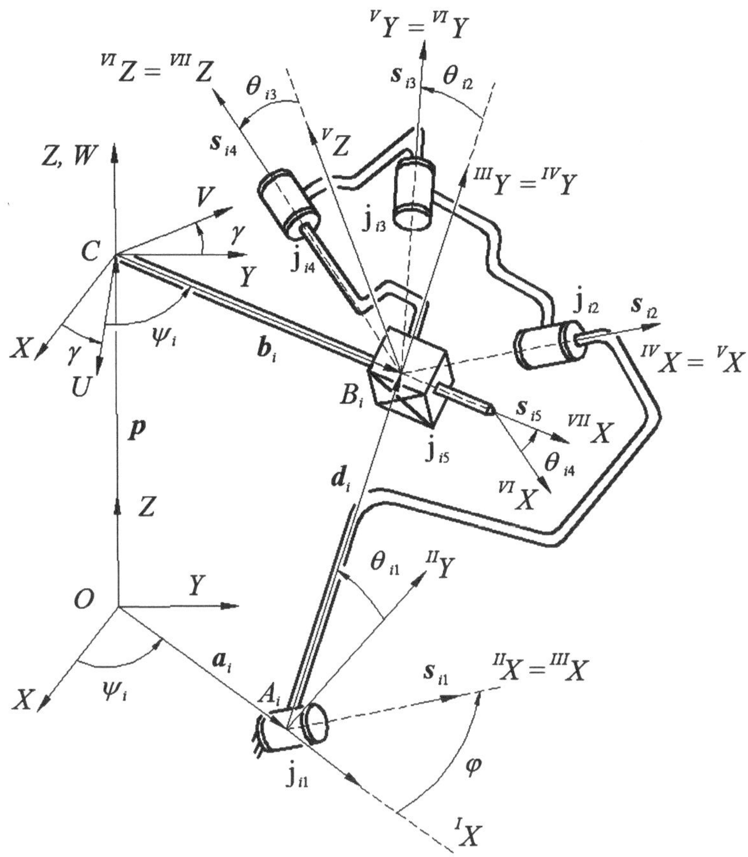


**Fig. 32.2** Leg $i$ of the 5-$\underline{R}SP$ parallel mechanism

$$\$_{i1} = \begin{bmatrix} \boldsymbol{s}_{i1} \\ \boldsymbol{a}_i \times \boldsymbol{s}_{i1} \end{bmatrix}, \quad \$_{i2} = \begin{bmatrix} \boldsymbol{s}_{i2} \\ (\boldsymbol{a}_i + \boldsymbol{d}_i) \times \boldsymbol{s}_{i2} \end{bmatrix}, \quad \$_{i3} = \begin{bmatrix} \boldsymbol{s}_{i3} \\ (\boldsymbol{a}_i + \boldsymbol{d}_i) \times \boldsymbol{s}_{i3} \end{bmatrix},$$
$$\$_{i4} = \begin{bmatrix} \boldsymbol{s}_{i4} \\ (\boldsymbol{a}_i + \boldsymbol{d}_i) \times \boldsymbol{s}_{i4} \end{bmatrix} \quad \text{and} \quad \$_{i5} = \begin{bmatrix} \boldsymbol{0} \\ \boldsymbol{s}_{i5} \end{bmatrix} \tag{32.1}$$

where

$$\begin{gathered} \boldsymbol{a}_i = [ac\psi_i, \quad as\psi_i, \quad 0]^{\mathrm{T}}, \quad \boldsymbol{s}_{i1} = [\mathrm{c}(\varphi + \psi_i), \quad \mathrm{s}(\varphi + \psi_i), \quad 0]^{\mathrm{T}}, \\ \boldsymbol{d}_i = [-dc\theta_{i1}\mathrm{s}(\varphi + \psi_i), \quad dc\theta_{i1}\mathrm{c}(\varphi + \psi_i), \quad ds\theta_{i1}]^{\mathrm{T}}, \quad \boldsymbol{s}_{i2} = [1, \quad 0, \quad 0]^{\mathrm{T}}, \\ \boldsymbol{s}_{i3} = [0, \quad 1, \quad 0]^{\mathrm{T}}, \quad \boldsymbol{s}_{i4} = [0, \quad 0, \quad 1]^{\mathrm{T}}, \text{ and } \quad \boldsymbol{s}_{i5} = [\mathrm{c}(\gamma + \chi_i), \quad \mathrm{s}(\gamma + \chi_i), \quad 0]^{\mathrm{T}} \end{gathered} \tag{32.2}$$

To simplify the task of calculating the screw system for this leg, the $S$ joint axes $\boldsymbol{s}_{i2}$, $\boldsymbol{s}_{i3}$, and $\boldsymbol{s}_{i4}$ are assumed to be always parallel to the $X$, $Y$, and $Z$-axes, respectively.

Substituting Eq. (32.2) into Eq. (32.1), and performing the vector operations, the motion screw system for leg $i$ can be written as

$$\{\$_i\} = \left\{ \begin{array}{l} \$_{i1} = [c(\varphi+\psi_i),\ \ s(\varphi+\psi_i),\ \ 0;\ \ 0,\ \ 0,\ \ as\varphi]^{\mathrm{T}}, \\ \$_{i2} = [1,\ \ 0,\ \ 0;\ \ 0,\ \ ds\theta_1,\ \ -dc\theta_1 c(\varphi+\psi_i) - as\psi_i]^{\mathrm{T}}, \\ \$_{i3} = [0,\ \ 1,\ \ 0;\ \ -ds\theta_1,\ \ 0,\ \ ac\psi_i - dc\theta_1 s(\varphi+\psi_i)]^{\mathrm{T}}, \\ \$_{i4} = \left[0,\ \ 0,\ \ 1; dc\theta_1 c(\varphi+\psi_i) + as\psi_i,\ dc\theta_1 s(\varphi+\psi_i) - ac\psi_i,\ \ 0\right]^{\mathrm{T}}, \\ \$_{i5} = [0,\ \ 0,\ \ 0;\ \ c(\gamma+\psi_i),\ \ s(\gamma+\psi_i),\ \ 0]^{\mathrm{T}} \end{array} \right\} \tag{32.3}$$

Note that the screws $\$_{i1}$, $\$_{i2}$, $\$_{i3}$, and $\$_{i4}$ are for revolute joints, therefore, these screws have zero pitch. Similarly, the screw $\$_{i5}$ is for a prismatic joint and, therefore, has an infinite pitch. Due to the geometry of the mechanism, the angular position of joint $j_{i1}$ is the same for all five legs and, therefore, $\theta_{i1}$ will be abbreviated as $\theta_1$ [17]. Although this mechanism can be only actuated with one joint $j_{i1}$, it is considered a simultaneous and synchronized actuation from the five joints $j_{i1}$. Each leg of the mechanism contains an independent set of screws. Since there are five screws per leg then there must be one reciprocal screw that complements the six screw system for each leg [13], namely

$$\$^r_{i1} = \left[-s(\gamma+\psi_i),\ \ c(\gamma+\psi_i),\ \ \tan\theta_1 c(\gamma-\varphi);\ \ P^r_i,\ Q^r_i,\ R^r_i\right]^{\mathrm{T}} \tag{32.4a}$$

where

$$\begin{aligned} P^r_i &= \tan\theta_1\left[dc\theta_1 s(\gamma-\varphi)s(\varphi+\psi_i) + ac(\gamma-\varphi)s\psi_i\right] \\ Q^r_i &= -\tan\theta_1\left[dc\theta_1 s(\gamma-\varphi)c(\varphi+\psi_i) + ac(\gamma-\varphi)c\psi_i\right] \\ R^r_i &= dc\theta_1 c\varphi s\gamma + c\gamma(a - dc\theta_1 s\varphi) \end{aligned} \tag{32.4b}$$

If Eq. (32.4a) is written for each leg of the parallel mechanism then the result is the platform constraint-screw system. Taking the reciprocal screw to this 5-system, the platform motion-screw for the mechanism can be written as

$$\$^f = [0,\ 0, \tan\theta_1 c(\gamma-\varphi);\ 0,\ 0,\ -dc\theta_1 c\varphi s\gamma - c\gamma(a - dc\theta_1 s\varphi)]^{\mathrm{T}} \tag{32.5}$$

This equation indicates that the platform has a single degree-of-freedom (i.e., a screw motion) along and about the $Z$-axis. It has been previously presented in [17] with the use of simulations that when the mechanism is assembled setting $\varphi = \pm 90°$ then $\gamma = 0°$. Note that in the configuration $\varphi = \pm 90°$ Eq. (32.5) reduces to

$$\$^f_1 = [0,\ 0,\ 0;\ 0,\ 0,\ dc\theta_1 - a]^{\mathrm{T}} \tag{32.6}$$

This equation shows that the platform will have a single translational degree-of-freedom along the $Z$-axis, which is consistent with the simulations of the mobility discussed in [17] for $\varphi = \pm 90°$. The following section presents the inverse and forward instantaneous kinematics of the 5-$\underline{R}SP$ parallel mechanism using screw theory. The section shows the determination of the Jacobian matrix with the use of the branch motion screw sets in order to solve the direct and inverse instantaneous kinematics of the mechanism.

## 32.4 Instantaneous Kinematics

The instantaneous twist $\$_p$ of the platform of a parallel mechanism can be expressed as a linear combination of $\ell$ instantaneous twists [6]

$$\$_p = \sum_{j=1}^{\ell} \dot{q}_{ij} \$_{ij} \tag{32.7}$$

where $\ell$ denotes the total number of joints per limb, and $\dot{q}_{ij}$ represents the scalar velocity of joint $\mathrm{j}_{ij}$. Each leg is comprised of a system of five motion-screws. Since $\$_{i1}$, $\$_{i2}$, $\$_{i3}$, and $\$_{i4}$ have zero pitch, and $\$_{i5}$ has an infinite pitch, the instantaneous twist of the platform can be written as

$$\$_p = \dot{\theta}_1 \$_{i1} + \dot{\theta}_{i2} \$_{i2} + \dot{\theta}_{i3} \$_{i3} + \dot{\theta}_{i4} \$_{i4} + \dot{b}_{i5} \$_{i5} \tag{32.8}$$

The Jacobian matrix can be obtained from screw theory by locking the active joint $\mathrm{j}_{i1}$ [7], hence the branch motion-screw system from Eq. (32.3) becomes a four-system when $\$_{i1}$ is eliminated and the branch constraint-screw system will have two reciprocal screws, namely

$$\$_i^r \begin{Bmatrix} \$_{i1}^r = \left[0,\ \ 0,\ \ 1;\ \ dc\theta_1 c(\varphi+\psi_i) + as\psi_i, dc\theta_1 s(\varphi+\psi_i) - ac\psi_i,\ \ 0\right]^{\mathrm{T}}, \\ \$_{i2}^r = \left[-s(\gamma+\psi_i),\ \ c(\gamma+\psi_i),\ \ 0;\ \ -ds\theta_1 c(\gamma+\psi_i), -ds\theta_1 s(\gamma+\psi_i),\ \ R_{i2}^r\right]^{\mathrm{T}} \end{Bmatrix}$$

$$\text{where} \qquad R_{i2}^r = dc\theta_1 c\varphi s\gamma + c\gamma(a - dc\theta_1 s\varphi) \tag{32.9}$$

The passive joints can be eliminated by applying the orthogonal product to each of the reciprocal screws in Eq. (32.9), that is

$$\$_{ij}^{r\mathrm{T}} \$_p = \$_{ij}^{r\mathrm{T}} \$_{i1} \dot{\theta}_1 \tag{32.10}$$

Performing the orthogonal product on the right side of Eq. (32.10), using the reciprocal screws form Eq. (32.9), gives

$$\$_{i1}^{r\mathrm{T}} \$_p = dc\theta_1 \dot{\theta}_1, \text{ and } \quad S_{i2}^{r\mathrm{T}} S_p = -dc(\gamma - \varphi) s\theta_1 \dot{\theta}_1 \tag{32.11}$$

For the five legs, Eq. (32.11) result in two independent linear systems where each system contains five scalar equations. The equations can be written in matrix form as

$$\mathbf{J}_{f=q} \$_p = \mathbf{J}_{g=q} \mathbf{\Theta} \quad (\mathrm{q} = 1, 2) \tag{32.12}$$

where $\mathbf{J}_{f=q}$ and $\mathbf{J}_{g=q}$ are commonly referred to as the direct kinematics Jacobian matrices and the inverse kinematics Jacobian matrices [7], respectively, and $\boldsymbol{\Theta}$ is the vector of the joint rates of the five rotational actuators. Note that the instantaneous twist $\$_p$ can be also expressed as

$$\$_p = \begin{bmatrix} \omega_x, & \omega_y, & \omega_z, & v_x, & v_y, & v_z \end{bmatrix}^{\mathrm{T}} \tag{32.13}$$

The direct kinematics Jacobian matrices can be written as

$$\mathbf{J}_{f=q} = \begin{bmatrix} \mathrm{A}_{1q} & \mathrm{B}_{1q} & \mathrm{C}_{1q} & \mathrm{D}_{1q} & \mathrm{E}_{1q} & \mathrm{F}_{1q} \\ \mathrm{A}_{2q} & \mathrm{B}_{2q} & \mathrm{C}_{2q} & \mathrm{D}_{2q} & \mathrm{E}_{2q} & \mathrm{F}_{2q} \\ \mathrm{A}_{3q} & \mathrm{B}_{3q} & \mathrm{C}_{3q} & \mathrm{D}_{3q} & \mathrm{E}_{3q} & \mathrm{F}_{3q} \\ \mathrm{A}_{4q} & \mathrm{B}_{4q} & \mathrm{C}_{4q} & \mathrm{D}_{4q} & \mathrm{E}_{4q} & \mathrm{F}_{4q} \\ \mathrm{A}_{5q} & \mathrm{B}_{5q} & \mathrm{C}_{5q} & \mathrm{D}_{5q} & \mathrm{E}_{5q} & \mathrm{F}_{5q} \end{bmatrix} \quad (\mathrm{q} = 1, 2) \tag{32.14a}$$

where

$$\begin{array}{l} \mathrm{A}_{i1} = dc\theta_1 \mathrm{c}(\varphi + \psi_i) + as\psi_i, \ \mathrm{B}_{i1} = dc\theta_1 \mathrm{s}(\varphi + \psi_i) - ac\psi_i, \ \mathrm{C}_{i1} = 0 \\ \mathrm{D}_{i1} = 0, \ \mathrm{E}_{i1} = 0, \text{ and } \ \mathrm{F}_{i1} = 1 \end{array} \tag{32.14b}$$

and

$$\begin{array}{l} \mathrm{A}_{i2} = -ds\theta_1 \mathrm{c}(\gamma + \psi_i), \ \mathrm{B}_{i2} = -ds\theta_1 \mathrm{s}(\gamma + \psi_i), \ \mathrm{D}_{i2} = -\mathrm{s}(\gamma + \psi_i), \\ \mathrm{C}_{i2} = dc\theta_1 \mathrm{c}\varphi \mathrm{s}\gamma + \mathrm{c}\gamma(a - dc\theta_1 \mathrm{s}\varphi), \ \mathrm{E}_{i2} = \mathrm{c}(\gamma + \psi_i), \text{ and } \ \mathrm{F}_{i2} = 0 \end{array} \tag{32.14c}$$

The inverse kinematics Jacobian matrices can be written as

$$\mathbf{J}_{g=1} = dc\theta_1 \mathbf{I} \tag{32.15a}$$

and

$$\mathbf{J}_{g=2} = -dc(\gamma - \varphi)s\theta_1 \mathbf{I} \tag{32.15b}$$

where $\mathbf{I}$ is the $5 \times 5$ identity matrix. The *forward instantaneous kinematic problem* is defined as follows: for a given set of joint rates of the actuators determine the general velocity vector of the platform (referred to as the platform twist velocity). This velocity can be written as

$$V = \mathbf{J}\boldsymbol{\Theta} \tag{32.16a}$$

where $\mathbf{J} = \mathbf{J}_{f=q}^{-1}\mathbf{J}_{g=q}$ is commonly referred to as the direct Jacobian matrix. The *inverse instantaneous kinematic problem*, on the other hand, is defined as follows: for a given platform twist velocity determine the joint rates of the actuators. Taking the inverse of Eqs. (32.16a) and (32.16b), the vector of the actuator joint rates is

$$\boldsymbol{\Theta} = \mathbf{J}^{-1} V \tag{32.16b}$$

where $\mathbf{J}^{-1} = \mathbf{J}_{g=q}^{-1}\mathbf{J}_{f=q}^{-1}$ is commonly referred to as the inverse Jacobian matrix [17]. Solving Eq. (32.14a) by adding the velocity contributions from each leg results inthe solution to the forward instantaneous kinematics problem as follows

$$V = [v_z \quad \omega_z]^{\mathrm{T}}, \quad \boldsymbol{\Theta} = \begin{bmatrix} \dot{\theta}_1 \\ \dot{\theta}_1 \end{bmatrix} \text{ and } \mathbf{J} = d\begin{bmatrix} \mathrm{c}\theta_1 & 0 \\ 0 & -\dfrac{\mathrm{c}(\gamma - \varphi)\mathrm{s}\theta_1}{d\mathrm{c}\theta_1\mathrm{c}\varphi\mathrm{s}\gamma + \mathrm{c}\gamma(a - d\mathrm{c}\theta_1\mathrm{s}\varphi)} \end{bmatrix} \tag{32.17}$$

The inverse instantaneous kinematic problem can be solved in a straightforward manner by taking the inverse of Eq. (32.17), that is

$$\begin{bmatrix} \dot{\theta}_1 \\ \dot{\theta}_1 \end{bmatrix} = \begin{bmatrix} \dfrac{1}{d\mathrm{c}\theta_1} & 0 \\ 0 & -\dfrac{d\mathrm{c}\theta_1\mathrm{c}\varphi\mathrm{s}\gamma + \mathrm{c}\gamma(a - d\mathrm{c}\theta_1\mathrm{s}\varphi)}{d\mathrm{c}(\gamma - \varphi)\mathrm{s}\theta_1} \end{bmatrix} \begin{bmatrix} v_z \\ \omega_z \end{bmatrix} \tag{32.18}$$

The following section will present a singularity analysis of the mechanism to determine the configurations in which the platform will gain or lose certain degrees-of-freedom.

## 32.5 Singularity Analysis

The singular positions of the mechanism can be obtained by identifying the configurations in which the determinant of the matrix **J** is zero. The determinant of Eq. (32.17) results in

$$|\mathbf{J}| = -d\frac{\mathrm{s}\theta_1\mathrm{c}\theta_1}{b\mathrm{c}(\gamma - \varphi) + ac\varphi\tan^2(\gamma - \varphi)} \tag{32.19}$$

Therefore, the determinant is zero when: (i) $\theta_1 = 90°$, (ii) $\theta_1 = 0°$, (iii) $\theta_1 = 180°$, (iv) $\varphi = 90°$, (v) $\gamma - \varphi = 90°$. Note that the determinant $|\mathbf{J}|$ approaches infinity $= \infty$ when (vi) $b\mathrm{c}(\gamma - \varphi) + ac\varphi\tan^2(\gamma - \varphi) = 0$. Singularity (i) occurs when all the legs are perpendicular to the base and parallel to each other, note that for this configuration $\gamma = 0°$. Singularities (ii) and (iii) occur when the platform is collapsed, i.e. the platform is folded onto the base when $p_z = 0$ and $\gamma = \mathrm{atan}\,[dc\varphi/(a - ds\varphi)]$. Singularity (iv) occurs when the joint axis $s_{i1}$ is tangential to the base circle, i.e. when $\varphi = 90°$. This assembly configuration has been identified in Sect. 32.3 to produce only translation of the platform along the central axis, and can be further proved when substituting $\varphi = 90°$ and $\gamma = 0°$ into the solution for the direct instantaneous kinematics. Singularity (v) occurs when $\gamma - \varphi = 90°$, and finally the singularity (vi) provides a condition that involves dimensional and assembly configuration parameters of the parallel mechanism in which the magnitude of the angular velocity $\omega_z$ of the platform becomes infinite, disregarding the input of the active joint velocities.

Investigating the singularity configuration (i), the branch motion-screw system for leg $i$ $\{\$_i\}$, expressed by Eq. (32.2), can be evaluated in the singular configuration (i) when $\theta_1 = 90°$ and $\gamma = 0°$, leading to the branch constraint-screw of leg $i$

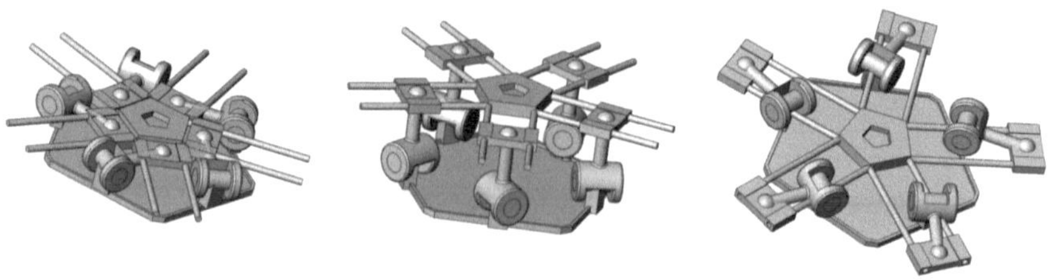

**(a)** **(b)** **(c)**

**Fig. 32.3** Singularity configurations

$$\$^r_{i1} = [0,\ \ 0,\ \ 1;\ \ as\psi_i,\ \ -ac\psi_i,\ \ 0]^{\mathrm{T}} \tag{32.20}$$

Note that the constraint imposed by the reciprocal screw is a force that is coaxial with the leg, i.e., parallel to the base and intersecting points $A_i$ and $B_i$. Writing the reciprocal screw for each leg of the parallel mechanism, result in the platform constraint-screw system, whose reciprocal screw system is the platform motion-screw system $\{\$^f\}$ for the mechanism in the singularity (i) configuration

$$\{\$^f\} = \begin{Bmatrix} \$^f_1 = [0,\ \ 0,\ \ 1;\ \ 0,\ 0,\ 0]^{\mathrm{T}}, \\ \$^f_2 = [0,\ \ 0,\ \ 0;\ \ 0,\ 1,\ 0]^{\mathrm{T}}, \\ \$^f_3 = [0,\ \ 0,\ \ 0;\ \ 1,\ 0,\ 0]^{\mathrm{T}} \end{Bmatrix} \tag{32.21}$$

This equation indicates that the platform has three degrees-of-freedom in the singular configuration (i) when $\theta_1 = 90°$ and $\gamma = 0°$, where the platform is able to rotate about the *Z*-axis and to experience translation along the *X*- and *Y*-axes. Figure 32.3a shows a CAD model of the mechanism in the singularity (i) configuration.

The singularity configuration (ii) is also investigated with the use of screw theory when the branch motion-screw system for leg *i* $\{\$_i\}$ from Eq. (32.2) is expressed for $\theta_1 = 0°$ and $\gamma = \mathrm{atan}\,[dc\varphi/(a - ds\varphi)]$, and the reciprocal screw to the branch motion-screw system of leg *i* for the singularity (ii) configuration is written as

$$\$^r_{i1} = [-\mathrm{s}(\gamma + \psi_i),\ \ \mathrm{c}(\gamma + \psi_i),\ \ 0;\ \ 0,\ 0,\ ac\gamma + ds(\gamma - \varphi)]^{\mathrm{T}} \tag{32.22}$$

where $\$^r_{i1}$ is a force perpendicular to the *P* joints of the platform and coplanar with the platform, intersecting the center of the *S* joints. The platform constraint-screw system $\{\$^r\}$ is formed by the reciprocal screws of all legs, and its reciprocal screw gives the platform motion-screw system $\{\$^f\}$ in the singularity (ii) pose, namely

$$\{\$^f\} = \begin{Bmatrix} S^f_1 = [0,\ \ 1,\ \ 0;\ \ 0,\ 0,\ 0]^{\mathrm{T}}, \\ \$^f_2 = [1,\ \ 0,\ \ 0;\ \ 0,\ 0,\ 0]^{\mathrm{T}}, \\ \$^f_3 = [0,\ \ 0,\ \ 0;\ \ 0,\ 0,\ 1]^{\mathrm{T}} \end{Bmatrix} \tag{32.23}$$

**Table 32.1** Simulation parameters and results for assembly configurations $\varphi = 30^\circ$ and $\varphi = 90^\circ$

| Parameters | | Initial | Final |
|---|---|---|---|
| $a = 150$ mm | $\theta_{11}, \theta_{12}, \theta_{13}, \theta_{14}, \theta_{15}$ (deg) | 24.14 | 162.14 |
| $\varphi = 30^\circ$ | $p_x, p_y$ (mm) | 0 | 0 |
| $d = 100$ mm | $p_z$ (mm) | 86.71 | 64.45 |
| Time = 5 s | $\alpha, \beta$ (deg) | 0 | 0 |
| $\dot{\theta}_{i1} = 4.6$ rpm | $\gamma$ (deg) | 37.17 | −22.64 |
| $a = 150$ mm | $\theta_{11}, \theta_{12}, \theta_{13}, \theta_{14}, \theta_{15}$ (deg) | 108.47 | 72.47 |
| $\varphi = 90^\circ$ | $p_x, p_y$ (mm) | 0 | 0 |
| $d = 100$ mm | $p_z$ (mm) | 94.85 | 95.36 |
| Time = 5 s | $\alpha, \beta$ (deg) | 0 | 0 |
| $\dot{\theta}_{i1} - 1.2$ rpm | $\gamma$ (deg) | 0 | 0 |

This equation shows that the platform has three DOF in the singular configuration (ii) when $\theta_1 = 0^\circ$, which apart from translating along the $Z$-axis, the platform rotates about the $X$- and $Y$-axes. Note that the singularity (iii) produces the same results as for the singularity (ii) from Eq. (32.23). Figure 32.3b and c illustrate the poses when the mechanism is subjected to singularity (ii) and singularity (iii) respectively.

Section 32.6 presents two numerical examples to demonstrate that the mobility and the solutions to the inverse and forward instantaneous kinematics presented in Sects. 32.3 and 32.4 are correct. The section also compares the results of the examples with a motion simulation using commercial software.

## 32.6 Numerical Example and Motion Simulations

***Simulation 1***. A CAD model of the mechanism was produced using a commercial motion simulation software package. The model uses the dimensional parameters $a = 150$ mm and $d = 100$ mm. Consider the two assembly configurations $\varphi = 30^\circ$, and $\varphi = 90^\circ$, the total simulation time and the angular velocity of the actuators $j_{i1}$ are shown in Table 32.1. The results of the simulations for the initial and final configurations of the mechanism are presented in the table.

The simulation plots for the assembly configuration $\varphi = 30^\circ$ are shown in Fig. 32.4. Figure 32.4a, c, and e show the $x$, $y$, and $z$ coordinates of the position, velocity, and acceleration of the centroid $C$ versus time in solid, dashed and dotted lines, respectively. These figures indicate that the platform can only translate along the Z-axis, that is, the positioning along the $X$- and $Y$-axes is constantly zero. Figure 32.4b, d, and f show the projected angles of rotation, the angular velocity, and the angular acceleration about the $X$-, $Y$-, and $Z$-axes in solid, dashed and dotted lines, respectively. These figures indicate that the platform can only rotate about the $Z$-axis. This agrees with the mobility analysis presented in Sect. 32.3,

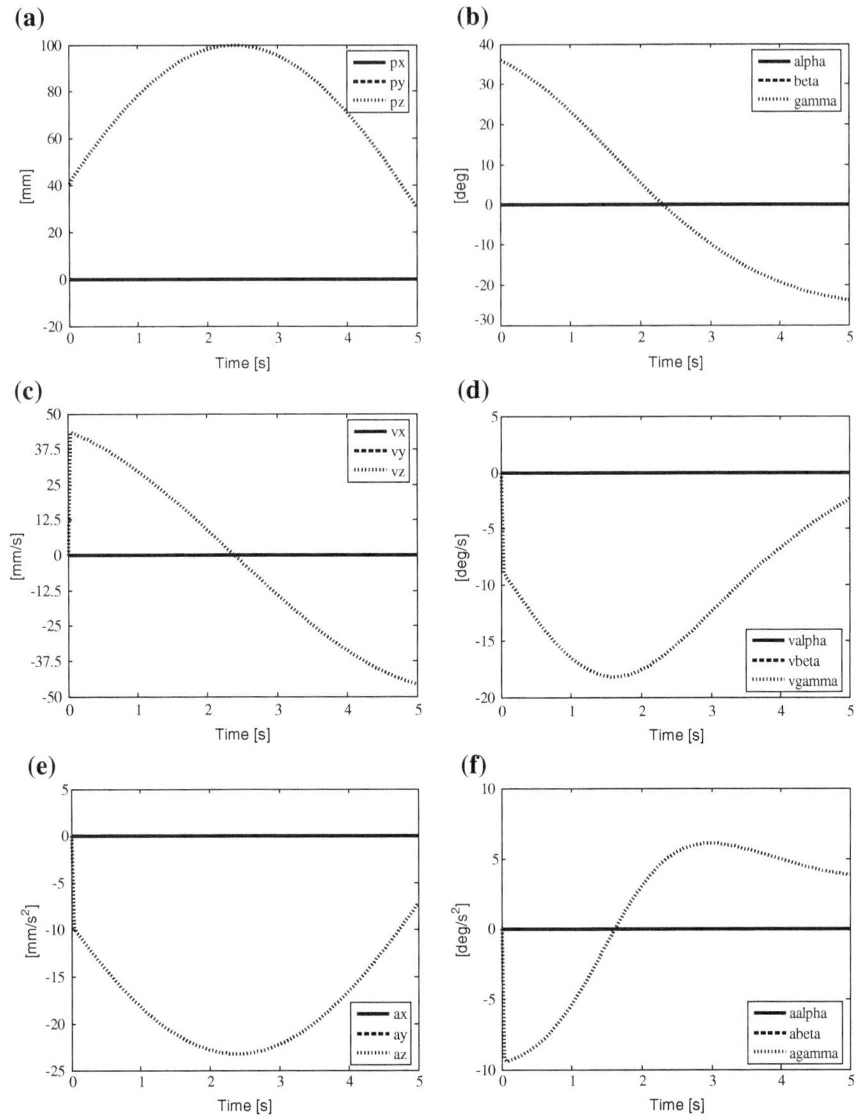


**Fig. 32.4** Position, velocity and acceleration simulation results for $\varphi = 30°$

that is, the rotation of the platform depends on the angle $\varphi$. The positioning simulation shows that the platform can only translate along the $Z$-axis when $\varphi = \pm 90°$ which agrees with Eq. (32.6), and produces identical results to the plots shown in Fig. 32.4a, c, and e. The simulation of the projected angles, angular velocity and angular acceleration of the platform in Fig. 32.5 show that the rotation about the $Z$-axis is eliminated.

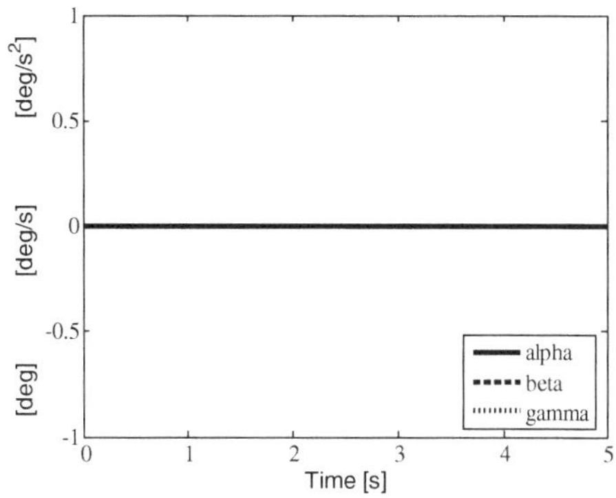


**Fig. 32.5** Angular position, velocity and acceleration simulation results of the mechanism for $\varphi = 90°$

**Table 32.2** Geometrical and simulation parameters for the direct instantaneous kinematics with assembly configuration $\varphi = 60°$

| Dimensional parameters | Simulation parameters |
|---|---|
| $a = 150$ mm | $a = 150$ mm |
| $\varphi = 60°$ | $0° \leq \theta 1 \leq 180°$ |
| $d = 100$ mm | Time $= 5$ s |
| $0° \leq \theta 1 \leq 180°$ | $\dot{\theta}_1 = 5.6$ rpm |

*Example 1* Consider the assembly configuration $\varphi = 60°$. The total simulation time and the angular velocity of the actuators are presented in Table 32.2, where $\dot{\theta}_1 = 5.6$ rpm (recall that due to the geometry of the mechanism, all the actuators have the same angular position, and velocity). The behavior of the calculated linear velocity of the platform centroid $C$ as a function of $\theta_1$ is shown in Fig. 32.6a.

Figure 32.6b shows the simulated linear velocity components of the centroid $C$ on the $X$-, $Y$-, and $Z$-axes, represented as plots in black, blue and green color, respectively. Furthermore, the translational velocity of $C$ along the $Z$-axis follows a cosinusoidal behavior and is identical to the computed velocity shown in Fig. 32.6a.

Substituting the parameters listed in Table 32.2 into Eq. (32.17) gives the angular velocity of the platform.

A plot of the angular velocity of the platform about the $Z$-axis as a function of the angular position of the active joint $\theta_1$ is shown in Fig. 32.6c. Figure 32.6d presents the plots produced by the simulation of the angular velocity components about the $X$-, $Y$-, and $Z$-axes in solid, dashed, and dotted plots, respectively.

The platform shows no angular velocity component about the $X$- and $Y$-axes as observed from Fig. 32.6c and d, which show identical velocity profiles. Note that the platform experiences rotation about the $Z$-axis, while the platform is always parallel to the base. The position and orientation simulations verify the results of the mobility analysis presented in Sect. 32.3.

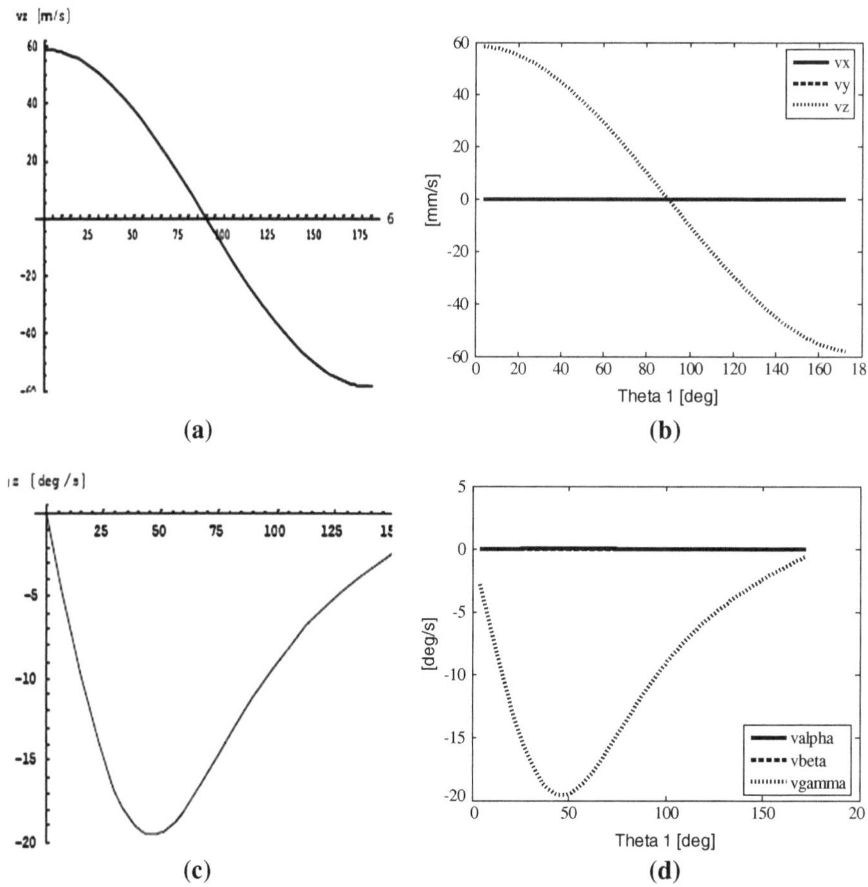


**Fig. 32.6** Calculated and simulated plots of the linear and angular velocity for $\varphi = 60°$

## 32.7 Conclusions and Suggestions for Future Research

This paper is a detailed kinematic analysis of a novel 5-$\underline{R}SP$ parallel mechanism with centralized motion by means of screw theory. First, the motion-screw sets comprised by each kinematic chain are obtained. Then the paper determines the platform constraint-screw set, which produces five independent constraints, leading a one degree-of-freedom motion consisting of a rotation about and a translation along an invariant axis. The motion-screw sets were then used to obtain the Jacobian matrix, which results in closed-form solutions for the inverse and forward instantaneous kinematics of the mechanism. The Jacobian matrix was analyzed for obtaining the singular configurations of the mechanism, namely: (i) when all the legs are perpendicular to the base ($\theta_1 = 90°$), (ii) and (iii) when the platform is collapsed ($\theta_1 = 0°$, or $\theta_1 = 180°$), and (iv) occurs when the joint axis $s_{i1}$ is tangential to the base circle, i.e. when $\varphi = 90°$. The paper evaluates the constraint and motion-screws of the platform in the singular configurations and observes that the platform can rotate about the $Z$-axis. To experience translation along the $X$- and $Y$-axes in the singular configuration (i) translate along the $Z$-axis; the platform will rotate about the $X$- and $Y$-axes in the singular configurations (ii) and (iii); and the platform will translate along the $Z$-axis in the configuration (iv). Numerical examples and motion simulations are included to validate the analytical results presented in the paper. A suggestion for future research is to perform a static and dynamic force analysis of the mechanism for practical industrial applications.

**Acknowledgments** This work was supported by the Grants of the Mexican Science and Technology Council (CONACYT) for a PhD degree program at King's College London, University of London, England.

## References

1. Ball RS (1900) A treatise on the theory of screws. Cambridge University Press, Cambridge
2. Von Mises R (1924) Motor calculus—a new theoretical device for mechanics. In Baker EJ, Wohlhart K (eds) University of Technology Graz, Austria
3. Brand L (1957) Vector and tensor analysis, sixth printing. Wiley, New York
4. Dimentberg FM (1965) The Screw Calculus and its Application in Mechanics. Izdat, Nauka, Moscow
5. Hunt KH (1978) Kinematic geometry of mechanisms. Oxford University Press, New York
6. Mohamed MG, Duffy J (1985) A direct determination of the instantaneous kinematics of fully parallel robot manipulators. J Mech Transm-T ASME 107(2):226–229
7. Joshi SA, Tsai LW (2002) Jacobian analysis of limited-DOF parallel manipulators. J Mech Des-T ASME 124(2):254–258
8. Tsai LW (1999) Robot analysis: the mechanics of serial and parallel manipulators. Wiley, New York
9. Davidson JK, Hunt KH (2004) Robots and screw theory: applications of kinematics and statics to robotics. Oxford University Press Inc., New York
10. Shai O, Pennock GR (2006) A study of the duality between planar kinematics and statics. J Mech Des-T ASME 128(3):587–598
11. Shai O, Pennock GR (2006) Extension of graph theory to the duality between static systems and mechanisms. J Mech Des-T ASME 128(1):179–191
12. Huang Z, Li QC (2002) General methodology for type synthesis of lower mobility symmetrical parallel manipulators and several novel manipulators. Int J Robot Res 21(2):131–146
13. Dai JS, Huang Z, Lipkin H (2006) Mobility of overconstrained parallel mechanisms. J Mech Des-T ASME 128(1):220–229
14. Huang Z, Wang J, Fang YF (2002) Analysis of instantaneous motions of deficient-rank 3-RPS parallel manipulators. Mech Mach Theory 37(2):229–240
15. Gosselin CM, Angeles J (1990) Singularity analysis of closed-loop kinematic chains. IEEE T Robotic Autom 6(3):281–290
16. Zhao JS, Feng ZJ, Khou K, Dong JX (2005) Analysis of the singularity of spatial parallel manipulator with terminal constraints. Mech Mach Theory 40(3):275–284
17. Rodriguez Leal E, Dai JS, Pennock GR (2011) Kinematic analysis of a 5-$\underline{R}SP$ parallel mechanism with centralized motion. Meccanica 46(1):221–237

# Chapter 33
# Experimental Modal Analysis for a 3-DOF PKM Module

**Jun Zhang, Hai-Wei Luo and Tian Huang**

**Abstract** To reveal the modal properties of a 3-dof PKM module, a wire-frame model is proposed and relative experimental modal test is conducted. The lower orders of natural frequencies, damping ratios are obtained and corresponding mode shapes are classified. The modal analysis reveals two categories of vibration modes for the PKM module. The first to the third orders of modes correspond to the overall vibration of the module while the fourth to the sixth orders of modes corresponding to pitch and yaw as well as their combination of the moving platform. The overall vibration modes of the module are expected to be eliminated through bolt-connection of the base and fixed frame. The moving platform vibration modes, however, rely on the weak stiffness of spherical joints. Thus, desirable dynamic performance of the PKM module demands delicate design for the spherical joint with high stiffness.

**Keywords** Experimental modal analysis · Parallel kinematic machine · Dynamic modeling

J. Zhang (✉)
School of Mechanical Engineering, Anhui University of Technology, Ma-xiang Road, Ma'an'shan City, Anhui, China
e-mail: junzhang@ahut.edu.cn

H.-W. Luo · T. Huang
School of Mechanical Engineering, Tianjin University, Tianjin City, China
e-mail: luohaiwei85@tju.edu.cn

T. Huang
e-mail: htiantju@public.tpt.tj.cn

J. S. Dai et al. (eds.), *Advances in Reconfigurable Mechanisms and Robots I*,
DOI: 10.1007/978-1-4471-4141-9_33, 

## 33.1 Introduction

A novel 3-dof parallel kinematic machine (PKM) module was proposed to satisfy high-speed machining of extra large component with complex geometries [1]. The PKM module stems from the Sprint Z3 Head but adopts a topological configuration of 3-R$\underline{P}$S parallel mechanism. Urged by the requirements for quick evaluation of dynamic performance, an analytical lumped-parameter dynamic model based on substructure synthesis was proposed and the distributions of lower natural frequencies throughout the entire workspace were predicted in a quick manner [2]. However, in this proposed model, several assumptions and approximations were made to simplify the dynamic modeling process. These simplified representations would degenerate the accuracy of prediction results. For example, the existence of clearance and friction in joints, as well as the deflection of the base connecting to the frame in real condition would change the dynamic characteristics of the PKM system and should not be neglected [3, 4]. In addition, further optimal design of structure and control demands explicit understand of dynamic characteristics of the machine system.

Motivated by above concerns, an experimental modal test and relative analyses is conducted on a prototype with the same parameters listed in Ref. [2]. With modal parameters distinguished by the testing, the previous analytical dynamic model can be modified with better prediction accuracy. On the other hand, the system dynamic characteristics can be predicted and analyzed based on modal truncation. With modal truncation, the sophisticated descriptions and approximations for component flexibility and joint nonlinearity can be avoided and make it much easier to achieve useful dynamic results.

## 33.2 Dynamic Equations for PKM and Its Transformation

The structure of the proposed PKM module is depicted in Fig. 33.1, which consists of a fixed base, a moving platform and three identical R$\underline{P}$S limbs [1].

With the combination of substructure synthesis and finite element method, the analytical governing equations of motion for the system can be derived as [2]

$$\tilde{M}\ddot{\tilde{U}} + \tilde{K}\tilde{U} = \tilde{F} \tag{33.1}$$

Herein, $\tilde{M}$ and $\tilde{K}$ are the global mass matrix and stiffness matrix, respectively; $\tilde{U}$ and $\tilde{F}$ are the global general coordinates vector and external load vector, respectively. It is worthy to point out that the matrices of $\tilde{M}$ and $\tilde{K}$ are time-varying and configuration-related.

According to the modal analysis theory, Eq. (33.1) can be transformed from physical coordinates into modal coordinates and vice versa. Generally, the PKM module moves in given postures at velocities of moderate amplitudes. Thus, the

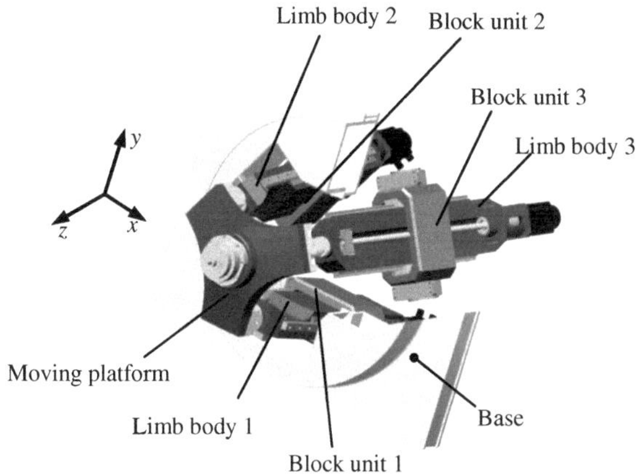


**Fig. 33.1** Structure of the PKM module

parallel mechanism can be considered as a linear system in steady state. And for a linear mechanical system, a structural damping assumption can be applied. Therefore, the Eq. (33.1) can be expressed as the following by adding a ratio damping matrix $c$

$$m\ddot{x}(t) + c\dot{x}(t) + kx(t) = f(t) \tag{33.2}$$

where $m$, $c$ and $k$ are mass matrix, damping matrix and stiffness matrix, respectively; $x(t)$ and $f(t)$ are general modal coordinates and external load vector, respectively.

The Laplace transform of the above equation is

$$\left(s^2m + sc + k\right)x(s) = F(s) \tag{33.3}$$

The system response is

$$x(s) = \left(s^2m + sc + k\right)^{-1}F(s) = H(s)F(s) \tag{33.4}$$

where $H(s)$ is transfer matrix of the system and can be written as

$$H(s) = \sum_{i=1}^{n}\left|\frac{\varphi_i\varphi_i^T}{m_i(s - s_i)} + \frac{\varphi_i^*\varphi_i^{*T}}{m_i^*(s - s_i^*)}\right| \tag{33.5}$$

Herein, $m_i$, $s_i$ and $\varphi_i$ denote the $i$th order modal mass, round frequency and mode vector, respectively; the superscript '*' represents the conjugation of a variable.

The element of $l$ row and $p$ column in the matrix $H(s)$ can be expressed as

$$H_{lp} = \sum_{i=1}^{n}\left|\frac{\varphi_{li}\varphi_{pi}^T}{m_i(s - s_i)} + \frac{\varphi_{li}^*\varphi_{pi}^{*T}}{m_i^*(s - s_i^*)}\right| \tag{33.6}$$

where $\varphi_{li}$ and $\varphi_{pi}$ denote the element of $l$ row and $p$ column in the $i$th order mode.

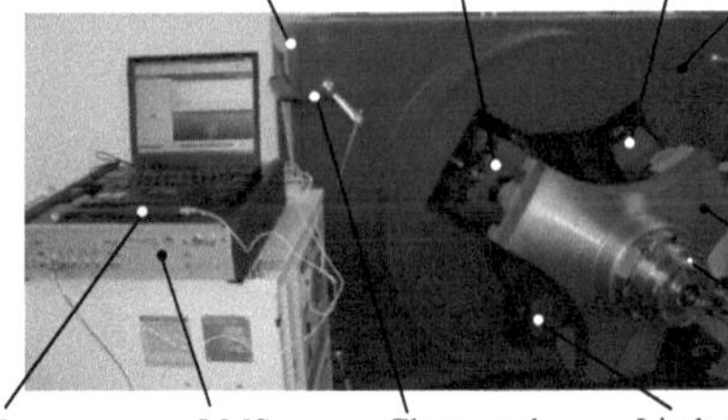


**Fig. 33.2** Testing rig for the PKM module

## 33.3 Experimental Modal Analysis for Typical Configuration

In an experimental modal analysis, by measuring applied excitations and corresponding dynamic responses, the transfer function from exciting points to recording points can be calculated, and homologous elements of transfer function in arbitrary row or column can be obtained. Then with the technique of modal identification and fitness, modal parameters of the system can be determined.

In this paper, the method of single input and single output (SISO) is adopted. And to facilitate the modal testing, the LMS Test.lab system (Version 9A) is applied. The experimental modal testing rig is demonstrated in Fig. 33.2.

Due to space limitation, the PKM module is connected to a fixed frame with four clamp rods at each corner of the base when conducting the test. This kind of connection, however, is quite different from real working condition in which the base of the PKM is bolt connected. Obviously, the stiffness of clamped connection is much smaller compared to bolt connection.

Before testing the modal properties of the system, a modal analysis model is needed. With the help of LMS Test.lab, a wire-frame model for the PKM module at any given configuration can be established. According to the structural features and kinematic relationships of the PKM module, a wire-frame model with 111 nodes is established in Fig. 33.3.

The wire-frame model consists of a fixed base, a moving platform, three identical limb bodies and block units. The connection between limb bodies and moving platform, i.e. the three identical spherical joints in real mechanism, is represented by three lines as shown in Fig. 33.3. For sake of clarity, the numbering of each node is not shown in Fig. 3. The detailed node numbering is listed in Table 33.1.

During the experimental testing, impulse excitation in three directions ($x$, $y$, $z$) is applied at the node of plat:1, which represents the installing position of the spindle. By recording the responses of each node in three corresponding directions ($x$, $y$, $z$), the transfer functions of the system can be obtained. To reduce the noise effect, the multi-average method is applied. The responses of each node (i.e. recording position) are measured for 5 times. Rectangular windows are added to input signals (hammer impulse) and exponential windows to response. The FRF of node plat:1 in three directions are demonstrated in Fig. 33.4.

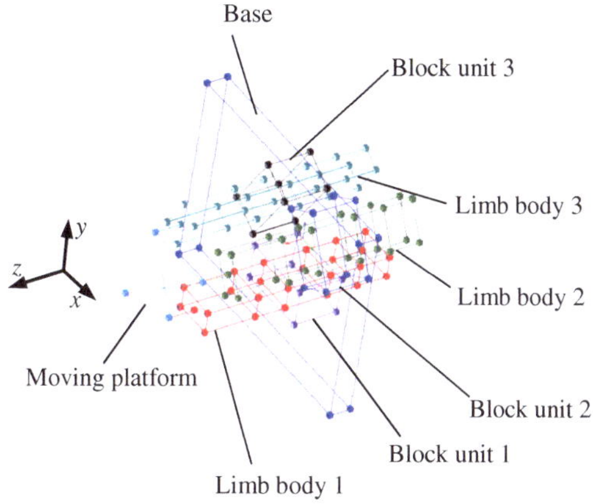


**Fig. 33.3** Wire-frame model of the PKM module

**Table 33.1** Node numbering of component in PKM module

| Component name | Node number | Component namc | Node number |
|---|---|---|---|
| Moving platform | Plat:1–plat:4 | Limb body 3 | L3_1–L3_25 |
| Fixed base | Base:1–base:8 | Block 1 | B1_1–B1_8 |
| Limb body 1 | L1_1–L1_25 | Block 2 | B2_1–B2_8 |
| Limb body 2 | L2_1–L2_25 | Block 3 | B3_1–B3_8 |

After averaging the transfer functions of all testing points, mode results of the PKM system can be identified and fitted in LMS Test.lab. The natural frequencies and corresponding mode shapes as well as damping ratios of the system are listed in Table 33.2 and the sketches for lower order natural modes are illustrated in Fig. 33.5.

From Table 33.2 and Fig. 33.5, it can be found that the first to the third orders of natural modes corresponding to the overall vibration of the PKM module. The physical explanation for the overall vibration modes relies on the 'weak' connection between the base and the fixed frame. As mentioned before, the base is clamped to the fixed frame, which results in a 'soft' supporting, i.e. small stiffness of the PKM module. Once the base and the fixed frame turns into bolt connection, the first three orders of mode are expected to be eliminated. On such circumstance, the present fourth to sixth orders of natural modes will become dominant. It can also be observed that these three natural frequencies are quite near in value and all correspond to the vibration of moving platform. The pitch and yaw as well as their combination vibrations of the moving platform imply that the stiffness of spherical joints is the 'bottleneck' in the whole parallel mechanism. The PKM module is most likely to vibrate when the spindle is machining.

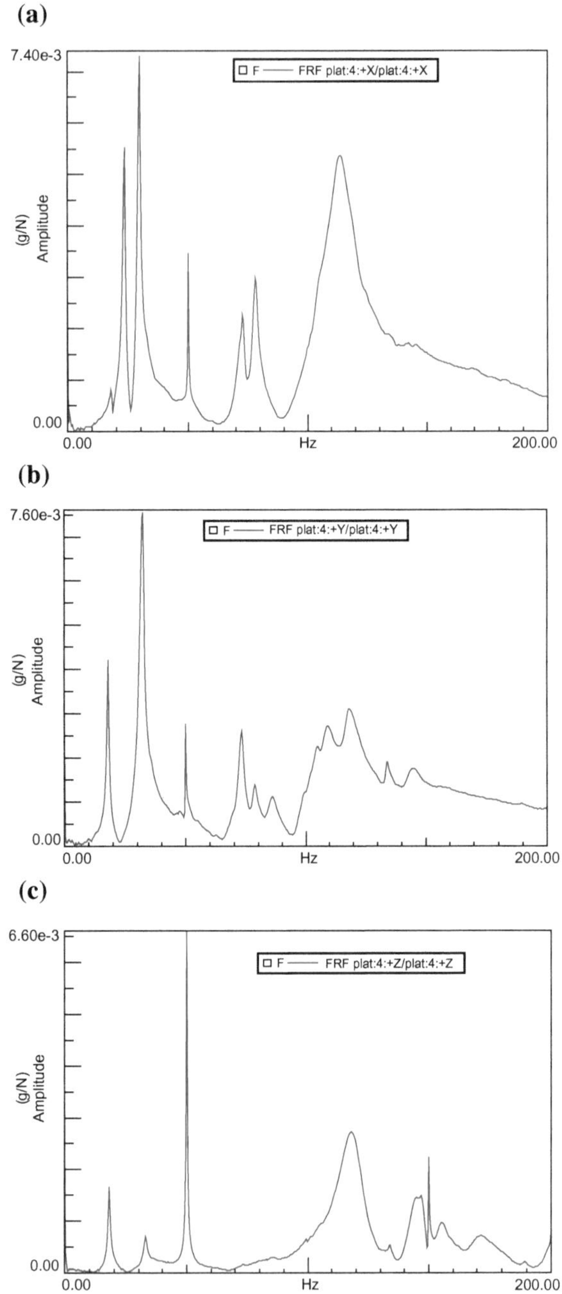


**Fig. 33.4** FRF of node plat:1 **a** FRF in *x* direction **b** FRF in *y* direction **c** FRF in *z* direction

**Table 33.2** Mode results of the PKM module

| Frequency/Hz | Damping/% | Description of mode shapes |
|---|---|---|
| $f_1 = 17.92$ | $C_1 = 1.15$ | Pitch of PKM module (Rotate about $x$ axis) |
| $f_2 = 18.02$ | $c_2 = 1.17$ | Translation in $z$ axis of PKM module |
| $f_3 = 23.22$ | $c_3 = 1.48$ | Yaw of PKM module (Rotate about $y$ axis) |
| $f_4 = 29.41$ | $c_4 = 1.27$ | Pitch of moving platform (Rotate about $x$ axis) |
| $f_5 = 32.00$ | $c_5 = 1.98$ | Yaw of moving platform (Rotate about $y$ axis) |
| $f_6 = 32.60$ | $c_6 = 1.61$ | Combining pitch and yaw of moving platform |

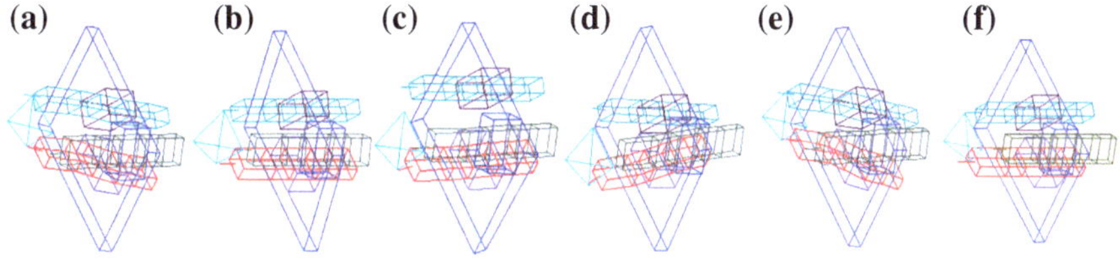


**Fig. 33.5** Lower order natural modes **a** 1st order **b** 2nd order **c** 3rd order **d** 4th order **e** 5th order **f** 6th order

## 33.4 Conclusions

In this paper, a wire-frame model for a 3-dof PKM module is established and an experimental modal test is implemented with SISO method. Based on the proposed model and its testing results, the modal properties such as lower orders of natural frequencies, damping ratios and corresponding mode shapes are revealed. The modal analyses results indicate that, except for the overall vibration modes which can be eliminated through bolt-connection of the base and fixed frame, the most likely occurred vibration modes for the PKM module are the pitch and yaw as well as their combination of the moving platform. These vibration modes of the moving platform imply that the spherical joint stiffness is the 'bottleneck' of the system and is of great concern during the design stage. To achieve a desirable dynamic performance, the spherical joint should be delicately designed for a high stiffness.

## References

1. Li YG, Liu HT, Zhao XM et al (2010) Design of a 3-DOF PKM module for large structural component machining. Mech Mach Theory 45:941–954
2. Zhang J, Li YG, Huang T (2010) Dynamic modeling and eigenvalue evaluation of a 3-DOF PKM module. Chinese J Mech Eng 23:166–173
3. Zhou ZL, Jeff X, Chris KM (2006) Modeling of a fully flexible 3PRS manipulator for vibration analysis. ASME J Mech Des 128:403–412
4. Shiau TN, Tsai YJ, Tsai MS (2008) Nonlinear dynamic analysis of a parallel mechanism with consideration of joint effects. Mech Mach Theory 43:491–505

# Chapter 34
# Analysis on Motion Characters for A 3-PRS Parallel Mechanism

**Zheng Gao, Rui Su, Jianting Zhao and Hongrui Wang**

**Abstract** The motion character of a special three degree-of-freedom (DOF) parallel mechanism is analyzed by screw theory. This mechanism's significant feature is that the axis of its prismatic pair is parallel to that of its revolute pair in the same limb. Firstly, by analysis, the mechanism's constraint screws are obtained, which leads to motions along the fixed-length links being restricted. Secondly, we get the mechanism's constraint screws' reciprocal productions, namely the mechanism's free motions, which are three screw motions. After that, a numerical example is presented to illustrate the analysis in detail. At last, we draw some conclusions and discuss the future work.

**Keywords** 3-PRS · Parallel mechanism · Screw theory · Motion character

---

Supported by Institutions Youth Science and Technology Research Fund of Education Department of Hebei Province(2010217)

---

Z. Gao (✉) · J. Zhao · H. Wang
Laboratory of Intelligent Robots, Hebei University, Baoding, China
e-mail: gaozheng1977@yahoo.com.cn

J. Zhao
e-mail: g_zheng@163.com

H. Wang
e-mail: hongrui@hbu.edu.cn

R. Su
Office of Committee Degrees, Hebei University, Baoding, China
e-mail: surui@hbu.edu.cn

J. S. Dai et al. (eds.), *Advances in Reconfigurable Mechanisms and Robots I*,
DOI: 10.1007/978-1-4471-4141-9_34, © Springer-Verlag London 2012

## 34.1 Introduction

Spatial parallel robot mechanism is one of the focuses of robot research field. Compared with serial robot, parallel robots have high rigidity, high payload capability, less cumulate error and good dynamic characteristic. The earliest spatial parallel was used for flight simulator, and then it is designed for parallel virtual machine tool, micro operation robot, sensor element, force feedback device and walking robot legs.

In recent years, new research has been focusing on new type and lower-mobility parallel robot. Because of its simple structure and low cost, lower-mobility parallel robot has wide application prospects in industry and other fields. One of the 3-DOF spatial parallel mechanism—3-PRS has induced great interests, it is applied on SPRINT Z3, telescope focusing and machine tool [1–8]. This kind of 3-PRS has a common characteristic, namely it can realize 2 rotational motion and a translational motion, the three spherical joints, which connecting the triangle mobile platform, move in three different Euler planes, its prismatic pair axis and revolute pair axis are perpendicular to each other [9]. However, 3-PRS parallel mechanism whose prismatic pair axis and revolute pair axis are parallel to each other has little study. Zhao [10] mentions a spatial parallel robot with one translational DOF and three rotational DOFs, which adding a PS limb in the center of 3-PRS parallel mechanism with its prismatic pair axis and revolute pair axis are parallel to each other.

Motion character is analyzed by screw theory for a 3-DOF 3-PRS parallel mechanism, whose prismatic pair axis and revolute pair axis are parallel to each other, this mechanism's mobile platform has three screw motions relative to the base platform.

## 34.2 Notation of Structure and Coordinates

3-PRS parallel mechanism is composed of a mobile platform, a base platform and three identical limbs, as it is shown In Fig. 34.1. $M_1M_2M_3$ constructs the base platform, which is an isosceles triangle with a circum circle radii $r_1$; $C_1C_2C_3$ denotes the mobile platform, which is an isosceles triangle, too, while its circum circle radii is $r_2$. The base isosceles triangle and the mobile one are similar triangles. In general, $r_2 < r_1$. $A_i (i = 1, 2, 3)$ are centers of the three sliders, which move in the directions of guides $M_1M_2$, $M_2M_3$ and $M_1M_3$; $B_i$ $(i = 1, 2, 3)$ are centers of the three revolute pairs, and $A_i$'s translational path is parallel to $B_i$'s revolute axis, $|A_iB_i| = a$ $C_i$ is center of the spherical pair, and $B_iC_i$ is rigid body and its length is $L$.

Base coordinate system $O - XYZ$ is fixed to the base platform; origin $O$ is located in the center of the circum circle $M_1M_2M_3$, and axis $X$ aligns with $OM_3$, axis $Z$ is upwards and perpendicular to base platform, while then axis $Y$ can be obtained by right-handed rule. Mobile coordinate system $o - xyz$ is fixed to the

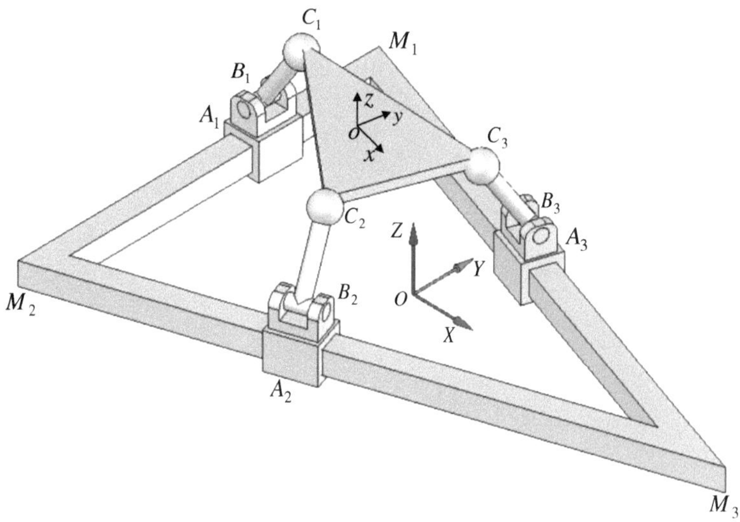


**Fig. 34.1** Structure of 3-PRS parallel mechanism and its coordinates

mobile platform, origin $o$ lies in the center of the circum circle $C_1C_2C_3$, axis $x$ is perpendicular to $C_1C_2$, at initial position it is in the same direction with axis $X$; axis $z$ is upwards and perpendicular to the mobile platform, while axis $y$ can be get by right-handed rule.

As shown in Fig. 34.2, in the fixed coordinate system $O-XYZ$, $M_i$'s coordinates are

$$\mathbf{M}_i = \begin{pmatrix} r_1 \cos \eta_i \\ r_1 \sin \eta_i \\ 0 \end{pmatrix}$$

where $i = 1, 2, 3$, and $\eta_i$ is an angel between $OM_i$ and axis $X$; by the geometry of the base platform, we have $\eta_1 = -\eta_2 = \eta$, $\eta_3 = 0$. For the mechanism can be structured, there should have the relation of $r_2 < |r_1 \cos \eta|$. Let $b = |r_1 \cos \eta|$, then $A_i$'s coordinates are as follows

$$\mathbf{A}_1 = \begin{pmatrix} -b \\ y_1 \\ 0 \end{pmatrix} \quad \mathbf{A}_2 = \begin{pmatrix} x_2 \\ y_2 \\ 0 \end{pmatrix} \quad \mathbf{A}_3 = \begin{pmatrix} x_3 \\ y_3 \\ 0 \end{pmatrix} \tag{34.1}$$

where $y_i$ is variable, and

$$\begin{aligned} x_2 &= ky_2 + r_1 \\ x_3 &= -ky_3 + r_1 \end{aligned} \tag{34.2}$$

$k$ in formula (34.2) is slope of line $M_2M_3$ in the fixed coordinate system $O-XYZ$, namely

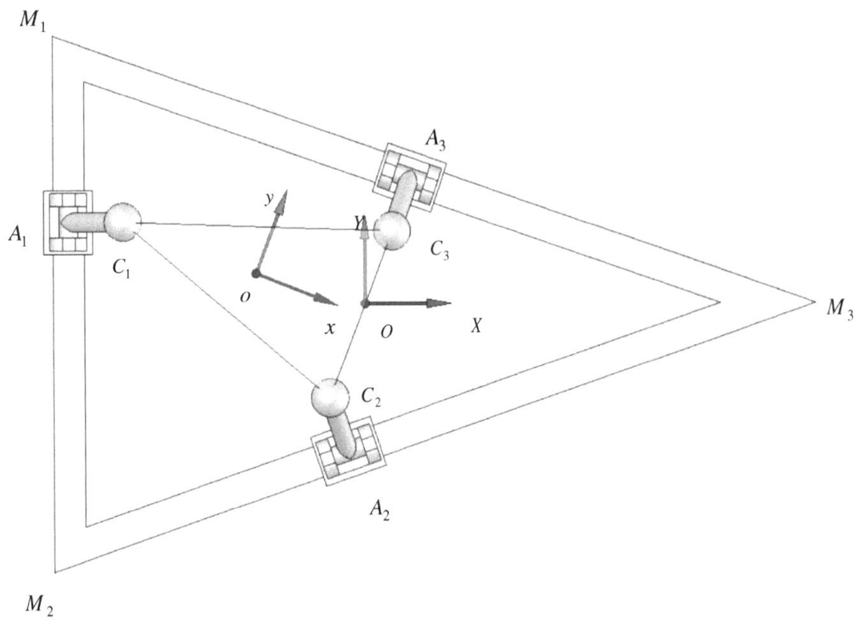


**Fig. 34.2** Top view of 3-PRS parallel mechanism

$$k = \frac{1 - \cos\eta}{\sin\eta} = \tan\alpha$$

where $\alpha \equiv M_1M_2M_3$.

According to $A_i$'s coordinates and the mechanism's geometry, $B_i$'s coordinates are obtained, namely

$$\mathbf{B}_1 = \begin{pmatrix} -b \\ y_1 \\ a \end{pmatrix} \quad \mathbf{B}_2 = \begin{pmatrix} x_2 \\ y_2 \\ a \end{pmatrix} \quad \mathbf{B}_3 = \begin{pmatrix} x_3 \\ y_3 \\ a \end{pmatrix} \tag{34.3}$$

And then $C_i$'s coordinate can be expressed as

$$\mathbf{C}_i = \mathbf{B}_i + L\mathbf{s}_{ibc} \tag{34.4}$$

where $\mathbf{s}_{ibc}$ is unit vector, which denotes $B_iC_i$'s direction, i.e.

$$\mathbf{s}_{ibc} = \begin{pmatrix} \cos\theta_i \cos\varphi_i \\ \cos\theta_i \sin\varphi_i \\ \sin\theta_i \end{pmatrix}$$

And $\theta_i$ $(i = 1, 2, 3)$ denotes the angle between $B_iC_i$ $(i = 1, 2, 3)$ and the base plane, $\varphi_i$ $(i = 1, 2, 3)$ denotes the angle between axis $X$ and $B_iC_i$'s projection on the base plane, as shown in Fig. 34.3.

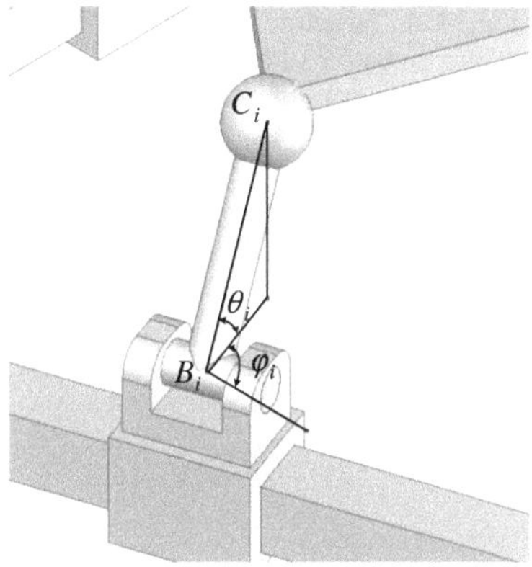


**Fig. 34.3** Angles between $B_iC_i$ and base plane

## 34.3 Motion Character Analysis for 3-PRS Parallel Mechanism

When we use kinematic screw to describe the motion of two rigid bodies connected by joints, the corresponding unit kinematic screw can also be used to express different type kinematic pair, such as revolute pair, prismatic pair and helical pair etc. Therefore, unit kinematic screw can also be called kinematic pair screw, which can be called KP screw for short [11].

According to the coordinate systems shown in Fig. 34.1 and the screws of each limb shown in Fig. 34.4, we get the following expressions:

$$\begin{cases} \$_{i1} = (0; \mathbf{s}_{i1}) \\ \$_{i2} = (\mathbf{s}_{i2}; \mathbf{r}_{bi} \times \mathbf{s}_{i2}) \\ \$_{i3} = (\mathbf{s}_{i3}; \mathbf{r}_{ci} \times \mathbf{s}_{i3}) \\ \$_{i4} = (\mathbf{s}_{i4}; \mathbf{r}_{ci} \times \mathbf{s}_{i4}) \\ \$_{i5} = (\mathbf{s}_{i5}; \mathbf{r}_{ci} \times \mathbf{s}_{i5}) \end{cases} \quad (34.5)$$

where $\mathbf{s}_{i1}$, $\mathbf{s}_{i2}$, $\mathbf{s}_{i4}$ express the direction vectors of corresponding kinematic pair axis respectively, which are parallel to each other; $\mathbf{s}_{i3}$ is in the direction of $B_iC_i$; $\mathbf{r}_{ai}$, $\mathbf{r}_{bi}$, $\mathbf{r}_{ci}$ are vectors from origin $O$ to points $A_i$, $B_i$, $C_i$. Due to mechanism's geometry, the following equations are obtained

$$\mathbf{s}_{i5} = \mathbf{s}_{i4} \times \mathbf{s}_{i3} \quad (34.6)$$

$$\mathbf{r}_{bi} = \mathbf{r}_{ai} + a\mathbf{s}_{iab} \quad (34.7)$$

$$\mathbf{r}_{ci} = \mathbf{r}_{bi} + L\mathbf{s}_{ibc} \quad (34.8)$$

$\mathbf{s}_{iab}$ and $\mathbf{s}_{ibc}$ denote the directions of $A_iB_i$ and $B_iC_i$, while $a$ and $L$ are lengths of $A_iB_i$ and $B_iC_i$ respectively. By linear combining, from formulas (34.5)–(34.8) acquired the following equivalent screw systems

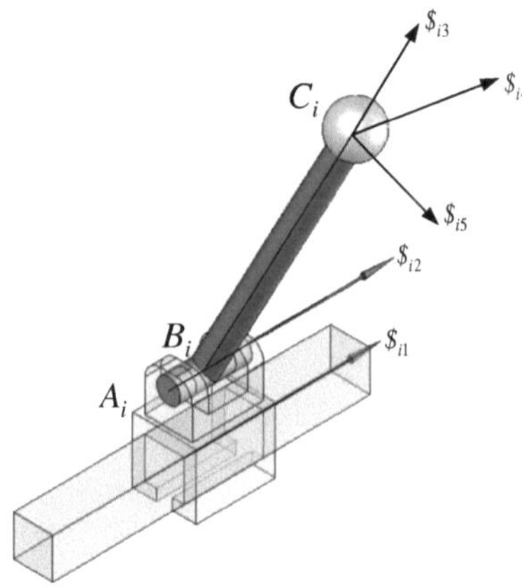


**Fig.34. 4** Structure of one of the limbs of 3-PRS and its screw distribution

$$\begin{cases} \mathbf{\$}_{ie1} = (0; \mathbf{s}_{i1}) \\ \mathbf{\$}_{ie2} = (\mathbf{s}_{i1}; \mathbf{r}_{bi} \times \mathbf{s}_{i2}) \\ \mathbf{\$}_{ie3} = (\mathbf{s}_{i3}; \mathbf{r}_{ci} \times \mathbf{s}_{i3}) \\ \mathbf{\$}_{ie4} = (0; L\mathbf{s}_{ibc} \times \mathbf{s}_{i4}) \\ \mathbf{\$}_{ie5} = (\mathbf{s}_{i5}; \mathbf{r}_{ci} \times \mathbf{s}_{i5}) \end{cases} \tag{34.9}$$

Take each point's coordinates into formula (34.9), and then obtain every kinematic screw expression of each limb, also each limb's constraint screw system can be gained more easily as follows

$$\mathbf{\$}_i^r = (\mathbf{s}_{ibc}; \mathbf{r}_{ci} \times \mathbf{s}_{ibc}) \tag{34.10}$$

The constraint screw is aligned with $\mathbf{\$}_{i3}$ and through the center of spherical pair, which points out that the motion of mobile platform is constrained in the direction of $B_iC_i$, as it is shown in Fig. 34.5.

The three linear independent reciprocal force screw sets construct the platform's constraint screw system, namely

$$\$^r = \begin{pmatrix} \$_1^r \\ \$_2^r \\ \$_3^r \end{pmatrix}$$

And $card(\{\$^r\}) = 3$, then the mechanism's DOF [12] is

$$M = 6 - card(\{\$^r\}) = 3$$

According to the reciprocal relations between kinematic screw system and constraint screw system, we can obtain mobile platform's kinematic system, that is

$$\$_i = (\mathbf{S}_i; \mathbf{S}_{0i}), \ i = 1, 2, 3 \tag{34.11}$$

The pitch of $\$_i$ can be written as

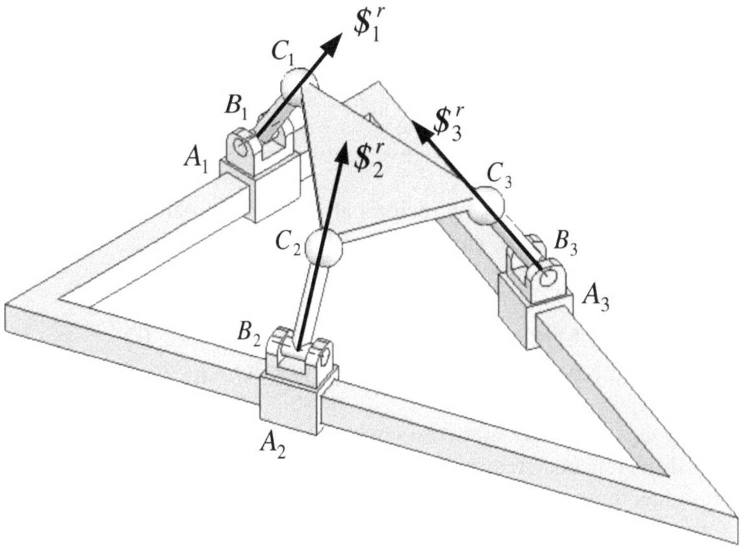


**Fig. 34.5** Constraint screws in limbs of the 3-PRS parallel mechanism

$$h_i = \frac{\mathbf{S}_i \cdot \mathbf{S}_{0i}}{\mathbf{S}_i \cdot \mathbf{S}_i} \tag{34.12}$$

And the projection of coordinate origin on each instantaneous axis is as following

$$\mathbf{r}_{op_i} = \frac{\mathbf{S}_i \times \mathbf{S}_{0i}}{\|\mathbf{S}_i\|^2} \tag{34.13}$$

## 34.4 Numerical Analysis

The following parameters are derived from the Solidworks model, as shown in Table 34.1.

Take these parameters above into formulas (34.1)–(34.10), we can achieve the terminal constraints of each kinematic link $B_iC_i$, and then finally obtain mobile platform $C_1C_2C_3$'s instantaneous terminal constraint matrix, as follows

$$\$^r = \begin{pmatrix} 0.69340 & 0 & 0.72055 & 0.41113 & 1.88978 & -0.39564 \\ -0.20733 & 0.60418 & 0.76940 & 0.47220 & -4.35004 & 3.54318 \\ -0.12953 & -0.37746 & 0.91692 & 1.17379 & 0.32826 & 0.30095 \end{pmatrix} \tag{34.14}$$

**Table 34.1** Geometry and kinematic parameters obtained from the Solidworks model

| $r_1$ | $r_2$ | $a$ | $L$ | $y_1$ | $y_2$ | $y_3$ |
|---|---|---|---|---|---|---|
| 2.82941 m | 1.13177 m | 0.405 m | 0.800 m | 0.57058 m | 0.93176 m | 1.13415 m |
| $\alpha$ | $\eta$ | $\mu$ | $\theta_1$ | $\theta_2$ | $\theta_3$ | |
| 71.06° | 142.11 | 37.89° | 59.32° | 50.30° | 66.48° | |

Then by formula (34.11), $\$^r$'s reciprocal production, namely the mobile platform $C_1C_2C_3$'s free motions, can be derived, which are

$$\begin{cases} \$_1 = k_1(0.49271 \quad -0.64015 \quad -0.79308 \quad 1 \quad 0 \quad 0) \\ \$_2 = k_2(0.51188 \quad -0.21716 \quad -0.50535 \quad 0 \quad 1 \quad 0) \\ \$_3 = k_3(-0.49913 \quad -0.40949 \quad -0.65337 \quad 0 \quad 0 \quad 1) \end{cases} \tag{34.15}$$

where $k_i$ $(i = 1, 2, 3)$ is any non-zero const. From Eq. (34.15) we can see that the platform's instantaneous free motions are three screw motions, and each translation motion has three parasitic orientation motions.

By formula (34.12) and Eq. (34.15), the free motions' pitches are derived, which are as follows

$$\begin{cases} h_1 = 0.38447 \\ h_2 = 1 \\ h_3 = -0.77441 \end{cases} \tag{34.16}$$

At last, by formulas (34.13) and (34.15), we can obtain the projection of coordinate origin in each instantaneous screw axis, as shown in the Eq. (34.17).

$$\mathbf{r}_{op_1} = \begin{pmatrix} 0 \\ -0.61885 \\ 0.49952 \end{pmatrix} \quad \mathbf{r}_{op_2} = \begin{pmatrix} 0.89513 \\ 0 \\ 0.90669 \end{pmatrix} \quad \mathbf{r}_{op_3} = \begin{pmatrix} -0.48535 \\ 0.59160 \\ 0 \end{pmatrix} \tag{34.17}$$

The analysis above explains that the mobile platform takes instantaneous screw motions relatively to the base platform. Each screw pitch and unit screw axis is shown in formula (34.16) and expression (34.17) respectively.

## 34.5 Conclusion and Future Work

Through motion character analysis for the special 3-PRS by screw theory, we obtain the following conclusions:

1. For the 3-PRS parallel mechanism, whose prismatic pair's axis paralleling to its revolute one, its constraint screws are in the directions of the fixed-length links and through the centers of spherical joints;
2. The mechanism's free motions are three screw motions. As for the given numerical example, each translation motion has three parasitic orientation motions.

For this specific mechanism, there is still much work to do, such as:

1. The mechanism's direct and inverse kinematics should be analyzed. However, because the mechanism's free motions are screw motions, therefore, the next work firstly have to do is to derive the analytical solutions of the mechanism's parasitic motion;
2. Finding out if there is any reasonable application for this mechanism, which is also the ultimate goal of mechanism research work.

## References


1. Wahl J (2000) Articulated tool head:Canada, CA 2349579 11 May 2000
2. Saenz Fernandez AJ, Collado Jimenez V, Gimenez Olazabal M (2002) Kinematical system for a movable platform of a machine:European, EP1245349 (A1), 10 Jan 2002
3. Lee KM, Shah DK (1988) Kinematic analysis of a three-degree-of-freedom in-parallel actuated manipulator. IEEE T Robotic Autom 4(3):354–360
4. Bürüncük K, Tokad Y (1999) On the kinematic of a 3-DOF stewart platform. J Robotic Syst 16(9):105–118
5. Carretero JA, Nahon MA, Podhorodeski RP (2000) Workspace analysis and optimization of a novel 3-DOF parallel manipulator. IEEE T Robotic Autom 15(4):178–188
6. Fan KC,Wang H,Zhao JW et a1 (2003) Sensitivity analysis of the 3-PRS parallel kinematic spindle platform of a serial-parallel machine tool. Int J Mach Tool Manu ,43:1561–1569
7. Pond GT,Carretero JA (2004) Kinematic analysis and workspace determination of the inclined PRS parallel manipulator. In: 15th CISM-IFToMM Symposium on Robot Design,Dynamics,and Control, Springer Quebec, pp 4–18
8. Liu X J,Pruschek P,Prjtschow G (2004) A new 3-dof parallel mechanism with full symmetrical structure and parasitic motions. In: International Conference on Intelligent Manipulation and Grasping, Grafica K s.n.c., Genoa, pp 389–394
9. Liu XJ, WU C, Wang JS, Bonev I (2008) Attitude description method of [PP]S type parallel robotic mechanisms. Chin J Mech Eng (Chinese Version) 44(10):19–23
10. Zhao JS, Feng ZJ, Chu FL (2009) Analytical theory of degrees of freedom for robot mechanisms. SciencePress, Beijing
11. Zhao TS (2000) Some theoretical issues on analysis and synthesis for spatial imperfect-dof parallel robot. Yan Shan University, Qinhuangdao
12. Yu JJ, Liu XJ, Ding XL, Dai JS (2008) Mathematical foundations of robot mechanism. China Machine Press, Beijing

# Chapter 35
# Effect of Different Terrain Parameters on Walking

**Shahram Mohseni-Vahed and Yun Qin**

**Abstract** This paper proposes an approach to analyse comprehend the effect of different uncertain ground impedance parameters on bipedal walking. A dynamic model of a rimless wheel in contact with an unpredictable visco-elastic terrain is presented. The mathematical model of the rimless wheel as it makes contact with an uneven ground has been developed based on Lagrangian dynamics. The uncertain ground impedance parameters induce structural visco-elasticity which is represented by a spring and damper pair in the horizontal direction and another in the vertical direction. A numerical simulation has been performed to investigate the behaviour of the proposed dynamic mathematical model. The results describe the effect of the interplay among the visco-elastic parameters at the ground-leg contact point of a legged walker and the emerging properties of interaction dynamics of walking on different impedance parameters. The results of simulation trials highlight the importance of the above interplay in the area of bipedal humanoid walking.

**Keywords** Dynamic mathematical model · Ground impedance parameters · Interaction dynamics of rimless wheel with ground

## 35.1 Introduction

Over the past several years, remarkable progress has been made in the practical implementation of legged mechanisms [1]. In early research walking robots were designed as static machines. This type of walking robot usually consists of four or

S. Mohseni-Vahed (✉) · Y. Qin
School of Natural and Mathematical Sciences, King's College London, Strand, London, WC2R 2LS, UK
e-mail: shahram.1.mohseni-vahed@kcl.ac.uk

J. S. Dai et al. (eds.), *Advances in Reconfigurable Mechanisms and Robots I*,
DOI: 10.1007/978-1-4471-4141-9_35, © Springer-Verlag London 2012 

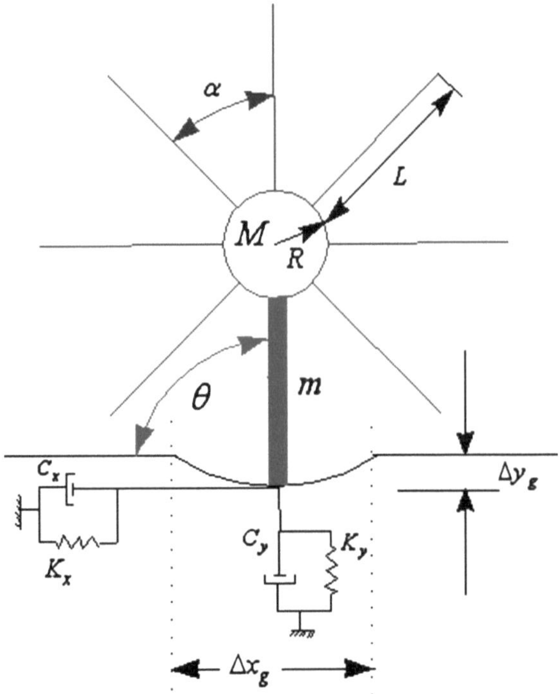


**Fig. 35.1** The schematic diagram of the ground contact model for a rimless wheel with a hub

six legs and relies on static equilibrium being maintained during motion over many steps [2]. Examples include the Odex series [3] and the Adaptive Suspension Vehicle [4].

Experimental results of recent walking robots such as Whegs, Rhex, and Mecharoach [5] apparently prove that the trial to trial variability of each robot's behavior cannot be explained by the neuromechanical models of their biological origins alone. Variability of walking is an indispensable property of walking [6] to deal with metastability of legged locomotion [7]. This evidence, as well as the results presented here, suggest that variability is an inevitable design consideration and is important to model and account for in legged locomotion.

The more complex case of dynamic walking machines has been introduced for humanoid robots, where the gait control is more successful and adaptable when actively applied. Dynamic control makes a robot potentially faster, due to compensating for inertia, gravity and dynamic discontinuities caused by collisions with the ground. In his demonstration, T. McGreey uses inertia and gravity to generate motion [2] in "passive walking" machines. This was important to show that optimization of dynamic walking should be achieved through energy efficiency utilizing the inertia of the swinging leg to help control the stance leg. Achieving stable dynamic walking on a bipedal robot is a difficult control problem because bipeds can only control the trajectory of their center of mass through the unilateral, intermittent, irregular force contacts with the ground.

A humanoid robot walking on two legs involves highly non-linear and multi-variable dynamics. It is highly suitable for general terrain types which can be uneven but risks tipping over. An important topic of recent research on humanoid robots is the balance and walking stability [8]. Balance in humanoids is best achieved through dynamic control and can be explained by the center of pressure (CoP) or zero moment point (ZMP) [9] at the base of the foot where it touches the ground. This point must be away from the edges of the foot and is the only means by which the robot maintains an upright stance and generates any movement. This only occurs during intermittent ground interactions on varying terrain, thus causing the complexity of legged locomotion [10]. Controlling stability is highly dependent on the ground type. In the presence of visco-elasticity balancing on one foot becomes more difficult due to the foot loosing friction and/or moving to an unpredictable position.

Walking robots typically have many degrees of freedom, which can cause a combinatorial explosion for control systems that attempt to optimize performance in every possible configuration of the robot. In order to analyze the walking motion it is better to use a simpler case, where the problem of remaining upright is negated, while the dynamics of the walking action on any terrain type is represented. Therefore this paper will discuss locomotive problems in terms of the rimless wheel [11]. The rimless wheel is a simple planar model of walking and consists of a central mass with several "spokes" extending radially outward. At any given moment one of the spokes is pinned to the ground, and the system follows the dynamics of a simple pendulum [12] with the foot (end of spoke) acting as a fixed point on irregular terrain types [12, 13].

Figure 35.1 shows the rimless wheel (central body mass with eight spokes) with the single leg in contact with an uneven terrain ground. The interaction dynamics resulting from the ground-leg contact can generally be represented by a spring and damper pair at each axis in the X and Y directions, simulating the ground impedance parameters pertaining to the structural visco-elasticity of the given ground type.

In this work, a dynamic mathematical model of the rimless wheel with the ground impedance parameters is utilized within an open loop system. Constant torque is applied at the minimum magnitude required to produce forward motion and not allow the body to fall backwards utilizing appropriate initial conditions for the starting step. The results of the simulation are integral to the design of control and balancing methods of humanoid robots during legged walking on uneven ground types.

## 35.2 Modeling Humanoid Locomotion in Terms of Environment Interaction

A humanoid robot shows linear and rotational stability when all acting forces and moments about the centre of mass (centroid) are summed to zero. Any force experienced by the robot must be counteracted by an opposing applied force in order to maintain stability (stop it from tipping). Physically, these opposing forces are created by the limbs, applying pressure from the feet onto the ground [14]. Assuming friction forces are of a high enough magnitude at the ground contact point, the foot becomes a "fixed joint" of the ground-robot system from which all opposing forces originate.

The ability to walk relies on maintaining this temporary "fixed joint" long enough to affect the robot's movements positively. If the friction between the foot and the ground contact point is insufficient or the ground deforms and the "fixed joint" is no longer "fixed" then control is adversely affected. Thus, considering the ground of the environment and the ground contact made by the feet is paramount to achieving stability of humanoid robots.

The simplest method to represent the bipedal motion of humanoid robots for testing purposes is to use a rimless wheel. This model is considered in this work since the sequence of steps made during walking and the interaction between the feet and the environment are the only major concerns when modelling such a system, and will be applicable to a subsection of more complex systems.

Passive walking using a rimless wheel accurately models the energy losses in the system accounting for the emergence of stochasticity, dominating other parameters such as variability in the initial launching velocity or variation of leg lengths. The RW provides the most basic and essential features of walking, in that it samples a discrete set of ground contact points from an infinite set of points on the ground profile.

Figure 35.1 depicts the kinematics and forces involved with passive locomotion of the RW. The interaction between the leg and ground is modelled as shown in Fig. 35.1, where $M$ is the mass of the robot's body and $m$ is the mass of a single leg, $L$ is the length of a single leg, $\theta$ is the angle of the leg from the horizontal plane, $\alpha$ is the angle between two consecutive legs of the rimless wheel, $X_g$is the deformation (visco-elastic displacement) of the ground surface caused by the robot's contact with the ground in the x-axis direction, $y_g$ is the deformation of the ground surface caused by the robot's contact with the ground perpendicular to the surface in the y-axis direction, $K_x, K_y$ are the ground stiffness coefficients, respectively, and $C_x, C_y$ are the viscosity at the contact point along and perpendicular to the surface, respectively. The advantage of considering the stiffness and viscosity along and perpendicular to the surface is that it allows understanding of the effect of energy loss during slip and slipping reversal along the surface as well as during compression and restitution perpendicular to the ground.

A schematic of the model is shown in Fig. 35.2. In this diagram the position and velocity of the hub at any given time is given by

$$CG = (-x_e + (L+R)\cos\theta)X + (-y_e + (L+R)\sin\theta)Y \tag{35.1}$$

$$v_{CG}^2 = \left(-\dot{x}_e + (L+R)\dot{\theta}\sin\theta\right)^2 + \left(-\dot{y}_e + (L+R)\dot{\theta}\cos\theta\right)^2 \tag{35.2}$$

The kinetic energy $(\tau)$ is given by:

$$KE = \tau = \frac{8}{12}m(L-R)^2\dot{\theta}^2 + \left(\frac{1}{2}MR^2\dot{\theta}^2\right) + \frac{1}{2}(M+8m)(v_{CG})^2 \tag{35.3}$$

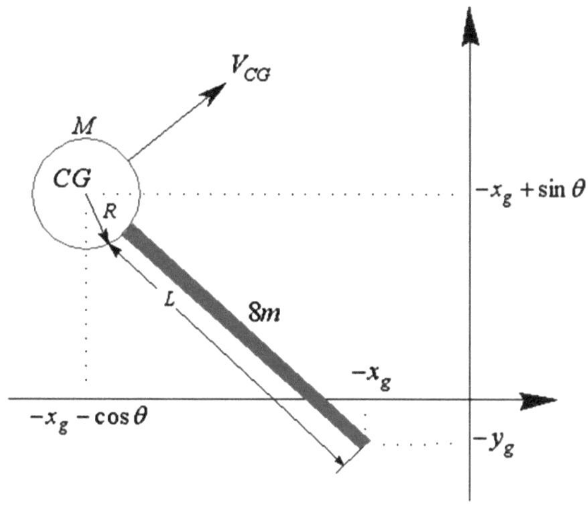


**Fig. 35.2** Diagram illustrates dynamic model of rimless wheel

The potential energy ($\vartheta$) is given by:

$$PE = \vartheta = (M + 8m)g((L + R)\sin\theta - y_e) + \frac{1}{2}k_x x_e^2 + \frac{1}{2}k_y y_e^2 \tag{35.4}$$

The total dissipation energy, including the energy due to friction in the joints ($D_{\text{joints}}$) between the motor and wheel can be written as:

$$D = D_x + D_y + D_{\text{joints}}$$

where

$$D_x = \frac{1}{2}C_x\dot{x}_e^2,\ D_y = \frac{1}{2}C_y\dot{y}_e^2 \quad \text{and} \quad D_{\text{joints}} = \frac{1}{2}b_\theta\dot{\theta}^2$$

Therefore the Lagrangian of the system is given by

$$\begin{aligned}\mathcal{L} &= \tau - \vartheta \\ &= \frac{8}{12}m(L - R)^2\dot{\theta}^2 + \frac{1}{2}MR^2\dot{\theta}^2 + \frac{1}{2}(M + 8m)[\left(-\dot{x}_e + (L + R)\dot{\theta}\sin\theta\right)^2 \\ &\quad + \left(-\dot{y}_e + (L + R)\dot{\theta}\cos\theta\right)^2 - [(M + 8m)g((L + R)\sin\theta - y_e) + \frac{1}{2}k_x x_e^2 + \frac{1}{2}k_y y_e^2\end{aligned} \tag{35.5}$$

The external force applied by the ground due to deformation along the ground direction is given by

$$\frac{d}{dt}\left(\frac{\partial\mathcal{L}}{\partial\dot{x}_g}\right) - \frac{\partial\mathcal{L}}{\partial x_g} + \frac{\partial D_{x_e}}{\partial\dot{x}_g} = 0 \tag{35.6}$$

$$(M+8m)\left(\ddot{x}_g+(L+R)\ddot{\theta}\sin\theta+(L+R)\dot{\theta}^2\cos\theta\right)-k_x x_g+C_x\dot{x}_g=0 \tag{35.7}$$

The external force applied by the ground due to deformation perpendicular to the ground direction is given by

$$\frac{d}{dt}\left(\frac{\partial\mathcal{L}}{\partial\dot{y}_g}\right)-\frac{\partial\mathcal{L}}{\partial y_g}+\frac{\partial D_{x_e}}{\partial\dot{y}_g}=0 \tag{35.8}$$

$$(M+8m)\left(\ddot{y}_g+(L+R)\ddot{\theta}\cos\theta-(L+R)\dot{\theta}^2\sin\theta\right)-(M+8m)g-k_x y_g+C_x\dot{y}_g=0$$

The externally applied torque is given by

$$\frac{d}{dt}\left(\frac{\partial\mathcal{L}}{\partial\dot{\theta}}\right)-\frac{\partial\mathcal{L}}{\partial\theta}+\frac{\partial D_{\text{joints}}}{\partial\dot{\theta}}=T \tag{35.9}$$

$$\begin{aligned}&(M+8m)\left(-\ddot{x}_g(L+R)\sin\theta+\dot{x}_g(L+R)\dot{\theta}\cos\theta+(L+R)^2\ddot{\theta}\right.\\&\left.-\ddot{y}_e(L+R)\cos\theta+\dot{y}_g(L+R)\dot{\theta}\sin\theta\right)\\&-(M+8m)g(L+R)\cos\theta)+\frac{4}{3}m(L-R)^2\ddot{\theta}+MR^2\ddot{\theta}+b_\theta\dot{\theta}=T\end{aligned}$$

## 35.3 Simulation of Interaction Dynamics of Legged Locomotion on Terrain

Simulation of interaction dynamics of legged locomotion on terrain Equations (35.7)–(35.9) can be re-arranged to form the nonlinear dynamic model which can be written compactly as

$$\overbrace{\begin{bmatrix}M+8m & 0 & (M+8m)(L+R)\sin\theta\\0 & M+8m & (M+8m)(L+R)\cos\theta\\-(M+8m)(L+R)\sin\theta & -(M+8m)(L+R)\cos\theta & (L+R)^2+\frac{4}{3}m(L-R)^2+MR^2\end{bmatrix}}^{\text{Inertia Matrix (I)}}\begin{bmatrix}\ddot{x}_e\\\ddot{y}_e\\\ddot{\theta}\end{bmatrix}$$

$$=\overbrace{\begin{bmatrix}((L+R)\dot{\theta}\cos\theta+k_x x_g-C_x\dot{x}_g)\\(L+R)\dot{\theta}\sin\theta+k_x y_g-C_x\dot{y}_g)\\(L+R)\dot{\theta}^2(\sin\theta-\cos\theta)-\dot{x}_g(L+R)\dot{\theta}(\cos\theta+\sin\theta)+(M+8m)g(L+R)\cos\theta-b_\theta\dot{\theta}+T\end{bmatrix}}^{\textit{Generalised force Vector}(F)} \tag{35.10}$$

The model can be rearranged into state-space form by defining the following state-variables:

$$X_1 = x_g, \quad X_3 = y_g, \quad X_5 = \theta_g$$

Thus

$$\begin{bmatrix} \dot{X}_2(tem(1)) \\ \dot{X}_4(tem(2)) \\ \dot{X}_6(tem(3)) \end{bmatrix} = I^{-1} \times F \overset{\text{yields}}{\rightarrow} \begin{matrix} X_2 = tem(1) = \dot{x}_g \\ X_4 = tem(2) = \dot{y}_g \\ X_6 = tem(3) = \dot{\theta}_g \end{matrix}$$

In order to analyze the effect of ground impedance parameters during the spinning of the rimless wheel a simulation has been conducted. The simulation involves an uncontrolled open loop system that has been set up with a zero initial condition for all variables for the first step k = 1 ($\dot{\theta}_{initial(1)} = 0, \dot{}_{ginitial(1)} = 0$ *and* $\dot{y}_{ginitial(1)} = 0$). The input torque applied to the system which is within the chosen range of impedance parameter values as shown in Eq. (35.10) below.

$$\begin{gathered} C_x \in (10 \text{ and } 100)[Ns/m] \text{ and } K_x \in (10 \; and \; 1000)[N/m] \\ C_y \in (10 \text{ and } 100)[Ns/m] \text{ and } K_y \in (10 \text{ and } 1000)[N/m] \\ \mathrm{M} = 0.3[\mathrm{kg}],\ \mathrm{m} = 0.2[\mathrm{kg}],\ \mathrm{L} = 0.2[\mathrm{m}],\ \mathrm{R} = 0.1[\mathrm{m}], \\ \mathrm{T} = 3[\mathrm{N.m}] \text{ and } g = 9.8[\mathrm{m/s^2}] \end{gathered} \tag{35.11}$$

<u>**Note:**</u> It should be mentioned that after the terminal values of the state variables at the end of the k-th step are used, as initial conditions for setting up the variables ($\dot{\theta}_{\mathrm{initial}(K+1)} = \dot{\theta}_{(K)}, \dot{x}_{\mathrm{ginitial}(K+1)} = (L + R)\dot{\theta}_{(K)}\sin\theta + \dot{x}_{g(K)}$) and $\dot{y}_{\mathrm{ginitial}(K+1)} = (L + R)\dot{\theta}_{(K)}\cos\theta + \dot{y}_{g(K)}$ for the next step (k + 1).

The results of the simulation illustrate how the ground impedance parameters affect the time duration of consecutive steps. In defining a strategy for balancing a humanoid robot for controlled stepping action, the monitoring and control of the variables corresponding to angular displacement (angle between legs), angular velocity, amount of slip (x-axis) or relocation in (y-axis) will be highly important Fig. 35.3.

## 35.4 Conclusion

The paper has analysed the effect of uncertain ground impedance parameters on bipedal walking. A dynamic passive model of a humanoid robot with many degrees of freedom is presented based on a rimless wheel. The mathematical model is derived from the energy components of the system using the Lagrangian technique. Using numerous simulations the effect of visco-elastic parameters related to the robot-ground interaction has been studied. The results clearly show that the variability of the terrain impedance across collisions during ground contact can lead to complex temporal variations in the speed and temporal step patterns of the walker.

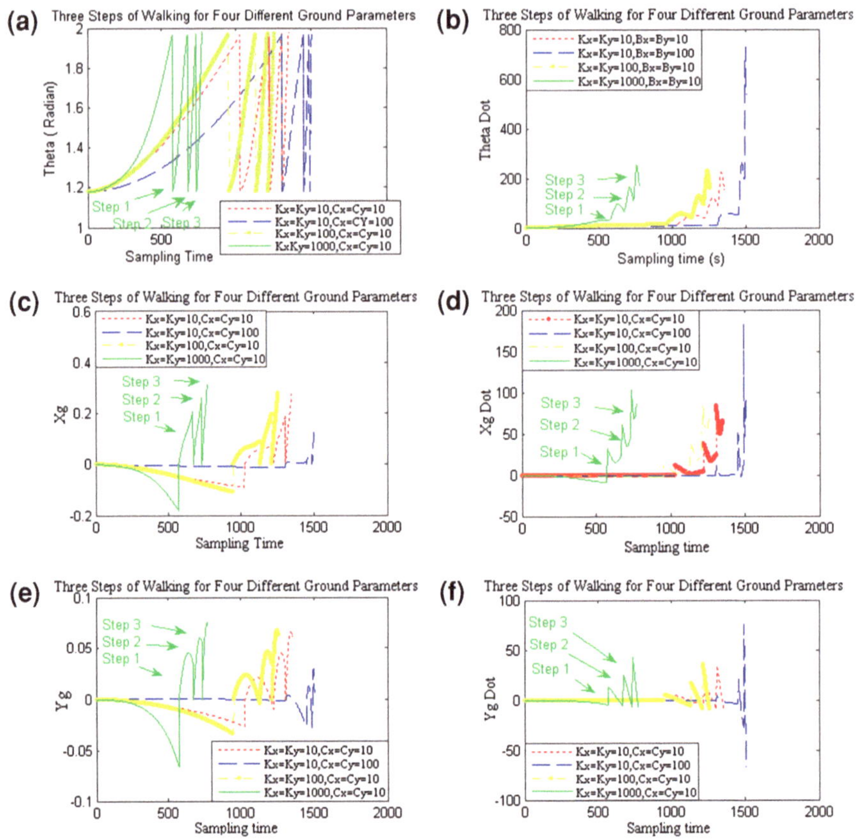


**Fig. 35.3** (**a** to **f**) show the behavior of the rimless wheel and the ground deformation pattern described by the variables $\theta, \dot{\theta}, xe, \dot{x}e, ye\ and\ \dot{y}e$ over three collisions between the feet and the ground with the impedance parameters set as stated earlier

Thus, if the controller is set-up for one set of ground impedance parameters, it may be inadequate for the next step if the ground impedance parameters change, possibly resulting in instabilities in the balance of the humanoid robot. Hence, either the feedback controller has to be robust to a pre-specified range of ground impedance parameters or some form of adaptation of the contact area similar to human walking. The design of a suitable mechanism of the foot contact area is a challenging task and will be a topic of future work.

## References


1. Paul M, Jamie M (2010) Optimal switching curve for a multiple link walking robot using nelder mead optimization, ndia ground vehicle systems. In: Engineering and Technology Symposium Robotic Systems Mini-Symposium, Dearborn
2. McGeer T (1990) Passive dynamic walking. Int J Robotics Res 9(2):61–82
3. Russell M (1983) Odex 1:the first functionoid Robot. Age 5:12–18
4. Waldron K (1986) Force and motion management in legged locomotion. IEEE J Robot Automat 2(4):214–220
5. Quin RD, Kingsley DA, Offi JT, Ritzmann, RE (2002) Improved mobility through abstracted biological principles. In: IEEE International Conference on Intelligent Robots and Systems (IROS), pp 2652–2657
6. Saranali U, Buehler M, Koditscek D (2001) Rhex a simple and highly mobile hexapod Robot. Int J Robotics Res, 20(7):616–631
7. Boggess MJ, Schroer RT, Quinn RD, Ritzmann RE (2004) Mechanized cockroach footpaths enable cockroach-like mobility. In: Proceedings of the International Conference on Information and Automation, vol. 3, pp 28712876
8. McGeer T, Palmer LH (1989) Wobbling, toppling, and forces of contact. Am J Phys, 57(12):1089–1097
9. Byl K, Tedrake R (2008) Metastable walking on stochastically rough terrain. In: Proceedings of Robotics: Science and Systems IV 6490–6495
10. Stephens B (2007) Integral control of humanoid balance. In: The Robotics Institute Carnegie Mellon University Pittsburgh, PA 15213, USA, bstephens@cmu.edu http://www.cs.cmu.edu/bstephe1
11. Jong Hyeon P (2001) Impedance control for biped robot locomotion. IEEE Trans Robotics Automat, 17(6):870–882
12. Wisse M (2004) Essentials of dynamic walking analysis and design of two-legged robots. In: PhD thesis, Universiteit Delft, 20 Sept 2004
13. Tedrake R, Zhang TW, Seung S (2001) Learning to walk in 20 minute. In: Brain and Cognitive Sciences Center for Bits and Atoms Massachusetts Inst. of Technology Cambridge
14. Narukawa T, Yokoyama K, Takahashi M, Yoshida K (2010) An experimental study of three-dimensional passive dynamic walking with flat feet and ankle springs.In: Cutting Edge Robotics, Keio University Japan
15. Lee S H. The reaction mass pendulum (RMP) model for humanoid robot gait and balance Control. In: Humanoid Robots, University of California, Los Angeles
16. Stephens B, Atkeson C () Modeling and control of periodic humanoid balance using the linear biped model. In: Carnegie Mellon University, Pittsburgh

# Part IV
# Bio-Reconfiguration Techniques and Biomedical Devices

# Chapter 36
# Bi-Behavioral Prosthetic Knee Enabled by a Metamorphic Compliant Mechanism

**Shannon Zirbel, Shane Curtis, Rachel Bradshaw, Luke Duffield, Greg Teichert, Nicholas Williams, Ron Rorrer, Spencer Magleby and Larry Howell**

**Abstract** Metamorphic mechanisms with two distinct behaviors were designed using compliant mechanism theory with a potential application as a prosthetic knee. The mechanism has discrete "locking" points to restrict rotation when under a compressive load. The designs use cross-axis flexural pivots, either in inversion or isolation, with engaging teeth to carry loads at distinct angles. Inverted compliant mechanisms function by inverting the mechanism so the compliant members are in tension when a compressive load is applied. Compliant mechanisms in isolation provide an alternative loading pattern which redirects the load to a passive rest. The mechanism incorporates teeth which engage during weight-bearing in flexion at up to 60° of flexion to lock the mechanism. When tension is applied to the device, the teeth are disengaged and the mechanism is allowed to rotate freely. The purpose of this design is to hold compressive loads both when un-flexed and flexed. The concept is applied in the preliminary design of a prosthetic knee joint. Proof-of-concept prototypes successfully demonstrate the metamorphic behavior.

**Keywords** Compliant mechanisms · Compression joint · Prosthetic knee · Cross-axis flexural pivot · Weight-bearing in flexion

---

S. Zirbel · S. Curtis · R. Bradshaw · L. Duffield · G. Teichert · S. Magleby · L. Howell (✉)
Brigham Young University, 435 CTB, Provo, UT 84602, USA
e-mail: lhowell@byu.edu

N. Williams · R. Rorrer
University of Colorado Denver, Denver, CO 80217-3364, USA

J. S. Dai et al. (eds.), *Advances in Reconfigurable Mechanisms and Robots I*,
DOI: 10.1007/978-1-4471-4141-9_36, 

## 36.1 Introduction

The objective of this work was to design a prosthetic knee using compliant mechanisms with metamorphic motion that enable two phases of motion: swing phase (free to rotate) and stance phase (locked and able to support high compressive loads). A compliant-mechanism-based prosthetic knee concept was previously described by Guerinot et al. [1]. This past work was used as a basis for our bi-behavioral prosthetic knee design.

## 36.2 Background

Two designs are presented in this work for a prosthetic knee using compliant mechanisms with metamorphic motion. A metamorphic mechanism [2, 3] auto-reconfigures when a specified input is changed. For the designs presented here, the behavior exhibited when the mechanism is under a compressive load is different from the behavior exhibited when the mechanism is under a tensile load.

The following background on current prosthetic knee designs and compliant mechanisms in compression provide context for this work.

### *36.2.1 Prosthetic Knee Design Considerations*

Gait has two distinct phases: swing and stance. To achieve efficient gait, the knee prosthetic should flex slightly on heel-strike (at the beginning of the stance phase) [4]. This is referred to as stance flexion, and is one of the primary metrics identified for this project. The following descriptions show how current prosthetic knee designs address considerations of stability, durability, and function.

The simplest type of prosthetic knee is a single pin-joint knee [5]. The pin joint is located slightly behind the load line for stability. It is often combined with a simple braking system which is activated when weight is applied. This brake provides enhanced stability in stance but would not provide sufficient resistance when the knee is loaded with a large amount of flexion. This design provides the least amount of stability, but is durable due to its simplicity.

The polycentric knee uses a four-bar mechanism to provide a high level of stability in stance [6, 7]. It is designed so that the instant center of rotation is behind the load line for the first 10–20° of rotation. An added benefit is that the overall leg length decreases during swing so that the amputee is less likely to trip. Some polycentric knees incorporate a mechanism to provide stance flexion; that is, it enables the knee to flex slightly at heel-strike to absorb shock and to enable a more natural gait. The polycentric knee has very good stability, while still being durable.

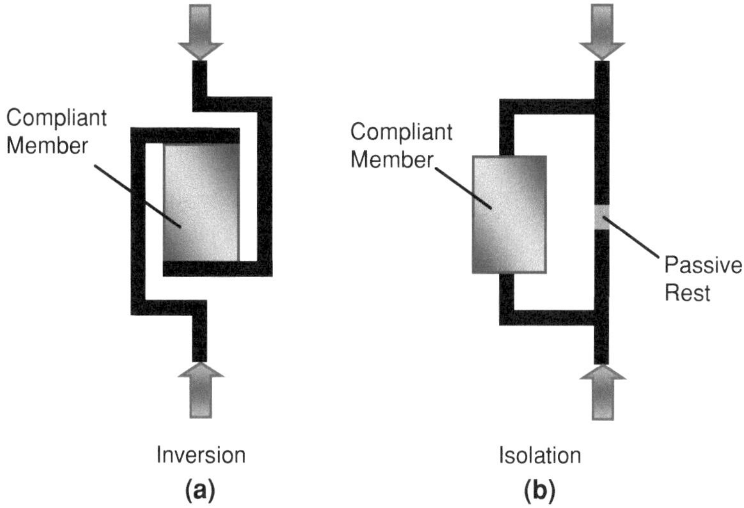


**Fig. 36.1** Compliant mechanisms can withstand compressive loads by **a** inverting the mechanism so the compressive load puts the compliant member into tension, or **b** isolating the compliant member away from the compressive load

Hydraulic and pneumatic knees provide damping to knee motion in addition to stable stance [6, 8]. Damping is desired to slow down the rotation of the knee as it approaches extreme flexion and extension. Increased damping may also be desired when weight is applied, enabling a controlled fall. These knees are less durable and require more maintenance than the pin joint or polycentric knees, but provide greater stability.

### *36.2.2 Compliant Mechanisms in Compression*

Compliant mechanisms can improve performance and reduce part count, cost, wear, and weight [1, 9]. The inherent flexibility of compliant mechanisms make them well-suited for knee prosthetics and other devices that undergo large rotations.

Because compliant members do not generally perform well in compression, Guerinot et al. developed two principles which provide viable alternatives to this challenge, referred to as isolation and inversion [1, 10]. These two principles work as demonstrated in Fig. 36.1. Compliant mechanisms in inversion function by inverting the mechanism so the compliant members are in tension when a compressive load is applied. Compliant mechanisms in isolation provide an alternative loading pattern which redirects the load to a passive rest. The metamorphic designs using isolation are behaviorally similar to contact-aided compliant mechanisms [11], which change their behavior when contact is established (as opposed to different behaviors in specific loading cases).

**Table 36.1** Primary needs ranked on a scale from 1 to 5, where 1 is most important

| Importance | Customer need |
|---|---|
| 1 | Support weight in flexion |
| 1 | Allow for natural gait/natural knee motion |
| 1 | Compliant |
| 1 | Supports load in stance |
| 1 | Safe |
| 1 | Extremely affordable |
| 1.5 | Has a long product life |
| 2.5 | Functional for various users |
| 3 | Light weight |
| 5 | Aesthetically pleasing |

## 36.3 Bi-Behavioral Need

The natural knee is actively controlled by muscles. This active control allows the knee to react differently in different phases of gait and during different activities. Musculature enables the knee to extend. Some current prosthetics use microprocessors to mimic the performance of natural knees, but the cost and maintenance required for these systems make such prosthetics unrealistic for many potential users.

To answer this challenge mechanically, the mechanism needs to perform at least two independent behaviors, suggesting a metamorphic design. To achieve the desired performance, the knee needs to (a) swing freely during the swing phase of gait and (b) support compressive loads, even when flexed. The prosthetic knee should swing freely for 130–140° to be functionally equivalent to other prosthetic knees. To maintain natural function, we also determined that the prosthetic knee should have stable "locking" positions at discrete angles of flexion for stability, referred to as *weight-bearing in flexion*. These two needs, free swing and weight-bearing in flexion, have competing objectives in the mechanical design. Other needs are listed in Table 36.1.

Based on our customer needs, we developed the functional specifications given in Table 36.2. To enable natural gait, all of these functions are necessary; however, some of them are conflicting in mechanical design. For clarity, the metrics are defined as follows: *swing completion* means that the knee returns to the extended position quickly enough to be ready for the next step; *stance flexion* is when the knee flexes slightly at heel-strike to absorb some of the impact force; and *weight-bearing in flexion* is the ability of the knee to support a load at discrete increments of flexion.

## 36.4 Metamorphic Designs

To provide the two behaviors of swing and weight-bearing in flexion, we designed a metamorphic compliant prosthetic knee. There are two iterations of the design: Design 1 employs the inversion principle of compliant mechanisms in compression

**Table 36.2** Functional specifications

| Quantitative metric | Units | Marginal | Ideal | Qualitative metric |
|---|---|---|---|---|
| Stable stance | deg | 0 | 0 | Can kneel |
| Swing (non-loaded flexion) | deg | 0–130 | 0–140 | Rigid in torsion |
| Swing completion | deg | 0 | 60–0[a] | Aligns in sagittal plane |
| Weight-bearing in flexion | deg | 0 | 0–45 | |
| Stance flexion | deg | 0 | 15 | |
| Load capability | $lb_f$ | 525 | 700 | |
| Weight | $lb_f$ | 4 | 2.5 | |
| Max knee height | in. | 7.5 | 7 | |
| Max knee lateral width | in. | 3 | 2.5 | |
| Max knee depth | in. | 4 | 3.5 | |
| Torsional strength | in.$-lb_f$ | 175 | 200 | |
| Lifetime | years | 3 | 5 | |
| Cycles at max flexion | cycles | 20,000 | Infinite | |

[a] Swing completion should begin at 60° and continue to 0°

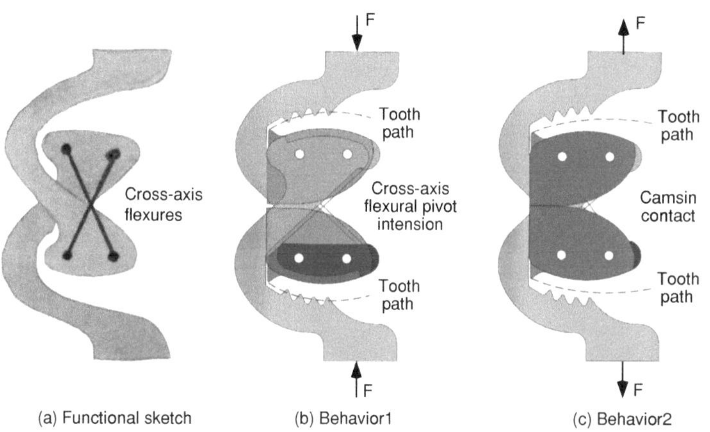


**Fig. 36.2** The two behaviors are illustrated in the context of Design 1. In Behavior 1, the cross-axis flexures are in tension when the compressive load is applied. The teeth (*in red*) engage and lock the mechanism. In Behavior 2, when a tensile load is applied from the weight of the shank, the cams come into contact and the cross-axis flexures are slack. The teeth do not engage so the prosthetic can rotate freely

and Design 2 employs the isolation principle. Both designs are bi-behavioral metamorphic mechanisms, allowing the knee to swing freely when no weight is applied, and to lock when a compressive load is applied.

A model of Design 1 is presented in Fig. 36.2. The device utilizes an inverted cross-axis flexural pivot [12–14] which is in tension when the device is under compression (Behavior 1), and a set of cams that are in contact when the device is in tension (Behavior 2). Design 1 employs the inversion principle, putting the cross-axis flexural pivot in tension when the device is under a load, as is the case during stance. The cams ensure that the knee does not collapse during the swing

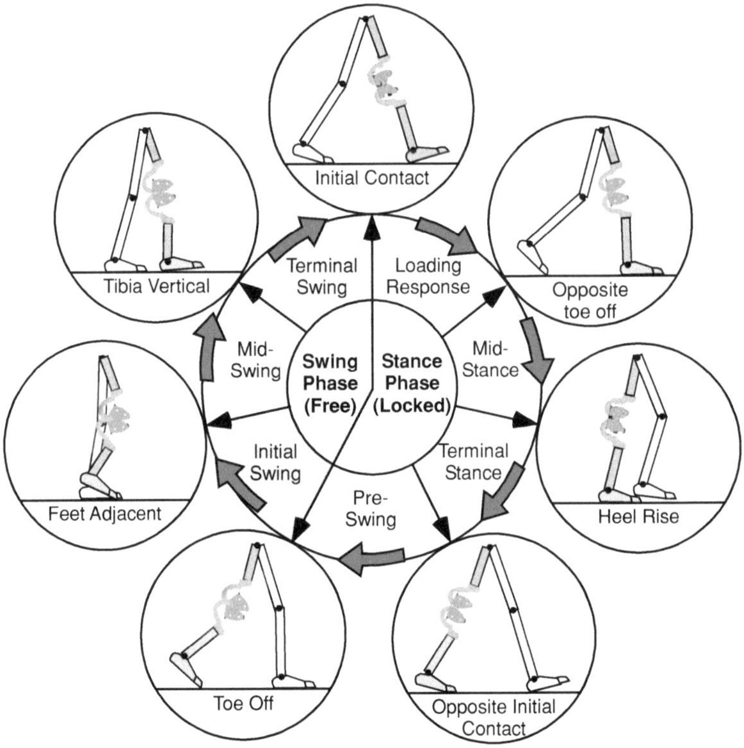


**Fig. 36.3** The figure, modified from [15], illustrates the bi-behavioral metamorphic motion of Design 1 during gait

phase of gait. The gap between the cams is minimal so that when the flexures are under a compressive load, there is no effect of instability from buckling. The flexures carry a negligible compressive load because the cams come into contact almost immediately when the prosthetic knee is under a tensile load. During Behavior 1, the teeth engage and lock the mechanism during stance or at discrete angles of flexion. This locking behavior is referred to as weight-bearing in flexion and increases the stability of the prosthetic knee. During Behavior 2, the teeth do not engage, enabling the device to swing freely. These behaviors, shown in different phases of gait, are illustrated in Fig. 36.3.

A model of Design 2 is shown in Fig. 36.4. This design uses a cross-axis flexural pivot and the isolation principle to withstand the compressive loads experienced during gait. The cross-axis flexures are shifted to the back of the knee to provide greater stability in stance. The flexures are also rotated to minimize the depth dimension of the knee.

Design 2 is also a bi-behavioral metamorphic mechanism. Behavior 1 occurs when a load is applied to the system. When weight is applied in extension (as in heel-strike or while standing), hard stops come into contact instead of the teeth.

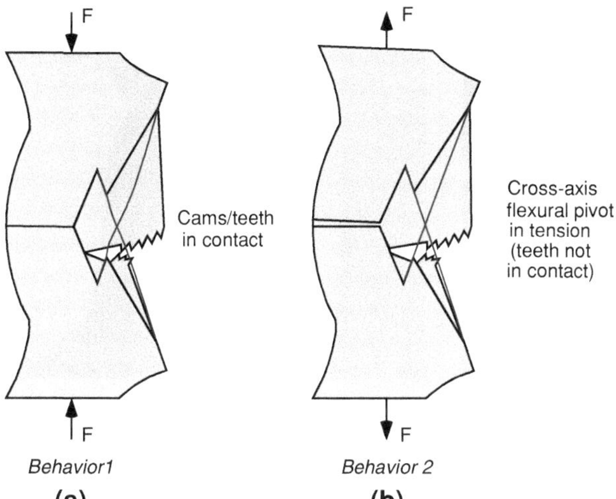


**Fig. 36.4** The two behaviors are illustrated in the context of Design 2. In Behavior 1, the cross-axis flexures are slack when the compressive load is applied. The hard stops come into contact and carry the compressive load. The teeth engage and lock the mechanism. In Behavior 2, when a tensile load is applied from the weight of the shank, the cross-axis flexures are in tension and carry the load. The teeth do not engage so the prosthetic can rotate freely

The hard stops take the compressive load, applying the isolation principle. When weight is applied between approximately 20 and 60° of flexion, the flexures deflect and the teeth engage to lock the knee. Behavior 2 occurs during the swing phase when the weight of the shank puts the flexures into tension. The contacting surfaces with the engaging teeth are designed to maintain a constant gap to allow up to 140 degrees of non-loaded flexion.

The flexures were designed using the pseudo-rigid-body model (PRBM) [9]. The geometry of the pseudo-rigid links defines the geometry of the elliptical contacting surface, or the path along which the teeth lie.

Due to the isolation principle, Design 2 minimizes the stresses in the flexures and is able to support large compressive loads while in extension. The design is such that the knee can be made from two unique parts, which reduces fabrication costs. Because the center of rotation is at the back of the knee, the knee is stable instance.

## 36.5 Prototypes and Demonstrations

The prototype of Design 1, shown in Fig. 36.5a, was built using a laser cutter. The majority of the prototype is acrylic, with the exception of the cross-axis flexures, which are made from polystyrene. Although not a complete prosthetic, this proof-of-concept prototype is beneficial for verifying the fundamental concepts encompassed

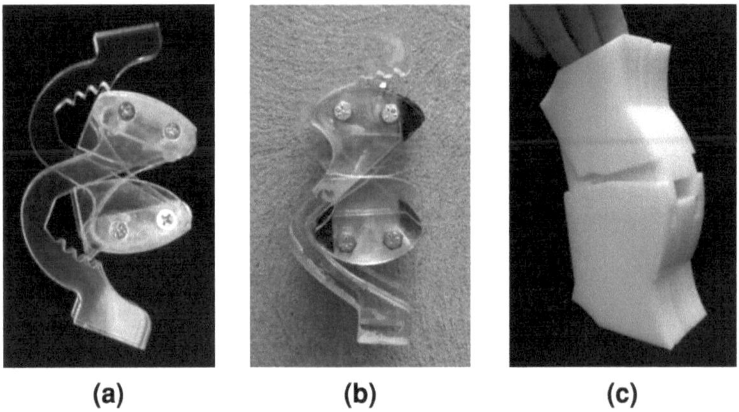


**Fig. 36.5** Prototypes of Designs 1 and 2. **a** The first prototype of Design 1 was made from acrylic with polystyrene flexures. **b** A second prototype of Design 1 was made from acrylic with black silicone at the tips of the cams. **c** The prototype of Design 2 was made from polypropylene

by this design. A second prototype of Design 1, shown in Fig. 36.5b, incorporated black silicone at the tips of the cams where the flexures are attached. The silicone tips provide sufficient compliance to facilitate tooth engagement. The position of the teeth was also modified to reduce bending loads on the teeth. The prototype of Design 2, shown in Fig. 36.5c, was made from layers of polypropylene cut on a CNC mill.

These prototypes successfully demonstrate the feasibility of achieving the two distinct behaviors needed for the desired prosthetic knee behavior. This is a foundational starting point for a detailed design of a prosthetic knee.

Device models included parametric finite element models (using ANSYS) of the cross-axis flexural pivot and pseudo-rigid-body models [9], including both the simple (pin joint) model and the four-bar model. Analytical models were also used to evaluate the stresses on the teeth. These models were used as design tools to determine the overall configuration and they may also be valuable in future detailed design and analysis. Careful design is required to select geometry and materials that will result in the desired behavior while maintaining stress constraints.

## 36.6 Discussion

The present work has provided two possible designs for a compliant prosthetic knee with metamorphic behavior. Both designs incorporate components which could lead to increased stability and performance for transfemoral amputees. These designs demonstrate metamorphic motion and are in the preliminary stages in the prosthetic design.

The metamorphic mechanism itself has potential applications in other areas where a locking function is needed when compressive loads are applied. This can be as a safety response, such as in safety and arming applications, or in space applications when high *g* loads are applied during launch.

**Acknowledgements** The authors gratefully acknowledge the support of BYU faculty members Matt Seeley, Mark Colton, and Anton Bowden for their insightful feedback in design reviews. We also express our appreciation to Lane Ferrin of Northwest Orthotics and Dale Price for their collaboration on this project.

## References

1. Guerinot AE, Magleby SP, Howell LL (2004) Preliminary design concepts for compliant mechanism prosthetic knee joints. Proceedings of ASME DETC, Salt Lake City
2. Dai J, Jones JR (2005) Matrix representation of topological changes in metamorphic mechanisms. J Mech Des 127:837–840
3. Dai J, Jones JR (1999) Mobility in metamorphic mechanisms of foldable/erectable kinds. J Mech Des 121:375–382
4. Kandel ER, Schwartz JH, Jessell TJ (2000) Principles of neural science, 4th edn. McGraw-Hill, New York
5. Wilson AB (1968) Recent advances in above-knee prosthetics. Artif Limbs 12:1–27
6. Radcliffe CW (1977) Above-knee prosthetics. Knud-Jansen Lecture, New York
7. Patil KM, Chakraborty JK (1991) Analysis of a new polycentric above-knee prosthesis with a pneumatic swing phase control. J Biomech 24:223–233
8. Johansson JL, Sherrill DM, Riley PO et al (2005) A clinical comparison of variable-damping and mechanically passive prosthetic knee devices. Am J Phys Med Rehab 84:563–575
9. Howell LL (2001) Compliant mechanisms. Wiley, New York
10. Guerinot AE, Magleby SP, Howell LL, Todd RH (2005) Compliant joint design principles for high compressive load situations. J Mech Des 127:774–781
11. Mankame ND, Ananthasuresh GK (2002) Contact aided compliant mechanisms: concept and preliminaries. Proceedings of ASME DETC conference, Montreal, Canada, pp 109–121
12. Jensen BD, Howell LL (2002) The modeling of cross-axis flexural pivots. Mech Mach Theory 37:461–476
13. Xu P, Jingjun Y, Guanghua Z (2010) An effective pseudo-rigid-body method for beam-based compliant mechanisms. Precis Eng 34:634–639
14. Hongzhe Z, Shusheng B (2010) Stiffness and stress characteristics of the generalized crossspring pivot. Mech Mach Theory 45:378–391
15. Whittle MW (2007) Gait analysis: an introduction, 4th edn. Butterworth-Heinemann, Oxford

# Chapter 37
# A Novel Actuator with Reconfigurable Stiffness for a Knee Exoskeleton: Design and Modeling

**Nikos C. Karavas, Nikos G. Tsagarakis, Jody Saglia and Darwin G. Galdwell**

**Abstract** This paper presents the design of a new inherently compliant actuator intended for the development of a knee exoskeleton. The proposed actuator has the ability to reconfigure the level of stiffness in order to achieve suitable torque-to-angular displacement profiles for different human tasks and users. The design specifications of the actuator have been obtained from motion capture and simulation data of sit-stand-sit motion cycle. The actuator functional principle and modeling are presented. Finally the mechatronic design of the actuator is described.

## 37.1 Introduction

During the last decades many researchers have developed different types of lower limb exoskeletons and active orthosis [1]. Applications for these devices can be categorized into three major groups: human performance augmenting exoskeletons [2, 3], assistive exoskeletons [4, 5] and rehabilitation exoskeletons [6, 7].

As exoskeletons are devices wearable by humans, safety of interaction preventing both harm and discomfort is mandatory. Conventional stiff actuation approach imposes on the exoskeletons significant limitations related to safety issues and the ability to interact with human. However, introducing inherent compliance to the actuation system decouples the inertia of the motor drive from the output link and make the output mechanical impedance to be low across the frequency spectrum [8]. Therefore, the robotic system improves its ability to

N. C. Karavas (✉) · N. G. Tsagarakis · J. Saglia · D. G. Galdwell
Department of Advanced Robotics, Istituto Italiano di Tecnologia,
via Morego 30, 16163 Genova, Italy
e-mail: nikolaos.karavas@iit.it

J. S. Dai et al. (eds.), *Advances in Reconfigurable Mechanisms and Robots I*,
DOI: 10.1007/978-1-4471-4141-9_37, 

intrinsically absorb impacts and enhances physical human-robot interaction (pHRI), which is a key aspect in functionality and adaptability of assistive robotic devices [9]. From an engineering point of view improvement of pHRI can be achieved by exploring high performance actuator technologies and dedicated control strategies. Furthermore, compliant robots can be more energy efficient than traditional stiff actuated systems, fact which is crucial in creating energetically autonomous exoskeletons for locomotion [10].

The primary goal of our work is to build a compliant lower body exoskeleton to be used as a lower limb assistive device, in order to gain the features explained above. First step along this path is the development of a compliant knee exoskeleton, which will provide motion assistance both to patients with impaired legs and healthy individuals for different motion tasks like walking, standing up-sitting down and squatting. The realization of such a compliant exoskeleton remains a challenging task requiring more compact, modular and high performance actuation units. This paper presents the design of an intrinsic compliant actuator with reconfigurable stiffness which has the ability to provide the required assistive torques to the knee joints of the wearer.

## 37.2 Human Leg Dynamic Behavior of Sit-Stand-Sit Movement

Understanding the biomechanics of human lower limbs is vital in the design of exoskeletons and orthotic devices for locomotion. Clinical Gait Analysis (CGA) data has been used as primary basis for the design specifications of the exoskeletons [1, 2]. The knee exoskeleton is intended to be a device which will assist humans in performing tasks of daily living i.e. walking, standing up and sitting down. Sit-stand-sit motion is a very frequent task and also one of the most demanding in terms of power, torque and range of motion [11, 12]. Thus, the specifications for the proposed actuator have been obtained from simulation and experimental data of the sit-stand-sit movement cycle.

In order to study the torques characteristics for standing up and sitting down in the sagittal plane, a simulation planar model (3 DOF) of a human leg has been developed. The anthropometric data on which the model is based has been obtained from [13]. It has been considered a male of 82.5 kg with 1.85 m height using the 50th percentile. Lumped mass models of the shank, thigh, pelvis and torso were considered according to the anthropometric data in [13]. To derive the torque characteristics of each joint of the human leg, parametric trajectories based on data of biomechanics literature firstly were used [11, 14–16]. Figure 37.2a shows the parametric trajectories which correspond to the standing-up motion and the resulting torques. In order to validate these results the following experiment took also place (Fig. 37.1): a healthy subject was instructed to repeatedly stand up and sit down without hand assistance (worst case/highest torques), while trajectories of hip, knee and ankle joints of both legs were recorded using the Vicon

**Fig. 37.1** Video frames of the standing up motion during the experiment

motion capture system. These trajectories, which correspond to the stand-sit-stand motion cycle, were used as input to the simulation model in order to obtain the torque characteristics (Fig. 37.2b). Note that Fig. 37.2 depicts also the ankle and hip trajectories to allow full comparison of the simulation results. However, for the actuator design only the knee trajectories have been considered.

In addition, in Fig. 37.2 it is possible to see that the task of standing up is performed approximately over the same duration both in parametric and Vicon trajectories. Therefore, it is reasonable to compare the two knee torque curves obtained with different joint input trajectories fact which results that the maximum torque of the knee joint for the task of standing up-sitting down is around 100 Nm (omitting the high peaks in Fig. 37.2b. This torque value, which can also be confirmed by biomechanics data in the literature [11, 14, 15], was used as torque specification for the knee actuator design. In addition, analysis of the torque-velocity requirements using these simulation results was conducted in order to determine the suitable selection of the pair DC motor-harmonic drive but this analysis is not described due to the paper length constraint. At this point authors are not aware of any study regarding the stiffness range of human knee joint during walking or sit-stand-sit movement. In [17] there is an attempt to specify the characteristic stiffness of the knee in flexion and extension modes. According to that, we adopt a suitable range of stiffness from 300 to 600 Nm/rad.

## 37.3 Actuation Design

### *37.3.1 Concept and Principle of Operation of CompAct-ARS*

In the Advanced Robotics Department of Italian Institute of Technology two compliant actuators have been developed recently. The CompAct-SEA (Serial Elastic Actuator) [18] and the CompAct-VSA (Variable Stiffness Actuator) [19]. Both actuators, as they are, cannot be directly applied for the development of a knee exoskeleton. The CompAct-SEA is a compact soft actuation unit which is used in

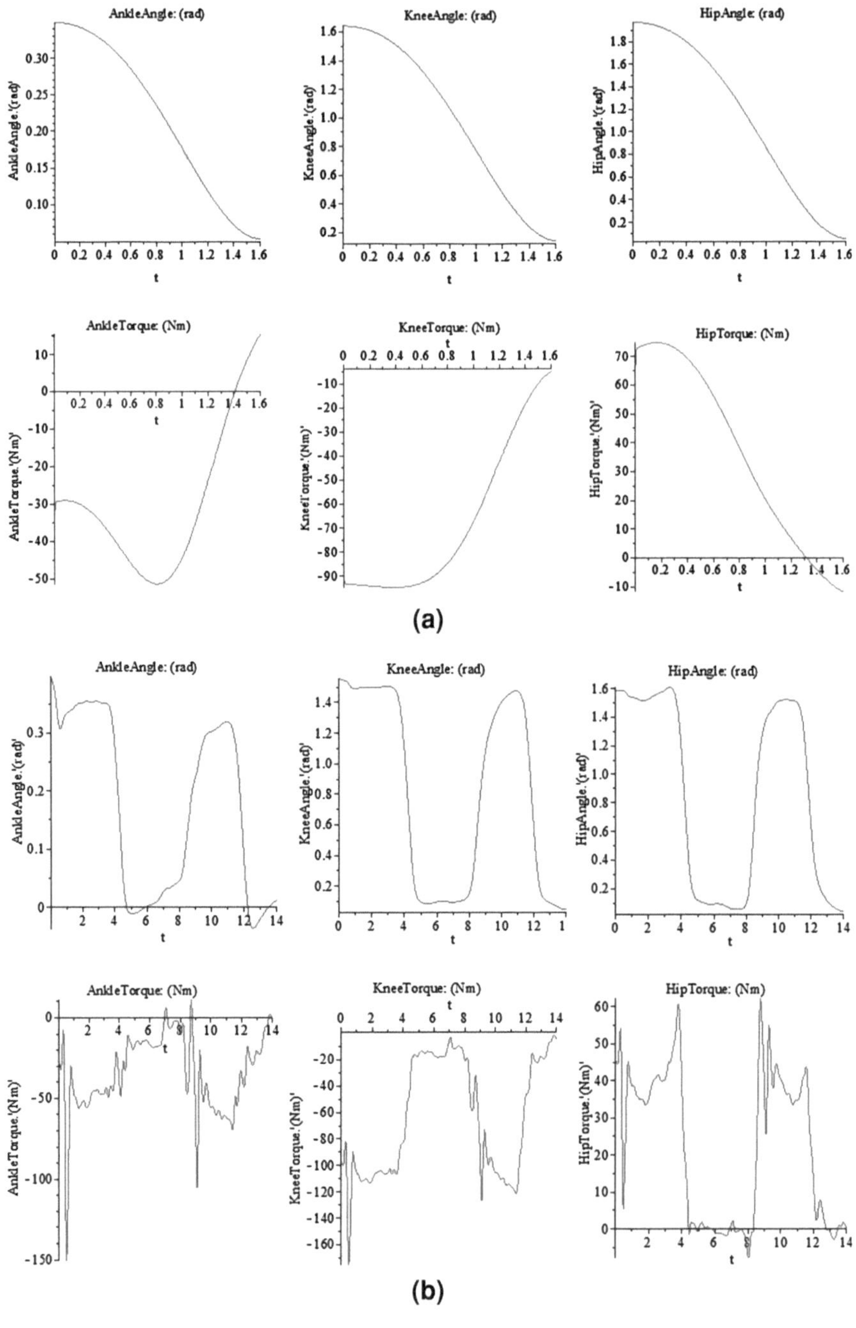


**Fig. 37.2** **a** Parametric trajectories and the corresponding torques of each joint. **b** Trajectories obtained from vicon and the corresponding torques of each joint

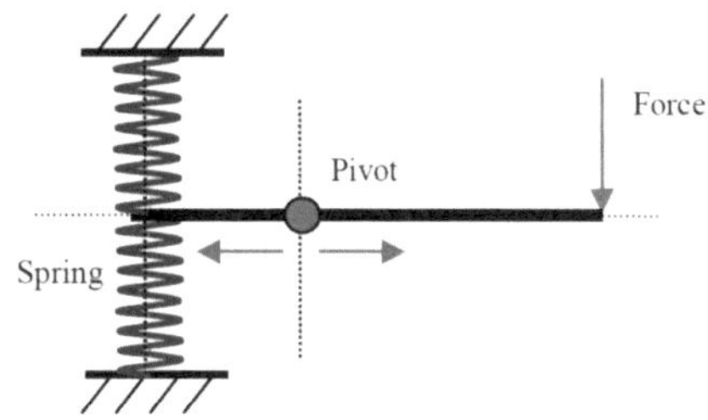


**Fig. 37.3** Schematic of the lever arm mechanism with variable pivot point

multi degree of freedom and small scale robotics systems such as the humanoid robot COMAN [20]. It shows high integration density and accompanied with proper control scheme is capable to vary the output impedance within a wide range. However, this actuator does not fulfill the elastic torque requirements of the actuation for the knee exoskeleton which is intended to be an assistive device (see Sect. 37.2). Even in a larger scale design, the CompAct-SEA cannot achieve the required angular deflection considering the desired stiffness and nominal torque.

On the other hand the CompAct-VSA, a novel realization of a variable stiffness actuator, can achieve wide stiffness range (theoretically from zero to infinite), high torque capacity and fast stiffness adjustment response. Although its compact size and relatively light weight, CompAct-VSA still cannot satisfy the dimensional and weight requirements of exoskeleton joints. As a consequence, a combination of both actuators functional principles is proposed in order to optimize the aforementioned actuation systems and to achieve the desired performance and compliant behavior. The proposed actuator is the CompAct-ARS (Actuator with Reconfigurable Stiffness) which extends the limitation of fixed compliance of CompAct-SEA and simultaneously reducing significantly the size and weight in comparison with the CompAct-VSA. The CompAct-ARS is an inherently compliant actuator with the ability to regulate off-line the level of stiffness in a wide as needed range. Thus, this hybrid concept of compliant actuator can satisfy the required performance (passive compliance, high torque capacity and feasible implementation for the exoskeleton joints).

The principle of function of the CompAct-ARS is based on CompAct-VSA principle [19], which uses a lever arm mechanism with a variable pivot axis (Fig. 37.3). The apparent stiffness at the lever can be adjusted by changing the pivot position while the location of the force and springs are kept fixed. The amplification ratio is the distance between pivot and the springs over the distance between pivot and the point at which force is applied. The main difference between CompAct-ARS and CompAct-VSA concerning their functional principles is that in our case the pivot position variation is manually achieved by tuning set screws and not by a second motor.

### *37.3.2 Modeling of CompAct-ARS*

The modeling of the CompAct-ARS is based on the schematic of the lever arm mechanism presented in Fig. 37.4. When an external torque $\tau_{ext}$ is applied at the output link, it rotates the lever arm with an angle $\phi$ with respect to the pivot $(P)$.

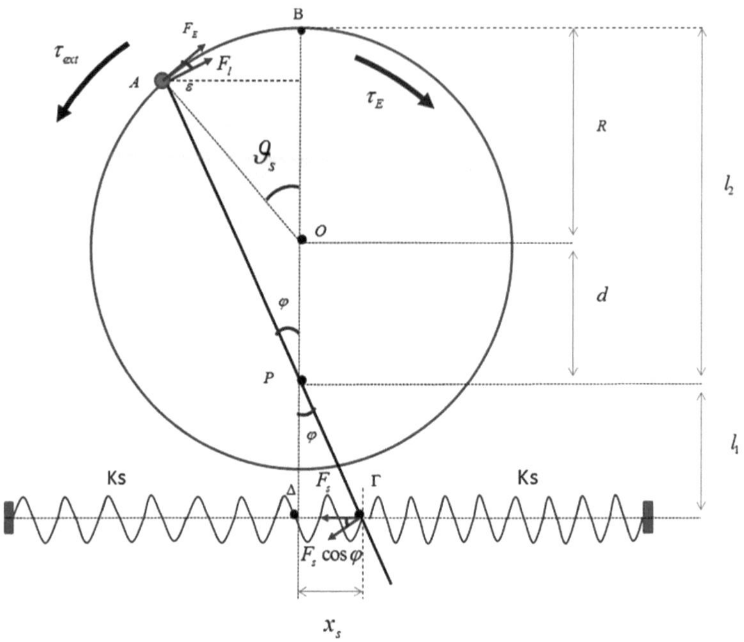


**Fig. 37.4** Schematic of the CompAct-ARS functional principle

This in turns displaces the springs with stiffness $k_s$ generating a force $F_s$ applied in $(\Gamma)$ given by

$$F_s = 2k_s l_1 \tan\phi \tag{37.1}$$

where $l_1$ is the distance form the pivot $(P)$ to point $(\Delta)$ which is the connection point between the springs and the lever at lever's equilibrium position. The perpendicular force to the lever arm is $F_s \cos\phi$. Due to the lever mechanism, a force

$$F_l = F_s \cos\phi \frac{P\Gamma}{AP} = 2k_s l_1 \tan\phi \cos\phi \frac{l_1}{AP\cos\phi} = 2k_s l_1^2 \tan\phi \frac{1}{AP} \tag{37.2}$$

is therefore applied in $(A)$, where $\frac{P\Gamma}{AP}$ is the amplification ratio and

$$AP = \sqrt{R^2 + d^2 + 2Rd\cos\theta_s} \tag{37.3}$$

where $R = AO$, $d$ is the distance from the pivot $(P)$ to the center of rotation of the joint $(O)$ and $\theta_s = q - \theta_M$ is the deflection of the elastic transmission defined as the difference between the angle of the output link $q$ and the angle of the motor $\theta_M$ after the reduction drive. The resultant force $F_E$ along the perpendicular to $AO$ is

$$F_E = \frac{F_l}{\cos\varepsilon} = \frac{F_l}{\cos(\theta_s - \phi)} = \frac{2k_s l_1^2 \tan\phi}{AP\cos(\theta_s - \phi)} \tag{37.4}$$

Since $\sin\phi = \frac{R}{AP}\sin\theta_s$ the (37.4) becomes

$$F_E = \frac{2k_s l_1^2 \sin\phi}{AP\cos\phi\cos(\theta_s - \phi)} = \frac{2k_s R l_1^2 \sin\theta_s}{AP^2\cos\phi\cos(\theta_s - \phi)} \tag{37.5}$$

As a consequence, the elastic torque $\tau_E = F_E R$ which, at the equilibrium, counterbalances the external torque is obtained such as

$$\tau_E = \frac{2k_s R^2 l_1^2 \sin\theta_s}{AP^2\cos\phi\cos(\theta_s - \phi)} \tag{37.6}$$

Assuming the deflection of the elastic transmission $\theta_s$ to be small and thus $\phi$ to be small the formulation of the elastic torque in (37.6) simplifies such as

$$\tau_E = \frac{2k_s R^2 l_1^2 \theta_s}{l_2^2} \tag{37.7}$$

Finally the stiffness of the elastic mechanism $K = \frac{\partial\tau_E}{\partial\theta_s}$ is formulated such as

$$K = \frac{2k_s R^2 l_1^2}{l_2^2} = 2k_s\alpha^2 R^2 \tag{37.8}$$

where $\alpha = \frac{l_1}{l_2}$ is the ratio of the lever arm.

### *37.3.3 Performance Analysis of CompAct-ARS*

Analysis of the CompAct-ARS performance was conducted through simulation on the basis of the model presented in Sect. 37.3.2. Initially, the behavior of the maximum angular deflection $\theta_s$ achievable by the elastic transmission has been analyzed. Basic feature of the functional principle of the actuator is that the maximum angular deflection $\theta_s$ is constrained by the maximum displacement $x_s$ of the pre-compressed springs and varies on the basis of the pivot position $l_1$. Figure 37.5a depicts the variation of the maximum allowable passive deflection as a function of the desired stiffness range. For comparison, the red line in Fig. 37.5a indicates the deflection angle which will be required for a given stiffness range in order to generate the desired maximum torque of 100 Nm estimated in Sect. 37.2.

The behavior of the elastic torque $\tau_E$, formulated as in (37.7), is further investigated through simulation and its variation with respect to the angular deflection is analyzed. From Fig. 37.5b it is possible to see that by increasing the spring rate the maximum elastic torque increases as well. However, the maximum spring compression decreases which results in reducing the storable potential energy. Thus, as a trade off the employed springs have a rate of 55 kN/m providing a maximum elastic torque of 80 Nm. This torque value is greater than knee joints need during gait cycle [21] and close to the maximum torque needed for standing

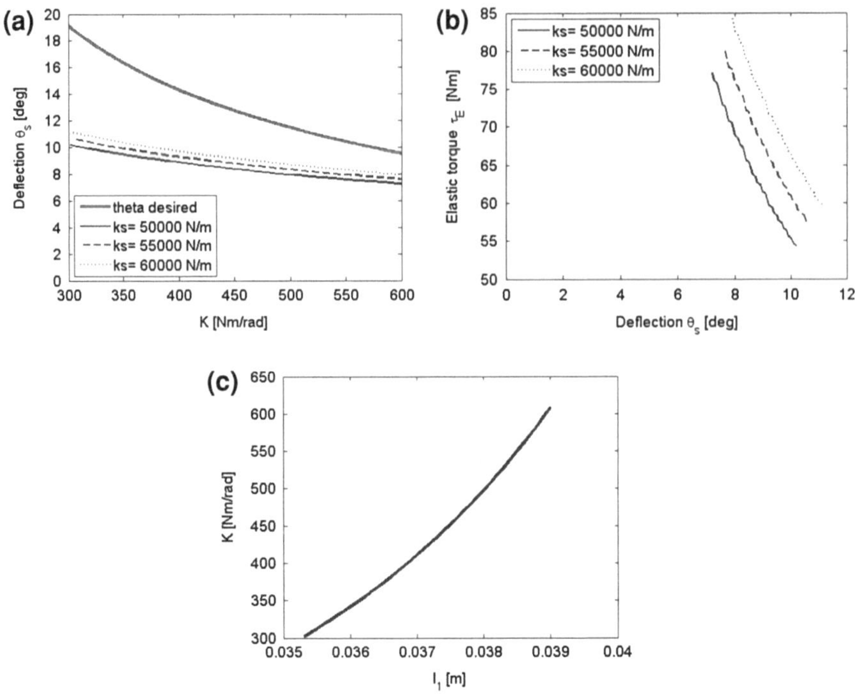


**Fig. 37.5** Theoretical specifications. **a** Angular deflection $\theta_s$ versus stiffness of the elastic mechanism for different values of the spring's rate. **b** Elastic torque $\tau_E$ versus deflection for stiffness range 300–600 Nm/rad. **c** Stiffness of the elastic transmission calculated as in (37.8)

up and sitting down motion. Finally the stiffness $K$ at the elastic transmission, formulated as in (37.8), is analyzed with respect to the variations of the pivot position $l_1$. In Fig 37.5c it is possible to notice that the desired stiffness range is achieved by a pivot position variation of 4 mm.

### *37.3.4 Mechanical Design*

The proposed actuator has been designed for physical implementation. Particular attention was paid to optimize the size, weight and the modularity of the mechanical assembly. As it is shown in Fig. 37.6c the CompAct-ARS is divided in two subassemblies: (1) the motor housing including the motor and the harmonic gear reduction drive and (2) the elastic module which contains the compliance elements and transfer the torque of the motor (after the gear reduction) to the output load. The actuator of the unit was formed by the combination of a Kollmorgen RBE1810 frameless brushless motor (peak torque of 1.53 Nm) with a

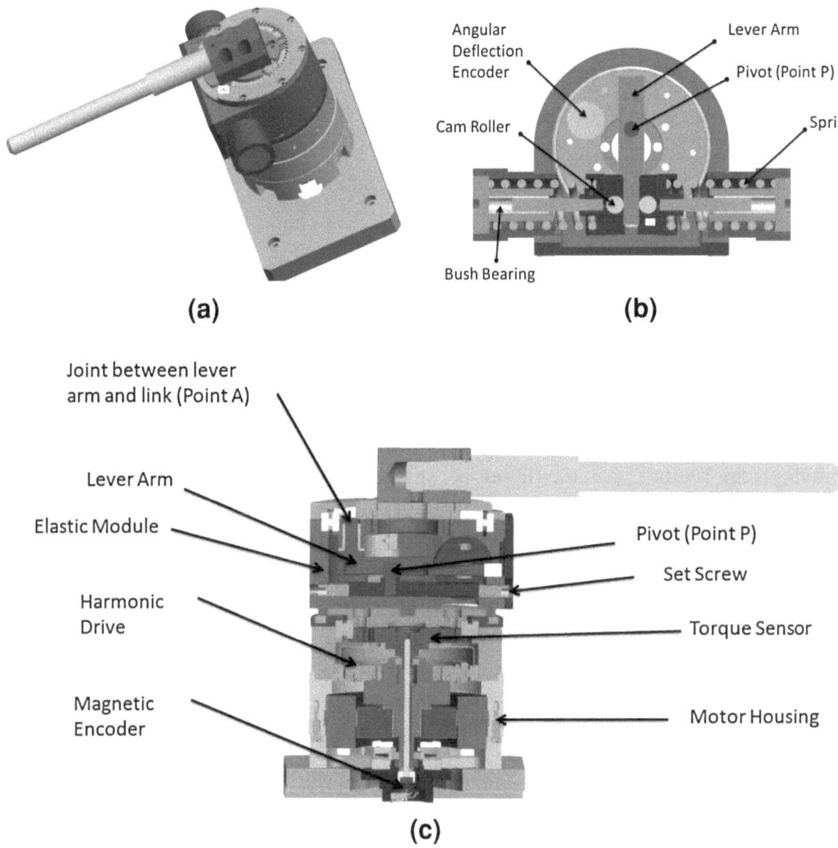


**Fig. 37.6** **a** CAD assembly of the CompAct-ARS. **b** Horizontal section view of the elastic module. **c** Section view of the CAD assembly

harmonic drive CSD 25 having a gear ratio 100:1. Moreover, the actuation module is provided with four position sensors and one torque sensor. In detail, an incremental optical encoder (MicroE optical encoder with 12 bit resolution) is mounted at the motor side, one absolute magnetic encoder (Austria Microsystems with 12 bit of resolution) monitors the position of the motor after the reduction drive and one additional optical incremental encoder (Avago Technologies encoder with 14 bit of resolution) measures the deflection angle of the link. A potentiometer also measures the position of the pivot point. In Fig. 37.6c are shown the two set screws which can be manually tuned in order to vary the position of the pivot point and therefore the level of stiffness. Figure 37.6b presents the elastic part and the employed springs which are pre-compressed half of the maximum displacement which is 20 mm. Finally in Table 37.1 the specifications of the actuator are introduced.

**Table 37.1** Specifications CompAct-ARS

| Description | Symbol | Value | Unit |
|---|---|---|---|
| Elastic torque (max) | $\tau_E$ | 80 | Nm |
| Elastic deflection (max) | $\theta_s$ | 11 | deg |
| Springs' rate | $k_s$ | 55 | kN/m |
| Stiffness range | $K$ | $300 \sim 600$ | Nm/rad |
| Diameter | $D$ | 90 | mm |
| Overall length | $L$ | 135 | mm |

## 37.4 Conclusion and Future Work

To summarize, a compliant actuator (CompaAct-ARS), which is intended to be adapted to form a knee exoskeleton, has been designed after the determination of the mechanical requirements using both simulation and experimental data. This actuator can reconfigure the level of stiffness in order to achieve the desired torque-to-angular displacement profiles for different motion tasks and different human operators. Moreover, a detailed analysis of the performance of the actuator regarding stiffness variation, elastic torque and angular deflection was conducted in order to evaluate the proposed design and select suitable actuator components (motor, reduction drive, elastic elements stiffness).

Future work will include the design of the structural parts of the knee exoskeleton in order to integrate the presented actuation module to the knee exoskeleton and evaluate its overall performance.

## References

1. Dollar AM, Herr H (2008) Lower extremity exoskeletons and active orthoses: challenges and state-of-the-art. Robot IEEE Trans 24(1):144–158
2. Zoss AB, Kazerooni H, Chu A (2006) Biomechanical design of the berkeley lower extremity exoskeleton (bleex). Mech IEEE/ASME Trans 11(2):128–138
3. Guizzo E, Goldstein H (2005) The rise of the body bots. Spect IEEE 42(10):50–56
4. Kawamoto H, Sankai Y (2002) Power assist system hal-3 for gait disorder person. In: Miesenberger K, Klaus J, Zagler W (Eds) Computers helping people with special needs, vol 2398. Lecture Notes in Computer Science, Springer, Berlin Heidelberg, pp 19–29
5. Neuhaus PD, Noorden JH, Craig TJ, Torres T, Kirschbaum J, Pratt JE (2011) Design and evaluation of mina: a robotic orthosis for paraplegics. Rehabilitation robotics (ICORR), 2011 IEEE international conference on, 29 June 2011–1 July 2011, pp 1–8
6. Veneman JF, Kruidhof R, Hekman EEG, Ekkelenkamp R, Van Asseldonk EHF, van der Kooij H (2007) Design and evaluation of the lopes exoskeleton robot for interactive gait rehabilitation. Neural Sys Rehab Eng IEEE Trans 15(3):379–386
7. Banala SK, Agrawal SK, Scholz JP (2007) Active leg exoskeleton (alex) for gait rehabilitation of motor-impaired patients. Rehabilitation robotics, 2007. ICORR 2007. IEEE 10th international conference on, June 2007, pp 401–407
8. Laffranchi M, Tsagarakis NG, Caldwell DG (2009) Safe human robot interaction via energy regulation control. Intelligent robots and systems, 2009. IROS 2009. IEEE/RSJ international conference on, oct 2009, pp 35–41

9. Beyl P (2010) Design and control of a knee exoskeleton powered by pleated pneumatic artificial muscles for robot-assisted gait rehabilitation. Ph.D. thesis, Vrije Universiteit Brussel
10. Van Ham R, Vanderborght B, Van Damme M, Verrelst B, Lefeber D (2007) Maccepa, the mechanically adjustable compliance and controllable equilibrium position actuator: design and implementation in a biped robot. Robot Auton Sys 55(10):761– 768
11. Kralj A, Jaeger RJ, Munih M (1990) Analysis of standing up and sitting down in humans: definitions and normative data presentation. J Biomech 23(11):1123–1138
12. Janssen WGM, Bussmann HBJ, Stam HJ (2002) Determinants of the sit-to-stand movement: a review. Phys Ther 82(9):866–879
13. NASA-STD-3000 man-systems integration standards, vol I
14. Riener R, Fuhr T (1998) Patient-driven control of fes-supported standing up: a simulation study. Rehab Eng IEEE Trans 6(2):113–124
15. Kagaya H, Shimada Y, Ebata K, Sato M, Sato K, Yukawa T, Obinata G (1995) Restoration and analysis of standing-up in complete paraplegia utilizing functional electrical stimulation. Arch Phys Med Rehab 76(9):876–881
16. Kerr KM, White JA, Barr DA, Mollan RAB (1997) Analysis of the sit-stand-sit movement cycle in normal subjects. Clin Biomech 12(4):236–245
17. Shamaei K, Dollar AM (2011) On the mechanics of the knee during the stance phase of the gait. Rehabilitation robotics (ICORR), 2011 IEEE international conference on, 29 June 2011–July 1 2011, pp 1–7
18. Tsagarakis NG, Laffranchi M, Vanderborght B, Caldwell DG (2009) A compact soft actuator unit for small scale human friendly robots. Robotics and automation, 2009. ICRA'09. IEEE international conference on. IEEE, 2009, pp 4356–4362
19. Tsagarakis NG, Sardellitti I, Caldwell DG (2011) A new variable stiffness actuator (compact-vsa): design and modelling. Intelligent robots and systems (IROS), 2011 IEEE/RSJ international conference on. IEEE, 2011, pp 378–383
20. Moro FL, Tsagarakis NG, Caldwell DG (20011) A human-like walking for the compliant humanoid coman based on com trajectory reconstruction from kinematic motion primitives. 11th IEEE-RAS International Conference on Humanoid Robots, pp 364–370, 2011
21. Winter DA (1991) Biomechanics and motor control of human gait: normal, elderly and pathological, 2nd edn. University of Waterloo Press, Waterloo

# Chapter 38
# Bio-Inspired Dynamic Model for a Reconfigurable Multiple Continuum Arm Robot

**Rongjie Kang, David T. Branson, Emanuele Guglielmino and Darwin G. Caldwell**

**Abstract** This paper proposes a dynamic model utilizing continuum arms to form a reconfigurable robotic system. The continuum arms are composed of serially-linked parallel mechanisms inspired by octopus arm anatomy. The kinematics and dynamics for a single continuum arm are formulated and then expanded to a multiple arm system using a modular modeling method. Simulation results show that this robotic system is capable of diverse locomotion patterns by changing the configuration of the arms.

**Keywords** Reconfigurable robot · Continuum arm · Parallel mechanism · Locomotion

## 38.1 Introduction

Conventional robots typically use fixed morphology designed for the performance of a specific task that are not easily adaptable to other tasks. To overcome this problem a novel type of robots, termed reconfigurable robots, have been developed recently. A reconfigurable robot is capable of changing its morphology, and can

R. Kang (✉) · D. T. Branson · E. Guglielmino · D. G. Caldwell
Department of Advanced Robotics, Istituto Italiano di Tecnologia,
Via Morego 30, 16163 Genoa, Italy
e-mail: rongjie.kang@iit.it

D. T. Branson
e-mail: david.branson@iit.it

E. Guglielmino
e-mail: emanuele.guglielmino@iit.it

D. G. Caldwell
e-mail: darwin.caldwell@iit.it

J. S. Dai et al. (eds.), *Advances in Reconfigurable Mechanisms and Robots I*,
DOI: 10.1007/978-1-4471-4141-9_38, © Springer-Verlag London 2012 

therefore adapt to variable tasks and environments [1–3]. However, the reconfigurable mechanism requires additional degree of freedom (DOF) and therefore increases the size, weight and complexity of the structure, especially for reconfigurable robots utilizing rigid components.

An octopus arm is an example of highly flexible manipulator in the natural world. It can elongate, shorten and bend at any point along its length [4]. Inspired by octopus arms, the concept of utilizing continuum arms in robotic systems has been proposed [5]. Continuum arms inherently have a large number of DOF and the capability to change their shapes, thus they are suitable for implementation in reconfigurable robots.

The investigation of continuum robotic arms requires kinematically and dynamically accurate modeling techniques. Chirikjian established a dynamic model for hyper-redundant manipulator using "backbone curves" [6]. Yekutieli et al. proposed a 2-dimensional dynamic model for an octopus arm [7]. Giri and Kang presented 3-dimensional dynamic models respectively for a continuum manipulator using parallel mechanisms [8, 9]. However, these are all models for a single arm system that is not reconfigurable. This paper presents a bio-inspired multiple continuum arm model for a reconfigurable robotic system based on previous work presented by Kang et al. [9].

The paper is organized as follows: in Sect. 38.2 the octopus arm anatomy is briefly introduced, in Sect. 38.3 the kinematics of the continuum arm is presented, in Sect. 38.4 the dynamic model for a single arm is created using a modular modeling method, in Sect. 38.5 the multiple continuum arm model is proposed, in Sect. 38.6 the simulation results for different configurations are presented, and in Sect. 38.7 the conclusions are presented.

## 38.2 Octopus Arm Inspiration

Octopus arm dexterity is based on an anatomical muscular structure composed of three muscle types (Fig. 38.1): longitudinal (axially running along the length of the arm), transversal (radial) and oblique muscles (diagonally surrounding the arm; not shown in the figure) [4]. The longitudinal and transversal muscles are divided into four groups spreading around the central connective tissue. These muscles generate antagonistic forces to enable local areas to elongate, shorten and bend while the oblique muscles produce torsion. Because we don't consider radial changes or twist motion at this time, transversal and oblique muscles are not included in the model.

## 38.3 Kinematic Model for a Single Arm

To model a continuum arm Kang et al. proposed to divide it into a finite number of segments with parallel actuation mechanisms (Fig. 38.2) [9]. Each single segment is composed of a fixed base, a moving platform, a central strut, and four

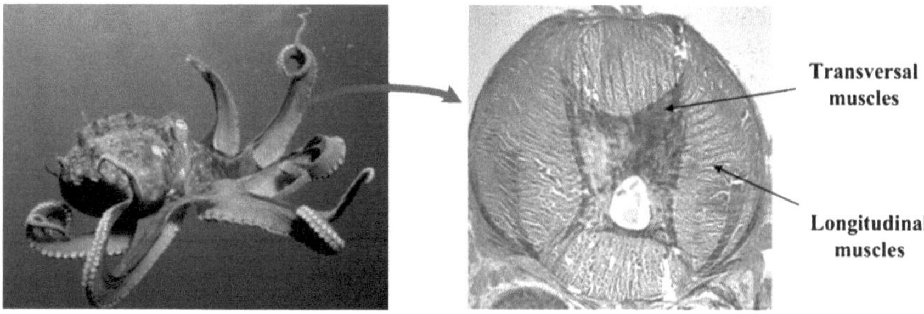


**Fig. 38.1** Octopus arm anatomy

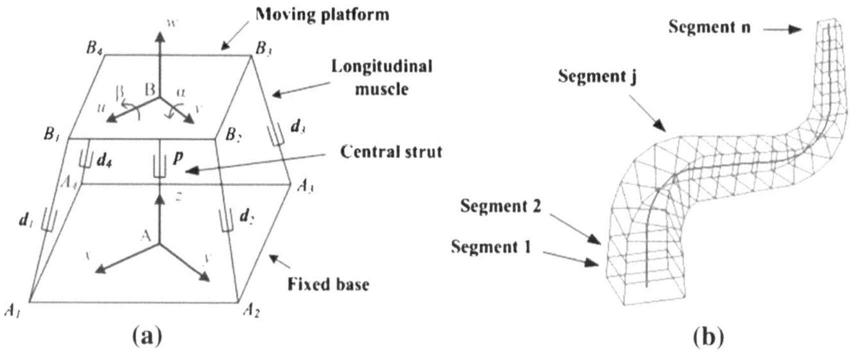


**Fig. 38.2** Geometry of a **a** single segment **b** multiple-segmented arm

longitudinal "muscles". The longitudinal muscles are considered equivalent to the linear piston/cylinder systems that can use any form of suitable actuators. They are attached to the moving platform with spherical joints, and to the base with universal joints. The central strut provides a kinematic constraint. It consists of a passive prismatic joint fixed to the centre of the base, and connects to the centre of the moving platform using a universal joint. The moving platform has two rotational DOF, about axes $u$ and $v$, and one translational DOF, along axis $z$.

Two Cartesian coordinate systems $A(x, y, z)$ and $B(u, v, w)$ are attached to the base and moving platform respectively. According to Fig. 38.2a, the kinematics of a single segment are described as:

$$\boldsymbol{d}_i = \boldsymbol{p} + {}^{A}\boldsymbol{R}_B {}^{B}\boldsymbol{b}_i - \boldsymbol{a}_i \tag{38.1}$$

$$L_i = \boldsymbol{d}_i^T \boldsymbol{d}_i \tag{38.2}$$

where $\boldsymbol{p} = \overrightarrow{AB}$ is the position vector of the centroid $B$, $\boldsymbol{b}_i = \overrightarrow{BB_i}$ is the position vector of $B_i$, and $\boldsymbol{a}_i = \overrightarrow{AA_i}$ is the position vector of $A_i$. ${}^{A}\boldsymbol{R}_B$ is the Euler rotation matrix, ${}^{B}\boldsymbol{b}_i$ is the position vector of $B_i$ and $L_i$ is the longitudinal muscle length. $\boldsymbol{p}$, $\boldsymbol{d}_i$, $\boldsymbol{a}_i$ are expressed in frame $A$ and ${}^{B}\boldsymbol{b}_i$ is expressed in frame $B$.

A homogeneous matrix ${}^{j-1}\boldsymbol{T}_j$ $(j = 1, 2, \ldots n)$ can then be defined to relate segment $j$ and $j-1$ as follows:

$$ {}^{j-1}\boldsymbol{T}_j = \begin{bmatrix} {}^{j-1}\mathbf{R}_j & {}^{j-1}\mathbf{p}_j \\ \mathbf{0} & \mathbf{1} \end{bmatrix} \tag{38.3} $$

$$ {}^{0}\boldsymbol{T}_n = {}^{0}\boldsymbol{T}_1 {}^{1}\boldsymbol{T}_2 \ldots {}^{n-1}\boldsymbol{T}_n \tag{38.4} $$

where $n$ is the total number of segments in a single arm.

## 38.4 Dynamic Model for a Single Arm

This section utilizes Newton's law and the modular modeling method to develop the dynamic model for a single arm based on previous work by Kang et al. [9].

### 38.4.1 Dynamics of a Single Segment

A single continuum arm is composed of multiple segments whose dynamic equations are formulated as follows:

#### 38.4.1.1 Longitudinal Muscle Dynamics

The longitudinal muscle is considered as a linear piston/cylinder system. The linear motion is:

$$ \xi_{ci} - f_{BLzi} = M_{pl}\ddot{L}_i + C_v\dot{L}_i + K_s(L_i - L_o) \tag{38.5} $$

where $\xi_{ci}$ is the control force, $f_{BLzi}$ the component force of joint $B_i$ acting on the piston and along the muscle $i$, $L_i$ the actual length of muscle $i$, $L_o$ the initial length of the muscle $i$, $M_{pl}$ the mass of the piston, $C_v$ the damping coefficient, and $K_s$ the stiffness coefficient.

The rotation of the longitudinal muscles is formulated in frame $A$ as:

$$ \dot{I}_i\boldsymbol{\omega}_{li} + I_i\dot{\boldsymbol{\omega}}_{li} = \boldsymbol{d}_i \times \boldsymbol{f}_{BLxyi} \tag{38.6} $$

where $\boldsymbol{f}_{BLxyi}$ is the component force of joint $B_i$ acting on the piston and perpendicular to muscle $i$, $I_i$ is the inertia moment of muscle $i$, and $\omega_{Li}$ is the corresponding angular velocity of muscle $i$.

To interlink several segments serially the reaction forces between the longitudinal muscles and the base, $\boldsymbol{F}_{ALi}$, are calculated by:

$$ \boldsymbol{F}_{ALi} = M_{cl}\dot{\boldsymbol{v}}_{li1} + M_{pl}\dot{\boldsymbol{v}}_{li2} - \boldsymbol{F}_{BLi} \tag{38.7} $$

where $\boldsymbol{F}_{ALi}$ is the joint force on $A_i$, $M_{cl}$ the mass of the cylinder, $\boldsymbol{F}_{BLi}$ is the joint force on $B_i$, $\boldsymbol{v}_{li1}$ is the velocity of the centre of mass of the cylinder, and $\boldsymbol{v}_{li2}$ is the velocity of the centre of mass of the piston. The calculation of $\boldsymbol{F}_{BLi}$, $\boldsymbol{v}_{li1}$ and $\boldsymbol{v}_{li2}$ is detailed in [9].

#### 38.4.1.2 Central Strut Dynamics

The linear motion along the central strut is:

$$f_{BCz} = M_{pl}\ddot{h} + C_v\dot{h} + K_s(h - h_o) \tag{38.8}$$

where $f_{BCz}$ is the component force of joint $B$ acting on the strut along its axis, $h$ is the actual height between point $A$ and $B$, and $h_o$ is the initial height.

The rotation of the central strut about point $A$ is:

$$\boldsymbol{T}_{AC} + \boldsymbol{T}_{BC} + \boldsymbol{p} \times \boldsymbol{F}_{BC} = \boldsymbol{0} \tag{38.9}$$

where $\boldsymbol{T}_{AC}$ is the reaction torque acting on the strut, $\boldsymbol{T}_{BC}$ is the constraint torque due to the universal joint $B$, and $\boldsymbol{p}$ is the position vector of $B$ where $\|\boldsymbol{p}\| = h$.

Applying Newton's law to the central strut yields:

$$\boldsymbol{F}_{AC} + \boldsymbol{F}_{BC} = M_{pl}[0 \quad 0 \quad \ddot{h}]^T \tag{38.10}$$

where $\boldsymbol{F}_{AC}$ is the reaction force acting on the strut, and $\boldsymbol{F}_{BC} = f_{BCz}\boldsymbol{p} + f_{BCxy}$ is the joint force of $B$ acting on the strut.

#### 38.4.1.3 Moving Platform Dynamics

The moving platform has one translational DOF along the axis of the central strut and two rotational DOF about axes $u$ and $v$ given by:

$$M_{plat}[0 \quad 0 \quad \ddot{h}]^T = \boldsymbol{F}_{CL1} + \boldsymbol{F}_{CL2} + \boldsymbol{F}_{CL3} + \boldsymbol{F}_{CL4} + \boldsymbol{F}_{CC} + \boldsymbol{F}_{ex} \tag{38.11}$$

$$\sum_{i=1}^{4} {}^{B}\boldsymbol{b}_i \times {}^{A}\boldsymbol{R}_B^T \boldsymbol{F}_{CLi} + {}^{B}T_{CC} + {}^{B}\boldsymbol{T}_{ex} = {}^{B}\boldsymbol{I}_{plat}[\ddot{\beta} \quad \ddot{\alpha} \quad 0]^T \tag{38.12}$$

where $\boldsymbol{F}_{CLi} = -\boldsymbol{F}_{BLi}$ is the joint force on $B_i$ acting on platform, $\boldsymbol{F}_{CC} = -\boldsymbol{F}_{BC}$ is the joint force on $B$ acting on the moving platform, and $\boldsymbol{F}_{ex}$ is the external force such as gravity, ${}^{B}\boldsymbol{T}_{CC}$ is the constraint torque of joint $B$ acting on the moving platform. ${}^{B}\boldsymbol{T}_{ex}$ is the external torque acting on the platform, and ${}^{B}\boldsymbol{I}_{plat}$ is the inertia matrix.

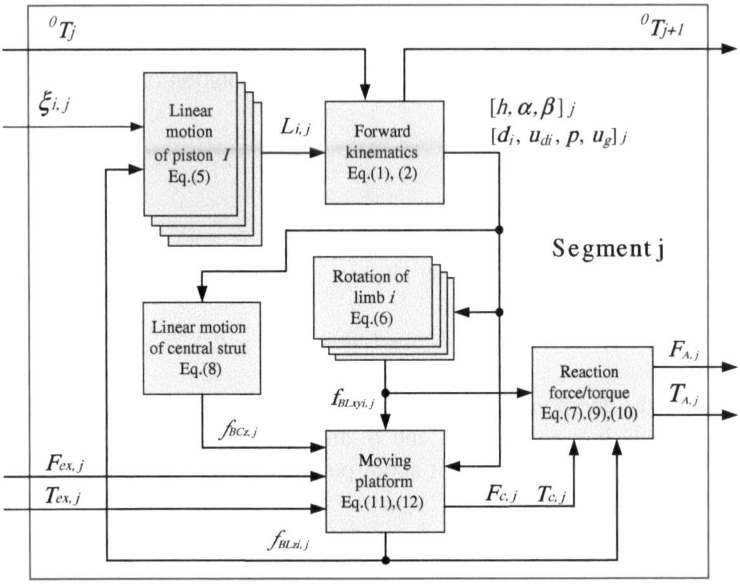


**Fig. 38.3** Modular dynamic model for a single segment

### *38.4.2 Modular Dynamic Model for a Single Arm*

A modular modeling method is then utilized to develop the full multiple segment arm model. Figure 38.3 shows the schematic of a single segment including all of the kinematic and dynamic equations mentioned previously. This schematic can be implemented in the MATLAB/Simulink environment. Each segment has a number of input/output ports used to transfer forces, torques and kinematic information to nearby segments.

To form the complete arm the individual segments are connected serially (Fig. 38.4). The dynamic forces and torques are solved from the arm tip to the base, while the kinematic transformation matrices are solved from the arm base to the arm tip.

## 38.5 Multiple Arm Robotic System

This section combines eight continuum arm models with a body to form the multiple arm system. A body coordinate system $C(x_c, y_c, z_c)$ is defined at the centroid of the robot body (Fig. 38.5). The base of each arm is described in frame $C$ by yaw angle, $\psi_m$, about axis $z_c$ and pitch angle, $\varphi_m$, about the $x_c - y_c$ plane where $m = 1, 2, \ldots, 8$ is the number of the arm.

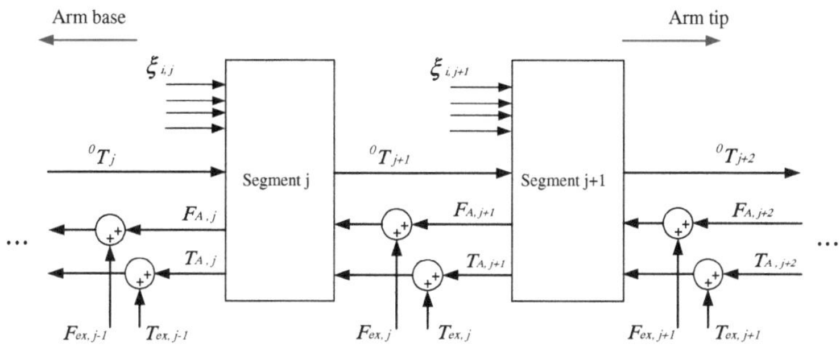


**Fig. 38.4** Modular dynamic model for multiple-segment arm

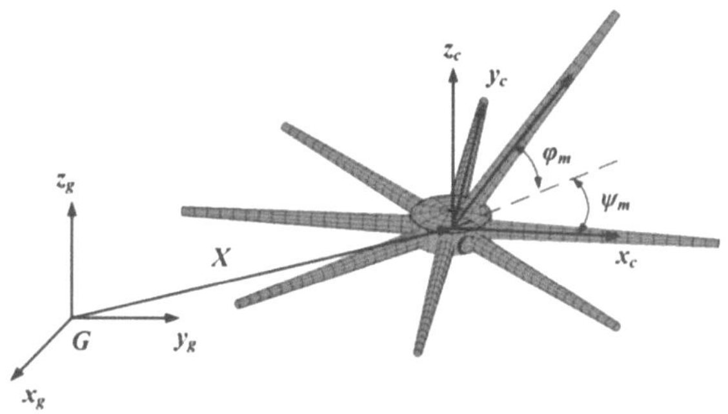


**Fig. 38.5** Geometry of multiple continuum arm model

The arm motions generate reaction forces acting on the body, and the body is considered to be rigid where its dynamics are formulated in frame $G$ as follows:

$$\sum_{m=1}^{8} \boldsymbol{F}_{A0,m} = M_b\ddot{\boldsymbol{X}} + C_b\dot{\boldsymbol{X}} \tag{38.13}$$

$$\sum_{m=1}^{8} \boldsymbol{T}_{A0,m} = \boldsymbol{I}_b\ddot{\boldsymbol{\Phi}} \tag{38.14}$$

where $M_b$ is the mass of the body, $\boldsymbol{X}$ the position vector of the body, $C_b$ is the damping coefficient, $\boldsymbol{I}_b$ is the inertia matrix of the body, $\boldsymbol{\Phi}$ is the Euler angle from frame $G$ to $C$, and $\boldsymbol{F}_{A0,\ m}$ and $\boldsymbol{T}_{A0,\ m}$ are the reaction forces and torques coming from the base segments of the arms. Thus, two additional homogeneous transformation matrices are applied to Eq. (38.4) for each arm to express the arm position and orientation in frame $G$ as:

$$^G\boldsymbol{T}_{n,m} = {}^G\boldsymbol{T}_C(\boldsymbol{X}, \boldsymbol{\Phi})^C\boldsymbol{T}_{0,m}(\varphi_m, \psi_m)^0\boldsymbol{T}_{1,m}{}^1\boldsymbol{T}_{2,m}\ldots{}^{n-1}\boldsymbol{T}_{n,m} \tag{38.15}$$

.

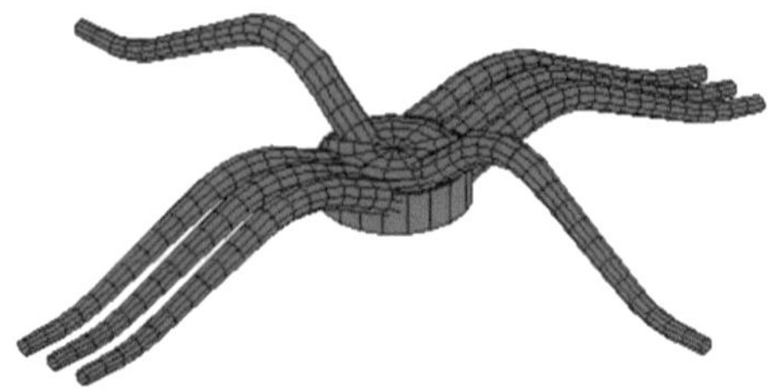

**Fig. 38.6** Crawler configuration

## 38.6 Arm Configurations

The proposed multiple continuum arm model is capable of changing its configuration to achieve diverse locomotion patterns. This section shows possible arm configurations for a reconfigurable robotic system. Locomotion simulations for the configurations were performed using a distributed hierarchical control architecture presented by Branson et al. [10]. This controller converts general motion commands for an arm into specific control inputs for each muscle.

### *38.6.1 Crawler Configuration*

In the crawler configuration, shown in Fig. 38.6, three arms are placed next to each other in front of the body and three arms are combined behind the body. These arms are used to push or pull the body in the desired direction, and the remaining arms are placed to either side of the body to provide direction control. This configuration is suitable for going through height and space limited spaces.

### *38.6.2 Quadruped Configuration*

To achieve a quadruped configuration the arms are combined into pairs with each pair forming a single leg (Fig. 38.7). This configuration allows for locomotions such as jumping, walking, running, and over obstacles.

### *38.6.3 Hexapod configuration*

In the hexapod configuration, shown in Fig. 38.8, six arms are used for locomotion and the remaining two arms are free. This configuration is capable of similar locomotions to the quadruped robot for jumping, walking, running, and movement over objects, while allowing for additional manipulation tasks to be performed by the two free arms.

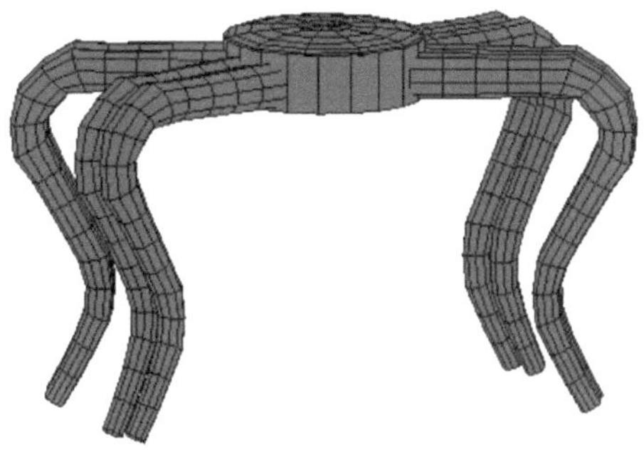

**Fig. 38.7** Quadruped configuration

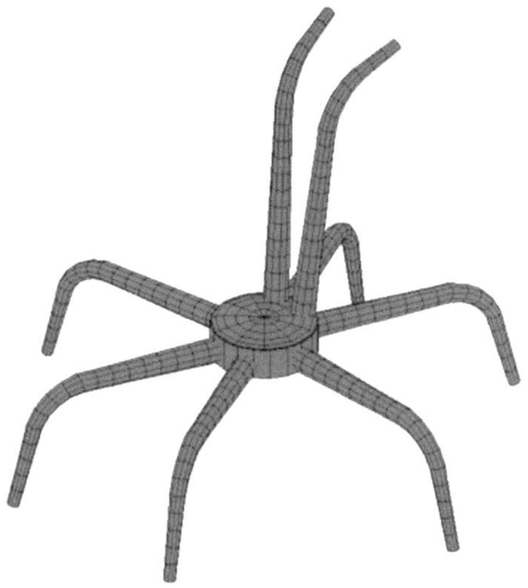

**Fig. 38.8** Hexapod configuration

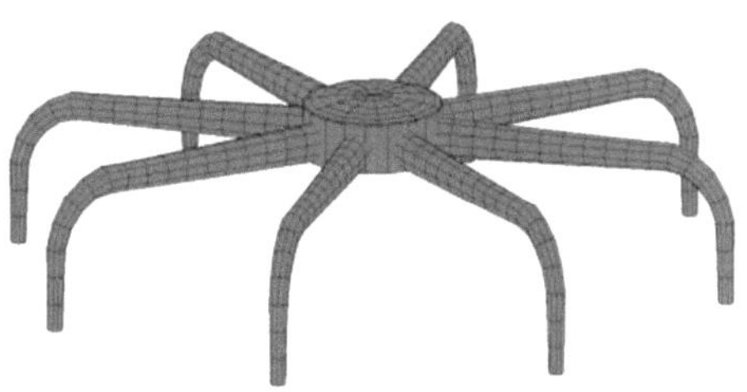

**Fig. 38.9** Octopod configuration

## *38.6.4 Octopod Configuration*

The octopod configuration is obtained by spreading the arms around the body evenly (Fig. 38.9). This configuration provides great mobility over rough terrains.

## 38.7 Conclusions

This paper applies the use of multiple continuum arms to a reconfigurable robotic system. Continuum arms are capable of elongating, shortening and bending at any point along their length. The use of continuum arms greatly enhances the flexibility of a reconfigurable robot and allows for diverse locomotion patterns. The presented system can be used to study morphology and control schemes for reconfigurable robotic systems based on continuum arms.

**Acknowledgments** This work was supported in part by the European Commission in the ICT-FET OCTOPUS Integrated Project, under grant agreement no. 231608.

## References

1. Dai JS, Zhang QX (2000) Metamorphic mechanisms and their configuration models. Chin J Mech Eng 13(3):212–218
2. Kutzer, M, Moses, MS, Brown, CY, Scheidt, DH, Chirikjian, GS et al (2010) Design of a new independently-mobile reconfigurable modular robot. In: IEEE international conference on robotics and automation, Anchorage, 2010, pp 2758–2764
3. Qi, Z, Wang, H, Huang, Z, Zhang, L (2009) Kinematics of a quadruped/biped reconfigurable walking robot with parallel leg mechanisms. In: 1st ASME/IFToMM international conference on reconfigurable mechanisms and robots, London, 2009, pp 558–564
4. Mather JA (1998) How do octopuses use their arms. J Comp Psycho 112(3):306–316
5. Walker, ID, Dawson, DM, Flash, T et al (2005) Continuum robot arms inspired by cephalopods. In: SPIE conference of unmanned ground vehicles, Orlando, 2005, pp 303–314
6. Chirikjian, GS (1993) A continuum approach to hyper-redundant manipulator dynamics. In: IEEE/RSJ international conference on intelligent robots and systems, Yokohama, 1993, pp 1059–1066
7. Yekutieli Y, Sagiv-Zohar R, Aharonov R, Engel Y, Hochner B, Flash T (2005) Dynamic model of the octopus arm. I. Biomechanics of the octopus arm reaching movement. J Neurophysiol 94:1443–1458
8. Giri, N, Walker, ID (2011) Three module lumped element model of a continuum arm section. In: IEEE/RSJ international conference on intelligent robots and systems, San Francisco, 2011, pp 4060–4065
9. Kang, RJ, Kazakidi, A, Guglielmino, E, Branson, D, Tsakiris, DP et al (2011) Dynamic model of a hyper-redundant, octopus-like manipulator for underwater applications. In: IEEE/RSJ international conference on intelligent robots and systems, San Francisco, 2011, pp 4054–4059
10. Branson, D, Kang, RJ, Guglielmino, E, Caldwell, D (2012) Control architecture for robots with continuum arms inspired by octopus *Vulgaris* neurophysiology. In: IEEE international conference on robotics and automation, St.Paul, (2012, in press)

# Chapter 39
# A Novel Reconfigurable Unit for High Dexterous Surgical Instrument

**Linan Zhang, Shuxin Wang, Jianmin Li, Xiaofei Wang, Chao He and Jinxing Qu**

**Abstract** Surgical instruments with multi-degree of freedom (DoFs) are widely used in single-port surgery and natural orifice transluminal endoscopic surgery. In this paper, a novel reconfigurable unit, which can be used as a basic component of a high dexterous surgical instrument (HDSI), was developed according to an isosceles trapezoid mechanism. Then the unit was improved and optimized to obtain a larger rotational range. The new unit can prevent driving cable from slackening during the moving. Two multi-joint prototypes, which are designed based on the proposed reconfigurable unit, are developed and experimented preliminarily. Experiment results show that the unit can avoid cable slack and is suitable to be used for HDSI.

**Keywords** Reconfigurable unit · Surgical instrument · Cable driving · Slack

L. Zhang · S. Wang (✉) · J. Li · X. Wang · C. He · J. Qu
Key Lab for Mechanism Theory and Equipment Design of Ministry of Education, Tianjin University, Tianjin, China
e-mail: shuxinw@tju.edu.cn

L. Zhang
e-mail: zla3452@tju.edu.cn

J. Li
e-mail: jianm.lee@gmail.com

X. Wang
e-mail: wangfifi@msn.com

C. He
e-mail: xiaoye0186@163.com

J. Qu
e-mail: tjuqujinxing@163.com

J. S. Dai et al. (eds.), *Advances in Reconfigurable Mechanisms and Robots I*,
DOI: 10.1007/978-1-4471-4141-9_39, © Springer-Verlag London 2012 

## 39.1 Introduction

High dexterous surgical instrument (HDSI) is widely applied in single-port surgery (SPS) and natural orifice transluminal endoscopic surgery (NOTES), which could be beneficial to reduce patient trauma, postoperative pain and recovery time [1–3]. To design a HDSI, reconfigurable unit, which is a cable driving modular joint, is often used as a basic element [4–6]. But cable slack is a key problem in the design of the unit [7, 8]. Cable in compression tend to buckle which would result in actuator backlash and low rigidity of the instrument [9].

Even ignoring cable's elastic elongation, cable slack still happen in some traditional units. Take a typical wrist as an example, as shown in Fig. 39.1a, the wrist is constructed by connecting several same traditional reconfigurable units end-to-end. Any two adjacent units can rotate around each other. As shown in Fig. 39.1b, $h$ and $d$ are the dimensional parameters of the traditional unit, $l_1$ and $l_2$ are the path lengths of driving cable located on each side of the unit, and $\theta$ is rotational angle of the unit. Then the total path length of driving cable in one unit can be obtained by

$$\begin{aligned} L &= l_1 + l_2 \\ &= \sqrt{(4h^2 - w^2)\cos^2(\theta/2) - 2hw\sin\theta + w^2} \\ &\quad + \sqrt{(4h^2 - w^2)\cos^2(\theta/2) + 2hw\sin\theta + w^2} \end{aligned} \tag{39.1}$$

According to Eq. (39.1), $L$ will change with rotational angel $\theta$. So the total path length of driving cable in a wrist changes with its motion. Therefore, cables in the wrist of Fig. 39.1a will go slack when the wrist moves.

Many research institutions have done their research on HDSIs which can prevent cable from slackening. Snake-like arm robot [10], flexible endoscopy instrument [11] and continuum manipulators [12] utilize elastic material, such as NiTi alloy, to avoid cable slack. However, comparing to an instrument which is based on module joint, these instruments needs more power to overcome the material deformation during moving. A Tensor arm [13] and an elephant's trunk manipulator [14] employ mechanism to compensate the path length change of driving cable, but it is too big and not properly for surgical application. In actuated joint section of endoluminal NOTES robotic system [15], path length change of cable is compensated by distributing cable crossly. However, the distributing way may increase friction between cable and instrument. Dexterous units are designed to keep path length unchangeable in some other robots [6, 8, 16, 17] but still under study. Only limited kinds of properly reconfigurable unit are developed, hence designing a new reconfigurable unit, which can prevent cable from slackening, is necessary to the design of HDSI.

In this paper, an isosceles trapezoid mechanism is used as a basic to design a reconfigurable unit for HDSI. The development of the unit is illustrated in Sect. 39.2. Section 39.3 introduces parameters optimization of the developed reconfigurable unit. Two prototypes of wrist based on the developed unit are presented in Sect. 39.4 to verify the availability of the unit. Conclusions are drawn in Sect. 39.5.

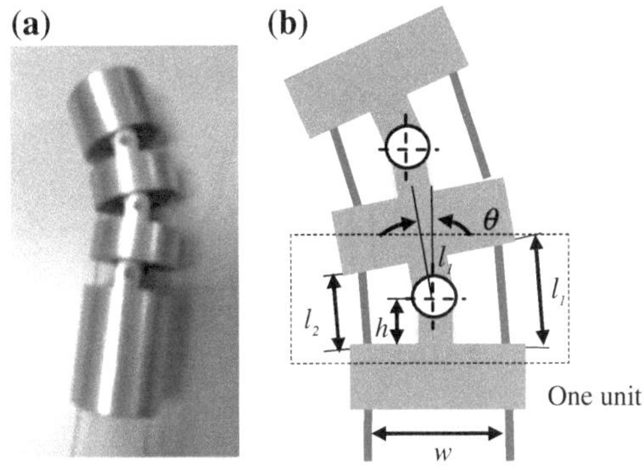


**Fig. 39.1** Wrist based on traditional unit. **a** Structure of the wrist. **b** Prototype of the wrist

## 39.2 Development

The isosceles trapezoid mechanism is constructed by links. As shown in Fig. 39.3a, it is composed of link *AB*, *CD*, *GH* and *MN*. The middle points of link *AB* and *CD, E* and *F*, are also the middle points of *MG* and *NH*, respectively. $AB = CD$, $MG = NH$, and $MN = GH$. So quadrilateral *NGMH* is constrained to be an isosceles trapezoid by link *MN* and *GH*. Two cables are distributed on both sides of the mechanism. Consequently, the two cables, links *AB* and *CD* construct an isosceles trapezoid *ABCD*. The total path length of the driven cables *L* is

$$L = BC + AD = 2EF \tag{39.2}$$

The length of *EF* changes with the movement of the mechanism, so the total path length of cable in the mechanism is not constant. In order to eliminate the length change of *EF*, link *GH* is replaced by link *EF*, as is shown in Fig. 39.2b. In addition, to extend rotational range, the unit of Fig. 39.2b is improved and the structure is shown in Fig. 39.2c. Link *MJ* is perpendicular to link *AB*, link *NK* is perpendicular to link *CD*.

Let $a$ be the length of link *AB* and *CD*, $b$ be the length of *EF*, $c$ be the length of *JK*, $d$ be the length of *ME* and *NF*, and both the distances from *M* to *J* and *N* to *K* are $e$. A well designed HDSI can keep two adjacent reconfigurable units coaxial when it is at initial position. So link *EF* is perpendicular to link *AB* and *CD* at initial position. The value of $c$ depends on $b$ and $d$ and it is

$$c = \sqrt{(b - 2e)^2 + 4d^2} \tag{39.3}$$

The equation of rotational angle $\theta$ is

$$\theta = 2\arctan\left(\left(-F \pm \sqrt{F^2 + E^2 - H^2}\right) \Big/ (H - E)\right) \tag{39.4}$$

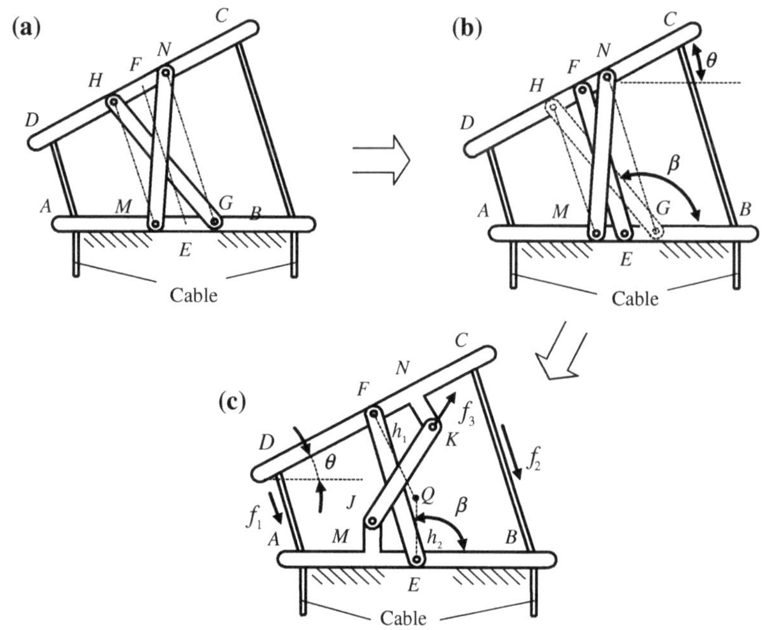


**Fig. 39.2** Development of reconfigurable unit. **a** isosceles trapezoid mechanism. **b** Reconfigurable unit

where $E = 2l^2 + 2bl\cos\beta$, $F = 2bl\sin\beta$, $H = b^2 + 2l^2 - c^2 + 2bl\cos\beta$, $l = \sqrt{d^2 + e^2}$. Then the total path length of cables in the unit is equal to the sum of length of $AD$ and $BC$.

$$L = \sqrt{A^2 + a^2/4 - Aa\cos B} + \sqrt{C^2 + a^2/4 - Ca\cos D} \tag{39.5}$$

where $A = \sqrt{a^2/4 + b^2 - ab\cos(\pi - \beta + \gamma)}$, $B = \pi/2 + \beta - \theta - \arcsin(a\sin(\theta - \beta)/(2A))$, $C = \sqrt{a^2/4 + b^2 - ab\cos(\beta - \gamma)}$, $D = \pi/2 - \beta + \theta - \arcsin(a\sin(\beta - \gamma)/(2C))$, $\gamma = \arctan e/d$.

The total path length changes of three units with rotational angle are shown in Fig. 39.3. The figure shows that the total path length of reconfigurable unit is almost constant within a certain rotational range, while the path lengths of the other two kinds of units change more quickly with rotational angle. Therefore, the developed reconfigurable unit is the best choice.

## 39.3 Parameters Optimization

For the developed reconfigurable unit, the total path length of cables is almost invariable within a certain rotational range. Large rotational range for each unit can increase the workspace and dexterity of the instrument, so it is necessary to

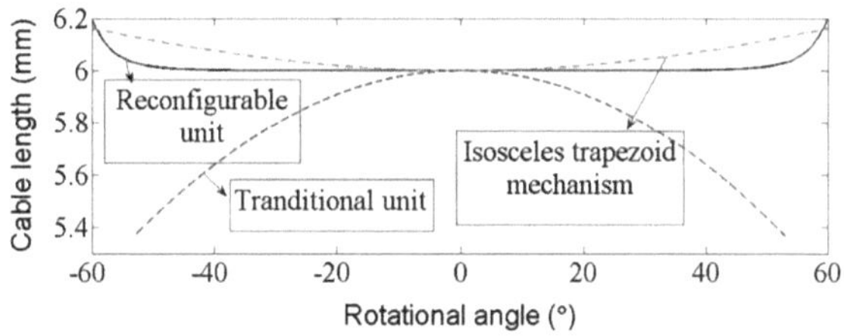


**Fig. 39.3** Total path length of cables change versus rotational angle ($a = 6$ mm, $b = 3$ mm, $d = 0.7$ mm, e = 0, $h = 1.5$ mm, $w = 6$ mm)

optimize the developed unit. During the procedure of optimization and improvement, the change of the total path length must be limited to an allowed extent. In generally, the elongation rate of cable is usually far less than 1/1000, so the allowed path change rate of cable is set as 1/1000 in this paper.

### *39.3.1 Parameter Analysis*

The rotational range of the designed reconfigurable unit varies with rotational range is

$$R = \theta_{\max} - \theta_{\min} \tag{39.6}$$

where $\theta_{\max}$ and $\theta_{\min}$ are the maximum and minimum value of $\theta$, respectively. $\theta$ can be easily calculated by Equation (39.4).

In actual application conditions, the smaller value of $a$ is, the less trauma patient would have. In this paper, reconfigurable unit is designed 7 mm in diameter. According to Eqs. (39.4) and (39.6), the rotational range $R$ is determined by $b$ and $d$. As shown in Fig. 39.4a, the rotational range $R$ increases with the increasing of $b$ and decreasing of $d$.

To maximize the rotational range of the unit, the variations of $\theta$ with the parameters $e$ and $d$ are optimized, as shown in Fig. 39.4a. The optimal rotational range is distributed along an arc line, so the value of $d$ can be chosen from a larger range comparing to the unit of Fig. 39.2b. When the value of $b$ changes, different arc lines of optimal rotational range can be obtained. Fig. 39.4b shows the optimized arc lines.

Large value of $b$ will result in low rigidity. With experience, the value of $b$ had better to be smaller than half of $a$. On the other hand, $b$ had better to be maximized to acquire a large rotational range. Therefore, the value of $b$ is set to be $b = a/2$. Furthermore, the larger the distance from point $E$ to $M$ ($H$ to $N$) is, the easier for manufacture and assembly. Keeping the value of $d$ and $e$ equal is a way to achieve this goal. Then $d$ and $e$ are determined according to the Fig. 39.8. The final optimized parameters are shown in Table 39.1. The values of $c$ in the units of Fig. 39.2b and c is determined by Eq. (39.3).

The curves of the total path length of cable versus rotational angle are shown in Fig. 39.5. The rotational range of primary unit in initial design is about 87°.

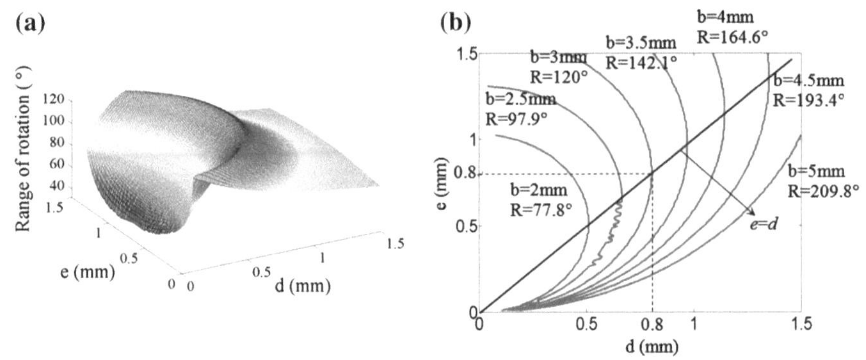


**Fig. 39.4** Parameters optimization. **a** Rotational range of the developed reconfigurable unit structure with the change of the value of $d$ and $e$ ($a = 6$ mm, $b = 3$ mm). **b** Optimal rotational range curves

**Table 39.1** The optimized parameters of the reconfigurable unit

| | $a$ (mm) | $b$ (mm) | $c$ (mm) | $d$ (mm) | $e$ (mm) |
|---|---|---|---|---|---|
| Reconfigurable unit | 6 | 3 | 2.126 | 0.8 | 0.8 |
| Primary reconfigurable unit | 6 | 3 | 3.4 | 0.8 | – |

Comparing to the initial design, the reconfigurable unit has larger rotational range, which is about 112°.

### 39.3.2 Statics Analysis

Actuator will exert force to each links of the reconfigurable unit when it is driven. As shown in Fig. 39.2c, link $MK$ is the weakest one of these links in engineering, so it need to decrease. $f_2$ and $f_1$ are driven force of cables. $f_3$ is the force exerted on link $MK$ and it can be obtained as

$$f_3 = \frac{d\sin(\pi - \beta)}{\cos(\alpha - \pi/2)}(f_2 - f_1) = k(f_2 - f_1) \tag{39.7}$$

where $k$ denotes influence coefficient of force $f_3$. $\beta$ and $\alpha$ can be calculated as

$$\beta = 2\arctan\left(\left(-N \pm \sqrt{N^2 + M^2 - P^2}\right)\Big/(P - M)\right) \tag{39.8}$$

$$\alpha = \arcsin((b\sin(\beta - \gamma) + l\sin\theta)/c) \tag{39.9}$$

where $M = -2bl - 2bl\cos\theta$, $N = 2bl\sin\theta$, $H = b^2 - c^2 + 2l^2 + 2l^2\cos\theta$.

The parameters of the two reconfigurable units are listed in Table 39.1. Comparing the influence coefficients of the two units, as shown in Fig. 39.6, it can be

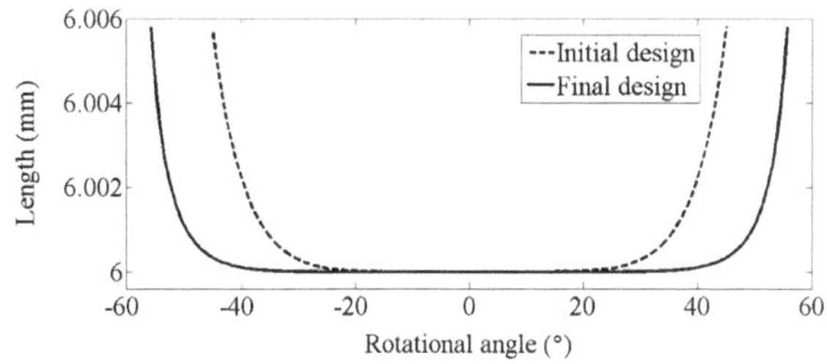


**Fig. 39.5** Total path length of reconfigurable unit versus rotational angle

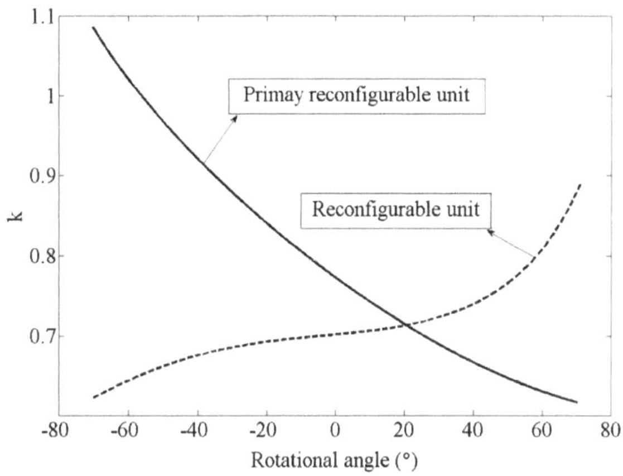


**Fig. 39.6** Influence coefficients of two reconfigurable units

seen that the influence coefficient of reconfigurable unit is smaller than that in primary reconfigurable unit, which demonstrates that the reconfigurable unit has better performance in statics.

## 39.4 Two Multi-Joint Prototypes Based on Reconfigurable Unit

### *39.4.1 Detailed Design*

According to the above analysis and optimization, a CAD model is developed, as shown in Fig. 39.7. Two ends of the middle link are installed on circular arc grooves of the end link and base link, respectively. Two constraint links, which are used to connect base link and end link, constraint end link to rotate around the base link. The end link is driven by cables which go through holes of the base link and the end link. These driving cables are not shown in the model for clarity.

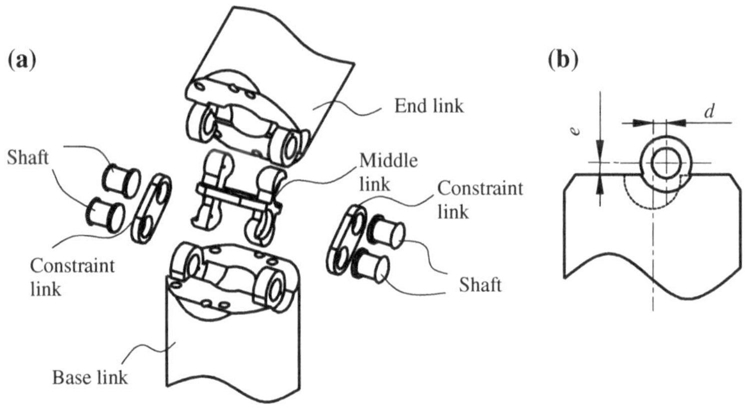


**Fig. 39.7** CAD model of the reconfigurable unit. **a** Reconfigurable unit. **b** Front view of the base link and end link

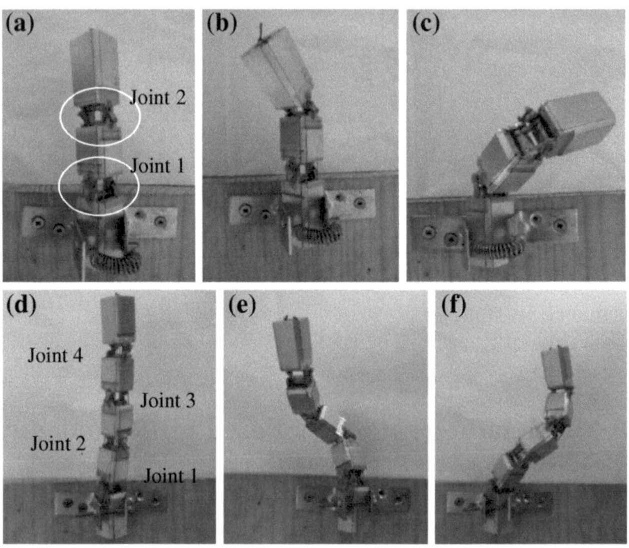


**Fig. 39.8** Flexible wrists constructed by reconfigurable unit. **a** Initial position of two-joint wrist. **b** Configuration when joint 2 is actuated. **c** Configuration when joint 1 is actuated. **d** Initial position of four-joint wrist. **e** Driving four-joint wrist to left side. **f** Driving four-joint wrist to right side

## *39.4.2 Prototypes and Preliminary Experiment*

According to the CAD model, a two-joint and four-joint wrists, as shown in Fig. 39.8, are constructed connecting the proposed reconfigurable units The size of the two wrists is magnified 5 times for easy manufacturing. Each joint is driven by two cables (0.6 mm in diameter). As shown in Figs. 39.8b and c, each joint in the wrist is driven separately once a time. In addition, four developed reconfigurable units construct a four-joint flexible wrist. By driving the cable, three positions of the wrist can be obtained, as shown in Figs. 39.8d, e, and f. The preliminary experiment results show the motion of joint $i$ ($i = 1, 2, 3, 4$) will only affect the rotations of joints from $(i + 1)$ to 4. For example, when only joint 2 is driven, joint 3 and 4 will rotate accordingly, while joint 1 won't. Both the two wrists can move dexterously and no cable slack occurs during the moving.

The novel reconfigurable unit developed in this paper plays an important role in HDSI. Firstly, the dexterity of instrument can be easily increased by increasing the number of reconfigurable units. Secondly, it is easy to manufacture, assemble and maintain the HDSI. Finally, the developed reconfigurable unit can keep path length of driving cable almost unchanged within a certain rotational range. The dexterous design may decrease the actuator backlash of HDSI and increase its rigidity. The motions of the two prototypes show that the novel reconfigurable unit is suitable to the design of HDSI.

## 39.5 Conclusion

In this paper, a novel reconfigurable unit, derived from an isosceles trapezoid mechanism, can keep total path length of driving cables constant within a certain rotational range. Under allowed elongation rate of cable, the rotational range is enlarged after optimizing the structure. Utilizing the reconfigurable unit, two prototypes, two-joint and four-joint flexible wrists, are developed to verify the usability of the reconfigurable for HDSI. The result shows that HDSIs with different DoFs can be easily constructed by using the reconfigurable unit, and these instruments will have high stiffness and low actuator backlash. The presented reconfigurable unit has a good perspective of application for HDSIs. The problem of coupled motion in these instruments needs to solve by control in future work.

**Acknowledgments** This work was supported by NSFC (National Science Foundation of China) under grant 50925520, The first author also likes to express his gratitude to Xiaogen Jiang from Birmingham University and Dongyang Xue from Tianjin University for their help.

## References

1. Bessler M, Stevens PD, Milone L et al (2007) Transvaginal laparoscopically assisted endoscopic cholecystectomy: a hybrid approach to natural orifice surgery. Gastrointest Endosc 66(6):1243–1245
2. Abbott, DJ, Becke, CR, Rothstein, I et al (2007) Design of an endoluminal NOTES robotic system. In: IEEE/RSJ international conference on intelligent robots and systems, IEEE Press, New York, 2007 pp 410–416

3. Low, SC, Tang, SWZ, Thant, M et al (2006) Master slave robotic system for therapeutic gastrointestinal endoscopic procedures. In: 28th annual international conference of the IEEE engineering in medicine and biology society, IEEE Press, New York, 2006, pp 3850–3853
4. Peris J, Reynaerts D, Van Brussel H (2001) A miniature manipulator for integration in a self-propelling endoscope. Sens Actuat A-Phys 92(1–3):343–349
5. Wei JQ, Akiko N, Michael H et al (2011) Multifunctional articulating surgical robot for NOTES. J Med Devices 5(2):1
6. Ma, RQ, Wu DM, Yan ZY et al (2010) Research and development of micro-instrument for laparoscopic minimally invasive surgical robotic system. In: 2010 IEEE international conference on robotics and biomimetic (ROBIO), IEEE Press, New York, 2010, pp 1223–1228
7. Noonan, DP, Vitiello, V, Shang, J et al (2011) A modular, mechatronic joint design for a flexible access platform for MIS. In: 2011 IEEE/RSJ international conference on intelligent robots and systems, IEEE Press, New York, 2011, pp 949–954
8. Aldrich, JB, Skelton, RE, Backlash-free motion control of robotic manipulators driven by tensegrity motor networks. In: Proceedings of the 45th IEEE conference on decision and control, IEEE Press, New York, 2006, pp 2300–2306
9. Camarillo DB, Milne CF, Carlson CR et al (2008) Mechanics moduleing of tendon-driven continuum manipulators. IEEE T Robot 24(6):1262–1273
10. Simaan N, Xu K, Wei W et al (2009) Design and integration of a telerobotic system for minimally invasive surgery of the throat. Int J Robot Res 28(9):1134–1153
11. Ott L, Nageotte F, Zanne P et al (2011) Robotic assistance to flexible endoscopy by physiological-motion tracking. IEEE T Robot 27(2):346–359
12. Bardou, B, Zanne, P, Nageotte, F et al (2010) Control of a multiple sections flexible endoscopic system. In: 2010 IEEE/RSJ international conference on intelligent robots and systems (IROS 2010), IEEE Press, New York, 2010, pp 2345–2350
13. Anderson, V, Horn, R (1967) Tensor arm manipulator design. ASME Trans 67-DE-57(2):1–12
14. Hannan MW, Walker ID (2003) Kinematics and the implementation of an elephant's trunk manipulator and other continuum style robots. J Robotic Syst 20(2):45–93
15. Abbott, DJ, Becke, C, Rothstein, RI et al (2007) Design of an endoluminal NOTES robotic system. In: 2007 IEEE/RSJ international conference on intelligent robots and systems, IEEE Press, New York, 2007, pp 410–416
16. Williams, MR, Creek, W (2011) Instrument wrist with cycloidal surfaces. U. S. patent, No. Us 2011/0152879 A1
17. Cooper, TG, Park, M (2008) Surgical instrument with parallel motion mechanism. U. S. patent, No. US 200810065102 A1

# Chapter 40
# Design and Kinematics Analysis of a Novel MR-Compatible Robot for Needle Insertion

**Chaochao Cheng, Shan Jiang, Jun Liu and Jinlong Lou**

**Abstract** The MR-compatible robot for needle insertion has attracted increasingly attentions due to high precision in minimally invasive surgery. A robotic system for prostate brachytherapy has been designed and is actuated by five ultrasonic motors to realize needle orientation and insertion. The application of cable transmission leads to a compact and dexterous mechanism, because all the motors are placed in the base and the moving part of the mechanism becomes as light and small as possible. What's more, placing the motors far away from the center of the scanner will eliminate artifacts and distortion of the MR images caused by the motors. The displacement analysis is carried out and the reachable workspace is obtained. Jacobian matrix is deduced in velocity mapping model for dimensional synthesis in further study. Configuration singularities are analyzed on the basis of Jacobian matrix.

**Keywords** Cable transmission · Ultrasonic motor · Kinematics · MRI

## 40.1 Introduction

Brachytherapy is considered to be an effective treatment for prostate cancer. Real-time imaging and adjusting needle orientation with robot assistance are two ways to improve the precision of the surgery. MRI can provide a clear image to

C. Cheng · S. Jiang (✉) · J. Lou
School of Mechanical Engineering, Center for Advanced Mechanisms and Robotics, Tianjin University, 300072 Tianjin, China
e-mail: shanjiang@tju.edu.cn

J. Liu
Department of Magnetic Resonance, Tianjin Union Medicine Centre, 300121 Tianjin, China
e-mail: cjr.liujun@vip.163.com

J. S. Dai et al. (eds.), *Advances in Reconfigurable Mechanisms and Robots I*,
DOI: 10.1007/978-1-4471-4141-9_40, 

distinguish tumor size, location and extension [1, 2]. Needle insertion with manipulator can achieve good performance by adjusting the needle orientation according to the tissue deformation timely. So MRI-guided robot is an ideal approach for prostate needle insertion surgeries [3]. Fischer et al. introduced a 4-DOF hybrid robot using MR-compatible cylinders for real-time control needle orientation under MRI guidance and performed insertion motion manually [4–6], Patriciua et al. reported a MR-compatible automatic robot for close radiotherapy driven by ultrasonic motor [7, 8]. To eliminate artifacts of the MR images caused by the motors, Larson et al. developed a robotic device to perform interventions in the breast [9]. The device is actuated by telescoping shafts with ultrasonic motors located more than 1 m far away from the breast.

In this paper, we report a description of the architecture of a novel robot system. Cable transmission applied in MRI-guided robot for the first time is introduced, which leads to a lighter and more compact mechanism. Displacement analysis is deduced and workspace is obtained. Velocity mapping model is established for Jacobian matrix, configuration singularities are analyzed at last.

## 40.2 Design of the Robot System

### *40.2.1 Requirements Analysis*

Different from conventional medical robot system, the MRI-guided needle insertion robot performs insertion surgery in high magnetic field and limited space. Firstly, most of MRI devices will generate a 0.5T–3T magnetic field which is the main limitation to a design of the robotic system. The material, actuators and sensors of the robotic system must be MR-compatible. The robotic system must not disturb magnetic field and worsen the quality of imaging. At the same time, MRI devices should not create a disturbance on performance of the robotic system to avoid hurting the patient and damaging robotic system itself.

The space in the MRI bore for prostate brachytherapy is another limitation. The confined size is shown in Fig. 40.1. The MRI bore diameter is approximately 0.6 m and the patient is placed on a sickbed in the supine position with legs spread and raised. In such limited space, the mechanism must be as compact as possible. In this paper, ultrasonic motors are used and cable transmission is adopted, which will be introduced in Sect. 40.2.3. It leads to a light, compact mechanism with the motors placed in the base.

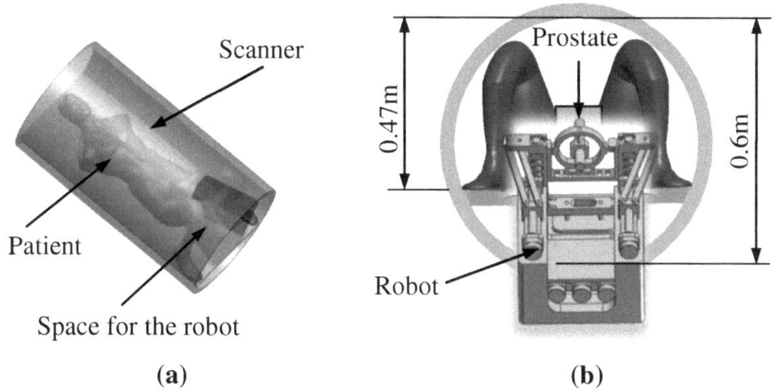


**Fig. 40.1** Space for prostate brachytherapy in MRI bore: **a** side view, **b** front view

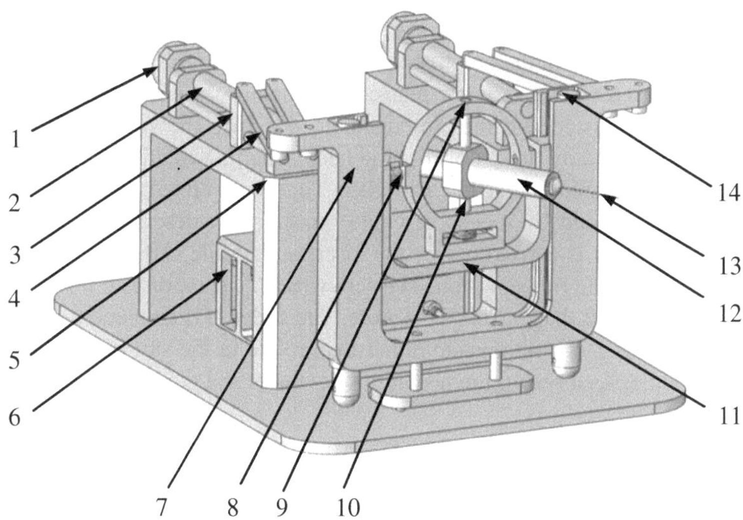


**Fig. 40.2** Detail of the prototype. 1-ultrasonic motor, 2-screw rod A, 3-nut platform, 4-link rods A, 5-base, 6-motors shelf, 7-moving platform, 8-the first rotating axis, 9-rotating head, 10-the second rotating axis, 11-lifting platform, 12-insertion module, 13-needle,14- screw rod B

## *40.2.2 Detailed Design of Prototype*

According to the requirements mentioned above, a novel prototype is designed. A 3D model of our prototype is shown in Fig. 40.2. The details of the motion are shown in Fig. 40.3.

The prototype mainly consists of four parts: the 2-dof parallel module (six-bar linkage), the vertical height coarse adjustment module, the 2-dof rotating head module and the insertion module. A detailed description of the prototype is given below.

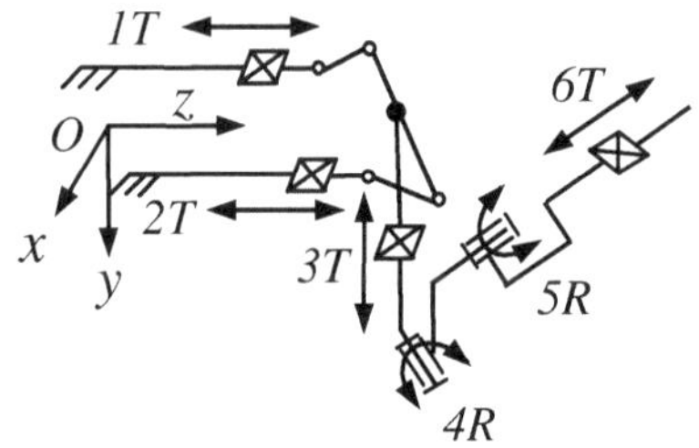


**Fig. 40.3** The equivalent kinematic diagram

As shown in Fig. 40.2, the ultrasonic motors on the base drive the screw rod A of the 2-dof parallel module, the moving platform is linked to the base by two symmetrical branched chains. The chain is composed of screw rod A, the nut platform and two parallel link rods A. The nut platform, the moving platform and link rods A form a parallelogram to ensure that the moving platform is parallel to the nut platform. The 2-dof parallel mechanism realizes rapid localization of the needle in horizontal plane, which is applied in MRI-guided robot for the first time.

The lifting platform in the height coarse adjustment module can move along the screw rod B fixed in the moving platform vertically. The height of the lifting platform is pre-adjusted and remains unchanged during the surgery. Rotation of the needle is adjusted by the 2-dof rotating head. The first rotation axis and second rotation axis are perpendicular to each other. The motors fixed in the motors shelf drive the rotating head by means of cable transmission described in Fig. 40.4b. The needle insertion is carried out by the insertion module shown in Fig. 40.4c. It is also driven by cable transmission. The screw rod *C*'s rotation is transformed into sliding of the slider B that carries the needle. The slider A is connected to the slider B by link rod B and it can translate rather than rotate in the slot of the base of the insertion module. As a result, the needle inserts under the guidance of slider A and slider B.

### *40.2.3 Cable Transmission*

In our prototype, the motors that drive the rotating head can't be fixed in the rotating head itself, so does the motor driving the needle insertion. Otherwise, the moving part will become too heavy for the motors fixed in the base to actuate. The rotating head and the insertion module should be as light and compact as possible to rotate and insert smoothly. But the ultrasonic motor is relatively large in size and heavy in weight. As a result, the ultrasonic motors are fixed in the base. Rotation and insertion are realized by cable transmission which is pioneering work in MRI-guided robot field.

Fig. 40.4a shows the structure of cable transmission in the motors shelf. For each dof, there should be two cables to transmit the bidirectional rotation, but only one cable is presented for simplicity. The shaft A used to circle the MRI

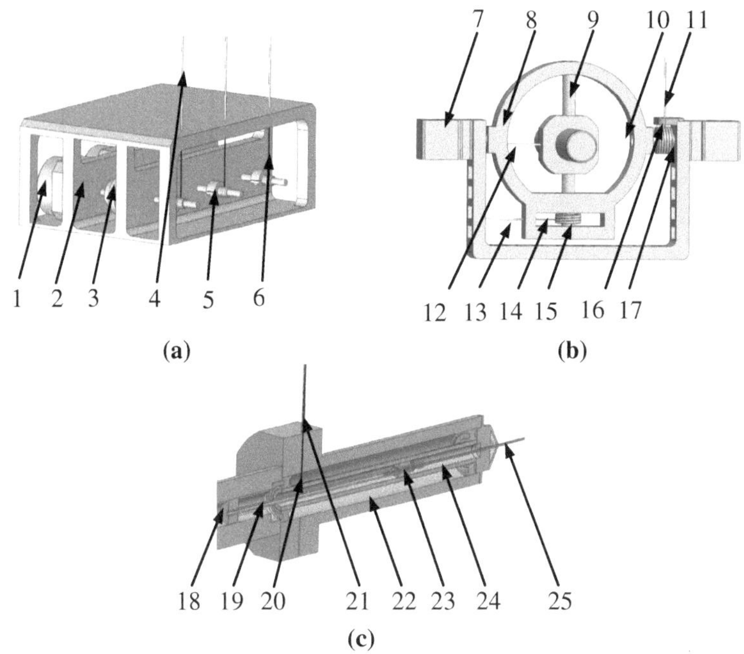


**Fig. 40.4** Detailed design of cable transmission: **a** motor shelf, **b** rotating head, **c** sectional drawing of needle insertion module. 1-ultrasonic motor, 2-motors shelf, 3-flexible coupling, 4-flexible cable sheath, 5-shaft A, 6-cable, 7-lifting platform, 8-rotating head, 9-the second rotation axis, 10-the first rotation axis, 11, 12, 13-cable sheath, 14, 16-cable, 15, 17-shaft B, 18-slider A, 19-link rod B, 20-cable, 21-cable sheath, 22-base of insertion module, 23-slider B, 24-screw rod C (shaft B), 25-needle

compatible cable is connected to the motors by flexible coupling. The same shaft B is connected to the rotating head shown in Fig. 40.4b and the insertion module shown in Fig. 40.4c. The cable is circled in shaft A on one side and shaft B on the other side. The cable is wrapped in the flexible cable sheath. The rotation of the motors can be transmitted to the end effecter by cable transmission precisely and the actual distance between shaft A and shaft B doesn't affect the transmission ratio theoretically, namely, the transmission ratio remains unchanged whether the flexible cable sheath is taut or loose.

Application of cable transmission has many advantages. The end effecter becomes lighter and more compact without load of the ultrasonic motors. In addition, the motors are permitted to be placed far away from the center of the MRI bore, which will lead to less damage to quality of imaging. The motors could even be placed outside of the borc if the precision of cable transmission is high enough. As a result, the range of choice of motors will be expanded from MR-compatible motors to all kinds of motors.

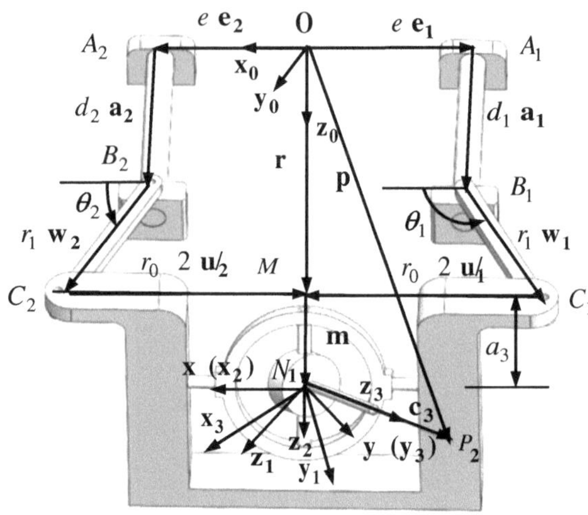


**Fig. 40.5** Coordinate systems of displacement analysis

## 40.3 Kinematics Analysis

### *40.3.1 Displacement*

Four coordinate systems are established as shown in Fig. 40.5. The $N - x_2y_2z_2$ coordinate system is obtained by rotating $N - x_1y_1z_1$ coordinate system around the first rotation axis of the rotating head by $\theta_4$, $R_1$ denotes the rotation matrix. The $N - x_3y_3z_3$ coordinate system is obtained by rotating $N - x_2y_2z_2$ coordinate system around the second rotation axis by $\theta_5$, $R_2$ denotes the rotation matrix. In the base coordinate system

$$\mathbf{p} = \mathbf{r} + \mathbf{m} + \mathbf{c} \tag{40.1}$$

where $\mathbf{c} = R\mathbf{c}_3$, $R = R_1R_2$, $\mathbf{c}_3 = [0, 0, a_5]^T$, $\mathbf{m} = [0, a_3, 0]^T$, $\mathbf{p} = [p_x, p_y, p_z]^T$, $d_i$ denotes the length of screw rod, $2e$ denotes the length between two motors in he base, $r_1$ denotes the length of link rod, $a_3$ denotes the height of rotating head, $r_0$ denotes the length of moving platform, $a_5$ denotes the length of needle.

$$a_5 = -(p_y - a_3)/\sin\theta_4\cos\theta_5 \tag{40.2}$$

As shown in Fig. 40.5, $\mathbf{a}_1 = [0, 1]^T$, $\mathbf{w}_1 = [\cos\theta_1, \sin\theta_1]^T$, $\mathbf{e}_1 = [0, 1]^T$, $\mathbf{u}_1 = [1, 0]^T$, $\mathbf{a}_2 = [0, -1]^T$, $\mathbf{w}_2 = [\cos\theta_2, \sin\theta_2]^T$, $\mathbf{e}_2 = [1, 0]^T$, $\mathbf{u}_2 = [-1, 0]^T$. We can derive the equation below according to Pythagorean Theorem,

$$d_1 = r_z - \sqrt{r_1^2 - [(r_0/2 - (e + r_x)]^2} \tag{40.3}$$

$$d_2 = r_z - \sqrt{r_1^2 - [(r_0/2 - (e - r_x)]^2} \tag{40.4}$$

$$\mathrm{r} = e\,\mathrm{e_i} + d_1\,\mathrm{a_i} + r_1\,\mathrm{w_i} + r_0\,\mathrm{u_i}\,/2 \tag{40.5}$$

Eq. (40.5) leads to

$$\mathrm{w_1} = \left[r_x - (r_0/2 - e), \sqrt{r_1^2 - [r_0/2 - e - r_x]^2}\right]^T / r_1 \tag{40.6}$$

$$\mathbf{w}_2 = \left[r_x + (r_0/2 - e), \sqrt{r_1^2 - [r_0/2 - e + r_x]^2}\right]^T / r_1 \tag{40.7}$$

### 40.3.2 Workspace

On the basis of displacement analysis, the boundary points of the workspace are determined by reduced dimension search and binary search. The reachable workspace is obtained as shown in Fig. 40.6a.

We can conclude from Fig. 40.6 that the shape of the reachable workspace of the robot is approximately a semi-ellipsoid (equatorial radius a = 0.3 m, b = 0.165 m, and polar radius c = 0.5 m), and its volume is 0.1038 $m^3$. The average size of the prostate is 0.05 × 0.035 × 0.04 m, and the volume of the prostate will expand by 25 percent under the impact of water-filled [10]. So the workspace satisfies requirements of the needle insertion surgery considering both the space in the MRI bore and the size of prostate. Given $\theta_4 = \pi/2$ and $\theta_5 = 0$, the reachable workspace is obtained shown in Fig. 40.6b and the section of the workspace is the same heart-shaped plane along y axis direction.

### 40.3.3 Velocity Mapping Model

As discussed in Sect. 40.3.1, differentiating Eq. (40.1) with respect to time yields

$$\dot{\mathbf{r}} = \dot{\mathbf{p}} - \dot{\mathbf{c}} = \dot{\mathbf{p}} - R\dot{\mathbf{c}}_3 - \dot{R}\mathbf{c}_3 \tag{40.8}$$

where $\dot{\mathbf{c}}_3 = \mathbf{v}_5 = [0, 0, v_5]^T$, $\dot{\mathbf{p}} = [vp_x, vp_y, vp_z]^T$, $\dot{\mathbf{r}} = [vr_x, 0, vr_z]^T$, we can get,

$$v_5 = -(vp_y + a_5\dot{\theta}_4 \cos\theta_4 \cos\theta_5 - a_5\dot{\theta}_5 \sin\theta_4 \sin\theta_5)/(\sin\theta_4 \cos\theta_5) \tag{40.9}$$

Differentiating Eq. (40.5) with respect to time yields, $\dot{\mathbf{r}} = \dot{d}_i\,\mathbf{a_i} + r_i\,\dot{\theta}_i\,\mathbf{Q}\,\mathbf{w_i}$ $(i = 1, 2)$, so we can get

$$\dot{d}_i = \frac{\mathbf{w_i}^T\dot{\mathbf{r}}}{\mathbf{w_i}^T\,\mathbf{a_i}} \tag{40.10}$$

Hence, the linear mapping relationship is established as follows,

$$\begin{bmatrix}\dot{d}_1 & \dot{d}_2 & \dot{\theta}_4 & \dot{\theta}_5 & v_5\end{bmatrix}^T = J \cdot \begin{bmatrix}vp_x & vp_y & vp_z & \dot{\theta}_4 & \dot{\theta}_5\end{bmatrix}^T \tag{40.11}$$

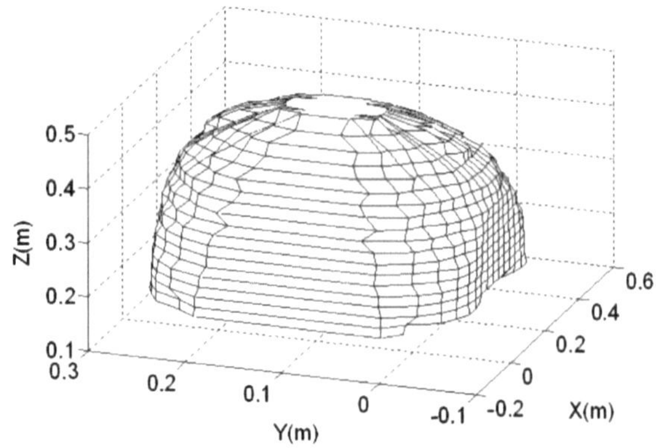


**(a)**

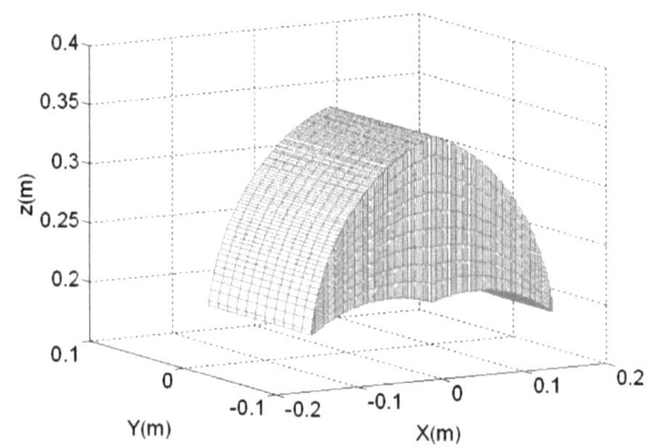


**(b)**

**Fig. 40.6** Workspace of the robot: **a** the reachable workspace, **b** the workspace when given $\theta_4 = \pi/2$ and $\theta_5 = 0$

$$J = \begin{bmatrix} J_{11} & J_{12} & 1 & J_{14} & J_{15} \\ J_{21} & J_{22} & 1 & J_{24} & J_{25} \\ 0 & 0 & 0 & 1 & 0 \\ 0 & 0 & 0 & 0 & 1 \\ 0 & J_{52} & 0 & J_{54} & J_{55} \end{bmatrix} \tag{40.12}$$

where J is the Jacobian matrix that establishes the linear mapping relationship between the velocity of the end effecter and the velocity of the driving links, $J_{11} = \frac{\cos\theta_1}{\sin\theta_1}$, $J_{12} = \frac{\sin\theta_1\cos\theta_4\cos\theta_5 + \cos\theta_1\sin\theta_5}{\sin\theta_1 s\theta_4 c\theta_5}$, $J_{14} = \frac{a_5\sin\theta_1 c\theta_5 + a_5\cos\theta_1 s\theta_5 c\theta_4}{\sin\theta_1 s\theta_4}$, $J_{15} = \frac{a_5\sin\theta_1 c\theta_5(c\theta_4 c\theta_5 - c\theta_4 s\theta_5) - a_5\cos\theta_1}{\sin\theta_1 c\theta_5}$, $J_{21} = \frac{\cos\theta_2}{\sin\theta_2}$, $J_{22} = \frac{\sin\theta_2 c\theta_4 c\theta_5 + \cos\theta_2 s\theta_5}{\sin\theta_2 s\theta_4 c\theta_5}$,

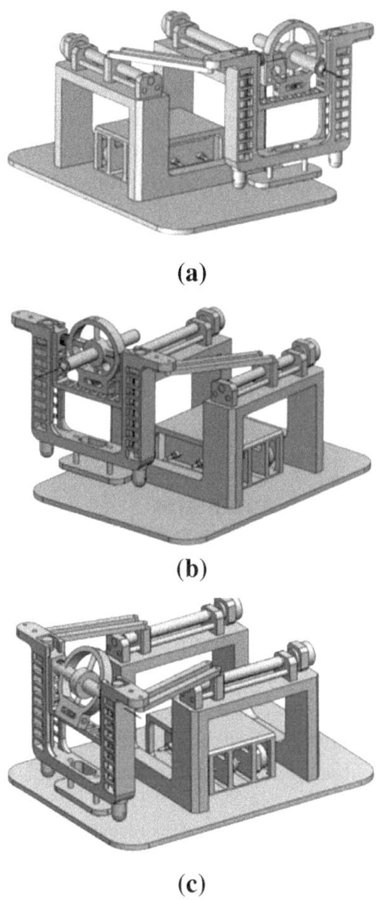


**Fig. 40.7** Singularity configuration of the prototype: **a** singularity configuration when $\sin\theta_1 = 0$, **b** singularity configuration when $\sin\theta_2 = 0$, **c** singularity configuration when $\cos\theta_5 = 0.11$

$$J_{24} = \frac{a_5 \sin\theta_2 c\theta_5 + a_5 \cos\theta_2 s\theta_5 c\theta_4}{\sin\theta_2 s\theta_4},\ J_{25} = \frac{a_5 \sin\theta_2 c\theta_5 (c\theta_4 c\theta_5 - c\theta_4 s\theta_5) - a_5 \cos\theta_2}{\sin\theta_2 c\theta_5},\ J_{52} = \frac{-1}{s\theta_4 c\theta_5},$$
$$J_{54} = \frac{-a_5 c\theta_4}{s\theta_4},\ J_{55} = \frac{a_5 s\theta_5}{c\theta_5}.$$

## *40.3.4 Configuration Singularities*

Configuration singularities study is very important for mechanical design and control. The algebraic meaning of singularities is that the determinant of the Jacobian matrix is zero or infinity. Singularity configuration is studied according to Jacobian matrix in this paper. They are shown in Fig. 40.7.

1. When $\sin\theta_1 = 0$, $\det(J) \to \infty$
   The right link rods coincide with the moving platform, the robot will lose the degree of freedom along the x axis direction.
2. When $\sin\theta_2 = 0$, $\det(J) \to \infty$
   The left link rods coincide with the moving platform, the robot will lose the degree of freedom along the x axis direction.
3. When $\cos\theta_5 = 0$, $\det(J) \to \infty$
   The needle coincides with the moving platform, the robot will lose the degree of freedom along the vertical direction.

The singularity configuration should be avoided by designing the parameters of the robot properly and restricting the rotation angles of the ultrasonic motors in the process of robot control.

## 40.4 Conclusion

A robot system of MR-compatible is designed to perform the prostate needle insertion surgery in limited space. Cable transmission is applied in MRI-guided robot for the first time, which leads to a compact, dexterous mechanism to perform needle orientation and insertion. Displacement analysis is deduced and the workspace calculated consequently satisfies the surgery requirements. Velocity mapping model is established for deducing the Jacobian matrix, on the basis of which configuration singularities are analyzed.

**Acknowledgment** This research is partly supported by National Science Foundation of China (Grant No. 51175373) and Education Program for New Century Excellent Talents (NCET-10-0625).

## References

1. Schoeppel SL, Ellis JH, LaVigne ML (1992) Magnetic resonance imaging during intracavitary gynecologic brachytherapy. Int J Radiat Oncol Biol Phys 23(1):169–174
2. Tardivon AA, Kinkel K, Lartigau E (1996) MR imaging during intracavitary brachytherapy of vaginal and cervical cancer: preliminary results. Radiographics. 16:1363–1370
3. Dai JS (2010) Surgical robotics and its development and progress. Robotica 28(161):161–161
4. Fischer GS, Iordachita II, DiMaio SP et al (2006) Design of a robot for transperineal prostate needle placement in MRI Scanner. In: IEEE international conference on mechatronics (ICM), pp 592–597
5. Fischer GS, DiMaio SP, Iordachita II, Fichtinger G (2007) Robotic assistant for transperineal prostate interventions in 3T closed MRI. MICCAI 10:425–433
6. Fischer GS, Iordachita I, Csoma C, Tokuda J, DiMaio SP, Tempany CM, Hata N, Fichtinger G (2008) MRI-Compatible pneumatic robot for transperineal prostate needle placement. IEEE/ASME Trans Mechatron 13(3):295–305

7. Patriciua A, Petrisor D, Muntener M, Mazilu D, Schar M, Stoianovici D (2007) Automatic brachytherapy seed placement under MRI guidance. IEEE Trans Biomed Eng 54(8):1499–1506
8. Muntenera M, Patriciua A, Petrisora D, Mazilua D, Baggaa H, Kavoussib L, Clearyc K, Stoianovicia D (2006) Magnetic resonance imaging compatible robotic system for fully automated brachytherapy seed placement. Urology 68(6):1313–1317
9. Larson BT, Erdman AG, Tsekos NV, Yacoub E, Tsekos PV, Koutlas IG (2004) Design of an MRI-compatible robotic stereotactic device for minimally invasive interventions in the breast. J Biomech Eng 126(4):458–465
10. Wallner K, Blasko J, Dattoli M (2001) Prostate brachytherapy made complicated, 2nd edn. Smart Medicine Press, Seattle

# Chapter 41
# A Flexible-Waist Quadruped Robot Imitating Infant Crawl

**Cheng Liu, Xiuli Zhang, Dongdong Li and Kunling Zhou**

**Abstract** Through investigating a 1-year-old human baby's body and crawl movement, an infant-sized quadruped robot "Babybot" is designed. The robot has a mechanism of two passive DOFs in its waist, namely the flexible waist. The biologically-inspired Central Pattern Generator (CPG) is employed for controlling Babybot's locomotion. The results of dynamic simulation show Babybot can crawl in a natural infant-like way and the flexible waist can improve the robot's crawling performance.

**Keywords** Quadruped robot · Infant crawl · Flexible waist · Rhythmic motion · Central pattern generator

## 41.1 Introduction

Infancy is an important stage in the growth of an individual. Before baby achieves bipedal walking, crawling movement plays a key role for infant locomotion. From the viewpoint of human evolution, upright walking human also evolved from the four-legged crawling animals. So study in infant crawl is beneficial for understanding both the development of human growth and evolution.

To study infant's development on cognitive and motion ability, there are some existing infant-like humanoid robots [1]: The "Infanoid" [2] has been developed mainly for imitating the infants social communication. "Pneuborn-7II" [3] puts emphasis on the development of crawling behavior through a pneumatic

C. Liu · X. Zhang (✉) · D. Li · K. Zhou
Beijing Jiaotong University, Beijing 100044, China
e-mail: zhangxl@bjtu.edu.cn

J. S. Dai et al. (eds.), *Advances in Reconfigurable Mechanisms and Robots I*,
DOI: 10.1007/978-1-4471-4141-9_41, © Springer-Verlag London 2012

musculoskeletal system. “iCub” [4], developed as a robotic platform for infant study, also has the ability of crawl. However, compared to the normal infant crawl, their crawling movements are not very flexible. Moreover, due to the large number of DOFs, these robots need to integrate complex mechanical structure and motion control system.

From the viewpoint of kinematics, infant’s crawling behavior is an unique motion that human moves in a quadruped gait with both hands and legs touching down on the ground. It means a bionic crawling infant robot could also be designed as a quadruped robot. Quadruped robots, usually inspired from animals, have much better mobility and stability than humanoid robot [5].

In this paper, we design a 10-DOFs quadruped robot as the foundational platform and improve it with a humanoid flexible-waist mechanism. We also use the Central Pattern Generator (CPG) to produce rhythmic crawling motion signals. With features from both quadruped and humanoid robot, the Babybot can crawl in a natural infant-like way.

## 41.2 Bionic Analysis

The infant we choose to study at is a 1-year-old boy. At that age, most infants have already been proficient in crawl. Under our model, an infant body can be divided into seven parts: head, main body, waist, two arms and two legs, which is very similar to the normal quadruped robot. The key point here is to understand the role of waist and infant baby’s special crawling gait.

### *41.2.1 Role of Human Waist*

The waist part plays an important role in crawling performance. On the basis of anatomy, human waist and hip have a structure as shown in Fig. 41.1.

During crawl, the skeletons of infant waist have a periodic rolling and yawing movements accompanying with the limbs’ pitching [8]. The waist can increase the motion range of infant’s hind body so that the legs propel body forward more efficiently [9]. Moreover, waist, with its compliant multi-DOF structure, also has the buffer effect when infant’s limbs touch down on the ground.

The important roles of waist have been largely ignored in former quadruped robotic studies. Compared to the multi-DOFs waist of humanoid robot [10], most quadruped robots have only rigid body which has no waist joint [11–13]. Researchers use more DOFs in legs to compensate the function of waist. However, it increases the complexity of robots. Moreover, when the speed increases or the robot fast turns around, the rigidness of trunk will cause uncoordinated motion.

There were several researchers noticing this and employing waist joint in their quadruped robots. “ELIRO-I“ [14] proves the waist-jointed walking robot is more

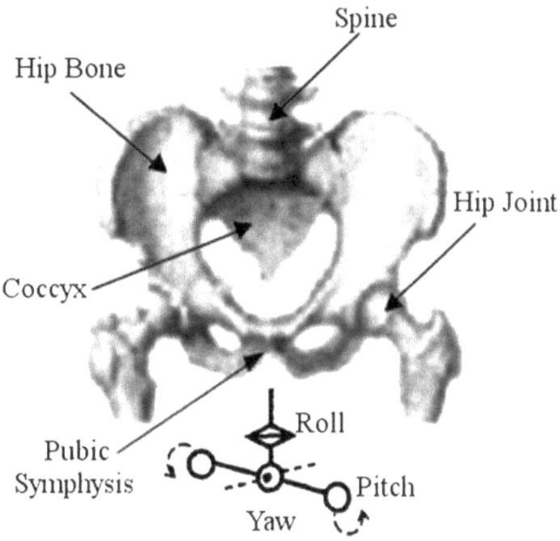


**Fig. 41.1** Skeleton structure and DOFs of human waist Waist is the part of the abdomen between the rib cage and legs. Besides the spine, two hipbones are combined by a cartilage, which is named pubic symphysis [6]. With this kind of structure, human waist can move in three directions: pitch, roll and yaw, providing human body with greater mobility and flexibility [7]

stable than a single rigid body robot as the waist could provide a larger Gait Stability Margin. Utku and Saranl's legged robot [15] also achieves a higher locomotion speeds and hopping height due to larger possible stride lengths resulting from a jointed spine's flexion. Researchers from Kyoto University and Doshisha University [16] found that changing the roll joint stiffness of their robot's body can even influence the robot's gait patterns. Their results show the potential of flexible waist for improving the quadruped robot's performance.

### *41.2.2 Infant Crawl Gait*

An infant's crawling behavior changes continuously along with its growth. Besides the crawling speed and efficiency, the crawling posture also changes with inter-related changes in infant's brains, bodies and motor experience [17].

The standard infant crawling gait is a kind of diagonal gait pattern in a sequence as left front limb and right hind limb, then right front limb and left hind limb, very similar to quadruped animals' walk. But a big difference here is that infant usually crawls on its knees, so the shanks tend to rely on the ground and have no propel movement.

Adolph et al. [18] had a detailed study on infant's crawling gaits. Generally, infants do not crawl in a perfect diagonal gait pattern. They lifted and placed their front limbs a few ahead than their hind limbs. This gait is shown in Fig. 41.2.

## 41.3 Robot Design

Unlike normal quadruped robot, our Babybot only uses six active DOFs locating on the left and right shoulders, elbows and hips to realize the crawling movement. The other four DOFs on knees and waist are designed as passive DOFs. This

**Fig. 41.2** Gait footprint of standard infant crawl

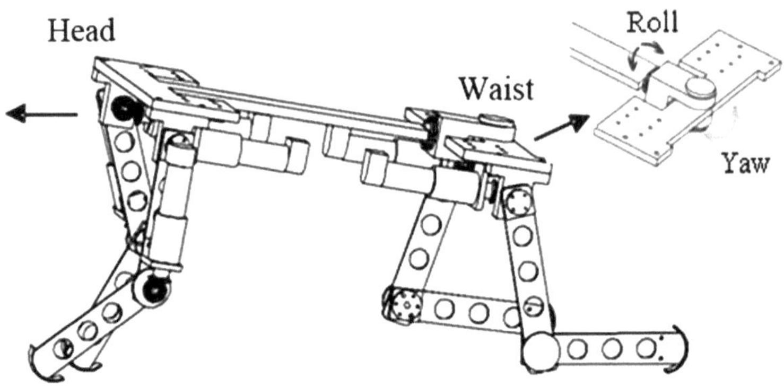


**Fig. 41.3** CAD model of the Babybot

design method reduces the mechanical complexity and makes the robot easier to control. All the joints on limbs rotate in pitch direction specifically. Six active DOFs are driven by the CPG signals. All passive DOFs are constrained by torsion springs along their axles.

### *41.3.1 Mechanism of Robots Waist*

Analysis in Sect. 41.2.1 shows that waist plays a very important role in crawling movement. To make the robot crawl more like an infant, a flexible waist is necessary

We design the waist with two DOFs respectively rotating in roll and yaw directions. Since infant moves forward mainly via its limbs, waist is rarely used to propel body forward. So we do not put actuators on the two rotation joints. Instead, two torsion springs are installed around the rotating axles to limit the motion and act as buffers. Figures 41.3 and 41.4 show the basic model of robot and its waist joint.

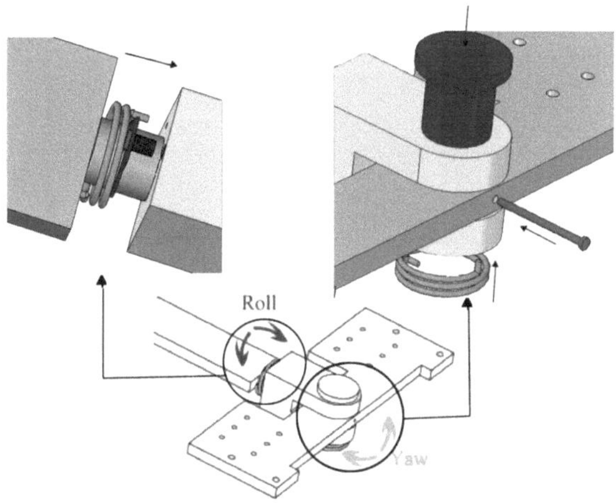


**Fig. 41.4** Mechanism of the flexible-waist

### *41.3.2 CPG Model for Imitating Infant Crawl*

Central Pattern Generator (CPG) is the neural networks located in the spine of vertebrates. It can automatically generate control signals for coordinating muscles' activities in a periodic phase [19, 20]. CPG models have been proved successful for applications in robotics [21].

We build a CPG model as Fig. 41.5 for controlling Babybot's crawling gaits.

We use sine function to model neural oscillator, so the CPG signals are represented as Formula (41.1). The configurations of parameters are presented in Table 41.1.

$$\begin{gathered}\theta_{ij}(t) = A_j \sin(2\pi t/T + \Phi_{ij}) \\ \theta_{iE}(t) = A_E \sin(2\pi t/T + \Phi_{iS} + \Phi_E) \\ i = L, R \\ j = S, H\end{gathered} \tag{41.1}$$

## 41.4 Dynamic Simulation

The Babybot's crawling simulation is operated in the Automatic Dynamic Analysis of Mechanical Systems (ADAMS) software. The total weight of Babybot is about 9.7 Kg, similar to a 1-year-old baby. The feet and the knees are set as rubber for compliant contacting with the ground. The simulation results (Fig. 41.6) show that Babybot successfully realizes the infant-like crawl.

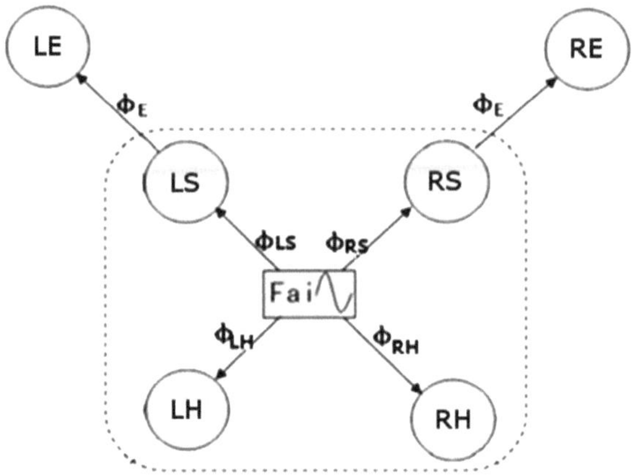


**Fig. 41.5** CPG model of robot six neural oscillators are coupled together with different phases. Each neural oscillator generates a periodic and continuous signal used for angle control. The left front shoulder is the reference. The right shoulder and two hip joints have phase difference $\Phi_{ij}$ with the reference. Each elbow joint has a fixed phase difference $\Phi_E$ with the shoulder on the same limb (*L* left, *R* right, *S* shoulder, *E* elbow, *H* hip)

**Table 41.1** CPG configurations

| Parameters | Meaning | Values |
|---|---|---|
| $\theta$ | Joint angle | Variable |
| T | Oscillating period | 2*s* |
| $A_S$ | Amplitude of shoulder joints | 20° |
| $A_H$ | Amplitude of hip joints | 15° |
| $A_E$ | Amplitude of elbow joints | 15° |
| $\Phi_{LS}$ | Phase of left shoulder | 0 |
| $\Phi_{RS}$ | Phase of right shoulder | $-\pi$ |
| $\Phi_{LH}$ | Phase of left hip | $-5\pi/4$ |
| $\Phi_{RH}$ | Phase of right hip | $-\pi/4$ |
| $\Phi_E$ | Phase difference between the elbow and shoulder | $\pi/2$ |

### *41.4.1 Comparative Study for the Flexible Waist*

To evaluate the flexible waist's effects on Babybot's crawling performance, we conducted comparative experiments between the robot with and without the flexible waist. From 0.5 to 8.5 s, when Babybot crawls steadily during four cycles, the robot's swing amplitude, the forward velocity, acceleration and the inner force applied to the waist joint are sampled for analysis. Figure 41.7 shows the comparisons and below are the analysis results.

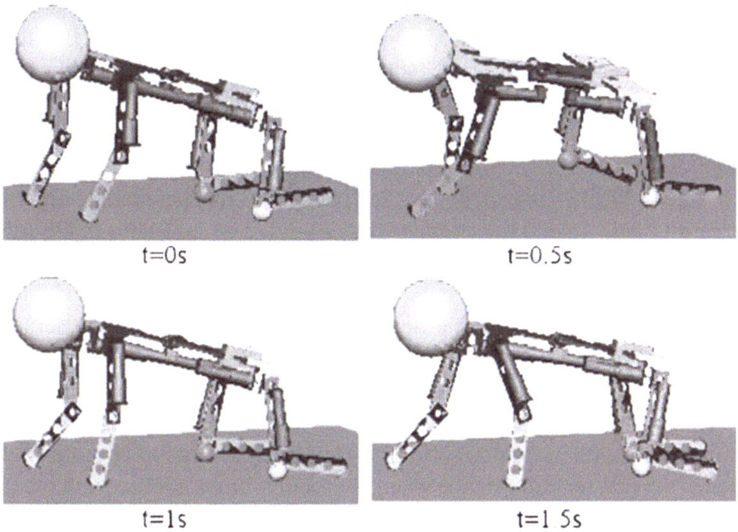


Fig. 41.6 Key frames from robot's crawling video

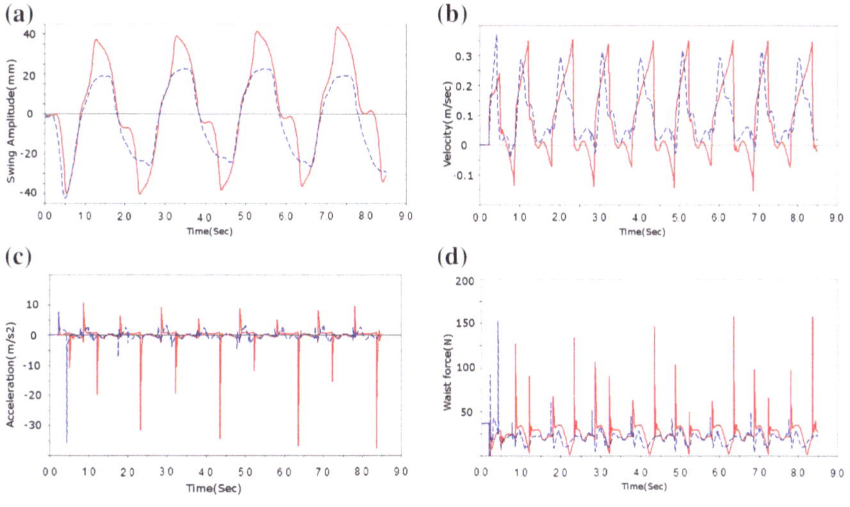


**Fig. 41.7** Crawling comparison results—Rigid waist robot blue—Flexible waist robot. **a** Swing amplitude. **b** Velocity. **c** Acceleration. **d** Force on waist joint

1. The flexible-waist robot swings in a smaller amplitude and more smoothly than the rigid-waist robot. It shows the flexible waist can reduce the useless sway of robot and make robot crawl more steadily and coordinated.

2. During four cycles, Babybot's velocity is changing periodically. The robot moves forward in most time, shown by the points above zero. But the robot moves backward in short time when the points below zero. This unnatural backward sliding impedes the robot's crawling velocity and cause abrasion in robot's hands and knees. Analyzing mathematically, the number of the points that below zero is 29.9 % of all points and the average velocity is 0.095 m/s for rigid waist; While change to the flexible waist, the percentage significantly decrease to 3.48 % and the average velocity increase a little to 0.098 m/s. This indicates that the flexible waist can decrease robot's disadvantageous backward moving.
3. When Babybot crawls with the flexible waist, both curves of acceleration and waist's inner force are smoother, with less and smaller peaks than rigid-waist crawling. In average, the inner force is 20.39 N with flexible waist while the number is 27.19 N with rigid waist, decreasing in 25 %. This proves that the flexible waist, with its torsion springs, can absorb reacting forces more effectively and provide greater compliance when the robot's limbs contact with the ground.

## 41.5 Conclusion

In this paper, we design a robot, "Babybot," based primarily on human infant. It has a quadruped-like DOF configuration, combining with a 2-DOF waist which is usually used in humanoid robot. We build a CPG model to produce position signals for robot's active joints. Thereby, the robot performs an infant-like crawling movement in dynamic simulation experiments.

Moreover, we conducted comparative experiments to evaluate the flexible waist's effects on Babybot's crawling performance. The results show that the bionic flexible waist can improve the robot's crawling performance with a smaller range of yaw swing, a smoother acceleration, a higher velocity with less draw backs and a gentler internal force. This kind mechanism of flexible waist can be a good approach for improving quadruped robot's performance in related areas.

**Acknowledgments** This work was supported by the National Natural Science Foundation of China (No. 50905012), and the Fundamental Research Funds for the Central Universities (No. 2009JBM089).

## References

1. Robot babies: cute or creepy? http://spectrum.ieee.org/slideshow/robotics/humanoids/robot-babies-cute-or-creepy/.. Cited 20 July 2011
2. Kozima H, Yano H (2001) A robot that learns to communicate with human caregivers. In: Proceedings of international workshop on epigenetic robotics, pp 47–52

3. Narioka K, Hosoda K (2011) Motor development of an pneumatic musculoskeletal infant robot. In: Proceedings of international conference on robotics and automation, pp 963–968, Shanghai, China
4. Righetti L, Ijspeert AJ (2006) Design methodologies for central pattern generators: an application to crawling humanoids. In: Proceedings of robotics: science and systems, pp 191–198
5. Hirose S, Kato K (2000) Study on quadruped walking robot in Tokyo Institute of Technology-past, present and future. In: Proceedings of IEEE international conference on robotics and automation, vol 1, pp 414–419, San Francisco, USA
6. Ogura Y, Aikawa H et al (2004) Development of a human-like walking robot having two 7-DOF legs and a 2-DOF waist. In: Proceedings of IEEE international conference on robotics and automation, pp 134–139, New Orleans, USA
7. Ogura Y, Aikawa H et al (2006) Development of a new humanoid robot WABIAN-2. In: Proceedings of IEEE international conference on robotics and automation, pp 76–81, Orlando, USA
8. BBCs science and educational film. First Steps. The Human Body, EP3
9. Li Y-J, Zhao T-J et al (2005) Development of humanoid robot and study of its waist mechanism. Mech Sci Technol 24(4):400–403 (in Chinese)
10. Xu W, Huang Q et al (2010) Mechanical design and balance control of a humanoid waist joint. In: Proceedings of IEEE international conference on automation and logistics, pp 249–254, Hong Kong and Macau, China
11. Raibert M, Blankespoor K et al (2008) Bigdog, the rough-terrain quaduped robot. In: Proceedings of international federation of automation control, pp 10822–10825, Seoul, Korea
12. Arikawa K, Hirose S (1996) Development of quadruped walking robot titan-viii. In: Proceedings of international conference on intelligent robots and systems, vol 1, pp 208–214, Osaka, Japan
13. Buehler M, Battaglia R et al (1998) SCOUT: a simple quadruped that walks, climbs, and runs. In: Proceedings of IEEE international conference on robotics and automation, vol 2, pp 1707–1712, Leuven, Belgium
14. Park SH, Kim DS, Lee Y-J (2005) Discontinuous spinning gait of a quadruped walking robot with waist-joint. In: Proceedings of international conference on intelligent robots and systems, pp 2744–2749, Edmonton, Canada
15. Culha U, Saranl U (2011) Quadrupedal bounding with an actuated spinal joint. In: Proceedings of international conference on robotics and automation, pp 1392–1397, Shanghai, China
16. Aoi S, Yamashita T et al (2010) Hysteresis in gait transition induced by changing waist joint stiffness of a quadruped robot driven by nonlinear oscillators with phase resetting. In: Proceedings of international conference on intelligent robots and systems, pp 1915–1920, Taipei
17. Tang S, Zhang Y (2009) Synopsis of the development of infant crawling movement. Contemp Preprimary Education 1:37–38 (in Chinese)
18. Adolph KE, Vereijken B, Denny MA (1998) Learning to crawl. Child Dev 69(5):1299–1312
19. Grillner S (1985) Neurobiological bases of rhythmic motor acts in vertebrates. Science 228(4696):143–149
20. Ijspeert AJ (2008) Central pattern generators for locomotion control in animals and robots: a review. Neural Netw 21(4):642–653
21. Zheng H, Zhang X, Li T, Duan G (2003) CPG-based methods for motion control of robot. High Technol Lett 7:64–68 (in Chinese)

# Chapter 42
# Design and Analysis of a Portable Reconfigurable Minimally Invasive Surgical Robot

**Chao He, Shuxin Wang, Xiaofei Wang, Anlin Zhang and Dongchun Liu**

**Abstract** Minimally invasive robotic surgery has many advantages over traditional open surgery and laparoscopic surgery, though current minimally invasive surgical (MIS) robots have significant drawbacks, including large volume and weight. This paper presents a new arrangement of a multi-degrees of freedom (DOF) MIS robot to maintain portability while incorporating a reconfigurable structure to achieve improved output force and stiffness. The performance and design of the reconfigurable structure are also optimized. Performance analysis shows this new reconfigurable robot can achieve a sufficiently large workspace and output forces for MIS.

**Keywords** Surgical robot · Reconfigurable structure · Minimally invasive surgical instrument · Force

## 42.1 Introduction

Compared to traditional surgical technologies such as open surgery and laparoscopic surgery, minimally invasive robotic surgery has many unique advantages in abdominal surgery, such as improved operation quality, greater operational comfort, shortened surgeon training time, reduced surgical trauma, and improved adaptability for complicated surgeries. These advantages have allowed minimally invasive surgical (MIS) robots to become widespread over the past decade.

Plenty of MIS robots have been developed in the field of robotics research including the Zeus robot [1], RTW robot [2], RobIn Heart (RIH) [3], da Vinci

C. He (✉) · S. Wang · X. Wang · A. Zhang · D. Liu
Key Lab for Mechanism Theory and Equipment Design,
Ministry of Education, Tianjin University, No. 92, Weijin Road, Tianjin, China
e-mail: xiaoye0186@163.com

J. S. Dai et al. (eds.), *Advances in Reconfigurable Mechanisms and Robots I*,
DOI: 10.1007/978-1-4471-4141-9_42, © Springer-Verlag London 2012 

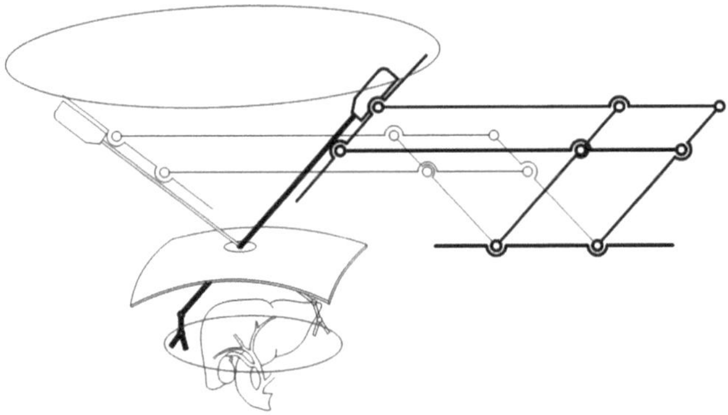

**Fig. 42.1** MIS robot containing an arm part and an instrument part

system [4] and the MicroHand A robot [5]. Due to the constraint at the incision point on the abdomen of the patient, these robots can be divided into two parts as Fig. 42.1 shows. The first part is apositioning arm outside the body. The second part is a multiple degrees of freedoms (DOFs) instrument used for dexterity enhancement of the whole robot.

In this kind of MIS robot, the robot arm must be arranged far away from the incision point for the purpose of avoiding conflict between the robot arm and the patient. Meanwhile, the small distance between the incision point and the operation target area in the MIS usually requires a large rotational range of motion around the incision point. These characteristics ultimately lead to a layout with significant size and weight. Furthermore, the big and heavy active arm requires a powerful passive positioning system which further increases the size and weight of the whole robot. Take the da Vinci robot for example, the height is 1.8 m and the weight is 533 kg [4]. This disadvantage of this kind of configuration has been pointed out in the literature, and much effort has been devoted to solving this problem.

One solution is adding degrees of freedoms of the instrument to endow the instrument with positioning ability to achieve a portable design. In [6], a dexterous instrument which can mimic the motion of a human arm is presented. In this robot, due to extra positioning ability of the instrument, the arm can be designed to be much smaller. However, in this kind of design, the increased moment arm between the instrument tip and the driven joint can cause a loss in the operating force of the instrument. Because of this, the diameter of this instrument is increased to 15 mm, which is not suitable for a MIS application.

Numerous snake-like robots have been developed to decrease the volume and weight of surgical robots. A lightweight snake-like robot for minimally invasive telesurgery is reported in [7]. The CardioArm robot [8] was developed for heart surgery. A Miniature Snake-Like robot is presented in [9]. Compared to traditional MIS robots, these snake-like robots and MIS devices have a significant improvement in volume and weight. However, one core weakness of the snake-like robots is their low output force. This disadvantage further leads to low performance in stiffness and accuracy [10].

This paper uses a reconfigurable structure to achieve a portable design and to improve the output force and stiffness of the robot. The remaining sections of this paper are organized as follows. In Sect. 42.2, the development of the robot will be presented. In Sect. 42.3, the performance of the robot will be analyzed. In Sect. 42.4, kinematics analysis of the robot will be discussed. In Sect. 42.5, simulation and validation will be given. In Sect. 42.6, conclusions to our study will be presented.

## 42.2 Description of the Robot

### *42.2.1 Overview of the Robot*

The presented robot system is a master–slave robotic system which contains a master unit and a slave unit. Similar to other typical master–slave robotic systems, during the operation the surgeon guides the master robot at the master unit side and the slave robot at the patient side mimics all movements of the master robot. This paper focuses on the design and analysis of a portable and lightweight slave robot for MIS.

As shown in Fig. 42.2, the slave unit is composed of a standard operating table and two slave arms. These two slave arms are kinematically identical to each other but have symmetrical structures. Each slave arm consists of two main parts: a 7-DOF passive arm and a (6 + 1)-DOF active robot.

The 7-DOF passive arm is used for passive positioning and preoperational adjustment of the active robot. It also provides enhancement of reachable workspace since the workspace of the active robot is relatively limited. The base of the passive arm is a locking mechanism which can be fixed on the operating table. It can also slide along the guide to provide a translational DOF. The main part of the passive arm is a lockable 6-DOF mechanism which contains two ball joints at both ends. Turning the handle can lock both ball joints at the same time. The length between the two ball joints is set to 210 mm to provide enough adjustment workspace for the active robot. Benefiting from the purely mechanical structure design, the passive arm is sterilizable and electrically safe.

During the operation all the joints of the passive arm are locked and the (6 + 1)-DOF active robot provides all the active movements.

### *42.2.2 The (6 + 1)-DOF Active Robot*

Figure 42.3 shows the scheme of the (6 + 1)-DOF active robot. The active robot can be divided to two parts. The first part is a 2-DOF driving part outside the abdomen which contains a translation DOF T1 and a rotation DOF R2. The second part is a (4 + 1)-DOF reconfigurable instrument inside the abdominal cavity which contains rotation DOFs R3, R4, R5, R6 and an open-close DOF O-C7.

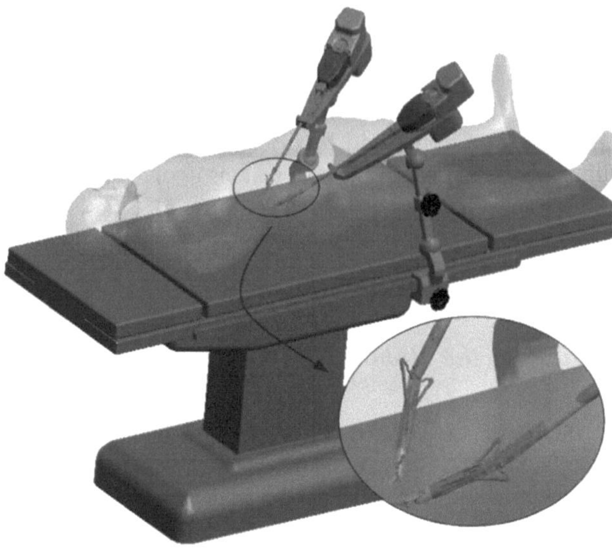

**Fig. 42.2** The slave unit applied to abdominal surgery

**Fig. 42.3** The (6 + 1) active robot

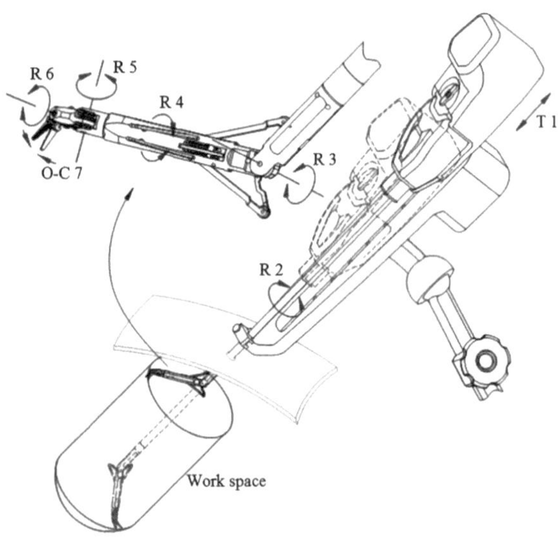


In these DOFs, only the R2 joint and the T1 joint move outside the abdomen. The R2 joint is a rotation about the instrument shaft and does not sweep out any volume. Meanwhile, the T1 joint is a translation along the linear guide which is mounted on the base of the active robot. Therefore, unlike other traditional MIS robots, the active robot does not rotate around the incision point when operation. This exceptional trait solves the problem of frequent conflict between robotic arms in other MIS robots. Furthermore, avoiding moving parts with a large range of motion enables the whole robot to adopt a compact, lightweight and miniature design. The length of the 2-DOF driving part outside the abdomen is 420 mm, the width is 100 mm, and the height is 200 mm.

All the DOFs R2, R3, R4, R5, R6 and O-C7 are driven by six independent closed-loop cables. Pre-stress mechanisms and motion transmission elements are mounted inside the quick-exchange interface. The reconfigurable instrument can be attached to and detached from the robot arm quickly.

In the active robot, DOFs T1, R2 and R3 provide positioning of the instrument, DOFs R4, R5 and R6 provide orientation of the instrument, and DOF O-C7 provides open-close control of the forceps or the scissors.

### *42.2.3 The Reconfigurable Instrument*

The reconfigurable instrument enables positioning and orientation of the instrument inside the abdominal cavity. The front part of the instrument, which contains R4, R5, R6 and O-C7, is the same cable-driven instrument as the MicroHand A robot, which has been tested in vivo experiments (see here [5] for more information).

As mentioned in Sect. 42.1, when adding extra DOFs to increase the instrument's positioning ability, the moment arm of the back joint, which is the length between the R6 joint and the R3 joint here, will increase significantly. However the diameter of the instrument is confined by the incision hole to decrease operative trauma, which means the radius of the driving wheel of the R3 joint is confined by the outer diameter of the instrument. The combination of these two factors causes a decrease in the output forces and stiffness of the instrument.

In this paper, the reconfigurable concept is introduced to improve the output force and stiffness of the instrument. Figure 42.4 shows the reconfigurable structure after unfolding. The body of the reconfigurable instrument S2 can rotate with respect to the instrument shaft S1 as DOF R3 shows. A pin connects bar L11 and S2 together which means L11 can rotate with respect to S2. Bar L21 is also rotationally connected to L11 by a pin. Pin P1 is welded at the lower end of L21 and can slide along groove G1. A cable is fixed at the joint connecting L11 and L12. The same mechanism is located in the other side of S2 symmetrically. A geometrically identical structure is placed in the quick-exchange interface to control folding and unfolding of the reconfigurable instrument.

As shown in Fig. 42.5, when unfolding the reconfigurable structure cables pull backward L11 and L12 and further pull backward P1 and P2. When P1 and P2 hit the ends of the grooves, the reconfigurable structure is locked and can be used for power transmission of the R3 DOF. During operation, the reconfigurable structure should be covered by a plastic cover to prevent tissue getting stuck in the retractable mechanism.

The folding procedure is the reverse of the unfolding procedure but the forces come from a conflict between L11 and L12, and a trocar. This conflict happens automatically when pulling the instrument out of the abdomen. The folded reconfigurable structure is shown in Fig. 42.6.

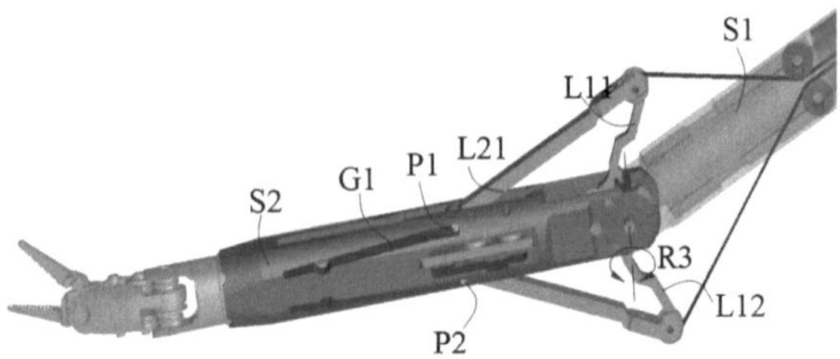


**Fig. 42.4** The unfolded reconfigurable structure

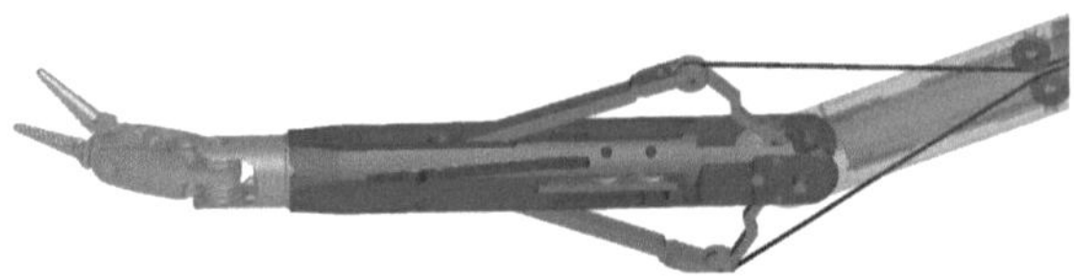

**Fig. 42.5** The reconfigurable structure unfolding

**Fig. 42.6** The folded reconfigurable structure

## 42.3 The Optimization of Structure Dimensions

After the unfolding, the whole reconfigurable structure is locked by the pre-tension of the cables as Fig. 42.7 shows, which means it can be considered as a fixed structure when cables drive the R3 DOF during operation. Therefore, this reconfigurable structure can be replaced by a line connecting point o and point p. this line is rigidly fixed on S2.

The driving force exerted on S2 is the tension of the cable $F_3$. Therefore, after drawing a line perpendicular to the cable axis through the point o, the length between point o and point p can be regarded as an equivalent driving radius $R_e$. This can be applied to calculate the driving toque $T_3$ of the R3 joint as

$$T_3 = F_3 \cdot R_e \tag{42.1}$$

Employing mathematical analysis to describe what is depicted in Fig. 42.7, the equivalent driving radius $R_e$ can be described as

$$\begin{aligned} l_{pk} = l_{op} \cdot \sin(\theta_i) - d_p \, l_{xk} = h_p - l_{op} \cdot \cos(\theta_i) \, l_{px} = \sqrt{(l_{pk})^2 + (l_{xk})^2} \\ \sin(\theta_{x1}) = r_p / l_{px} \tan(\theta_{x2}) = l_{pk} / l_{xk} \, \theta_x = \theta_{x1} + \theta_{x2} \\ l_p = \sqrt{(h_p - r_p \cdot \sin(\theta_x))^2 + (d_p - r_p \cdot \cos(\theta_x))^2} \, l_c = l_{px} \cdot \cos(\theta_{x1}) \\ \cos(\theta_m) = \left((l_c)^2 + (l_p)^2 - (l_{op})^2\right) / 2l_c \cdot l_p \, R_e = \sin(\theta_m) \cdot l_p \end{aligned} \tag{42.2}$$

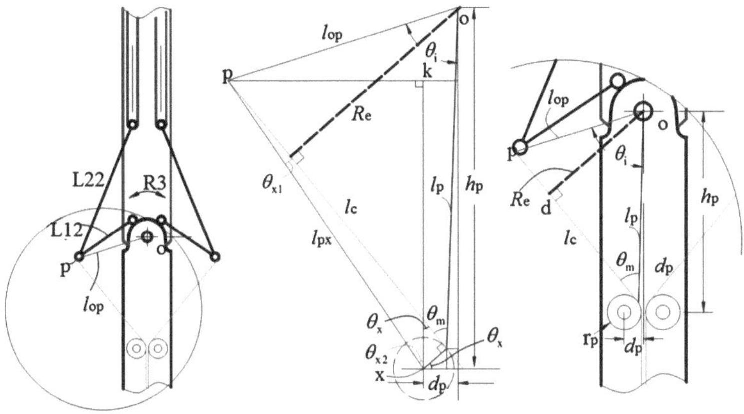


**Fig. 42.7** The diagram of the reconfigurable structure

in which $r_p$ is the radius of the transmission pulley and is set to 2.25 mm because a smaller pulley may wear the cable passing though it, $l_{op}$, $\theta_i$, $d_p$ and $h_p$ are structure variables and need to be optimized.

First, consider $d_p$ and $h_p$ which present the position of the transmission pulley. Here the initial value of $l_{op}$ is set in advance to 20 mm. A set of values of $\theta_i$ varying from 10°to 150° is set as well. Then, changes of $R_e$ with respect to $d_p$ and $h_p$ can be determined and analyzed. Calculation results when $\theta_i$ are equal to 20, 50, 80 and 150 degrees are given in Fig. 42.8.

Obviously, $d_p$ grows, the value of $R_e$ always increases. Meanwhile, $d_p$ is restricted by the outer diameter of S1 which is 10 mm. Therefore the biggest value of $d_p$ the structure design can afford, is chosen as 2.25 mm. If $\theta_i$ exceeds 50 °, the best value of $h_p$ is around 30 mm. after that, $R_e$ increases slowly. However, too large a value of $h_p$ causes conflict between the cable and the trocar. Hence, it is set to be 30 mm. Then, changes of $R_e$ with respect to $l_{op}$, and $\theta_i$ can be calculated and are given in Fig. 42.9.

It can be seen that $R_e$ increases according to the growth of $l_{op}$. However, $l_{op}$ stands for the whole width of the configurable structure and a smaller value of $l_{op}$ can avoid conflict between the robot and organs. Therefore, 18 mm was chosen for $R_e$. Considering $\theta_i$ separately, the impact of $\theta_i$ on $R_e$ can be calculated and plotted in Fig. 42.10.

Since $R_e$ decreases faster on the left side of the highest point, 85°was chosen for $\theta_i$, to make sure $R_e$ maintains a higher value when R3 rotates to each sides.

Now, all the four structure variables have been optimized and are given in Table 42.1. These key structure variables are used for detailed structure design.

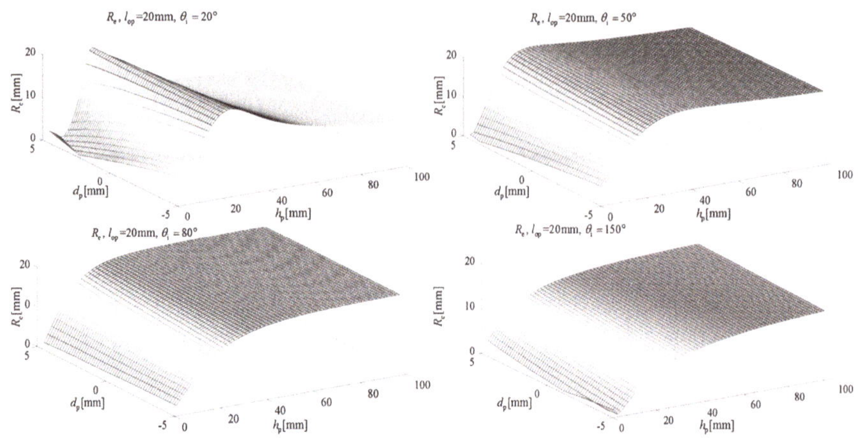


**Fig. 42.8** Calculation results of $R_e$

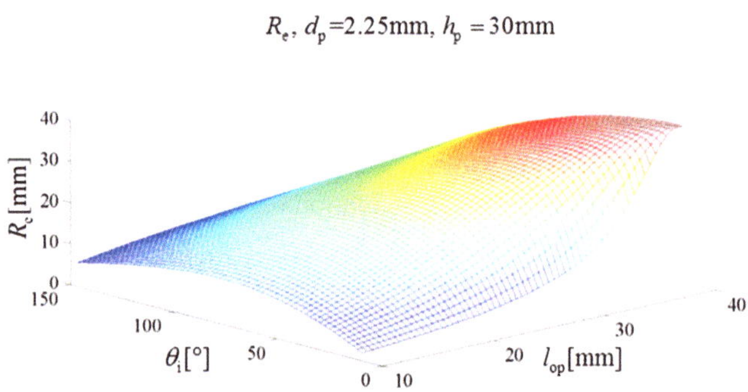


**Fig. 42.9** Calculation results of $R_e$ with respect to $l_{op}$, and $\theta_i$

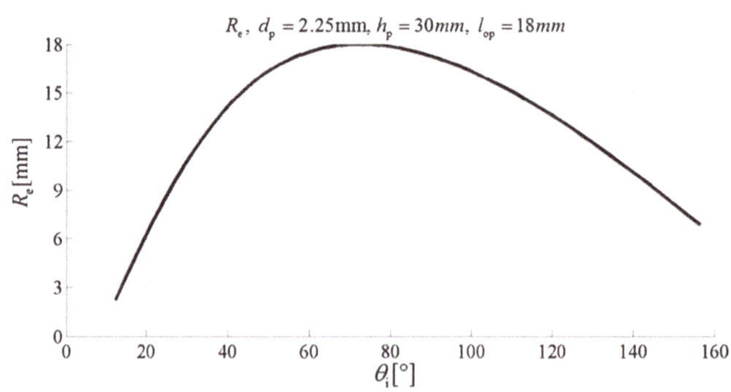


**Fig. 42.10** The impact of $\theta_i$ on $R_e$

**Table 42.1** Optimized structure variables

| $l_{op}$ | $\theta_i$ | $d_p$ | $h_p$ |
|---|---|---|---|
| 18 mm | 85° | 2.25 mm | 30 mm |

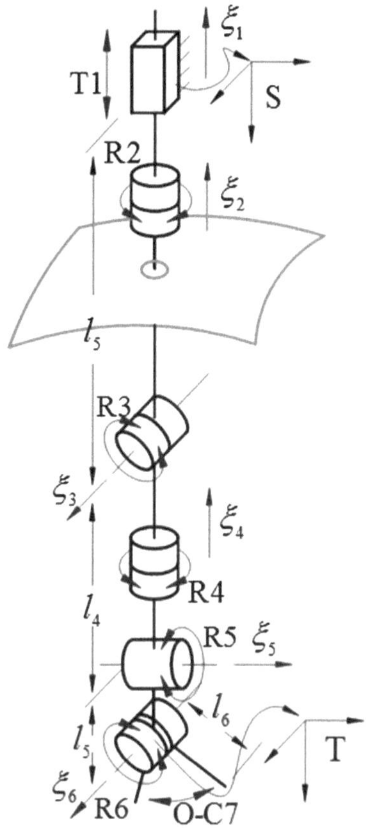


**Fig. 42.11** The zero configuration of the active robot

## 42.4 Kinematics Analysis

The initial twists of all six DOFs, a spatial frame S and a body frame T [11] are established in the following kinematics analysis and shown in Fig. 42.11. The O-C7 DOF is ignored since it is irrelevant to the pose of frame S.

Therefore, the forward kinematics of the active robot are given by

$$\mathbf{g}_{st}(\theta) = e^{\hat{\xi}_1\theta_1} e^{\hat{\xi}_2\theta_2} e^{\hat{\xi}_3\theta_3} e^{\hat{\xi}_4\theta_4} e^{\hat{\xi}_5\theta_5} e^{\hat{\xi}_6\theta_6} \mathbf{g}_{st}(0) \tag{42.3}$$

where $\theta = [\theta_1\ \theta_2\ \theta_3\ \theta_4\ \theta_5\ \theta_6]^T$ is the joint variable of the active robot, $\xi_i$ is the initial twist of the $i$th joint, and $\mathbf{g}_{st}(0)$ is the initial configuration of T to S.

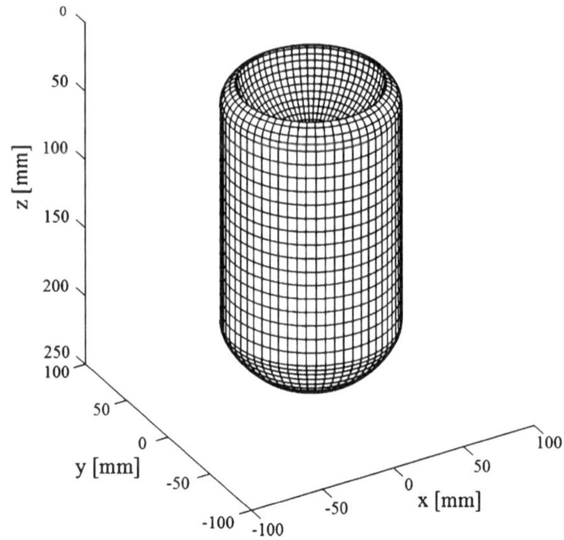


**Fig. 42.12** The reachable workspace of the active robot

## 42.5 Performance Specifications

### *42.5.1 Workspace*

Based on the forward kinematics, the reachable workspace of the active robot part can be derived and is plotted in Fig. 42.12.

The workspace is an approximate cylinder with an outer diameter of 165 mm and a height of 300 mm. It shows the robot has enough range of motion in the z direction but has a relatively small range of motion in x and y directions only. However, the workspace can already cover a typical MIS operation area. Take gallbladder removal for example, the size of the gallbladder is less than 40 mm in diameter and 100 mm in length [12]. The workspace can also be adjusted using the passive arm to reach the entire abdominal cavity.

### *42.5.2 Output Force*

The rotation range of R3 is ±60°. Therefore, according to Fig. 42.10, $R_e$ varies from 8.8 mm to 18 mm. Meanwhile, for a 10 mm traditional MIS robotic instrument, the maximum radius of its driving wheel is 5 mm. That means the reconfigurable structure improves the output torque of the R3 joint 0.76–2.8 times. According to the strength of the adapted 0.5 mm stainless steel cable, which is safe at 150 N, the output force of the R3 joint at the instrument tip is 14.8–30.3 N. This force is big enough to be used in MIS [13].

R4, R5, R6 and R7 joints are same as joints in the 'MicroHand A' instrument. They have already been shown to have enough output force for MIS [5]. T1 and R2 have enough space to enhance the strength of transmission part since they are outside the abdomen. Therefore they also have enough output force.

## 42.6 Conclusions

In this paper, a new portable reconfigurable MIS robot was presented. Five DOFs of this robot were set up inside the abdominal cavity to avoid large range of motion of the arm part outside the abdominal wall. This feature allows the robot to be both portable and lightweight. The volume of the active arm of the robot is 420 mm in length, 100 mm in width, and 200 mm in height.

The reconfigurable structure which drives the R3 DOF was optimized. Four key structure dimensions were gained though this optimization process to be used for the detailed design.

Performance specifications about the workspace and output force were discussed. Relative analysis shows the robot to have enough workspace for MIS surgery. Compared to a traditional tendon-driven joint, the reconfigurable structure improves the output torque of the R3 joint 0.76–2.8 times. This system benefits from a reasonable DOF arrangement and a reconfigurable structure, and all the DOF have enough output forces for MIS.

**Acknowledgments** This research is supported by the National Natural Science Foundation of China (NSFC) under Grant No. 50925520. The authors would like to thank Kevin Olds for his help in editing the English manuscript.

## References

1. Steven EB, Moji G (2003) Transforming a surgical robot for human telesurgery. J IEEE Trans Robot Autom 19:818–824
2. Çavusoglu MC, Williams W, Tendick F, Sastry SS (2003) Robotics for telesurgery: second generation berkeley/UCSF laparoscopic telesurgical workstation and looking towards the future applications. Ind Robot Int J 30:22–29
3. Podsedkowski L (2005) RobIn Heart 0, 1, and 3—mechanical construction development. J Bull Pol Ac Tech 53:79–85
4. http://www.intuitivesurgical.com
5. Sang H, Wang S, Li J, He C, Zhang L, Wang X (2011) Control design and implementation of a novel master–slave surgery robot system, MicroHand A. Int J Med Robot 7:334–347
6. Song HS, Kim KY, Lee JJ (2009) Development of the dexterous manipulator and the force sensor for minimally invasive surgery. In: Proceedings of the 4th international conference on autonomous robots and agents. IEEE Press, Wellington, pp 524–528.
7. Simaan N, Taylor RH, Flint P (2004) High dexterity snake-like robotic slaves for minimally invasive telesurgery of the upper airway. MICCAI 2004(3217):17–24
8. Takeyoshi O, Degani A, Schwartzman D, Zubiate B, McGarvey J (2009) A highly articulated robotic surgical system for minimally invasive surgery. J Ann Thorac Surg 87:1253–1256

9. Webster RJ, Okamura AM, Cowan NJ (2006) Toward active cannulas: miniature snake-like surgical robots. In: IEEE/RSJ international conference on intelligent robots and systems, Beijing, pp 2857–2863
10. Simaan N, Taylor RH (2005) Devices, systems and methods for minimally invasive surgery of the throat and other portions of mammalian body. US Patent, 20050059960
11. Murray RM, Li ZX, Sastry SS (1994) A mathematical introduction to robotic manipulation. CRC Press, Boca Raton
12. Westat Inc. (1988) Third national health and nutrition examination survey gallbladder ultrasonography procedure manual. Rockville
13. Madhanir AJ, Niemeyer G, Salisbury JK Jr (1998) The black falcon: a teleoperated surgical instrument for minimally invasive surgery. In: IEEE/RSJ international conference on intelligent robots and systems, Victoria, pp 936–944

# Chapter 43
# A Passive Robotic Platform for Three-Dimensional Scanning of Ex Vivo Soft Tissue

**Jichun Li, Jelizaveta Zirjakova, Wei Yao, Kaspar Althoefer, Prokar Dasgupta and Lakmal D Seneviratne**

**Abstract** This paper presents a novel portable passive robotic platform for three-dimensional scanning (3DS) of soft tissue, capable to evaluate mechanical properties and geometry in ex vivo condition. The platform comprises six degrees of freedom (DOF) passive robotic arm (Phantom Omni), a data acquisition system and a set of stiffness probes for force and stiffness measurement. The performance of the developed platform was validated by sliding indentation and uniaxial tissue indentation measurements on silicone phantoms, porcine organs and human prostates. The results show that the platform can perform effective measurements of soft tissue mechanical properties and help surgeons to identify embedded tumours.

**Keywords** Robotic platform · Soft tissue · 3D scanning

J. Li (✉) · J. Zirjakova · W. Yao · K. Althoefer · L. D Seneviratne
Centre for Robotics Research, King's College London, London, WC2R 2LS, UK
e-mail: jichun.li@kcl.ac.uk

J. Zirjakova
e-mail: jelizaveta.zirjakova@kcl.ac.uk

W. Yao
e-mail: wei.yao@kcl.ac.uk

K. Althoefer
e-mail: kaspar.althoefer@kcl.ac.uk

L. D Seneviratne
e-mail: lakmal.seneviratne@kcl.ac.uk

P. Dasgupta
Guy's Hospital, Centre for Urology, King's College London, London, SE1 9RT, UK
e-mail: prokarurol@gmail.com

L. D Seneviratne
Khalifa University of Science, Technology and Research, Abu Dhabi, UAE

J. S. Dai et al. (eds.), *Advances in Reconfigurable Mechanisms and Robots I*,
DOI: 10.1007/978-1-4471-4141-9_43, 

## 43.1 Introduction

Quantitative assessment of the mechanical properties of soft tissues ex vivo is very important in both clinical and research fields. Firstly, mechanical property is an import measure for surgeons to indentify the cancer since a malignant tumour is typically stiffer than the healthy tissue [1]. Therefore, two-dimensional (2D) or three-dimensional (3D) mechanical imaging is used as an efficient way to indentify tumours. Secondly, validation of a novel surgical device is usually implemented in several stages. Tests performed on ex vivo organs of animals and human being is an important and necessary step for the evaluation of the performance of the device. Only when the devise shows its feasibility in ex vivo conditions, it is possible to test it during the real surgery.

Medical robots play a growing and critical role in modern medical field. Da Vinci™ Surgical System and Zeus™ Surgical System are the most well-known medical robots. Using them, medical professionals can make smaller incisions which results to decreased blood loss for patients, less pain, and shorter hospitalization period. However, an inherent problem with robot-assisted MIS system is the complete lack of haptic or tactile feedback.

A variety of master–slave robotic systems attached with medical probes, capable of providing haptic feedback through kinaesthetic sensing have been developed [2–4]. For such systems driven by industrial robot, the accurate control of the motion could be achieved. It is relatively easy solution for laboratory experiments, as the path of the probe can be pre-programmed and the performance is evaluated under the same conditions. However, it is an expensive and difficult alternative for surgeons. Moreover, it does not give the possibility to detect organ curvatures and easily change the measurement path, in order to compose an accurate 3D representation. In addition, moving parts of the probes could reduce the reliability and could be difficult to fabricate in mass production. From the other side, many hand-held devices are developed to measure the mechanical feedback of soft tissue [5–7]. Such systems showed good accuracy and feasibility. However, they were designed to carry out local stiffness measurement and not capable of generating 3D stiffness representation.

Imaging modalities such as magnetic resonance imaging (MRI) or computer tomography (CT) are used to identify tissue abnormalities. However, these techniques are expensive and require long processing time. The significant limitation is that sometimes they are not able to distinguish between prostate tumor and edema fluid, especially in the case of small size formation. The optical coherence tomography (OCT) is another imaging technique which can instantly visualize sub-surface abnormalities with micrometer resolution. However the penetration depth of OCT is limited to 1–2 mm below the tissue surface [8]. The elastography is a technique for generating images of tissue elasticity by propagating mechanical waves through the investigated tissues [9]. This method is effective for detecting tumors. The limitation is that the elasticity images are difficulty to generate in real time.

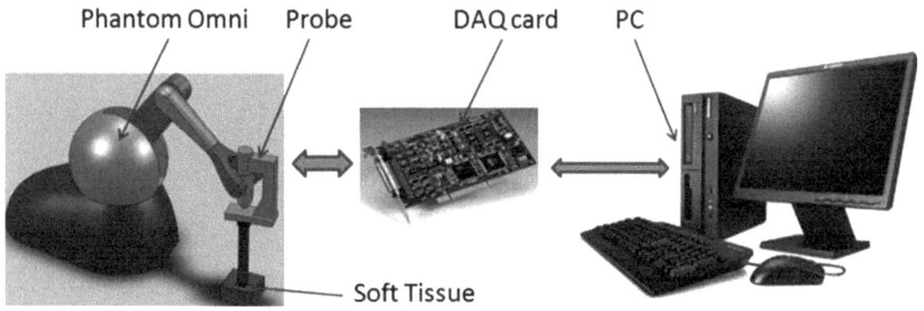


**Fig. 43.1** Schematic model of the robotic platform—frame assignment

Taking into consideration the limitations of the aforementioned devices, we have proposed a novel portable passive robotic platform for 3DS of ex vivo soft tissue by measuring the mechanical properties and geometry. Advantages of the system include low cost, mobility, ease-of-use, and minimal training required.

This paper is organized as follows. In Sect. 43.2, the prototype and the basic working principle of the system is described. In Sect. 43.3, we discuss experiments for 3D scanning of the soft tissue based on sliding indentation and unixal indentation tests. Finally, discussions and conclusions are drawn in Sect. 43.4.

## 43.2 Development of the Platform

### *43.2.1 Overview of the Platform*

The schematic overview of the device is shown on Fig. 43.1. The portable robotic platform is composed of a passive robotic arm (Phantom Omni) [10], a set of stiffness probes for force and stiffness measurement, a data acquisition card and a PC.

The Phantom Omni haptic device (SensAble technologies) was used for many applications due to its properties and availability to public, for example in [11], [12]. However, to our knowledge, it has not been used broadly combined with the probe for soft tissue stiffness examination.

Phantom Omni is a robotic device with range of motion set by hand wrist movements and with nominal positional resolution of 0.055 mm [10]. Three potentiometers and three encoders read the outputs of the device—six variable angles—three joint angles for translation and three gimbal angles for rotation. The user is able to set any necessary trajectory of the probe, thus, making soft tissue examination more effective. It is possible to capture the curvature of a soft tissue and to detect hard formations more accurately.

Combining the geometry information from the Phantom Omni device and the information about the mechanical properties of soft tissue from the probe, a three-dimensional scanning of soft tissue can be obtained.

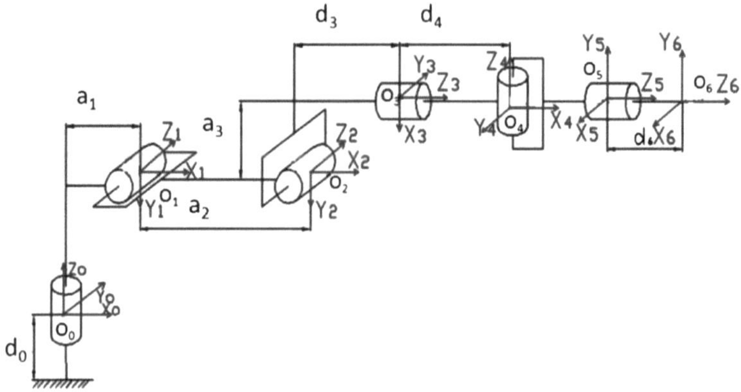


**Fig. 43.2** Kinematics model of the robotic platform

The indentation probe for force and stiffness measurements is attached to the stylus of the device. Experimental data from the probe are acquired using the National Instruments LabView 8.0 software package and associated PCI Data Acquisition Card (NI PCI 6034E, which is a 16-bit DAQ card).

## 43.2.2 Kinematics Analysis

To represent the trajectory of the probe, it is necessary to calculate 3D coordinates of the end effecter. By applying forward kinematics equations the position of end point can be calculated. The commonly used Denavit–Hratenberg algorithm was used to define the position of each link in space. Coordinate frame was assigned to each joint as shown on Fig. 43.2 Phantom Omni kinematics relationship can be found looking onto each frame's relative movement. The homogenous transformation matrixes $A_i^{i-1}$were used, where $i$ define the number of the frame, base frame is assigned to sequence number zero: For the robotic platform there are six frames, thus $i = 6$ and six transformation matrixes can be defined. After each transformation matrix is calculated, the resultant

$$A_i^{i-1} = \begin{pmatrix} cos\theta_i & -sin\theta_i cos\alpha_i & sin\theta_i sin\alpha_i & a_i cos\theta_i \\ sin\theta_i & cos\theta_i cos\alpha_i & -cos\theta_i sin\alpha_i & a_i sin\theta_i \\ 0 & sin\alpha_i & cos\alpha_i & d_i \\ 0 & 0 & 0 & 1 \end{pmatrix} \quad (43.1)$$

$\theta_i$—rotation angle about z axis, $\alpha_i$—angle about common normal, $d_i$—offset along z axis, $a_i$—offset along common normal.

Homogeneous matrix $A_6^0$ is obtained. $A_6^0$ defines the position and orientation of the probe tip in absolute coordinates.

### 43.2.3 Stiffness Probe

The set of stiffness probes for force and stiffness measurement includes two probing devices, which are discussed hereafter. In the work we have chosen to use two measurement devices due to the different, but still important advantages of both.

We propose a novel design of a stiffness probe Pa which can measure the indentation force and indentation depth simultaneously. As shown in Fig. 43.3a, the proposed probe is composed of a force sensor Nano 17 (SI-25-0.25, ATI Industrial Automation) and an image acquisition unit capable of obtaining images from the contact area. The image acquisition unit consists of a digital camera with 12 mm diameter and a probe with transparent round head with a diameter of 10 mm. Generally, when the probe indents towards the target soft tissue, a circular area will appear around the point where the indentation depth happens due to the imparted force and the round-shaped head of the probe. When increasing the indentation depth, the diameter of the circular indentation area increases. The camera captures the change of contact area due to indentation during probing. The contact area as acquired from the camera images then can be converted into indentation depth estimates. Knowing the indentation force and depth, we can obtain the stiffness of the target tissue. To the best of our knowledge, there has not been any work that considers measuring the indentation depth by capturing the contact area. This probe provides user with accurate measurements of soft tissue mechanical properties. Moreover, due to attached camera, it can be used as an endoscope during MIS. However, it is not very suitable for rapid tissue scanning, what is often needed for surgical conditions.

The second probing device, Pb, Fig. 43.3b, is designed to slide with the indentation over the tissue. This probe can slide over the soft tissue with relatively good speed. Therefore, allowing quickly generating 3D stiffness representation. The indentation ball is fixed in the plastics connector, which is attached directly to the force sensor, in order to obtain most precise sensor readings. The indentation depth is designed to be a constant value of 1, 2, 3, 4 and 5 mm. The probe is controlled by user while performing the quick scanning.

## 43.3 Validation Experiments

### 43.3.1 Three-Dimensional Scanning by Sliding Indentation Test

Mechanical imaging is a new technique used for soft tissue examination [13]. It has got many advantages over the ultrasound imaging, which is widely used in medical applications. However, non-uniform structures and speckles on the prostate surface create noise in ultrasound images and reduce reliability of the representation. The prostate is connected to other organs and further segmentation is needed to extract its image from the surrounding. As a result, full prostate boundary is not easy to detect without prior contour specification, Fig. 43.4.

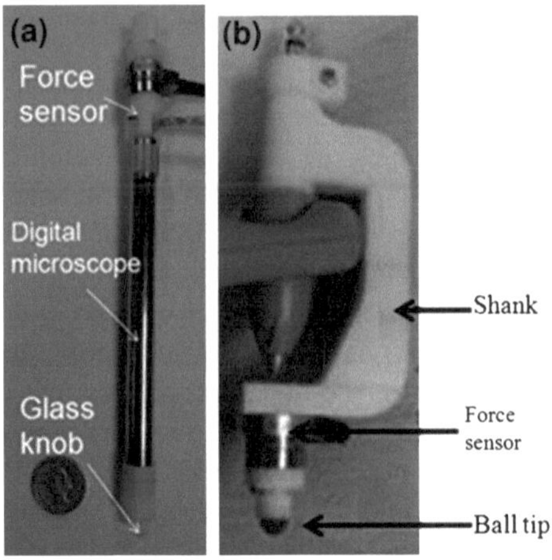


**Fig. 43.3** A stiffness probe, Pa, based on vision and force modalities with accurate measurements (**a**) and a stiffness probe, Pb, with constant indentation depth (**b**)

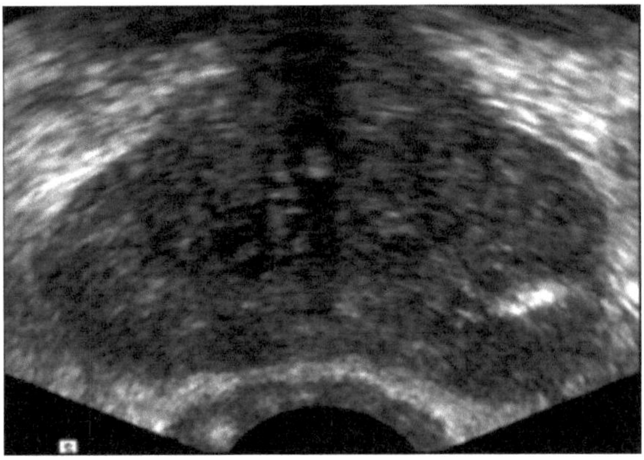

**Fig. 43.4** Prostate specimen ultrasound image example (E J Halpern, 2006)

To validate the performance of tissue abnormality identification by 3DS method, sliding indentation tests were performed on an ex vivo human prostate, obtained after surgery. The developed probe, Pb, was slid over the surface of the prostate. During this process, the forces and trajectory are recorded. Then, force feedbacks at each sampled location are combined together to generate a colour-coded mechanical image, Fig. 43.5a. Since the image indicates the stiffness distribution within the soft tissue, the location of the tumours can be identified. It can be seen from Fig. 43.5a that thanks to the kinematics of the robotic device is possible to examine small specimens of soft tissue, taking into account organ curvature.

To show the accuracy of the platform, an experienced surgeon was asked to palpate the specimen. The detected area is shown on Fig. 43.5b. It can be seen, that detected tumour location corresponds to the stiff area detected by palpation.

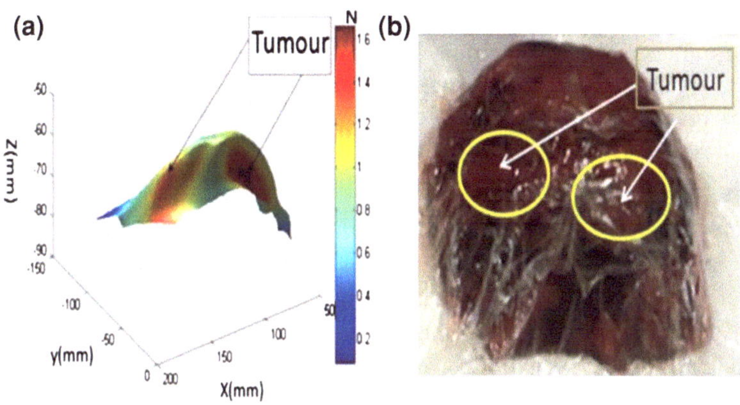


**Fig. 43.5** 3D mechanical image of the examined prostate (**a**) and stiffer area detected by manual palpation (**b**)

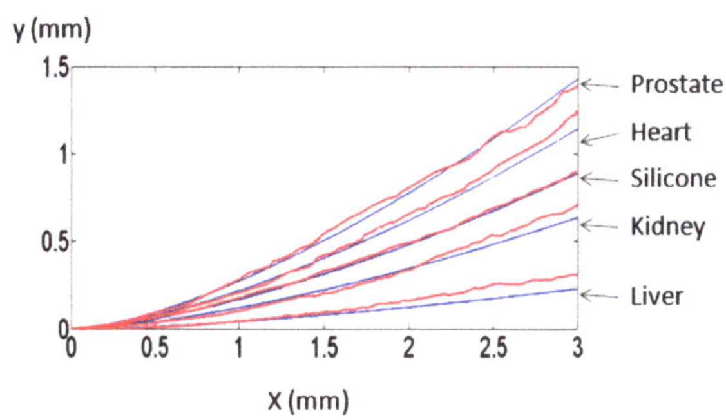


**Fig. 43.6** The relation of tissue reaction force with the indentation depth of the stiffness probe; the blue solid lines are the estimations using Eq. 43.2, the red solid lines are the experimental measurements

## *43.3.2 Uniaxial Indentation Tests*

Tissue stiffness can be represented using the tissue's elastic modulus through the measurements of tissue reaction force and indentation depth [14]. The elastic modulus of tissue can be estimated as:

$$E = \frac{3F(1+v)}{8\mathrm{d}_i\sqrt{\mathrm{r}d_i}}, \tag{43.2}$$

where $F$ is the tissue reaction force normal to tissue surface, $v$ is the Poisson ratio ($v = 0.5$), $E$ is the elastic modulus, $d_i$ is the indentation depth and $r$ is the radius of the sphere head of indenter.

In order to validate the above mentioned assumption, indentation tests were carried out on silicone phantom, porcine liver, porcine kidney and archived human prostate using the developed probe, Pa.

Figure 43.6 shows the results of the experiments. It was found that Eq. 43.2 fits the measurement well. According to Eq. 43.2, the elastic modulus of the liver, kidney, silicone, heart and prostate are estimated as 7.9, 21.8, 30.4, 39.2 and 49 Kpa respectively. The estimated results show good agreement with existing literatures [14–17].

## 43.4 Conclusions

In our work we have described the working principle and results of the passive robotic platform for ex vivo tissue diagnosis used for three-dimensional scanning. The platform consists of Phantom device and set of stiffness probes. It was shown, that the setup is suitable for quantitative assessment of the mechanical properties of soft tissue ex vivo specimen. Thanks to both positional information and stiffness, three-dimensional and 2D stiffness maps were created. The platform is specially created for soft tissue ex vivo examination and allows user to validate the feasibility of the probe before in vivo tests.

The experimental results show good feasibility of the setup to generate 3D stiffness maps for ex vivo tissue examination and to detect embedded tumours.

Uniaxial indentation tests, using probe Pa, were performed with the help of passive robotic platform. The relation between tissue reaction force and the indentation depth was found experimentally and was close to theoretical estimation.

## References


1. Murphy DA, Miller JS, Langford DA (2006) Endoscopic robotic mitral valve surgery. J Thorac Cardiovasc Surg 132:776–781
2. Rosen J, Hannaford B (1999) Force controlled and teleoperated endoscopic grasper for minimally invasive surgery—experimental performance evaluation. IEEE Trans Bio Med Eng 46:1212–1221
3. Liu H, Li J, Song X et al (2011) Rolling indentation probe for tissue abnormality identification during minimally invasive surgery. IEEE Trans Robot 27:450–460
4. Trejos AL, Jayander J, Perri MT et al (2008) Experimental evaluation of robot-assisted tactile sensing for minimally invasive surgery. In: 2nd IEEE RAS/EMBS international conference on biomedical robotics and biomechantronics, pp 971–976, IEEE press, Scottsdale
5. Rosen J, Hannaford B (2001) Markov modeling of minimally invasive surgery based on tool/ tissue interaction and force/torque signatures for evaluating surgical skills. IEEE Trans Bio Med Eng 48:579–591
6. Tholey G, Pillarisetti A, Green W et al (2004) Design, development, and testing of an automated laparoscopic grasper with 3-D force measurement capability. In: Stephane C, Dimitris M (eds) Medical simulation. LNCS, vol 3078. Springer, Berlin, pp 38–48
7. Peng P, Sezen AS, Rajamani R, Erdman AG (2010) Novel MEMS stiffness sensor for force and elasticity measurements. Sens Actuator 158:10–17
8. Zysk AM, Nguyen FT, Oldenburg AL, Marks DL, Boart SA (2007) Optical coherence tomography: a review of clinical development from bench to bedside. J Biomed Opt 12:051403

9. Ophir J, Alam SK, Garra B, Kallel F, Konofaqou E, Krouskop T, Varqhese T (1999) Elastography: ultrasonic estimation and imaging of the elastic properties of tissues. Proc Inst Mech Eng H 213:203–233
10. Phantom Omni® haptic device. http://www.sensable.com/haptic-phantom-omni.htm#techspecs
11. Hayn H, Schwarzmann D (2010) A haptically enhanced operational concept for a hydraulic excavator. In: Mehrdad HZ (ed) Advances in Haptics, pp 199–220. InTech Press
12. Sankaranarayanan G, King H, Ko S et al (2010) Portable surgery master station for mobile robotic telesurgery. In: 1st International ICST conference on robot communication and coordination, pp 1–8, Athens
13. Egorov V, Ayrapetyan S, Sarvazyan AP (2006) Prostate mechanical imaging: 3-D image composition and feature calculations. IEEE Trans Med Imag 25:1329–1340
14. Liu H, Noonan D, Seneviratne LD, Dasgupta P, Althoefer K (2010) Rolling indentation for tissue abnormality localization during minimally invasive surgery. IEEE Trans Biomed Eng 57:404–414
15. Hannaford B, Trujillo J, Sinanan M, Moreyra M, Rosen J, Brown J, Leuschke R (1998) MacFarla: computerized endoscopic surgical grasper. In: Westwood JD, Westwood SW, Haluck RS, Hoffman HM, Mogel GT, Phillips R, Robb RA, Vosburgh KG (eds) Medicine meets virtual reality. IOS Press, Amsterdam, pp 265–271
16. Samur E, Sedef M, Basdogan C, Avtan L, Duzgun O (2007) A robotic indenter for minimally invasive measurement and characterization of soft tissue response. Med Image Anal 11:361–373
17. Krouskop TA, Wheeler TM, Kallel F, Garra BS, Hall T (1998) Elastic moduli of breast and prostate tissues under compression. Ultrason Imaging 20:260–274

# Chapter 44
# Design and Fabrication of DNA Origami Mechanisms and Machines

**Hai-Jun Su, Carlos Ernesto Castro, Alexander Edison Marras and Michael Hudoba**

**Abstract** The goal of this paper is to introduce scaffolded DNA origami as a viable approach to the design of nanoscale mechanisms and machines. Resembling concepts of links and joints in macro scale mechanisms and machines, we propose the concept of *DNA Origami Mechanisms and Machines (DOMM)* that are comprised of multiple links connected by joints. Realization of nanoscale machines would pave the way for novel devices and processes with potential to revolutionize medicine, manufacturing, and environmental sensing. The realization of nanoscale machines and robots will enable scientists to manipulate and assemble nano objects in a more precise, efficient and convenient way at the molecular scale. For example, DNA nanomachinery could potentially be used for nano manufacturing, molecular transport in bioreactors, targeting cancer cells for drug delivery, or even repairing damaged tissue. As a proof of concept, we build a nanoscale spatial Bennett 4-bar mechanism that can be completely folded and unfolded with a specified kinematic motion path. The links comprise a 16 double stranded DNA (dsDNA) helices bundled in a 4 by 4 square cross-section yielding a high mechanical stiffness. The joints (in this case hinges) are designed using single strand DNA (ssDNA) connections between the links. This DOMM was designed within caDNAno, a recently developed computer-aided DNA origami design software, and then fabricated via a molecular self-assembly process. The resulting structure was imaged by transmission electron microscopy to identify structural conformations. Our results show that the designed DNA origami Bennett mechanism closely follows the kinematics of their rigid body counterparts.

H.-J. Su (✉) · C. E. Castro · A. E. Marras · M. Hudoba
Department of Mechanical and Aerospace Engineering,
The Ohio State University, Columbus, USA
e-mail: su.298@osu.edu

C. E. Castro
e-mail: castro.39@osu.edu

J. S. Dai et al. (eds.), *Advances in Reconfigurable Mechanisms and Robots I*,
DOI: 10.1007/978-1-4471-4141-9_44, © Springer-Verlag London 2012 

This research has the potential of opening a new era of design, analysis and manufacture of nanomechanisms, nanomachines and nanorobots.

## 44.1 Introduction

A mechanism, or linkage, is a mechanical device that transforms an input motion to some desirable pattern. They are typically comprised of a collection of interconnected components, individually called links that are physically connected with joints [1]. A machine typically contains mechanisms that are designed to transform forces and power [2]. For centuries, we have been designing and manufacturing mechanisms and machines with various natural (biological) and engineered materials at macro, and more recently at meso, micro and nano scales.

Nanomachines are devices in the size range of nanometers. They are expected to lead breakthroughs in numerous applications in technology and life sciences. In recent years, several authors [3] in the kinematics and mechanisms field have worked on analysis and design of protein based nanomachines. They argued that a practical and viable approach to design and fabricate nanomachines is to use polypeptide chains as building blocks. Chirikjian et al. developed a statistical kinematics approach for computing folding and unfolding of proteins [4]. Kazerouninan models protein structures with kinematic linkage models and developed a Matlab program called ProtoFold for predicting protein kinematics [5, 6]. On the other hand, Mavroidis [7–9] worked on design and analysis of bio-nanorobots with proteins and DNA.

While proteins provide an attractive material for nano-construction due to the huge range of naturally occurring protein-based molecular machinery with a wide variety of 3D geometries and mechanical behavior, designing specific synthetic protein structures has proven difficult owing to the multitude of complex amino acid interactions that govern protein folding [10]. To date, completely synthetically designed protein structures that have been realized in the lab are limited to relatively small ($\sim$nm) and simple structures [10–13]. DNA on the other hand, self-assembles primarily by well-understood Watson and Crick base pairing and base stacking interactions [14, 15]. Nadrian Seeman's pioneering work in the 1980 and 1990s founded the field of structural DNA nanotechnology [16–19]. More recently the development of scaffolded DNA origami has enabled the contraction of nanoscale objects with unprecedented 3D structural complexity by self-assembly [20–22]. Figure 44.1 shows several examples of nano-structures constructed using scaffolded DNA origami [20, 21, 23–27] (figure adapted from [22]).

The DNA origami self-assembly process is driven by DNA base-pairing between a long ($\sim$7000–8000) bases single-stranded DNA (ssDNA) "scaffold" and many shorter ($\sim$30–50 bases) ssDNA "staple" strands. The scaffold, which must have a fully known base sequence, is generally derived from the M13MP18 bacteriophage virus genome [20–22]. The staple strand base sequences are

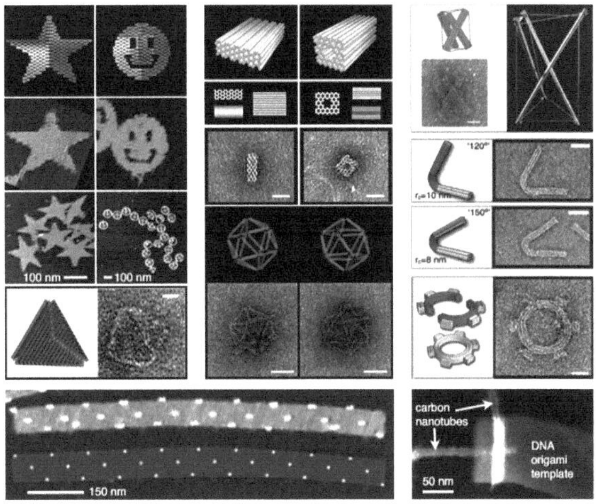


**Fig. 44.1** Examples of 2 and 3D DNA origami structures [20, 21 23–27]. Objects on the *bottom row* have been functionalized with streptavidin molecules (*bottom left*) [24] and carbon nanotubes (*bottom right*) [25] in a pre-programmed arrangement. Scale bars are 20 nm unless otherwise specified

designed to be piece-wise complementary to sections of the scaffold that may be distant in primary sequence. That is, one staple strand may bind parts of the scaffold that are separated by several thousand bases. During self-assembly, the scaffold must fold into a compact structure to spatially collocate sections of the scaffold that are complementary to a single staple strand. The loss of entropy and repulsive charge interactions (DNA is inherently negatively charged) are outweighed by the cumulative base pairing and base stacking energy [15] of many staple strands ($\sim$150–200) thereby stabilizing the scaffold into close-packed bundles of double-stranded DNA (dsDNA). Structures are designed so that these bundles arrange into the desired three-dimensional structure.

Owing to its promise as a viable method for functional nano construction, a series of computer aided engineering (CAE) tools have been developed to facilitate DNA origami design [22, 28, 29]. In keeping with current research, these computational tools have been geared towards the design of static 2D and 3D structures. The goal of this paper is to introduce scaffolded DNA origami as a viable approach to the design of nanoscale mechanisms and machines comprised of multiple components with directed motion. We introduce the design of DNA origami objects with directed motion and discuss the implementation of these moving parts into nanoscale mechanisms and machines. We construct a nanoscale spatial Bennett 4-bar mechanism to demonstrate the design and fabrication process.

## 44.2 Design and Fabrication Process

A linkage or kinematic mechanism is a collection of interconnected components, individually called "links". The physical connections between links are called "joints." Here we explore using the concepts of links and joints to design

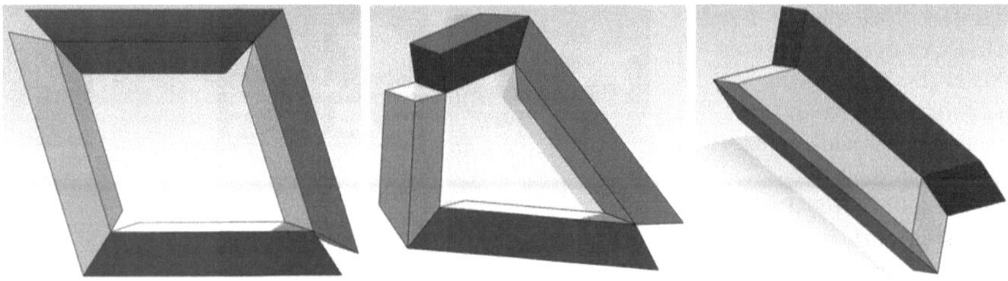

**Fig. 44.2** The Bennett 4R linkage is at the open (*unfolded*) position (*left*), pop-up position (*middle*) and the closed (*folded*) position. All links have the same length

nanoscale DNA origami machines and mechanisms (DOMM). The design of DNA origami mechanisms follows a top-down process described as follows. First, we design a conceptual mechanism that satisfies the desired kinematic function. This step essentially follows the classical design process of mechanisms which includes number synthesis, type synthesis and kinematic analysis or synthesis. Second, we design links with a sufficiently high stiffness by utilizing scaffolded DNA origami. This step is done in existing computer-aided DNA origami design software. Third, we design flexible connections between the links to act as joints with specified degrees of freedom. Finally, the design is fabricated by molecular self-assembly and evaluated by transmission electron microscopy (TEM).

### 44.2.1 Kinematic Design of a Foldable Bennett 4-Bar Linkage

Our goal is to design a spatial DOMM that can be folded and unfolded in 3D by a single actuator. Here we choose the spatial Bennett 4R (4 revolute joints) linkage due it is capability of generating a folding/unfolding motion in space. It is well known that this linkage has a mobility of one if its link dimensions satisfy the so called Bennett condition [30]. In particular, we consider the equilateral case of Bennett linkages, i.e. all links have the same length. This is also what is called the "alternative form" of Bennett 4R linkages by Chen [31]. It has been proven that this alternative form of Bennett linkage can be unfolded to an open position and folded completely into a bundle without any physical interference. Figure 44.2 shows the linkage in three representative configurations.

### 44.2.2 Design of Rigid Links with DNA Origami

Our next step is to design the individual links with scaffolded DNA origami. The first step of the DNA origami design process is to approximate the desired geometry with 2 nm cylinders that fill space in either a honeycomb- [21] or

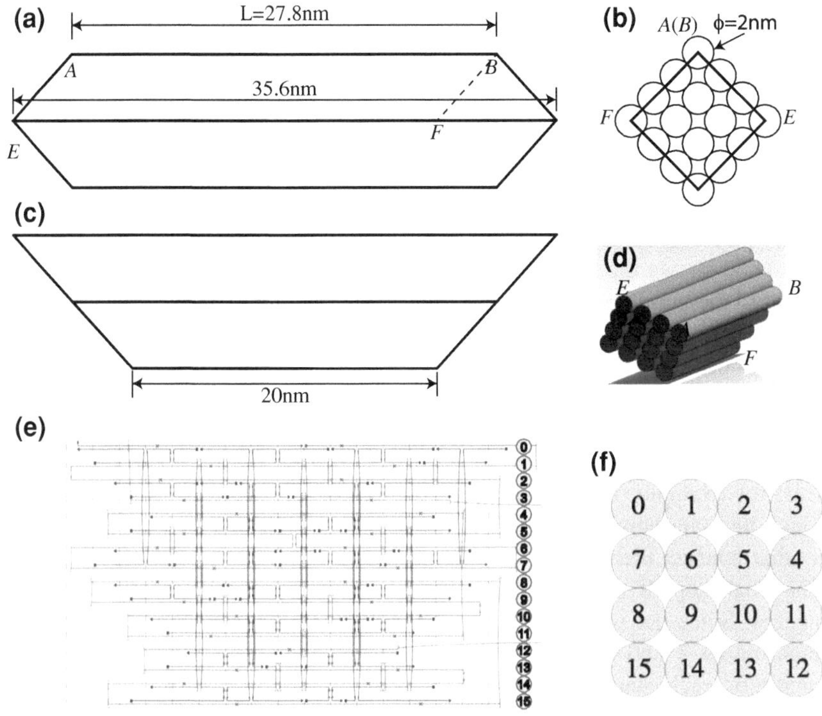


**Fig. 44.3** Design of rigid links with bundles of 4 by 4 dsDNA helices. Each dsDNA helix has a diameter of 2 nm. **a** front view, **b** side view, **c** top view, **d** 3D CAD model, **e** scaffold and staple pattern of a DNA link, **f** cross section of scaffold pattern

square-packed [32] cross-sectional geometry. The geometry of these links must satisfy the bennett linkage kinematic constraints. In the case of an equilateral bennett linkage, the cross-section must be square. Another important consideration is the link stiffness. Here we chose a square cross-section with 16 dsDNA helices bundled in a 4 by 4 arrangement as shown in Fig. 44.3b. This satisfies both kinematic and stiffness constraints. The desired link geometry is depicted in Fig. 44.3a–c with dimensions. We use different length cylinders to form a jagged edge that approximates the desired angled edge in order to minimize physical interference upon folding of the Bennett linkage.

The second step of the design process is populating the cylinder-based structure with scaffold and staple bases. For this step, it is useful to consider the cylinders to be made out of constituent DNA base pair (bp) building blocks. The long scaffold continuously weaves throughout the entire object generally contributing one base to every bp comprising the structure. The staples contribute the other base to each bp in the structure. Staples run along a single cylinder/helix for several bp and then cross over to a neighboring cylinder/helix making a "u-turn" in the process.

The resulting u-turn cross-over geometry is termed a "Holliday Juntion," and occurs in several physiological DNA structures [33]. Several of these Holliday junction cross-overs are spaced out along the length of the object. Collectively, these connections stabilize the bundle geometry and provide mechanical integrity.

The computer aided DNA origami design software caDNAno was employed to design the links for this application. caDNAno enables fast and flexible design optimization of DNA origami scaffold and staple routing in a convenient graphical user interface that maps a three-dimensional structure onto a 2D blueprint [29]. Figure 44.3e shows the scaffold and staple design in a 2D blueprint for a single link next to the numbered square cross-section in Fig. 44.3f. In the blueprint, each row represents one dsDNA helix in the cross-section. The scaffold, shown in light blue, winds throughout the entire structure, and the staples, shown in red, make the large majority of cross-overs that stabilize the structure (there are usually also a handful of scaffold cross-overs that provide some stability). caDNAno projects the link structure onto the 2D blueprint in a row by row manner. Therefore, some helices, for example, helix 0 and helix 7, appear separated in the 2D blueprint although they are neighbors in the 3D structure as indicated by the staple cross-overs that connect them.

The final step of the design process is to determine the staple strand base sequences. Since the scaffold and staples should follow Watson and Crick DNA base-paring (A-T and G-C pairs) and the sequence of the scaffold is known, the sequence of each staple can simply be read off of the scaffold/staple routing scheme according to which section of the scaffold they bind to. caDNAno automates this process saving a great deal of time.

## *44.2.3 Design of DNA Origami Hinges*

After designing the DNA origami links as shown in Fig. 44.3, the next step in the Bennett linkage design is to connect the links with hinges. First of all, from Figs. 44.2 and 44.3 we identify the location of kinematic joints as the edges *AE* and *BF* of the cross section of links. To form the hinges, we use ssDNA cross-overs between links at multiple locations spread out along the appropriate edges. One approach would be to make these connections with staple cross-overs, which would require that separate link structures find each other by random diffusion in solution. While this is a feasible strategy, the coupling efficiency of links would limit the successful assembly of Bennett linkages. Furthermore, the possibility would exist that more or less than four links could be coupled together. Therefore, we take the approach of using scaffold cross-overs. Essentially we will fold four different links out of the same scaffold, and then route the scaffold to form ssDNA connections at the hinge locations. Using scaffold based hinges ensures that connections between links are made with 100 % efficiency and practically eliminates the possibility of forming linkages with more or less than four links.

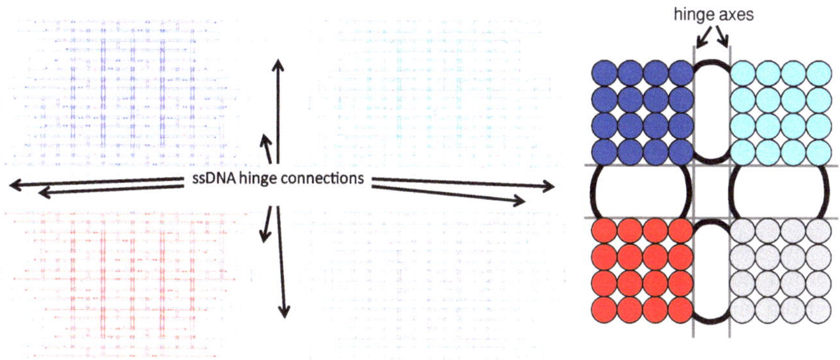


**Fig. 44.4** Design of hinge joints and the Bennett mechanism with scaffolded DNA origami

Figure 44.4a shows the design blueprint for all four links joined by ssDNA connections. Figure 44.4b depicts a cross-sectional view of the four links denoting the positions of the hinge connections. The solid black lines indicate connections that are made on the front side of the links, and dotted black lines indicate connections that are made on the back side of the links. Two elements were taken into account when designing the ssDNA hinge connections. First, owing to the helical nature of DNA, the scaffold backbone rotates about the long axis as it travels along the length of each cylinder. The ssDNA connections between links were placed at a position where the scaffold was pointing in the appropriate direction. For example, in the bottom most hinge connection shown in Fig. 44.4b between the red and gray links, the connection was placed at a position where the scaffold was pointing to the right. Second, for every link, the hinge axes converge at a point slightly outside the link cross-section as shown in Fig. 44.4b. For example, for the red link, the hinge axes converge at a point just above and to the right of the top right cylinder. Since the ssDNA connections are placed slightly removed from that point, a hinge with zero length would result in interference when the bennett linkage folds up. Therefore, the hinge connections were designed to be 3 ssDNA bases long ($\sim$1.5 nm). This also ensures flexibility of the hinge.

### 44.2.4 Fabrication Process and Prototypes

The fabrication of DNA origami structures is done by a molecular self-assembly process which is driven by Watson and Crick DNA base-pairing [14] between the long scaffold molecule and many shorter staple molecules. The culmination of the design process described above is a list of staple sequences that when mixed with the scaffold at the appropriate solution conditions, will form the desired structure in order to maximize base pairing. The equilateral bennett linkage was assembled with a 7,560 base long ssDNA scaffold derived from M13MP18 bacteriophage and

179 synthetic ssDNA staples (MWG eurofins) ranging in length from 20 to 48 bases. The self-assembly reaction was performed in a buffer solution containing 22 mm $MgCl^2$, 5 mm TRIS, 5 mm NaCl, and 1 mm EDTA [22]. The scaffold was put into the folding reaction at a concentration of 20 nm and the staples at 200 nm (each of the 179 staples is at 200 nm). The 10-fold excess of staples is used to ensure that the scaffold gets completely populated with full length staples (a small fraction of incomplete staples may result from the staple synthesis process).

The folding reaction was prepared in a volume of 50 μl and subjected to a thermal annealing ramp where the reaction is heated to initially melt staple-scaffold interactions and then very slowly cooled. The structure folding results are usually sensitive to cooling rate, and it is often helpful to try different cooling rates as well as non-constant cooling rates to optimize folding results. Here, the folding reaction was heated up to 65 °C and then slowly cooled down to room temperature over a time scale of 65 h and then stored at 4 °C. Typical successful folding reaction yields vary from ~5 to 40 %. Well-folded structures can be purified away from misfolded structures, aggregated structures, and excess staples using agarose gel electrophoresis. Purified well-folded structures can then be imaged by negative stain TEM. Protocols for gel purification and TEM are given in [22].

Figure 44.5 shows a TEM image of the DNA origami Bennett mechanism prototypes. Since the Bennett linkage has mobility one, and no actuators were used in our design, DNA origami mechanisms can be at any configuration constrained by the kinematics of Bennett linkages (to be discussed in the following section). For instance, Fig. 44.5a and d show that mechanisms are in the completely folded and the completely open position respectively. And Fig. 44.5b shows that the mechanism in a near folded position while (c) shows a pop-up configuration.

## 44.3 Kinematic Analysis of the Bennett Linkage

In this section, we will analyze the kinematics of the Bennett linkage. Although many authors have worked on this subject [34], all the existing work is based on the twist angles and normal distance between two adjacent joint axes. However these parameters cannot be directly measured from physical prototypes of DNA origami mechanisms. Therefore we would like derive the kinematic constraint equation based on the angles between edges on two adjacent links which can be easily measured from TEM images of physical prototypes.

### *44.3.1 The kinematic Constraint Equation*

Figure 44.6a shows the schematic view of the spatial Bennett 4R linkage at the open position. To do kinematic analysis of this linkage, we build the global

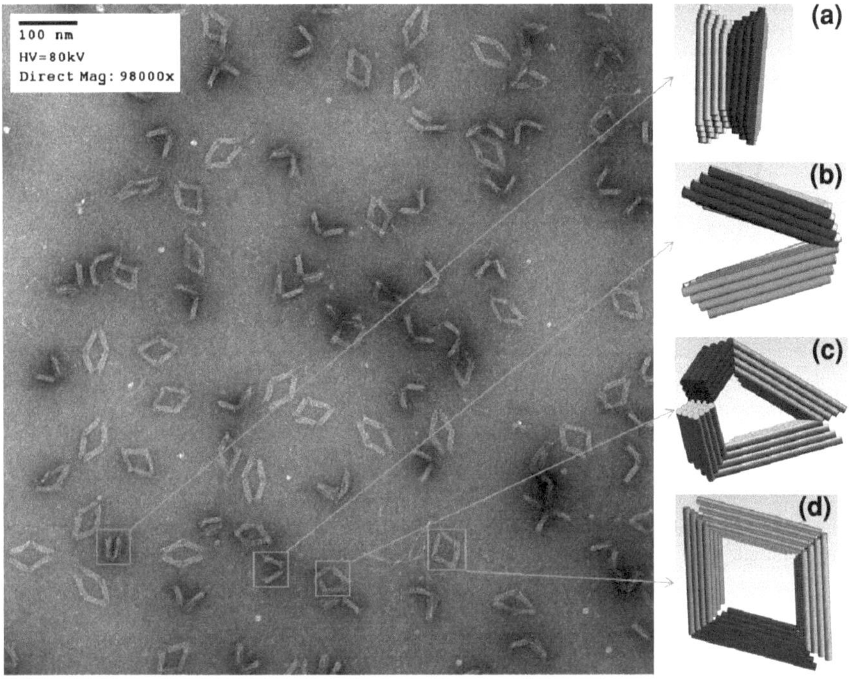


**Fig. 44.5** Prototypes of DNA origami Bennett mechanisms showing **a** in the completely folded configuration, **b** near the folded position, **c** in a pop-up position and **d** in the completely open position

coordinate frame at the corner $A$ of one link. The $x$ axis is chosen along one edge $AB$ of the link. The $xy$ plane is chosen to be the plane formed by edges $AB$ and $AD$ when the linkage is at the fully open position.

The length of links $AB$, $BC$, $CD$ and $DA$ are all defined by $l$. Please note that this $l$ is not the link length along the common normal of two adjacent axes. However these are the actual dimensions which we can measure from TEM images of DNA origami mechanisms. The four joint axes are along edges $AE$, $BF$, $CG$ and $DH$ defined by unit vectors $\mathbf{S}_i\ (i = 1, \ldots, 4)$. And the four joint angles are defined by $\theta_i$.

We use kinematic analysis to derive the relationship between these angles. Since it is known that $\theta_1 = \theta_3$ and $\theta_2 = \theta_4$, we only have to derive the relationship between $\theta_1$ and $\theta_2$. This is done as follows. First, the link $DA$ is rotating about axis $\mathbf{S}_1$ by an angle $\theta_1$. Hence the coordinates of point $\mathbf{D}$ only depend on $\theta_1$. Similarly the coordinates of point $\mathbf{C}$ depend only on $\theta_2$. That is,

$$\mathbf{D} = \left[e^{\theta_1 \mathbf{S}_1}\right]\begin{Bmatrix} l \\ 0 \\ 0 \end{Bmatrix}, \quad \mathbf{C} = \mathbf{B} + \left[e^{\theta_2 \mathbf{S}_2}\right]\begin{Bmatrix} -l \\ 0 \\ 0 \end{Bmatrix}, \tag{44.1}$$

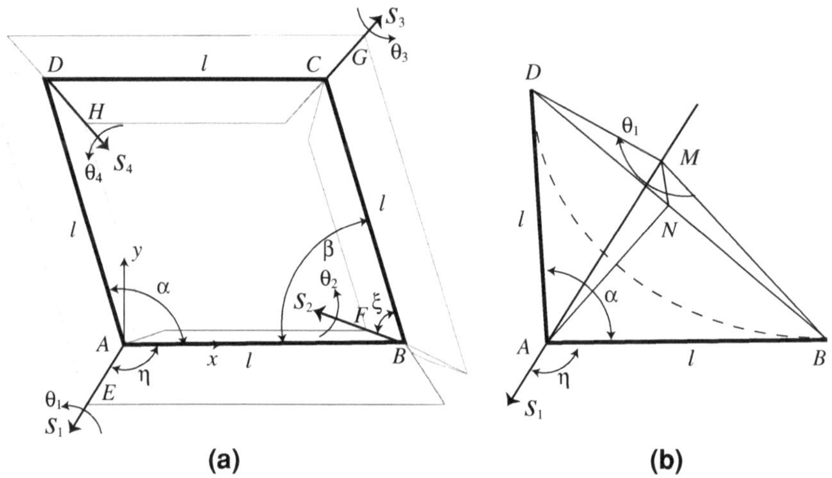


**Fig. 44.6** **a** Schematic of a spatial Bennett 4R linkage. **b** Relationship between angle $\alpha$ and rotation angle $\theta_1$

**Table 44.1** Dimensions of the DNA Bennett linkage

| $l = 27.8$ nm, $\eta = 123^\circ$, $\xi = 57^\circ$ |
|---|
| $\mathbf{S}_1 = (-0.544984, -0.734471, -0.404406)$, |
| $\mathbf{S}_2 = (-0.544984, 0.404406, -0.734471)$ |

where $\left[e^{\theta_i \mathbf{S}_i}\right]$ are the 3 by 3 matrix defining the rotation about an arbitrary axis $\mathbf{S}_1$ for $\theta_i$. The values of $\mathbf{S}_i$ are given in Table 44.1

The kinematic constraint equation is obtained by equating the distance between points **D** and **C** to length $l$, i.e.

$$(\mathbf{D} - \mathbf{C}) \cdot (\mathbf{D} - \mathbf{C}) - l^2 = 0, \tag{44.2}$$

which we solve for $\theta_2$ for any given value of $\theta_1$ or vice versa.

## *44.3.2 Calculating Angles Between Edges on Links*

Angles $\theta_1$ and $\theta_2$ cannot be directly measured from experimental TEM images. Therefore we would like to calculate angles $\alpha$ and $\beta$ which are the angles between links *BA* and *DA* and *AB* and *CB* respectively.

By observing the geometry shown in Fig. 44.6b, we have

$$|NB| = |MB| \sin(\theta_1/2), \quad \text{where } |NB| = l \sin\left(\frac{\alpha}{2}\right), \ |MB| = l \sin \eta. \tag{44.3}$$

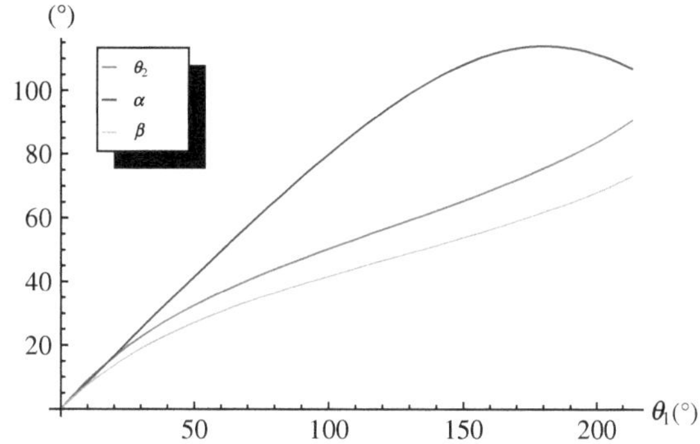


**Fig. 44.7** Kinematic analysis of the Bennett linkage. The linkage is completely folded at $\theta_1 = 0$ and completely unfolded (*open*) at $\theta_1 = 213.4\,^\circ$.

We obtain the relationship between $\theta_1$ and $\alpha$ as

$$\sin\left(\frac{\alpha}{2}\right) = \sin\eta \sin\left(\frac{\theta_1}{2}\right). \tag{44.4}$$

Similarly we can also calculate $\beta$ from $\theta_2$ as

$$\sin\left(\frac{\beta}{2}\right) = \sin\xi \sin\left(\frac{\theta_2}{2}\right). \tag{44.5}$$

By substituting the design parameters shown in Table 44.1, we can plot $\theta_2$, $\alpha$ and $\beta$ versus the input angle $\theta_1$ as in Fig. 44.7. See Chen and You [35, 36] for similar work on kinematic analysis of the alternative form of Bennett linkages.

## 44.4 Kinematic Validation of Prototypes

Our design process is based on the assumption that the stiffness of the links is much higher than that of joints. Also we assume that the rotational stiffness of the joint in the axial direction is much lower than in other directions. Since currently the stiffness of the joints and links in prototypes cannot be determined physically, the only way to verify the results is to check if the angles between links of the DNA origami prototype follow the kinematics of the Bennett linkage described in the previous section.

TEM images provide a 2D projection of the 3D geometry of structures deposited on a surface. Here we assume the DNA mechanism lands so that the plane formed by the bottom edges of two adjacent links, e.g. the *xy* plane formed by edges *AB* and *BC*, is flat on the imaging surface when the linkage is at the open position as shown in Fig. 44.6a. In keeping with this assumption, we only study the cases when the angle $\beta$ is sufficiently large. We measure the angles between links from the TEM images, denoted by $\alpha'$ and $\beta'$. Please note these are angles $\alpha$ and $\beta$ projected on the *xy* plane. Since the mechanism lands on the edges *AB* and

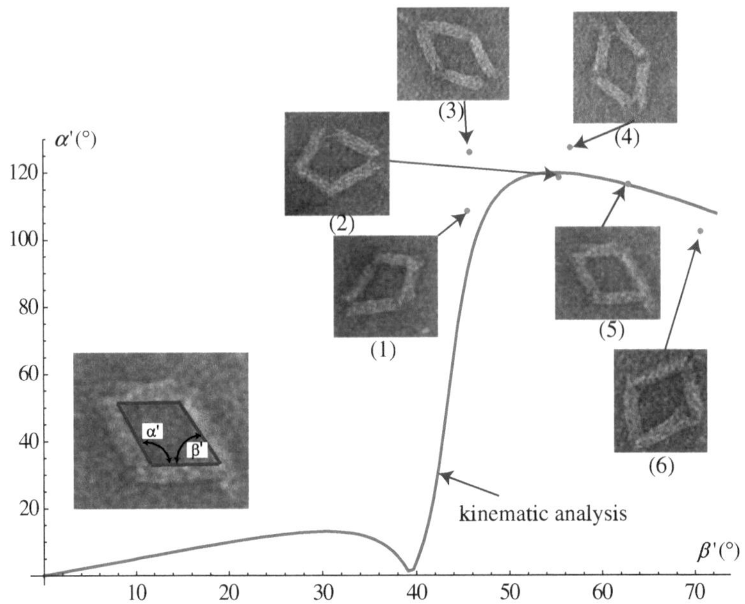


**Fig. 44.8** Kinematic validation of physical prototypes. The *curve* is obtained by the kinematic analysis. And *dots* represent sample prototypes

$BC$, we have $\beta' = \beta$. By following geometry of the Bennett linkage, we can also obtain the projected angle $\alpha'$. These angles are determined as follows.

$$\beta' = \beta, \quad \alpha' = \cos^{-1}\left(\frac{\mathbf{D}' \cdot \mathbf{B}}{l^2}\right), \tag{44.6}$$

where $\mathbf{B} = (l, 0, 0)$. And $\mathbf{D}' = (D_x, D_y, 0)$ is the projection point of $\mathbf{D}$ onto the $xy$ plane which can be derived from the kinematics of Bennett linkage. We plot $\alpha'$ versus $\beta'$ as the curve in Fig. 44.8.

Figure 44.8 shows the measured angles $\alpha'$ and $\beta'$ from six sample prototypes which are selected such that $\alpha'$ and $\beta'$ are sufficiently large. It can be seen that when the prototype is in the open position (see (5) in Fig. 44.8), $\alpha' + \beta'$ is close to 180°.

## 44.5 Conclusions and Future Work

In this paper, we have introduced the concept of DNA Origami Mechanisms and Machines (DOMM). These DOMM are formed with bundles of double stranded DNA helices (relatively rigid) resembling links which are joined with single stranded DNA connections (relatively compliant) resembling hinges. The design and fabrication process of a DNA origami Bennett 4-bar mechanism was presented. We showed that physical prototypes follow the analytical kinematics of rigid body mechanisms fairly well. We believe this process is applicable to general DOMM. Currently we are working on the design of other kinds of kinematic joints as well as various actuators to control the motion of DOMM. To the kinematics and mechanisms community, this research has the potential of opening a new era of design, analysis and manufacture of nanomechanisms, nanomachines and nanorobots.

## References

1. McCarthy JM (2000) Geometric design of linkages. Springer, New York
2. Erdman AG, Sandor GN, Kota S (2001) Mechanism design analysis and synthesis, 4th edn. Prentice Hall, Upper Saddle River
3. Chirikjian GS, Kazerounian K, Mavroidis C (2005) Analysis and design of protein based nanodevices: challenges and opportunities in mechanical design. J Mech Des 127(4):695–698
4. Chirikjian GS (2001) Conformational statistics of macromolecules using generalized convolution. Comput Theor Polym Sci 11(2):143–153
5. Kazerounian K (2004) From mechanisms and robotics to protein conformation and drug design. J Mech Des 126(1):40–45
6. Kazerounian K, Latif K, Alvarado C (2005) Protofold: a successive kinetostatic compliance method for protein conformation prediction. J Mech Des 127(4):712–717
7. Mavroidis C, Dubey A (2004) Bio-Nanorobotics: state of the art and future challenges. In: Yarmush ML (ed) The Biomedical Engineering Handbook, 3rd edn. CRC Press LLC
8. Hamdi M, Ferreira A (2009) Multiscale design and modeling of protein-based nanomechanisms for nanorobotics. Int J Robot Res 28(4):436–449
9. Sharma G, Mavroidis C, Rege K, Yarmush ML, Budil D (2009) Computational studies of a protein-based nanoactuator for nanogripping applications. Int J Robot Res 28(4):421–435
10. Dill KA, Ozkan SB, Shell MS, Weikl TR (2008) The protein folding problem. Ann Rev Biophys 37:289–316
11. Kuhlman B, Dantas G, Ireton G, Varani G, Stoddard B, Baker D (2003) Design of a novel globular protein fold with atomic-level accuracy. Science 302(5649):1364–1368
12. Baker D (2006) Prediction and design of macromolecular structures and interactions. Philos Trans R Soc B Biol Sci 361(1467):459–463
13. Leaver-Fay A, Jacak R, Stranges PB, Kuhlman B (2011) A generic program for multistate protein design. PLos ONE 6(7):e20937
14. Watson J, Crick F (1953) Molecular structure of nucleic acids: a structure for deoxyribose nucleic acid. Nature 171(4356):737–738
15. Yakovchuk P, Protozanova E, Frank-Kamenetskii MD (2006) Base-stacking and base-pairing contributions into thermal stability of the DNA double helix. Nucleic Acids Res 34(2):564–574
16. Seeman NC (1982) Nucleic acid junctions and lattices. J Theor Biol 99(2):237–247
17. Chen JH, Seeman NC (1991) Synthesis from DNA of a molecule with the connectivity of a cube. Nature 350(6319):631–633
18. Li X, Yang X, Qi J, Seeman NC (1996) Antiparallel DNA double crossover molecules as components for nanoconstruction. J Am Chem Soc 118(26):6131–6140
19. Seeman NC (2003) DNA in a material world. Nature 421(6921):427–431
20. Rothemund PWK (2006) Folding DNA to create nanoscale shapes and patterns. Nature 440(7082):297–302

21. Douglas SM, Dietz H, Liedl T, Hogberg B, Graf F, Shih WM (2009) Self-assembly of DNA into nanoscale three-dimensional shapes. Nature 459(7245):414–418
22. Castro CE, Kilchherr F, Kim D-N, Shiao EL, Wauer T, Wortmann P, Bathe M, Dietz H (2011) A primer to scaffolded DNA origami. Nat Methods 8(3):221–229
23. Ke Y, Sharma J, Liu M, Jahn K, Liu Y, Yan H (2009) Scaffolded DNA origami of a DNA tetrahedron molecular container. Nano Lett 9(6):2445–2447 (PMID: 19419184)
24. Jungmann R, Scheible M, Kuzyk A, Pardatscher G, Castro CE, Simmel FC (2011) DNA origami-based nanoribbons: assembly, length distribution, and twist. Nanotechnology 22(27):275–301
25. Maune HT, Han S-P, Barish RD, Bockrath M, Iii WAG, Rothemund PWK, Winfree E (2010) Self-assembly of carbon nanotubes into two-dimensional geometries using DNA origami templates. Nat Nanotechnol 5(1):61–66
26. Liedl T, Hogberg B, Tytell J, Ingber DE, Shih WM (2010) Self-assembly of three-dimensional prestressed tensegrity structures from DNA. Nat Nanotechnol 5(7):520–524
27. Dietz H, Douglas SM, Shih WM (2009) Folding DNA into twisted and curved nanoscale shapes. Science 325(5941):725–730
28. Andersen ES, Dong M, Nielsen MM, Jahn K, Lind-Thomsen A, Mamdouh W, Gothelf KV, Besenbacher F, Kjems J (2008) DNA origami design of dolphin-shaped structures with flexible tails. ACS Nano 2(6):1213–1218
29. Douglas SM, Marblestone AH, Teerapittayanon S, Vazquez A, Church GM, Shih WM (2009) Rapid prototyping of 3d DNA-origami shapes with cadnano. Nucleic Acids Res 37(15):5001–5006
30. Bennett GT (1914) The skew isogram mechanism. Proc London Math Soc s2-13(1):151–173
31. Chen Y (2009) Design of Structural Mechanisms. D.Phil. Dissertation, University of Oxford, UK
32. Ke Y, Douglas SM, Liu M, Sharma J, Cheng A, Leung A, Liu Y, Shih WM, Yan H (2009) Multilayer DNA origami packed on a square lattice. J Am Chem Soc 131(43):15903–15908
33. Lilley D (1997) All change at holliday junction. Proc Natl Acad Sci U S A 94(18):9513–9515
34. Perez A, McCarthy J (2002) Bennett's linkage and the cylindroid. Mech Mach Theor 37(11):1245–1260
35. Chen Y, You Z (2006) Square deployable frames for space applications. Part 1: Theory. Proc Inst Mech Eng [G] 220(4):347–354.
36. Chen Y, You Z (2007) Square deployable frames for space applications. Part 2: Realization. Proc Inst Mech Eng [G] 221(1):37–45

# Chapter 45
# Multibody Modelling Applied to Origami Carton Folding

**Ferdinando Cannella, Jian S. Dai and Daniele Clari**

**Abstract** This paper presents an application of the virtual prototyping on the reconfigurable mechanisms and in particular, in this study, is shown a multibody origami carton folding model. The aim of this work is to reproduce via numerical model the D-RAPS reconfigurable multifinger robot that is in use for this kind of application. Thanks to the trajectory matching, the authors reach model validation only by comparing analytical and numerical results. In addition, this model permits to investigate the carton folding deeper than via experimental tests. In fact, as example, the contact forces between the mechanism finger and carton panel are computed.

**Keywords** Reconfigurable mechanism · Origami carton folding · Multibody simulation

## 45.1 Introduction

In the highly competitive consumer market, new carton product models come out frequently with fancy shapes, attractive styles and various closing methods, to form tempting containers for consumer products, mostly in small production

---

F. Cannella (✉)
Dipartimento di Advanced Robotics, Fondazione Istituto Italiano di Tecnologia, via Morego 30, Genova, Italy
e-mail: ferdinando.cannella@iit.it

J. S. Dai · D. Clari
Department of Mechanical Engineering, King's College London, University of London, Strand, London C2R 2LS, UK
e-mail: jian.dai@kcl.ac.uk

D. Clari
e-mail: danieleclari@email.it

J. S. Dai et al. (eds.), *Advances in Reconfigurable Mechanisms and Robots I*,
DOI: 10.1007/978-1-4471-4141-9_45,  

batches. Packaging industry which creates these cartons has problems in automation to the extent that manual effort is still required. This not only affects the quality, but also leads to wrist labor injuries due to the constant twisting motion in various carton folding processes. Examples of manual operation are commonly found in prestigious personal care product packaging including cosmetics and perfumery packaging, [1]. Historically, manual methods have been used because they are able to adapt to different types of carton with changing styles. This changing styles takes about 10 % of the work order and is called upon as assembly lines to take the order of promotional products [2]. Though manual workers are used for diverse and assorted cartons, dedicated machines are used for simple cartons and are specifically designed for one specific type of cartons. Moreover, in the confectionary market, when the carton shape is elaborate, it is possible to compare these masterpieces to the origami, according to the complicated procedure to fold them. The main characteristic of these cartons is that they are not fold by glue, but by tuck-in operation; this requires that the cartonboard are joint each other trough flaps and slots. Considering the complexity of this operation, every single carton shape has different folding procedures [3]. They require, in most cases, a change of more than 40 points [4] to fit into the same type of carton of different dimensions, which means one specific type of cartons requires one packaging line. Capital expenditure increases with the changeover [5] from one type to another type of carton folding assembly lines. Thus the flexibility is lost due to these limitations and the associated cost in the changeover. So in the last decades the challenge for companies and research centers was to develop a manipulation planning strategy for a new generation of packing devices, which can deal with different types of cartons of reasonable dimensions without any manual adjustment. Automation with reconfigurability of this kind of production remains a challenge to technology in current carton packaging machines and raises an issue on how to automatically handle and manipulate complex and diverse cartons of different product models. One of the solutions was the Dexterous Reconfigurable Assembly and Packaging System (D-RAPS), developed and used as carton folding test rig [2, 4]. Thanks to this robot the complex analytical theories [6] and algorithmic models [7, 8] on the folding process created by the computational geometry community [9] could be proved. In order to improve this device and using it for carton folding investigation, the authors implemented D-RAPS into MSC.ADAMS program. In this way all multibody software advantages were available for studying the virtual test rig model. The aim of this work is to show these advantages compared to the previous method.

## 45.2 Carton Folding

The manipulation process of carton folding has to be investigated in order to develop algorithms for the identification of folding sequences and corresponding desired trajectories of packaging cartons; moreover this investigation is necessary for creating a manipulation planning strategy and to develop a generic manipulation planning system for carton folding applications in multifinger robotic packaging. As far as analysis is concerned, it is essential to simplify the complex process into a simple one based on mathematics. In this section the problem of complex carton folding process is reduced to conventional mechanism analysis based on graph theory and robotics [10]. The challenge can be formulated in this way: given geometric parameters, initial positions and orientations, target position and orientation of various cartons, a system of multiple cooperative robotic fingers in reachable workspace generates feasible sequences and corresponding trajectories for cartons. These cartons could be modeled as articulated objects in the packing process. Modelling is a useful tool to describe the carton folding, so a concept of an equivalent mechanism is proposed [11]: carton creases, which are dominant element in carton folding, as revolute joints and carton panels, which do not need to bend during carton folding process, as mechanism links. This study has been undertaken using screw theory, matrix theory or graph theory [2, 12]. Moreover the assembly sequence depends on mechanical characteristic as the creases [13], panels [14] and relative position in the final configuration. Computer-aided process planning forms a vital link in the realization of a complexly automated manufacturing environment [15, 16]. Some authors addressed a temporary framework for assembly sequence representation and analysis, and showed that assembly sequence properties can be formulated and rigorously provided via mechanical theorem [17]. According to the aim of this work, the carton is modeled as equivalent mechanism with joint and links; afterward the kinematics (in particular the trajectories) is computed as shown in the next section.

## 45.3 Case Study

The origami carton chosen as case study is shown in Fig. 45.1a in flattened state, b folded state. Firstly carton model and its mobility must be modeled as an equivalent mechanism.

The carton is formed by a combination of different geometric shapes from flattened to final complete pack. It is possible to figure out that the equivalent mechanism in open loop, as shown in the Fig. 45.2a, changes its mobility during the folding, for instance, when the panel 15 is attached to the panel 14 the structure is different, as shown in the Fig. 45.2b. The mobility was analyzed by using the Grübler–Kutzbach criterion [18].

$$M_e = m_c + m_o = b\,(n_c - g_c - 1) + g_c + g_o \tag{45.1}$$

where $m_c$ is the mobility in closed loops, $m_o$ is the mobility in open loops and is equal to number of joints, $g_o$, in open loops, $b$ the order of the screw system formed by joint axes of a closed loop, $n_c$ the number of links in a closed loop and $g_c$ the number of joints in a closed loop. The application of this equation to the

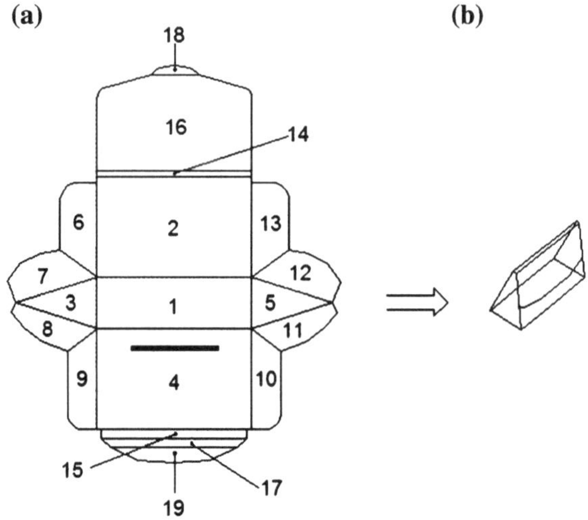


**Fig. 45.1** Origami carton in **a** flattened state, **b** folded state

equivalent mechanism model shown in Fig. 45.2a gives as flattened mobility result $M_f = 21$. In the folding configuration state there is a mixture between closed and open loop: the closed loop involves panels 14, 2, 1, and 4, as shown in Fig. 45.2b, which behave as a four bar sub-mechanism and the mobility is $m_c = 1$; the open loop has $m_o = g_o = 14$ (where $m_c$ is the mobility in closed loops, $m_o$ is the mobility in open loops and $g_o$ number of joints in an open loop). Hence the mobility from flattened to the erectable state passes from $M_f = 21$ to $M_e = 1 + 14 = 15$. So it can be considered a new mechanism. In order to compute the carton folding trajectories Eq. (45.2) is used [1]. It is also the trajectory that the multiple robotic finger tracks. The corresponding simulation is shown in the following section, where these trajectories will be applied in the physical carton folding with dexterous reconfigurable assembly packaging system.

$$\begin{aligned} \boldsymbol{T}_j = [&\boldsymbol{T}^1_{1g(6,7)}, \boldsymbol{T}^1_{2g(8,9)}, \boldsymbol{T}^1_{3g(10,11)}, \boldsymbol{T}^1_{4g(12,13)}, \boldsymbol{T}^2_{5q(1,3)}, \boldsymbol{T}^2_{6q(1,5)}, \boldsymbol{T}^3_{7q(1,4)}, \\ &\boldsymbol{T}^4_{8q(1,2)}, \boldsymbol{T}^5_{9q(4,17)}, \boldsymbol{T}^5_{10q(17,19)}, \boldsymbol{T}^6_{11q(2,16)}, \boldsymbol{T}^7_{12q(16,18)}]^T \end{aligned} \tag{45.2}$$

where $\mathbf{T}_{k,\,j}$ 4 × 4 transformation matrix by translating from link $j$ to link $k$. Then the motion and the configuration of the carton have been analyzed in order to find the relationships between the carton motion and mechanism motion. The configuration formula of adjacent panels is based on configuration transformation analysis of the carton motion.

Then a hereditary connectivity matrix was composed to identify the folding sequence of a carton and its corresponding folding trajectory, in terms of the adjacency matrix and the configuration transformation of the carton motion [1]. The folding trajectories in the simulation represent configuration transformation between the initial configuration, process configurations and folded configuration. The simulated test case carton in shown in the Fig. 45.3 in its flattened, half

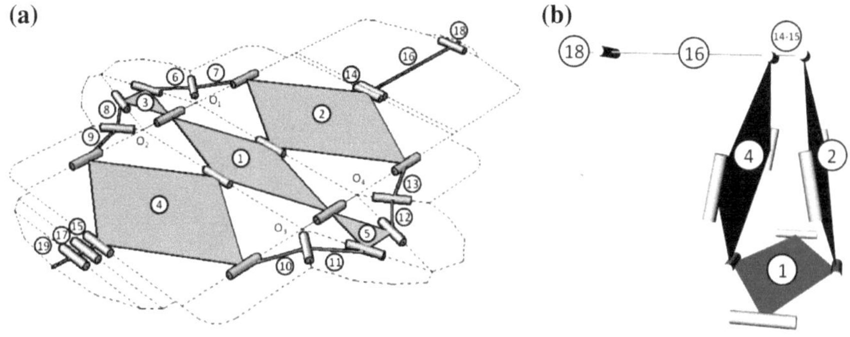


**Fig. 45.2** **a** Equivalent mechanism open-loop. **b** Equivalent mechanism in closed-loop

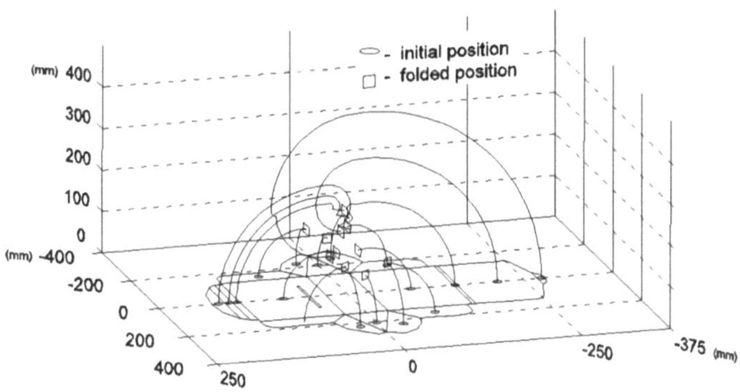


**Fig. 45.3** Carton folding trajectories

erected and folded configuration. These trajectories were used by the authors for studying the finger displacements in the D-RAPS robot [2, 19].

## 45.4 Experimental Test Rig

The D-RAPS robot is a Dexterous Reconfigurable Assembly and Packaging System, compound by dexterous fingers. It was built in order to demonstrate that the process time could be reduced not only during the set up, but even during the carton folding. The manual process of packaging was investigated in order to discover the degree of dexterity required and the procedure for folding the chosen carton. In order to simulate the packaging processes, kinematic model of carton as well as packaging machine components (fingers, pushers and the turn-table) need to be developed and integrated together [19]. The kinematic model of the carton has been well studied [20, 21], so in this work only the contact between the fingers

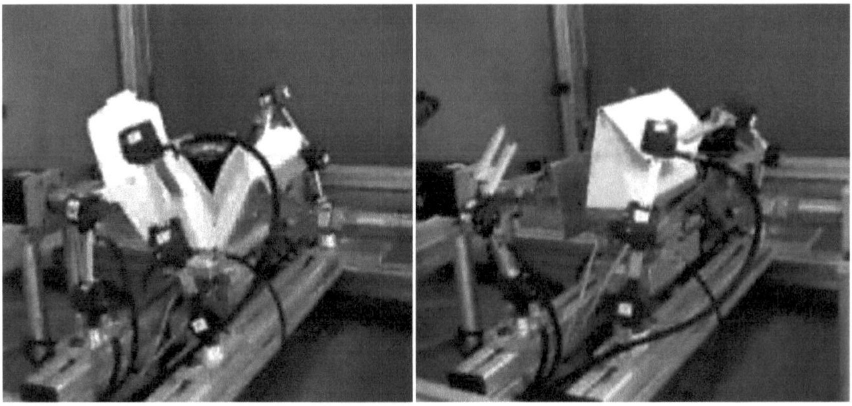

**Fig. 45.4** Two actions of the experimental carton folding process

and the carton was presented. This condition is reached locating various contact points on the carton and recording the displacement of these points as the folding of the carton takes place. The contact points on the carton can be identified by geometrical interpretation of the folding sequences [22]. Moreover, following the previous section, it is possible to compute the trajectories of the single finger [23]. These data are directly used to drive the fingers.

Contact points are then used to find joint displacement for each fingers' joints. A simulation provides many valuable information that is used for experimental packaging-machine design as well as its control [24]: position of the fingers, motion-data, trajectories. The data-file generated by the simulation was fed to the controller after ensuring that both the model and the machine have geometric and configuration equivalence, as shown in Fig. 45.4. That permitted to establish the capability of the packaging system in performing erection and folding of cartons.

## 45.5 Multibody Simulation

The multi body model of D-RAPS is built in order to simulate carton folding with numerical software. The MSC.ADAMS was used to reach this aim. First of all the carton model, as shown in the Fig. 45.5a, and the reconfigurable mechanism are built, as shown in the Fig. 45.5b. The origami carton was divided into two parts: the panels and the creases. The panels and the creases have a complex behaviour [13, 14] and they are a challenge for the virtual prototyping modelling [3, 15]; considering that the comparison is with rigid panel models [1, 10, 23], their compliance is not take in account.

The D-RAPS mathematic model is composed by 22 different CAD parts (built in PRO-E cad environment) assembled in ADAMS environment. As the physical robot, the design has four fingers: two with three degrees of freedom and two with

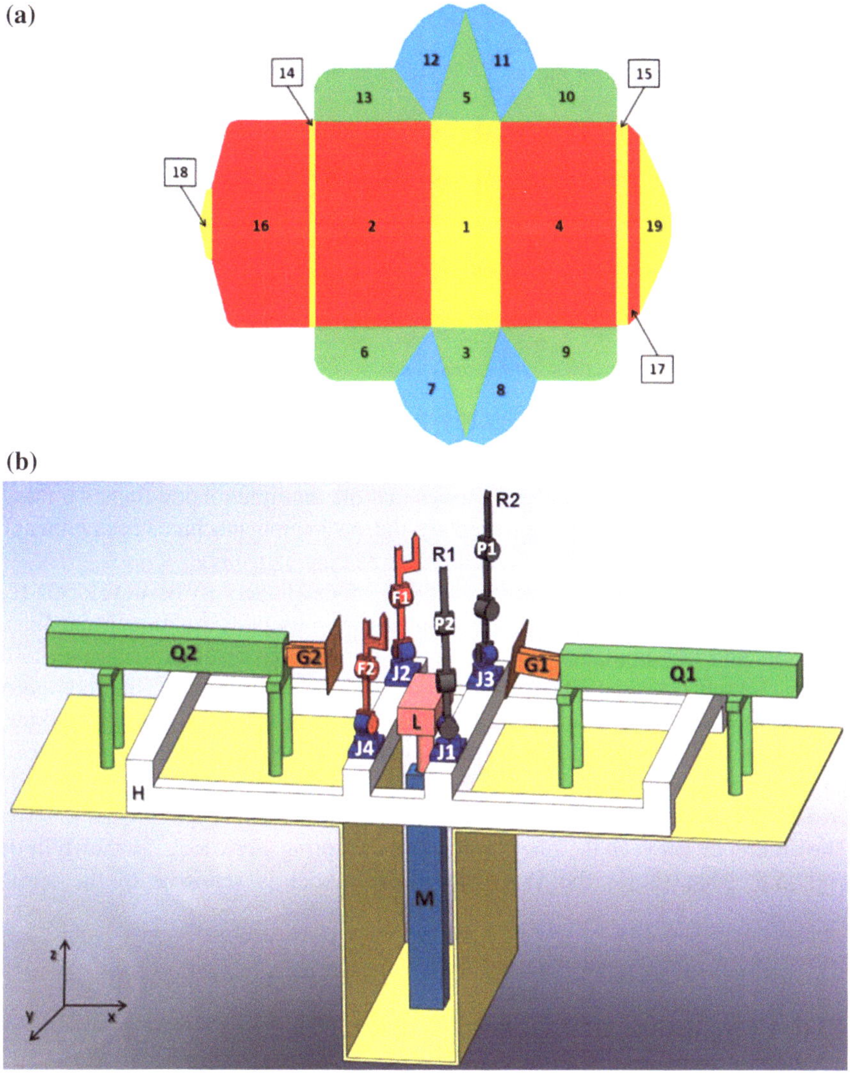


**Fig. 45.5** MSC.ADAMS **a** (*top*) carton model, **b** (*down*) D-RAPS multibody model

two degrees of freedom each. The three-degree-of-freedom fingers provide a yaw motion at the base and pitch motions on the following two joints forming a Y–P–P configuration. The two-degree-of-freedom fingers have only pitch motions allowing it to move on a planar surface. Differently to the physical model, the virtual one has sensor in each component: sensors for measuring forces on the fingers during the contact with panels, for measuring the torque supplied by motors or absorbed by springs during the folding. Moreover all the mechanical properties of the model can be changed in few seconds. The carton panel trajectories obtained

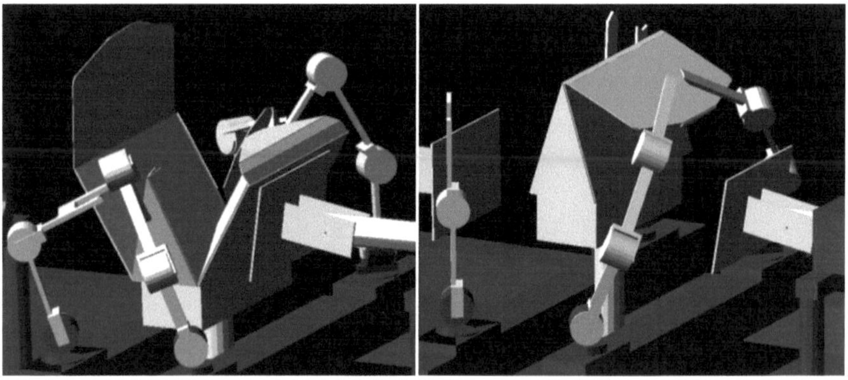

**Fig. 45.6** Two actions of carton folding process simulation

from the equation are also the trajectories that the multiple robotic fingers track. If the carton folding is to be achieved by the packaging machine, each movable element of the machine needs to be connected cinematically.

This requires inverse kinematic solution of the fingers to be developed [2]. Then, the obtained joint angles are input to the multibody rotational and linear actuators and the simulation were done. First simulation result is the sequence of carton folding, as shown in the Fig. 45.6. Second result is the computed displacements of panels, as shown in the Fig. 45.7. The physical finger R1 (shown in the Fig. 45.5b) displacements are shown and compared with the numerical finger R1 displacements; the agreement can be considered very good. Then the creases are modeled with nonlinear spring [13] and another simulation was run. The difference between the linear and nonlinear spring were done, as shown in the Fig. 45.8. That shows the D-RAPS virtual model is sensitive to the spring properties.

## 45.6 Conclusions and Future Work

There are perfect coincidences between the carton folding sequence of the physical model, shown in the Fig. 45.4, and the numerical one, shown in the Fig. 45.7. That means the motor angle input is well interpolated by the MSC.ADAMS solver. In fact the given trajectories of fingers from the theory and these ones computed, shown in the Fig. 45.7, are almost coincident, the error is less than 1 %, due to the computational approximation. The last comparison is between the contact forces measured with linear crease stiffness and nonlinear crease stiffness: the Fig. 45.8 shows that with linear spring it is necessary apply more force in order to begin and finish the folding, but less peak force.

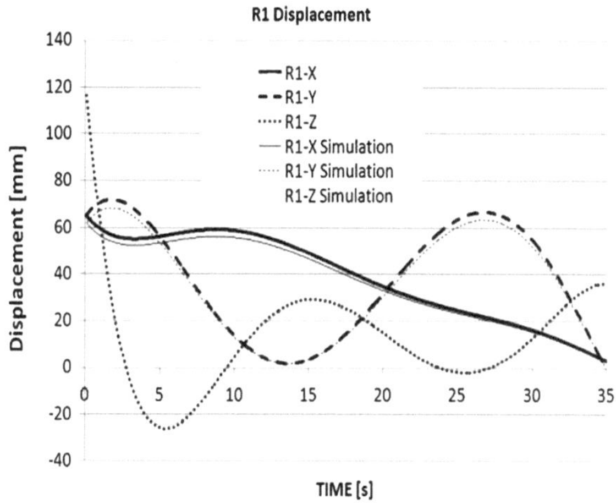


**Fig. 45.7** Finger R1 displacement, X, Y and Z mean along X, Y and Z axes respectively

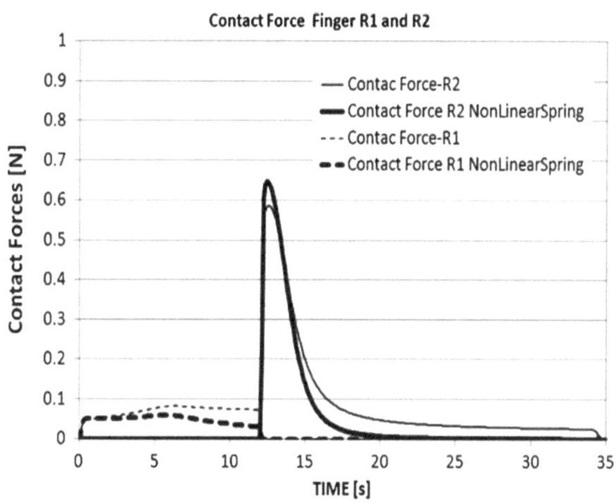


**Fig. 45.8** Finger R1 and R2 contact force with linear and nonlinear spring (simulation)

Considering that is easy to change the mechanical properties of D-RAPS virtual model or to set the motor parameters, that lead to have the following advantages: equations derived from the theory can be quickly proved; the difference between different mechanical properties (more complex crease stiffness behaviour and panel compliance) or carton configuration or actuator parameter can be appreciate; measurement of torque, force, displacement, speed, acceleration, stress and strain can be recorded; when the inverse kinematics is too complex to solve in an analytical way, it is possible to compute it and verify the results comparing with result of the direct one. Moreover new carton folding kinematics can be optimized in order to find the best sizes, without any physical prototyping; that means saving time and increasing the quality of research results. Cause of that MSC.ADAMS was used: for having not only the kinematics, but even the dynamics of the carton folding.

## References


1. Liu HH, Dai JS (2002) Carton manipulation analysis using configuration transformation. J Mech Eng Sci 416:543–555
2. Dubey VN, Dai JS (2007) Complex carton packaging with dexterous robot hands. In: Huat LK (ed) Industrial robotics: programming, simulation and application, pp 583–594, InTech
3. Cannella F, Dai JS (2009) Origami-carton tuck-in with a reconfigurable linkage. In: ReMaR 2009, IEEE Press, London, pp 512–520
4. Dai JS (1996) Survey and business case study of the dexterous reconfigurable assembly and packaging system (D-RAPS). In: Science and technology report, No PS960321. Unilever Research, UK
5. Dai JS (1996) Conceptual design of the dexterous reconfigurable assembly and packaging system (D-RAPS). In: Science and technology report, No PS960326, Unilever Research, UK
6. Ambekar AG, Agrawal VP (1987) Identification of kinematic chains, mechanism, path generators and function generators using min codes. J Mech Mach 22(5):463–471
7. Kawahata F (1995) Fantasy origami. Gallery Origami House, Tokyo
8. Kawasaki T (1989) On the relation between mountain-creases and valley-creases of a flat origami. In: Proceedings of the 1st international meeting on origami science and technology, pp 229–237
9. Biedl T, Demaine A, Lubiw E, O'Rourke J, Overmars M, Robbins S, Whitesides S (1998) Unifolding some classes of orthogonal polyedra. In: Proceeings of the 10th canada conference on computing and geometry, pp 70–71
10. Liu H, Dai JS (2002) Manipulation analysis of carton-folding task by using GGPs and CCPs. In: Proceedings of the 2002 IEEE international symposium on intelligent control, pp 637–641, IEEE Press, Vancouver
11. Dai JS, Rees JJ (1998) Mobility ub metamorphic mechanism of foldable/erectable kinds. In: 25th ASME Biennial mechanisms and robotics conference, Atlanta
12. Guang S, Nancy MA (2001) A motion planning approach to folding from paper craft to protein folding. In: IEEE robotics and automation conference
13. Cannella F, Dai JS (2006) Crease stiffness and panel compliance of carton folds and their integration in modelling. J Mech Eng Sci Proc IMechE 220(6):847–855
14. Cannella F, Dai JS (2008) Stiffness characteristics of carton folds for packaging. Trans ASME J Mech Des 130(2):022305
15. Yao W, Cannella F, Dai JS (2011) Automatic folding of cartons using a reconfigurable robotic system. Rob Comp Integr Manuf 27:604–613
16. Lee HR, Gemmill DD (2001) Improved method of assembly sequence determination for automatic assembly system. Eur J Oper Res 131(3):611–621
17. Seow KT, Devanathan R (1994) A temporal framework for assembly sequence representation and analysis. IEEE Trans Rob Autom 10(2):220–229
18. Davis TH (1968) An extension of manulescu's calssifation of planar kinematic cjains and mechanism of monility m > 1, using graph theory. In J Mech 3:87–100
19. Dubey VN, Crowder RM (2003) Designing a dexterous reconfigurable packaging system for flexible automation. In: ASME design engineering technical conference, DETC2003/DAC-48812, Chicago
20. Dubey VN, Dai JS, Stamp KJ et al (1999) Kinematic simulation of a metamorphic mechanism. In Tenth world congress on the theory of machine and mechanisms (IFToMM), pp 98–103
21. Dai JS, Jones RJ (1999) Mobility in metamorphic mechanisms of foldable/erectable kinds. ASME J Mech Des 21(3):375–382

22. Dubey VN, Dai JS (2001) Modeling and kinematic simulation of a mechanism extracted from a cardboard fold. In J Eng Simul 2(3):3–10
23. Liu H, Dai JS (2003) An approach to carton-folding trajectory planning using dual robotic fingers. Rob Auton Syst 42:47–73
24. Dubey VN, Dai JS, Stamp KJ (2000) Knowledge-based carton specification system. Science and Technology Report, Unilever Research, UK
25. Cannella F, Dai JS (2008) Stiffness characteristics of carton folds for packaging. Tran ASME J Mech Des 130(2):022305

# Part V
# Analysis and Design of Reconfigurable Robots

# Chapter 46
# Hardware Design and Testing of ModRED: A Modular Self-Reconfigurable Robot System

**S. G. M. Hossain, Carl A. Nelson and Prithviraj Dasgupta**

**Abstract** Unstructured environments are challenging for conventional robots, and modular self-reconfigurable robots (MSRs) can be deployed to overcome this challenge. The goal of thc current work was to develop a flexible, cost effective multi-module robot system capable of self-reconfiguration and achieving various gaits in unstructured environments. This paper discusses the communication aspects of the Modular Robot for Exploration and Discovery (ModRED) robot system from a hardware perspective. To ensure enhanced flexibility and local autonomy as well as better reconfiguration, each robot module is built with four independent degrees of freedom, and a novel docking interface provides interconnection of modules. The prototyping effort is described with emphasis on the implementation of inter-module communication. The electronic hardware layout and control system are described, and the communication system is outlined. Finally, some preliminary testing of the developed prototype is presented.

**Keywords** Modular self-reconfigurable robots · Distributed control · Radio communication

S. G. M. Hossain · C. A. Nelson (✉)
University of Nebraska-Lincoln, Lincoln, NE, USA
e-mail: cnelson5@unl.edu

S. G. M. Hossain
e-mail: smgmamur@yahoo.com

P. Dasgupta
University of Nebraska, Omaha, NE, USA
e-mail: pdasgupta@mail.unomaha.edu

J. S. Dai et al. (eds.), *Advances in Reconfigurable Mechanisms and Robots I*,
DOI: 10.1007/978-1-4471-4141-9_46, © Springer-Verlag London 2012

## 46.1 Introduction

Since their inception in the 1990s, a significant amount of effort has been spent addressing the challenges of developing and improving modular self-reconfigurable robots (MSRs). These robots are different from conventional robots in terms of their design and control architecture, and the reason for such variation stems from the difference of their applications compared to conventional robots. In the case of well defined, structured environments such as industrial settings, robots are used for performing specific tasks and can perform effectively and efficiently. However, in scenarios requiring multitasking, it is more feasible to use robots having a modular design without compromising the quality of performance. Furthermore, in cases where the environment is unstructured or dynamically changing (uncertain), conventional robots offer limited performance and MSRs are better suited because of their capability to be programmed as "agent-based" systems and their ability to adapt to such environments. An overview of the basic principles and characteristics of MSRs can be found in [1, 2]. Both of these papers also discuss the state of the art and challenges involved for further improvement in terms of hardware design, controls and communication among the modules and self-reconfiguration planning. In this paper we will primarily focus on the inter-module communication issues for the development of our chain-type MSR system, which will be followed by some early gait-related achievements of the developed prototype leveraging the communication between cooperating modules.

### *46.1.1 MSR Design State of the Art*

Performance of a self-reconfigurable robot is highly dependent on its mechanical and electronic control design. To date, a number of different designs have been developed and evaluated [1]. In a previous study [3], we presented some design-related challenges such as dexterity versus self reconfiguration, actuator size versus power, weight versus actuated DOF, etc., and our approach to address them. A number of researchers have pursued their own approaches to solve similar problems for advancing the state of MSR hardware design. A central part of the hardware design of self-reconfigurable robots relates to the geometrical arrangement of actuator axes and interconnections [2]. As of 2007 [1], the highest number of active modules connected in a system was Polybot with 56 modules, developed by Yim et al. [4], whereas Butler et al. were capable of simulating 2.2 million modules in computer simulation [5]. The strongest actuation also was generated by Polybot, lifting five modules in a cantilevered configuration [4], whereas the most robust self-reconfiguration was achieved by the MTRAN II system with 14 modules [6]. Dexterity of MSRs varies; there are systems such as Polybot [7] and Molecube [8] with one DOF, and CONRO [9], Polypod [10] and MTRAN III [11] with two DOF. Superbot [12], which has combined features of the ATRON,

CONRO and MTRAN robots, offers more flexibility with three DOF. More recently, iMobot [13] demonstrated advantages of even higher DOF (four controllable DOF) that ensured improved local autonomy to a single module. Based on this literature, it appears that tradeoffs exist between lower and higher per-module DOF designs, related to complexity and flexibility of lower and higher extents respectively. None of the MSRs surveyed mix prismatic and revolute DOFs in their architecture. Docking between modules is another important consideration which is key to reconfiguration ability and variable morphology of these robots. Types of docking connectors vary from magnetic devices to mechanical locks and use a variety of actuator types [14, 15]. These kinematic and docking issues are all coupled to selection of communication, actuation, and sensor components in MSR design.

### *46.1.2 Previous Work in MSR Communication*

MSRs can work in a distributed manner, requiring local autonomy of the modules that leads to global emergent behavior, and this requires intercommunication for various purposes such as reconfiguration and maneuvering. The robot modules also need to exchange information, sense other modules or obstacles and have appropriately designed navigation systems. MTRAN III [16] modules demonstrated a centralized communication using a Bluetooth interface. Superbot [17] modules included a master–slave controller including an RF receiver for distal communication up to a range of one meter and an SPI bus through the docking interfaces for communicating while the modules are connected to each other. MTRAN II [18] used electrical communication using the module surfaces for power supply, global communication (RS 485) and local communication between neighboring modules. For detecting and distance-sensing of adjacent modules, Superbot and Kilobot [17, 19] used an IR sensor-receiver array. It is obvious that an MSR system having higher local autonomy for the modules will need higher ranges for inter-module communication. Sensor prices can also limit scalability as the overall expense rises significantly for multiple modules. To develop an effective and low-cost communication system for the MSR modules, tradeoffs between sensor performance and expense are important points to consider. In the following sections, the aspects of the ModRED MSR system that address the aforementioned issues are presented.

### *46.1.3 Development of ModRED*

The motivation to build ModRED (Modular Robot for Exploration and Discovery) was to develop a novel and inexpensive robot system with dexterity and distributed

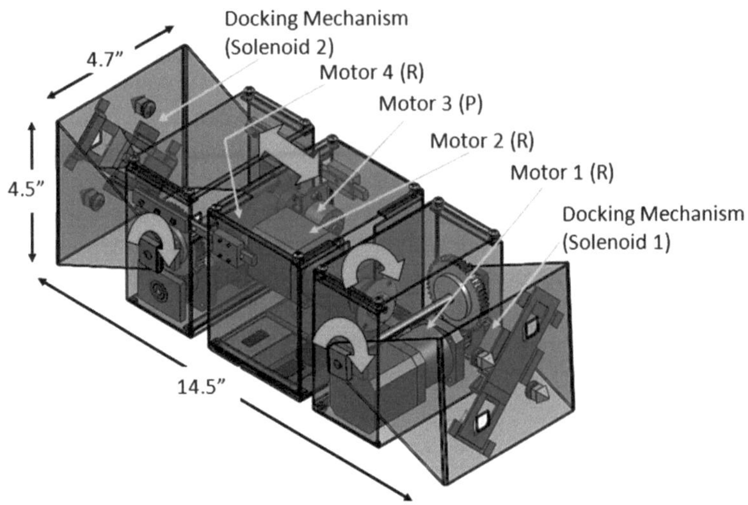


**Fig. 46.1** Dimensions and actuators for a ModRED module. The CAD model shows inner components including motors, transmissions and docking mechanisms

control architecture. The application included exploration of unstructured environments characterized by uncertainty such as extra-terrestrial environments including the moon or Mars. Improvement of the dexterity of an individual robot module as well as the entire robot system was expected to be achieved by developing four-DOF robot modules in a novel kinematic arrangement including a prismatic DOF. The communication system design was based on an inexpensive yet high-range RF system with an accompanying sensor array. For effective and computationally inexpensive reconfiguration and maneuvering, a game theory and multi-agent based coordination technique was chosen [20]. In this paper, we discuss the mechanical design and communication system of the robot. The following sections cover topics from design to prototype development and achievable gaits of the ModRED system.

## 46.2 ModRED Mechanical Design

The mechanical design of ModRED modules was based on four independent per-module degrees of freedom. An architecture with higher per-module DOF was targeted to ensure enhanced flexibility of a single module and more kinematically capable configurations involving only a few modules. The combination of rotational and prismatic DOF is a novel and promising concept that can lead towards this target. Figure 46.1 shows the RRPR kinematic structure of a ModRED module and its primary components.

## 46.3 Communication and Electronics

An MSR system requires electronics to control the maneuvering of individual modules as well as to sense the presence of other modules, exchange information and perform as an MSR system by generating multi-module gaits. As a distributed system, communication is extremely important for decision making and reconfiguration. ModRED modules use a microcontroller-based computational approach to manage the output to actuators, input from sensors and the communication between modules. Figure 46.2 presents a schematic of the electronic system for a module.

For computation and control purposes, an Arduino Fio (ATmega328P) microcontroller is used for each module, characterized by 8 MHz clock speed, 1 kB of EEPROM and 32 kB of flash memory. Eight analog input pins and 14 digital I/O pins are available with an operating voltage of 3.3 V. Supply voltage is permitted from 3.35–12 V. Lithium-polymer rechargeable batteries (3.7 V, 1000 mAh) are used as power supply. These batteries are lightweight and compact, appropriate for use in a mobile robot. Each of the robot modules is actuated by four bipolar, four-wire stepper motors of which three are rotary steppers with gearbox reduction of 60:1 and step-angle of 1.8°. The other motor is a linear stepper actuator with 0.0417 mm of travel per step. To control these motors, stepper motor drivers are used, which require high/low pulses from the microcontroller to change the direction of rotation and a PWM input to energize the coils for running the motors. The drivers are capable of supplying up to 750 mA per phase and provide permanent 8-step microstepping. For the docking mechanism, latching solenoids are used to minimize the power drawn. With 12 V supply, the solenoid latches and maintains its position without any power supply thereafter. To open the latch, a reverse voltage of 6.5–8.5 V is required which activates a spring to take the solenoid back to its original position. To supply bidirectional voltage to the solenoids, L298 N H-bridges are used. Table 46.1 summarizes some important aspects of the hardware architecture for a ModRED module.

### *46.3.1 Navigation and Sensing*

Sensing is required for an MSR module to locate another module and to find the interfacing orientation properly for docking. First, the module detects another module using proximity sensors, and then to ensure proper interfacing, a tactile (bump) sensor or a combination of tactile and proximity sensors is needed. The robot modules are equipped with two infrared (IR) sensors for proximity sensing, with a range of 4–30 cm and a calibrated output analog voltage of 2.5–0.4 V. To ensure successful docking, bump switches are incorporated in the front face of the docking bracket. These sensors can also be used for obstacle detection purposes. For navigation, a 6-bit dual axis Hitachi HM55B compass module is used. This

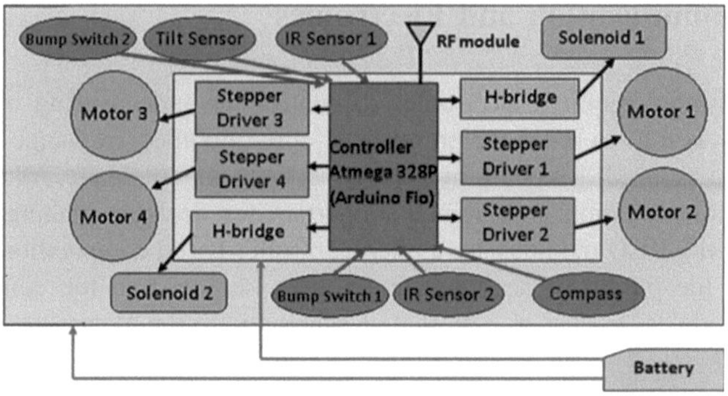


**Fig. 46.2** Schematic of electronic hardware of a ModRED module. *Yellow* inner area: processing and control units; outer *green* area: contains sensors and actuators

**Table 46.1** Some important hardware features of a ModRED module

| Computation | Arduino Fio (Atmel ATmega328P) |
|---|---|
| Communication | Wireless—2.4 GHz RF (120 m range) |
| Sensors | Infrared (range: 4–30 cm, one mounted on the front face of each of the two docking plates, two more mounted on the two sides of the central segment) |
| | Tilt switch (mercury based, one sensor mounted inside each module) |
| | Bump switch (one mounted on the front face of each of the two docking plates) |
| Navigation | Hitachi HM55B compass module (6 bit, dual axis, one sensor per module) |
| Power | Lithium-polymer battery |
| Motor | Bipolar four-wire stepper |

sensor is capable of detecting biaxial direction with a resolution of 64 directions (increments of 5.625°). For navigating through unstructured environments, elevation is another important parameter. We use a simple mercury-based tilt switch for preliminary detection of the inclination information of the robot module. The application of the sensors and navigation system can be visualized in Fig. 46.3. This shows an unstructured surface where two modules are navigating, sensing each other and identifying forward or backward inclination.

### 46.3.2 Wireless Communication

The wireless communication is achieved using an XBee modem that can be connected directly to the Arduino Fio microcontroller board. The microcontroller reads the RF input as serial data. The XBee modem includes a chip antenna of 2.4 GHz RF and 120 m (unobstructed) range with low (1 mW) transmitting power.

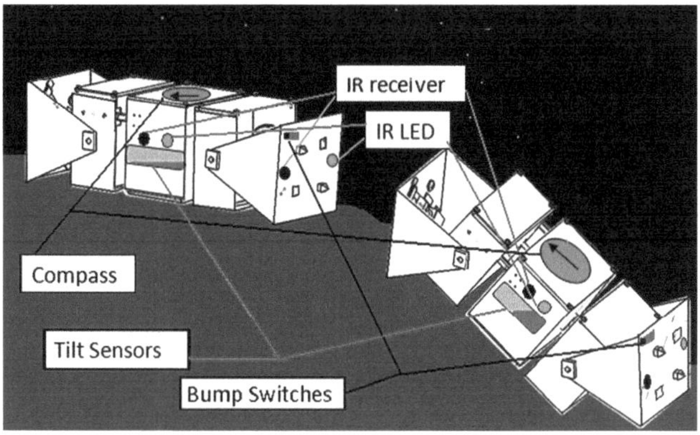


**Fig. 46.3** Two ModRED modules maneuvering through unstructured terrain. The robot modules communicate and sense using the various sensors provided

## 46.4 Development of Gaits Using the Prototype

To maneuver through unstructured environments, a modular robot should be capable of generating various types of gaits so that it can overcome obstacles and navigate more optimally. Higher per-module DOF is an advantage for ModRED modules to develop different types of locomotion gaits. Rolling, steering and inchworm motions are developed both by individual and multiple modules. All these gait developments with sequences of actuations have been discussed in detail in [3]. The communication between modules plays an important role to synchronize the actuators' motion sequences to achieve functional gait. Some of these gaits were demonstrated [21] by computer simulation of ModRED modules. For testing of the gaits, two MSR modules have been prototyped and tested, equipped with electronic components necessary to develop such gaits as previously enumerated. The single-module inchworm and single-module steering and rolling gaits have been validated as demonstrated in Fig. 46.4. Plans to test the other gaits discussed in [3] and to evaluate the reconfiguration capabilities are underway.

## 46.5 Conclusion

This paper presents fundamental aspects of ModRED MSR development. The communication design utilizes a wide-range radio frequency system, ensuring that the robot modules can locate and communicate with each other even over large distances, lending itself to unstructured planetary surfaces or disaster affected zones. The other sensors are also chosen to meet the requirements of the robot system to reconfigure and develop various gaits. Two modules are developed to implement node-to-node communication and sensing in its simplest form. In future

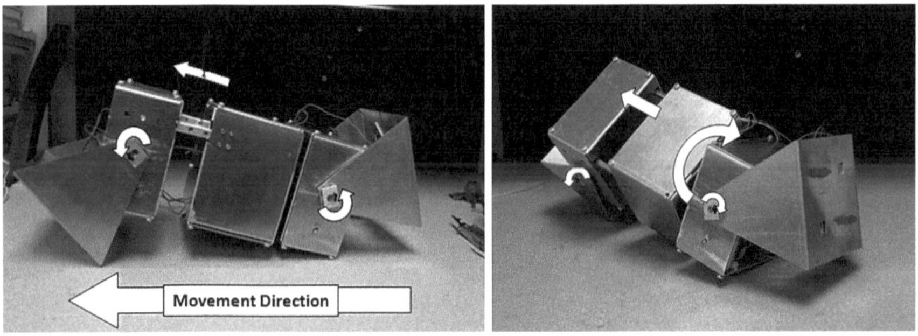


**Fig. 46.4** ModRED prototype: (*left*) demonstrating single-module inchworm and (*right*) single-module steering and rolling gaits

work, the expectation is to build and test an increased number of modules which will use more complicated communication protocols and multi-agent system configurations. This will enable the robots to achieve more novel gaits and techniques for reconfiguration. Servo actuators may also be beneficial in this respect in conjunction with position feedback control. Such improvements will lead toward building a flexible, robust and fully self-reconfigurable robot system to advance the state of MSR technology.

**Acknowledgment** We gratefully acknowledge support from NASA Nebraska Space Grant and NASA EPSCoR.

## References

1. Yim, M, Shen W, Salemi B, Rus D, Moll M, Lipson H, Klavins E, Chirikjian G (2007): Modular self-reconfigurable robot systems—challenges and opportunities for the future. In: IEEE Robotics and Automation Magazine, March 2007, pp 43–52
2. Murata S, Kurokawa H (2007) Self-reconfigurable robots—shape changing cellular robots can exceed conventional robot flexibility. In: IEEE Robotics and Automation Magazine, March 2007, pp 71–78
3. Chu, KD, Hossain SGM, Nelson C (2011) Design of a four DOF modular self-reconfigurable robot with novel gaits. In: Proceedings of ASME international design engineering technical conferences (IDETC/CIE). Washington, DC, August 2011
4. Yim M, Zhang Y, Duff DG (2002) Modular robots. IEEE Spectr 39(2):30–34
5. Butler Z, Kotay K, Rus D, Tomita K (2004) Generic decentralized locomotion control for lattice-based self-reconfiguring robots. Int J Robot Res 23(9):919–938
6. Murata S, Yoshida E, Kamimura A, Kurokawa H, Tomita K, Kokaji S (2002) M-TRAN: self reconfigurable modular robotic system. IEEE-ASME T Mech 7(4):431–441
7. Yim M, Zhang Y, Roufas K, Duff D (2003) Eldershaw C connecting and disconnecting for chain self-reconfiguration with polybot. IEEE-ASME T Mech 7(4):442–451
8. Zykov V, Chan A, Lipson H (2007) Molecbues: an open-source modular robotic kit. In: IEEE IROS 2007 self-reconfigurable robotics workshop
9. Castano A, Behar A, Will P (2002) The conro modules for reconfigurable robots. IEEE-ASME T Mech 7(4):403–409

10. Yim M (1994) Locomotion gaits with polypod. In: Video proceedings of the ieee international conference on robotics and automation, San Diego
11. Kurokawa H, Tomita K, Kamimura A, Kokaji S, Hasuo T, Murata S (2008) Distributed self reconfiguration of M-TRAN III modular robotic system. Int J Robot Res 27(3–4):373–386
12. Salemi B, Moll M, Shen W (2006) Superbot: a deployable, multi-functional, and modular self-reconfigurable robotic system. In: IEEE/rsj international conference on intelligent robots and systems, pp 3636–3641
13. Ryland G (2010) Design of iMobot, an intelligent reconfigurable mobile robot with novel locomotion. In: 2010 IEEE international conference on robotics and automation, anchorage. Alaska, May 2010, pp 60–65
14. Kurokawa H, Tomita K, Kamimura A, Kokaji S, Hasuo T, Murata S (2007) Self-reconfigurable modular robot M-TRAN: distributed control and communication. In: Proceedings of the 1st international conference on robot communication and coordination. IEEE Press, Piscataway
15. Shen W, Kovac R, Rubenstein M (2009) SINGO: a single-end-operative and genderless connector for self-reconfiguration, self-assembly and self-healing. In: 2009 IEEE international conference on robotics and automation, Kobe, May 2009
16. Murata S, Kakomura K, Kurokawa H (2007) Toward a scalable modular robotic system—navigation, docking, and integration of M-TRAN. IEEE Robot Autom Mag 14(4):56–63
17. Shen W, Chiu H, Rubenstein M, Salemi B (2008) Rolling and climbing by the multifunctional superbot reconfigurable robotic system. In: Proceedings of space technology and applications international forum (STAIF-08), AIP conference proceedings No 969, American Institute of Physics, Melville, February 2008 pp 839–848
18. Kurokawa H, Kamimura A, Yoshida E, Tomita K, Murata S, Kokaji S (2002) Self-reconfigurable modular robot (M-TRAN) and its motion design. In: 7th international conference on control, automation, robotics and vision (ICARCV'02), Singapore, December 2002
19. Rubenste, M, Hoff N, Nagpal R: Kilobot (2011) A low cost scalable robot system for collective behaviors. In: Technical report TR-06-11, Harvard University, June 2011
20. Ramaekers Z, Dasgupta P, Ufimtsev V, Hossain SGM, Nelson C (2011) Self-reconfiguration in modular robots using coalition games with uncertainty. In: AAAI workshop on autonomous action planning for autonomous mobile robots (PAMR), pp 57–63
21. Nelson C, Chu K, Dasgupta P (2010) ModRED: a modular self-reconfigurable robot for autonomous extra-terrestrial exploration and discovery. In: Planetary rovers workshop (IEEE international conference on robotics and automation), anchorage, AK

# Chapter 47
# Structures and Characteristics in Reconfigurable Modular Robots

**B. Madhevan and M. Sreekumar**

**Abstract** Reconfigurable robot is a mechatronic device that can change its own shape. Such robots are not only designed to perform a specific task but also to execute multiple tasks by changing their shape and size. In order to change their own shape, reconfigurable robots are built from multiple identical modules that can rearrange themselves to form a large variety of structures. Based on module specifications, the potential ability of the complete robot is determined. Further, module specifications in turn determine the functional reconfiguration of these robots which increases the adaptability against the change of environment. Hence, accurate movements of module are needed, but this may be extremely difficult due to the errors inherent within the control system. In this paper, structures and characteristics in self-reconfigurable robots towards change in environment are investigated. A comprehensive survey of available literature is done to identify key challenges and opportunities in this area.

**Keywords** Self-reconfigurable robots · Modular robots · Mobile robots · Robot architecture · Adaptability · Obstacle avoidance

B. Madhevan · M. Sreekumar (✉)
Indian Institute of Information Technology Design and Manufacturing (IIITD&M)
Kancheepuram, Off Vandalur Kelambakkam Road, Melakottaiyur, Chennai 600048, India
e-mail: msk@iiitdm.ac.in

B. Madhevan
e-mail: bmadhevan@iiitdm.ac.in

J. S. Dai et al. (eds.), *Advances in Reconfigurable Mechanisms and Robots I*,
DOI: 10.1007/978-1-4471-4141-9_47, © Springer-Verlag London 2012 

## 47.1 Introduction

Traditionally, robots use sense-think-act cycle of control. The idea is to map the sensor input onto an internal representation of the environment [1]. Using this representation, an appropriate action is planned and then executed. This approach is suitable for known environments, but problem arises with real world environments at many levels due to non synchronization of the internal model with the real world. The planning problem in the internal model turned out to be more difficult. Even if the planning is successful, the outcome is not as expected because the environment changes when the robot is thinking. Within the 1980s, subsumption architecture based robots are developed in which complex intelligent behaviour is decomposed into many simple behaviour modules, which is in turn organized into layers. During 1990s, behavior based robots are developed and this breakthrough introduced a new way of structuring the control systems for robots. A behavior-based control system is composed of many task achieving behaviors. Instead of mapping sensor input onto an internal representation, simple calculations are performed to figure out whether a specific response to the environment is needed. The notion is that, real world problems are solved by correctly structuring the control system [2].

More recently, new ideas started to flow in the field of robotics from the area of artificial intelligence. Self-reconfigurable robots are related to multi-robot systems in the sense that these systems also consist of groups of more or less independent robots [3]. The difference is that in multi-robot systems, the robots do not interact physically. On the other hand, in self-reconfigurable robots the modules are physically coupled. Amorphous computing systems are built from a massive number of small elements to power these modules which are made to communicate locally. This line of research is still pursued and the hope is that the problems will reduce with improved sensor technology and increased computational power.

Apart from the introductory section, this paper is presented as follows. Section 47.2 discuss about hierarchy of modular robots which explain different architectures in modular robots. Section 47.3 recalls various types of self-reconfigurable robots and in Sect. 47.4 hardware issues of self-reconfigurable robots are discussed. Section 47.5 review control strategies for self-reconfigurable robots. Future research aspects are presented in Sect. 47.6 which is followed by conclusion and references.

## 47.2 Hierarchy of Modular Robots

The hierarchy of modular robots is shown in Fig. 47.1. Modular Robots is a robot which consists of several modules. If a module breaks, it can be identified and replaced in a relatively short time, whereas in non-modular robots the entire robot will be replaced. In Reconfigurable modular robots, the modules are independent and have high degree of homogeneity [4–7]. The reconfigurable modular robot is

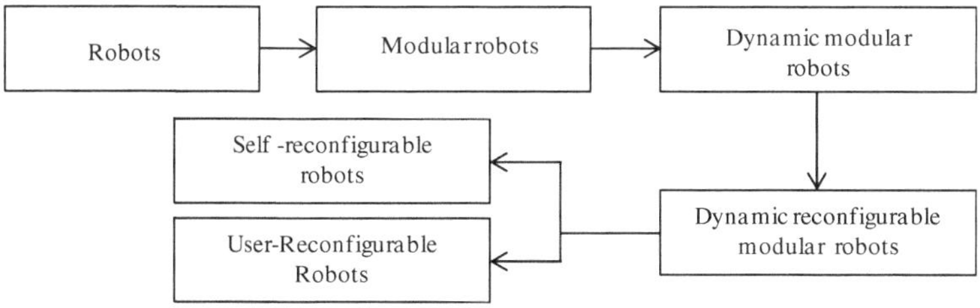


**Fig. 47.1** Hierarchy of modular robots

**Table 47.1** Comparison of lattice, chain, and hybrid robots

| Types | Reconfigure | Motion generation | No of actuators | No of connectors | Homogeneity | Symmetry |
|---|---|---|---|---|---|---|
| Lattice | Easy | Hard | Many | Many | Homogeneous | Isotropic |
| Chain | Hard | Easy | Few | Few | Heterogeneous | Anisotropic |
| Hybrid | Easy | Easy | Few | Few | Homogeneous | Anisotropic |

assumed to have a fixed shape at run-time, but in off-line it can be reconfigured into different shapes and sizes. The possibility of changing the shape and size of the robot increases robot's versatility.

Dynamically reconfigurable modular robots can be reconfigured at run-time. This happens because the modules have functionality to detect detachments and attachments of modules at run-time, and potentially to change behavior depending on the new configuration. Designing the module morphology and control software for such a system is challenging (Table 47.1).

## 47.3 Taxonomy of Reconfigurable Modular Robots

(Fig. 47.2, Table 47.2)

## 47.4 Hardware Issues in Self-Reconfigurable Robots

The most important part of the hardware design for self-reconfigurable robots is the fundamental properties of a module, especially the shape of the module and the geometrical arrangement of actuator axes and connection. The module should be made more functional without further complication of the hardware.

In order to find such a solution, systematic design is required. Another important issue is the reliability of the hardware [20–23]. Major factors restricting the size of the self-reconfigurable robots system are mechanical reliability,

**Fig. 47.2** A few models of modular robots

**Table 47.2** Development of self-reconfigurable robots [8–19]

| System | Class | DOF | Shape | Author | Year |
|---|---|---|---|---|---|
| Self replicator | Mobile | – | – | John Von Neumann | 1950 |
| Mechanical SR | Mobile | – | – | Lionel Penrose | 1959 |
| CEBOT | Mobile | Various | – | Fukuda et al. (Tsukuba) | 1988 |
| Polypod | Chain | 2 | 3D | Yim (Stanford) | 1993 |
| Metamorphic | Lattice | 6 | 2D | Chirikjian (Caltech) | 1993 |
| Fracta | Lattice | 3 | 2D | Murata (MEL) | 1994 |
| Fractal Robots | Lattice | 3 | 3D | Michael(UK) | 1995 |
| Tetrobot | Chain | 1 | 3D | Hamline et al. (RPI) | 1996 |
| ANAT Robot | Chain | – | 3D | Charles Khairallah (CA) | 1997 |
| 3D Fracta | Lattice | 6 | 3D | Murata et al. (MEL) | 1998 |
| Molecule | Lattice | 4 | 3D | Kotay and Rus (Dartmouth) | 1998 |
| CONRO | Chain | 2 | 3D | Will and Shen (USC/ISI) | 1998 |
| PolyBot | Chain | 1 | 3D | Yim et al. (PARC) | 1998 |
| TeleCube | Lattice | 6 | 3D | Suh et al. (PARC) | 1998 |
| Vertical | Lattice | – | 2D | Hosakawa et al. (Riken) | 1998 |
| Crystalline | Lattice | 4 | 2D | Vona and Rus (Dartmouth) | 1999 |
| I-Cube | Lattice | – | 3D | Unsal (CMU) | 1999 |
| M-TRAN I | Hybrid | 2 | 3D | Murata et al.(AIST) | 1999 |
| Pneumatic | Lattice | – | 2D | Inou et al. (TiTech) | 2002 |
| Uni Rover | Mobile | 2 | 2D | Hirose et al. (TiTech) | 2002 |
| M-TRAN II | Hybrid | 2 | 3D | Murata et al. (AIST) | 2002 |
| Atron | Lattice | 1 | 3D | Stoy et al. (U.S Denmark) | 2003 |
| S-bot | Mobile | 3 | 2D | Mondada et al. (EPFL) | 2003 |
| Stochastic | Lattice | 0 | 3D | White, Kopanski, Lipson (Cornell) | 2004 |
| Superbot | Hybrid | 3 | 3D | Shen et al. (USC/ISI) | 2004 |
| Y1 Modules | Chain | 1 | 3D | Gonzalez-Gomez et al. (UAM) | 2004 |
| M-TRAN III | Hybrid | 2 | 3D | Kurokawa et al. (AIST) | 2005 |
| AMOEBA-I | Mobile | 7 | 3D | Liu JG et al. (SIA) | 2005 |
| Catom | Lattice | 0 | 2D | Goldstein et al. (CMU) | 2005 |
| Stochastic-3D | Lattice | 0 | 3D | Zykov, Lipson (Cornell) | 2005 |
| Molecubes | Chain | 1 | 3D | Zykov, Mytilinaios, Lipson (Cornell) | 2005 |
| Prog. Parts | Lattice | 0 | 2D | Klavins (U. Washington) | 2005 |
| Miche | Lattice | 0 | 3D | Rus et al. (MIT) | 2006 |
| GZ-I Modules | Chain | 1 | 3D | Zhang and Gonzalez-Gomez (U. Hamburg, UAM) | 2006 |
| Odin | Hybrid | 3 | 3D | Lyder et al., Modular Robotics Research (USD) | 2008 |
| Evolve | Chain | 2 | 3D | Chang Fanxi, Francis (NUS) | 2008 |
| Roombots | Hybird | 3 | 3D | Sproewitz, Moeckel, Ijspeert, BioroboticsLabaratory (EPFL) | 2009 |

electrical reliability, and high cost. Electrical reliability and high cost can be resolved by mass production. However, the mechanical reliability of the mechanism cannot be predicted until the system is actually constructed. Table 47.3 explains about the characteristics in self-reconfigurable robots. Few of them are

**Table 47.3** Electrical and mechanical characteristics of self-reconfigurable robots

| Names of robot | Electrical characteristics | | Mechanical characteristics | |
|---|---|---|---|---|
| | CPU | Power | Dimension | Attachment method |
| USC/ISI, CONTRO | Yes | Yes | 3D | Mechanical, SMA[a] |
| Stanford polypod | Yes | No | 3D | Mechanical |
| PARC polybot | Yes | No | 3D | Mechanical, SMA[a] |
| JHU metamorphic | Yes | No | 2D | Mechanical |
| Dartmouth, crystalline | Yes | Yes | 2D | Mechanical |
| MEL, fractum | Yes | No | 2D | Electro magnets |
| MEL, micro-unit | Yes | No | 2D | Mechanical |
| RIKEN, vertical | No | Yes | 2D | Permanent magnets |
| PARC, telecube | n/a | n/a | 3D | Switching permanent |
| MEL, 3D-Unit | No | No | 3D | Mechanical |
| MEL, MTRAN | Yes | Mo | 3D | Permanent magnets SMA |
| Dartmouth, molecule | Yes | No | 3D | Mechanical |
| CMU, I-cubes | Yes | Yes | 3D | Mechanical |

[a] Shape memory alloy

- Dexterity of individual modules defines the flexibility of the robot.
- Motion precision, power and torque define the efficiency of the modules.
- Limits on mechanical and electrical field robustness.
- Limits on strength and precision for bonding modules.

## 47.5 Classification of Control Systems

The relationship between the control system and the modules of the robot determines the type of control system to be developed for reconfigurable robots. Such control systems can be divided into three main categories as shown in Fig. 47.3a.

Locomotion is an act or the power of moving robot's module from place to place. Self-reconfigurable robots have the advantage of adapting their shape for changing the locomotion pattern to fit the environment [24, 25]. Locomotion pattern can be classified into two main categories as shown in Fig. 47.3b.

### *47.5.1 Open Control Platform*

This section presents an open control platform (OCP) which provides interaction among modules and supports dynamic reconfiguration and customization of modules in real time [26–29]. Its primary goals are

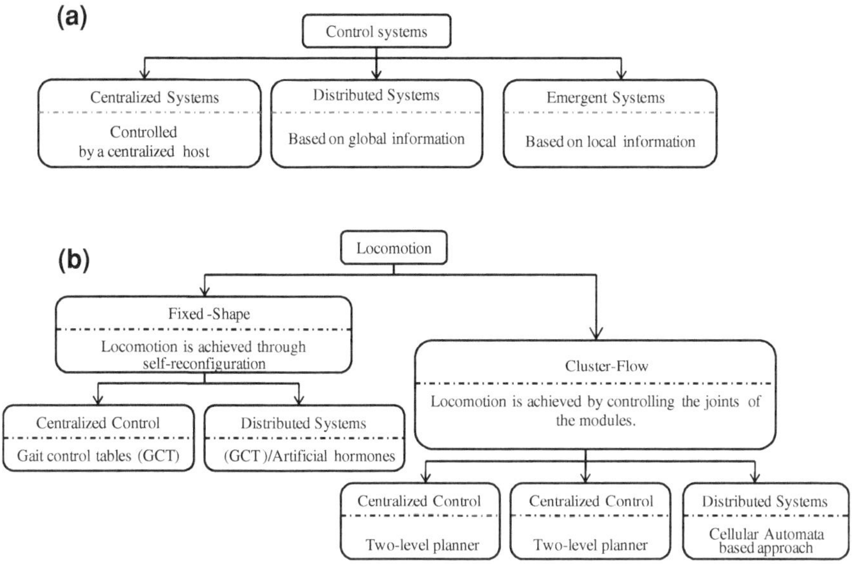


**Fig. 47.3** **a** Control structure for self-reconfigurable robots. **b** Locomotion systems for self-reconfigurable robots

- To accommodate rapidly changing application requirements,
- Easily incorporate new technology,
- Maintain viability in unpredictable and changing environments.

The challenge addressed by the OCP is to deal with the complexity of integrating these modules in a unknown environment where the robot has to take spot decisions. Open Control Platform Design: The OCP consists of multiple layers of application programmer interfaces (APIs) that increase in abstraction and become more domain specific at the higher layers as shown in Figs. 47.4 and 47.5.

Each layer builds on the modules defined in lower layers. In the bottommost "core" layer, the OCP leverages from and extends new advances in real-time distributed object computing which allow distributed, heterogeneous modules to communicate asynchronously in real time. It also supports highly decoupled interaction between the distributed modules of the system, which tends to localize architectural or configuration changes so that they can be made quickly and with high reliability. The middle "reconfigurable controls" layer provides abstractions for integrating and reconfiguring control system modules; the abstractions bridge the gap between the controls domain and the core distribution substrate. The abstract interface is based on familiar controls engineering concepts such as block diagram modules, input output ports, measurement and command signals. The third "hybrid controls" layer supports reconfiguration management by making reconfiguration strategies and rationale for reconfiguration decisions explicit and reusable. It contains generic patterns of integration and reconfiguration that are found in hybrid, reconfigurable control systems.

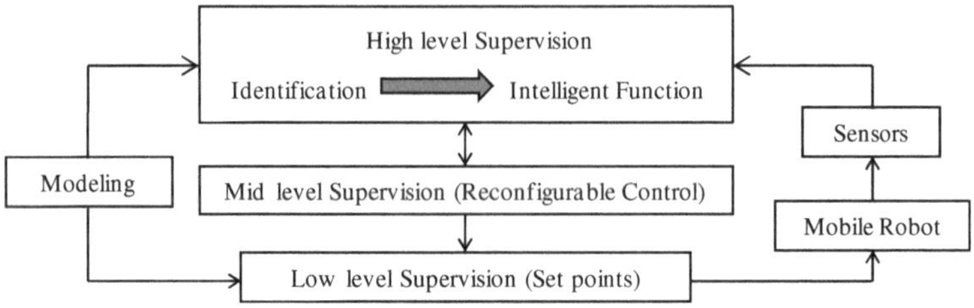


**Fig. 47.4** Control architecture for self-reconfigurable robots

**Fig. 47.5** Open control platform for self-reconfigurable robots

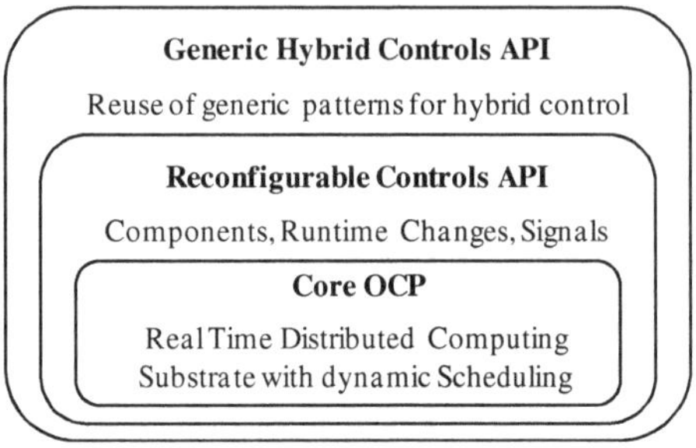


## 47.6 Discussion

Self-reconfigurable robots will move robotics into new areas and avenues of application. In addition to traditional mass production environments, self-reconfigurable robots may become useful in real-world environments. These environments are characterized by being unstructured, complex, and dynamic. Self-reconfigurable robots have an advantage over fixed shape robots in these environments because of their special abilities. These abilities include versatility, robustness, adaptability, and scale extensibility. Because of these abilities, self-reconfigurable robots can easily explore in an unknown environment. The robots must move over the surface, despite its condition. It must not die even in difficult times; a self-reconfigurable robot can change its shape and escape from such situations. It is also capable of repairing itself. Various tasks for these robots will be found in orbit, in the deep sea, and in areas disaster-stricken by radiation or toxic chemicals. Most of the existing path planning algorithms concentrates on avoiding moving and stationary obstacles within the workspace of the robot. When moving obstacles are present, the problem becomes more complex for self-re-configurable robots to identify their path. In order to realize the potential of self-reconfigurable robots more research is needed in the areas of hardware and control. The research on control needs to be focused on real-world environments.

## 47.7 Conclusions

The structures and characteristics of self-reconfigurable robots are investigated and presented in this paper. The structure of self-reconfigurable robot was composed of basic motion modules, articulate modules and connection modules. Each module has uniform standard interface. The mechanical configuration was realized through the reassemble of different modules. Focus of future research will be on motion coordination for self reconfigurable robots that is accurately controllable, so that it could map and utilize all information obtained from sensors regarding the obstacles and environment.

## References

1. Von Neumann J (1966) Theory of self-reproducing automata. University of Illinois Press, Urbana
2. Jones R, Cherry A, Farrall S (1994) Application of intelligent control in automotive vehicles. In: International conference of control, pp 159–164
3. Grob R (2006) Autonomous self-assembly in swarm-bots. IEEE Trans Robot 22(6): 1115–1130
4. Rus D, Vona M (2001) Crystalline robots: self-reconfiguration with compressible unit modules. Auton Robot 10(1):107–124
5. Park M, Chitta S, Teichman A, Yim M (2008) Automatic configuration recognition in modular robots. Int J Robot Res 27(3–4):403–421
6. Lau HYK, Ko AWY, Lau TL (2008) The design of a representation and analysis method for modular self-reconfigurable robots. Robot Cim-Int Manuf 24:258–269
7. Walter J, Welch J, Amato N (2002) Concurrent metamorphosis of hexagonal robot chains into simple connected configurations. IEEE T Robot Autom 18(6):945–956
8. Fitch R, Rus D, Vona MA (2000) Basis for self-repair using crystalline modules. In: Proceedings of intelligent autonomous systems conference, pp 903–909
9. Støy K, Shen W-M, Will PM (2002) Using role-based control to produce locomotion in chain-type self-reconfigurable robots. IEEE/ASME Trans Mechatron 7(4):407–410
10. Yim M, White P, Park M, Sastra J (2004) Modular self-reconfigurable robots. Encyclopedia of Complexity and System Science. Springer Magazine, NewYork, pp 19–32
11. Yim M, Shen W-M, Salemi B, Rus D et al (2007) Modular self-reconfigurable robot systems. IEEE T Robot Autom 14(1):43–52
12. Murata S, Kurokawa H (2007) Self-reconfigurable robots. IEEE T Robot Autom 71–78
13. Butler Z, Fitch R, Rus D (2002) distributed control for unit-compressible robots: goal-recognition, locomotion, and splitting. IEEE/ASME Trans Mechatron (special issues on self reconfigurable robots) 7(4):418–403
14. Castano A, Behar A, Will PM (2002) The conro modules for reconfigurable robots. IEEE/ASME Trans Mechatron 7(4):403–409
15. Unsal C, Kiliccote H, Kohsla P (2001) A modular self-reconfigurable bipartite robotic system: implementation and motion planning. Auton Robot 10(1):23–40
16. Yim M, Zhang Y, Roufas K et al (2002) Connecting and disconnecting for chain self-reconfiguration with polybot. IEEE/ASME Trans Mechatron 7:442–451
17. Murata S, Yoshida E, Kamimura A et al (2002) M-TRAN: self-reconfigurable modular robotic system. IEEE/ASME Trans Mechatron 7(4):431–441

18. Ostergaard EH (2004) Distributed control of the ATRON self-reconfigurable robot. Ph.D. thesis, University of Southern Denmark
19. Kurokawaa H (2006) Self-reconfigurable M-TRAN structures and walker generation. Robot Auton Syst 54:142–149
20. Murata S, Yoshida E, Kurokawa H et al (2001) Self-repairing mechanical system. Auton Robot 10(1):7–21
21. Posadas JL (2008) Agent-based distributed architecture for mobile robot control. Eng Appl Artif Intell 21:805–823
22. Qiao L, Sato M, Abe K et al (1995) Environment recognition in power train control. In: Proceedings of IEEE, pp 730–734
23. Murphy TF, Yurkovich S, Chen SC (1996) Intelligent control for paper machine moisture control. In: Proceedings of IEEE conference control applications, pp 826–833
24. Stoy K, Shen WM, Will P (2003) A simple approach to the control of locomotion in self-reconfigurable robots. Robot Auton Syst 44:191–199
25. Stoy K (2004) Emergent control of self-reconfigurable robots. Ph.D. thesis, University of Southern Denmark
26. Wills L, Kannan S, Sander S et al (2001) An open platform for reconfigurable control. In: IEEE control systems magazine, pp 49–64
27. Wills L, Kannan S, Heck B et al (2000) An open software infrastructure for reconfigurable control systems. In: Proceedings of American control conference, pp 2799–2803
28. Wills L, Sander S, Kannan S et al (2000) An open control platform for reconfigurable, distributed, hierarchical control systems. In: Proceedings of digital avionics systems, pp 1–8
29. Kannan S, Restrepo C, Yavrucuk I et al (1999) Control algorithm and flight simulation integration using the open control platform for unmanned aerial vehicles. Proc Digit Avionics Syst Conf 6 (A):1–10

# Chapter 48
# Typical 3 + 3 Gait Motion Analysis of a Radial Symmetrical Six-Legged Robot Based on Metamorphic Theory

**Xilun Ding and Kun Xu**

**Abstract** The locomotion of multi-legged robots has some characters similar to metamorphic mechanisms. Assuming the constraints of the feet on the ground with hinges, the supporting legs and the hexapod body are taken as a parallel mechanism, and each transporting leg is regarded as a usual serial manipulator. In different period there may been different equivalent mechanisms of multi-legged robot. The locomotion of hexapod robot can be considered as a series of varying hybrid serial-parallel mechanisms. Four typical 3 + 3 gaits of a radial symmetrical six-legged robot are studied in this paper based on metamorphic theory. A new static stability workspace is proposed to establish the static stability of four gaits, and a new method to calculate the stride length of multi-legged robots is present by analyzing the relationship between the workspaces of two adjacent equivalent parallel mechanisms in one gait cycle.

**Keywords** Six-legged robot · Metamorphic · Serial-parallel mechanism · Static stability · Stride length

## 48.1 Introduction

Metamorphic mechanism was proposed by Dai and Rees at 25th ASME Biennial Mechanisms and Robotics Conference in 1998 [1]. This kind of mechanisms can

X. Ding · K. Xu (✉)
Robotics Institute, Beihang University, 10083, Beijing, China
e-mail: xk_sea@me.buaa.edu.cn

X. Ding
e-mail: xlding@buaa.edu.cn

J. S. Dai et al. (eds.), *Advances in Reconfigurable Mechanisms and Robots I*,
DOI: 10.1007/978-1-4471-4141-9_48, © Springer-Verlag London 2012 

change their topological configuration and shapes, their effective links number and their DOFS can hence be varied [1, 2]. Different from traditional mechanism which only has an unchanged topological configuration, the metamorphic mechanisms have various topological configurations. A certain configuration of metamorphic mechanism has its intrinsic effective links and DOFS. Metamorphic mechanisms have multiple configurations and they can change their configuration to provide suitable effective links and DOFS to adapt to different tasks.

Multi-legged robots display much better adaptability to rough terrain than wheeled robots and tracked robots because they do not need continuous support on the ground [3]. This type of robots includes biped robots, quadruped robots, hexapod robots and more legged ones. In nature most arthropods have six legs so they can easily maintain static stability while they are moving. Hexapod robots show good robustness in case of leg faults [4–8]. For these reasons, the hexapod robot has attracted considerable attention in research community. The key problem of multi-legged robots mostly focuses on kinematics and dynamics analyses and gait planning.

Gait is a key factor for multi-legged robots to move on the ground, and depends on the robot body and its structure of legs. Typical hexapod robots can be divided into bilateral symmetrical rectangular hexapods and radial symmetrical hexagonal ones. Rectangular hexapods have a rectangular body with six legs distributed symmetrically on two sides [9, 10]. Hexagonal hexapods have a circular or hexagonal body with six legs evenly distributed around [4]. 3 + 3 gait [5], 4 + 2 gait [6, 7], 5 + 1 gait [8] are common periodic gaits of hexapods. Kugushev and Jaroshevskij [11] proposed a free gait to adapt different terrain. McGhee et al. [12]. and other researchers [13, 14] also went on studying free gaits of hexapods. Because of the special structure of hexagonal robots, Wang et al. [15, 16] proposed insect-wave tripod gait, mammal kick-off tripod gait and mixed tripod gait, they also studied its fault-tolerance when one or two legs were damaged.

When the multi-legged robot is moving on the ground, the system composed of the ground, supporting legs and the robot body can be seen as a parallel mechanism. Ding Xilun and Xu Kun [17] designed a novel metamorphic wheel-legged robot and analyzed its DOF from this view. Wang et al. [15, 16] investigated the stride size and the stability margin of hexapod robot using this method. But most of them neglected the influence of different foothold points on walking performance of the robot. In different periods while multi-legged robot moves on the ground, it can have different equivalent mechanisms. Hence it is not suitable to take the system as a single specific mechanism. The locomotion of hexapod robot can be considered as a series of moving hybrid serial-parallel mechanisms.

Because of the similarities between the multi-legged robots and metamorphic mechanisms, by taking NOROS robot (Fig. 48.1) built in space robot lab of BUAA, based on metamorphic theory a new method was proposed to analyze the movement performance of multi-legged robots in this paper. One gait cycle is divided into several stages. Assuming the constraints of the feet on the ground with hinges, in specific stage, the ground, supporting legs and the robot body form a parallel mechanism, each swinging leg is taken as a series manipulator. In different

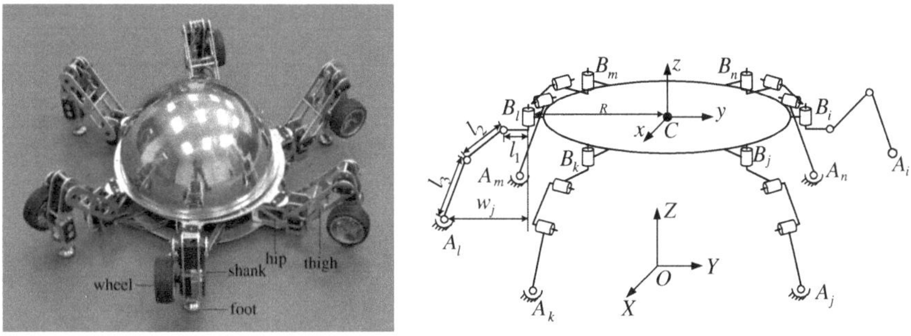


**Fig. 48.1** NOROS robot and its one equivalent mechanism

periods the equivalent mechanisms of the system may not same. The walking movements of multi-legged robot are the motion and transition process of these equivalent mechanisms.

## 48.2 Typical 3 + 3 Gaits

The body of the NOROS robot is composed of a hemispherical shell and a double-deck chassis. Each leg consists of five elements: foot, wheel, shank, thigh and hip, as shown in Fig. 48.1. Shank, thigh and hip are linked together by two parallel revolute joints, knee and hip joint. The leg is connected to the chassis by waist joint whose rotating axis is perpendicular to the chassis. Main length parameters of small prototype of NOROS robot, which will be used for subsequent analysis and simulation, are listed in Table 48.1.

Wang et al. proposed 3 kinds typical 3 + 3 tripod gaits: insect-wave gait, mammal kick-off gait and mixed gait [16]. In fact there is another insect-wave gait, so here the insect-wave gait mentioned in [16] is named insect-wave gait I and the other is named insect-wave gait II. The aforementioned 4 kinds of typical 3 + 3 tripod gaits can be divided into 2 groups according to their initial standing posture. Insect-wave gait I and mammal kick-off gait have the same or similar initial standing posture with six legs distribute parallel, as seen in Fig. 48.2. Their supporting triangles are called first type supporting triangle. Their moving directions form an angle of 90°. As can be seen from the Fig. 48.2, the robot can walk along 2 directions using each gait. Mixed gait and insect-wave gait II have the same or similar initial standing posture with six legs even distribution, as seen in Fig. 48.4. Their supporting triangles are called second type supporting triangle. The moving direction of mixed gait can along the direction of any leg and the moving direction of insect-wave gait II can along the angular bisector of two adjacent legs. So the robot can move along 6 directions using these two gaits (Fig. 48.3).

**Table 48.1** Main length parameters of small prototype of NOROS

| | Body | Hip | Thigh | Shank |
|---|---|---|---|---|
| Length(mm) | $R = 150$ | $l_1 = 50$ | $l_2 = 120$ | $l_3 = 130$ |

**Fig. 48.2** Initial standing posture with legs distribute parallel

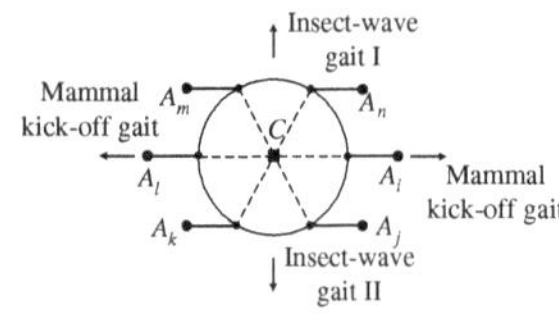


**Fig. 48.3** Initial standing posture with legs distribute divergently

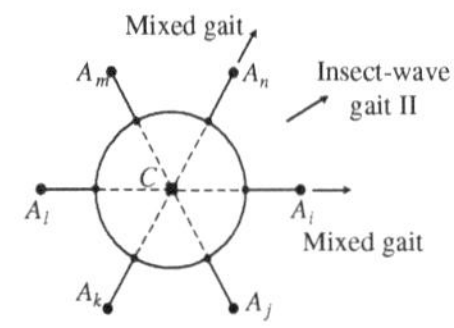


At the same location the robot can change its six legs footholds to adjust its standing posture and switch its walking gait. Thus the radial symmetrical six-legged robot can walk along any direction, realizing zero turning radius. So the radial symmetrical six-legged robot is omni directional mobile robot.

The locomotion of hexapod robot is considered as a series of varying hybrid serial-parallel mechanism. One gait cycle can be divided into several different moving stages. Figures 48.4, 48.5, 48.6 and 48.7 show the motion sequence of four gaits. "•" means supporting foot, "O" means transporting foot, and "→" denote the walking direction of robot. In the figures phase ① means the initial standing posture, phase ② and phase ③ are one moving stage in which there are three legs lifting and swing, and other three legs supporting and kicking, so the robot body moves forward part of one stride. Phase ④ indicates one switching period in which the transporting legs contact ground and the supporting legs do not lift. Phase ⑤ and phase ⑥ are one moving stage in which the transporting legs in previous moving stage become supporting legs and the supporting legs in previous moving stage become transporting legs. Similar to phase ④, Phase ⑦ is one switching period. Phase ⑧ and phase ⑨ are one moving stage in which the transporting legs and supporting legs change to each other. Phases ①–⑥ compose one starting gait cycle and phases ④–⑨ form one gait cycle in continuous walking. The two gait cycles are not identical.

As can be seen in Figures 48.4, 48.5, 49.6 and 49.7, in moving stages there are 3 transporting legs and 3 supporting legs, so the equivalent mechanism is a hybrid serial-parallel mechanism with 3 parallel branches and 3 serial branches. In initial standing phase and switching period there are 6 supporting legs, so the equivalent mechanism is a parallel mechanism with 6 parallel branches. In static stability

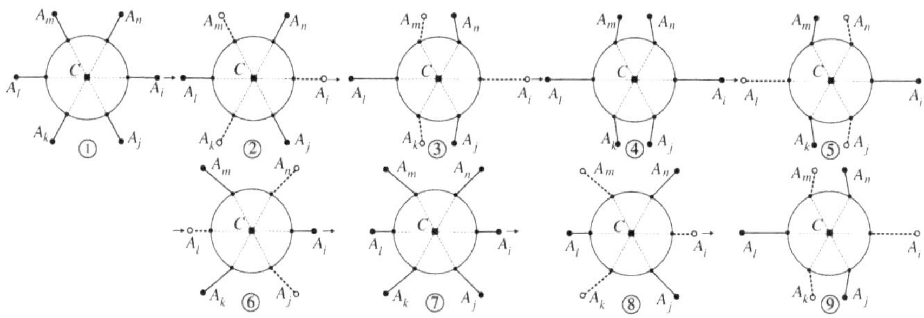

**Fig. 48.4** Sequence diagram of mixed tripod gait

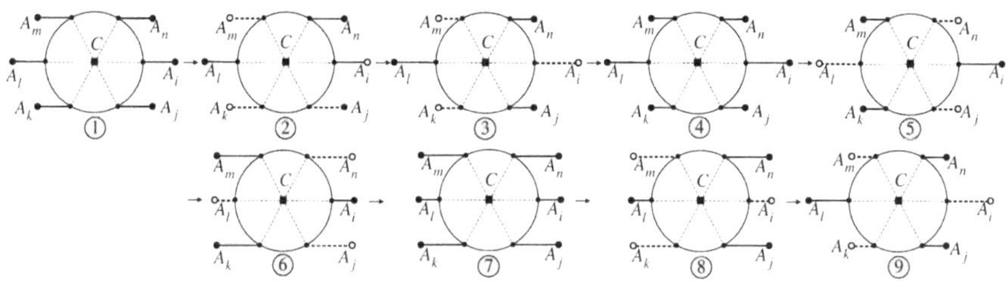

**Fig. 48.5** Sequence diagram of mammal kick-off tripod gait

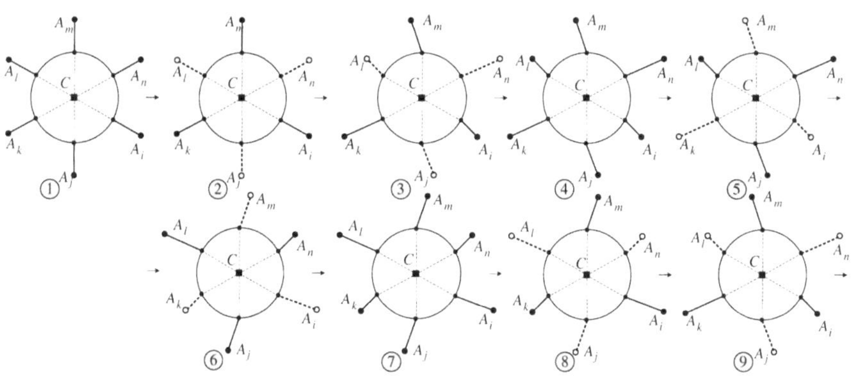

**Fig. 48.6** Sequence diagram of insect-wave tripod gait I

3 + 3 tripod gaits there are two type equivalent mechanisms: hybrid serial-parallel mechanism with 3 parallel branches and 3 serial branches and parallel mechanism with 6 parallel branches, as seen in Figs. 48.8 and 48.9. Because of the different footholds the supporting polygons may not same, so there are some dimension difference among the same equivalent mechanisms.

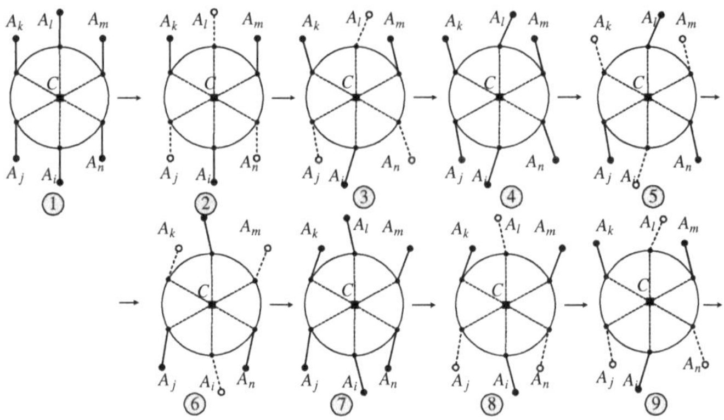

**Fig. 48.7** Sequence diagram of insect-wave tripod gait II

**Fig. 48.8** Equivalent mechanism with 6 parallel branches

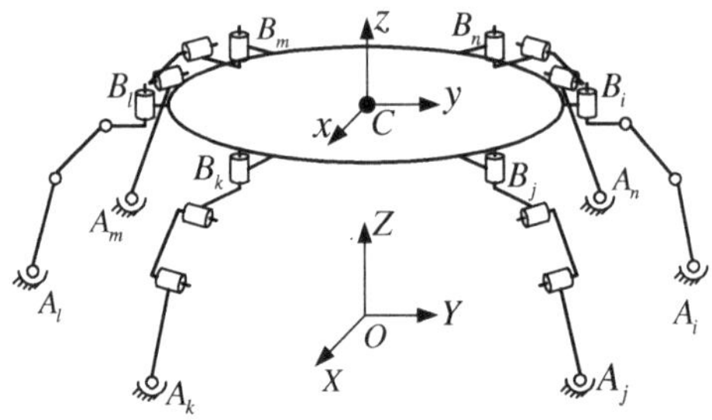

**Fig. 48.9** Equivalent mechanism with 3 parallel branches and 3 serial branches

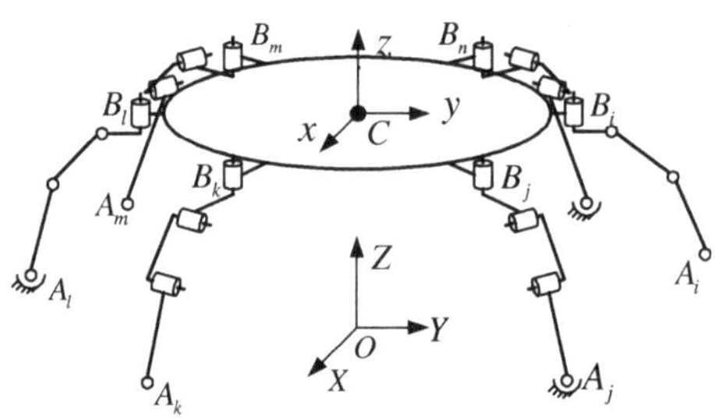

## 48.3 Workspace of Equivalent Mechanisms

Workspace is an important index for parallel mechanism. Geometrical methods and numerical methods are main two types to analyze the workspace of parallel mechanism [18–22]. Here we used the geometrical method to find the constant orientation workspace of robot body. Because of the structure limit, there are certain ranges of knee joint, hip joint and waist joint, as shown in Table 48.2.

**Table 48.2** Rotation range of leg joints

| Knee joint | Hip joint | Waist joint |
|---|---|---|
| $-140^\circ \leq \alpha \leq 40^\circ$ | $-90^\circ \leq \beta \leq 100^\circ$ | $-65^\circ \leq \gamma \leq 65^\circ$ |

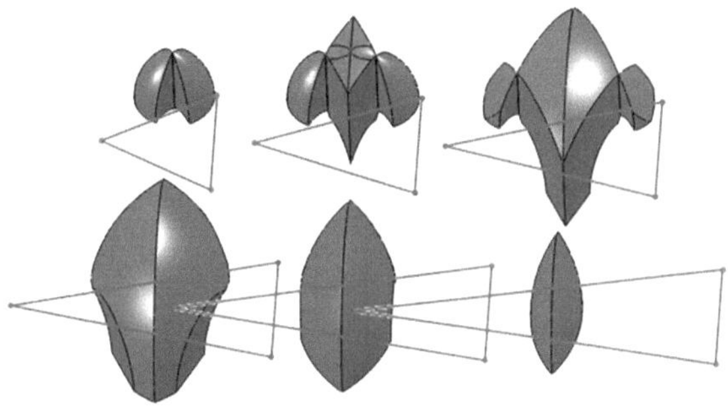

**Fig. 48.10** Constant orientation workspaces of parallel mechanism having first type supporting triangle

By using the forward kinematics established in previous studies we can get the workspace of the supporting leg which has certain foothold point.

We assume that supporting triangle of mixed gait and insect-wave gait II is equilateral triangle and the supporting triangle of mammal kick-off gait and insect-wave gait I is isosceles triangle, and each supporting leg has equal foothold offset. Once the supporting triangle change, the performance of the equivalent mechanism also change. Figures 48.10 and 48.11 show the constant orientation workspaces while the footholds offset $w = 0, 50, 100, 150, 200, 250$ mm, respectively.

Several static stability evaluation criteria have been defined in the history research of multi-legged robots. The frequently used criteria are the stability margin, the longitudinal stability margin and the crab longitudinal stability margin. The static stability margin is the smallest distance from the projection of center of gravity to the edges of the supporting polygon. According to the definition of static stability margin we assume a virtual ball whose radius is stability margin $s_m$ and it locates at the robotic center of gravity. When the vertical projection of this ball completely locates in the supporting triangle, the robot movement satisfies the static stability margin $s_m$. So the vertical projection trajectory of the robotic center of gravity while the vertical projection of the virtual ball tangent to the supporting triangle form the static stability supporting triangle, as seen in Fig. 48.12.

The part workspace whose vertical projection locates in the static stability supporting triangle is named static stability workspace, which truly represents the performance of the mechanism. The robotic center of gravity must be in the static stability workspace while the robot walks on the ground using static stability gaits. Figures 48.13, 48.14, 48.15 and 48.16 show the static stability workspace while the static stability margin $s_m = 0$ mm and $s_m = 30$ mm.

Insect-wave gait I and mammal kick-off gait have first type supporting triangle. Figure 48.17 shows the volume curves of reachable constant orientation workspace, 0 mm static stability margin constant orientation workspace and 30 mm static stability margin constant orientation workspace. The three curves have the

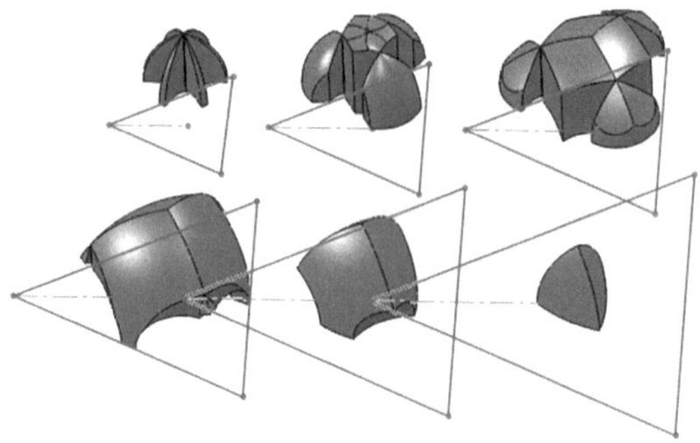

**Fig. 48.11** Constant orientation workspaces of parallel mechanism having second type supporting triangle

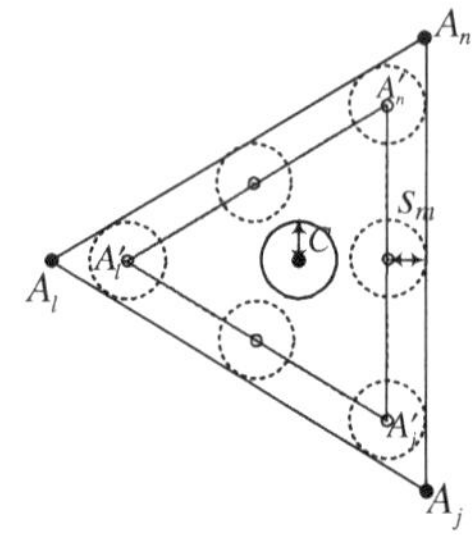


**Fig. 48.12** Supporting triangle and static stability supporting triangle

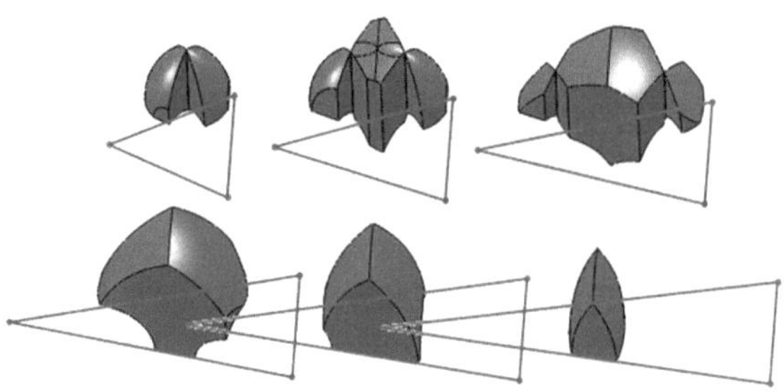

**Fig. 48.13** Static stability workspace of parallel mechanism having first type supporting triangle while $s_m = 0$ mm

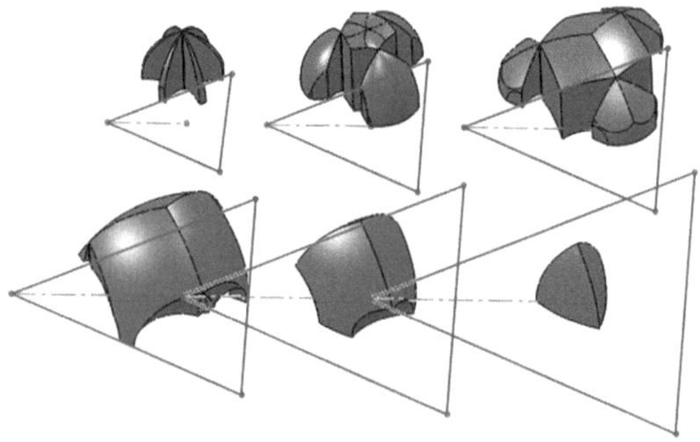

**Fig. 48.14** Static stability workspace of parallel mechanism having second type supporting triangle while $s_m = 0$ mm

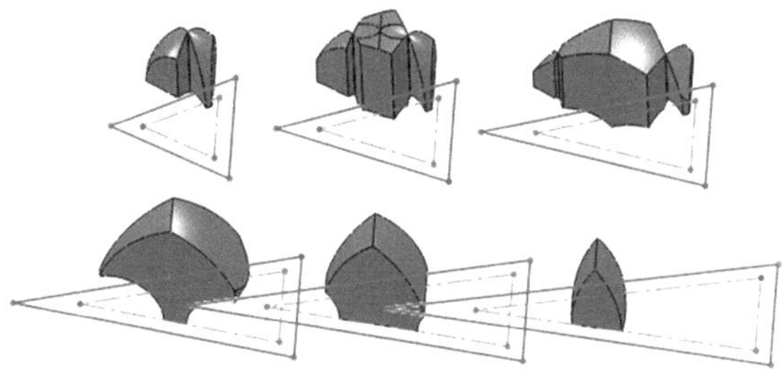

**Fig. 48.15** Static stability workspace of parallel mechanism having first type supporting triangle while $s_m = 30$ mm

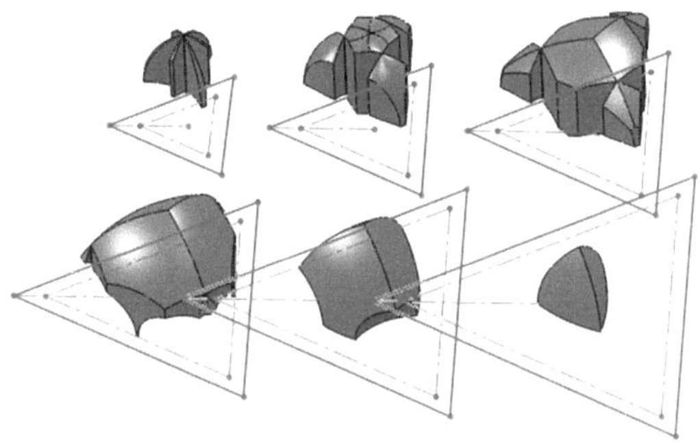

**Fig. 48.16** Static stability workspace of parallel mechanism having second type supporting triangle while $s_m = 30$ mm

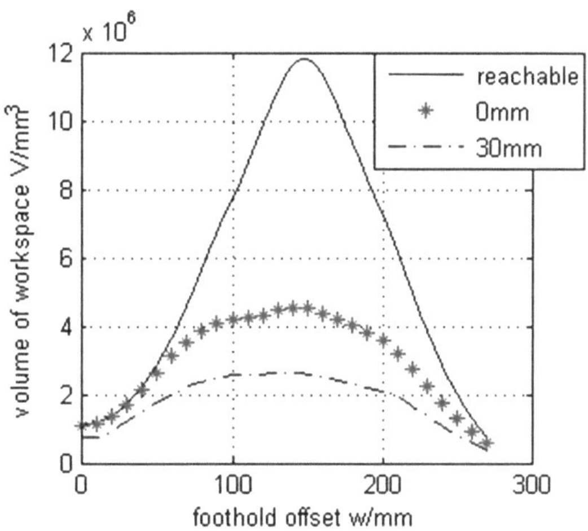


**Fig. 48.17** Workspace volume curves of parallel mechanism having first type supporting triangle

same variation trend. At the beginning the curves ascend with the increase of foothold offset. After the curves get their peak values they descend with the increase of foothold offset. As seen in figure, the three curves are not close to each other. This suggests that there is considerable reachable workspace whose vertical projection dose not locate in supporting triangle.

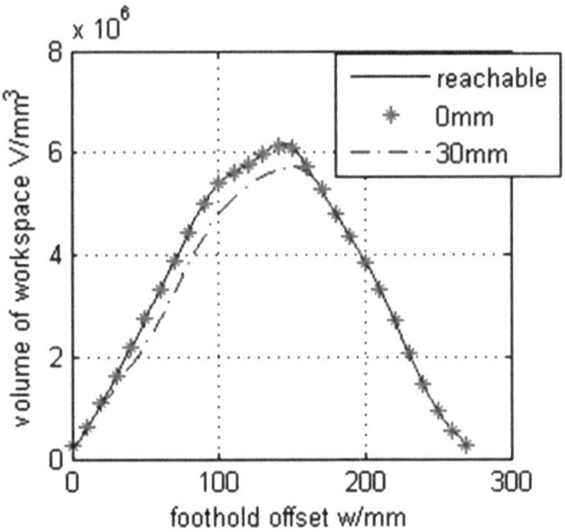


**Fig. 48.18** Workspace volume curves of parallel mechanism having second type supporting triangle

Mixed gait and insect-wave gait II have second type supporting triangle. Figure 48.18 shows the volume curves of reachable constant orientation workspace, 0mm static stability margin constant orientation workspace and 30 mm static stability margin constant orientation workspace. The three curves have the same variation trend. At the beginning the curves ascend with the increase of foothold offset. After the curves get their peak values they descend with the increase of foothold offset. The three curves are very close to each other. This suggests that vertical projection of reachable workspace mostly locates in supporting triangle.

In Figures 48.19, 48.20 and 48.21, although the equivalent mechanism of mammal kick-off gait and insect-wave gait I has larger workspace than the equivalent mechanism of mixed gait and insect-wave gait II, their workspace satisfied the same static stability margin are smaller. From this point of view, the second type supporting triangle has bigger static stability margin than the first type supporting triangle.

## 48.4 Stride Size of Four Gaits

The relationship between the workspace of two adjacent moving stage mechanisms is determined by the robot's movement. On the other hand performance of the robot can be analyzed and motion planning can be made by using the relationship between the workspace of two adjacent moving stage mechanisms.

If there are not any intersections between the workspaces of two adjacent moving stage mechanisms, as seen in Figure 48.22a, there must be not switch period, so robot must be airborne with six foot off the ground. In other words the robot jumps on the ground. In this case the robot can not walk using static stability gaits. Instead, if there are some intersections between the workspace of two

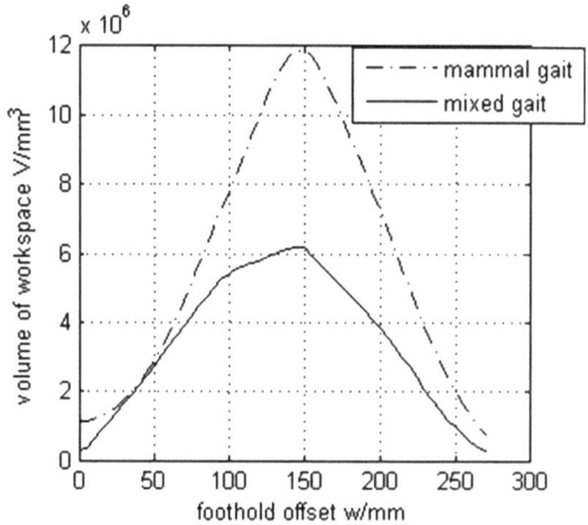


**Fig. 48.19** Reachable workspace volume curves of two type parallel mechanisms

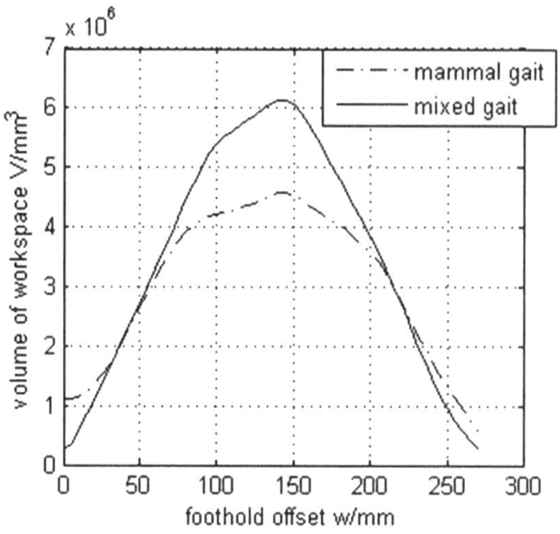


**Fig. 48.20** 0 mm static stability margin workspace of two type parallel mechanisms

adjacent moving stage mechanisms, as seen in Fig. 48.22b, there can be switch period and the robot can walk using static stability gaits. Thus, to the four gaits aforementioned there must be some intersections between the workspace of two adjacent moving stage mechanisms. The intersections are workspace of switch period mechanism.

Figures 48.23, 48.24, 48.25 and 48.26 represent the vertical sections of workspaces. Using geometric method to analyze the vertical section we can find the regions where the hexapod can move on the ground by static stability gaits. The region between line 1 and line 2 is the area in which the robot can walk on the ground using static stability gaits. If we want to calculate the maximum stride size we must determine the height of hexapod body and the boundary of two workspaces are exactly connect to each other. Using this method we can deduce the

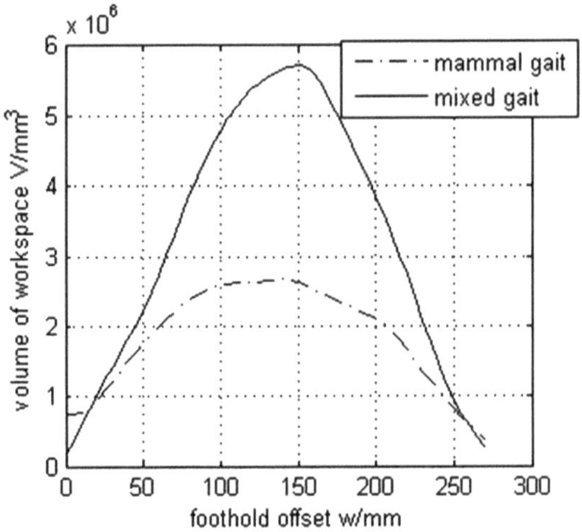


**Fig. 48.21** 30 mm static stability margin workspace of two type parallel mechanisms

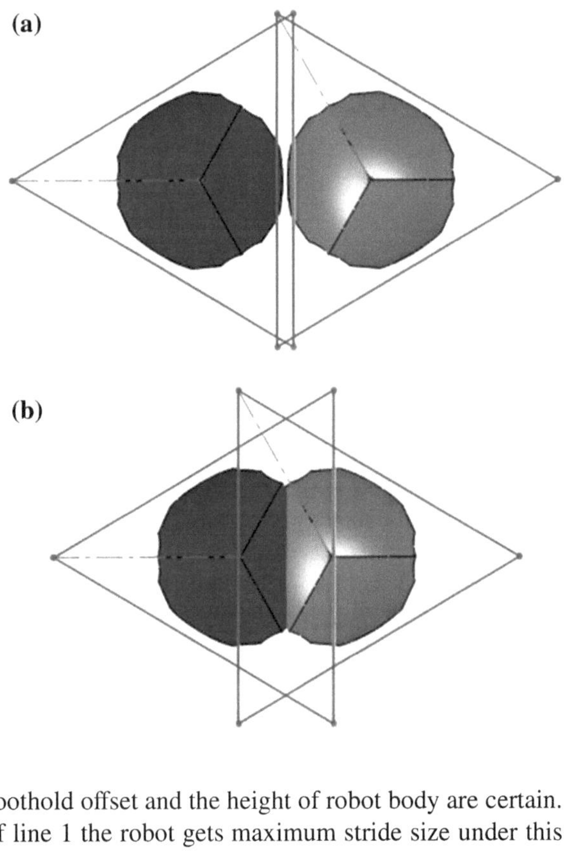


**Fig. 48.22** Relationship between the workspaces of two adjacent parallel mechanisms of mixed tripod gait

maximum stride size if the foothold offset and the height of robot body are certain. In above figures, at height of line 1 the robot gets maximum stride size under this height.

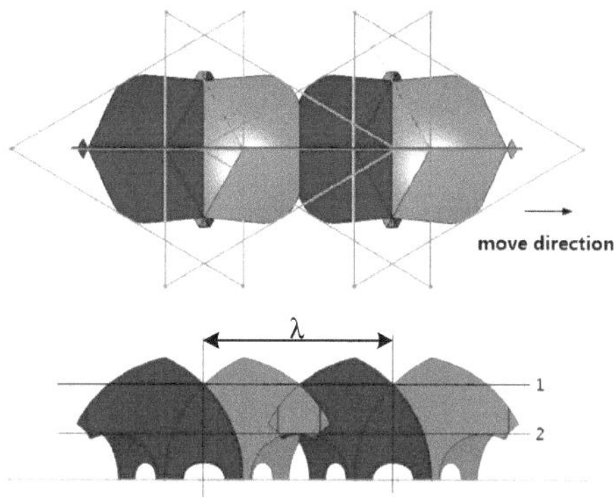


**Fig. 48.23** Equivalent mechanism workspaces of two adjacent gait cycle in mixed gait

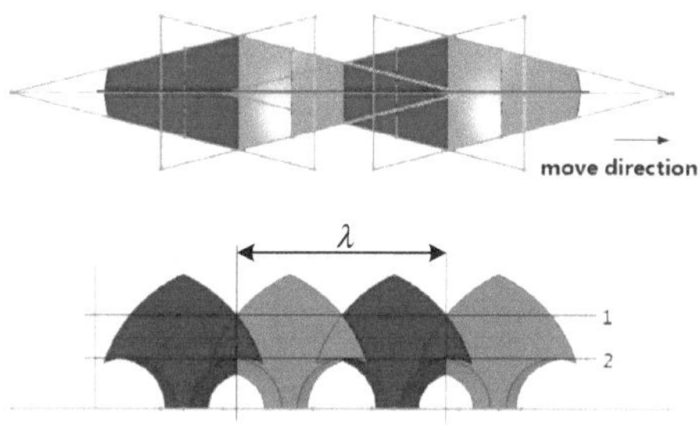


**Fig. 48.24** Equivalent mechanism workspaces of two adjacent gait cycle in mammal kick-off gait

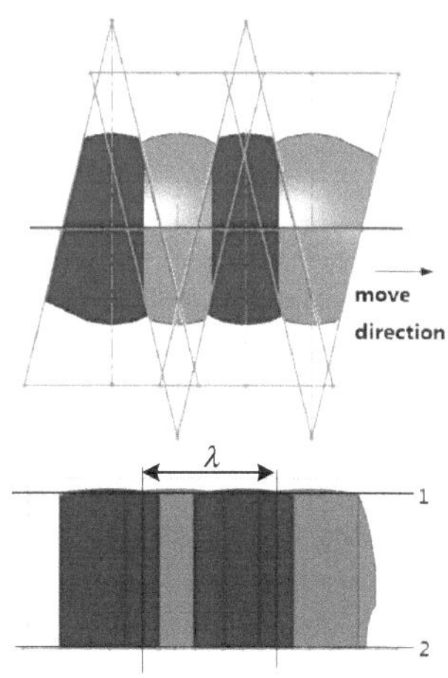


**Fig. 48.25** Equivalent mechanism workspaces of two successive gait cycle in insect-wave gait I

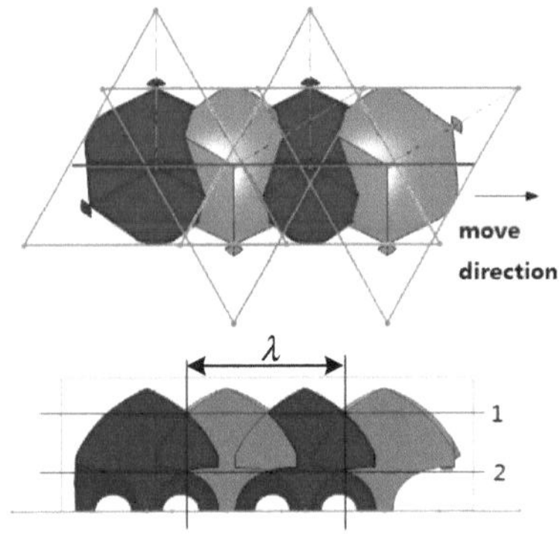


**Fig. 48.26** Equivalent mechanism workspaces of two successive gait cycle in insect-wave gait II

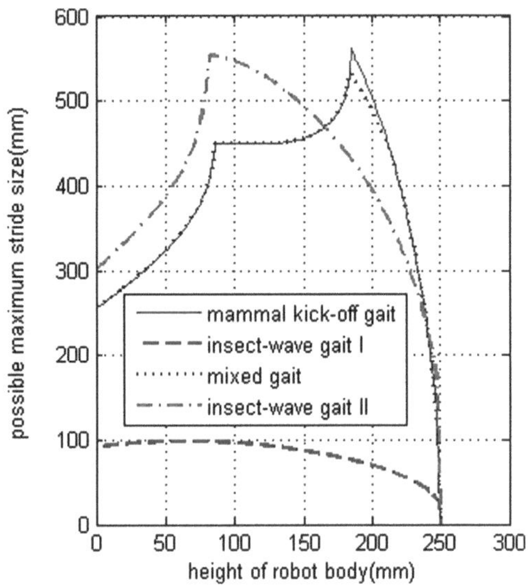


**Fig. 48.27** Possible maximum stride size of four gaits

According to above method and constrains, we can deduce the maximum stride size while the foothold offset and the height of robot body are certain. Further more we can get the possible maximum stride size while robot body is at certain height.

Figure 48.27 shows the possible maximum stride size of four gaits. In the figure the stride size of insect-wave gait I is the smallest, so the robot are not fit to walking using this gait. From the figure we can see that the maximum stride size of insect-wave gait II is the largest while the height of robot body $h < 165$ mm. This result agrees with the natural phenomena: the body of most insects is near to the surface while they are walking. When the robot body $h > 165$ mm mammal kick-off gait and mixed gait have larger stride size.

## 48.5 Conclusions

A new method has been proposed to determine the stride size of the hexapod robot. In this method one gait cycle is divided into several different periods. In one particular period the system is taken as a specific mechanism. If the hexapod robot realizes moving on the ground by using static stability gaits, the workspaces of different equivalent parallel mechanisms in neighboring moving periods must have intersection. The maximum stride size can be obtained by analyzing the workspaces. This method is more suitable and precise than the other methods.

Static stability workspace is defined to establish the static stability of gaits. Simulation results show that mixed gait and insect-wave gait II which have second type supporting triangles have better static stability.

**Acknowledgments** The authors would like to acknowledge Project (51125020) supported by National Natural Science Foundation of China (NSFC).

## References

1. Dai JS, Rees Jones J (1998) "Mobility in metamorphic mechanisms of foldable/erectable kinds." In: Proceedings 25th ASME biennial mechanisms conference, 1998, ASME: DETC98/MECH5902(1998)
2. Dai JS, Rees JJ (1999) Mobility in metamorphic mechanisms of foldable/erectable kinds. ASME Trans J Mech Des 121(3):375–382
3. Preumont A, Alexandre P, Ghuys D (1991) "Gait analysis and implementation of a six leg walking machine," 911CAR, 5th international conference on advanced robotics 'robots in unstructured environments,' Pisa, Italy, vol 2. pp 941–945, 19–22 Jun 1991
4. Lee TT, Liao CM, Chen TK (1988) On the stability properties of hexapod tripod gait. IEEE J Robotic Autom 4(4):427–434
5. Yang JM, Kim JH (1998) Fault-tolerant locomotion of the hexapod robot. IEEE Trans Syst Man Cybern Part B 28(1):109–116
6. Yang JM, Kim JH (1998) "A strategy of optimal fault tolerant gait for the hexapod robot in crab walking," IEEE international conference on robotics and automation. Leuven, Belgium, vol 2. pp 1695–1700, 16–20 May 1998
7. Preumont A, Alexandre P, Ghuys D (1991) "Gait analysis and implementation of a six leg walking machine," 911CAR, 5th international conference on advanced robotics 'robots in unstructured environments,' Pisa, Italy, vol 2. pp 941–945, 19–22 Jun 1991
8. Kaneko M, Abe M, Tanie K (1985) A hexapod walking machine with decoupled freedoms. IEEE J Robotic Autom 1(4):183–190
9. Lee WJ, Orin DE (1988) Omnidirectional supervisory control of a multilegged vehicle using periodic gaits. IEEE J Robotic Autom 4(6):635–642
10. Kugushev EI, Jaroshevshij VS (1975) "Problem of selecting a gait for an integrated locomotion robot," Process fourth international conference. Artificial intelligence, Tbilisi, Georgian SSR, USSR, pp 789–793, Sept 1975
11. McGhee RB, Iswandhi GI (1979) Adaptive locomotion of a multilegged robot over rough terrain. IEEE Trans Syst Man Cybern 9(4):176–182
12. Porta JM, Celaya E (2004) "Reactive free gait generation to follow arbitrary trajectories with a hexapod robot." Robotic Auton Syst 47:187–201

13. Erden MS, Leblebicioglu K (2008) Free gait generation with reinforcement learning for a six-legged robot. Robotic Auton Syst 56:199–212
14. Wang ZY, Ding XL, Rovetta A (2009) Analysis of typical locomotion of a symmetric hexapod robot. Robotica V28(6):893–907
15. Ding XL, Wang ZY, Rovetta A (2010) Typical gaits and motion analysis of a hexagonal symmetrical hexapod robot. Robots V32(6):759–765
16. Kamikawa K, Arai T et al. (2004) "Omni-directional gait of multi-legged rescue robot," IEEE international conference on robotics & automation, New Orleans, pp 2171–2176
17. Merlet JP, Gosselin CM, Mouly N (1998) Workspaces of planar parallel manipulators. Mech Mach Theory 33:7–20
18. Gosselin C (1990) Determination of the workspace of 6-DOF parallel manipulators. J Mech Des 112(3):331–336
19. Haug EJ, Luh CM, Adkins FA, Wang JY (1996) Numerical algorithms for mapping boundaries of manipulator workspaces. J Mech Des 118(2):228–234
20. Snyman JA, Plessis LJD, Duffy J (2000) An optimization approach to the determination of the boundaries of manipulator workspaces. J Mech Des 122(4):447–456
21. Plessis LJD, Snyman JA (2001) A numerical method for the determination of dextrous workspaces of gough–stewart platforms. Int J Numer Meth Eng 52(4):345–369
22. Garcia E, Estremera J, Santos P (2002) A comparative study of the stability margins for walking machines. Robotica 20:595–606

# Chapter 49
# An Automatic Dynamics Generation Method for Reconfigurable Modular Robot

**Wenbin Gao and Hongguang Wang**

**Abstract** A method of automatic dynamics generation for modular robots is presented on modular level. The robot's link parameters are got from the modules' parameters and the Assemble Incidence Matrix (AIM) describing the module types, assembled orientations and sequence of a given robot configuration. Adjoint matrices are adopted to describe the forward mapping of velocities, accelerations from frames on the $(i-1)$ link to that of the $i$th link, as well as the dual adjoint matrices are taken to describe the backward mapping of moments and forces between frames on the $i$ and the $(i-1)$th links. A mathematically consistent recursive approach for dynamics of modular robot is got from the Newton–Euler formula in Lie Group form.

**Keywords** Reconfigurable modular robot · Dynamics · Lie group

## 49.1 Introduction

Reconfigurable Modular Robot System (RMMS) is comprised by a serious of standardized modules with different dimensions and certain assembling styles, such as link modules, joint modules and gripper modules. Joint modules are all-in-one

W. Gao (✉) · H. Wang
State Key Laboratory of Robotics, Shenyang Institute of Automation, Chinese Academy of Sciences, 110016 Shenyang, China
e-mail: gaowenbin@sia.cn

H. Wang
e-mail: hgwang@sia.cn

W. Gao
Graduate School of Chinese Academy of Sciences, 100049 Beijing, China

J. S. Dai et al. (eds.), *Advances in Reconfigurable Mechanisms and Robots I*,
DOI: 10.1007/978-1-4471-4141-9_49, © Springer-Verlag London 2012

components which can generate rotary or translational motions and communicate with the upper controller. Manipulators with different DOFs or configurations can be rapidly constructed by choosing proper modules and then connecting them together. The closed-form dynamics model is widely applied in robot design, calibration and motion optimization. However, it's impractical to derive the dynamics manually for every configuration of modular robot, because the number of possible robot geometries is very huge. So this method that can generate the dynamics model of a given configurations automatically is important in module robot's researches and applications [1–4].

The researches of dynamics of modular robots mostly are based on the Lagrangian and Newton–Euler formulation. Wang [5] and [6] proposed the dynamics constructing method for modular robots based on Lagrangian formula. In both of their works, the given configurations are regarded as normal manipulators, which cannot be adopted to automatically generate the dynamics. Fei [7] and [8, 9] gave the recursive methods based on Newton–Euler formula, both of their methods can automatically generate the dynamics. Chen's method which is expressed in the Lie Group form is a more compact approach.

In this paper, it introduces a Modular Reconfigurable Robot Experiment System (MRRES) and presents a modular level method for automatic dynamics generation of modular robots based on Newton–Euler formulation in the form of Lie Group and Lie Algebra. The links' dynamics parameters of a given modular robot are obtained automatically from the information of the AIM and module parameters. Compare with Chen's method, our method is a complete modular level approach.

## 49.2 MRRES and AIM

The MRRES is developed by the State Key Laboratory of Robotics in Shenyang Institute of Automation, China. It takes a distributed control system with an Industrial Personal Computer as the main controller and DSPs in the joint modules. The data communication is based on CAN bus. Figure 49.1 shows the module library and a 6-DOF configuration.

As shown in Fig. 49.2a, both the revolute and prismatic joints are 1-DOF modules. In a joint module, the input part connecting to the lower part of a given manipulator is named as part I, and the output part connecting to the gripper direction is named as part II. Both part I and II have their own input ports and output ports. The Cartesian coordinates are set on the input port (subscript 0) and output port (subscript 1) respectively to describe the position and orientation of the output port relative to input port. The output frame of part I and the input frame of part II are superposed when the joint is at zero position. Their $z$-axes are collinear with the joint axis and their origins are set on the midpoint of the joint axis. As shown in Fig. 49.2b–d, the Cartesian coordinate is set on each link modules, based modules and gripper are alike, to describe the position and orientation of the output port relative to input port.

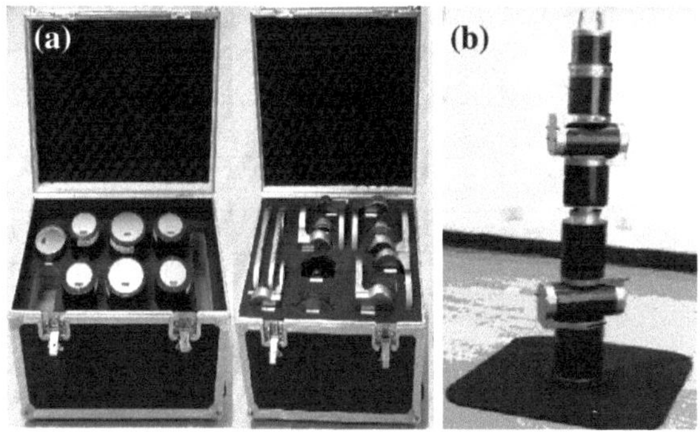


**Fig. 49.1** Module library and a 6-DOF configuration. **a** Module library. **b** A 6-DOF configuration

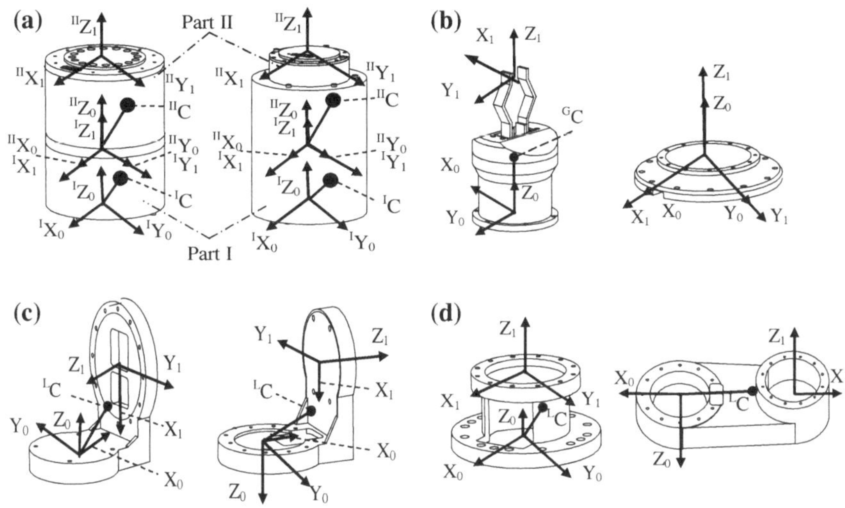

**Fig. 49.2** Cartesian coordinates on modules. **a** Revolute joint and Prismatic joint. **b** Gripper and base modules. **c** A and B styles of right-angled links. **d** Linear link and parallel link

AIM is taken to represent module types, assembled orientations and sequence of a given configuration of modular robots. The AIM of the 6-DOF configuration shown in Fig. 49.1 is given in Table 49.1. The first row describes the joint modules types and connecting sequence from base to gripper. $J_R$, $J_P$, $J_G$ are the representations of the revolute joint, prismatic joint and gripper respectively. The first column describes the possible types of the link modules. $L_C$, $L_P$, $L_R$ are the representations of linear link, parallel link, A and B styles of right-angled links separately. The other part of Table 49.1 excluding the first row and first column above mentioned is the AIM. Each element of AIM contains three numbers. The first one represents the orientation of the two assembled modules and the 1, 2, 3, 4 mean that the input port of a module turning 0, 90, 180, 270° about the $z$-axis of output port of former module in a clockwise direction separately; The second and third one are the numbers of the given types of joint and link modules respectively.

**Table 49.1** AIM of the configuration shown in Fig. 49.2

| | $J_R$ | $J_R$ | $J_P$ | $J_R$ | $J_R$ | $J_R$ | $J_G$ |
|---|---|---|---|---|---|---|---|
| B | (1,1,1) | | | | | | |
| $L_C$ | ↓ | | (1,1,1) → | (1,3,1) ↓ | | (1,5,2) → | (2,1,2) |
| $^AL_R$ | (2,2,1) → | (1,2,1) ↓ | ↑ | (3,3,2) → | (1,4,2) ↓ | ↑ | |
| $^BL_R$ | | (3,2,1) → | (1,1,1) | | (3,4,2) → | (1,5,2) | |
| $L_P$ | | | | | | | |

Taking the third row and the fourth column of AIM for example, $G_{34}$ (3, 3, 2) means that the number 2 of A style right-angled links turning 180° in a clockwise direction relative to the output port of the number 3 revolute joint.

## 49.3 Kinemics of Modular Robot

Based on the assembly information from AIM, the link's information of robot can be got from the parameters of link and joint modules (In the following, links are all mean the robot's links after being assembled). As shown in Fig. 49.3, the part II of joint module ($j-1$), link module ($j-1$) and part I of joint module $j$ consist the ($i-1$)th link of the robot. And the ($i-1$)th joint is formed by joint module ($j-1$), the body's fixed frame of the ($i-1$)th link is superposed with input frame of part II of joint module ($j-1$).

When all joints are at zero position, the link fixed coordinate transformation from the $i$ link to the ($i-1$)th link is

$$\boldsymbol{M}_{i-1,i} = \boldsymbol{T}_{j-1,j} = \boldsymbol{T}^{\mathrm{II}}_{j-1} {}_J\boldsymbol{T}^{\mathrm{L}}_{j-1} \boldsymbol{T}^{\mathrm{L}}_{j-1} {}_{\mathrm{L}}\boldsymbol{T}^{\mathrm{J}}_{j} {}_I\boldsymbol{T}^{\mathrm{II}}_{j} = \begin{bmatrix} \boldsymbol{R}_i & \boldsymbol{p}_i \\ 0 & 1 \end{bmatrix} \quad i,j = 1\cdots n+1 \quad (49.1)$$

Where, $\boldsymbol{T}_{j-1,j} \in \mathrm{SE}(3)$is the homogeneous transformation matrix from the input port of part II of joint module $j$ to that of joint module ($j-1$); $\boldsymbol{T}^{\mathrm{II}}_{j-1} \in \mathrm{SE}(3)$ is the homogeneous transformation matrix from the output frame to the input frame of part II of joint module ($j-1$); ${}_{\mathrm{J}}\boldsymbol{T}^{\mathrm{L}}_{j-1} \in \mathrm{SE}(3)$ is the homogeneous matrix describing the orientation of link module ($j-1$) relative to joint module ($j-1$); $\boldsymbol{T}^{\mathrm{L}}_{j-1} \in \mathrm{SE}(3)$ is the homogeneous transformation matrix from the output frame to the input frame of link module ($j-1$); ${}_{\mathrm{L}}\boldsymbol{T}^{\mathrm{J}}_{j} \in \mathrm{SE}(3)$ is the homogeneous matrix describing the orientation of joint module $j$ relative to link module ($j-1$); ${}_{\mathrm{I}}\boldsymbol{T}^{\mathrm{II}}_{j} \in \mathrm{SE}(3)$ is the homogeneous transformation matrix from the input frame of part II of joint module $j$ to the input frame of part I.

The kinemics of sub-assembly (Fig. 49.3) from frame $i$ to frame ($i-1$) is as following [10, 11]

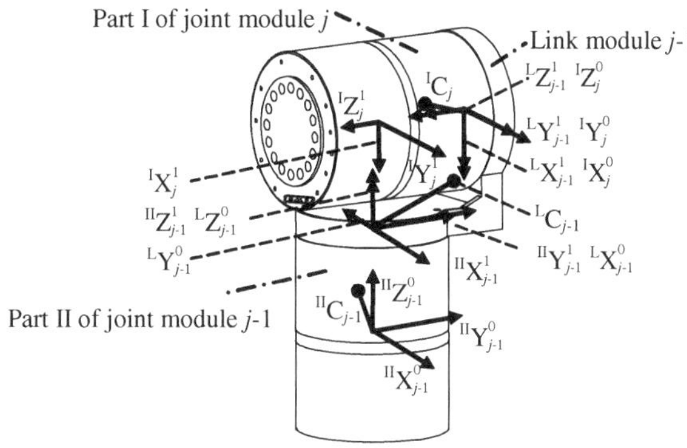


**Fig. 49.3** Sub-assembly of a given configuration

$$f_{i-1,i} = \boldsymbol{M}_{i-1,i} e^{S_i x_i} \tag{49.2}$$

where, $\boldsymbol{S}_i \in se(3)$ is the $i$th joint screw written in the body-fixed frame of the $i$th link and $x_i \in \boldsymbol{R}$ is the current position of the $i$th joint relative to a specified zero position. $\boldsymbol{S}_i$ can also be expressed in a 6-dimensional vector, $\boldsymbol{S}_i = [\boldsymbol{w}_i^{\mathrm{T}} \ \boldsymbol{u}_i^{\mathrm{T}}]^{\mathrm{T}} \in \boldsymbol{R}^{6\times 1}$, termed twist coordinate. For revolute joint $\boldsymbol{S}_i = [\boldsymbol{w}_i^{\mathrm{T}} \ 0]^{\mathrm{T}}$, where $\boldsymbol{w}_i = [0\,0\,1]$ is the unit directional vector of the joint $i$ expressed in frame $i$. For prismatic joint $\boldsymbol{S}_i = [0 \ \ \boldsymbol{u}_i]^{\mathrm{T}}$, where $\boldsymbol{u}_i = [0\,0\,1]$ is the unit directional vector of joint $i$ relative to frame $i$.

According to the Eq. (49.2), the transformation of the frame fixed on the $i$th link to the inertial frame can be got from combining a series of sequential matrices, as follows

$$f_i = f_{0,1} f_{1,2} \cdots f_{i-1,i} \tag{49.3}$$

Then, the kinemics of a given modular robot's configuration can be modeled alike [11]

$$f_{0G} = f_{0,1}\, f_{1,2} \cdots f_{i-1,i}\, f_{i,G} \tag{49.4}$$

$f_{i,G}$ is the transformation matrix from the frame of gripper to that of the $i$th link.

## 49.4 Dynamics of Modular Robot

### *49.4.1 Dynamics Parameters of Sub-Assembly*

Frames, which are parallel to the input frames of link modules or each part of joint modules, are set on the centers of masses (Marked by C in Fig. 49.2) respectively. The vector from the origin of input frame of module to the center of mass is named as $\boldsymbol{r}$. And the homogeneous matrix of the frame on the center of mass relative to the input frame is

$$\boldsymbol{T}_j^{k\mathrm{C}} = \begin{bmatrix} \boldsymbol{I} & \boldsymbol{r}^{k\mathrm{C}} \\ 0 & 1 \end{bmatrix} \quad k = \mathrm{I, II, L} \tag{49.5}$$

where, L means a link module and I and II mean the two parts of a joint module respectively.

Taking a sub-assembly for example (As shown in Fig. 49.3), the frame on the center of mass of link module ($j-1$) which is expressed in the ($i-1$)th link frame (superposed with input frame of part II of link module ($j-1$)) is

$$\boldsymbol{T}_{i-1}^{\mathrm{LC}} = \boldsymbol{T}_{j-1}^{\mathrm{II}}\,{}_{J}\boldsymbol{T}_{j-1}^{\mathrm{L}}\boldsymbol{T}_{j-1}^{\mathrm{LC}} = \begin{bmatrix} \boldsymbol{R}_{i-1}^{\mathrm{LC}} & \boldsymbol{r}_{i-1}^{\mathrm{LC}} \\ 0 & 1 \end{bmatrix} \tag{49.6}$$

$\boldsymbol{T}_{j-1}^{\mathrm{II}}$, ${}_{\mathrm{J}}\boldsymbol{T}_{j-1}^{\mathrm{L}}$ and $\boldsymbol{T}_{j}^{\mathrm{LC}}$ are defined in Eqs. (49.1) and (49.5).

The frame on the center of mass in part I of joint module $j$ which is expressed in the ($i-1$)th link frame is

$$\boldsymbol{T}_{i-1}^{\mathrm{IC}} = \boldsymbol{T}_{j-1}^{\mathrm{II}}\,{}_{J}\boldsymbol{T}_{j-1}^{\mathrm{L}}\boldsymbol{T}_{j-1}^{\mathrm{L}}\boldsymbol{T}_{j}^{\mathrm{IC}} = \begin{bmatrix} \boldsymbol{R}_{i-1}^{\mathrm{IC}} & \boldsymbol{r}_{i-1}^{\mathrm{IC}} \\ 0 & 1 \end{bmatrix} \tag{49.7}$$

The definitions of $\boldsymbol{T}_{j-1}^{\mathrm{II}}$, ${}_{\mathrm{J}}\boldsymbol{T}_{j-1}^{\mathrm{L}}$, $\boldsymbol{T}_{j-1}^{\mathrm{L}}$ and $\boldsymbol{T}_{j}^{\mathrm{IC}}$ are in Eqs. (49.1) and (49.5).

The links' dynamics parameters are the basic informations for constructing the dynamics equations. Taking ($i-1$)th link for example, its total mass is

$$\mathrm{m}_{i-1} = \mathrm{m}_{j-1}^{\mathrm{II}} + \mathrm{m}_{j-1}^{\mathrm{L}} + \mathrm{m}_{j}^{\mathrm{I}} \tag{49.8}$$

$\mathrm{m}_{j-1}^{\mathrm{II}}$, $\mathrm{m}_{j-1}^{\mathrm{L}}$, and $\mathrm{m}_{j}^{\mathrm{I}}$ are the masses of part II of joint module ($j-1$), link module ($j-1$) and part I of joint module $j$ respectively.

The product of the mass of the ($i-1$)th link and the vector from the origin of the ($i-1$)th frame to that of the center of mass is

$$\mathrm{m}_{i-1}[\boldsymbol{r}_{i-1}] = \mathrm{m}_{j-1}^{\mathrm{II}}\left[\boldsymbol{r}_{i-1}^{\mathrm{IIC}}\right] + \mathrm{m}_{j-1}^{\mathrm{L}}\left[\boldsymbol{r}_{i-1}^{\mathrm{LC}}\right] + \mathrm{m}_{j}^{\mathrm{I}}\left[\boldsymbol{r}_{i}^{\mathrm{IC}}\right] \tag{49.9}$$

Taking $\boldsymbol{r}_{i-1}^{\mathrm{IC}}$ (As defined in Eq. (49.7)) which is the vector from the origin of the ($i-1$)th frame to the center of mass of the part I of joint module $j$ for example. If $\boldsymbol{r}_{i-1}^{\mathrm{IC}} = [\,r_1 \quad r_2 \quad r_3\,]$, $[\boldsymbol{r}_{i-1}^{\mathrm{IC}}]$ is its $3 \times 3$ real screw-symmetric form belonged to the Lie Algebra of so (49.3), as follows

$$\left[\boldsymbol{r}_{i-1}^{\mathrm{IC}}\right] = \begin{bmatrix} 0 & -r_3 & r_2 \\ r_3 & 0 & -r_1 \\ -r_2 & r_1 & 0 \end{bmatrix} \tag{49.10}$$

Other three dimensional vectors expressed in the brackets are alike.

The inertia of the ($i-1$)th link relative to the ($i-1$)th frame is [12]

$$\begin{aligned} \boldsymbol{I}_{i-1} &= \boldsymbol{I}_{i-1}^{\mathrm{C}} - \mathrm{m}_{i-1}[\boldsymbol{r}_{i-1}]^2 \\ &= \boldsymbol{I}_{j-1}^{\mathrm{II}} - \mathrm{m}_{j-1}^{\mathrm{II}}\left[\boldsymbol{r}_{i-1}^{IIC}\right]^2 + \boldsymbol{R}_{i-i}^{\mathrm{LC}}\boldsymbol{I}_{j-1}^{\mathrm{L}}\left(\boldsymbol{R}_{i-1}^{LC}\right)^{\mathrm{T}} - \mathrm{m}_{j-1}^{\mathrm{L}}\left[\boldsymbol{r}_{i-1}^{\mathrm{LC}}\right]^2 \\ &\quad + \boldsymbol{R}_{i-1}^{\mathrm{IC}}\boldsymbol{I}_{j}^{\mathrm{I}}\left(\boldsymbol{R}_{i-1}^{\mathrm{IC}}\right)^{\mathrm{T}} - \mathrm{m}_{j}^{\mathrm{I}}\left[\boldsymbol{r}_{i-1}^{\mathrm{IC}}\right]^2 \end{aligned} \tag{49.11}$$

$I_{i-1}^{\mathrm{C}}$ is the inertia relative to a frame on the $(i-1)$th link's center of mass which is parallel to the $(i-1)$th frame; $r_{i-1}$ is the vector from the origin of the $(i-1)$th frame to its center of mass. $I_{j-1}^{\mathrm{II}}$, $I_{j-1}^{\mathrm{L}}$ and $I_{j}^{\mathrm{I}}$ are the inertias relative to the frames at the centers of mass which are parallel to the input frames of part II of joint module $(j-1)$, link module $(j-1)$ and part I of joint module $j$, respectively. $R_{i-1}^{\mathrm{LC}}$ and $R_{i-1}^{\mathrm{IC}}$(As defined in Eqs. (49.6) and (49.7)) stand for the orientation matrixes of the frames on the centers of mass of link module $(j-1)$ and part I of joint module $j$ relative to the frame of the $(i-1)$th link, respectively.

## 49.4.2 Dynamics Based on Lie Group

This method of automatic generation of dynamics is got from Newton–Euler formula based on the Lie Group and Lie Algebra [13]. The parameters of robot's links have been got from the modules' parameters in Sects. 49.3 and 49.4.1. So the link parameters used in the recursive method for dynamics are all mean the combined parameters.

### 49.4.2.1 Forward Recursion for Velocities and Accelerations

***Initialization:***

$$V_0 = 0, \quad \dot{V}_0 = [0 \quad 0 \quad g]^{\mathrm{T}} \tag{49.12}$$

***Forward recursion: for i = 1 to n, do***

$$V_i = \mathrm{Ad}_{f_{i-1,i}^{-1}}(V_{i-1}) + S_i\dot{x}_i \tag{49.13}$$

$$\dot{V}_i = S_i\ddot{x}_i + \mathrm{Ad}_{f_{i-1,i}^{-1}}\left(\dot{V}_{i-1}\right) + \mathrm{ad}_{\mathrm{Ad}_{f_{i-1,i}^{-1}}(V_{i-1})}S_i\dot{x}_i \tag{49.14}$$

- $V_i = [\omega_i^{\mathrm{T}} \ v_i^{\mathrm{T}}]^{\mathrm{T}}$ is the six-dimensional generalized velocity of the $i$th frame, which is expressed in the $i$th frame (The following three and six dimensional vectors are all expressed in the $i$th frame, except for special notes). And $v_i$ represents the velocity of the $i$th frame, $\omega_i$ represents the angular velocity of the $i$th frame. If $f_i = f_{0,1}f_{1,2}\cdots f_{i-1,i}$ is used to describe the location of the $i$th frame relative to the inertial reference frame, then $V_i = f_i^{-1}\dot{f}_i$. $\dot{V}_i = [\dot{\omega}_i^{\mathrm{T}} \ \dot{v}_i^{\mathrm{T}}]^{\mathrm{T}}$is the six-dimensional generalized acceleration of the $i$th frame. $\dot{v}_i$ represents acceleration of the $i$th frame; $\dot{\omega}_i$ represents angular acceleration of the $i$th frame.
- $\mathrm{Ad}_{f_{i-1,i}^{-1}}$ which is used to transform $V_{i-1}$ and $\dot{V}_{i-1}$ from the $(i-1)$ frame to the $i$th frame is the adjoint matrix. $f_{i-1,i}$ is defined in Eq. (49.2), if $f_{i-1,i}$ is in the form as $f_{i-1,i}^{-1} = \begin{bmatrix} R & p \\ 0 & 1 \end{bmatrix}$. Then $\mathrm{Ad}_{f_{i-1,i}^{-1}} = \begin{bmatrix} R & 0 \\ [p]R & R \end{bmatrix}$.

- $\mathrm{ad}_{\mathrm{Ad}_{f_{i-1,i}^{-1}}(V_{i-1})}\boldsymbol{S}_i\dot{x}_i$ indicates $\boldsymbol{S}_i\dot{x}_i$s linear mapping via Lie bracket. $\mathrm{Ad}_{f_{i-1,i}^{-1}}(\boldsymbol{V}_{i-1})$ and $\boldsymbol{S}_i\dot{x}_i$ are elements of Lie Algebra. If $\mathrm{Ad}_{f_{i-1,i}^{-1}}(V_{i-1}) = x = \begin{pmatrix}\omega_1^{\mathrm{T}} & v_1^{\mathrm{T}}\end{pmatrix}^{\mathrm{T}}$ and $S_i\dot{x}_i = y = \begin{pmatrix}\omega_2^{\mathrm{T}} & v_2^{\mathrm{T}}\end{pmatrix}^{\mathrm{T}}$. Then, $\mathrm{ad}_x\boldsymbol{y} = (\boldsymbol{\omega}_1 \times \boldsymbol{\omega}_2, \boldsymbol{\omega}_1 \times \boldsymbol{v}_2 - \boldsymbol{\omega}_2 \times \boldsymbol{v}_1)$. The matrix representation is $\mathrm{ad}_{\boldsymbol{x}}\boldsymbol{y} = \begin{bmatrix}[\boldsymbol{\omega}_1] & 0 \\ [\boldsymbol{v}_1] & [\boldsymbol{\omega}_1]\end{bmatrix}\begin{bmatrix}\boldsymbol{\omega}_2 \\ \boldsymbol{v}_2\end{bmatrix}$.

#### 49.4.2.2 Backward Recursion of Forces and Moments

The equation of motion for a rigid body by Newton–Euler method is expressed as following

$$\boldsymbol{F}_i^{\mathrm{I}} = \begin{bmatrix}\boldsymbol{m}_i^{\mathrm{I}} \\ \boldsymbol{f}_i^{\mathrm{I}}\end{bmatrix} = \boldsymbol{J}_i\dot{\boldsymbol{V}}_i - \mathrm{ad}_{\boldsymbol{V}_i}^*(\boldsymbol{J}_i\boldsymbol{V}_i) \tag{49.15}$$

- $\boldsymbol{F}_i^{\mathrm{I}}$ is the general inertial force of the $i$th link. $m_i^{\mathrm{I}}$ and $f_i^{\mathrm{I}}$ are inertial moment and inertial force respectively.
- $\boldsymbol{J}_i = \begin{bmatrix}\boldsymbol{I}_i^{\mathrm{C}} - m_i[\boldsymbol{r}_i]^2 & m_i[\boldsymbol{r}_i] \\ -m_i[\boldsymbol{r}_i] & m_i \cdot \boldsymbol{1}\end{bmatrix} \in \boldsymbol{R}^{6\times 6}$ is a symmetric positive definite matrix. $\boldsymbol{I}_i^{\mathrm{C}} - m_i[\boldsymbol{r}_i]^2$ and $m_i[\boldsymbol{r}_i]$ are defined in Eqs. (49.9) and (49.11).
- If $\mathrm{ad}_{V_i} = x = (\omega^{\mathrm{T}}\ \ v^{\mathrm{T}})^{\mathrm{T}}$, then $\mathrm{ad}_{\boldsymbol{V}_i}^* = \begin{bmatrix}-[\boldsymbol{\omega}_1] & -[\boldsymbol{v}_1] \\ 0 & -[\boldsymbol{\omega}_1]\end{bmatrix}$ is the dual operator of $\mathrm{ad}_{V_i}$.

***Backward recursion: for i = n to1 do***

$$\boldsymbol{F}_i = \mathrm{Ad}_{f_{i,i+1}^{-1}}^*(\boldsymbol{F}_{i+1}) + \boldsymbol{J}_i\dot{\boldsymbol{V}}_i - \mathrm{ad}_{\boldsymbol{V}_i}^*(\boldsymbol{J}_i\boldsymbol{V}_i) \tag{49.16}$$

$$\tau_i = \boldsymbol{S}_i^{\mathrm{T}}\boldsymbol{F}_i \tag{49.17}$$

- $\boldsymbol{F}_i$ is the total generalized force transmitted from the $(i-1)$ link to the $i$th link through joint $i$, the first three components are the moment vector.
- $\mathrm{Ad}_{f_{i,i+1}^{-1}}^* = \begin{bmatrix}\boldsymbol{R}^{\mathrm{T}} & \boldsymbol{R}^{\mathrm{T}}[\boldsymbol{p}]^{\mathrm{T}} \\ \boldsymbol{0} & \boldsymbol{R}^{\mathrm{T}}\end{bmatrix}$ is the dual operator of $\mathrm{Ad}_{f_{i,i+1}^{-1}}$ (Eq. (49.13)).
- $\tau_i$ is the $i$th actuator's torque or force.

## 49.5 Process of Automatic Generation for Dynamics

The process of the automatic generation of the dynamics for module robot is as following

(1) Giving the AIM of a modular robot's configuration;
(2) Getting the kinematics parameters of each link which is composed by parts of two joint modules and a link module;
(3) Getting the kinematics transformation between adjacent sub-assembles;
(4) Getting the dynamic parameters of each link;
(5) Generating dynamic equations by Lie Group and Lie Algebra formulation based on the Newton–Euler recursive method.

## 49.6 Conclusions

In this paper, a method for automatic generation of dynamics for modular robot is presented. The configuration information of a given manipulator about geometry and DOFs is described by a AIM. The robot's link parameters are got from the assembled modules' kinematics and dynamics parameters based on the assembling information from AIM. Adjoint matrix is adopted to describe the forward mapping of velocities and accelerations from frame of the $(i-1)$ link to that of $i$th. Meanwhile taking dual adjoint matrixes to describe the backward mapping from frame of the $i$ to that of $(i-1)$th. And then a clear and elegant expression of closed-form dynamic equation is generated based on the recursive Newton–Euler algorithm in the form of Lie Groups and Lie Algebra. This automatic dynamics generation method can be applied to modular robot's configuration and motion optimization, calibration and optimal control and so on.

## References

1. Wen-Hong Z, Tom L, Patrick B (2007) Modular robot manipulators with preloadable modules. Proceedings of the IEEE international conference on Mechatronics and Automation, Harbin, China, pp 7–12
2. Matsumaru T (1995) Design and control of the modular robot system: TOMMS. Proc IEEE Int Conf Robotic Autom 2:2125–2131
3. Chen I-M, Gao Y (2001) Closed-form inverse kinematics solver for reconfigurable robots. Proceedings of the IEEE international conference on robotics and automation, Seoul, Korea, pp 21–26
4. Wang J, Li Y, Zhao X (2010) Inverse kinematics and control of a7-DOF redundant manipulator. Int J Adv Rob Syst 7(4):1–9
5. Wang J, Li Y (2009) Analysis on the interaction between the nonholonomic mobile modular robot and the environment. Proceedings of the IEEE international conference on robotics and biomimetics, Guilin, China, pp 86–91
6. Wang W, Zhao J, Gao Y (2008) Dynamics analysis of reconfigurable robots based on screw theory. Chin J Mech Eng 44(11):99–104
7. Fei Y, Zhao X, Xu WL (1998) Kinematics and dynamics of reconfigurable modular robots systems. Proc IEEE Int Conf Man Cybern 4:3335–3341

8. Chen I-M, Yang G (1997) Automatic generation of dynamics for modular robots with hybrid geometry. Proceedings of the IEEE international conference on robotics and automation, New Mexico, USA, pp 2288–2293
9. Chen I-M, Yang G (1998) Automatic model generation for modular reconfigurable robot dynamics. ASME J Dyn Syst Meas Control 120(3):346–352
10. Frank C (1994) Park: computational aspects of the product-of-exponentials formula for robot kinematics. IEEE Trans Autom Control 39(3):643–647
11. Murry RM, Li Z, Sastry SS (1994) A mathematic introduction to robotic manipulation. CRC Press, Boca Raton
12. Liu Y, Hong J, Yang H (1983) Dynamics of multi rigid body systems. Higher Education Press, Beijing
13. Park FC, Bobrow JE (1994) A recursive algorithm for robot dynamics using lie groups. Proceedings of the IEEE international conference on robotics and automation, California, USA, pp 1535–1540

# Chapter 50
# A Kind of Architecture and Key Technologies for Developing Modular Robot

**Hongxing Wei, Jingtao Lei and Tianmiao Wang**

**Abstract** Modular robot technology is of great significance for developing service robots and promoting its industrialization. This paper presents a kind of modular robot architecture which is composed of the Robot Operating System (ROS), standard bus and modular functional components with standard interfaces. The robot has advantages of openness, compatibility, separated function of hardware and software. With this robot, the paper presents the challenges for developing modular robots and explores the uniform software platform for developing modular robots with ROS and integration technology. With this development, this paper presents a way for integrating the functional components of software and hardware into various robots.

**Keywords** Modular robot · Architecture · ROS · Functional component · Integrated development environment

## 50.1 Introduction

Generally, robots are classified into two categories that are industrial-type and service-type according to their functionalities. Service robot is developed to provide different services to various consumers.

---

H. Wei · T. Wang
School of Mechanical Engineering and Automation,
Beihang University, Beijing, China
e-mail: whx1630@163.com; itm@buaa.edu.cn

J. Lei (✉)
School of Mechanical Engineering and Automation,
Shanghai University, Shanghai, People's Republic of China
e-mail: jtlei2000@163.com

J. S. Dai et al. (eds.), *Advances in Reconfigurable Mechanisms and Robots I*,
DOI: 10.1007/978-1-4471-4141-9_50, © Springer-Verlag London 2012

The industrial-type robots have some characteristics such as single purpose, work in unchanged environment. In contrast to industrial robots, service-type robots have characteristics: multiple purposes, working in complex and different environments, interacting with people directly, completing a variety of complex operations.

It is much more complex to develope service robot than industrial robot. Developing service robots need highly flexible design method so that service robots can be developed quickly to meet various functional needs.

The modular design method is flexible. The most advantages is that hardware or software modules can be reused and integrated into different robots according to various consumers' functional needs.

Developing modular robots has become a hot topic in recent years. The core of robotics modularity technology is developing Robot operating system (ROS), a software platform and functional components with standard interfaces for robots. Under the management of ROS, a number of homogeneous or heterogeneous components are integrated into various robots by the software platform.

First of all, the architecture of modular robot should be determined. Some architecture of modular robot has been proposed. Literature [1] presented a layered architecture and distributed control system. Literature [2, 3] proposed a network-based robot controller, the modular open software architecture was adopted. The architecture of modular robot and the design method based on configuration has been studied [4–7]. Literature [8–12] focus on developing modular robot architecture based on robot middleware technology in Japan. The driver, sensor and controller are considered as the basic elements of robot systems, which are defined as functional components. A variety of modular components with standard interfaces are integrated to different robots by middleware. These architectures had their advantages and researched on robot modular technology from different perspective, such as machinery, software and so on.

The architecture of modular robot should have the versatility and universality. Some work has been carried out by us. The architecture of service robots and middleware technology has been studied, modular robot based on the separated software and hardware has been developed [13, 14]. A kind of architecture with better universality of modular robot is presented in this paper, which is composed of the ROS, standard bus structure and modular functional components of hardware and software. The concept of machinery bus and the module attribute with standard interface are proposed.

## 50.2 The Architecture of Modular Robot

The architecture of modular robot should reflect the relationship among components and its functions distribution. It should determine the information flow relationship and logical calculation structure of one or more intelligent robotics system. In a word, the architecture of robot should describe clearly the working principle and its modular layer of hardware and software.

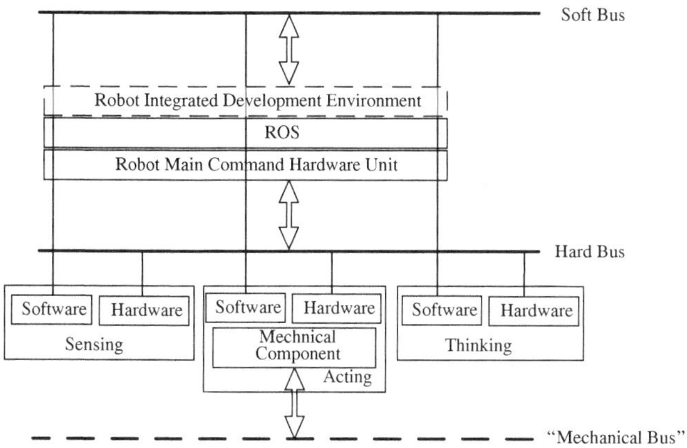


**Fig. 50.1** The architecture frame of modular robot (I)

It is difficult to determine a general architecture for developing modular robot. Because structures and functions of robots are much more complex than that of computers. And the configurations of robots are various.

It is necessary to extract common features from different configurations of robots, and present the general architecture. Firstly, robot is a kind of automated machine with intelligent ability similar to human or biology, such as perception, planning, acting and collaboration. Secondly, from the point of developing modular robots, the differences of external structure of robots should not be considered first. The common characteristics, which are basic elements of various robots, should be considered. These basic mechanical or electrical elements should be designed as modular functional components with standard interfaces.

In general, modular robot architecture has the advantages of openness, and modularization and standardization of functional components so that they have characteristics of compatibility and reusability.

According to two points above, a kind of general architecture of modular robot with separated software and hardware was proposed as shown in Fig. 50.1. It is composed of ROS, standard buses and modular functional components of software and hardware, which work together to achieve basic functions of perception, acting and thinking.

The hierarchical architecture of modular robot is as shown in Fig. 50.2. The hierarchy model from bottom to up is composed of hardware layer, the hardware bus interface, the device driver layer, the robot main command unit, the software bus interface, software functional module layer, application layer and user interface layer. The main command unit of modular robot is composed of the robot control unit hardware, ROS and Integrated Development Environment (IDE). Under IDE for developing modular robots, the physical structure, the inner topological structure, internal control structure and data exchange routing among functional components can be established.

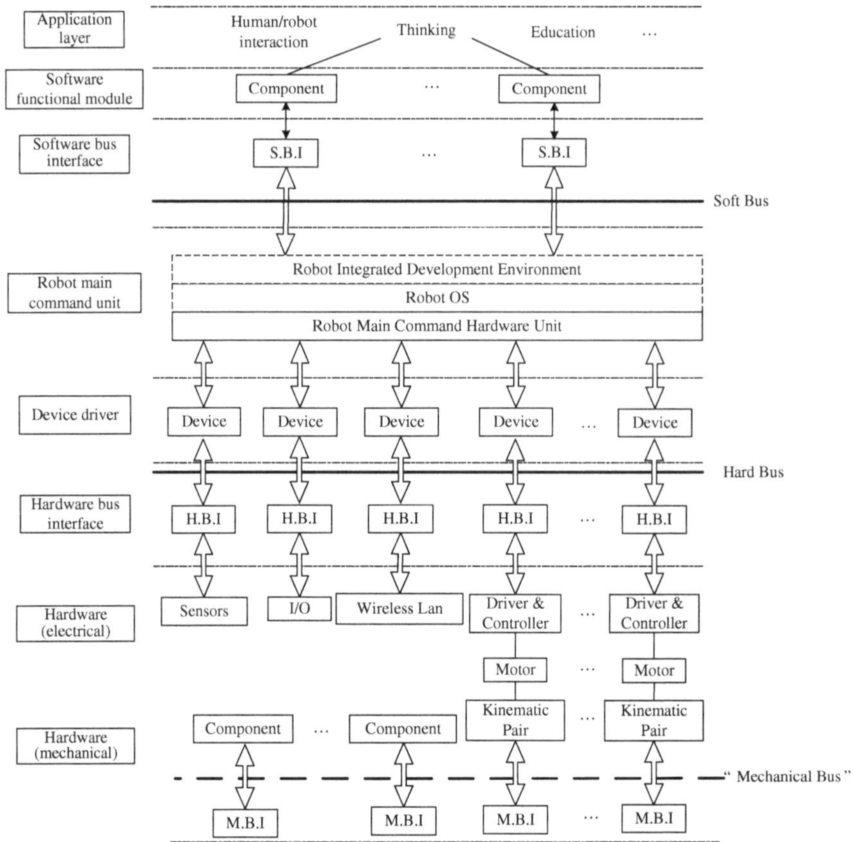


**Fig. 50.2** The hierarchical architecture of modular robot (II)

The architecture of modular robot clearly describes the topology relationship between robot's structure and function. The robot is composed of various software and hardware functional components which are developed with the characteristics of functional independence. This method reflects the overall concept design of general modular robots.

## 50.3 The Key Technologies for Developing Modular Robot

### *50.3.1 The Robot Operating System*

The ROS is used to organize, control and manage effectively robot's various resources so as to provide convenience for high-level applications. It is more difficult to develop ROS for modular robot than that of computer. Because computer only deal with specific mathematical tasks, but service robots have to operate much more complex tasks.

## *50.3.2 Module Functional Component Library*

### 50.3.2.1 Module Division and Module Granularity

The configuration and functions diversity of robot lead to the difficulty of functional components division, such as, the module granularity, the relationships among layers of modules, simplified interface between the modules and so on.

The robot should be divided into reasonable layer number according to the complexity of robot. The more layers of the modular unit, the smaller granularity, the more assembly interface, more assembly time and lower accuracy, and difficult to create and manage complex, thus increasing cost. Generally, the motor, sensor with single function are looked as independent module. And various algorithm units, mobile robot chassis unit, the operating arm and other complex modules are looked as modular units.

### 50.3.2.2 Modular Functional Component Library

Various granularity modules of mechanical and electrical consist of the functional components library. When we integrate robots by selecting various components from library, the attribute information of functional components should be defined clearly to be managed in IDE.

For example, logic description of the attribute information [15] for mechanical functional components is shown in Fig. 50.3.

## *50.3.3 The Integration Technology in Graphical Development Environment*

Generally, robots are designed for specific tasks, various software platforms have been developed for different robot, and these software platforms are incompatible with one another. It is necessary to develop general software platform for developing modular robot.

### 50.3.3.1 Graphical Integrated Development Environment

The graphical software platform IDE shown in Fig. 50.4 for developing modular robot was studied. The hardware and software functional components with

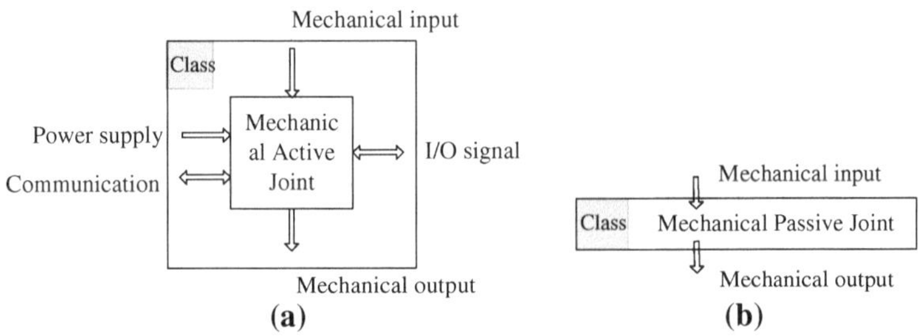


**Fig. 50.3** The attribute information of mechanical functional components. **a** Active joint module. **b** Passive joint module

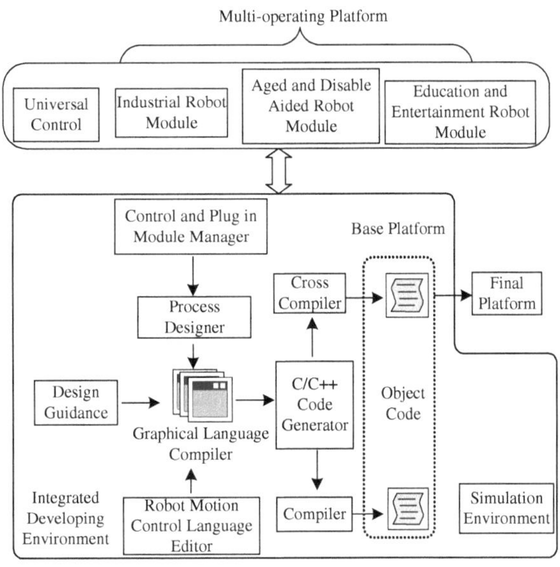


**Fig. 50.4** Integrated development environment (IDE)

relevant functions can be integrated into various robots. The functional components have standard interface, commonality and interchangeability. In IDE, we can not only complete the properties definition of the functional components, then define the connection relationship between components, but also finish automatic modeling of kinematics and dynamics. At last, the simulation and optimization of various robot configurations can be analyzed.

#### 50.3.3.2 Automated Kinematics and Dynamics Modeling

The kinematics and dynamics analysis for modular robot is different from the robot with fixed configuration. It should be generated automatically when the robot configurations are changed.

For simple configuration such as series or multi-branching series, the kinematics automatic modeling is obtained by the POE formulas of screw theory according to the Assembly Incidence Matrix. For complex configuration such as parallel or hybrid, the finite element division method based on constraint conditions of the kinematics-pair can be used to automatically obtain the kinematics model [16, 17].

#### 50.3.3.3 Topological Configuration Optimization

Different configurations of robots can be integrated with joint modules, link modules and other modular functional components. The optimal configuration that meets the given task can be determined by configuration optimization. The automated modeling technology of kinematics and dynamics is the basis of topological configuration optimization. The optimal or sub-optimal configuration of modular robot can be determined under the IDE.

## 50.4 Conclusion

A kind of general architecture for developing modular robot is presented, which is composed of ROS, standard buses and modular functional components with standard interfaces. The architecture has some advantages of openness, compatibility, separated function of hardware and software. A uniform software platform IDE for developing modular robots is presented. The modular functional components of software & hardware with standard interfaces are integrated into various robots according to different needs.

**Acknowledgments** This work was supported by the National High-Tech 863 Research and Development Program of China.

## References

1. Ly DN, Regenstein K, Asfour T et al (2004) A modular and distributed embedded control architecture for humanoid robots. IEEE/RSJ Int Conf Intell Robots Syst 3(28):2775–2780
2. Cho Y-J, You B-J, Oh S-R et al (1999) A compact/open network-based controller incorporating modular software architecture for a humanoid robot. Intell Robotic Syst 25:341–355
3. Ly DN, Regenstein K, Asfour T et al (2004) A modular and distributed embedded control architecture for humanoid robots. IEEE/RSJ Int Conf Intell Robots Syst 3:2775–2780
4. Bi, ZM, Zhang WJ, Lang, SYT (2002) Modular robot system architecture. In: 7th international conference on control, automation, robotics and vision, vol 2, pp 1054–1059
5. Bi, ZM, Lang SYT (2005) General modular robot architecture and configuration design. In: Proceedings of the IEEE international conference on mechatronics and automation, vol 1. Niagara Falls, pp 268–273

6. Larizza P, Murciano G, Pappagallo L et al (2006) A new generation of modular robots. IEEE Int Symp Ind Electron 4:3367–3371
7. Pozna C (2007) Modular robots design concepts and research directions. In: 5th international symposium on intelligent systems and informatics. Subotica, pp 113–118
8. Ikezoe A, Nakamoto H, Nagase M (2006) Development of RT-middleware for image recognition module. In: SICE-ICASE international joint conference, pp 2036–2041
9. Tetsuo K, Makoto M (2006) Robot middleware and its standardization. In: OMG.SICE-ICASE international joint conference, pp 2028–2031
10. Toru Y, Eri S (2007) Humatronics and RT-middleware. advances in information processing and protection. Springer, US. ISBN 978-0-387-73136-0, pp 3–12
11. RT middleware developed for realizing open robot architecture, for creation of new industry of service robots (2005) http://www.is.aist.go.jp/rt/
12. Jia S, Hada Y, Gakuhari H et al (2006) Intelligent home service robotic system based on robot middleware technology. In: IEEE/RSJ international conference on intelligent robots and systems
13. Wei H, Li S, Zou Y, Yang L, Wang T (2008) A middleware based control architecture for modular robot systems. In: IEEE/ASME international conference on mechtronic and embedded systems and applications. Beijing, pp 327–332
14. Tao Y, Wei H, Wang T (2008) A multi-layered interaction architecture for elderly companion robot. In: International conference on intelligent robotics and applications. Wuhan, pp 538–546
15. Larizza P, Murciano G, Pappagallo L et al (2006) A new generation of modular robots. In: IEEE international symposium on industrial electronics, vol 4, pp 3367–3371
16. Pozna C (2007) The modular robots kinematics. Acta Polytech Hung 4(2):5–18
17. Bi ZM, Gruver WA, Zhang WJ et al (2006) Automated modeling of modular robotic configurations. Robotics Auton Syst 54(12):1015–1025

# Chapter 51
# Metamorphic Mechanism Analysis of a Chinese Massage Robot End-Effector

**Lv-Zhong Ma, Peng Fei, Jun Zhang and Guanghong Zhao**

**Abstract** This paper introduces a Chinese massage robot metamorphic end-effector composed by parallel and serial mechanism, the numbers of whose driving elements, driven elements and kinematic pairs are variable, which can work in four different ways of Chinese medical massage techniques. In this paper, the principles of motions, mechanism composition, motion characteristics, metamorphic methods and configuration variable changes of four common massage techniques are analyzed. This metamorphic end-effector has been used in a National High Technology Research and Development Supported Program in China.

**Keywords** Metamorphic mechanism · Chinese massage · Robot · End-effector

## 51.1 Introduction

Metamorphic mechanism is a hot topic in nowadays mechanism theory study. It is characterized by an instantaneous change in configurations, which enable it afford variable DOF or variable number of components, so that the mechanism can perform different tasks according to the requirements. Chinese massage consists a whole set of complex body movements, which in general include swinging, rubbing, fluttering, pressing, striking and movable joint methods etc. when classified according to the movement types of the massage techniques. When classified by

L.-Z. Ma (✉) · P. Fei · J. Zhang · G. Zhao
Institute of Intellectual Machinery and Robot, Jiangsu University, 212013, Zhenjiang, China
e-mail: mlz@ujs.edu.cn

P. Fei
e-mail: feipeng@ujs.edu.cn

J. S. Dai et al. (eds.), *Advances in Reconfigurable Mechanisms and Robots I*,
DOI: 10.1007/978-1-4471-4141-9_51, © Springer-Verlag London 2012 

**Fig. 51.1** Common massage manipulations

the orientations of the massage forces of each technique, there are vertical force, planar force, rotation force, symmetrical force resultant, counter force and flexion extension movement joint, and so on. In actual practice, massage owns character of variability of motion forms and force orientations, so a metamorphic mechanism can be used in the massage end-effector to meet the multifunction and movement control requirements. This paper presents a metamorphic method which can achieve fluttering, pressing, kneading and rolling techniques in detail discussion. With only one metamorphic body the massage end-effector can meet a variety of massage techniques requirements.

## 51.2 An Analysis about the Principles of Traditional Chinese Massage

Techniques are the key to Chinese massage, and play a decisive role of the treatment effects. Common massage techniques are shown in Fig. 51.1, from left to right: rolling, fluttering, pressing and kneading method [1]. These are the single basic techniques, which also include rubbing, wiping, striking and pinching methods. Single-style manipulations are often combined in a definite way to perform certain different duplex type techniques, which can be employed in practical therapy.

Among them, the kneading method is a soft flexible surface massage up and down, right and left or in rotation way. Pressing is the way using the finger, palm or gadget to press the operated position in certain rhythm, frequently works together with kneading method. Other methods can be found in the relevant articles. Their kinematics characteristics can be described as Table 51.1 [2]. In the case of rolling method, forces are both needed in $X$, and $Z$-axis direction and a rotation around the $X$-axis should be there altogether. How to meet the kinematics and dynamics requirements of manipulation is the basic starting point to design the massage robot. Meanwhile, the speed of the robot movement and rotation should try to be uniformed and smooth, to prevent harm to humans [3].

It can be seen, for each massage techniques, the principle of its movement can be achieved using the corresponding reciprocating or rotary motion mechanism, such as connecting rod, crank rocker, etc., but to achieve each of these massage techniques with an independent body, it is difficult to work as the way that

**Table 51.1** Kinematics and dynamics characteristics of massage manipulations

| Manipulation | Rolling | Fluttering | Pressing | Kneading |
|---|---|---|---|---|
| DOF | $\begin{bmatrix} x & y & z \\ \alpha & \bullet & \bullet \end{bmatrix}$ | $\begin{bmatrix} x & y & z \\ \bullet & \bullet & \bullet \end{bmatrix}$ | $\begin{bmatrix} x & y & z \\ \bullet & \bullet & \bullet \end{bmatrix}$ | $\begin{bmatrix} x & y & z \\ \bullet & \bullet & \gamma \end{bmatrix}$ |
| Force exerted | $F_Z, F_X, M_X$ | $F_Z$ | $F_Z$ | $F_Z, M_Z$ |
| Frequency(BPM) | 120–160 | 400–600 | 100–200 | 120–160 |

*Note* In the number of degrees of freedom, "·" represent the non-desired output is a constant

massagist acts in. It is even more difficult to meet the requirements of flexibility because different persons like different massage intensity, frequency and amplitude. Furthermore if you want one robot to possess multiple functions, its mechanical body must include several independent kinematic chains, which will lead the system too complex and difficult to be optimized, and its accuracy and reliability will decline, so the difficulty of design and manufactory has increased. Therefore, in the following paragraph we will use the idea from metamorphic mechanism, with the advanced modern control theory integrated, by changing the body's movable parts, original movement and the number of movable parts, to propose a metamorphic mechanism used in the end-effector of massage robot to meet its multi-functional, biomimetic and kinetic requirements,

## 51.3 Mechanism Design of the Metamorphic Massage End-Effector and Its Functional Analysis

According to the demands of high frequency, high stiffness, high precision, low inertia requirements of Chinese massage techniques and considering characteristics of series and parallel mechanism, a metamorphic mechanism composed of a 2-DOF 1P1R parallel mechanical body [4] and a 4R crank rocker is chosen as executive body to accomplish the massage action, which is shown in Fig. 51.2.

The upper platform of the parallel is fixed and there are four branched-chains connected with it. Two of them are driver branched-chains P-R-R, by driving the prismatic pair $P_2$, $P_4$, movement along the *Z*-axis and rotation around the *X*-axis of the under platform can be achieved. Where prismatic pair $P_1$, $P_3$ located are follower branched-chains, the role is played when the movable platform only move up and down, a follower branched-chain can improve the parallel mechanism stiffness, and when the two active branched-chain singly take the reciprocating motion up and down (in order to meet the massage force strength requirements), the follower branched-chains can satisfy the up and down movement, also act as the pivot of the movable platform swinging. By driving the rotation pair $R_5$ in the component crank, a four rotation pair buildup crank rocker fixed on the under platform can achieve the rotation around the *Z*-axis of the end-effector. This body structure are high-stiffness, the movable platform of which can swing at large angle, whose movements are also sensitive and rapid, so that the requirements of massage operating characteristics can be met.

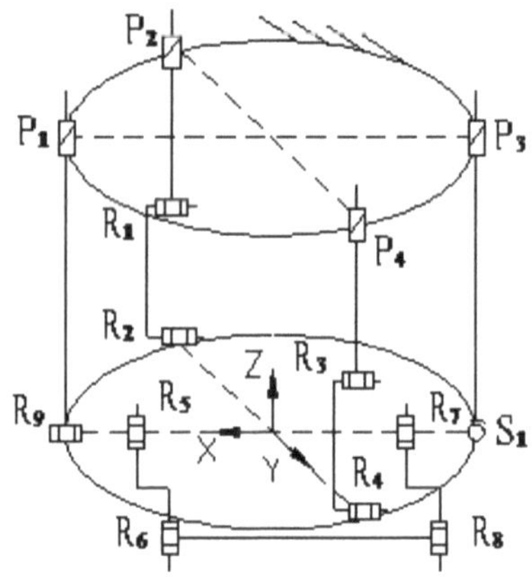


**Fig. 51.2** Mechanism diagram of the end actuator

When the end-effector execute the technique of pressing, fluttering or patting (patting is the massage manipulation swinging arm up and down, using hollow palm area to flap at the treatment position smoothly and rhythmically, exerting forces briefly and uniformly), all rotation pairs $R_1$, $R_2$,...$R_6$, $R_7$, $R_8$ are motion-restricted, so the DOF of the end-effector is 1, and the number of the components is 2. The movable platform at the lower position of the end-effector move along the *Z*-axis by controlling the electric cylinders $P_2$, $P_4$ synchronously operating in certain rules, to achieve the above-mentioned three manipulations. When the end-effector act as rolling method, $P_2$, $P_4$ synchronously operate in opposite orientations, at this moment $R_1$, $R_2$, $R_3$, $R_4$ work in relative rotation, and the DOF of the end-effector change to 2, number of the components change to 8. Topology and the DOF of the end-effector both changed, so the criteria of the metamorphic mechanism are satisfied [5]. And in the same way when the four cranks builtup crank rocker begins to move, with the changes of the DOF and the number of the components, it is also a metamorphic process. Meanwhile, the robot arm above can move the end-effector along *X*, *Y* and *Z*-axis, targeting the end-effector's location. So the massage robot can achieve the four massage techniques shown as Table 51.1. Summary of above mentioned, this end-effector is a metamorphic mechanism whose numbers of components and kinematic pairs are changeable and controllable. In Sect. 51.3 we will put up a comprehensive analysis of mechanism about its configuration changes.

## 51.4 Configuration Description of the Metamorphic Process

Take the transition from pressing to rolling for example, when in pressing method function, the end-effector can simply be described as Fig. 51.3, and its configuration transformation matrix is

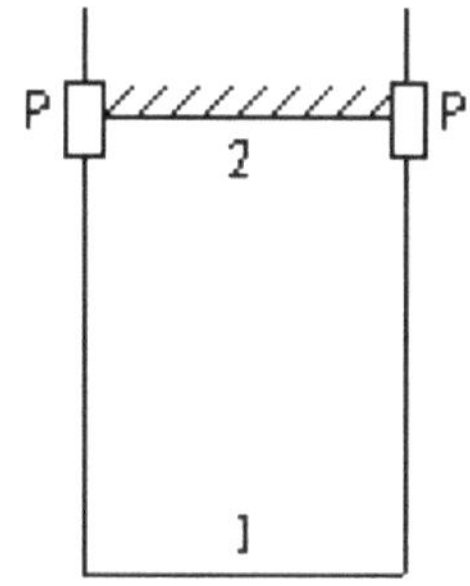


**Fig. 51.3** Mechanism diagram of the pressing method

$$A_1 = \begin{bmatrix} L_1 & P \\ 1 & L_2 \end{bmatrix}, \tag{51.1}$$

$$DOF = 3(n-1) - \sum_{f=1}^{2} fP_f = 3 \times (2-1) - 2 = 1 \tag{52.2}$$

When turning to the configuration of rolling method, its mechanism diagram is shown in Fig. 51.4.

Where R (S) is the projection of $S_1$, $R_5$ in the plane formed by $R_2$, $R_4$, $P_2$, $P_4$, it is overlapped in the visual angle perpendicular to the plane when the mechanism is in the rolling configuration. Therefore, both pairs in connection with the branched-chains where they located can be simplified to a floating support frame, that is, frame 7 (8) float according to comparison of massage force sensor values and the preset value, downward while massage forces too weak, upward when the force value is too large, and keeping still if massage force in appropriate value. And its configuration transformation matrix is

$$A_2 = \begin{bmatrix} L_1 & R & 0 & 0 & 0 & P & 0 & 0 \\ 1 & L_2 & R & 0 & 0 & 0 & 0 & 0 \\ 1 & 1 & L_3 & R & 0 & 0 & R & S \\ 0 & 1 & 0 & L_4 & R & 0 & 0 & 0 \\ 1 & 0 & 0 & 0 & L_5 & P & 0 & 0 \\ 0 & 1 & 0 & 0 & 1 & L_6 & P & P \\ 0 & 0 & 0 & 1 & 0 & 0 & L_7 & 0 \\ 0 & 0 & 0 & 0 & 0 & 0 & 3 & L_8 \end{bmatrix} \tag{51.3}$$

$$DOF = 3(n-1) - \sum_{f=1}^{2} fP_f = 3 \times (8-1) - 2 \times 9 - 1 = 2 \tag{51.4}$$

The metamorphic process can be described as metamorphic equation, $A_2 = B\,A_1\,B^{\mathrm{T}}$, where $B$ and $B^{\mathrm{T}}$ respectively represent metamorphic matrix and its transpose. There are already literatures existed giving detailed analysis about them [6], so considering the length of this paper, further discussion will not be extended about them.

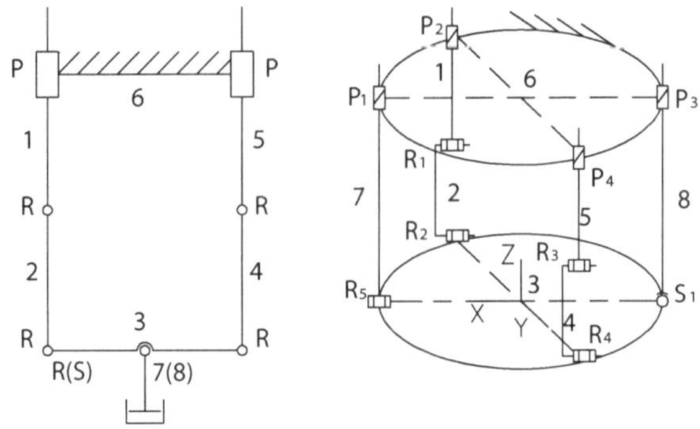


**Fig. 51.4** Mechanism diagram of the rolling method

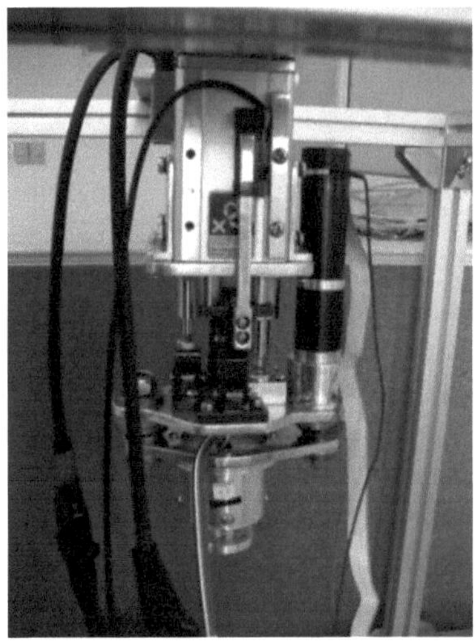

**Fig. 51.5** The metamorphic end-effector practical prototype instrument

## 51.5 Conclusion

Research on the metamorphic mechanism discussed in this paper is the key part of the National High Technology Research and Development Supported Program: *the massage robot*. A practical prototype instrument has been produced, shown in Fig. 51.5. Through the prototype performance test, the massage can achieve the desired effect.

Two Chinese national invention patents [7, 8] are also applied for and has been authorized with the program. In the mechanical design stage, research group members use a variety of design methods and mechanism principles for selection and optimization analysis of the prototype. This article carries out further analysis from the perspective of metamorphic mechanism.

## References

1. Luo C (2004) Applied chinese medical massage. Sichuan Science and Technology Press, Chengdu, pp 82–98
2. Yu S, Ma L, Guo Z (2005) Analyses of the kinematics and dynamics characteristics of the Chinese medical massage manipulations [J]. J Shandong Uni Technol 19(3):82–85
3. Yu S (2006) Design theory of traditional Chinese medical massage manipulato based on parallel mechanisms [D]. School of Mechanical Engineering of Jiangsu University, Zhenjiang, pp 49–51
4. Yin X, Xu Y, Ma L et al (2010) Choosing optimum and kinematical analysis of end—effectors of Chinese massage robot [J]. China Mech Eng 21(24):2899–2903
5. Li D, Zhang Z, Dai J, Zhang K (2010) Overview and prospects of metamorphic mechanism. Chin J Mech Eng 46(13):14–21
6. Wang D, Dai J (2007) Theoretical foundation of metamorphic mechanism and its synthesis [J]. Chin J Mech Eng 43(8):32–42
7. Ma L, Yu S et al. (2006) A new class of series-parallel massage robot [P]. CN200510094463.X, 08 Mar 2006
8. Ma L, Chen X et al. (2010) One kind of massage robot actuator [P]. CN201010220907.0, 24 Nov 2010

# Chapter 52
# Roller Skating Analysis of a Novel Quadruped Robot with Extendable Body

**Xilun Ding and Hao Chen**

**Abstract** This paper proposes a skating locomotion mode of a novel quadruped robot with extendable body and passive wheels. The kinematic model and skating principle have been analyzed according to the character of the robot's special structure. The gaits of straight skating and steering skating are discussed in detail. Lastly, we make some simulations to verify the skating principle and the kinematic analysis above.

**Keywords** Skater robot · Quadruped robot · Kinematics · Gait planning

## 52.1 Introduction

Robots with passive wheels and legs are a new type of wheel-legged robots. Compared to active wheel-legged robots, the passive wheel-legged robot has the advantages of good mobility and energy efficiency without attaching additional drive, steering, brakes and other devices. The research of passive wheeled robots began in 1990s. In 1994, Shigeo Hirose and Hiroki Takeuchi from Tokyo Institute of Technology proposed the initial concept of skater robots with passive wheels [1]. In 1996, they established a physical skating model for a quadruped robot under some assumptions [2]. In 1999, Gen Endo and Shigeo Hirose successfully developed a prototype of skater robots called Roller-Walker by installing a

X. Ding · H. Chen (✉)
Institute of Robotics, Beihang University, Beijing, China
e-mail: howard_chen@me.buaa.edu.cn

X. Ding
e-mail: xlding@buaa.edu.cn


DOI: 10.1007/978-1-4471-4141-9_52, 

rotational ankle mechanism on the quadruped robot Titan VIII [3]. In 2000, Gen Endo and Shigeo Hirose improved Roller-Walker which achieved straight roller-walk, rotational roller-walk, steering roller-walk and crab roller-walk [4]. In 2008, Gen Endo and Shigeo Hirose demonstrated that adjusting leg trajectory could provide the Roller-Walker with similar control scheme as a usual automatic transmission of a car [5]. In 2006, Xu Geli and Lv Tiansheng from Shanghai Jiao Tong University succeeded in designing a biped ice-skater robot on the basis of quadruped skater robots [6].

Body extendable quadruped robot (BEQR) is a novel wheel-legged robot with extendable body designed by the Space Robot Laboratory of Beihang University in 2007 [7]. This paper uses this robot as a platform to research its skating locomotion with passive wheels. A new skating mode is proposed according to the robot's special structure. The gaits of straight skating and steering skating will be discussed in details. Some simulation experiments help to verify the analysis above.

## 52.2 Characteristics of the BEQR

Figure 52.1 shows the structure of the robot BEQR. The body is constituted of two separated parts which connect each other by a pair of cross bars. One end of the bar connects the corresponding part of body by revolute joint, and the other end of the bar connects the opposite part of body by prismatic joint. Thanks to this mechanism, the body of BEQR is capable to extend or shrink according to the angle between the cross bars.

Four mechanical legs with the same structure are installed on both sides of the robot's body symmetrically. The leg is constituted of hip, thigh, knee, calf, ankle and foot (passive wheel) from top to bottom, imitating four-legged mammals. Each leg has four active degrees of freedom, including two in the hip, one in the knee and one in the ankle. The foot is a freely rotating wheel which has a passive degree of freedom. When the robot is walking, its swinging leg can be seen as serial mechanism, while the entire body can be seen as parallel mechanism with three or four serial chains. So overall, this robot can be regarded as a spatial series and parallel hybrid mechanism.

## 52.3 Kinematic Analysis

The BEQR can move on the flat ground by a special gait called gourd step in addition to walk. Gourd step is a figure skating technology, as shown in Fig. 52.2. The skater always keeps her feet on the ground, and repeats opening and drawing back the legs, adjusting the forward direction of feet at the same time. The reaction force from the ground pushes the skater forward.

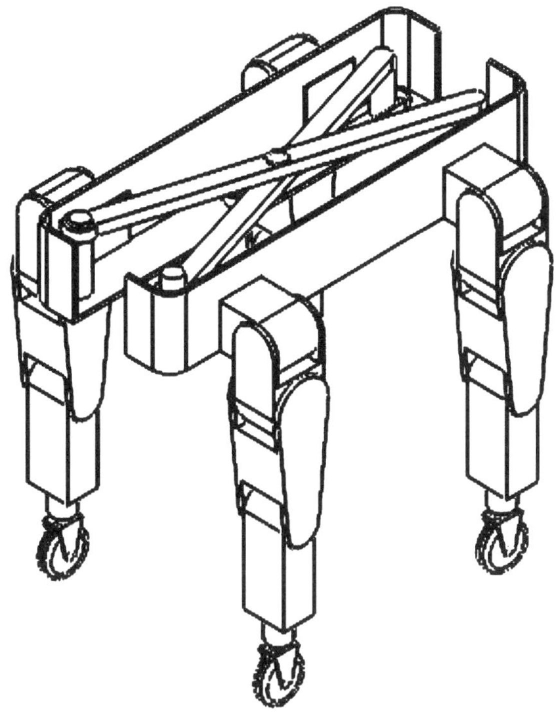

**Fig. 52.1** Structure of BEQR

The BEQR can skate on the flat ground by using similar action. The front and hind pairs of legs open and draw back periodically. When the legs open, the forward directions of wheels rotate outward, so the ground will produce reaction forces perpendicular to the forward direction on the left and right legs. The lateral components of the reaction forces offset each other, and the superposition of longitudinal components push the robot forward. As shown in Fig. 52.3, P represents the forward direction of wheels, N represents the normal direction perpendicular to P, $\alpha$ represents the angle between P and $y$axis of the inertial coordinate system, $L$ represents the distance between left and right wheels, $F_n$ represents the reaction force acting on the wheels.

When the legs drawback, the forward directions of left and right wheels rotate inward. The kinematics analysis is similar to the condition when legs open, as shown in Fig. 52.4.

The angle $\alpha$ is an important parameter determining the wheel's rolling forward direction which depends on the joint angles of the leg. We establish a coordinate system $O - XYZ$ fixed on the ground, another coordinate system $O' - xyz$ fixed on the wheel, as shown in Fig. 52.5. The angle $\alpha$ satisfy

$$\tan\alpha = \frac{x \cdot Y}{x \cdot X} \tag{52.1}$$

Fig. 52.2 Gourd step

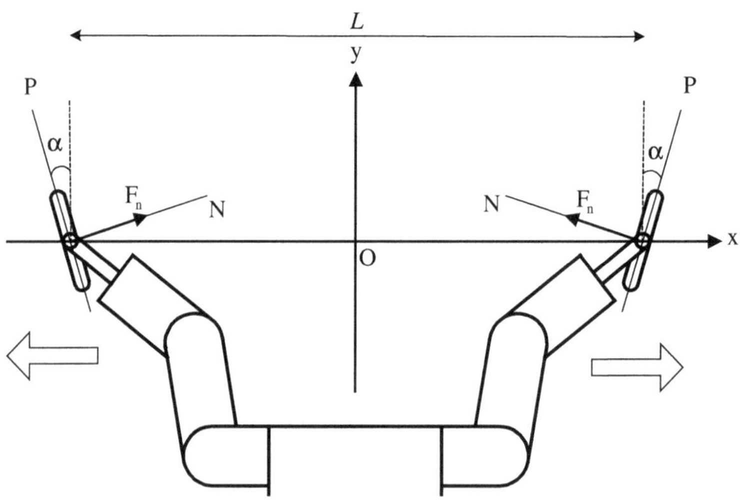


Fig. 52.3 Extending legs

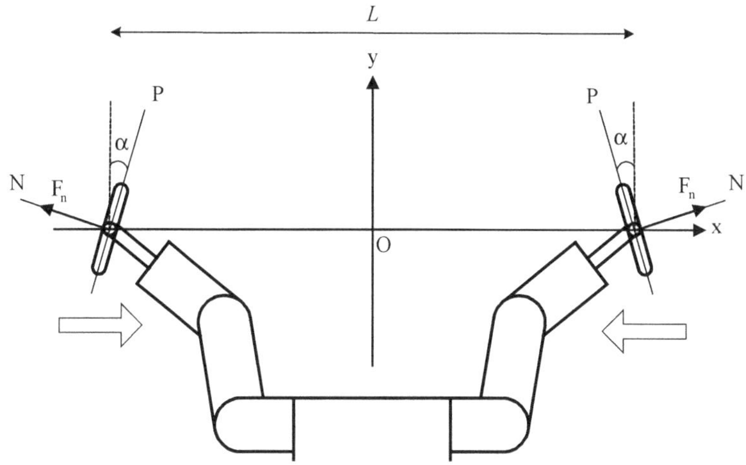


**Fig. 52.4** Shrinking legs

**Fig. 52.5** Forward direction of wheels

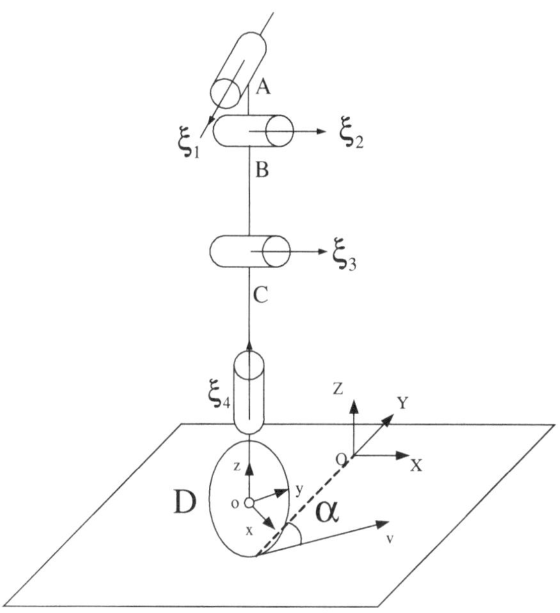


$X$ and $Y$ are unit vectors along corresponding axes of space coordinate system $O - XYZ$, $x$ can be found by solving the forward kinematic problem of legs. The product of exponentials 8 of the leg is

$$g_{st}(\boldsymbol{\theta}) = g_0 \cdot e^{\hat{\xi}_1\theta_1} \cdot e^{\hat{\xi}_2\theta_2} \cdot e^{\hat{\xi}_3\theta_3} \cdot e^{\hat{\xi}_4\theta_4} \tag{52.2}$$

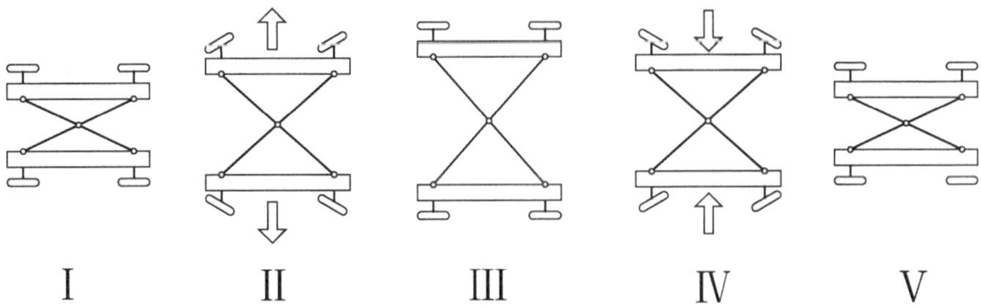


**Fig. 52.6** Action sequence of straight skating

Here, $g_{st}(\boldsymbol{\theta})$ is the transformation matrix from the wheel coordinate system $O' - xyz$ to the space coordinate system $O - XYZ$, $g_0$ is the transformation matrix from the hip joint to the space coordinate system $O - XYZ$ depending on the position and orientation of the robot's body, $\xi_i(i = 1,2,3,4)$ is motion twist of each joint, $\theta_i(i = 1,2,3,4)$ is the joint angle of each joint. Eq. (52.1) can be expressed as

$$\tan\alpha = \frac{i_1 \cdot g_{st}(\boldsymbol{\theta}) \cdot i_2^T}{i_1 \cdot g_{st}(\boldsymbol{\theta}) \cdot i_1^T} \tag{52.3}$$

Here, $i_1$ and $i_2$ are constant vectors, $i_1 = [1 \quad 0 \quad 0 \quad 0]$, $i_2 = [0 \quad 1 \quad 0 \quad 0]$ Though the forward direction of wheel cannot be controlled directly, it's convenient to calculate the required joint angles corresponding to given angle $\alpha$ by using Eq. (52.3).

## 52.4 Straight Skating

BEQR is able to achieve gourd step through swinging its legs as common quadruped robots. Furthermore, it can also extend and shrink its body to achieve gourd step skating because of the characteristic of its variable width. The action sequence of which is shown in Fig. 52.6.

Assuming the wheels don't sliding completely on the ground, we can derive the kinematics formula of the skating process, given by

$$\begin{cases} \dot{y} = \frac{\cot\alpha}{2} \cdot \dot{L} \\ \ddot{y} = \frac{4\mathbf{F}_n \sin\alpha}{M} \end{cases} \tag{52.4}$$

Here $y$ represents the displacement of forward direction, $L$ represents the distance between left and right wheels, M represents the total mass of the robot. If the trajectories of $\alpha$ and $L$ are planned in advance, then the skating velocity of the robot can be calculated through this formula. Generally, the trajectories of $\alpha$ and $L$ are set as,

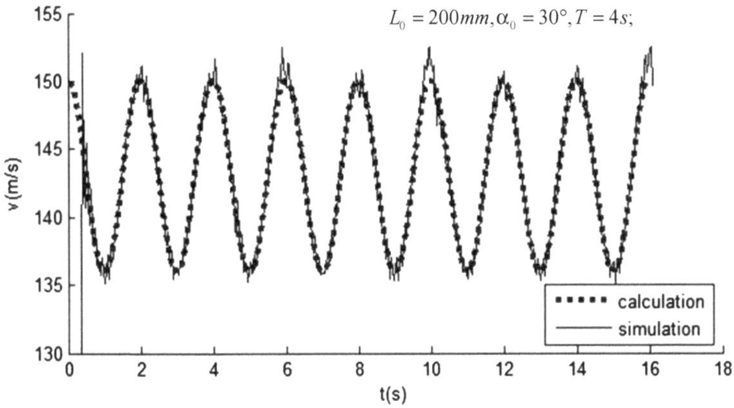


**Fig. 52.7** Velocity of straight skating

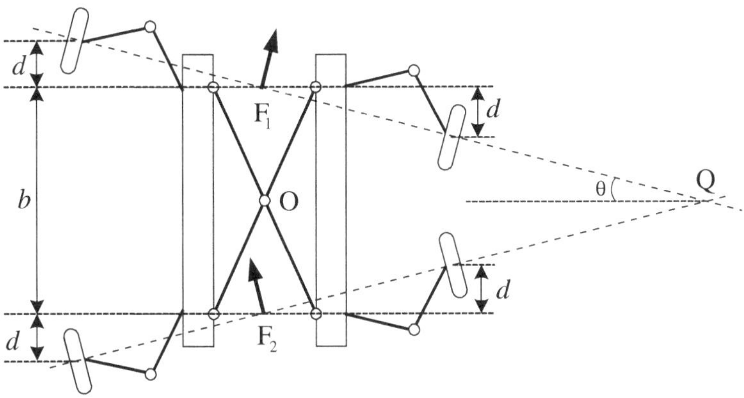


**Fig. 52.8** Steering skating

$$L = L_{offset} + \frac{L_0}{2}\left(1 - \cos\left(2\pi\frac{t}{T}\right)\right)$$

$$\alpha = \alpha_0 \sin\left(2\pi\frac{t}{T}\right)$$

Here, $L_0$ and $\alpha_0$ are the amplitudes of the prismatic motion and swing motion of the wheels, $L_{offset}$ is a constant depend on the structural parameters of the robot, $T$ is the period of cyclic motion.

Figure 52.7 demonstrates the velocity time diagram of the robot's skating by extending its body. It can be seen that the velocity curve of simulation is approximately sinusoidal, which is basically coincident with our calculation result. There are some errors at the peaks and valleys of the sinusoidal curve, because the

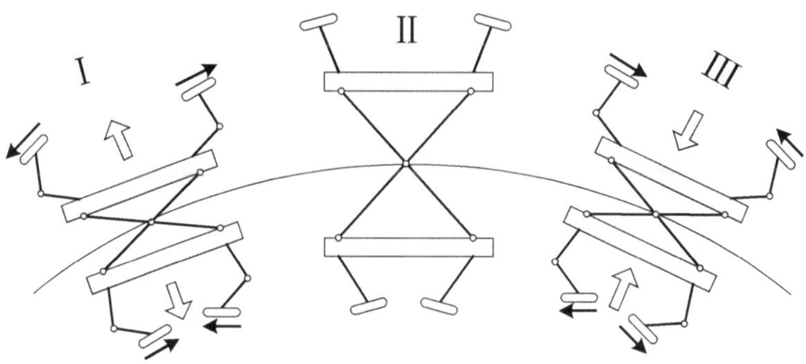


**Fig. 52.9** Action sequence of steering skating

robot moves to singular positions at these points, the ground cannot offer enough friction to push the robot, and the wheels slide on the ground, so that Formula (52.4) becomes invalid.

## 52.5 Steering Skating

Similar with straight skating, the robot repeats changing the distances the forward directions of its feet to achieve steering skating. When steering skating, the trajectory lines of the pairs of front and hind feet are not parallel with each other, which is the difference from straight skating, as shown in Fig. 52.8.

In Fig. 52.8, Point O represents the geometric center of the robots, dashed lines represent the trajectory lines of feet on the ground, $\theta$ is the angle between the dashed line and the transverse direction, $F_1$ is propellant force of the front legs, $F_2$ is propellant force of the front legs, $b$ is the distance between front legs and hind legs. According the above analysis, the propellant forces of legs are perpendicular to the corresponding trajectory lines relatively. If the trajectory lines are not parallel, the propellant forces will not pass through the geometric center of the robot, which produces a torque to drive the robot rotate. The rotation center Q is the intersection of two trajectories. The turning radius $R$ is given by

$$R = \frac{b}{2} \cdot \cot \theta \tag{52.5}$$

Since the robot can only change the width of its body, it needs to move the feet forward and backward to achieve steering skating. The distance that the feet move is denoted by $d$ shown in Fig. 52.7. In order to keep the steering angle $\theta$ constant, the parameter $d$ should satisfy

$$d = L \cdot \tan \theta \tag{52.6}$$

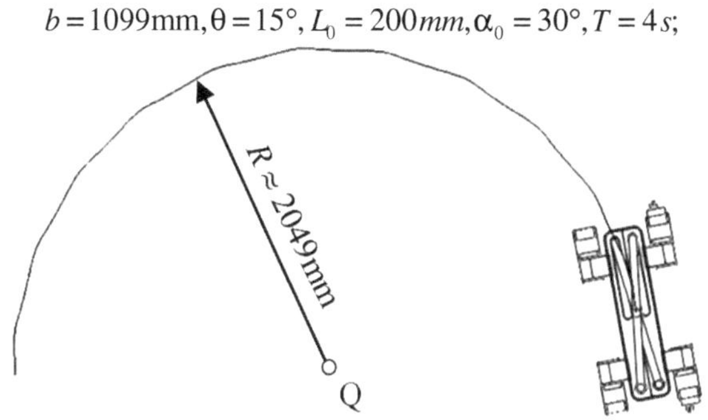


**Fig. 52.10** Body's trajectory of steering skating

Let's take turning right for a example. First, the robot extends its body and moves the legs in the corresponding directions. When the width of body reaches the largest value, it begins to shrink and move the legs in the opposite direction. Then the robot repeats these actions until finishing turning. The action sequence of steering skating is shown in Fig. 52.9.

A simulation result of steering skating is demonstrated in Fig. 52.10. It can be seen that the trajectory of the robot's geometric center goes along an approximate arc. The radius of the arc is 2049 mm, while the theoretical radius calculated by (52.5) is 2051 mm. the error is no more than one thousandth. The simulation result proves our kinematic analysis of steering skating.

## 52.6 Conclusion

In this paper, we develop a new locomotion mode for the quadruped robot BEQR. The robot is able to skate on flat ground by a special gait called gourd step. Unlike other skater robots, BEQR skates by extending and shrinking the body due to its variable structure. Through this method, BEQR has the abilities of straight skating and steering skating on the ground. Generally, these two abilities are enough to meet basic requirements of locomotion. Then we analyze the kinematic model of the robot and give the kinematic formula of skating. Some simulation experiments are taken to prove our works mentioned above.

BEQR is designed as a planetary exploring robot that can move on all terrains. Thanks to its character of variable structure, BEQR has many kinds of locomotion modes and it can choose appropriate modes according to different environments. In the future research, our work will concentrate on structure optimization and energy saving in order to make the robot move more controllably and efficiently.

## References

1. Hirose S, Takeuchi H (1995) Roller-walker: a proposal of new leg-wheel hybrid mobile robot. In: Proceedings of ICAR'95
2. Hirose S, Takeuchi H (1996) Study on roller-walk (basic characteristics and its control). In: Proceedings of 1996 IEEE international conference on robotics and automation, vol 4, pp 3265–3270, 22–28 April 1996
3. Endo G, Hirose S (1999) Study on roller-walker (system integration and basic experiments). In: Proceedings of 1999 IEEE international conference on robotics and automation, vol 3, pp 2032–2037, 10–15 May 1999
4. Endo G, Hirose S (2000) Study on roller-walker (multi-mode steering control and self-contained locomotion). In: Proceedings of ICRA '00. IEEE international conference on robotics and automation, vol 3, pp 2808–2814, 24–28 April 2000
5. Endo G, Hirose S (2008) Study on roller-walker (Adaptation of characteristics of the propulsion by a leg trajectory). Intelligent Robots and Systems, IROS 2008, 1532–1537
6. Xu Z, Lv T, Xu Z, Li J (2006) Principle of biped ice-skater robot and its kinematic analysis. Ch Mech Eng 3(17):230–233
7. Ding X, Xu K (2009) Design and analysis of a novel metamorphic wheel-legged rover mechanism. J Cent South Univ 40(Suppl 1):91–101

# Chapter 53
# Performance Analysis of a Quadruped/ Biped Reconfigurable Walking Robot with Parallel Leg Mechanism

**Hongbo Wang, Xing Hu, Lingfeng Sang, Zhen Xu and Jianjun Wang**

**Abstract** Combining the modularization and reconfigurable ideas, a quadruped/ biped reconfigurable walking robot with parallel leg mechanism is presented. According to design requirements, structural parameters of quadruped/biped walking robot are determined and numeric area of them and the relationship between them are analyzed. The requirements of walking velocity and bearing capacity of quadruped/biped walking robot are taken as measurable indexes to analyze the dexterousness, global minimum velocity index and bearing capacity index, and obtain the optimum numeric area of every parameter, which lay the theoretical foundation for manufacturing of prototype.

**Keywords** Reconfigurable · Parallel mechanism · Performance analysis

H. Wang (✉) · X. Hu · L. Sang · Z. Xu · J. Wang
College of Mechanical Engineering, Yanshan University,
Qinhuangdao, China
e-mail: hongbo_w@ysu.edu.cn

X. Hu
e-mail: huxing918@163.com

L. Sang
e-mail: sanglingfeng@163.com

Z. Xu
e-mail: yuchen735@163.com

J. Wang
e-mail: wjl337@163.com

J. S. Dai et al. (eds.), *Advances in Reconfigurable Mechanisms and Robots I*,
DOI: 10.1007/978-1-4471-4141-9_53, © Springer-Verlag London 2012 

## 53.1 Introduction

With the increasing of the number of the elderly and disabled, the robots for the elderly and disabled have been a research hotspot for the scholars at home and abroad [1]. Among them, the primary function of walking robot for the elderly and disabled is to help them walk freely in complex ground conditions.

The walking robots fall into three categories: legged type, wheel type and legged-wheel type putted forward by many scholars in recent years. The leg mechanism is the key part of walking robot, which decides the motion performance of walking robot [2, 3]. At present, the main forms of walking robot's leg mechanism are convergence-divergence mechanism [4], four bar linkage mechanism, parallel mechanism (PM), parallel bar mechanism, multi-joint serial mechanism and buffered virtual spring leg mechanism. Most of walking robots for the elderly and disabled use serial mechanism as their leg mechanism, such as: "Hubo FX-1" robot [5] made by Humanoid Robot Research Center of South Korea, Hyperion4 developed by Chiba Institute of Technology [6] and the walking chair prototype for lower limb disabled launched by Tokyo Institute of Technology [7]. In PM as leg mechanism study, some prototypes have been already developed. In 1992, Para-walker series robots [8] were researched by Shigeo Hirose who is a professor in Tokyo Institute of Technology and applied PM with six degree of freedom (DOF) into the walking robot firstly. WL-15 [9] which was the first biped walking robot with parallel leg mechanism in the world was developed by Waseda University in 2001, and WL-16 [10], WL-16R [11], WL-16RII [12], WL-16RIII [13], WL-16RIV [14] were developed later. Among them, WL-16RIV also belongs to walking robot for the elderly and disabled. Although there is no prototype of walking robot with parallel leg mechanism for the elderly and disabled in China, the mechanism theory for biped walking robot with parallel leg mechanism was studied by Hebei University of Technology [15]. Yanshan University has also obtained some achievement in theoretical research about the quadruped walking robot with parallel leg mechanism since 2007 [16].

Compared with serial mechanism, it is found that PM has greater stiffness, more stable structure and no accumulation error; it is easy to solve inverse kinematics and to be convenient for the online real-time calculation of the robot. The bearing capacity is the biggest advantage of PM. Therefore, PM is the best choice for leg mechanism of walking robot for the elderly and disabled.

Based on the above analysis, this paper presents a quadruped/biped reconfigurable walking robot with parallel leg mechanism for the elderly and disabled. The 3-SPU PM is selected as the basic leg mechanism. The robot can realize not only quadruped walking, but also biped walking.

In the mechanical design process, in addition to the selection of mechanical configuration, the synthesis of dimension parameters is also very important [17]. For the selected mechanism, the combination approaches of its mechanical dimensions are countless in terms of theory. To judge which one is the best choice, many scholars put forward mechanism performance index to quantize mechanical

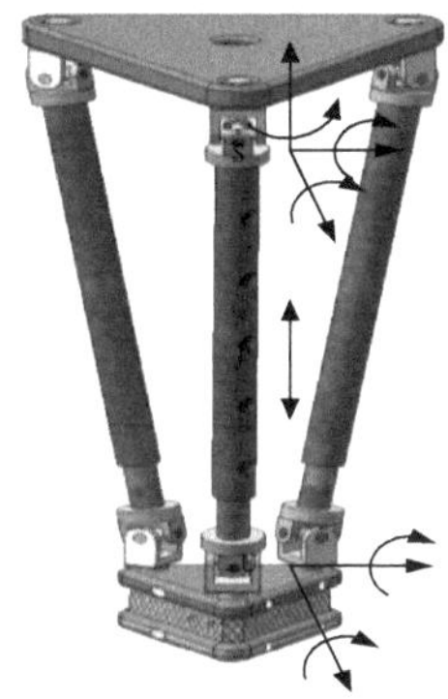

**Fig. 53.1** 3-SPU PM

performance and optimize mechanism design such as robot's dexterousness, isotropy of Jacobian matrix, velocity, acceleration, bearing capacity, inertia force, stiffness and precision and so on [18]. For the research of mechanism performance, Jacobian condition is often used as measurable index, but Jacobian matrix has a relationship with position and orientation of mechanism. In a position and orientation, the corresponding Jacobian matrix is a constant matrix, which is inappropriate to evaluate performance of mechanism in the whole workspace. The global performance index which is putted forward by Gossolin can solve the performance index atlas of the robot with each kind of specific mechanical dimensions in the whole workspace [19, 20].

Based on the robot's walking velocity and bearing capacity, the indexes of the dexterousness, global minimum velocity and bearing capacity for quadruped/biped walking robot are applied in this paper to obtain the performance atlases and the change rules between the dimension and performance index of the body mechanism of quadruped walking robot in the static walking and the leg mechanism of biped walking robot in the walking.

## 53.2 The Design Idea of the Robot

Based on the modularization and reconfigurable ideas, the modular structure design is made for leg mechanism of the walking robot. The structure of four legs for the robot is the same, and each leg is a module. The 3-SPU PM, as shown in Fig. 53.1, is the basic leg mechanism. The configuration design of the 3-SPU PM is described in literature [21].

Under the condition of quadruped walking, the vertical rotation joint hinged with the moving platform is locked, and the parallel leg mechanism changes into 3-UPU (as shown in Fig. 53.2), which has three translational DOF. Four 3-UPU PMs with 3-DOF are used as four legs. When the front foot and hind foot on both sides are combined respectively, and the vertical rotation joint hinged with the moving platform is unlocked, then a biped walking robot is formed as shown in Fig. 53.3.

**Fig. 53.2** Quadruped robot

**Fig. 53.3** Biped robot

## 53.3 The Structural Parameters of the Walking Robot

The structural dimension parameters of the walking robot include the circumradius $R, r$ of the upper platform and the lower platform of 3-SPU PM which is basic leg mechanism, the layout parameters $m$ of four 3-UPU PMs on the seat, and combination parameters $d_1, d_2$ of the lower platform.

### *53.3.1 The Circumradius of Upper and Lower Platforms of Basic Leg Mechanism*

The basic leg mechanism of the walking robot is 3-SPU PM, whose the upper platform and the lower platform are equilateral triangles. In this research, the circumradius of two equilateral triangles are selected as structural parameters and the standard size of wheelchair is used to restrain the structural dimensions of walking robot. According to requirement of parameter design in human engineering, the base of seat is designed as square with a size of 690 mm $\times$ 690 mm. When the layout dimension $m$ of the upper platform is minimum value, $R$ can reach the maximum 199.19 mm. Similarly, $r_{max} = 200$ mm, and $R \geq r$ is required.

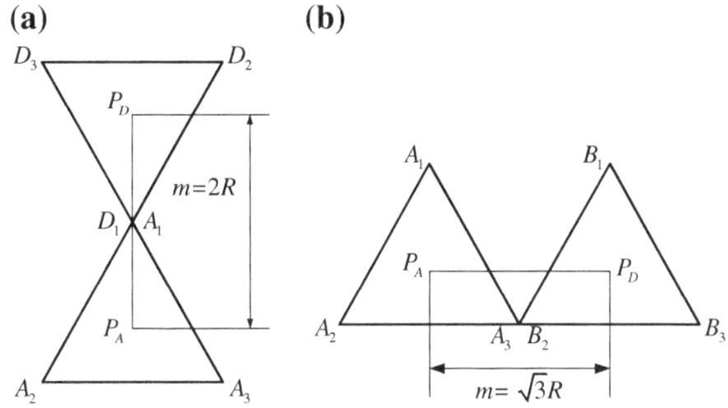


**Fig. 53.4** The layout limit case of the quadruped walking robot

## *53.3.2 The Layout Parameter of the Upper Platform*

Four 3-UPU PMs are arranged symmetrically around the base of the walking robot on which the layout parameter of the upper platform is $m$. The robot is used for manned, therefore, the upper structure of the robot is the same as a chair. The layout parameter $m$ represents the distance between the center points of the two upper platforms which are arranged on the same side of the two leg mechanisms. As a result, there are two limit cases, which are as shown in Fig. 53.4a,b.

By analyzing the geometric relations on the above diagram, the following relationship can be obtained:

$$m \geq 2R, m \geq \sqrt{3}R, m + \sqrt{3}R \leq 690 \tag{53.1}$$

## *53.3.3 The Combination Parameters of the Lower Platforms*

Figure 53.5 shows the two limit locations when the leg mechanisms are combined: $d_2 = 0$ and $d_2 = 3r$. In the practical process of combination, $0 < d_2 < 3r$ is required. The combination diagram of the lower platform is shown in Fig. 53.6. The distance between $E_2E_3$ and $H_1H_2$, which are two adjacent edges in triangle, is recorded as $d$, In addition, $d \geq 0\,mm$ and $|E_1H| \geq \sqrt{3}r/2$ are required.

According to the geometric relations on the Figs. 53.5 and 53.6, the following relationship is got:

$$0 < d_2 < 3r, 0 < d_2 < \sqrt{3}d_1, \frac{\sqrt{3}}{2}r < d_1, d_1 + \sqrt{3}d_2 \geq \sqrt{3}r \tag{53.2}$$

.

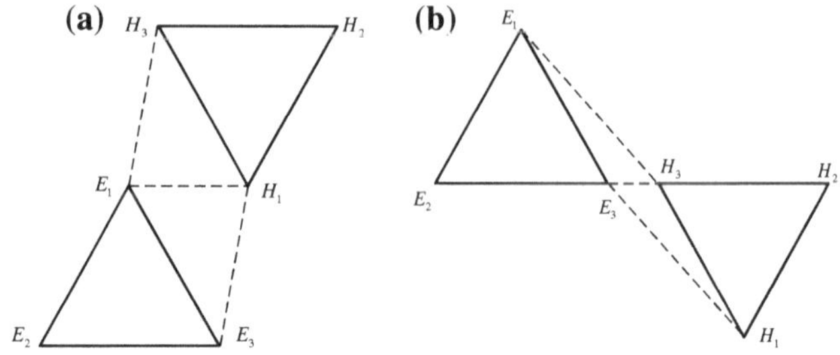


**Fig. 53.5** The two limit locations diagram. **a** $d_2 = 0$. **b** $d_2 = 3r$

**Fig. 53.6** The combination layout diagram of the lower platforms

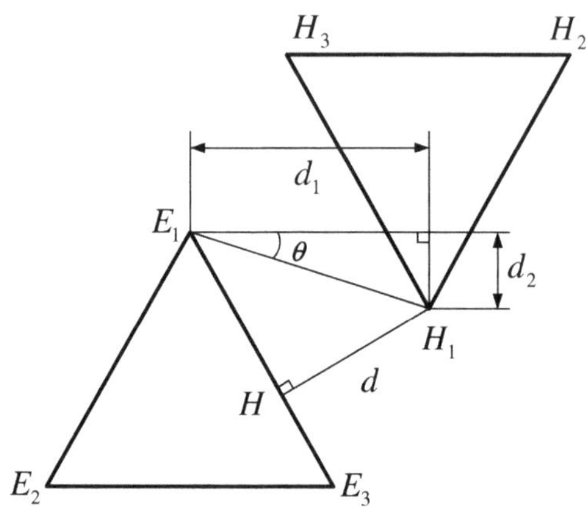


## 53.4 The Performance Analysis of the Quadruped Walking Robot

This section mainly discusses the performance analysis of body mechanism in the static walking. When the quadruped walking robot (as shown in Fig. 53.7) is static walking at any time, the swing leg is 3-UPU PM, the standing leg is 9-UPU PM. In literature [22], the kinematics and dynamics of standing legs and swing leg are analyzed in details. Its structural dimension parameters include circumcircle radius $R, r$ of the upper and lower platforms of 3-UPU parallel leg mechanism and layout parameters $m$ of four 3-UPU PMs on the seat.

According to the requirements of walking velocity and bearing capacity, the performance atlases are drawn along with the determined indexes, which are global condition performance, global minimum velocity and bearing capacity. Among the indexes, the global minimum force is selected as bearing capacity index.

(1) When $m = 350$ mm, $R = 110$–$170$ mm, $r = 40$–$100$ mm, the performance atlases of the mechanism are obtained as shown in Fig. 53.8a–c.

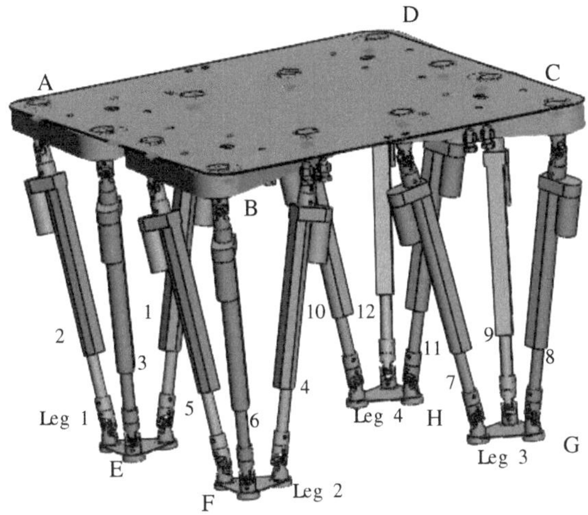


**Fig. 53.7** Quadruped walking robot

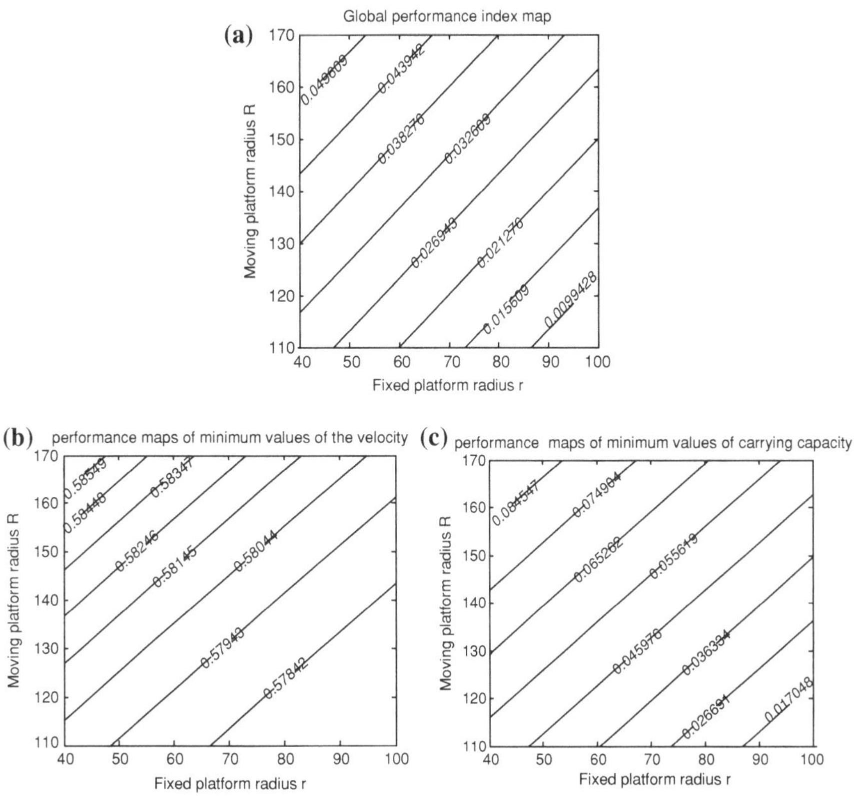


**Fig. 53.8** **a** Global performance index map. **b**, **c** The performance atlases of the quadruped walking

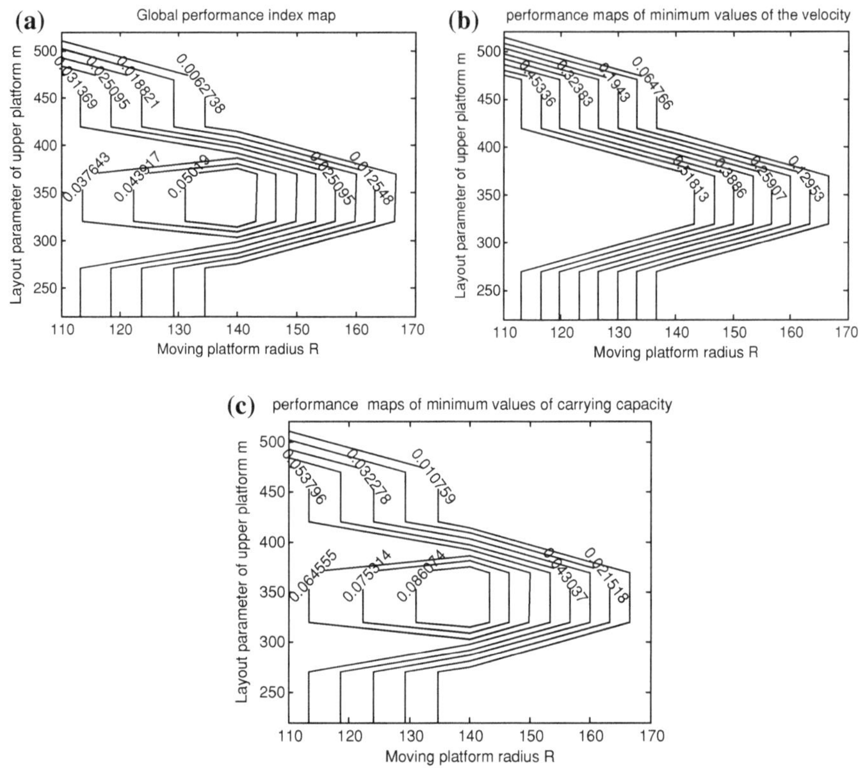


**Fig. 53.9** The performance maps of the quadruped walking robot

(2) When $r = 60$ mm, $m = 220$–$560$ mm, $R = 110$–$170$ mm, the performance atlases of the mechanism can be obtained as shown from Fig. 53.9a–c.

We can see from Fig. 53.8a–c: when circumcircle radius of the upper platform become bigger and circumcircle radius of the lower platform become smaller, three performance indexes value all become greater. This result is the same as the analysis of 3-UPU PM.

We can see from Fig. 53.9a–c: the three performance atlases are symmetrical about $r = 350$ mm. When $m$ value is in the range of 300–400 mm and $R$ value is in the range of 130–150 mm, the three performance indexes can be maximum.

## 53.5 The Performance Analysis of the Biped Walking Robot

The structure of the biped walking robot is shown in Fig. 53.10. Its structural parameters include circumcircle radius $R$, $r$ of 3-SPU PM, layout parameters $m$ of the four 3-SPU PMs on the seat, the combination parameters $d_1$, $d_2$ of the lower platform.

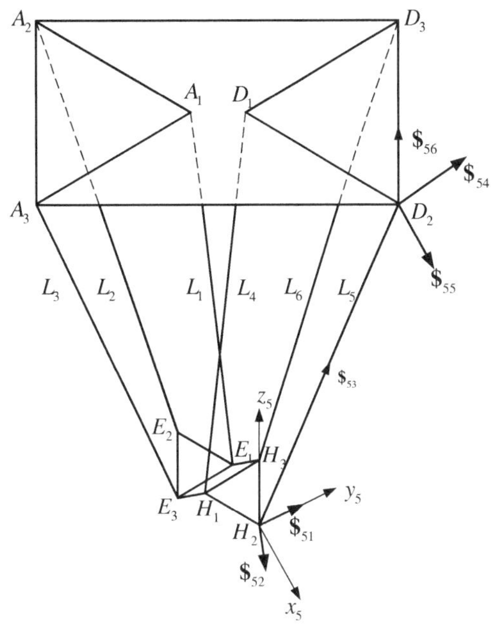


**Fig. 53.10** Diagram of leg mechanism

According to the requirements of walking velocity and bearing capacity, the performance atlases are drawn along with the determined indexes, which are global linear velocity, global minimum velocity and bearing capacity. Among the indexes, the global minimum force is selected as bearing capacity index.

Since the robot has five structural dimensions in total, it can be composed of ten groups of mechanisms, one of which is analyzed as follows:

When $m = 350$ mm, $R = 145$ mm, $r = 60$ mm, $d_l = 50$–$80$ mm, $d_2 = 35$–$80$ mm, performance atlases of the mechanism are obtained as shown in Fig. 53.11.

From Fig. 53.11a–d, the following conclusion can be obtained:

(1) Performance index of linear velocity has a better performance when $d_l$ and $d_2$ are relatively larger.
(2) When $d_2$ is larger and $d_l$ is smaller, performance index of force and bearing capacity are better. When $d_2$ is smaller and $d_l$ is larger, global minimum linear velocity is better which is opposite to the changes to performance index of force and bearing capacity.

Taking all the factors into account, the conclusion can be drawn: when $R$ is in the range from 130 to 150 mm, $r$ is in the range from 40 to 55 mm, $d_l$ is in the range from 50 to 60 mm, $d_2$ is in the range from 65 to 80 mm and $m = 350$ mm, the synthetic performance of the mechanism is preferable.

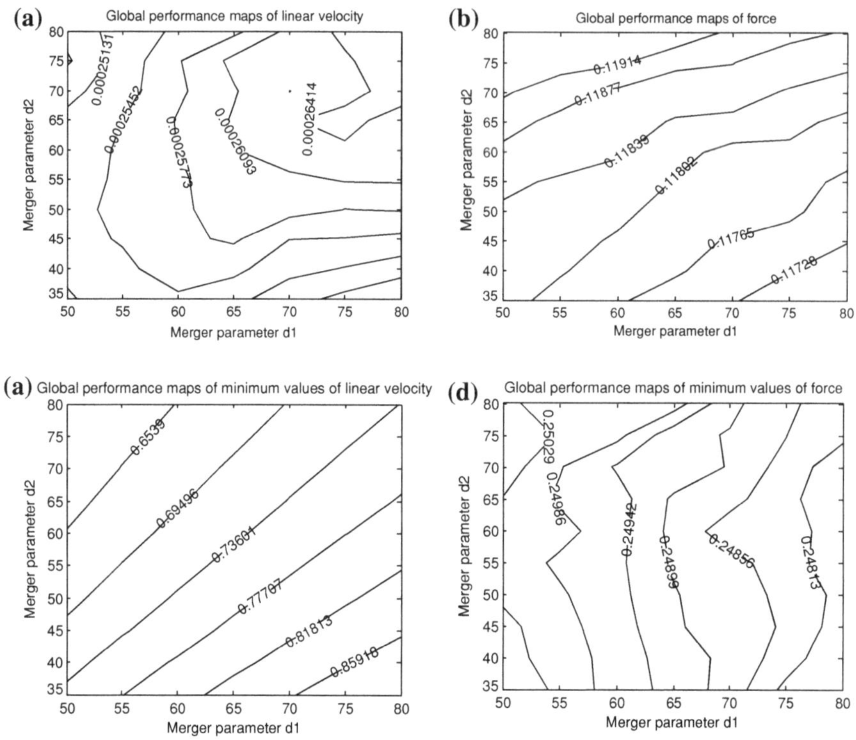


**Fig. 53.11** The performance atlases of the biped walking robot

## 53.6 Conclusion

(1) According to walking environment and the required number of degrees of freedom in walking process, quadruped/biped reconfigurable parallel leg walking robot is proposed and 3-SPU PM is selected as basic leg mechanism.
(2) According to the mechanical model, the required structural parameters are determined, the performance index of the body mechanism in quadruped static walking and the leg mechanism in biped walking are analyzed. The change rules between mechanical dimensions and performance index are studied. As a result, the size range with a preferable synthetic performance of the mechanism is obtained. These efforts lay a theoretical foundation for manufacturing of the prototype.

**Acknowledgments** This work was supported by the National Natural Science Foundation of China under Grant No. 61075099.

## References


1. Dong YH, Hou CY, Tian SL (2010) The logic control simulation of service robot for the elderly based on MATLAB/stateflow. Appl Sci Technol 37(11):10–13
2. Fan SC, Han SK, Gu HR (2007) The stability of biped walking robot parallel leg mechanism. J Hebei Univ Technol 33(1):27–31
3. Wang JD, Lu KY, Xu SF, Lei YY (2009) Research situation and prospect on quadruped walking robot. Manuf Autom 31(2):4–6
4. Cha XF, Zhang RP (1995) Research on the kinematics of leg mechanism for multilegged walking robots. J Southeast Univ 25(2):103–107.
5. Lee J, Kim JY, Park IW et al (2006) Development of a humanoid robot platform HUBO FX-1. In: International joint conference, Busan, South Korea, pp 1190–1194
6. Yoneda K (2007) Light weight quadruped with nine actuators. J Robotics Mechatron 19: 160–165
7. Wu Y, Nakamura H, Takeda Y et al (2007) Development of a power assist system of a walking chair based on human arm characteristics. J Adv Mech Des Syst Manuf 1(1):141–154
8. Ota Y, Yoneda K, Ito F et al (2001) Design and control of 6-DOF mechanism for twin-frame mobile robot. Auton Robots 10(3):297–316
9. Carbone G, Sugahara Y, Lim HO et al (2003) Stiffness performances estimation for biped locomotor WL-15. In: The IEEE/ASME international conference on advanced intelligent mechatronics, Kobe, Japan, pp 956–961
10. Sugahara Y et al (2004) Realization of dynamic human-carrying walking by a biped locomotor. In: Proceedings of the 2004 IEEE international conference on robotics and automation, New Orleans, USA, pp 3055–3060
11. Sugahara Y, Mikuriya Y, Hashimoto K et al (2005) Walking control method of biped locomotors on in-clined plane. In: The 2005 IEEE international conference on robotics and automation, Barcelona, Spain, pp 1977–1982
12. Sugahara Y, Carbone G, Hashimoto K et al (2007) Experimental stiffness measurement of WL-16RII biped walking vehicle during walking operation. J Robotics Mechatron 19(3):272–280
13. Hashimoto K, Sawato T, Hayashi A et al (2008) Static and dynamic disturbance compensation control for a biped walking vehicle. In: The 2nd biennial IEEE/RAS-EMBS international conference on biomedical robotics and biomechatronics, Scottsdale, USA, pp 457–462
14. Hashimoto K et al (2007) Unknown disturbance compensation control for a biped walking vehicle. In: Proceedings of the 2007 IEEE/RSJ international conference on intelligent robots and systems, San Diego, USA, pp 2204–2209
15. Han SK (2004) Biped foot robot of parallel leg mechanism of kinematics analysis and gait planning. Master dissertation, Hebei University of Technology
16. Wang HB, Qi ZY, Hu ZW, Huang Z (2009) Application of parallel leg mechanisms in quadruped/biped reconfigurable walking robot. Chin J Mech Eng 45:24–30
17. Wei XF (2007) Global performance analysis and optimization of the 6-DOF hybrid-chain robot mechanism. Master dissertation, Nanjing University of Science and Technology, pp 26–40
18. Guo XJ (2010) Performance index theory and simulation of mechanism. Science Press, Beijing, pp 59–83
19. Gosselin C, Angeles JA (1991) New performance index for the kinematics optimization of robotic manipulators. Trans ASME 113:220–226
20. Zhang LJ (2007) Performance analysis and dimension optimization of 2-DOF parallel manipulators. Doctoral dissertation, Yanshan University, pp 87–115

21. Wang HB, Xu GL, Zhang DF, Hu X (2010) Kinematics modeling and simulation of quadruped/biped walking robot with parallel leg mechanism for the elderly and the disabled. J Yanshan Univ 34:508–515
22. Xu GL (2011) Design and analysis of a quadruped walking robot with parallel leg mechanism for the elderly and the disabled. Master dissertation, Yanshan University, pp 30–38

# Chapter 54
# Intelligent Fingertip Sensing for Contact Information Identification

**Hongbin Liu, Xiaojing Song, Joao Bimbo, Kaspar Althoefer and Lakmal Senerivatne**

**Abstract** In this paper we propose a new iterative algorithm for effective contact information identification, allowing a fingertip equipped with 6-axis force/torque sensor to accurately estimate contact information, including the contact location on the fingertip, the direction and the magnitude of the friction and normal forces, the local torque generated at the surface. The proposed algorithm is highly computational efficient and achieves an update rate of 833 Hz. The accuracy of the proposed algorithm has been validated experimentally. The results show that the algorithm provides precise estimation for all the identified contact properties.

**Keywords** Contact information identification · Force/torque sensing · Grasping

## 54.1 Introduction

For many applications such as health care, robotic surgery and planetary exploration, robotic hands or grippers are often required to interact with the environment based on the force/tactile feedback especially when vision is occluded or lost [1, 2]. In such case, the precise real-time identification of the contact information between the finger and the object, such as contact locations, normal grasping force, and friction force is essential to achieve high performance control of grasping and manipulation [1]. In addition, the accurate estimation of contact information is important for force/tactile based object properties exploration such as pose estimation [3, 4], object shape recognition, surface features detection [5],

H. Liu (✉) · X. Song · J. Bimbo · K. Althoefer · L. Senerivatne
King's College London, Strand, London WC2R 2LS, UK
e-mail: hongbin.liu@kcl.ac.uk

J. S. Dai et al. (eds.), *Advances in Reconfigurable Mechanisms and Robots I*,
DOI: 10.1007/978-1-4471-4141-9_54, © Springer-Verlag London 2012

object surface texture classification [6, 7] as well as tumour localization during minimally invasive surgery [2, 8].

The human finger has sophisticated contact sensing mechanisms, permitting humans to be extremely proficient in manipulating object and recognizing surface properties based on the sense of touch alone [9]. To narrow the gap between the contact sensing capability of human and that of robots, extensive researches have been carried out to date [10]. One promising approach for identify the contact information is to use force/torque sensors, since this type of sensor is relative easy to construct and mount into the fingertip. The initial analytical solution to allow a robot fingertip identify the locations of point contacts based on force/torque sensing was proposed in [11, 12] and implemented in [13]. In extension to earlier works, the "intrinsic contact sensing" method was proposed in [14]. This method can effectively identify the contact location and local torque using force/torque measurements and is easy to implement. This method has been applied in [15] for the contact sensing of the feet of a pipe crawling robot and achieved good performance. In [16, 17], this method was applied to identify the surface shape properties using a soft fingertip. The limitation of the "intrinsic contact sensing" method is that it is strictly constrained to low friction contacts. An iterative method for identifying contact information was also proposed in [14]. While this method has no restriction on the contact friction level, the algorithm needs large number of iterations to converge.

In the view of the limitation of previous works, we propose a new iterative algorithm for contact information identification based on 6-axis force/torque sensing and the Levenberg–Marquardt (LM) optimization algorithm. The developed algorithm allowing a fingertip to provide rich and precise contact information, including the contact location on the fingertip, the direction and the magnitude of the friction and normal forces, the local torque generated at the surface. Validation tests show that the root mean square error (RMSE) of contact location identification is 0.266 mm, RMSE error for the estimation of normal force, friction force and the local torque are 0.0403 N, 0.0497 N, and 0.792 N mm respectively. Moreover, comparing to the iterative method proposed in [14], the computational efficiency of the new algorithm is significantly increased. To identify the contact information, the algorithm of [14] needs more than thousands iterations while the new algorithm requires less than ten iterations. In addition, the developed algorithm requires only a single 6-six force/torque sensor and a fingertip with known surface equation. Therefore the proposed algorithm is easy to be implemented and has enormous potential in robotic grasping applications.

## 54.2 The Design of Fingertip

To study the contact information identification, a fingertip equipped with a 6-DoFs force/torque sensor (ATI Nano17, resolution = 0.005 N) was constructed. The fingertip is connected to the middle phalange of the Shadow$^{TM}$ robotic hand as shown in Fig. 54.1.

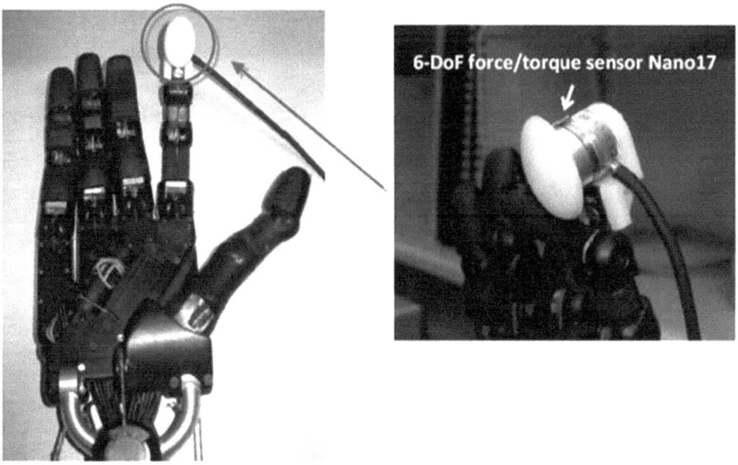


**Fig. 54.1** The design of the fingertip for the intelligent contact sensing

**Fig. 54.2** The force/torque sensing diagram of the ellipsoidal fingertip

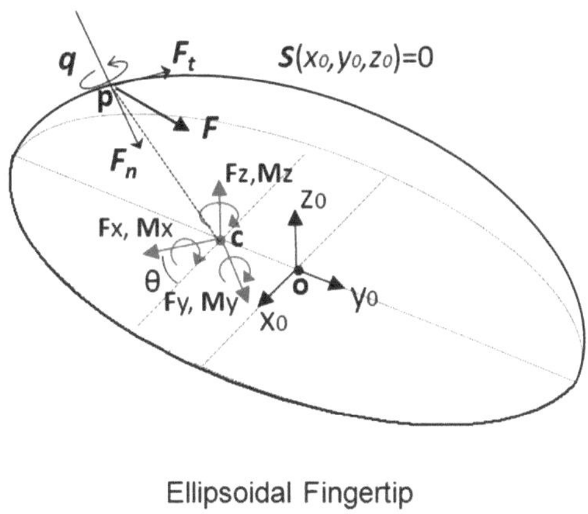


The fingertip is an ellipsoidal surface with radius $a = 9$ mm, $b = 14$ mm and $c = 6$ mm. Point **o** is the center of the ellipsoid as shown in Fig. 54.2. Assume the coordinates of the contact point is $\boldsymbol{p_o} = [\mathrm{x_0}, \mathrm{y_0}, \mathrm{z_0}]^{\mathrm{T}}$ in the **o** frame, since $\boldsymbol{p_o}$ is on the ellipsoid surface, it holds that:

$$S(x_0, y_0, z_0) = \frac{x_0^2}{a^2} + \frac{y_0^2}{b^2} + \frac{z_0^2}{c^2} = 1 \tag{54.1}$$

Let **c** to be the origin of the force/torque sensor, and the coordinate of the contact location in frame **c** is $\boldsymbol{p_c} = [x, y, z]^{\mathrm{T}}$. The homogenous transformation between $\boldsymbol{p}_o\boldsymbol{p}_c$ is:

$$\begin{bmatrix} \boldsymbol{p}_o \\ 1 \end{bmatrix} = \begin{bmatrix} \boldsymbol{R}^{\mathrm{T}} & -\boldsymbol{R}_{oc}^{\mathrm{T}}\overrightarrow{oc} \end{bmatrix} \begin{bmatrix} \boldsymbol{p}_c \\ 1 \end{bmatrix} \tag{54.2}$$

where $\overrightarrow{oc}$ is $[d_x\ d_y\ d_z]^{\mathrm{T}}$ in the sensor's frame, $\boldsymbol{R}_{oc}$ is the rotation matrix,

$$\boldsymbol{R}_{oc} = \begin{bmatrix} \cos\theta & -\sin\theta & 1 \\ \sin\theta & \cos\theta & 0 \\ 0 & 0 & 0 \end{bmatrix}$$

Applying the Eq. 54.2, the relationship between $\boldsymbol{p}_o$ and $\boldsymbol{p}_c$ is:

$$\begin{cases} x_0(x,y,z) = \cos\theta x + \sin\theta y - \cos\theta d_x - \sin\theta d_y \\ y_0(x,y,z) = -\sin\theta x + \cos\theta y + \ \sin\theta d_x - \cos\theta d_y \\ z_0(x,y,z) = z - d_z \end{cases} \tag{54.3}$$

Substitute Eq. 54.3 into Eq. 54.1, the surface equation with respect to the sensor frame $S(x,y,z)$ can be obtained. As shown in Fig. 54.2, the surface equation of the fingertip force is indicated using function $S(x, y, z)$, $\boldsymbol{F} = [f_x, f_y, f_z]^T$ $\boldsymbol{M} = [\mathrm{m_x}\ \mathrm{m_y}\ \mathrm{m_z}]^{\mathrm{T}}$ are the measurements of the force/torque sensor. It can be assumed that the local torque $\boldsymbol{q}$ always perpendicular to the surface [14]. Hence local torque $\boldsymbol{q}$ at a contact location $\boldsymbol{p_c} = [x,\ y,\ z]^{\mathrm{T}}$ can be described as $k\nabla S(\boldsymbol{p}_c)$.

$$\begin{cases} \boldsymbol{p_c} \times \boldsymbol{F} + \boldsymbol{q} = \boldsymbol{M} \\ S(x,y,z) = 0 \end{cases} \tag{54.4}$$

It can be seen from above that the contact location and the local torque can be described using a vector $\mathbf{x} = [x\ y\ z\ k]^{\mathrm{T}} = [\boldsymbol{p}_{\mathrm{c}}\ k]^{\mathrm{T}}$. The problem of identifying $\mathbf{x}$ is equivalent to solving $\boldsymbol{g}(\mathbf{x}) = 0$:

$$g(\boldsymbol{x}) = \begin{bmatrix} k\dfrac{\partial S}{\partial x} - f_y z + f_z y - m_x \\ k\dfrac{\partial S}{\partial y} - f_z x + f_x z - m_y \\ k\dfrac{\partial S}{\partial z} - f_x y + f_y x - m_z \\ S(x\ y\ z) \end{bmatrix} = 0, \tag{54.5}$$

## 54.3 Levenberg–Marquardt Method for Solving g(x)

It is assumed that the fingertip is undeformable and there is only one contact location. To solve $\boldsymbol{g}(\mathbf{x}) = 0$ the contact location $\boldsymbol{p}_{\mathrm{c}}$, and $k$, the gradient descent algorithm was proposed in [14]. With the gradient descent algorithm, the contact location can be iteratively updated by using the following rule:

$$\mathbf{x}_{k+1} = \mathbf{x}_k - \lambda \mathbf{J}(\mathbf{x})^T \mathbf{g}(\mathbf{x}), \tag{54.6}$$

where $\mathbf{J}(\mathbf{x})$ is Jacobian matrix of $\boldsymbol{g}(\mathbf{x})$. The $\mathbf{J}(\mathbf{x})$ is computed as:

$$\mathbf{J}(\mathbf{x}) = \frac{\partial \mathbf{g}}{\partial \mathbf{x}} = \begin{bmatrix} 2k\left(\frac{\cos\theta^2}{a^2} + \frac{\sin\theta^2}{b^2}\right) & 2k\left(\frac{\cos\theta\sin\theta(b^2-a^2)}{a^2b^2}\right) + f_z & -f_y & \frac{\partial S}{\partial x} \\ 2k\left(\frac{\cos\theta\sin\theta(b^2-a^2)}{a^2b^2}\right) - f_z & 2k\left(\frac{\cos\theta^2}{a^2} + \frac{\sin\theta^2}{b^2}\right) & f_x & \frac{\partial S}{\partial y} \\ f_y & -f_x & \frac{2k}{c^2} & \frac{\partial S}{\partial z} \\ \frac{\partial S}{\partial x} & \frac{\partial S}{\partial y} & \frac{\partial S}{\partial z} & 0 \end{bmatrix}$$

The parameter $\lambda$ denotes the learning rate. The limitation of the gradient descent method is that, to guarantee the method converge, the learning rate $\lambda$ has to be set to a small value (0.01). Thus the algorithm needs large numbers of iterations to compute. Furthermore an initial guess which is far away from the solution often significantly slows down the converging speed of the gradient descent method. To reduce the computational cost and increase the algorithm's robustness to the initial guess, we proposed the use of LM method for solving the problem. First, a positive definite function $\chi^2$ is defined as Eq. 54.7:

$$\chi^2 = \frac{1}{2}\mathbf{g}(\mathbf{x})^T \mathbf{g}(\mathbf{x}) \tag{54.7}$$

Estimating parameter vector $\mathbf{x} = [x\ y\ z\ k]^{\mathrm{T}}$ is thus the problem of minimizing $\chi^2$. The derivative of the function $\chi^2$ with respect to estimated parameters is:

$$\frac{\partial \chi^2}{\partial \mathbf{x}} = \mathbf{g}(\mathbf{x})^T \frac{\partial \mathbf{g}(\mathbf{x})}{\partial \mathbf{x}} = \mathbf{g}(\mathbf{x})^T \mathbf{J}(\mathbf{x}) \tag{54.8}$$

Applying the LM method, parameter vector $\mathbf{x}$ can be iteratively estimated using the following rule:

$$\begin{aligned} \mathbf{h} &= -\left[\mathbf{J}^T\mathbf{J} + \lambda\,\mathrm{diag}\left(\mathbf{J}^T\mathbf{J}\right)\right]^{-1}\mathbf{J}^T\mathbf{g}(\mathbf{x}) \\ \mathbf{x}_{k+1} &= \mathbf{x}_k + \mathbf{h}, \end{aligned} \tag{54.9}$$

where $\mathbf{h}$ is the perturbed step. Large value of parameter $\lambda$ leads to gradient descent update while small value of parameter $\lambda$ leads to Gauss–Newton update. When the current estimate is far from its real value, then the LM updating uses large $\lambda$; when the current estimate becomes close to its real value, then the value of $\lambda$ is adaptively reduced. The condition of adaptively changing $\lambda$ is [18]:

$$\chi(\mathbf{x}_k)^2 - \chi(\mathbf{x}_{k+1})^2 > \varepsilon_1 \mathbf{h}^T[\lambda\mathbf{h} - \mathbf{J}^T\mathbf{g}(\mathbf{x})]$$

If the condition holds, then $\lambda$ is reduced by a factor of ten; otherwise, $\lambda$ is increased by a factor of ten. $\varepsilon_1$ is a small positive number and was set to $10^{-2}$. Based on the estimation of contact location, the friction force and normal force are readily obtained without strict requirements on the orientation of the finger relative to object surfaces. Define $\boldsymbol{Q} = \nabla S(\boldsymbol{p}_c)$ as the normal vector of the contact location, Fig. 54.2, the inner product of the normal vector and the friction force $\boldsymbol{F_t}$ is zero, i.e.

$$\boldsymbol{Q}^T \cdot \boldsymbol{F}_t = \boldsymbol{Q}^T \cdot (\boldsymbol{F} - \boldsymbol{F}_n) = 0$$

Let normal force $\boldsymbol{F_n} = a\boldsymbol{Q}$, the scalar $a$ can be solved as:

$$a = \frac{\boldsymbol{Q}^T\boldsymbol{F}}{\boldsymbol{Q}^T\boldsymbol{Q}} \tag{54.10}$$

Thus, given a contact location $\boldsymbol{p}_\mathrm{c} = [x, y, z]^\mathrm{T}$, the normal force are:

$$\boldsymbol{F}_n = a\boldsymbol{Q}$$

Let $\theta$ to be the angle between the normal force and the resultant force, then it holds: $\cos\theta = \frac{F_n}{F}$. The friction force can be then computed as:

$$\boldsymbol{F}_t = \frac{\sqrt{1 - \cos\theta^2}}{\cos\theta}\boldsymbol{F}_n$$

## 54.4 Comparison of Computation Costs

The computation costs of contact information identification of using both the gradient descent (GD)-based algorithm and the proposed LM-based algorithm are compared. It was found that the computational efficiency of using the LM algorithm is far superior to the use of GD method. To identify the contact information, the LM algorithm needs less than ten iterations while the GD method requires more than thousands iterations. It was observed that when the initial guess is close to the solution, the GD needs 4,800 iterations to converge, while LM needs only four iterations, Fig. 54.3; when the initial guess is far from the solution, the GD needs more than 16,000 iterations to converge, while LM needs only nine iterations, Fig. 54.4. Using a standard PC (2.40 GHz Intel® Core™ 2 Duo processor and 2 GB RAM), the average computation time for GD is 245 ms (4.1 Hz), for LM is 1.2 ms (833 Hz).

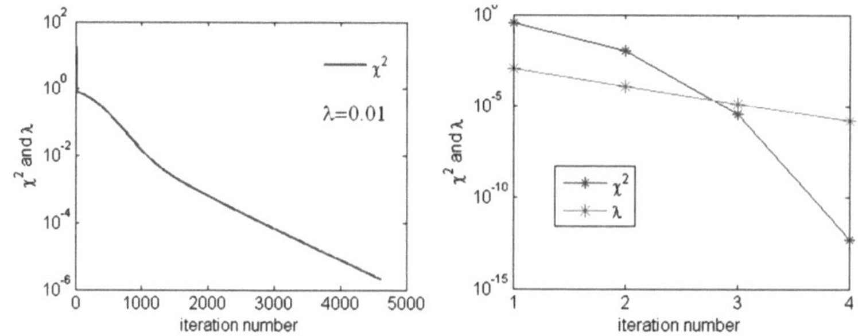


**Fig. 54.3** The convergence history for gradient descent method for Levenberg–Marquardt method with good initial guess

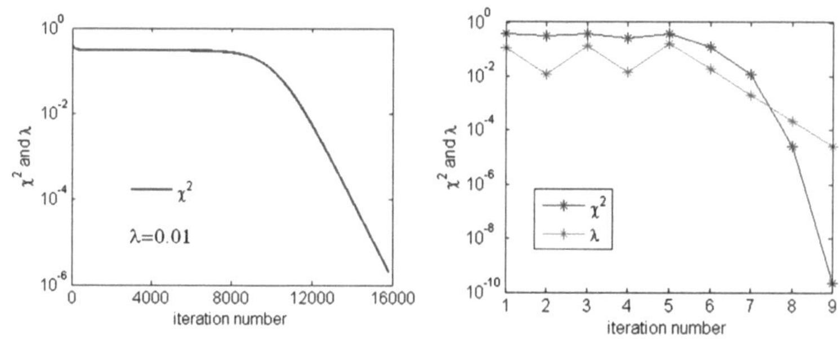


**Fig. 54.4** The convergence history for gradient descent method for Levenberg–Marquardt method with bad initial guess

# 54.5 Algorithm Validation

## *54.5.1 Accuracy of Contact Location Identification*

Validation tests have been carried out in order to validate the accuracy of the LM based contact sensing algorithm. For validating the accuracy of the contact location identification, 28 locations were marked on the fingertip surface, Fig. 54.5a. During tests, a thin rod with a semispherical tip was used to touch each marked location ten times. The identified contact locations were compared with the ground-truth values as shown in Fig. 54.5b. The results show that the identified contact locations match the ground-truth location in very good agreement. The root mean square error (RMSE) the identified contact location is 0.266 mm.

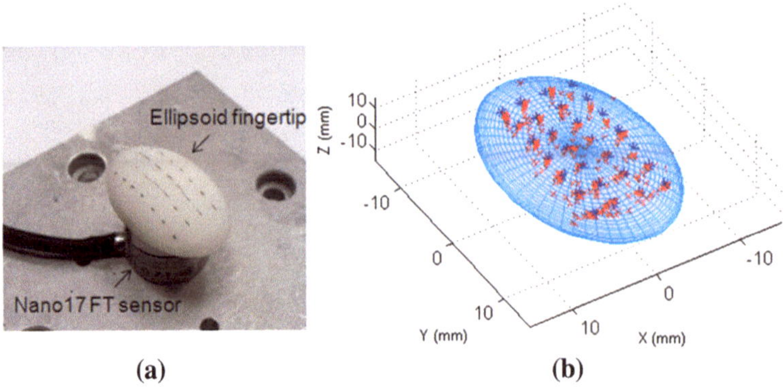


**Fig. 54.5** The test setup for validating the accuracy of contact location identification (**a**), and the validation results (**b**); the *red dots* are the identified contact locations; *blue* * signs indicate the ground-truth values

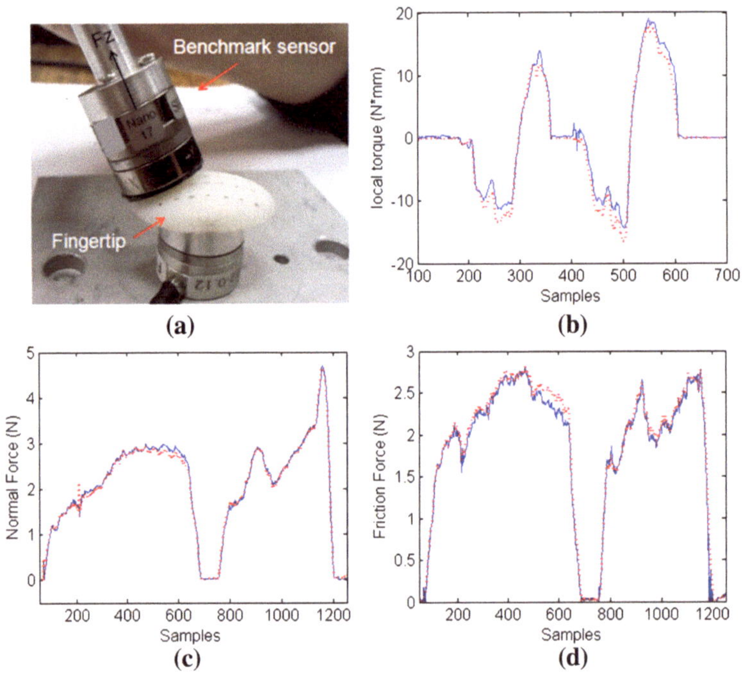


**Fig. 54.6** The test rig (**a**) for validation of estimation of the local torque, (**b**), the normal force (**c**) and the friction force (**d**); the *solid blue line* is the estimation, the *red dashed line* is the benchmark value

### 54.5.2 *Accuracy of Normal Force, Friction Force and Local Torque Estimation*

To validate the estimation accuracy of the estimation error for normal force, friction force and the local torque, a test rig was set up as show in Fig. 54.6a. Another Nano17 sensor was used as the benchmark sensor. The benchmark sensor has a flat end-surface which is perpendicular to its z axis; a thin layer of rubber was attached to increase the surface friction coefficient. When the benchmark sensor is in contact with the fingertip surface, the z axis of the benchmark sensor is coincident with the surface normal direction. Thus the torque measured around the z axis of the benchmark sensor, $m_z$, is the local torque generated at the finger surface; force measured along z axis of the benchmark sensor, $f_z$, is the normal force; the resultant force of $f_x$ and $f_y$ of the benchmark sensor is the friction force.

For validating the local torque, the benchmark sensor interacted with fingertip at different locations and applied local torques with different directions. The magnitude of the identified local torque was compared with the benchmark measurements. It was found the RMSE error of local torque identification is 0.792 N mm. For validating the normal and friction force, the benchmark sensor slides over the finger surface while changing the orientation. The magnitudes of the estimated normal and friction forces were compared with the benchmark. It was found that the RMSE error of normal force estimation is 0.0403N, RMSE error of friction force estimation is 0.0497 N.

The above tests demonstrate that developed algorithm is highly computational efficient and provides precise estimation for the contact location on the fingertip, the direction and the magnitude of the friction and normal forces, the local torque generated at the surface. The proposed algorithm is easy to be implemented and has broad applications for the contact sensing in the field of robotics.

**Acknowledgments** The research leading to these results has been supported by the HANDLE project, which has received funding from the European Community's Seventh Framework Programme (FP7/2007-2013) under grant agreement ICT 231640.

## References

1. Bicchi A (2000) Hands for dexterous manipulation and robust grasping: a difficult road toward simplicity. IEEE Trans Robotics Autom 16(6):652–662
2. Liu H, Li J, Song X, Seneviratne L, Althoefer K (2011) Rolling indentation probe for tissue abnormality identification during minimally invasive surgery. IEEE Trans Robotics 27(3):450–460
3. Jia Y-B, Erdmann M (1999) Pose and motion from contact. Int J Robotics Res 18:466–487
4. Lefebvre T, Bruyninckx H, Schutter JD (2005) Polyhedral contact formation identification for autonomous compliant motion: exact nonlinear bayesian filtering. IEEE TransRobotics 21(1):124–129
5. Okamura AM, Cutkosky MR (2001) Feature detection for haptic exploration with robotic fingers. Int J Robotics Res 20:925

6. Jamali N, Sammut C (2011) Majority voting: material classification by tactile sensing using surface texture. IEEE Trans Robotics 27(3):508–521
7. Liu H, Song X, Nanayakkara T, Althoefer K, Seneviratne L (2011) Friction estimation based object surface classification for intelligent manipulation. In: IEEE ICRA 2011 workshop on autonomous grasping, Shanghai
8. Liu H, Noonan DP, Challacombe BJ, Dasgupta P, Seneviratne LD, Althoefer K (2010) Rolling mechanical imaging for tissue abnormality localization during minimally invasive surgery. IEEE Trans Biomed Eng 57(2):404–414
9. Lederman SJ, Klatzky RL (1990) Haptic classification of common objects: knowledge-driven exploration. Cognit Psychol 22:421–459
10. Dahiya RS, Metta G, Valle M, Sandini G (2010) Tactile sensing—from humans to humanoids. IEEE Trans Robotics 26(1):1–20
11. Salisbury JK (1984) Interpretation of contact geometries from force measurements. In: Brady M, Paul R (eds) Proceedings of 1st international symposium on robotics research. MIT Press, Cambridge, pp 565–577
12. Brock DL, Chiu S (1985) Environment perceptions of an articulated robot hand using contact sensors. In: Proceedings of ASME winter annual meeting, Miami, pp 228–235
13. Tsujimura T, Yabuta T (1988) Object detection by tactile sensing method employing force/torque information. IEEE Trans Robotics Autom 5(4):444–450
14. Bicchi A, Salisbury JK, Brock DL (1993) Contact sensing from force and torque measurements. Int J Robotics Res 12(3):249–262
15. Gálvez JA, de Santos González P, Pfeiffer F (2001) Intrinsic tactile sensing for the optimization of force distribution in a pipe crawling robot. IEEE/ASME Trans Mechatron 6(1):26–35
16. Murakami K, Hasegawa T (2005) Tactile sensing of edge direction of an object with a soft fingertip contact. In: Proceedings of the IEEE ICRA, pp 240–247
17. Yamada T, Tanaka A, Yamada M, Yamamoto H, Funahashi Y (2010) Autonomous sensing strategy for parameter identification of contact conditions by active force sensing. In: IEEE international conference on robotics and biomimetics, pp 839–844
18. Madsen K, Nielsen HB, Tingleff O (2004) Methods for non-linear least square problems. Department of Mathematical Modelling, Technical University of Denmark, Denmark

# Chapter 55
# Conceptual Design and Kinematic Analysis of a Five-Fingered Anthropomorphic Robotic Hand

**Pramod Kumar Parida, B. B. Biswal and Rabindra Narayan Mahapatra**

**Abstract** Handling of objects with irregular shapes and that of flexible/soft objects by ordinary robot grippers is difficult. It is required that various objects with different shapes or sizes could be grasped and manipulated by one robot hand mechanism for the sake of factory automation and labour saving. Dexterous grippers will be the appropriate solution to such problems. Corresponding to such needs, we have developed an articulated mechanical hand with five fingers and 25 degrees-of-freedom which has an improved grasp capability. Since the developed hand is possible to envelope and grasp an object mechanically, it can be used easily and widely in the factory and for medical rehabilitation purpose. This work presents the conceptual design and the kinematic analysis of such a hand.

**Keywords** Multi-fingered gripper · Anthropomorphic · Kinematics · Simulation

## 55.1 Introduction

Hand is one of the most complex organs of the human body from the dexterity and kinematic point of view. Most of the studies in this area refer to the categorization and study of six grasps: cylindrical, fingertip, hook, palmar, spherical and lateral,

P. K. Parida
National Institute of Technology, Rourkela, Odisha, India
e-mail: pramodparida71@yahoo.com

B. B. Biswal (✉) · R. N. Mahapatra
ITER, SOA University, Bhubaneswar, Odisha, India
e-mail: bbbiswal@nitrkl.ac.in

R. N. Mahapatra
e-mail: rnm26@rediffmail.com

J. S. Dai et al. (eds.), *Advances in Reconfigurable Mechanisms and Robots I*,
DOI: 10.1007/978-1-4471-4141-9_55, 

as in [1, 2]. Dexterity is the first requirement for the robot hand. The multi-fingered robot hand acts as a multipurpose gripping device for various tasks. Some important multi-fingered hands are Utah/MIT hand [3], DLR hand [4], Shadow Dexterous Hand [5], Robonaut hand [6], NAIST-Hand [7] and Gifu hand [8]. Such a hand provides a promising base for supplanting human hand in execution of tedious, complicated and dangerous tasks, more precisely than a human hand, especially in situations such as manufacturing, space, undersea etc. These robot hands are helpful for the patients who are partially paralyzed due to neurological or orthopedic damages [9]. All the hands developed by previous researchers have limited flexibility due to less number of degrees-of-freedom (DoFs) incorporated. In order to enhance the flexibility of the hand the present work considers the anatomy of the human hand and considers 25 DoFs and the two DoFs in the wrist are only considered to be fixed. The human hand consists of connected parts composing kinematical chains so that hand motion is highly articulated. All fingers in the model have the same essential structure. Each finger has five DoFs: one DoF corresponding to the part of carpometacarpal articulation considered as belonging to the respective finger, two DoFs corresponding to metacarpophalangeal articulation, one DoF corresponding to proximal-interphalangeal (PIP) articulation and one DoF corresponding to DIP. The thumb has a different structure: three DoFs corresponding to the carpometacarpal (CMC) articulation, two DoFs corresponding to metacarpophalangeal (MCP) articulation, and one DoF corresponding to the interphalangeal (IP) articulation.

The aim of the present study is to obtain a human hand model, as natural as possible, that is capable of realizing various tasks in 3-D environment. The motion is studied by representing the active space as a complex surface (reach envelope). The intersections between this active space and the global reference frame planes represent the fingertips' trajectories. The model correctness is appreciated by comparing these trajectories to the real ones.

## 55.2 Modeling of Hand

The multi-fingered robot hand acts as a multipurpose gripping device for various tasks. Since it is designed to substitute the human hands, most anthropomorphic robot hands duplicate the shape and function of human hands. The size of the hand is a significant part in the research. Multi-fingered robot hand can be directly attached to the end of an industrial robot arm or play a role in the prosthetic applications. The structure of the fingers of human hands is almost the same and independent, as shown in Fig. 55.1. The finger segments in human hand give us the inspiration to design an independently driven finger segment to construct a whole finger. The segmental lengths of the thumb and fingers are taken proportionately to HL and HB with a fixed wrist.

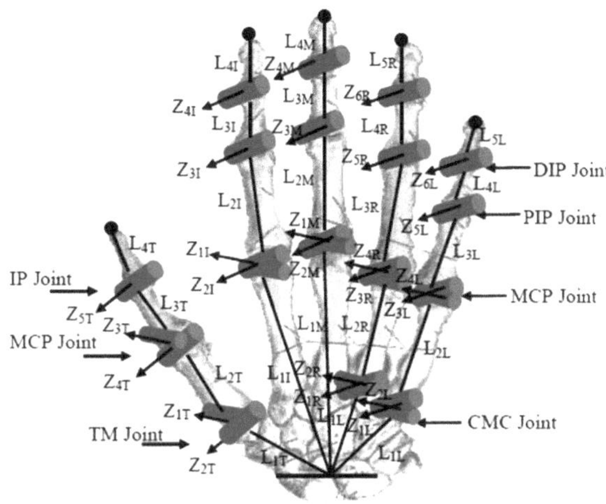


**Fig. 55.1** Kinematical model of human hand

### *55.2.1 Kinematic Model*

Typically the hand mechanism is approximated to have 27 DoFs, which consists of 25 DoFs at different joints of the fingers and 02 DoFs at wrist. In the present study the wrist is considered as a fixed origin. Hence, only 25 DoFs are considered. The thumb is modeled with five DoFs. The index and middle fingers are modeled with four DoFs each. The ring and little fingers are modeled with six DoFs each considering two DOFs each at CMC joint for palm arch. The Trapeziometacarpal (TM) joint, all five Mecapophalangeal (MCP) joints and two CMC joints are considered with two rotational axes each for both abduction–adduction and flexion–extension. The Distal-Interphalangeal (DIP) joints on the other four fingers possess one DoF each for the flexion–extension rotational axes. Figure 55.1 illustrate the human hand model while the thumb and other fingers' parameters are tabulated in Tables 55.1 and 55.2 respectively.

### *55.2.2 Anthropometry Data and Joint Limits*

The estimated measurement of the members of the hand is given in Tables 55.4 and 55.5, where HL is Hand Length and HB is Hand Breadth [10]. The joints limits of human hand are given in Tables 55.6, 55.10 [11].

### *55.2.3 Locating the Finger Tip*

A kinematic model is developed to calculate the fingertip position. Given the joint angles, the fingertip position in the palm frame is calculated by the kinematic

**Table 55.1** DH table of thumb

| Link (i) | Link twist angle ($\alpha_{i-1}$) | Link length ($a_{i-1}$) | Joint distance ($d_i$) | Joint angle ($\theta_i$) |
|---|---|---|---|---|
| 1T | 90° | $L_{1T}$ | 0 | $\theta_{1T}$ |
| 2T | −90° | 0 | 0 | $\theta_{2T}$ |
| 3T | 90° | $L_{2T}$ | 0 | $\theta_{3T}$ |
| 4T | −90° | 0 | 0 | $\theta_{4T}$ |
| 5T | 0° | $L_{3T}$ | 0 | $\theta_{5T}$ |
| $\varepsilon_T$ | 0° | $L_{4T}$ | 0 | 0 |

**Table 55.2** DH table of index and middle fingers

| Link (i) | Link twist angle ($\alpha_{i-1}$) | Link length ($a_{i-1}$) | Joint distance ($d_i$) | Joint angle ($\theta_i$) |
|---|---|---|---|---|
| 1F | 90° | $L_{1F}$ | 0 | $\theta_{1F}$ |
| 2F | −90° | 0 | 0 | $\theta_{2F}$ |
| 3F | 0° | $L_{2F}$ | 0 | $\theta_{3F}$ |
| 4F | 0° | $L_{3F}$ | 0 | $\theta_{4F}$ |
| $\varepsilon_F$ | 0° | $L_{5F}$ | 0 | 0 |

**Table 55.3** DH table of ring and little fingers

| Link (i) | Link twist angle ($\alpha_{i-1}$) | Link length ($a_{i-1}$) | Joint Distance ($d_i$) | Joint angle ($\theta_i$) |
|---|---|---|---|---|
| 1F | 90° | $L_{1F}$ | 0 | $\theta_{1F}$ |
| 2F | −90° | 0 | 0 | $\theta_{2F}$ |
| 3F | 90° | $L_{2F}$ | 0 | $\theta_{3F}$ |
| 4F | −90° | 0 | 0 | $\theta_{4F}$ |
| 5F | 0° | $L_{3F}$ | 0 | $\theta_{5F}$ |
| 6F | 0° | $L_{4F}$ | 0 | $\theta_{6F}$ |
| $\varepsilon_F$ | 0° | $L_{5F}$ | 0 | 0 |

**Table 55.4** Segment length for metacarpal bones

| Finger | Metacarpal bones | Link |
|---|---|---|
| Thumb | $0.251*HL$ | $L_{2T}$ |
| Index | $\sqrt{(0.374 * HL)^2+(0.126 * HB)^2}$ | $L_{2I}$ |
| Middle | $0.373*HL$ | $L_{2M}$ |
| Ring | $\sqrt{(0.336 * HL)^2+(0.077 * HB)^2}$ | $L_{2R}$ |
| Little | $\sqrt{(0.295 * HL)^2+(0.179 * HB)^2}$ | $L_{2L}$ |

**Table 55.5** Segment length for phalangeals

| Finger | Proximal | Link | Middle | Link | Distal | Link |
|---|---|---|---|---|---|---|
| Thumb | $0.196*HL$ | $L_{3T}$ | – | – | $0.158*HL$ | $L_{4T}$ |
| Index | $0.265*HL$ | $L_{3I}$ | $0.143*HL$ | $L_{4I}$ | $0.097*HL$ | $L_{5I}$ |
| Middle | $0.277*HL$ | $L_{3M}$ | $0.170*HL$ | $L_{4M}$ | $0.108*HL$ | $L_{5M}$ |
| Ring | $0.259*HL$ | $L_{3R}$ | $0.165*HL$ | $L_{4R}$ | $0.107*HL$ | $L_{5R}$ |
| Little | $0.206*HL$ | $L_{3L}$ | $0.117*HL$ | $L_{4L}$ | $0.093*HL$ | $L_{5L}$ |

**Table 55.6** Joint limits of thumb

| Joints | Rotations | $\theta_i$ | $\theta_{min}$ | $\theta_{max}$ |
|---|---|---|---|---|
| TM | Abduction–adduction | $\theta_{1T}$ | 0 | $\pi/3$ |
| | Flexion–extension | $\theta_{2T}$ | $-5\pi/36$ | $7\pi/36$ |
| MCP | Abduction–adduction | $\theta_{3T}$ | 0 | $\pi/3$ |
| | Flexion–extension | $\theta_{4T}$ | $-\pi/18$ | $11\pi/36$ |
| IP | Flexion–extension | $\theta_{5T}$ | $-\pi/12$ | $4\pi/9$ |

**Table 55.7** Joint limits of index finger

| Joints | Rotations | $\theta_i$ | $\theta_{min}$ | $\theta_{max}$ |
|---|---|---|---|---|
| MCP | Abduction–adduction | $\theta_{1I}$ | $-\pi/6$ | $\pi/6$ |
| | Flexion–extension | $\theta_{2I}$ | $-\pi/18$ | $\pi/2$ |
| PIP | Flexion–extension | $\theta_{3I}$ | 0 | $\pi/2$ |
| DIP | Flexion–extension | $\theta_{4I}$ | 0 | $\pi/3$ |

model. The DH method is implemented to determine the DH parameters for all the fingers which are tabulated in Tables 55.1 and 55.2. The middle and ring fingers are kinematically identical to the index finger. They all have four DOFs. The coordinate systems are located along each joint; a global coordinate system for hand is located in the wrist as shown in Fig. 55.1. Assuring the transfer from a reference frame to the next one the general expression of the matrix can be written as follows:

$$^{i-1}T_i = \begin{bmatrix} \cos q_i & -\sin q_i \cos \alpha_i & \sin q_i \sin \alpha_i & L_i \cos q_i \\ \sin q_i & \cos q_i \cos \alpha_i & -\cos q_i \sin \alpha_i & L_i \sin q_i \\ 0 & \sin \alpha_i & \cos \alpha_i & d_i \\ 0 & 0 & 0 & 1 \end{bmatrix} \tag{55.1}$$

By multiplying the corresponding transfer matrices written for every finger, the kinematical equations describing the fingertip motion with respect to the general coordinate system can be determined as:

$$P_x = (L_4\, c_{234} + L_3\, c_{23} + L_2\, c_2)\, c_1 \tag{55.2}$$

$$P_y = (L_4\, c_{234} + L_3\, c_{23} + L_2\, c_2)\, s_1 \tag{55.3}$$

$$P_z = L_4\, s_{234} + L_3\, s_{23} + L_2\, s_2 \tag{55.4}$$

## 55.3 Motion Study Through Simulation

It is now possible to develop a model using Eqs. 55.2, 55.3 and 55.4. A computer program using these equations in MATLAB-7.1 is developed to capture the motion of the fingers. Every joint variable range as per Table 55.3, 55.4, 55.5, 55.6, 55.7,

**Table 55.8** Joint limits for middle finger

| Joints | Rotations | $\theta_i$ | $\theta_{min}$ | $\theta_{max}$ |
|---|---|---|---|---|
| MCP | Abduction–adduction | $\theta_{1M}$ | $-2\pi/45$ | $7\pi/36$ |
| | Flexion–extension | $\theta_{2M}$ | 0 | $4\pi/9$ |
| PIP | Flexion–extension | $\theta_{3M}$ | 0 | $5\pi/9$ |
| DIP | Flexion–extension | $\theta_{4M}$ | $-\pi/18$ | $\pi/2$ |

**Table 55.9** Joint limits for ring finger

| Joints | Rotations | $\theta_i$ | $\theta_{min}$ | $\theta_{max}$ |
|---|---|---|---|---|
| CMC | Abduction–Adduction | $\theta_{1R}$ | 0 | $\pi/18$ |
| | Flexion–Extension | $\theta_{2R}$ | $\pi/90$ | $\pi/18$ |
| MCP | Abduction–Adduction | $\theta_{3R}$ | $-14\pi/180$ | $\pi/9$ |
| | Flexion–Extension | $\theta_{4R}$ | 0 | $4\pi/9$ |
| PIP | Flexion–Extension | $\theta_{5R}$ | 0 | $5\pi/9$ |
| DIP | Flexion–Extension | $\theta_{6R}$ | $-\pi/6$ | $\pi/2$ |

**Table 55.10** Joint limits for little finger

| Joints | Rotations | $\theta_i$ | $\theta_{min}$ | $\theta_{max}$ |
|---|---|---|---|---|
| CMC | Abduction–adduction | $\theta_{1L}$ | 0 | $\pi/12$ |
| | Flexion–extension | $\theta_{2L}$ | $\pi/36$ | $\pi/12$ |
| MCP | Abduction–adduction | $\theta_{3L}$ | $-19\pi/180$ | $11\pi/60$ |
| | Flexion–extension | $\theta_{4L}$ | 0 | $4\pi/9$ |
| PIP | Flexion–extension | $\theta_{5L}$ | 0 | $5\pi/9$ |
| DIP | Flexion–extension | $\theta_{6L}$ | $-\pi/6$ | $\pi/2$ |

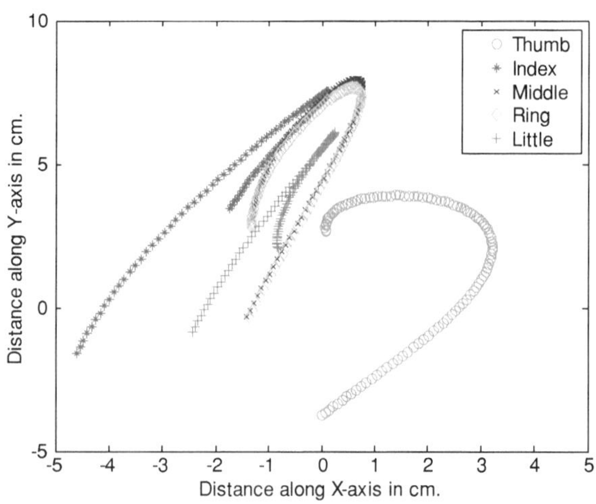


**Fig. 55.2** Profile of fingertips in the X–Y plane

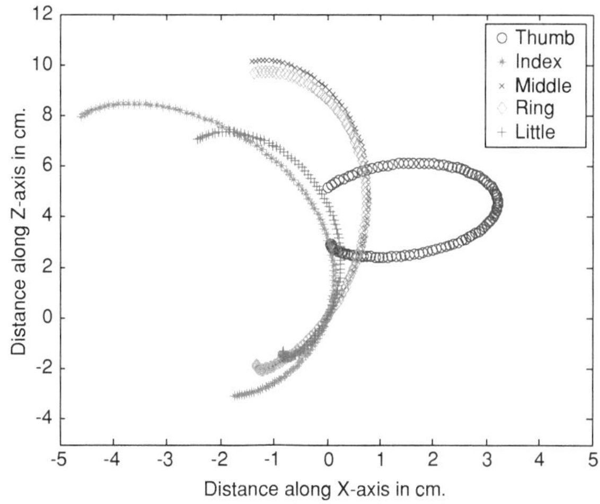


**Fig. 55.3** Profile of fingertips in the X–Z plane

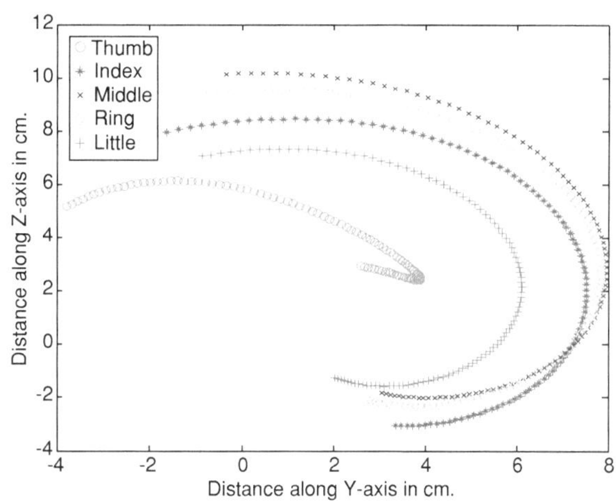


**Fig. 55.4** Profile of fingertips in the Y–Z plane

55.8 is divided to an appropriate number of intervals in order to have enough fingertips positions to give confident images about the spatial trajectories of these points. By connecting these positions and the complex surface bordering the active hand model workspace is obtained. The complex surface could be used to verify the model correctness from the motion point of view, and to plan the hand motion by avoiding the collisions between its active workspace and obstacles in the neighborhood.

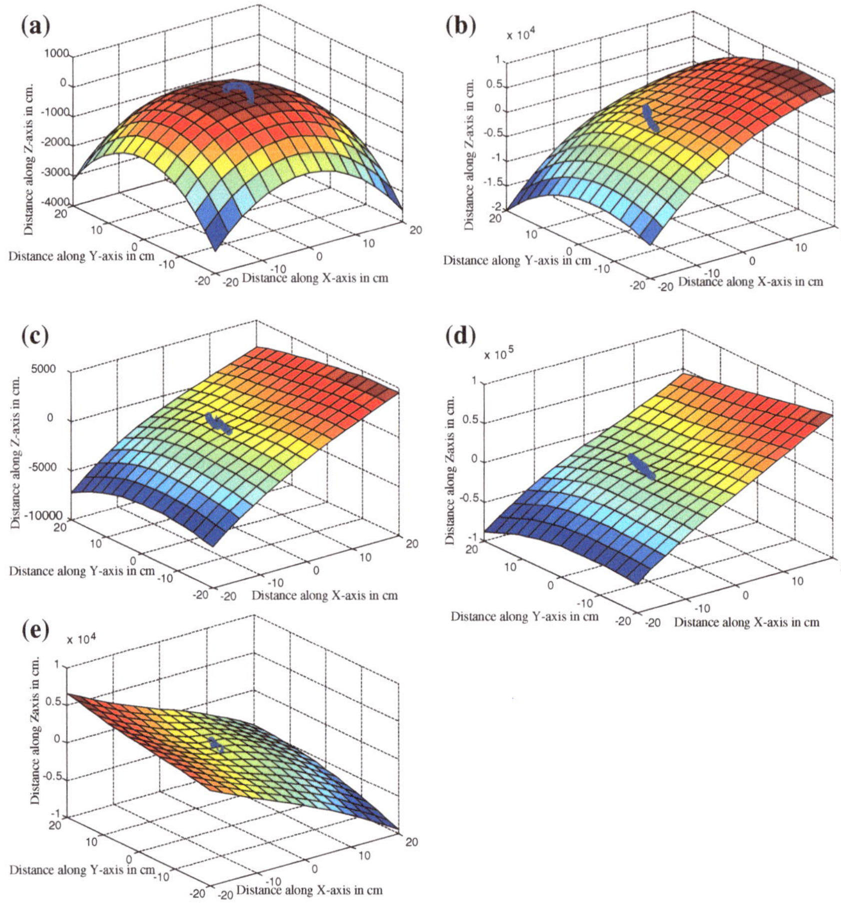


**Fig. 55.5** Area profiles of the finger tips **a** profile of thumb, **b** profile of index finger, **c** profile of middle finger, **d** profile of ring finger, and **e** profile of little finger

## 55.4 Results

Using the Eqs. 55.2, 55.3 and 55.4 along with the parametric data of human fingers presented in Tables 55.1, 55.2, 55.3, 55.4, 55.5, 55.6, 55.7, 55.8 the complex surface described by each finger tip is generated. In all the cases each angular range is divided into equal divisions. The simulation is realized using MATLAB. The profile of the independent finger tips are generated spatially. However, for the purpose of understanding and simplicity, these are presented in X–Y, X–Z and Y–Z planes in Figs. 55.2, 55.3 and 55.4 respectively. The area profiles of the thumb, index middle, ring and little finger tips in the 3D plane are presented in Fig. 55.5 respectively.

## 55.5 Conclusion

The present work aims at developing a kinematic model of a five-fingered dexterous robotic hand with 25 degrees-of-freedom which may find its potential applications in industries and other work places for manipulation of irregular and that of soft objects. The conceptual design has been done keeping human hand's anatomy in mind so that it has the flexibility close to the human hand and the kinematic behaviour is similar to that of the human hand. The model considers five fingers that are essential for grasping and manipulating objects securely. The joints, links and other kinematic parameters are chosen in such a way that they represent those of a human hand. The simulation result is very encouraging for the prototype development of the hand. The kinematic simulation is carried out to estimate the work volume and assess kinematic constraints of the conceptualized hand. It is expected that the proposed hand can be used for orthopaedic rehabilitation of human hand.

## References

1. Schlesinger G (1919) Der Mechanische Aufbau and der Kunstlichen Glieder. In: Borchardt M et al. (eds) Ersatzglieder und Arbeitshilfen fur Kriedgsbeschadigte und Unfallverletzte. Springer, Berlin, pp 321–699
2. Taylor CL, Schwarz RJ (1955) The anatomy and mechanics of the human hand. Artif Limbs 2:22–35
3. Jacobsen SC, Wood JE, Knutti DF, Biggers KB (1984) The UTAH/M.I.T dexterous hand: work in progress. Int J Robot Res 3(4):21–50
4. Butterfass J, Grebenstein M, Liu H, Hirzinger G (2003) DLR-Hand II: next generation of a dexterous robot hand. In: IEEE international conference on robotics and automation, Taipei, Taiwan, pp 109–114
5. Walkler R (2004) Developments in dexterous hands for advanced robotic applications. In: The sixth biannual world automation congress, Seville, Spain, pp 123–128
6. Lovchik CS, Diftler MA (1999) The robonaut hand: a dexterous robot hand for space. In: IEEE international conference on robotics and automation, Detroit, USA, pp 907–912
7. Jun U, Ishida Y, Kondo M, Ogasawara T (2005) Development of the NAIST-hand with vision-based tactile fingertip sensor. In: IEEE international conference on robotics and automation, Barcelona, Spain, pp 2332–2337
8. Kawasaki H, Komatsu T, Uchiyama T (2002) Dexterous anthropomorphic robot hand with distributed tactile sensor: Gifu hand II. IEEE/ASME Trans Mechatron 7(3):296–303
9. Cooper RA, Ohnabe H, Hobson DA (eds) (2007) An introduction to rehabilitation engineering. CRC Press, Taylor & Francis group, Boca Raton
10. Buchcholz B, Armstrong T, Goldste S (1992) Anthropometric data for describing the kinematics of the human hand. Ergonomics 35(3):261–273
11. Parasuraman S, Zhen C (2009) Development of robot assisted hand stroke rehabilitation system. In: International conference on computer and automation engineering, March, pp. 70–74

# Chapter 56
# Dynamic Analysis of Cable-Driven Humanoid Arm Based on Lagrange's Equation

**Jianhua Wang, Xiang Cui, Weihai Chen and Yan Jin**

**Abstract** Human beings have flexible arms and legs that can ensure them conduct various challenging operations to work out daily requirements. Therefore, Humanoid mechanism design become hot research areas recently. To mimic the skeleton structure and driving scheme of a human arm, a 7-DOF cable-driven mechanism is determined to develop. Cable-driven mechanism allows actuators don't have to be placed on joints that makes it possesses several advantages such as low weight, high dexterity, and large reachable workspace compared with traditional configurations. The forward and inverse displacement analysis issues are conducted. In this paper, dynamic analysis of the cable-driven humanoid arm (CDHA) is conducted based on Lagrange's method. The dynamic model is established by making use of twist-product-of-exponential formula and the cable tension distribution analysis is also addressed. At last, both the dynamic model and the algorithm are validated through several simulations in MATLAB.

**Keywords** Lagrangian dynamics · Humanoid robot · Cable-driven · Screw theory

---

This work is supported by the National Natural Science Foundation of China under Grant No. 50875011 and No. 50975017.

---

J. Wang · X. Cui (✉) · W. Chen
Beihang University, Beijing, People's Republic of China
e-mail: c11251987@yahoo.com.cn

Y. Jin
Queens University Belfast, Belfast, UK
e-mail: y.jin@qub.ac.uk

J. S. Dai et al. (eds.), *Advances in Reconfigurable Mechanisms and Robots I*,
DOI: 10.1007/978-1-4471-4141-9_56, © Springer-Verlag London 2012 

## 56.1 Introduction

The researches of humanoid robots become popular because of the clear superiority of biological characters: dexterity, high stiffness and fault-tolerance. They start playing a good role in manufacturing, rehabilitation, military affairs, entertainments, and so on. MIT Artificial Intelligence Laboratory developed a cable-driven humanoid hand for the famous humanoid robot named Cog [1]. Harbin engineering university constructed a cable-driven parallel rehabilitation manipulator for pelvis recovery [2]. Although kinematic analysis had been theoretically studied, there was little dynamic modeling research being carried out for these cable-driven mechanisms. However, a precise dynamic model is crucial to precision control for cable-driven mechanism.

Based on the skeleton structure of the human arm, an S-R-S serial configuration is worked out [3]. Human muscle is mimicked by the scheme of cable-driven since the driving cables are functionally similar to muscle fiber. The CDHA possesses seven degrees of freedom, which is the minimum number of DOF with capacity of fault tolerance. Using two paralell cable-driven mechanism to control spherical joints and the whole becomes a hybrid serial-parallel manipulator with three decoupled modules, as shown in Fig. 56.1. Kinematic analysis had been done by Weihai Chen et al. [4] already based on local POE formulas.

A robot is usually a multi-body system, and the corresponding dynamic analysis is often with regard to a complex system. Currently, there exist several dynamic modeling methods, e.g., the Newton-Euler law [5, 6], Lagrange's approach [7, 8], Kane's method [9] and some [10, 11]. However, the difficulty of use of the various methods differs. Lagrange's equation method provides a method for holonomic system with disregarding all interactive and constraint forces that do not do work. The need to differentiate scalar energy functions usually makes it complex computationally. This is not much of a problem for small multi-body systems and this method may be a good choice. Kane's method is initially proposed by Kane [12] at Stanford University and is improved to a computer-oriented multi-body modeling method. Actually, Lagrange's equation is a special case of Kane's equation to a holonomic system. Taking into consideration the CDHA's structure, Lagrange's method is adopted for efficiency.

The objective of this paper is to build up a dynamic model of the CDHA. To this end, two main steps are needed, i.e., the dynamic modeling of equivalent serial structure and the cable tension redistribution analysis. Because of the negligible quantity of the cables' energy, a serial equivalent model without cables can be built to simply the CDHA's dynamic model. Therefore, the dynamic modeling work is lightened and become friendly to computing process. Then the tension of cables can be calculated based on tension distribution algorithm. The cables' tension solution is redundant and some valuation indices are set up on account of its characters.

The organization of the paper is as follows: Sect. 56.2 outlines the physical structure of CDHA and gives an entire description of this kind of manipulator. Section 56.3 presents the dynamic model of CDHA with Lagrange's method.

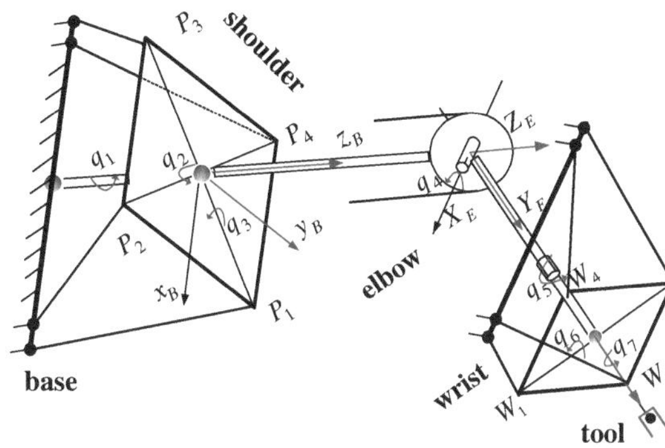


**Fig. 56.1** Schemetic diagram of CDHA ($P_1, P_2, P_3$ and $P_4$ are the attachment points of the platform whose center is the shoulder joints, and it's the same for $W_1, W_2, W_3$ and $W_4$)

The torques of revolute joints are worked out and the cable tension analysis is then conducted. General solution space of tensions including both special solutions and null-space solutions are obtained. Then simulation validations are conducted in Sect. 56.4. Finally, Sect. 56.5 summarizes the paper.

## 56.2 Sketch of CDHA

The prototype of CDHA is shown in Fig. 56.2.

The physical structure is composed of the base and the CDHA. The base is a fixed part and all motors and control system are assembled there. The configuration of CDHA can be simplified as a equivalent S-R-S configuration consisting of a 3-DOF shoulder assembly, a 1-DOF elbow assembly, and a 3-DOF wrist assembly. The shoulder and the wrist contain a similar parallel structure of four-cable-driven spherical joint and every cable is driven by a motor for redundant control. The elbow module is a two-cable-driven revolute joint and the two cables are driven by a single winch with cables twining around it in opposite directions. For the introduction of brake cables, the upper 3 modules are fully decoupled. In all, the whole is a cable-driven hybrid serial-parallel manipulator with three decoupled modules.

The control system of the prototype includes the following components: encoders that feedback angles of the rotational joints, a cable tension testing system, driving mechanisms of motors, and the motion control system consisting of PMAC multi-axis control card and a computer. All these are placed on the base except encoders.

## 56.3 Dynamic Modeling of CDHA

For simplification, some assumptions are listed. The inertia of cables is negligible so that dynamic modeling can firstly be operated without the effect of cables. Then CDHA's dynamic model comes down to the model of a 7-R serial configuration with seven links. Futher on, for the inertia of link 1, 2, 5, 6 are far less than link 3,

**Fig. 56.2** The prototype of CDHA. (It consists two main parts: the physical structure and control system. The axes of revolute joints are along with directions of the *red lines* and the *blue lines* stand for cables.)

4, 7, so the former are regarded as zero. Consequently, the dynamic model is specified to an equivalent S-R-S model composed of link 3, 4, 7. The dynamic analysis includes two parts as: (1) The dynamic modeling of 7-R serial configuration; (2) The cable tension analysis of the decoupled modules.

## *56.3.1 Dynamic Modeling of Serial Configuration*

Lagrange's method [13] is an efficient method to robot systems, which gives a visualized symbolic expression of dynamic model, as shown in (56.1):

$$\mathbf{M}(\mathbf{q})\ddot{\mathbf{q}} + \mathbf{C}(\mathbf{q}, \dot{\mathbf{q}})\dot{\mathbf{q}} + \mathbf{G}(\mathbf{q}, \dot{\mathbf{q}}) = \tau \tag{56.1}$$

where $\mathbf{q} = [q_1, q_2, \ldots, q_7]^T$ denotes the generalized coordinate, $\xi = [\xi_1, \xi_2, \ldots, \xi_7]^T$ denotes the corresponding forces or torques, $\mathbf{M}$ denotes the inertia matrix, $\mathbf{C}$ denotes Coriolis matrix, and $\mathbf{G}$ includes the gravity and other forces acting on the joints. With the use of screw theory, the coefficient matrixes can be acquired in a simplified form for ease of computation [14, 15].

The kinematic variables are acquired by inverse kinematic analysis [4]. Firstly, for the 7-R serial configuration, a linear trajectory of end-effector is planned and a serial of pose are obtained by means of interpolation algorithm [17]. Then we can get the values of $q$, velocities $\dot{q}$ and accelerations $\ddot{q}$ of joints from inverse kinematics.

There is no need to compute the body Jacobin of link 1, 2, 5, 6 and it is assumed that $\mathbf{J}_i^b(\mathbf{q}) = \mathbf{0}_{6\times7}$ when $i = 1, 2, 5, 6$. Then bodies Jacobins of link $i$ are obtained as:

$$\mathbf{J}_i^b(\mathbf{q}) = \begin{bmatrix} \xi_1^+ & \cdots & \xi_i^+ & \mathbf{0} & \mathbf{0} & \mathbf{0} & \mathbf{0} \end{bmatrix}, \tag{56.2}$$

where the columns are: $\xi_j^+ = Ad^{-1}_{(e^{\hat{\xi}_j q_j} \cdots e^{\hat{\xi}_i q_i}\mathbf{T}_i(\mathbf{0}))} \xi_j \ (j = 1, \ldots, i)$.

Using the notation of adjoint transformation symbol **A** [15, 16], the body Jacobin can be expressed in an explicit form [14]. Assuming $\mathrm{B}_l, \mathrm{B}_l'$ as the inertia of the *lth* link relative to body and base frame, $\mathbf{T}_l(\mathbf{q})$ is the pose matrix, the inertia matrix and Coriolis matrix are obtained:

$$\mathrm{B}_l' = Ad^T(\mathbf{T}_l^{-1}(\mathbf{0})) \cdot \mathrm{B}_l \cdot Ad(\mathbf{T}_l^{-1}(\mathbf{0})), \tag{56.3}$$

$$\mathbf{M}_{ij}(\mathbf{q}) = \sum_{l=\max(i,j)}^{7} (\mathbf{A}_{li}\xi_i)^T \mathrm{B}_l{}' \mathbf{A}_{lj}\xi_j \tag{56.4}$$

$$\mathbf{C}_{ij}(\mathbf{q}, \dot{\mathbf{q}}) = \frac{1}{2}\sum_{k=1}^{7} (\mathbf{R}(i,j,k) + \mathbf{R}(i,k,j) - \mathbf{R}(k,j,i))\dot{\mathbf{q}}_k \tag{56.5}$$

where $\mathbf{R}(i,j,k)$ is the differential $\partial \mathbf{M}_{ij}(\mathbf{q})/\partial \mathbf{q}_k$:

$$R(i,j,k) = \sum_{l=\max(i,j)}^{7} [\mathbf{A}_{k-1,i}\xi_i, \xi_k]^T \mathbf{A}_{lk}{}^T \mathrm{B}_l{}' \mathbf{A}_{lj}\xi_j + \sum_{l=\max(i,j)}^{7} \xi_i^T \mathbf{A}_{li}{}^T \mathrm{B}_l{}' \mathbf{A}_{lk}[\mathbf{A}_{k-1,j}\xi_j, \xi_k] \tag{56.6}$$

Taking no account of other conservative forces such as frictional forces, **G** is computed based on geopotential. The displacement component of gravity direction $h_i(\mathbf{q})$ is derived from instantaneous kinematics. The Y-axis of global coordinate system is parallel with the vertical direction. $h_i(\mathbf{q})$ can be then derived from $\mathbf{T}_i(\mathbf{q})$ based on the structure of transformation matrix. Assuming two vectors $\mathbf{E}_1 = [0,1,0,0]$ and $\mathbf{E}_2 = [0;0;0;1]$, then the potential energy is obtained as:

$$\mathbf{P}(\mathbf{q}) = \sum_{i}^{3,4,7} m_i g \mathbf{E}_1 \cdot \mathbf{T}_i(\mathbf{q}) \cdot \mathbf{E}_2 \tag{56.7}$$

Then the expression of $\mathbf{G}(\mathbf{q})$ can be solved as follows:

$$\mathbf{G}(\mathbf{q}) = \begin{bmatrix} \frac{\partial \mathbf{P}(\mathbf{q})}{\partial q_1} & \frac{\partial \mathbf{P}(\mathbf{q})}{\partial q_2} & \cdots & \frac{\partial \mathbf{P}(\mathbf{q})}{\partial q_7} \end{bmatrix}^T \tag{56.8}$$

To sum up,the equivalent torques acting on revolute joints can be solved as:

$$\boldsymbol{\tau} = \begin{bmatrix} \tau_1 & \tau_2 & \cdots & \tau_7 \end{bmatrix} \tag{56.9}$$

### *56.3.2 Cable Tension Analysis*

Cables tensions are computed based on torques of the dynamic model. The three modules, i.e., shoulder module, elbow module and wrist module are decoupled. The shoulder and wrist module is alike, which is a structure of spherical joint driven by four cables. For the similarity, shoulder module is taken for example of

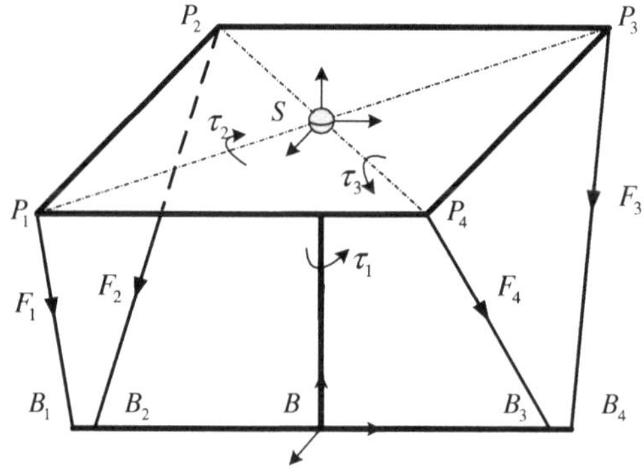


**Fig. 56.3** Kinetostatic equilibrium of shoulder module. (The directions of three torques $\tau_1, \tau_2$ and $\tau_3$ are along with the axes of revolute joints.)

analysis. The elbow module possesses an explicit revolute joint driven by two cables.

#### 56.3.2.1 Shoulder Module Analysis

The physical structure of shoulder module is shown in Fig. 56.3. The matrix of distances between rotation center and the upper cables' joints is $\mathbf{SP} = \left[ \overrightarrow{\mathbf{SP}_1} \quad \overrightarrow{\mathbf{SP}_2} \quad \overrightarrow{\mathbf{SP}_3} \quad \overrightarrow{\mathbf{SP}_4} \right]$. The tensions unit direction matrix, i.e., the direction matrix of cables $[\mathbf{l}_1 \quad \mathbf{l}_2 \quad \mathbf{l}_3 \quad \mathbf{l}_4]$ is given as $\mathbf{l}_i = \overrightarrow{\mathbf{P}_i\mathbf{B}_i}/l_i$, where $l_i$ represent the length of cable marked $i$. It can be solved by kinematic analysis. Then the Jacobin matrix can be expressed as:

$$\mathbf{J}_s = \left[ \overrightarrow{\mathbf{SP}_1} \times \mathbf{l}_1 \quad \overrightarrow{\mathbf{SP}_2} \times \mathbf{l}_2 \quad \overrightarrow{\mathbf{SP}_3} \times \mathbf{l}_3 \quad \overrightarrow{\mathbf{SP}_4} \times \mathbf{l}_4 \right] \tag{56.10}$$

Based on Kinetostatic the equilibrium analysis, an equation is built as:

$$\mathbf{J}_s \cdot \mathbf{F} - \mathbf{S}' = \mathbf{0}, \tag{56.11}$$

where $\mathbf{F} = [F_1 \quad F_2 \quad F_3 \quad F_4]^T$ denotes the cable tension and $\mathbf{S}'$ denotes torques solved by dynamic model in the global coordinate system.

Assuming a wrench $\mathbf{S} = [0 \quad 0 \quad 0 \quad \tau_2 \quad \tau_3 \quad \tau_1]^T$, then compute it in global coordinate system with adjoint transformation. If the adjoint matrix is $Ad(\mathbf{T}_3)$, $\mathbf{S}'$ is gotten as:

$$\mathbf{S}' = \left[ Ad^T(\mathbf{T}_3) \right]^{-1} \mathbf{S} \tag{56.12}$$

Without loss of generality, the cable tension vector can be solved as:

$$\mathbf{F} = \mathbf{J}_s{}^+\mathbf{S}' + (\mathbf{I} - \mathbf{J}_s{}^+\mathbf{J}_s)\mathbf{f} \quad (\mathbf{f} \in \mathbf{R}^{4\times 1}), \tag{56.13}$$

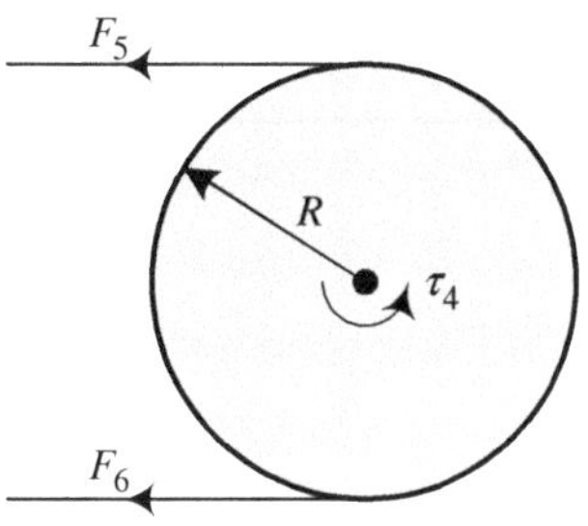


**Fig. 56.4** Kinetomatic equilibrium of elbow module. (The driven radius is $R$ and $F_5, F_6$ must be balanced with $\tau_4$ of the dynamic model.)

where $\mathbf{J}_s^+$ is Moore-Pennose pseudoinverse of $\mathbf{J}_s$, and $\mathbf{f}$ is a free vector.

#### 56.3.2.2 Elbow Module Analysis

The physical structure of the elbow module is shown in Fig. 56.4.

If the setting minimum of tension is $F_m$ and when $F_6$ is equal to the value, the solution of elbow module can be expressed as followings:

$$\begin{cases} F_6 = F_{\mathrm{m}} \\ F_5 = \tau_4/R + F_{\mathrm{m}} \end{cases} \tag{56.14}$$

## 56.4 Numerical Simulations

In this section, computer simulation is carried out on the CDHA by means of MATLAB. It is to demonstrate the validity and reasonability of the dynamic analysis. The mass values and inertial tensor matrixes of links 3, 4 and 7 are obtained by means of ADAMS, as shown in Table 56.1 as:

Then the inertia matrix can be constructed:$\mathrm{B}_i = \begin{bmatrix} m_i\mathbf{I}_3 & \mathbf{0} \\ \mathbf{0} & \Im_i \end{bmatrix} \quad (i = 3, 4, 7).$

To conduct this simulation, three procedures are employed.

1. A linear trajectory of end-effector from point [−59.29, −389.95, 219.84] to point [−144.27, −361.25, 414.62] is programmed using trapezoidal speed-time planning as shown in Fig. 56.5 with a rate of 60 mm/sec and an acceleration of 40 mm/sec$^2$. Kinematic parameters are obtained by inverse kinematics [17]. The values, velocities and accelerations of the revolute joints are also shown in Fig. 56.5.
2. Dynamic modeling of the serial configuration of CDHA is conducted based on Lagrange's method. With use of screw theory, the torques of the seven revolute joints is computed. Here we take no account of external load.
3. Cable tension analysis of the three decoupled modules is performed by means of tension distribution algorithm. The minimum of cable tension is set as 5N,

**Table 56.1** Mass and inertia

| Body | Mass (Kg) | Inertial tensor (Kg mm) |
|---|---|---|
| Link 3 | 0.959 | $\begin{bmatrix} 10581.75 & 0 & 0 \\ 0 & 10357.98 & 0 \\ 0 & 0 & 4891.39 \end{bmatrix}$ |
| Link 4 | 0.472 | $\begin{bmatrix} 2802.28 & 0 & 0 \\ 0 & 1556.47 & 0 \\ 0 & 0 & 1358.71 \end{bmatrix}$ |
| Link 7 | 0.685 | $\begin{bmatrix} 1637.14 & 0 & 0 \\ 0 & 1306.15 & 0 \\ 0 & 0 & 1227.75 \end{bmatrix}$ |

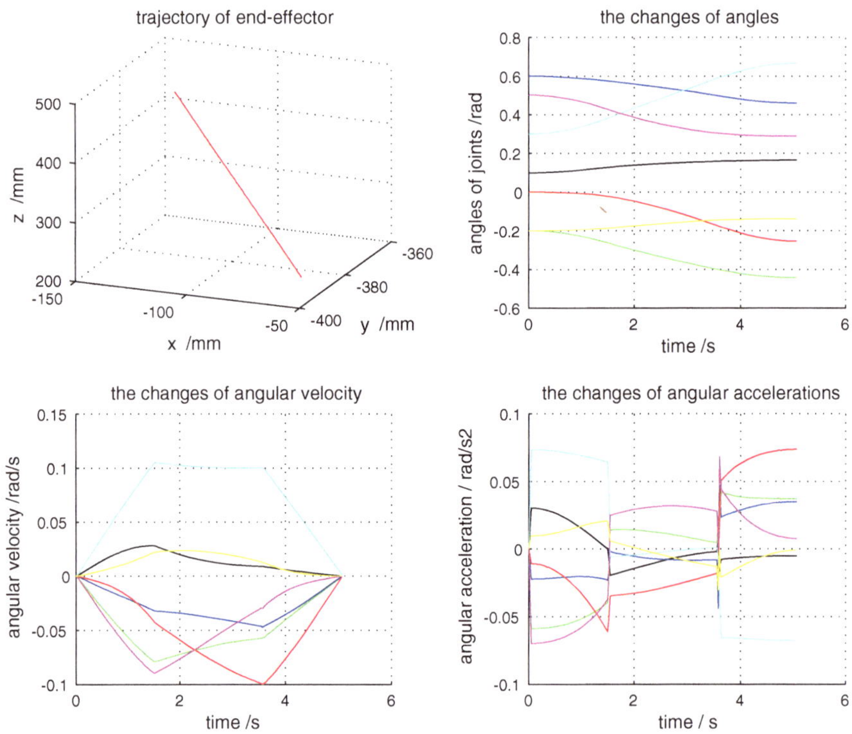


**Fig. 56.5** The solutions of kinematic parameters. (The trajectory of end-effector, the variation of joints' values, velocities and acceleration.)

which is decided by the choice of free vector **f**. The tensions curves of the three decoupled model are shown in Fig. 56.6.

According to the simulation above, the dynamic modeling of CDHA is proper and the curves of cable tension is smooth enough. The range of tensions is within

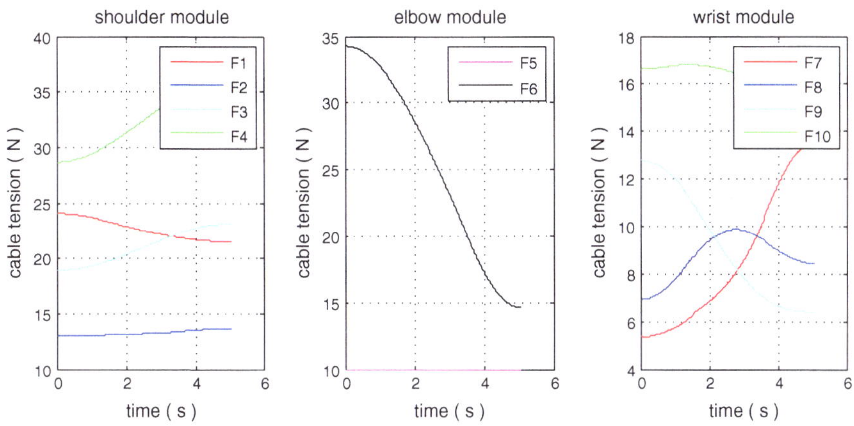


**Fig. 56.6** The cables tensions of the three modules

acceptable limit, which is suitable for manipulator's property and cable's stress limit.

## 56.5 Conclusion

For this special structure of CDHA, there are two main steps in order to conduct the dynamic analysis. Firstly, the CDHA is evolved to a serial configuration through an equivalent transformation based on its dynamics characteristics. Then, the cable tension analysis is conducted. The dynamic model based on Lagrange's method is built by means of screw theory. Compared with the other methods, this way is more efficient and the result of the derivation is a second-order vector differential equation which is friendly to design controllers. With regard to the redundant cable-driven mechanism, an algorithm for cable tension redistribution is developed. The following simulation is illustrated to verify the paper's train of thought, which lays down a good theory foundation for further research on effective control and performance improvement of the CDHA. What's more, this type of cable-driven manipulator is not a fixed configuration and free combination of the two kinds of cable-driven module,i.e., shoulder module and elbow module, can be used for cable-driven reconfigurable modular robot.

**Acknowledgments** This work is supported by the National Natural Science Foundation of China under Grant No. 50875011 and No. 50975017.

## References


1. Oky M (1997) The mechanisms in a humanoid robot hand. Auton Robot 4(2):199–209
2. Zhang L-X, Wang K-Y (2007) Research on the controllability of a wire-driven parallel rehabilitation manipulator for exercising the pelvis. J Harbin Eng Univ 28(7):790–794
3. Chen Q, Chen W, Rong L, Zhang J (2010) Mechanism design and tension analysis of a cable-driven humanoid-arm. Chin J Chem Eng 46(13):83–90
4. Chen W, Chen Q, Zhang J (2007) Inverse kinematic analysis for cable-driven humanoid arm. Chin J Chem Eng 43(4):12–20
5. Khalil W, Guegan S (2004) Inverse and direct dynamic modeling of Gough-Stewart robots. IEEE Trans Robotic, 20(4):754-761.
6. Dasgupta B, Mruthyunjaya TS (1998) Closed-form dynamic equations of the general stewart platform through the newton-euler approach. Mech Mach Theory 33(7):993–1012
7. Vakil M, Fotouhi R, Nikiforuk PN, Salmasi H (2008) A constrained Lagrange formulation of multilink planar flexible manipulator. J Vib Acoust 130(3):1–16
8. Lee KM, Shah DK (1988) Dynamic analysis of a three degrees of freedom in-parallel actuated manipulator. IEEE Trans Robotic Autom 4(3):361–367
9. Everett LJ (1989) An extension of Kane's method for deriving equations of motion of flexible manipulators. IEEE international conference on robotics and automation, 1989, Scottsdale, USA, pp 716–721
10. Khalil W, Ibrahim O (2004) General solution for the dynamic modeling of parallel robots. IEEE international conference on robotics and automation, 2004, pp 3665–3670
11. Lopes AM, Almeida F (2009) The generalized momentum approach to the dynamic modeling of a 6-DOF manipulator. Multibody Syst Dyn 21(2):123–146
12. Kane TR (1961) Dynamics of nonholonomic systems. J Appl Mech-Trans ASME 28(4):574–578
13. Tsai W (1999) Robot analysis: the mechanism of serial and parallel manipulators. Wiley, New York
14. Murray RM, Li Z, Sastry SS (1994) A Mathematical introduction to robotic manipulation. CRC Press, Boca Raton
15. Brockett RW, Stokes A, Park FC (1993) Geometrical formulation of the dynamical equations describing kinematic chains. IEEE international conference on robotics and automation,1993, Atlanta, USA, pp 6370–642
16. Park FC, Bobrow JE, Ploen SR (1995) A Lie group formulationof robot dynamics. Int J Robotics Res 14(6):609–618
17. Chen W, Chen IM, Lim WK, Yang GL (1995) Cartesian coordinate control for redundant modular robots. Proceedings of IEEE international conference on systems, man and cybernetics, Nashville, USA, 2000, pp 3253–3258

# Chapter 57
# Reconfiguration of System for Waste Collection with Robotics Methodologies

**Alberto Rovetta**

**Abstract** The collection of waste can be improved by adopting new methods of monitoring and new strategies for economic and efficient service. Over the past 200 years there have been no particular developments. Today you can reconfigure the systems for collecting, with the methods of robotic autonomous systems. It uses sensors to monitor the containers. Wireless communications and Internet are adopted to control the collection process. Statistical evaluations are performed to optimize the costs. Green waste Project strongly has reconfigured the process of waste collecting, with technical advantages and economic improvements.

**Keywords** Reconfigurable waste collection · Management of the waste collection system

## 57.1 Introduction

For thousands of years, the waste material is thrown out and considered as a mass to be deleted. Over the centuries, the collection of waste has always been open loop. It occurred in practice without the knowledge of the amount of waste and the possibility of recovering the material, rather than destroy it [1–12].

If the collection method is reconfigured, a more economical and rational result may be obtained.

A. Rovetta (✉)
Laboratorio di Robotica ARIAL, Dipartimento di Meccanica,
Politecnico di Milano, Via Lamasa 1, 20156 Milan, Italy
e-mail: alberto.rovetta@polimi.it

J. S. Dai et al. (eds.), *Advances in Reconfigurable Mechanisms and Robots I*,
DOI: 10.1007/978-1-4471-4141-9_57, © Springer-Verlag London 2012

Greenwaste Project was founded in 2004 with the Project Cleanwings between Italy and China with the support of the Ministries of Environment. Wastenergy Project continued, from 2008 to 2009, and led to a presentation in EXPO 2010 in Shanghai. BURBA Project is on development now in Europe and China. Greenwaste Program has revolutionized the way people think and carry out the collection of waste. Greenwaste has high level of knowledge and quality of action. Methodologies of robotics were used, transferred with the reconfiguration in the processes of waste collection, which today are, above all, manuals. Now they are highly developing with autonomous robotics and innovative automation methodologies.

## 57.2 Features Needed for the Reconfiguration

The robotic science also applies to the collection of waste, to have a better quality of life, for cleaning and material recovery. The materials can be recycled cost-effectively if the collection is done in a rational manner and without defects. The reconfiguration of systems of waste collection is modeled on the principles of robotics applied to autonomous systems.

The main concepts are:

1. sensorization of individual containers in any number they are
2. transmission of data from a single container to a central data collection. The center manages the selection process of collection and transmits control commands to the collection trucks
3. intelligent management of all the behaviors of the container. Timing and costs of collection are optimized and the layout of the container on the land is reconfigured.

The reconfiguration process changes the structure of knowledge on the behavior of the container. The containers are filled with the passage of time for action of users and are emptied with a certain periodicity. The first reconfiguration concept is the rationality in a different way of collection. Instead of having a permanently programmed collection, the collection shall be implemented according to the weight of material in the container. The logic is to act on containers and trucks only when the collection is necessary.

The reasons are many and varied: excess material escaping from the container, fermentation of organic materials, and sewage smells. Greenwaste eliminates the disturbing vision for the citizens of the spill and loss of material around the container with worsening environmental conditions.

The reconfiguration requires the use of robotic methods. The first step concerns the reconfiguration of sensorization of containers.

**Fig. 57.1** First "intelligent" bin

## 57.3 Sensorization

Sensorization is performed with different systems.

The container of Fig. 57.1 is equipped with a camera inside, which controls the posted material. The camera transmits images to an on-board processor that processes them. A sensor measures the weight collected in the container. It transmits it via GPRS or ZigBee to a collection center. The data is transmitted to the Control Center for the collection.

The reconfiguration makes the selection of active systems and their self-emptying. The ancient collection operated daily or periodically, whatever the condition of the container, without any verification of real costs. With the reconfiguration we are switching from an uncertain situation to a situation known and under control.

Another sensor measures the humidity and transmits the data. According to the value of the moisture, you can predict the decay of organic material collected in the bin. It is possible to reconfigure the order of collection; it becomes urgent for reasons of hygiene.

A sensor measures the temperature inside the container. It manages the temperature signals to inform the control center if cases under abnormal temperature occur in service. In addition, the temperature and humidity together enable to know what kind of putrefaction could take place, and thus can trigger the order of collection and discharge, immediately, without waiting for long periods.

Figure 57.2 represents the bin fully equipped with intelligent sensors and control units. The box is an intelligent actor of the system. The control box contains a microprocessor, a battery power supply, a memory unit, and sensors of weight and volume, a small camera that looks inside the box, a temperature sensor, a humidity sensor.

**Fig. 57.2** Position of sensors

**Fig. 57.3** Manual waste collection

All data collected substantially reconfigure the relationship between collection units and the environment. The information on environment are numerous and significant. The reconfiguration has raised the level of communication between the environment and the waste collection. In the near future Greenwaste will also propose the measurement of all parameters for measuring air quality and traffic circulation through the surveys that the electronic box can do. Reconfiguration will be a landmark, because the control box will become a fundamental part of urban environment, in every case free and documentable.

The reconfiguration of the electronic box is a decisive action to achieve the objectives of the Project Greenwaste. The main term is the human action, i.e. the operator must move the ecological box, place it, empty it, make sure it is empty, replace it and move to a new location.

The Fig. 57.3 shows the operator in action, and with difficulty moving the bin to place it. Greenwaste reconfigured human action, which has become the control, and not just manual operation. The operator now moves out of the means of transport, placing the box in only the first phase of installation, and emptying bins according to the data set, watching the computer data, transmitted via Internet. It can use all

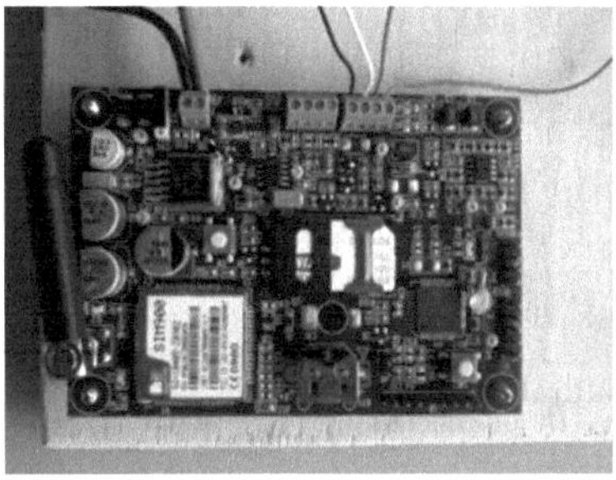

**Fig. 57.4** Board

highly automated systems and improve performance. It can use mobile phones too. The reconfiguration is needed for the operational project Greenwaste.

Figure 57.3 shows a bin "intelligent" in the first reconfiguration. It contains basically the weight sensor, which measures the weight of the bin to detect when it becomes filled and to be emptied. The design is innovative. The reconfiguration requires a technical adding that changes the nature of the electronic box. The electronic box is reconfigured and become an active element of the system. The reconfiguration of the container becomes an active element in the system Greenwaste.

The sensorization is the basis for the entire system to be reconfigured. It is possible to add other sensors; and to connect multiple containers side by side. For big cities like Beijing, a statistical analysis to 20 million users can support to schedule the waste collection. It is possible to apply the same methodology for similar territories.

## 57.4 Transmission of Data

Data can be collected by hand, as is used by many cities, Municipalities, users. The manual collection can be sent to the Technical Department of the Municipality.

The changes in the container and in the collection are related to the results of data processing, in statistical form, with multivariate analysis.

Management via Internet allows real-time results. The reconfiguration of the transmission is also in relation to changes in techniques and technology of today. Today, the GPRS transmission has taken a very low cost and always decreasing with the passage of time.

Now reconfiguration suggested leaving the solution of ZigBee for many reasons. They are the inability to act on significant distances; passivity in the face of great obstacles and/or of dense material, like concrete; relatively high cost of boards and systems development.

GPRS is the right choice now with both 2G and 3G. Each provider of communication has its own characteristics and Greenwaste must adapt to these.

The control board (Fig. 57.4), in the 2011 version, contains a GPRS module for data transmission. The economic cost of ZigBee modules and new capabilities of GPRS networks have created the system with GPRS. The reconfiguration has not

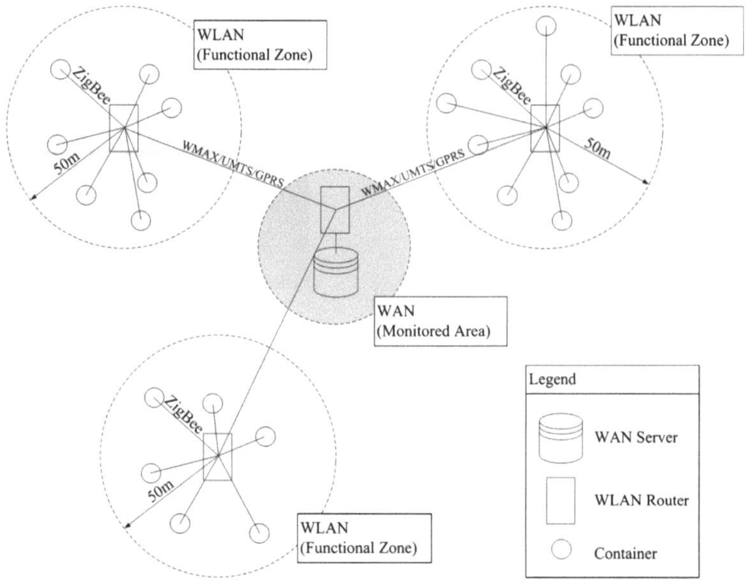


**Fig. 57.5** Star connection

been painless. We had to redesign the firmware, repeat the software, re-creating a control board efficient and totally reliable. The advantage of the reconfiguration was also linked to the flexibility of the board, which was designed for each implementation. It will also serve to detect different types of sensory signals. Modules can also be interfaced with special features, simply making the appropriate interface.

The star scheme (Fig. 57.5) is an effect of the reconfiguration of the system. Whether Greenwaste work is in "islands", or the bins are arranged "in line", it is possible to establish a hierarchy in the collection. It relies on the management of a local collection of data. The complexity grows: interface modules must be created and an efficient network without any weakness must be built. Until now, the vastness of the project has preferred to focus on broadcasting one-to-one, with a single link for each bin. Greenwaste is ready to flexible reconfiguration for a more economical action with the same efficiency, if it will be possible in many cases.

The map drawing (Fig. 57.6) is the actual description of a working situation, which is now developed by the project Greenwaste. Generally it is agreed to divide an area into Zones, easier to control. The zones are represented with a map, with the representation of streets, squares, houses, and the provision of bins represented by an icon. A large monitor for controlling the situation is used.

Computer monitors, mobile phones, tablets for local and remote control are used. Handhelds connected to Internet to deliver dynamics in the area present a great efficiency (Fig. 57.7).

Managers of the collection process are always connected to each other (Fig. 57.8) and work via Internet and satellite networks. The costs today are very

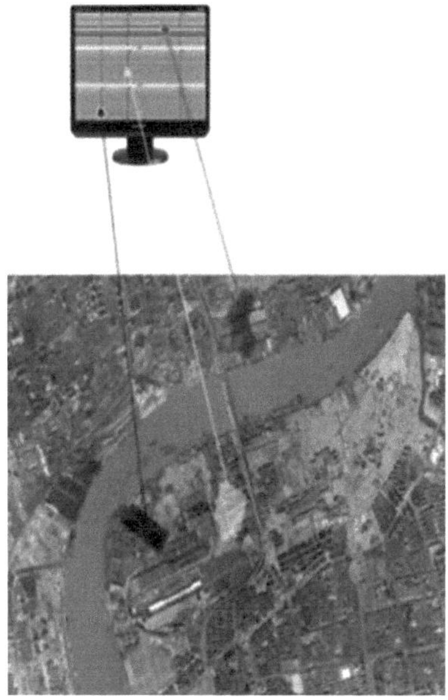

Fig. 57.6 Mapping and zones

Fig. 57.7 Internet use

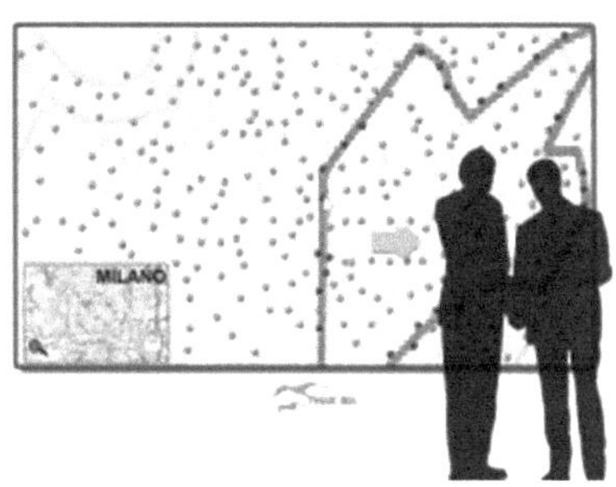


Fig. 57.8 ITC results

limited. Networks are used for data transmission. Generally special discounts from companies that provide services networks are offered.

## 57.5 Data Management

The Greenwaste Project uses a server that collects and manages all data in a single center. The user is always connected to the server, and requires the necessary data. The data are transmitted in an archive, in other words a file that contains the codes, the values of the weight, volume, battery level, the energy level of the cell where the container is placed, the date by day, hour, minute, second of every launch GPRS. The program shows the flags that indicate the status transitions. Each user may use its graphical interface. The results are always excellent. The reconfiguration provides a snapshot image of the collection container and the results of the collection.

The savings in the path length is the logical first step for the total achieved savings, step by step, from the use of waste collection, suggested by Greenwaste Project.

## 57.6 The Reconfiguration

Reconfiguration is also the transformation of the way of thinking regarding the collection of waste. The waste is often recycled. The recycled waste assumes a second life, a second economic value. Economic values are very high. Greenwaste changes the way you approach. The "thing", placed in the waste, will always retain its value.

From this point of view, waste collection is reconfigured as "the collection of values." Waste recovery is a satisfying result, without going beyond the very nature of things (Fig. 57.9).

A second important point of reconfiguration is the educational aspect. The Engineering provides technical values, and takes values even more complete if it considers the most important aspects of life. The education of a child shall include knowledge of and respect for the end use of things. Must be maintained respect for their value, even after use. Waste collection will be appreciated as a way to make more active the economic balance daily, between use, consumption and expenditure.

The collection of waste with Greenwaste offers savings with very high economic values. The citizen has the highest respect for his city and its environment, as the farmer and the mountain have for their land. This reconfiguration is a strong spring for cultural progress, even in the collection of waste (Fig. 57.10).

### *57.6.1 Reconfiguration with the Methods of Robotics*

A subsequent reconfiguration is born from the need to have a local application on a real reality. It has been applied with the weight sensor in Beijing, Chaoyang district. This data is sent over the Internet to a central control: the collection of waste has been transformed completely (Fig. 57.11).

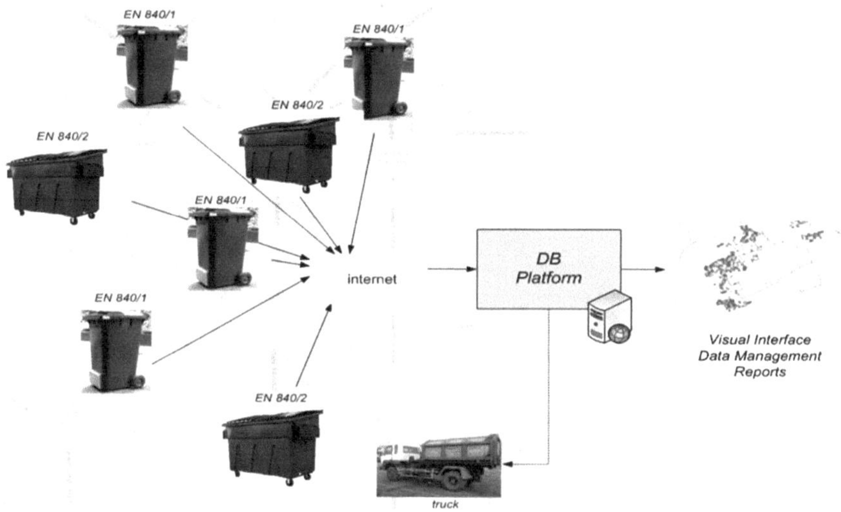


Fig. 57.9 Flow chart

## 57.6.2 Sensors

The weight sensor is formed, as PIC, with a strain gauge sensor, suitable for the measurement of bending. The arrangement of the sensors and the choice is influenced by the type of container and the system design, in accordance with the designer of the manufacturer. The project selected four sensors arranged at the vertices of a square. A second version has a single sensor, for very high loads.

There is a reconfiguration to make the system as small as possible. It saves the volumes, weights, costs of assembly and transport costs of the system.

The volume sensor indicates whether the bin is full or empty. It measures visually if the full line is blocked or free. The combination of measures of weight and volume increases significantly the effectiveness and safety of the measurement. The reconfiguration of the plot of the data allows a very accurate and valid diagnostic, in front of each type of material that is thrown in the trash. The density is a very significant parameter.

## 57.6.3 Application of Greenwaste in Beijing, 2011–2012

The provision of bins in Beijing as needed represents an enormous effort of reconfiguration. In Fig. 57.12, the final island of the three bins. The reconfiguration is the first mechanical structure. There are three metal structures, high, to protect the containers, anchored to the floor. One block is used to prevent removal or tampering. Container can be any type, provided its size enabling it to enter structure. Then the system is suitable for each container and becomes versatile. This feature is measuring

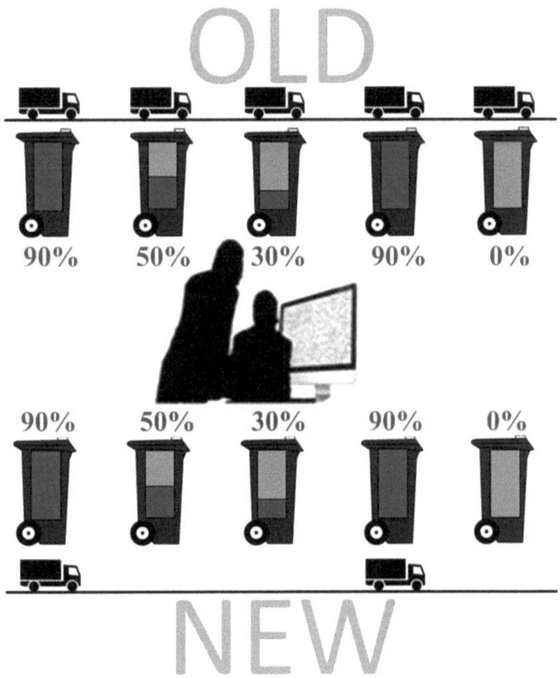


Fig. 57.10 Team work with "white collars"

weight, with all the electronic processing and transmission. The reconfiguration is also in the economic relationship with the bin. Greenwaste system takes the existing containers. It does not require them any change. This absence of changes in the bins is a great economic richness of the new system Greenwaste.

### *57.6.4 Reconfiguration for Expo 2010 Shanghai*

The bins in EXPO 2010 (Fig. 57.13) have shown the sign of the validity of the reconfiguration of the waste collection. 31 bins were taken to EXPO in collaboration with the company Aoto, producer, and were adopted. Security requirements have limited the duration outside. The waste collection system with remote transmission, with solar energy, has been the symbol of a project devoted to the opportunities of this technology. The presentation was made at the level of the European Union. The presentation in China on October 20th, 2010 in Environment Forum in Shanghai, expressed the great potential of the reconfiguration of the bins. The beauty and elegance of the shapes of the two containers, side by side, joined the innovation freshness with powerful technology.

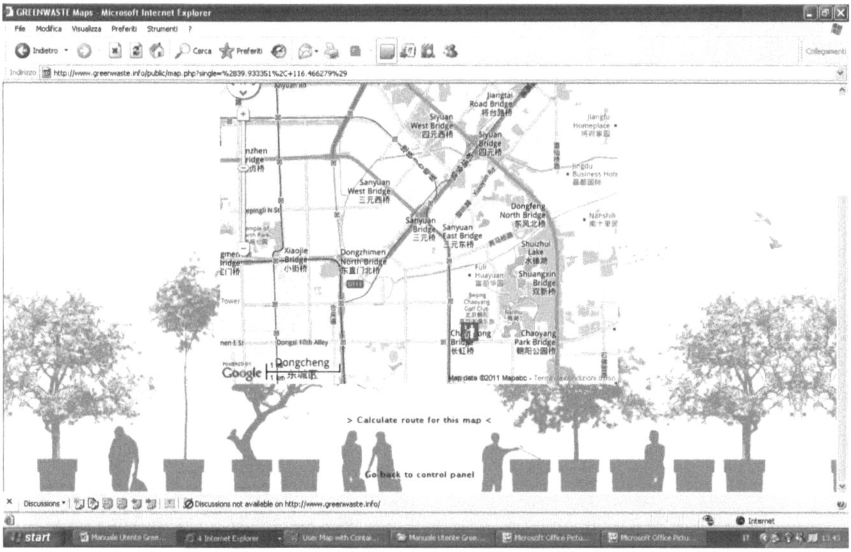


Fig. 57.11 Path selection

Fig. 57.12 PIC, Beijing

## 57.7 Software

The software of Greenwaste has many different aspects (Figs. 57.14, 57.15).

The technical aspect meticulously describes all data, at any time and performs an analysis as radiographic of the reliable system.

The user manual is accompanied by suitable use. The reconfiguration is to add intelligent working knowledge of the system. It seems a small step, but it is

**Fig. 57.13** EXPO 2010, Shanghai

revolutionary. It eliminates the printed paper to be delivered to the operator and submit the data on his mobile phone. The software is the interpreter of data. The details of the software are presented in the "User Manual of Greenwaste". The diagnostic is complete. The mapping is inside Greenwaste. The software also includes the calculation of time and space for the location of waste collection.

It was also set up a software that manages the economic point of view for the collection. It foreshadows the optimal choice of the points in which place the bins.

## 57.8 Design

Reconfiguration is the best design to make use of the new form of containers, the most acceptable on different territories. It is possible to modify the materials: steel, plastic, fiberglass, aluminum, and composite systems. Results are beautiful, high-impact projects, as was the case for EXPO 2010 in Shanghai. The pair of cylindrical containers entered the history of the perfect collection of waste.

Systems have made PIC square, cylindrical, cubic, a simple under posed box, with cameras and multi sensors complex.

The reconfiguration has also invested heavily in the interface design of the Project Greenwaste. The main need is to be clear, transparent, and easy to understand. Greenwaste has to be intelligible even by persons of the highest culture and minimal culture. The clear, colorful icons are the best form of post op. Everyone recognizes the icon. All implicit messages are sent immediately and without delay due to

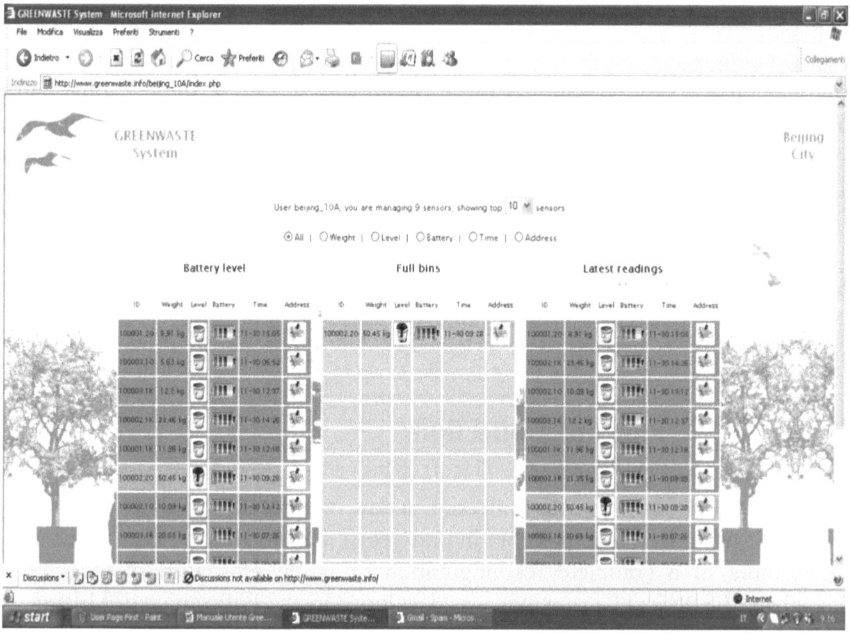


**Fig. 57.14** Software of Greenwaste

**Fig. 57.15** Bins categories

interpretation. In particular, a unique icon makes clear communication. In Greenwaste reconfiguration of the project is given much weight to the interaction between objects, people and goals.

Software is made in 5 languages (English, Italian, Chinese, Japanese, and French, by now). In the GUI there is no doubt the power of icons. Here is a simple icon that indicates that the bin is filled. Although color plays an important role, we used own methods of traffic lights, common throughout the world: green, yellow, and red. In this way the empty container is colored green on the monitor. The full container is colored red. Red recalls the urgency of the collection, and calls to collect and empty the bin with the best of times, with the most appropriate cost.

This has been the target of Greenwaste and it is active in all reconfigurations. It always draws a maximum respect for the environment, quality of life and love for the human being who lives in the environment and reaches his wellness.

## 57.9 Construction

The methodology of assembly is linear. However, it may be subject to reconfiguration if highly automated machines and robots are adopted. Here is an example of the assembly module, which has been delivered to operators. It is always subject to reconfiguration, after evaluation of the time, cost, and the redesign of many components, year after year.

An evolution occurred from 2010 to 2011 bins with PIC. The measuring system is both internal and external to the container. All systems are patented and are an example of active reconfiguration project.

## 57.10 The Strategy of Collection

The reconfiguration of waste collection today requires a complete and general process. It has never been possible in past years. The diagram of Fig. 57.16 gives an example of a complete reconfiguration of waste collection in the city of Shanghai. The project Cleanwings was developed by the Politecnico di Milano and the SJTU Shanghai Jiaotong University in collaboration with EPB of Shanghai Environmental Protection Bureau.

The collection allows getting smart and planned economy to a final stage to the incinerator, landfill and recovery of recyclable material. The wet material can be recovered and reused as a fertilizer compound as an alternative.

The sense of robotic application deals with the closed loop, which involves now, with Greenwaste project, the waste collection process. Sensors give their values to the control, and the control actuates the motion process, which is made by trucks and operators. The system changes for the effect of the closed loop, like in robotics, and the data exert the double action of getting information and of giving commands. The software improves and increases in its quality according to the experience of collection, and some intelligence in the waste collection sequence is introduced by the elements in the mechanical and electronic chain. A robotic exploitation is also the GPRS use, which enlarges to communication of a single element to all the environmental word. The methodologies of tests and developments are very similar to the methodologies of robotic autonomous systems.

## 57.11 Summary and Conclusions

The reconfiguration of the process of waste collecting has transformed the basic manual action to collect the waste. It now uses a procedure with Greenwaste refined reporting and monitoring, smart metering for cost control, planning of the

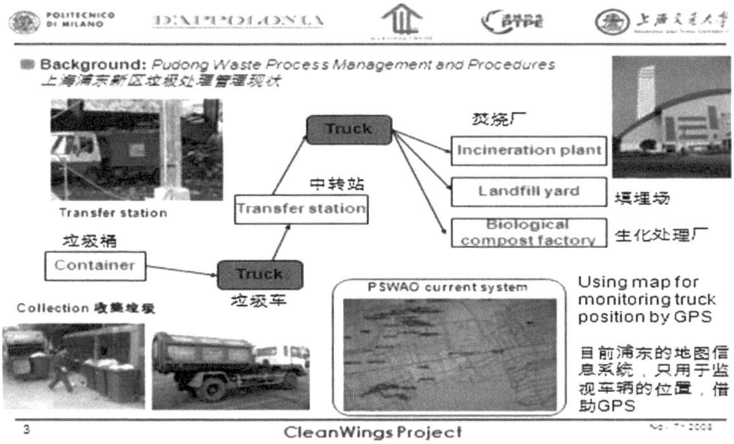


**Fig. 57.16** Complete diagram of waste flow

collection. The results are revealing in terms of economic, technical and scientific progresses. Culture and sensitivity to the process of collecting permit an evolution towards a better quality of life.

**Acknowledgments** The work was developed within Cleanwings Project, funded by the Ministries of Environment of Italy and China, European—Chinese Project WASTENERGY, European Project Burba (on development).

The author thanks Eng. A. Giusti, F. Vicentini, I. Zangrandi, A. Caravati, D. Ninfa, G. Rama, M. Marino and Dr S. Suevo, A. Desantis for collaboration in projects.

# References

1. Rovetta A, Fan X (2009) Early detection and evaluation of waste through sensorized containers for a collection monitoring application. Waste Manag 29:2939–2949
2. Vicentini F, Giusti A, Rovetta A, et al (2009) Sensorized waste collection container for content estimation and collection optimization, waste management, 2008. Report of Wastenergy Project, FP7, November 2009
3. Frosch RA (1996) Toward the end of waste: reflections on a new ecology for industry. Daedalus 125:199–212
4. Yan HS, Liu NT (2000) Finite-state-machine representations for mechanisms and chains with variable topologies. In: Proceedings of ASME design engineering technical conference, Baltimore, DETC2000/MECH-14054
5. Kramer J, Magee J (1985) Dynamic configuration for distributed systems. IEEE Trans Softw Eng SE-11(4):424–435
6. Zhang L, Dai JS (2008) Genome reconfiguration of metamorphic manipulators based on lie group theory. In: Proceedings of ASME international design engineering technical conferences & computers and engineering technical conferences, Brooklyn, DETC2008-49906
7. Rooker M, Ebenhofer G, Strasser T (2009) Reconfigurable control in distributed automation system. In: ASME/IFToMM international conference, pp 705–714
8. Shick PL (2007) Topology: point-set and geometric. Wiley, Hoboken

9. PISA Consortium (2006) PISA—flexible assembly systems through workplace-sharing and time-sharing human–machine cooperation, funded European Integrated Project, contract no. NMP2-CT-2006-026697. www.pisa-ip.org. Accessed December 2006
10. Koren Y et al (1999) Reconfigurable manufacturing systems. Annals CIRP 48(2):527–540
11. Ziadat AH, Mott H (2005) Assessing solid waste recycling opportunities for closed campuses. Manag Environ Quality Int J 16(3):250–256
12. User Manual for Greenwaste Project (2011) Report, Department of Mechanics, Politecnico di Milano, December 2011

# Part VI
# Control of Reconfigurable Robots

# Chapter 58
# Study on Decentralized Control of Reconfigurable Manipulator Based on Third-Order ESO

**Yanli Du, Yanfeng Qiao, Zhiqian Wang, Mujun Xie and Yuanchun Li**

**Abstract** This paper presents a decentralized controller based on third-order extended state observers (ESO) for reconfigurable manipulator with interconnection terms. Coupling interconnection terms of the joints and modeling uncertainties are estimated and compensated real-timely by using third-order ESO. Using particle swarm optimization (PSO), the parameters of ESO were adaptively adjusted. The controller is applied to the trajectory tracking control of 4-DOF reconfigurable manipulator. Simulation results show the effectiveness of the proposed method in treatment of the interconnection terms.

**Keywords** Reconfigurable manipulator · Decentralized control · ESO · PSO

---

Y. Du · Y. Li
State Key Laboratory of Automobile Dynamic Simulation, Jilin University, 130012 Changchun, China

M. Xie · Y. Li (✉)
Department of Control Engineering, Changchun University of Technology, 130012 Changchun, China
e-mail: liyc@mail.ccut.edu.cn

Y. Du
College of Electrical Information Engineering, Beihua University, 132021 Jilin, China

Y. Qiao · Z. Wang
Changchun Institute of Optics, Fine Mechanics and Physics, Chinese Academy of Sciences, 130033 Changchun, China

J. S. Dai et al. (eds.), *Advances in Reconfigurable Mechanisms and Robots I*,
DOI: 10.1007/978-1-4471-4141-9_58, 

## 58.1 Introduction

Application of decentralized control method in reconfigurable manipulator is more and more extensive recently. The dynamic system of reconfigurable manipulator is represented as a set of nonlinear interconnected subsystems. Then, local controller is designed for each subsystem so as to inhibit interconnection effect. Reference [1] developed a adaptive fuzzy tracking controller for a class of uncertain nonlinear MIMO systems based on the state observer. Fuzzy system was used to approximate subsystem dynamic model and interconnection terms; Ref. [2] designed a robust decentralized adaptive output-feedback controller for a class of large-scale systems with dynamic and static interconnections by using the MT-filter and backstepping design method; Ref. [3] investigated the stability and stabilization problems for the fuzzy large-scale system in which the system is composed of a number of Takagi–Sugeno fuzzy model subsystems with interconnections; In Ref. [4], the neural network was to approximate the uncertainties terms and the interconnection weights of the neural network can be tuned online; In Ref. [5], a new efficient estimation approach was proposed to estimate the perturbation from the dynamics of the sliding surface to compensate for model uncertainties, unknown external disturbances, and time-varying parameters; Ref. [6] presented a new approach to robust quadratic stabilization of nonlinear systems within the framework of Linear Matrix Inequalities (LMI) that recognized the matching conditions by returning a feedback gain matrix for any prescribed bound on the interconnection terms; In Ref. [7], a novel dynamic surface direct adaptive control scheme has been presented by using multilayer neural networks for a class of strict-feedback uncertain SISO nonlinear systems. The controller has been made to be free from singularity problem by introducing integral-type Lyapunov function; Ref. [8] designed a dynamic surface controller for uncertain, mismatched nonlinear systems in the strict feedback form based on neural network; Ref. [9] developed a decentralized NN controller for large-scale uncertain robotic systems. NN was used to approximate unknown functions to include the case of both local dynamic uncertainty and nonlinear interconnections in each subsystem.

Interconnection terms were approximated mostly by using NN or fuzzy system in the above decentralized controllers. Adjustable parameters of these methods were being multiple and design of its adaptive law being difficult, so develop of the decentralized control was restricted. This paper presents a decentralized control method based on third-order ESO for reconfigurable manipulator with interconnection terms. Position, velocity, coupling interconnection terms and modeling uncertainties of each joint are estimated by using third-order ESO. Using PSO, the parameters of ESO were adaptively adjusted.

## 58.2 Problem Description

Reconfigurable manipulator is the opening kinematics chain that is composed of joint modules and link modules. By using Newton–Euler equation, dynamic model of n-DOF reconfigurable manipulator is stated as:

$$M(q)\ddot{q} + C(q,\dot{q})\dot{q} + G(q) = \tau \tag{58.1}$$

where $q \in R^n$ is the joint position; $M(q) \in R^{n\times n}$, the inertia matrix; $C(q,\dot{q})\dot{q} \in R^n$ is the vector of centrifugal and Coriolis force; $G(q) \in R^n$, the gravity term; $\tau \in R^n$ is the control input.

Every joint of the robotic system is served as a subsystem and its dynamic model could be described as:

$$M_i(q_i)\ddot{q}_i + C_i(q_i,\dot{q}_i)\dot{q}_i + G_i(q_i) + Z_i(q,\dot{q},\ddot{q}) = \tau_i \tag{58.2}$$

where

$$\begin{aligned} Z_i(q,\dot{q},\ddot{q}) = & \left\{ \sum_{j=1,\, j\neq i}^{n} M_{ij}(q)\ddot{q}_j + [M_{ii}(q) - M_i(q_i)]\ddot{q}_i \right\} \\ & + \left\{ \sum_{j=1,\, j\neq i}^{n} C_{ij}(q,\dot{q})\ddot{q}_j + [C_{ii}(q,\dot{q}) - C_i(q_i,\dot{q}_i)]\dot{q}_i \right\} \\ & + [\bar{G}_i(q) - G_i(q)] \end{aligned} \tag{58.3}$$

Let $x_i = [x_{i1}, x_{i2}] = [q_i, \dot{q}_i]^T (i = 1, 2, \ldots, n)$, then Eq. (58.2) can be written as:

$$\begin{cases} \dot{x}_{i1} = x_{i2} \\ \dot{x}_{i2} = f_i(q_i,\dot{q}_i) + h_i(q,\dot{q},\ddot{q}) + g_i(q_i)\tau_i \\ y_i = x_{i1} \end{cases} \tag{58.4}$$

where

$$\begin{aligned} f_i(q_i,\dot{q}_i) &= M_i^{-1}(q_i)[-C_i(q_i,\dot{q}_i)\dot{q}_i - G_i(q_i)] \\ g_i(q_i) &= M_i^{-1}(q_i) \\ h_i(q,\dot{q},\ddot{q}) &= -M_i^{-1}(q_i)Z_i(q,\dot{q},\ddot{q}) \end{aligned} \tag{58.5}$$

The major task of this paper is to design controller for each subsystem and makes it track the desired trajectory.

## 58.3 Design of Decentralized Controller Based on Third-Order ESO

### 58.3.1 Derivation of Decentralized Control Law

In Eq. (58.4), let $x_{i3} = f_i(q_i, \dot{q}_i) + h_i(q, \dot{q}, \ddot{q})$ as extended state, then Eq. (58.4) is changed into:

$$\begin{cases} \dot{x}_{i1} = x_{i2} \\ \dot{x}_{i2} = x_{i3} + g_i(q_i)\tau_i \\ \dot{x}_{i3} = d_i \\ y_i = x_{i1} \end{cases} \tag{58.6}$$

where $d_i$ is the differential coefficient of $x_{i3}$. The forms of third-order ESO can be expressed as follows:

$$\begin{cases} e_i(t) = z_{i1} - y_i \\ \dot{z}_{i1} = z_{i2} - \beta_{i1} K_{i1}[e_i(t)] \\ \dot{z}_{i2} = z_{i3} - \beta_{i2} K_{i2}[e_i(t)] + g_i(q_i)\tau_i \\ \dot{z}_{i3} = -\beta_{i3} K_{i3}[e_i(t)] \end{cases} \tag{58.7}$$

where $z_{ij}(t)$ $(i = 1, 2, \ldots, n,\ j = 1, 2, \ldots, m)$ is the estimated value of output state of system by using ESO, $\beta_{ij}$ $(i = 1, 2, \ldots, n,\ j = 1, 2, \ldots, m)$ is the observer parameter, which can be written as:

$$z_{i1} \to x_{i1}, z_{i2} \to x_{i2}, \ldots, z_{im} \to x_{im}$$

$K_{ij}(\cdot)$ $(i = 1, 2, \ldots, n,\ j = 1, 2, \ldots, m)$ defined as a saturated continuous nonlinear function:

$$K_{ij}(e_i, \alpha, \delta) = \begin{cases} |e_i|^{\alpha} sign(e_i) & |e_i| > \delta \\ \dfrac{e_i}{\delta^{1-\alpha}} & |e_i| \le \delta \end{cases} \tag{58.8}$$

In Eq. (58.8), $0 < \alpha < 1, \delta > 0$. Stability analysis of third-order ESO is given by Ref. [10], and points out that third-order ESO can make state estimation error limit in a very small range.

Let the tracking error in the $i$th subsystem be:

$$e_{i1} = z_{i1} - y_{ri} \tag{58.9}$$

where $y_{ri}$ is the desired trajectory of $i$th subsystem and its filtered error as:

$$s_i = \dot{e}_{i1} + \lambda_i e_{i1} \tag{58.10}$$

where $\lambda_i > 0$. Notice that if $s_i$ is ultimately bounded by $|s_i(t)| \le \psi_i$, then $e_{i1}$ and $\dot{e}_{i1}$ are ultimately bounded by $|e_{i1}(t)| \le \psi_i/\lambda_i$ and $|\dot{e}_{i1}(t)| \le 2\psi_i$, respectively.

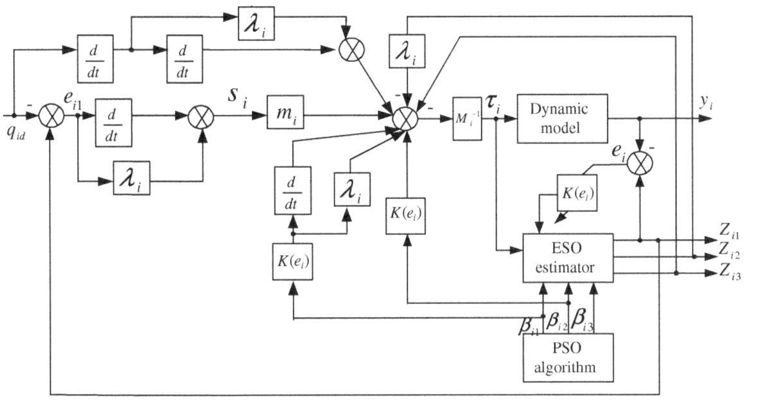


**Fig. 58.1** Block diagram of the $i$th subsystem

Therefore, our objective is to design a control law to ensure the filtered error $s_i$ to be ultimately bounded.

The dynamics of $s_i$ is obtained from Eqs. (58.7) and (58.10) as:

$$\dot{s}_i = f_i(z_i, K_i) + b_i(y_{R,i}) + g_i(q_i)\tau_i \tag{58.11}$$

where $f_i(z_i, K_i) = z_{i3} - \beta_{i2}K_{i2}[e_i(t)] - \beta_{i1}\dot{K}_{i1}[e_i(t)] + \lambda_i(z_{i2} - \beta_{i1}K_{i1}[e_i(t)]), b_i(y_{R,i}) - \ddot{y}_{ri} - \lambda_i\dot{y}_{ri}, z_i = [z_{i2}, z_{i3}], K_i = \{K_{i1}[e_i(t)], K_{i2}[e_i(t)], \dot{K}_{i1}[e_i(t)]\}$ and $y_{R,i} = [\dot{y}_{ri}, \ddot{y}_{ri}]$.

According to Eq. (58.11), we propose a control law as follows:

$$\tau_i = \frac{1}{g_i(q_i)}(-f_i(z_i, K_i) - b_i(y_{R,i}) - m_i s_i) \tag{58.12}$$

where $m_i > 0$. The closed-loop system is obtained by substituting Eq. (58.12) into Eq. (58.11):

$$\dot{s}_i = -m_i s_i \tag{58.13}$$

This guarantees the filtered tracking error $s_i(t) \to 0$ exponentially. Therefore, it follows from Eq. (58.10) that $e_{i1}(t) \to 0$ and $\dot{e}_{i1}(t) \to 0$ exponentially.

**Theorem** *Consider subsystem dynamic model of reconfigurable manipulator as Eq. (58.2), if decentralized control law is designed as Eq. (58.12), then trajectory tracking error of reconfigurable manipulator system would approached zero asymptotically.*

Structure diagram of each subsystem is shown in Fig. 58.1.

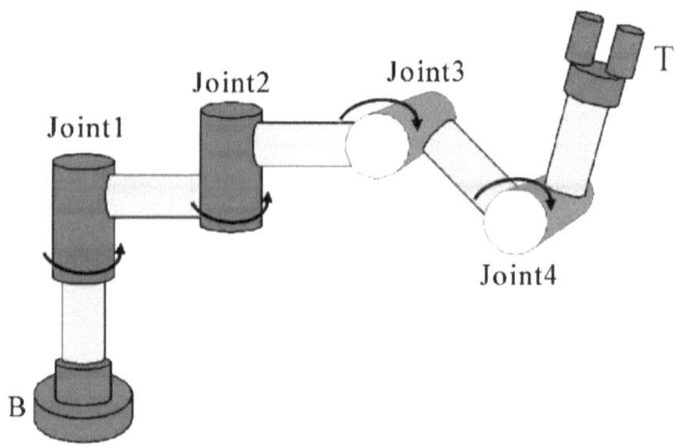


**Fig. 58.2** 4-DOF reconfigurable manipulator

### 58.3.2 Training Process of PSO Algorithm for ESO Parameters $\beta_{i1}$, $\beta_{i2}$, $\beta_{i3}$

ESO parameters $\beta_{i1}, \beta_{i2}, \beta_{i3}$ can be adaptively adjusted by using PSO algorithm, training process is as follows [11]:

(1) Initialize positions and associated velocity of all individuals, scale of the particle swarm, the inertia weight $w$, acceleration coefficient $c_1$,$c_2$, maximum evolutional generation maxgen etc.;
(2) Evaluate the fitness value of all individuals according to the objective function;
(3) Compare current fitness value of every individual with the fitness value corresponding to itself best position. If the current fitness value is better, then update current best position of every individual;
(4) Compare current fitness value of every individual with the fitness value corresponding to global best position. If the current fitness value is better, then update global best position;
(5) Update position and velocity of every individual.

If the best position of $i$ particle is $X_{pbest}$, the position of best particle is $X_{gbest}$, the current position of $i$ particle is $X_i$, then the velocity and position of every particle is updated according to following Eq. (58.14):

$$\begin{aligned} V_i(gen+1) &= w * V_i(gen) + c_1 * rand() * (X_{pbest} - X_i) \\ &\quad + c_2 * rand() * (X_{gbest} - X_i) \\ &\quad If\ |V_i| > V_{\max},\quad V_i = rand() \\ X_i(gen+1) &= X_i(gen) + V_i(gen+1) \end{aligned} \tag{58.14}$$

where $gen$ is the current evolutional generation and maximum velocity is set at $V_{\max}$.

(6) Repeat step (2)–(6) until better fitness value is reached or predefined number of iterations is completed.

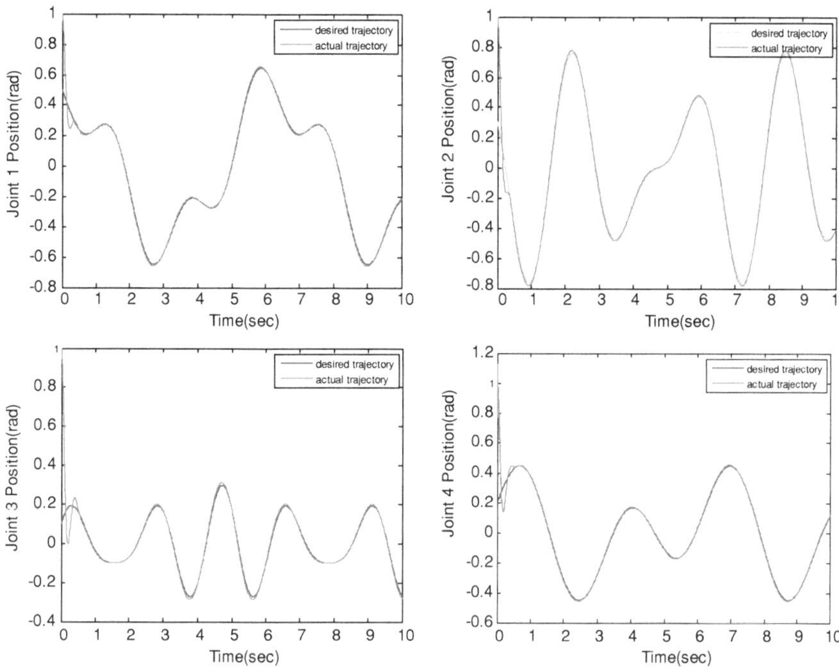


**Fig. 58.3** Trajectory tracking curve of each joint

## 58.4 Simulation

In order to testify the effectiveness of the proposed method, it was applied to 4-DOF reconfigurable manipulator in Fig. 58.2. Modular parameters and dynamic modeling process may be referred to Ref. [12].

The desired trajectory is:

$$q_1^d = 0.5\cos(t) - 0.2\sin(3t), \quad q_2^d = 0.3\cos(3t) - 0.5\sin(2t)$$
$$q_3^d = 0.2\sin(3t) + 0.1\cos(4t), \quad q_4^d = 0.3\sin(2t) + 0.2\cos(t)$$

The initial position was set at $q_1(0) = q_2(0) = q_3(0) = q_4(0) = 1$, the initial velocity was set at $\dot{q}_1(0) = \dot{q}_2(0) = \dot{q}_3(0) = \dot{q}_4(0) = 0$, the initial state of ESO was set at:

$$z_{11}(0) = z_{21}(0) = z_{31}(0) = z_{41}(0) = 1$$
$$z_{12}(0) = z_{22}(0) = z_{32}(0) = z_{42}(0) = 0$$
$$z_{13}(0) = z_{23}(0) = z_{33}(0) = z_{43}(0) = 0$$

Using control law in Eq. (58.12), ESO parameters are $\alpha = 0.5$, $\delta = 0.01$, $\lambda_i = 2$, $m_i = 500$, the scale of particle swarm is set at 20, maximum evolutional generation maxgen $= 30$, $c_1 = c_2 = 2$, $w_{\max} = 1$, $w_{\min} = 0.4$, $V_{\max} = 1.2$.

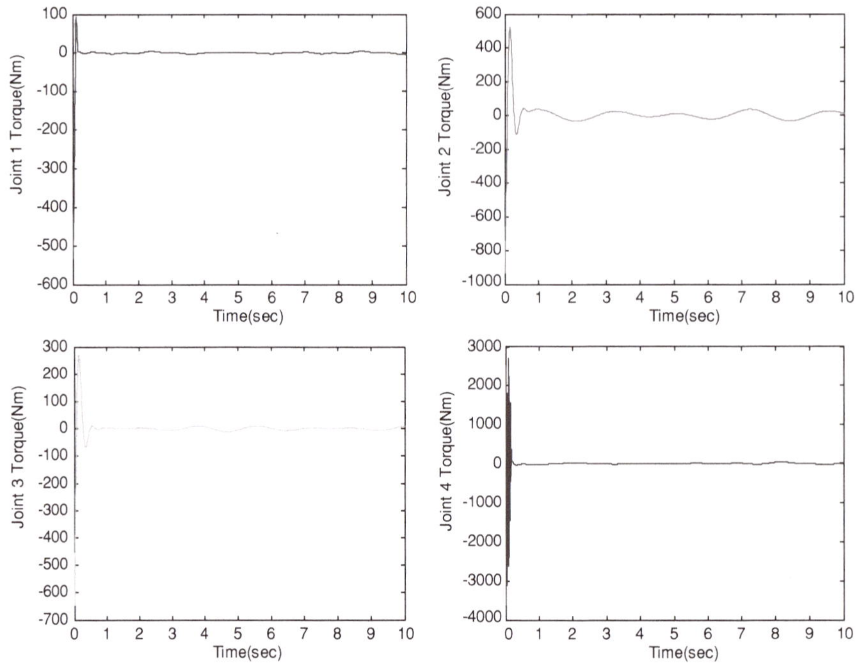


**Fig. 58.4** Control torque curve of each joint

**Fig. 58.5** Trajectory tracking curve of the end-effector

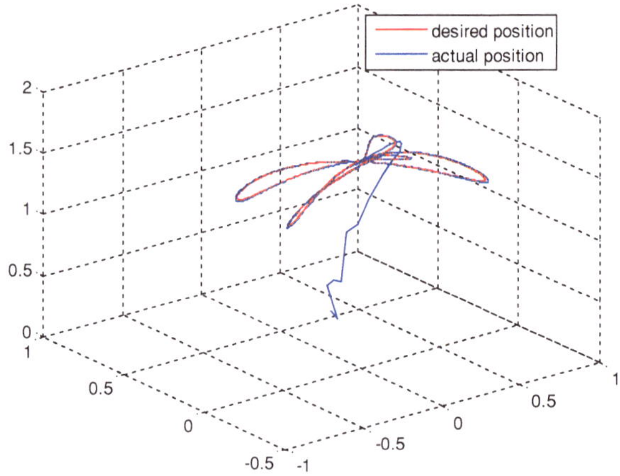


Trajectory tracking curves are shown in Fig. 58.3, control torque curves are shown in Fig. 58.4.

The results shown in Fig. 58.3 indicate that tracking errors of four joints was occurred on the initial stage owing to lack of dynamic model knowledge of sub-system. About 0.5 s later, actual trajectory can track well desired trajectory, it is proved that proposed decentralized control method can ensure stability of whole system and trajectory tracking performance.

Trajectory tracking curve of the end-effector is shown in Fig. 58.5.

## 58.5 Conclusion

In this paper, a decentralized control method based on third-order ESO for reconfigurable manipulator with interconnection terms is presented. Reconfigurable manipulator system is described as a set of every coupling subsystem, and the state and interconnection terms of each joint is estimated by using third-order ESO. Using PSO algorithm, the parameters of ESO were adaptively adjusted. Because ESO can estimate real-timely coupling interconnection terms, so this paper doesn't care about the problem whether or not interconnection terms are bounded, meanwhile, stability proving process of the decentralized controller is simplified. The effectiveness of the proposed decentralized control method is verified by simulating 4-DOF reconfigurable manipulator.

**Acknowledgments** This work was supported by the National Natural Science Foundation of China (No. 60974010) and the Scientific and Technological Development Plan Project in Jilin province of China (No. 20110705).

## References

1. Liu Y, Tong S, Li T (2011) Observer-based adaptive fuzzy tracking control for a class of uncertain nonlinear MIMO systems. Fuzzy Set Syst 64(2):25–44
2. Wu Z, Xie X, Jing Y, Zhang S (2004) Robust decentralized adaptive stabilization for large-scale systems with dynamic and static interconnections. Control Decis 19(8):862–866
3. Wang W-J, Luoh L (2004) Stability and stabilization of fuzzy large-scale systems. IEEE Trans Fuzzy Syst 12(3):309–315
4. Peng J, Wang Y, Yu H (2007) Neural network-based robust tracking control for nonholonomic mobile robot. Springer, Berlin, pp 804–812
5. Lin H-C, Lee C-S (2000) Predictability of back propagation and discrete Hopfield neural networks in harmonic compensation systems. Power Syst Technol 1:339–344
6. Siljak DD, Stipanovic DM (2000) Robust stabilization of nonlinear systems: the LMI approach. J Appl Math Decis Sci 6(5):461–493
7. Zhang TP, Ge SS (2007) Direct adaptive NN control of nonlinear systems in strict-feedback form using dynamic surface control [C]. In: IEEE international symposium on intelligent control, Singapore, pp 315–320
8. Swaroop D, Hedrick JK, Yip PP et al (2000) Dynamic surface control for a class of nonlinear systems. IEEE Trans Automatic Control 45(10):1893–1899
9. Tan KK, Huang S, Lee TH (2009) Decentralized adaptive controller design of large-scale uncertain robotic systems. Automatica 45:161–166
10. Shao L-W, Liao X-Z, Xia Y-Q, Han J-Q (2008) Stability analysis and synthesis of third order discrete extended state observer. Inform Control 37(2):135–139

11. Bingul Z, Karahan O (2011) Dynamic identification of Staubli RX-60 robot using PSO and LS methods. Expert Syst Appl 38(4):4136–4149
12. Zhu M (2006) Study on the kinematics, dynamics and control method of reconfigurable modular robots. College of Communication Engineering, Jilin University, Changchun

# Chapter 59
# A Combined Backstepping Terminal Sliding Mode Algorithm Based Decentralized Control Scheme for Reconfigurable Manipulators

**Bo Zhao, Zhiqian Wang, Yanfeng Qiao, Keping Liu and Yuanchun Li**

**Abstract** A decentralized control scheme based on a combined backstepping terminal sliding mode algorithm for reconfigurable manipulators is proposed. Based on Lyapunov stability theory, backstepping technique and terminal sliding mode are utilized in the first and second order of the subsystem respectively, the unknown terms and interconnection term are approximated or compensated by neural networks whose weights are updated with adaptive laws. For the serious chatter of the controller with the linear sliding mode, the terminal sliding mode replaced the linear sliding mode. In contrast, the method improves the convergence rate and the tracking accuracy, the control signal is smoother. Finally, the simulation results show the effectiveness of proposed scheme for different configurations with no need to modify any parameters.

**Keywords** Reconfigurable manipulators · Decentralized control · Backstepping technique · Terminal sliding mode · Neural networks

B. Zhao · Y. Li (✉)
State Key Laboratory of Automobile Dynamic Simulation, Jilin University,
Changchun, 130022, China
e-mail: liyc@mail.ccut.edu.cn

B. Zhao · K. Liu · Y. Li
Department of Control Engineering, Changchun University of Technology,
Changchun, 130012, China

Z. Wang · Y. Qiao
Changchun Institute of Optics, Fine Mechanics and Physics, Chinese Academy of Sciences,
Changchun, 130033, China

J. S. Dai et al. (eds.), *Advances in Reconfigurable Mechanisms and Robots I*,
DOI: 10.1007/978-1-4471-4141-9_59, 

## 59.1 Introduction

With the rapid development of modern industry, traditional robots which are limited by their fixed structure seem impossible to undertake variety of unknown or changeable tasks ranging from rescue and nuclear environment to space station. However, reconfigurable manipulators or robots that consist of interchangeable link and joint modules with standardized connecting interfaces are able to fulfill the tasks excellently by changing their kinematics and dynamics parameters due to the different environment [1].

The control strategies of reconfigurable manipulators have received great deal of attention in recent years. A neurofuzzy control scheme based on PD control of accurate model was developed [2] to identify and compensate structured and unstructured uncertainties for reconfigurable manipulators. Then Li et al. [3] proposed a computing torque control based neurofuzzy compensator for reconfigurable manipulators to compensate structured and unstructured uncertainties. A distributed adaptive sliding mode controller was proposed [4] for modular and reconfigurable robots. A robot dynamics system can be divided into some subsystems by a decomposing algorithm, and then an adaptive sliding mode controller is designed for each subsystem to handle the modeling uncertainty. In Ref. [5], a modular distributed control technique for modular and reconfigurable robots was proposed that is based on joint torque sensing, a modular and reconfigurable robot is stabilized joint by joint, and modules can be added or removed without the need to adjust control parameters of the other modules of the robot. All those methods were centralized control and distributed control. In practical engineering applications, they are probably difficult to be applied due to the high computation costs and structural complexities of centralized control and the time delay in the information exchange process among the subsystems of distributed control. Compared with centralized control and distributed control, decentralized control scheme can not only effectively reduce the computational burden of centralized control structure, but also avoid the communication time delay in distributed control. Therefore, more and more decentralized control systems have been proposed. Zhu and Li [6] proposed a stable decentralized adaptive fuzzy sliding mode control scheme for reconfigurable modular manipulators to satisfy the concept of modular software. A fuzzy logic system was introduced to approximate the unknown dynamics, then the effect of interconnection term and fuzzy approximation error is removed by employing an adaptive sliding mode controller. Backstepping adaptive sliding mode control based on hierarchical sliding mode control theory is presented [7] for a 2-DOF horizontal underactuated manipulator and gets better adaptability and control results. In Ref. [8], a decentralized adaptive iterative learning control algorithm for reconfigurable manipulators is proposed. Learning controller was conducted by a neural network learning component and a robust learning component to adaptively compensate the unknown dynamic functions and interconnections.

In this paper, a decentralized control scheme based on a combined backstepping terminal sliding mode algorithm for reconfigurable manipulators is proposed. In order to simplify the design of controller, backstepping technique [9] is utilized in the first step, then based on Lyapunov stability theory, the terminal sliding mode control which not only has strong robustness and anti-disturbance ability [10, 11] but also can smooth the control signal [12] is used in the second step. In each subsystem, the unknown term, uncertainty term and interconnection term should be approximated or compensated by the neural networks. The simulation results show high tracking performance and smoother control signal.

## 59.2 Problem Description

The dynamics of a reconfigurable manipulator with $n$ degrees of freedom is described, by using Newton–Euler formulation, as [6]

$$M(q)\ddot{q} + C(q,\dot{q})\dot{q} + G(q) = u \tag{59.1}$$

where $q \in R^n$ is the vector of joint displacements, $M(q) \in R^{n\times n}$ is the inertia matrix, $C(q,\dot{q}) \in R^n$ is the Coriolis and centripetal force, $G(q) \in R^n$ is the gravity term, and $u \in R^n$ is the applied joint torque.

In the literatures of the decentralized control, each joint is considered as a subsystem of the entire manipulator system interconnected by coupling torque. By separating terms only depending on local variables $(q_i, \dot{q}_i, \ddot{q}_i)$ from those terms of other joint variables, each subsystem dynamical model can be formulated in joint space as

$$M_i(q_i)\ddot{q}_i + C_i(q_i,\dot{q}_i)\dot{q}_i + G_i(q_i) + Z_i(q,\dot{q},\ddot{q}) = u_i \tag{59.2}$$

with

$$\begin{aligned} Z_i(q,\dot{q},\ddot{q}) &= \left\{ \sum_{j=1, j\neq i}^{n} M_{ij}(q)\ddot{q}_j + [M_{ii}(q) - M_i(q_i)]\ddot{q}_i \right\} \\ &+ \left\{ \sum_{j=1, j\neq i}^{n} C_{ij}(q,\dot{q})\dot{q}_j + [C_{ii}(q,\dot{q}) - C_i(q_i,\dot{q}_i)]\dot{q}_i \right\} + [\bar{G}_i(q) - G_i(q_i)] \end{aligned}$$

where $q_i, \dot{q}_i, \ddot{q}_i, \overline{G}_i(q), F_i(q_i,\dot{q}_i)$ and $\bar{u}_i$ are the $i$th element of the vectors $q, \dot{q}, \ddot{q}, G(q), F(q,\dot{q})$ and $u$, respectively. $M_{ij}(q)$ and$C_{ij}(q,\dot{q})$ are the $ij$th element of the matrices $M(q)$ and$C(q,\dot{q})$, respectively. Let

$$x_i = \begin{bmatrix} x_{i1} \\ x_{i2} \end{bmatrix} = \begin{bmatrix} q_i \\ \dot{q}_i \end{bmatrix} \quad i = 1, 2, \ldots, n \tag{59.3}$$

Each subsystem motion equation may be presented by the following state equation

$$S_i : \begin{cases} \dot{x}_{i1} = x_{i2} \\ \dot{x}_{i2} = f_i(q_i, \dot{q}_i) + g_i(q_i)u_i + h_i(q, \dot{q}, \ddot{q}) \\ y_i = x_{i1} \end{cases} \tag{59.4}$$

where $x_i$ is the state vector of subsystem $S_i$, $y_i$ is the output of subsystem$S_i$, and

$$\begin{aligned} f_i(q_i, \dot{q}_i) &= M_i^{-1}(q_i)[-C_i(q_i, \dot{q}_i)\dot{q}_i - G_i(q_i)] \\ g_i(q_i) &= M_i^{-1}(q_i) \\ h_i(q, \dot{q}, \ddot{q}) &= -M_i^{-1}(q_i)Z_i(q, \dot{q}, \ddot{q}) \end{aligned}$$

The control objective is to design a decentralized controller which combined backstepping technique and terminal sliding mode for (59.2) to make the recon-figu-reable manipulators follow a desired trajectory, make the control signal smoother and guarantee boundness of all variables of the closed-loop system.

## 59.3 Combined Backstepping Terminal Sliding Mode Decentralized Controller Design

In this section, a combined backstepping terminal sliding mode decentralized control design approach is given. The detailed design procedures are described as followed.

**Assumption 1** The desired trajectories $y_{id}, \dot{y}_{id}$ and $\ddot{y}_{id}$ are bounded.

*Step 1:* Define the state tracking error of the subsystem as

$$\begin{cases} e_{i1} = x_{i1} - x_{i1d} \\ e_{i2} = x_{i2} - \alpha_i \end{cases} \tag{59.5}$$

where $\alpha_i$ is the desired virtual control as

$$\alpha_i = \dot{x}_{i1d} - c_i e_{i1} \tag{59.6}$$

where the $c_i$ is a positive constant. Consider (59.5), one can obtain that

$$\begin{cases} \dot{e}_{i1} = \dot{x}_{i1} - \dot{x}_{i1d} = e_{i2} - c_i e_{i1} \\ \dot{e}_{i2} = \dot{x}_{i2} - \dot{\alpha}_i = f_i(q_i, \dot{q}_i) + g_i(q_i)u_i + h_i(q, \dot{q}, \ddot{q}) - \dot{\alpha}_i \end{cases} \tag{59.7}$$

Choose following Lyapunov function for the first order of subsystem

$$V_{i1} = \frac{1}{2}e_{i1}^2 \tag{59.8}$$

Its derivative along the trajectory is

$$\dot{V}_{i1} = e_{i1}\dot{e}_{i1} = e_{i1}e_{i2} - c_i e_{i1}^2 \tag{59.9}$$

*Step 2:* Introduce the teminal sliding mode for the second order of the subsystems as

$$s_i = e_{i2} + \beta_i e_{i1}^{q_i/p_i} \tag{59.10}$$

where $p_i$ and $q_i$ are positive odds, and $1 < {q_i}/{p_i} < 2$, $\beta_i$ is also a positive constant. And its time derivative is

$$\dot{s}_i = \dot{e}_{i2} + \beta_i \frac{q_i}{p_i} e_{i1}^{q_i/p_i - 1}\dot{e}_{i1} = f_i(q_i, \dot{q}_i) + g_i(q_i)u_i + h_i(q, \dot{q}, \ddot{q}) - \dot{\alpha}_i + \beta_i \frac{q_i}{p_i} e_{i1}^{q_i/p_i - 1}\dot{e}_{i1} \tag{59.11}$$

Then use the RBF neural networks to approximate the unknown term and uncertainty term $f_i(q_i, \dot{q}_i)$, $g_i(q_i)$

$$f_i(q_i, \dot{q}_i, W_{if}) = W_{if}^T \Phi_{if}(q_i, \dot{q}_i) + \varepsilon_{if} \quad \|\varepsilon_{if}\| \le \varepsilon_1 \tag{59.12}$$

$$g_i(q_i, W_{ig}) = W_{ig}^T \Phi_{ig}(q_i) + \varepsilon_{ig} \|\varepsilon_{ig}\| \le \varepsilon_2 \tag{59.13}$$

where the $W_{if}$ and $W_{ig}$ are the ideal neural network weights, $\Phi(\cdot)$ is the neural network basis function and $\varepsilon_{if}$ and $\varepsilon_{ig}$ is the neural network approximation errors, $\varepsilon_1$ and $\varepsilon_2$ are known constants. Define $\hat{W}_{if}$ and $\hat{W}_{ig}$ as the estimations of $W_{if}$ and $W_{ig}$, respectively. $\hat{f}_i(q_i, \dot{q}_i, W_{if})$ is estimation value of $f_i(q_i, \dot{q}_i)$, and $\hat{g}_i(q_i, \hat{W}_{ig})$ is estimation value of $g_i(q_i)$. $\hat{f}_i(q_i, \dot{q}_i, W_{if})$ and $\hat{g}_i(q_i, \hat{W}_{ig})$ can be expressed as

$$\hat{f}_i(q_i, \dot{q}_i, \hat{W}_{if}) = \hat{W}_{if}^T \hat{\Phi}_{if}(q_i, \dot{q}_i) \tag{59.14}$$

$$\hat{g}_i(q_i, \hat{W}_{ig}) = \hat{W}_{ig}^T \hat{\Phi}_{ig}(q_i) \tag{59.15}$$

Define the estimation errors as $\tilde{W}_{if} = W_{if} - \hat{W}_{if}$ and $\tilde{W}_{ig} = W_{ig} - \hat{W}_{ig}$. Therefore

$$f_i(q_i, \dot{q}_i, W_{if}) - \hat{f}_i(q_i, \dot{q}_i, \hat{W}_{if}) = \tilde{W}_{if}^T \hat{\Phi}_{if}(q_i, \dot{q}_i) + W_{if}^T \tilde{\Phi}_{if}(q_i, \dot{q}_i) + \varepsilon_{if} \tag{59.16}$$

$$g_i(q_i, W_{ig}) - \hat{g}_i(q_i, \hat{W}_{ig}) = \tilde{W}_{ig}^T \hat{\Phi}_{ig}(q_i) + W_{ig}^T \tilde{\Phi}_{ig}(q_i) + \varepsilon_{ig} \tag{59.17}$$

where the neural output error with Gaussian activation $\tilde{\Phi}(\cdot)$ is given by

$$\tilde{\Phi}_{if}(q_i, \dot{q}_i) = \Phi_{if}(q_i, \dot{q}_i) - \hat{\Phi}_{if}(q_i, \dot{q}_i) \tag{59.18}$$

$$\tilde{\Phi}_{ig}(q_i) = \Phi_{ig}(q_i) - \hat{\Phi}_{ig}(q_i) \tag{59.19}$$

Next, the neural network $\hat{p}_i(|s_i|, \hat{W}_{ip})$ which can be expressed as following function is proposed to compensate the interconnection term.

$$\hat{p}_i(|s_i|, \hat{W}_{ip}) = \hat{W}_{ip}^T \hat{\Phi}_{ip}(|s_i|) \tag{59.20}$$

where $\hat{W}_{ip}^T$ is the estimation of $W_{ip}^T$, and the weight estimation error is $\tilde{W}_{ip}^T = W_{ip}^T - \hat{W}_{ip}^T$.

**Assumption 2** The interconnection term $h_i(q, \dot{q}, \ddot{q})$ is bounded by

$$|h_i(q, \dot{q}, \ddot{q})| \leq \sum_{j=1}^{n} d_{ij} S_j \tag{59.21}$$

where $d_{ij} \geq 0, S_j = 1 + |s_j| + |s_j|^2$. By defining $p_i(|s_j|) = n \max_{ij}\{d_{ij}\} S_i$. Define approximation error

$$w_{i1} = W_{if}^T \tilde{\Phi}_{if}(q_i, \dot{q}_i) + W_{ig}^T \tilde{\Phi}_{ig}(q_i) u_i + \varepsilon_{if} + \varepsilon_{ig} u_i \tag{59.22}$$

$$w_{i2} = p_i(|s_i|) - W_{ip}^T \hat{\Phi}_{ip}(|s_i|) \tag{59.23}$$

$$w_i = |w_{i1}| + |w_{i2}| \tag{59.24}$$

The decentralized controller is designed as

$$\begin{aligned} u_i = & -[\hat{f}_i(q_i, \dot{q}_i, \hat{W}_{if}) + \operatorname{sgn}(s_i)\hat{p}_i(|s_i|, \hat{W}_{ip}) - \dot{\alpha}_i \\ & + \beta_i \frac{q_i}{p_i} e_{i1}^{q_i/p_i - 1} \dot{e}_{i1} + e_{i1} + l_{i1}\operatorname{sgn}(s_i) \\ & + l_{i2} s_i] / \hat{g}_i(q_i, \hat{W}_{ig}) \end{aligned} \tag{59.25}$$

**Theorem** *Consider the subsystem* (59.2) *with assumptions* 1–2, *the combined backstepping terminal sliding mode algorithm based decentralized control scheme is designed as* (59.25) *guarantees that all the variables of the closed-loop system are bounded. With the adaptive laws as*

$$\dot{\hat{W}}_{if} = \Gamma_{if} s_i \hat{\Phi}_{if}(q_i, \dot{q}_i) \tag{59.26}$$

$$\dot{\hat{W}}_{ig} = \Gamma_{ig} s_i \hat{\Phi}_{ig}(q_i) u_i \tag{59.27}$$

$$\dot{\hat{W}}_{ip} = \Gamma_{ip} |s_i| \hat{\Phi}_{ip}(|s_i|) \tag{59.28}$$

where $\Gamma_{if}$, $\Gamma_{ig}$, $\Gamma_{ip}$ are positive constants.

*Proof* Choose following Lyapunov function for the second order of subsystem

$$V_{i2} = \frac{1}{2} s_i^2 + \frac{1}{2} \tilde{W}_{if}^T \Gamma_{if}^{-1} \tilde{W}_{if} + \frac{1}{2} \tilde{W}_{ig}^T \Gamma_{ig}^{-1} \tilde{W}_{ig} + \frac{1}{2} \tilde{W}_{ip}^T \Gamma_{ip}^{-1} \tilde{W}_{ip} \tag{59.29}$$

Its time derivative is

$$\dot{V}_{i2} = s_i\dot{s}_i - \tilde{W}_{if}^T\Gamma_{if}^{-1}\dot{\hat{W}}_{if} - \tilde{W}_{ig}^T\Gamma_{ig}^{-1}\dot{\hat{W}}_{ig} - \tilde{W}_{ip}^T\Gamma_{ip}^{-1}\dot{\hat{W}}_{ip} \tag{59.30}$$

Therefore the derivative of subsystem's Lyapunov function is

$$\begin{aligned}
\dot{V}_i &= \dot{V}_{i1} + \dot{V}_{i2} \\
&= e_{i1}e_{i2} - c_ie_{i1}^2 + s_i[f_i(q_i,\dot{q}_i) + g_i(q_i)u_i + h_i(q,\dot{q},\ddot{q}) \\
&\quad - \tilde{W}_{if}^T\Gamma_{if}^{-1}\dot{\hat{W}}_{if} - \tilde{W}_{ig}^T\Gamma_{ig}^{-1}\dot{\hat{W}}_{ig} - \tilde{W}_{ip}^T\Gamma_{ip}^{-1}\dot{\hat{W}}_{ip} \\
&= e_{i1}e_{i2} - c_ie_{i1}^2 + s_i[f_i(q_i,\dot{q}_i) + \hat{g}_i(q_i)u_i + (g_i(q_i) - \hat{g}_i(q_i))u_i + h_i(q,\dot{q},\ddot{q}) - \dot{\alpha}_i \\
&\quad + \beta_i\frac{q_i}{p_i}e_{i1}^{q_i/p_i-1}\dot{e}_{i1}] - \tilde{W}_{if}^T\Gamma_{if}^{-1}\dot{\hat{W}}_{if} - \tilde{W}_{ig}^T\Gamma_{ig}^{-1}\dot{\hat{W}}_{ig} - \tilde{W}_{ip}^T\Gamma_{ip}^{-1}\dot{\hat{W}}_{ip}
\end{aligned} \tag{59.31}$$

Substituting (59.25) into (59.31), one can obtain that

$$\begin{aligned}
\dot{V}_i &= -c_ie_{i1}^2 - \beta e_{i1}^{q_i/p_i+1} - l_{i1}|s_i| - l_{i2}s_i^2 + s_i[\tilde{W}_{if}^T\hat{\Phi}_{if}(q_i,\dot{q}_i) + \tilde{W}_{ig}^T\hat{\Phi}_{ig}(q_i)u_i + h_i \\
&\quad - \operatorname{sgn}(s_i)\hat{W}_{ip}^T\hat{\Phi}_{ip}(|s_i|) + w_{i1}] - \tilde{W}_{if}^T\Gamma_{if}^{-1}\dot{\hat{W}}_{if} - \tilde{W}_{ig}^T\Gamma_{ig}^{-1}\dot{\hat{W}}_{ig} - \tilde{W}_{ip}^T\Gamma_{ip}^{-1}\dot{\hat{W}}_{ip} \\
&= -c_ie_{i1}^2 - \beta e_{i1}^{q_i/p_i+1} - l_{i1}|s_i| - l_{i2}s_i^2 + \tilde{W}_{if}^T(s_i\hat{\Phi}_{if}(q_i,\dot{q}_i) - \Gamma_{if}^{-1}\dot{\hat{W}}_{if}) + \tilde{W}_{ig}^T(s_i\hat{\Phi}_{ig}(q_i)u_i \\
&\quad - \Gamma_{ig}^{-1}\dot{\hat{W}}_{ig}) + s_iw_{i1} + s_i(h_i - \operatorname{sgn}(s_i)\hat{W}_{ip}^T\hat{\Phi}_{ip}(|s_i|)) - \tilde{W}_{ip}^T\Gamma_{ip}^{-1}\dot{\hat{W}}_{ip}
\end{aligned} \tag{59.32}$$

From (59.26), (59.27), yields

$$\begin{aligned}
\dot{V} &\le \sum_{i=1}^{n}\left(-c_ie_{i1}^2 - \beta e_{i1}^{q_i/p_i+1} - l_{i1}|s_i| - l_{i2}s_i^2 + s_iw_{i1} + |s_i|\sum_{j=1}^{n}d_{ij}S_j - |s_i|\hat{W}_{ip}^T\hat{\Phi}_{ip}(|s_i|) - \tilde{W}_{ip}^T\Gamma_{ip}^{-1}\dot{\hat{W}}_{ip}\right) \\
&\le \sum_{i=1}^{n}\left(-c_ie_{i1}^2 - \beta e_{i1}^{q_i/p_i+1} - l_{i1}|s_i| - l_{i2}s_i^2 + s_iw_{i1} - |s_i|\hat{W}_{ip}^T\hat{\Phi}_{ip}(|s_i|) - \tilde{W}_{ip}^T\Gamma_{ip}^{-1}\dot{\hat{W}}_{ip}\right) \\
&\quad + \max_{ij}\{d_{ij}\}\sum_{j=1}^{n}|s_i|\sum_{j=1}^{n}S_j
\end{aligned} \tag{59.33}$$

Notice that $|s_i| \le |s_j| \Leftrightarrow S_i \le S_j$, using Chebyshev inequality, one can obtain that

$$\sum_{i=1}^{n}|s_i|\sum_{j=1}^{n}S_j \le n\sum_{i=1}^{n}|s_i|S_i \tag{59.34}$$

Combine (59.33) and (59.34),

$$\begin{aligned}
\dot{V} &\le \sum_{i=1}^{n}\left(-c_ie_{i1}^2 - \beta e_{i1}^{q_i/p_i+1} - l_{i1}|s_i| - l_{i2}s_i^2 + s_iw_{i1} - |s_i|\hat{W}_{ip}^T\hat{\Phi}_{ip}(|s_i|) + |s_i|n\max_{ij}\{d_{ij}\}S_i - \tilde{W}_{ip}^T\Gamma_{ip}^{-1}\dot{\hat{W}}_{ip}\right) \\
&= \sum_{i=1}^{n}\left(-c_ie_{i1}^2 - \beta e_{i1}^{q_i/p_i+1} - l_{i1}|s_i| - l_{i2}s_i^2 + s_iw_{i1} - |s_i|\hat{W}_{ip}^T\hat{\Phi}_{ip}(|s_i|) + |s_i|p_i(|s_i|) - \tilde{W}_{ip}^T\Gamma_{ip}^{-1}\dot{\hat{W}}_{ip}\right)
\end{aligned} \tag{59.35}$$

Substituting (59.20) into (59.35),

$$\dot{V} \leq \sum_{i=1}^{n} \left( -c_i e_{i1}^2 - \beta e_{i1}^{q_i/p_i+1} - l_{i1}|s_i| - l_{i2}s_i^2 + s_i w_{i1} + |s_i| w_{i2} + \tilde{W}_{ip}^T (|s_i| \hat{\Phi}_{ip}(|s_i|) - \Gamma_{ip}^{-1} \dot{\hat{W}}_{ip}) \right) \tag{59.36}$$

From (59.24) and (59.28),

$$\dot{V} \leq \sum_{i=1}^{n} \left( -c_i e_{i1}^2 - \beta e_{i1}^{q_i/p_i+1} - l_{i1}|s_i| - l_{i2}s_i^2 + |s_i| w_i \right) \tag{59.37}$$

Denoting $c = [c_1, c_2, \ldots, c_n]^T, \beta = [\beta_1, \beta_2, \ldots, \beta_n]^T, s = [s_1, s_2, \ldots, s_n]^T, p = [p_1, p_2, \ldots, p_n]^T$, $q = [q_1, q_2, \ldots, q_n]^T$, $l_1 = [l_{11}, l_{12}, \ldots, l_{1n}]^T$, $l_2 = [l_{21}, l_{22}, \ldots, l_{2n}]^T$ and $w = [w_1, w_2, \ldots, w_n]^T$, yields

$$\dot{V} = -ce_1^2 - \beta e_{i1}^{q_i/p_i+1} - |s|(l_1 + l_2|s| - w) \tag{59.38}$$

It is clear that the time derivative of $V$ is negative outside a compact set $\Omega_s$ defined by

$$\Omega_s = \left\{ s : 0 \leq |s| \leq \frac{w - l_1}{l_2} \right\} \tag{59.39}$$

## 59.4 Simulation Results

In order to verify the effectiveness and reliability of the proposed decentralized control scheme, two different configurations of reconfigurable manipulators as shown in Fig. 59.1 are simulated, whose dynamic model can be described by (59.2).

The dynamic model of the configuration a is defined by

$$M(q) = \begin{bmatrix} 0.36\cos(q_2) + 0.6066 & 0.18\cos(q_2) + 0.1233 \\ 0.18\cos(q_2) + 0.1233 & 0.1233 \end{bmatrix}$$

$$C(q, \dot{q}) = \begin{bmatrix} -0.36\sin(q_2)\dot{q}_2 & -0.18\sin(q_2)\dot{q}_2 \\ 0.18\sin(q_2)(\dot{q}_1 - \dot{q}_2) & 0.18\sin(q_2)\dot{q}_1 \end{bmatrix}$$

$$G(q) = \begin{bmatrix} -5.88\ \sin(q_1 + q_2) - 17.64\ \sin(q_1) \\ -5.88\ \sin(q_1 + q_2) \end{bmatrix}$$

The desired trajectories of joint modules are $q_d = [q_{1d}, q_{2d}]^T$ with $q_{1d} = 0.2\sin(3t) + 0.1\cos(4t)$ and $q_{2d} = 0.3\sin(2t) + 0.2\cos(t)$. The initial positions of joint modules are $q_1(0) = q_2(0) = 1$, and the initial velocities of joints are zeros.

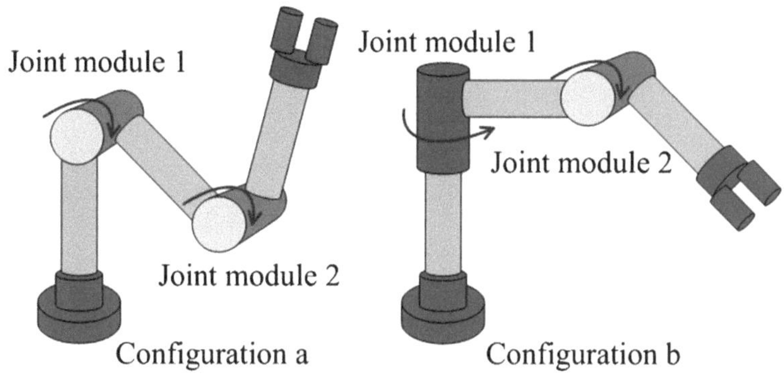


**Fig. 59.1** Configurations for simulation

For each subsystem, the RBF neural networks are used to approximate $f_i(q_i, \dot{q}_i)$, $g_i(q_i)$ and $h_i(q, \dot{q}, \ddot{q})$. The neural network Gaussian basis function is as:

$$\Phi_j = \exp(-\|X - c_j\|^2/2\sigma^2),\ j = 1, 2, \ldots, m \tag{59.40}$$

Where $c_j, b_j$ are the radial basis function neural networks centers and widths.

According to the proposed scheme in [7], the linear sliding mode and the combined decentralized controller for reconfigurable manipulators should be

$$s_i = e_{i2} + k_i e_{i1} \tag{59.41}$$

$$\begin{aligned} u_i = & -[\hat{f}_i(q_i, \dot{q}_i, \hat{W}_{if}) + \text{sgn}(s_i)\hat{p}_i(|s_i|, \hat{W}_{ip}) - \dot{\alpha}_i + k_i \dot{e}_{i1} + \frac{1}{k_i} e_{i2} \\ & + l_{i1}\text{sgn}(s_i) + l_{i2}s_i]/\hat{g}_i(q_i, \hat{W}_{ig}) \end{aligned} \tag{59.42}$$

The controller parameters are taken as $c = [8, 6]^T$, $k = [5, 1]^T$, $l_1 = [8, 8]^T$, $l_2 = [5, 5]^T$, $\Gamma_{if} = 0.002$, $\Gamma_{ig} = 0.002$, $\Gamma_{if} = 500$. The tracking performances (TP) of two joints are shown as Fig. 59.2, the tracking errors appear at the beginning of process due to the lack of knowledge about the dynamical model of subsystem. Nearly one second, the actual trajectories and the desired trajectories almost overlap with one another. Thus, the tracking performance is satisfied. However, the control torque (CT) has serious chatter, so we employ terminal sliding mode to improve it.

The control law with terminal sliding mode as (59.25), and the controller parameters are taken as $\Gamma_{if} = 0.002$, $\Gamma_{ig} = 0.002$, $\Gamma_{if} = 500$, $c = [20, 10]^T$, $\beta = [4, 4]^T$, $l_1 = [0.001, 10]^T$, $l_2 = [0.001, 10]^T$, $p = [3, 5]^T$, $q = [5, 7]^T$. Figure 59.3 show high tracking performance and the CT is greatly improved. Therefore the simulation results illustrate the effectiveness of the proposed decentralized control.

Now consider configuration b, the dynamic model of the configuration a is defined by

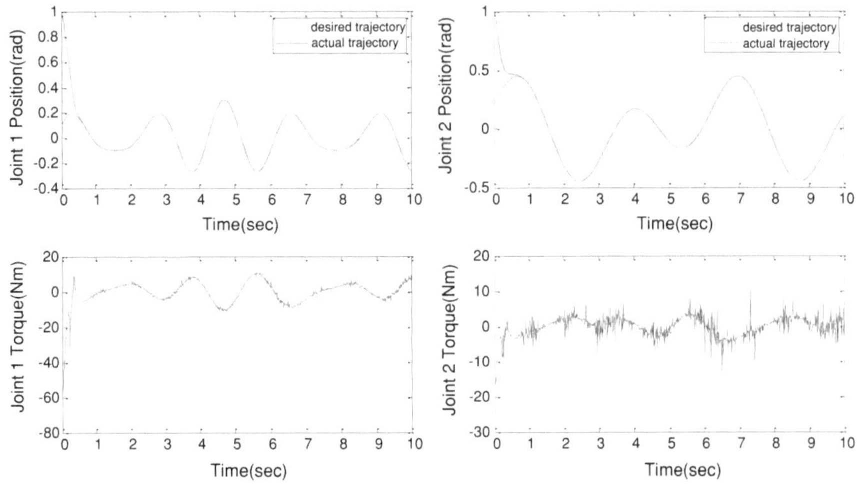


**Fig. 59.2** TP and CT with linear sliding mode of configuration a

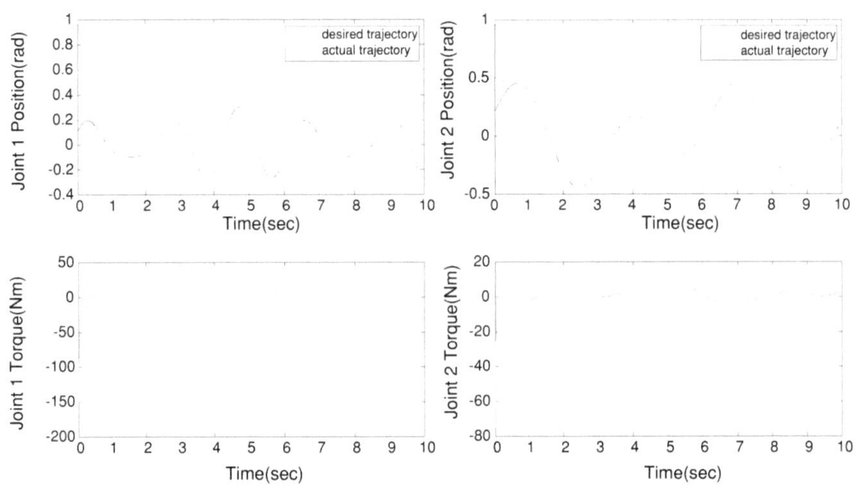


**Fig. 59.3** TP and CT with terminal sliding mode of configuration a

$$M(q) = \begin{bmatrix} 0.17 - 0.1166\cos^2(q_2) & -0.06\cos(q_2) \\ -0.06\cos(q_2) & 0.1233 \end{bmatrix}$$

$$C(q,\dot{q}) = \begin{bmatrix} 0.1166\sin(2q_2)\dot{q}_2 & 0.06\sin(q_2)\dot{q}_2 \\ 0.06\sin(q_2)\dot{q}_2 - 0.0583\sin(2q_2)\dot{q}_1 & -0.06\sin(q_2)\dot{q}_1 \end{bmatrix}$$

$$G(q) = \begin{bmatrix} 0 \\ -5.88\ \cos(q_2) \end{bmatrix}$$

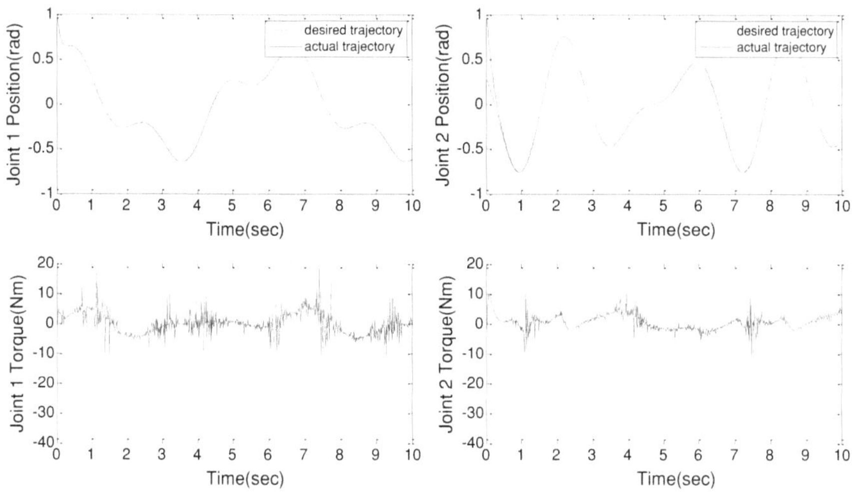


**Fig. 59.4** TP and CT with linear sliding mode of configuration b

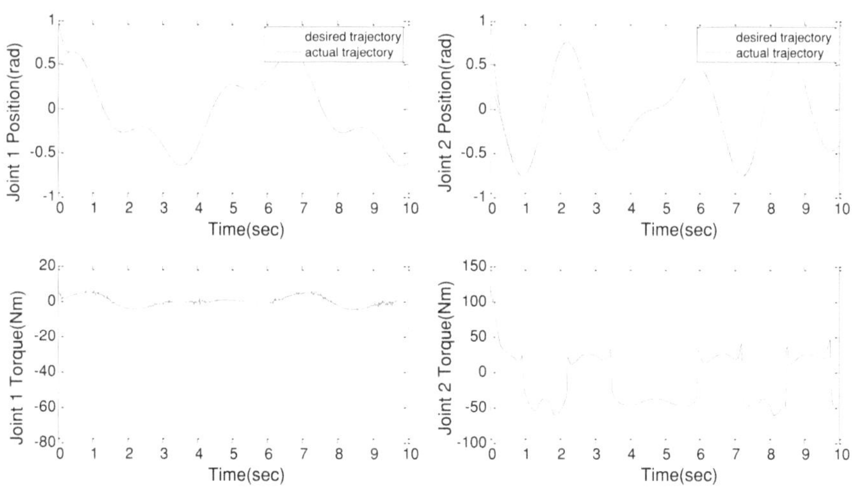


**Fig. 59.5** TP and CT with terminal sliding mode of configuration b

$q_{1d} = 0.5\cos(t) + 0.2\sin(3t)$ and $q_{2d} = 0.3\cos(3t) - 0.5\sin(2t)$ are desired trajectories of configuration b, other parameters are the same as configuration a. Figures 59.4 and 59.5 show the simulation results of configuration b under the controller with linear sliding mode and terminal sliding mode respectively. And the same conclusion is obtained.

## 59.5 Conclusion

In this paper, a stable combined backstepping terminal sliding mode based decentralized controller for reconfigurable manipulators is presented. Application of backstepping technique simplifies the controller design, guarantees globally asymptotically stable of the closed-loop system. Then a terminal sliding mode which can smooth the control signal is used. The unknown terms and interconnection term are approximated by neural networks whose weights are updated with adaptive laws. The comparing simulation verified that the controller with terminal sliding mode not only makes the control signal smoother, but also improves the convergence rate and the tracking accuracy.

**Acknowledgments** This work is supported by National Natural Science Foundation of China under Grant 60974010 and the Scientific and Technological Development Plan Project in Jilin Province of China under Grant 20110705.

## References

1. Paredis CJJ, Brown HB, Khosla PK (1997) A rapidly deployable manipulator system. Robot Auton Syst 21:289–304
2. Li Y, Zhu M, Li Y (2007) Neurofuzzy compensation control for reconfigurable manipulator. J Jilin University (Eng Technol Edn) 37:206–211
3. Li Y, Zhu MC, Li YC (2007) Simulation on robust neurofuzzy compensator for reconfigurable manipulator motion control. J Syst Simul 19:5169–5174
4. Zhu M, Li Y, Li Y (2007) Distributed adaptive sliding mode control of modular and reconfigurable robots. In: Proceedings of the 2007 IEEE international conference on mechatronics and automation, IEEE Press, New York, pp 2622–2627
5. Liu G, Abdul S, Goldenberg AA (2008) Distributed control of modular and reconfigurable robot with torque sensing. Robotica 26:75–84
6. Zhu M, Li Y (2010) Decentralized adaptive fuzzy sliding mode control for reconfigurable modular manipulators. Int J Robust Nonlin 20:472–488
7. Lin Z, Duan G, Song S (2009) Backstepping adaptive sliding mode control for horizontal underactuated manipulators. J Robot 31:131–136
8. Zhu L, Dong B, Zhao B, Li Y (2011) Decentralized adaptive iterative learning control for reconfigurable manipulators. In: Proceedings of the 30th Chinese control conference, IEEE Computer Society, Washington, DC, pp 3990–3995
9. Lin FJ, Shen PH, Hsu SP (2002) Adaptive backstepping sliding mode control for linear induction motor drive. IEE proc: Electr power appli 149:184–194
10. Swaroop D, Hedrick JK, Yip PP, Gerdes JC (2000) Dynamic surface control for a class of nonlinear systems. IEEE Trans Automat Control 45:1893–1899
11. Liu Y, Li Y (2005) Sliding mode adaptive neural-network control for nonholonomic mobile modular manipulators. J Intell Robot Syst 44:230–224
12. Li Z, Changsheng J, Yanli D (2009) A robust and adaptive terminal sliding mode control based on backstepping. J Control Theory Appl 26:678–682

# Chapter 60
# Underactuated Gripper That Is Able to Convert from Precision to Power Grasp by a Variable Transmission Ratio

**Stefan A. J. Spanjer, Ravi Balasubramanian, Aaron M. Dollar and Just L. Herder**

**Abstract** This paper investigates the possibility of converting grasped objects from precision grasps to power grasps using a variable transmission ratio for an underactuated finger. Reconfiguration happens when the precision grasp converts to a power grasp, because the number of contact points changes which changes the topology of the grasp. To this effect, a variable radius pulley was designed. A simulation study is presented to analyse grasping behaviour and a potential energy method is used to predict the equilibrium positions of finger and object. With this method stable and unstable equilibrium positions are determined. This is followed by an experiment to verify the theory. This paper is a first step in dextrous manipulation with large movements of objects using underactuated fingers.

**Keywords** Underactuated · Convert · Precision grasp · Power grasp · Variable transmission ratio

---

S. A. J. Spanjer (✉) · J. L. Herder
Laboratory of Mechanical Automation and Mechatronics, Faculty of Engineering Technology, University of Twente, Enschede, The Netherlands
e-mail: s.a.j.spanjer@student.utwente.nl

J. L. Herder
e-mail: j.l.herder@utwente.nl

R. Balasubramanian
School of Mechanical, Industrial & Manufacturing Engineering, Oregon State University, Corvallis, OR, USA
e-mail: ravi.balasubramanian@oregonstate.edu

A. M. Dollar
Department of Mechanical Engineering and Materials Science, School of Engineering and Applied Science, Yale University, New Haven, CT, USA
e-mail: aaron.dollar@yale.edu

J. S. Dai et al. (eds.), *Advances in Reconfigurable Mechanisms and Robots I*,
DOI: 10.1007/978-1-4471-4141-9_60, 

## 60.1 Introduction

There exists much research on underactuated robotic hands that show adaptive behavior and therefore these are able to pick up different objects [1–4]. Some underactuated hands are able to pick up objects with an enveloping grasp (power grasp) or with a precision grasp depending on the position of the object with respect to the hand and geometry parameters of the hand [2, 5]. With the so-called equilibrium point method the type of grasp can be easily predicted [2].

For any grasp type, robustness is an important aspect. Robustness is defined here as the ability to resist external forces [6].

The precision grasp is necessary for a successful grasp when objects are small or hard to pick up. The robustness of a precision grasp is less than the robustness of a power grasp, because the number of contact points is less. In [7] the robustness of the precision grasp is improved by actively varying the transmission ratio of an underactuated finger after an object is grasped. The transmission ratio is changed such that the required friction force between finger and object is reduced to zero. This way, the equilibrium margin for unknown disturbances is maximized. Still, for high accelerations of hand and object the power grasp is preferred because of its greater robustness. Ideally a hand should be able to convert from precision grasp to power grasp after an object is grasped, because then both small and large objects can be picked up and the desired robustness can be achieved.

The objective of this paper is to present a way to convert from precision to power grasp. This is achieved by varying the transmission ratio in an underactuated finger after an object is grasped (Fig. 60.1), which leads to motion of the finger and object. The principle of the variable transmission ratio is the same as in [7], but now this ratio is changed in order to break equilibrium and move the object into the hand: by changing the radius of the proximal pulley of a finger with a tendon-pulley mechanism the equilibrium point can be controllably shifted outwards (Fig. 60.1). As a consequence the distal link rotates counterclockwise and brings the object into the palm of the hand until equilibrium is reached again. For simplicity, we assume symmetry between the two fingers that hold the object. The object is able to move freely in y-direction while the x-direction is prohibited. The principle of potential energy is used to estimate stable and unstable positions of the object in the hand [8]. Therefore, the theory used in this paper is also applicable to other mechanisms for underactuation, for example four bar mechanisms.

The structure of the paper is as follows. We describe the model of grasp transition and the experiment set-up to validate this model in Sects. 60.2 and 60.3, respectively. In Sect. 60.4 we describe the simulation results and show with the experiment that the finger converts from precision to power grasp. In Sects. 60.5 and 60.6 the discussion and conclusion follow respectively.

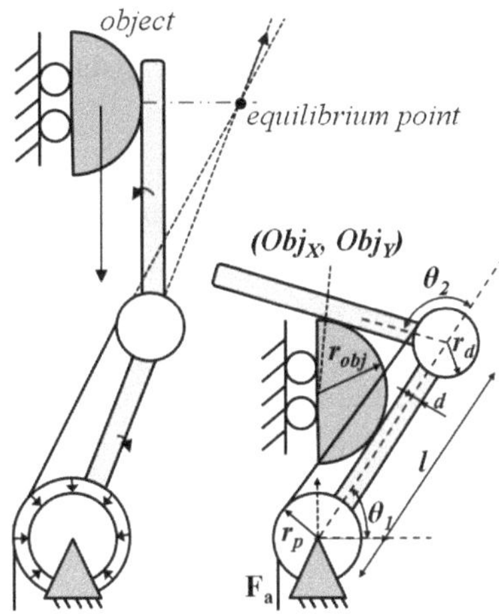


**Fig. 60.1** Transition from precision to power grasp

## 60.2 Model of Grasp Transition

This section consists of three parts: background about the equilibrium point, a simulation model of the grasping behavior and the potential energy method.

In the model the object is fixed: for different fixed positions the equilibrium of the finger and subsequently the potential energy are estimated. The derivative of the potential energy is equal to the resultant force on the object. This force is zero at points where the derivative is zero. Therefore these points are the positions of the object where the object and finger are in force equilibrium in case the object is not fixed. In the experiment the object is able to move in y-direction (Figs. 60.1 and 60.3) and the force equilibrium positions along this line are validated.

The finger model is based on the model in [7] and consists of a finger and an object at different fixed positions. The finger is tendon driven and consists of two phalanges with length $l = 0.1$ m and thickness $2d = 0.015$ m, a proximal pulley with adjustable radius $r_p$ and a distal pulley with radius $r_d = 0.01$ m (Fig. 60.1). It has two torsional springs at the joints with stiffnesses $K_p = 0.001$Nm and $K_d = 0.005$ Nm and its rest position angles are $\theta_{1,ini} = 25^\circ$ and $\theta_{2,ini} = 45^\circ$, respectively [9]. The object is a cylinder with radius $r_{obj} = 1.5$ cm and its position is given by $Obj = \left(x_{obj}, y_{obj}\right)^T$.

### 60.2.1 Equilibrium Point

In Fig. 60.1 the equilibrium point is drawn. This is the intersection of the tendon line and the proximal phalanx line. The sliding direction and rotation of the distal phalanx on the object can be easily verified with use of the equilibrium point. If the contact force produces a clockwise moment around the equilibrium point, then the distal phalanx will rotate clockwise and vice versa. There is equilibrium if the contact force points at the equilibrium point. In this paper the projection of the equilibrium point on the distal phalanx is called the projected equilibrium point.

## 60.2.2 Grasping Process

The finger contacts the object initially with the proximal or distal phalanx. Subsequently the finger slides along the object until equilibrium is reached or contact is lost. In this section the grasping process is described. Friction is not included in the model.

### 60.2.2.1 Identifying Contact Mode

Proximal Contact

The following equations are used in case the proximal link makes first contact with the object:

$$a_1\left\{\begin{array}{c}\cos\left(\theta_{1,ini}+\delta\theta_1\right)\\ \sin\left(\theta_{1,ini}+\delta\theta_1\right)\end{array}\right\}-\left(r_{obj}+d\right)\left\{\begin{array}{c}\cos\left(\theta_{1,ini}+\delta\theta_1-\pi/2\right)\\ \sin\left(\theta_{1,ini}+\delta\theta_1-\pi/2\right)\end{array}\right\}-\left\{\begin{array}{c}x_{obj}\\ y_{obj}\end{array}\right\}=\mathbf{0} \tag{60.1}$$

where $\delta\theta_1$ is the angle between object and proximal link and $a_1$ is the contact position on the proximal link. The required actuation force to make contact is calculated as follows:

$$F_a=\frac{K_p\delta\theta_1}{r_p} \tag{60.2}$$

After the proximal phalanx has made contact with the object, the additional actuation force is also calculated with (60.2) where $\delta\theta_1$ and $r_p$ are replaced by the angle between object and distal phalanx and $r_d$ respectively. This also results in a contact force exerted by the object on the proximal link $F_{c,p}$. The force exerted by the object on the distal link $F_{c,d}$ is equal to zero, because the distal phalanx just touches the object. Now the finger has contact with the object at two points. The equations used for a further increase of the actuation force are described in Sect. 60.2.2.3.

Distal Contact

If the distal link makes contact first with the object instead of the proximal link, then the following equations are used:

$$\delta\theta_1-\delta\theta_2\frac{K_d}{K_p}\frac{r_p}{r_d}=0 \tag{60.3}$$

$$l\begin{Bmatrix} \cos(\theta_{1,ini}+\delta\theta_1) \\ \sin(\theta_{1,ini}+\delta\theta_1) \end{Bmatrix} + a_2\begin{Bmatrix} \cos(\theta_{12,ini}+\delta\theta_{12}) \\ \sin(\theta_{12,ini}+\delta\theta_{12}) \end{Bmatrix} - (r_{obj}+d)\begin{Bmatrix} \cos(\theta_{12,ini}+\delta\theta_{12}-\pi/2) \\ \sin(\theta_{12,ini}+\delta\theta_{12}-\pi/2) \end{Bmatrix} - \mathbf{Obj} = \mathbf{0} \tag{60.4}$$

where $\theta_{12,ini} = \theta_{1,ini} + \theta_{2,ini}$ and $\delta\theta_{12} = \delta\theta_1 + \delta\theta_2$ and $a_2$ is the contact location on the distal link. The required actuation force to make contact is calculated with (60.2). The equations used for a further increase of the actuations force are described in Sect. 60.2.2.3.

#### 60.2.2.2 Contact Kinematics

The contact constraints are derived according to [10]. From analysis it appears that for double contact (proximal and distal) the constraints result in $\Delta a_2 = \Delta\theta_1 = \Delta\theta_2 = 0$. The contact constraints for single contact are as follows:

$$-\Delta a_2 - \frac{\Delta\theta_1(r_{obj}+d)l\cos(\theta_2)}{a_2} - \Delta\theta_1 l\sin(\theta_2) = 0 \tag{60.5}$$

$$-\Delta\theta_2 - \frac{\Delta\theta_1 l\cos(\theta_2)}{a_2} - \Delta\theta_1 = 0 \tag{60.6}$$

$d$ is the thickness of the finger and $\Delta a$ is the difference in contact location on the distal link.

#### 60.2.2.3 Static Equilibrium During Contact

Distal Contact only

The quasi-static method is equivalent to the quasi-static procedure in [7]. The following equations are used:

$$-\boldsymbol{J}_c^T\boldsymbol{\Delta F}_c - \boldsymbol{K\Delta\theta} + \boldsymbol{J}_a^T\Delta F_a = 0 \tag{60.7}$$

$\boldsymbol{K} = diag(K_p, K_d)$ is the stiffness matrix and the matrix $\boldsymbol{J_a} = (r_p, r_d)$ maps an increase in the actuation force $\Delta F_a$ to joint torques. The force exerted by the object on the finger increases with $\boldsymbol{\Delta F_c} = (\Delta F_{c,p}, \Delta F_{c,d})^T$. The matrix $\boldsymbol{J_c}$ transfers the contact forces to joint torques [2]:

$$\boldsymbol{J_c} = \begin{bmatrix} a_1 & 0 \\ a_2 + l\cos\theta_2 & a_2 \end{bmatrix} \tag{60.8}$$

$varvec\Delta\theta = (\Delta\theta_1, \Delta\theta_2)^T$ is the vector containing the difference in angles as a result of an increment of the actuation force $\Delta F_a$. Because there is only contact with the distal phalanx, the increment of the force on the proximal link $\Delta F_{c,p}$ is equal to zero (the total force $F_{c,p}$ is also equal to zero). Together with (60.7) the contact constraint Eqs. (60.5) and (60.6) are used in the quasi-static approach. After every iteration $a_2$, $\theta_1$, $\theta_2$ and $J_c$ are updated. The iteration procedure is stopped when the increments $\Delta a_2$, $\Delta\theta_1$ and $\Delta\theta_2$ are very small which means that the finger has reached an equilibrium state or the procedure is stopped when contact is lost $(a_2 > l)$.

The equilibrium point can be used to investigate how the finger will slide along the object. If $F_{c,d}$ produces a clockwise moment about the equilibrium point, then the distal phalanx rotates clockwise and slides until $F_{c,d}$ points in the direction of the equilibrium point or until contact is lost. This is also true in case $F_{c,d}$ produces a counterclockwise moment, except that the finger rotates counterclockwise.

Two Point Contact (Proximal and Distal)

A quasi-static method is used to calculate the contact forces as a consequence of a further increase of the actuation force. Eq. (60.7) is used again, but now without the product $\mathbf{K}\boldsymbol{\Delta\theta}$. The unknowns are $\Delta F_{c,p}$ and $\Delta F_{c,d}$. After every iteration the total forces $F_{c,p}$ and $F_{c,d}$ are updated. There are 2 options: $F_{c,p}$ increases (stable) or decreases (unstable). The finger does not move as long as $F_{c,p}$ is positive. The quasi-static procedure is stopped when $F_{c,p}$ is zero or $F_{c,p}$ and $F_{c,d}$ are sufficiently large. When $F_{c,p}$ is zero and $F_a$ is increased, the proximal phalanx loses contact with the object and the finger shows caging behavior [11]: the proximal phalanx rotates clockwise and the distal phalanx rotates counterclockwise. Single contact by the distal link is already modeled in Sect. 60.2.2.3: Eq. (60.7) is used in combination with the contact constraints in (60.5) and (60.6).

The equilibrium point can also be used to see if caging behavior will occur [2]. If $F_{c,d}$ produces a clockwise moment around the equilibrium point, then the finger is in a stable mode (Fig. 60.2). Caging behavior occurs when $F_{c,d}$ produces a counterclockwise moment around the equilibrium point.

### *60.2.3 Potential Energy*

The gradient of the potential energy is equal to the resultant force on the object. The partial derivative of the potential energy with respect to degree of freedom $y$ is the force in $y$ -direction. With the potential energy method the potential energy as a function of the object position can be obtained. For every object position the model is simulated to determine the equilibrium state of the finger. Every time the potential energy of this equilibrium is calculated:

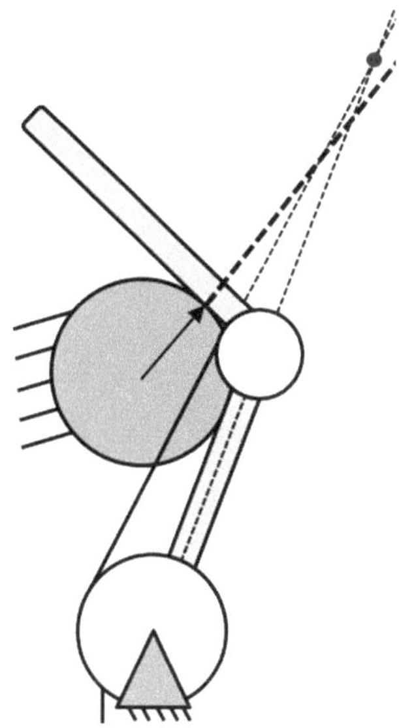

**Fig. 60.2** Unstable mode: force direction leads to caging

$$E_P = F_a \Delta h + \frac{1}{2} K_p \left(\theta_1 - \theta_{1,ini}\right)^2 + \frac{1}{2} K_d \left(\theta_2 - \theta_{2,ini}\right)^2 \tag{60.9}$$

The total actuation force $F_a$ is considered constant for different positions. $\theta_{1,ini}$ and $\theta_{2,ini}$ are the rest position angles at the proximal and distal joint respectively; at this position the torsion springs perform no work. $\Delta h$ is the change in tendon length:

$$\Delta h = r_p \left(\theta_1 - \theta_{1,ini}\right) + r_d \left(\theta_2 - \theta_{2,ini}\right) \tag{60.10}$$

For every object position the potential energy is calculated. The potential energy can be displayed as a function of the y-position of the object. Different energy curves can be obtained for different radii of the proximal pulley.

## 60.3 Experiment Set-Up

An experiment was set up to verify that the finger and object can convert from precision to power grasp by changing the transmission ratio. The experiment was also used to validate the end position of the object which is equal to the position of the minimum of the potential energy curve. The parameters are the same as in the model, only the joint stiffnesses are different: $K_p = 0.015$ Nm and $K_d = 0.048$ Nm.

The setup of the experiment is shown in Fig. 60.3. The same acrylic finger is used as in [7] where music-wire springs are mounted at the proximal and distal joint. The constant actuation force is supplied by a 1 kg weight.

Fig. 60.3 Experiment set up

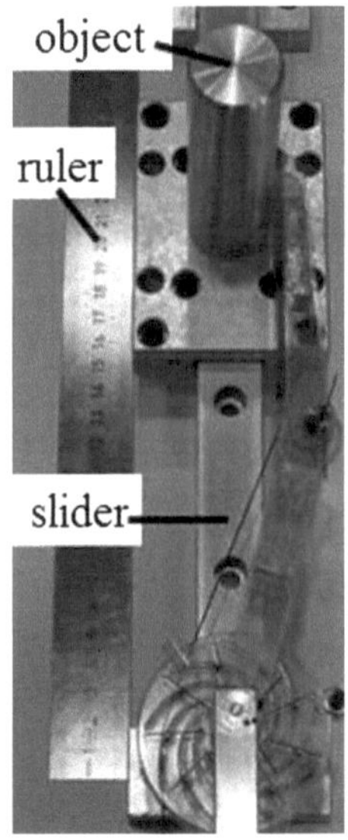


Fig. 60.4 Variable radius

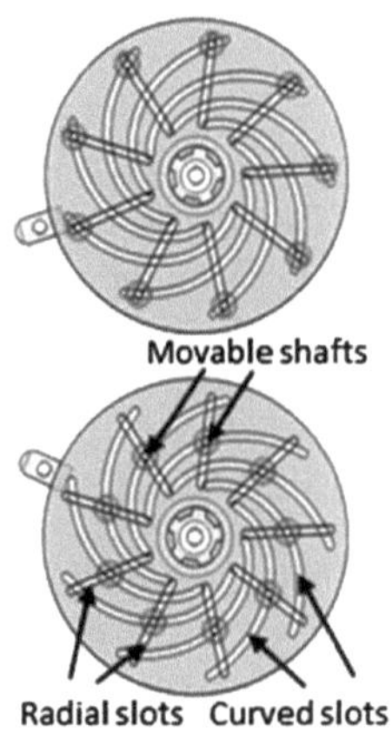


The radius of the proximal pulley is adjustable by rotating a top and bottom plate with respect to one another (Fig. 60.4). Shafts around which the tendon is routed (i.e. the effective pulley) are located between the straight slots in the bottom plate and the curved slots in the top plate. As the plates rotate with respect to one another, the shafts move radially, changing the effective pulley radius. On the shafts small pulleys are mounted. The logarithmic spiral slots are designed to have a relatively large angle with the radial slots such that friction for adjustment of the pins is low but high enough to make the adjustment system nonbackdrivable. The whole system is connected to the proximal shaft with bearings, so it can rotate freely.

The start position of the finger in the experiment is the precision mode: the finger touches the object by the distal link and is in an equilibrium state (Fig. 60.3). The y-position of the object depends on the radius of the proximal pulley, so that the

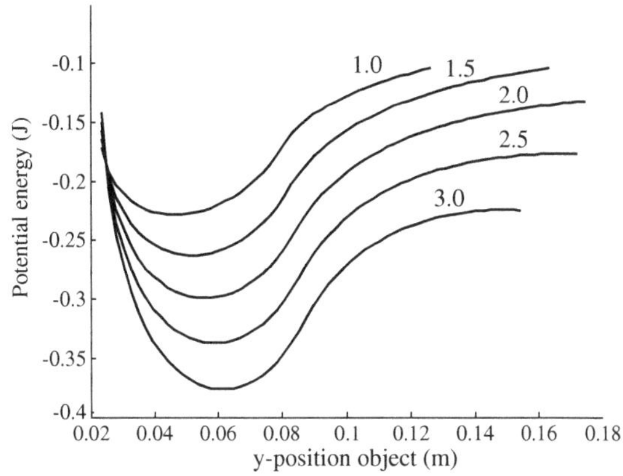


**Fig. 60.5** Potential energy curves for different radii (cm)

object is located at the level of the equilibrium point. In the next step the radius is lowered until the object starts to move. The end position of the object will be compared with the expected minimum of the energy curve of that radius.

## 60.4 Results

### *60.4.1 Simulation Results*

Figure 60.5 shows potential energy curves for different radii of the proximal pulley. In case a curve has no extreme, it was not possible to establish contact. It is clear that the curve with $r_p = 0.03$ m has a minimum and a maximum. The stable minimum is the position of the object when finger and object are in power grasp mode. Small disturbances will lead to other positions, but the object will return at the point with minimum energy. The same curve shows also an unstable maximum at $y = 0.14$ m. When only the distal phalanx has contact with the object, equilibrium will be reached when the contact force does not have a component in y-direction. When the radius is changed, the derivative at $y = 0.14$ m deviates from zero which means a force develops which tends to move the object. Decrease of the radius results in a more distal location of the equilibrium point. As a result, the distal phalanx rotates into a power grasp.

### *60.4.2 Experimental Results*

From the experiment it is clear that the object moves to the proximal phalanx. Figure 60.6 shows different trials of the experiment. The start- and end positions of the object together with the start- and end radii of the proximal pulley are depicted. The expected start- and end positions from the simulation results are also shown.

## 60.5 Discussion

The potential energy method is a simple method to estimate stable and unstable equilibrium positions. Simulation results predict equilibrium positions well.

In Fig. 60.6 the differences between simulation and experiment in precision grasp (average of 19 mm) are larger than the differences in power grasp (average of 3.4 mm) for a finger with link length 100 mm and an object with radius 15 mm. This is probably due to friction, which is not modeled and has more influence in a precision grasp.

In finger object interaction without friction, the equilibrium of the precision grasp is lost for an infinitesimal small proximal radius change. When friction is included, the equilibrium point has to be shifted outside the friction cone. The required radius change to convert from precision to power grasp increases for a larger coefficient of friction between finger and object.

The principle of conversion between precision and power grasp is analyzed and experimented with one finger. Because of symmetry, we assume the theory is also valid for a gripper with two identical fingers. Then the object is also able to move in x-direction and this could lead to asymmetric grasps. The result of radii change for an asymmetric precision grasp is that the fingers convert at different times, because the equilibrium points and force directions per finger are different. When the first finger converts, the object moves and the other finger(s) follow.

The underactuated finger needs an extra actuator for the radius change of the pulley. An advantage of this with respect to fully actuated hands is that the finger shows adaptive behavior and no sensors are needed in the fingers to measure the applied forces.

The potential energy method and the theory of transition between precision and power grasp is demonstrated for a finger with a pulley-tendon mechanism. These methods are also applicable for other types of underactuated fingers, e.g. the equilibrium point of a four bar mechanism could be shifted through a slider that has replaced one of the bars.

From Fig. 60.5 it appears that the minimum of the potential energy curve with a radius of 10 mm is at the same level as the maximum of the curve with a radius of 30 mm. A higher radius results in an even lower maximum. Other actuation forces also result in different curves. In this paper we investigated the method to convert from precision to power, but maybe it is also possible to convert from power to precision. It would be very interesting to investigate this dexterous manipulation and to see if it is possible to reposition the object at every position in the hand.

The radii of the pulleys in Hirose's two-fingered soft gripper with 10 phalanges per finger are designed such that objects can be grasped with uniform pressure along the whole finger [1]. If we replace every pulley by a pulley with variable radius in the soft gripper, it may be possible to pick an object with the gripper and move it all the way back to the proximal link by controlling the different radii.

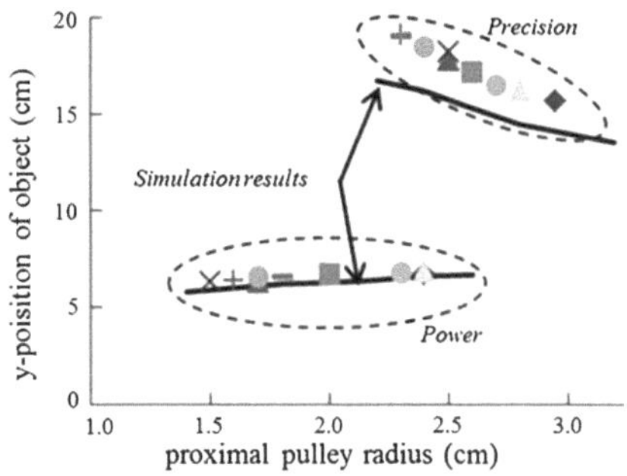


**Fig. 60.6** Experimental results

## 60.6 Conclusion

In this paper we showed an underactuated finger that is able to convert from precision to power grasp by a change of the transmission ratio. With the potential energy method the start position (precision grasp) and the end position (power grasp) of the object are predicted and these are verified with an experiment. The differences between simulations and experiment are about 19 mm for precision grasps and 3.4 mm for power grasps for a finger with link length 100 mm and an object with radius 15 mm.

## References

1. Hirose S, Umetani Y (1978) The development of soft gripper for the versatile robot hand. Mech Mach Theory 13:351–359
2. Birglen L, Laliberté T, Gosselin C (2008) Underactuated robotic hands. Springer, Berlin
3. Dollar AM, Howe RD (2010) The highly adaptive SDM hand: design and performance evaluation. Int J Robotics Res 29(5):585–597
4. Meijneke C, Kragten GA, and Wisse M (2011) Design and performance assessment of an underactuated hand for industrial applications. IFToMM/ASME International workshop on underactuated grasping, pp 9–15
5. Delft hand 3, http://compliantmechanisms.3me.tudelft.nl/mw/index.php/Video_DH3
6. Kragten GA, Herder JL (2010) The ability of underactuated hands to grasp and hold objects. Mech Mach Theory 45:408–425
7. Spanjer SAJ (2011) Improved grasp robustness through variable transmission ratios in underactuated fingers. MSc Dissertation, University of Twente
8. Kragten GA (2011) Underactuated hands: fundamentals, performance analysis and design. PhD Dissertation, Delft University of Technology
9. Dollar AM, Howe RD (2005) Towards grasping in unstructured environments: grasper compliance and configuration optimization. Adv Robotics 19(5):523–543
10. Montana DJ (1988) The kinematics of contact and grasp. Int J Robotics Res 7(3): 17–32
11. Balasubramanian R, Belter JT,and Dollar AM (2010) External disturbances and coupling mechanisms in underactuated hands. International design engineering technical conferences and computers and information in engineering conference

# Chapter 61
# Transparency Analysis of a Force Sensorless Master-Slave Control by Force Feedback Based Virtual Impedance Controller with Time Delay

**Ryosuke Horie, Kiyotoshi Komuta and Toshiyuki Murakami**

**Abstract** In this paper, an analysis of a master-slave control by force feedback based virtual impedance controller with time delay is conducted. Especially, the relationship between time delay and transparency is focused on. The proposed method consists of force feedback based virtual impedance controller of master manipulator and perfect tracking position controller of slave manipulator. From the controller analysis, it is confirmed that the relationship of "Reproducibility" and "Operationality" is simple in the proposed method and this makes clear parameter design of the controller as compared with conventional method. This is one of the remarkable points. Moreover, the proposed controller can be flexibly applied to many kinds of time-delayed teleoperation systems because models of a slave robot and communication time delay are not required. The validity of proposed controller is verified by some simulations.

**Keywords** Master-slave system · Force feedback · Virtual impedance control · Time delay · Reproducibility · Operationality

R. Horie (✉) · K. Komuta · T. Murakami
Department of System Design Engineering, Keio University, Yokohama, Japan
e-mail: ryosuke@sum.sd.keio.ac.jp

K. Komuta
e-mail: komuta@sum.sd.keio.ac.jp

T. Murakami
e-mail: mura@sd.keio.ac.jp

J. S. Dai et al. (eds.), *Advances in Reconfigurable Mechanisms and Robots I*,
DOI: 10.1007/978-1-4471-4141-9_61, © Springer-Verlag London 2012

## 61.1 Introduction

Haptics is one of key technologies to develop a new market in robot applications. Many researchers have investigated haptics in recent years. Bilateral teleoperation with master and slave robot is one of the haptic technologies. Bilateral control enables human operators to feel haptic information occured at remote environment. This technique is expected to be used in many situations such as space, surgery, nuclear reactor, tele-communication, and so on. Various researches concerning bilateral control have been done in last several decades. Hannaford constructed the ideal relationship between master and slave system based on hybrid matrix [1] and this relationship is formulated as "Transparency" [2]. Then, acceleration control on bilateral teleoperation is achieved by disturbance observer (DOB) [3, 4]. This enables to improve robustness and transparency of bilateral control. In addition, reaction torque observer (RTOB) has been implemented to enable force feedback without force/torque sensors [5]. In recent years, bilateral teleoperation has applied in multiple ways, for example, multi-degree-of-freedom (MDOF) bilateral system [6], bilateral teleoperation using two-wheel mobile manipulator [7] and so on. Also, evaluation indices are defined in order to evaluate the controller quantitatively in bilateral control systems [8]. The target goals of bilateral teleoperation are considered as following two points. One is a reproduction of environmental impedance in master side. The other is a realization of small operational force. Corresponding to two goals, indices are defined as "Reproducibility" and "Operationality".

- In teleoperation systems, communication time delay between master and slave robots is also a serious problem. To solve this problem, various researches have been done. One of the most effective stabilization methods of time delayed control systems is Smith predictor [9]. The method uses model of control object and time delay. Since the method is simple and easy to implement, many researches about the method have been studied so far. However, Smith predictor has a serious defect. Since time delay model of the system is needed for implementation of Smith predictor, it is difficult to implement in case that time delay model is not accurate. Particularly in case that it is difficult to obtain accurate time delay model (like time-varying delay case), the performance and stability of Smith predictor seriously deteriorates. In these circumstances, the communication disturbance observer (CDOB) was proposed to estimate unpredicatable and varying time delay as network disturbance (ND) and to compensate time delay effect [10]. However, the performance is greatly affected by order rearrengement of transmitted data due to network conditions, since CDOB requires slave models including integral elements.
- In this paper, a master-slave control by force feedback based virtual impedance controller is proposed. This proposed controller will be effective if the system includes communication delay because the transmission information of proposed controller is fewer than that of conventional controller. Moreover, models of a slave robot and communication time delay are not required. To make clear

the difference between conventional controller and proposed controller, performance analysis is conducted by using "Reproducibility" and "Operationality" in this paper.

## 61.2 Master-Slave Control with Time Delay

First, an ideal bilateral control is simply shown. Second, two types of conventional controllers are explained. In addition, the proposed controller is described.

### 61.2.1 Bilateral Control

Acceleration based force sensorless bilateral controller is constructed based on Disturbance observer (DOB) and Reaction torque observer (RTOB). The purpose of bilateral control is to realize a tracking control of position in master and slave systems, and to reproduce reaction force occurred in each system. An ideal bilateral teleoperation system would satisfy the following conditions. Here, subscripts $m$ and $s$ denote master and slave, respectively.

$$F_m + F_s = 0 \tag{61.1}$$

$$x_m - x_s = 0 \tag{61.2}$$

Eqs. (61.1) and (61.2) illustrate realization of "law of action and reaction" and ideal position tracking of master and slave.

### 61.2.2 General 4ch Sensorless Bilateral Controller

Figure 61.1 shows an overview of block diagram of general 4ch bilateral controller with time delay. This controller is well known as one of the best controller in bilateral system. In Fig. 61.1, control parameters are set as follows.

$$C_1 = C_4 = C_m = C_s = C_p(s) \tag{61.3}$$

$$C_2 = C_3 = C_5 = C_6 = C_f \tag{61.4}$$

Therefore, the acceleration references of master and slave are expressed as follows.

$$\ddot{x}_m^{ref} = C_p(x_s e^{-T_2 s} - x_m) + C_f(\hat{F}_s e^{-T_2 s} + \hat{F}_m) \tag{61.5}$$

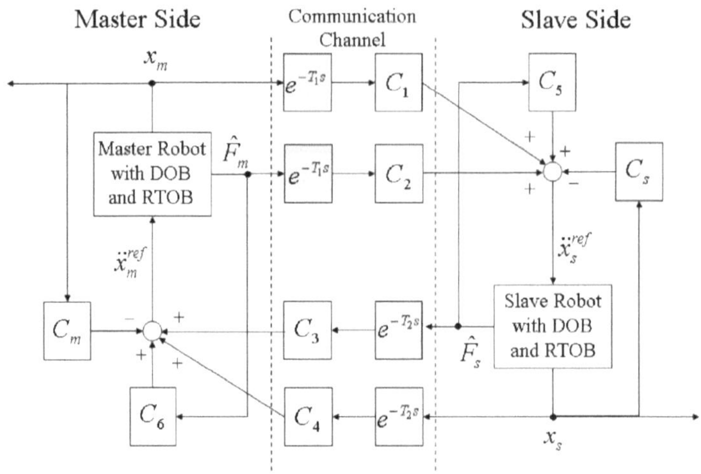


**Fig. 61.1** Block diagram of conventional bilateral controller with time delay

$$\ddot{x}_s^{ref} = C_p(x_m e^{-T_1 s} - x_s) + C_f(\hat{F}_m e^{-T_1 s} + \hat{F}_s) \tag{61.6}$$

Equations (61.5) and (61.6) mean that position and force control are unified through acceleration dimension. Here, DOB estimates disturbances and the estimated disturbance is fed back to achieve robust bilateral control [4]. Furthermore, RTOB provides an estimation of external force without torque/force sensor and brings torque/force sensorless control [5]. If communication delay is zero, 4ch controller achieves high transparency. However, in case of the system includes the communication delay, 4ch controller includes the effect of time delay in position feedback loop and a force feedback loop like (61.5) and (61.6). Therefore, transparency is deteriorated by the time delay effect.

### *61.2.3 3ch Bilateral Controller*

In Fig. 61.1, control parameters are set as follows.

$$C_1 = C_s = C_p(s) \tag{61.7}$$

$$C_4 = C_m = 0 \tag{61.8}$$

$$C_2 = C_3 = C_5 = C_6 = C_f \tag{61.9}$$

The difference between 4 and 3ch controller is that the master controller does not utilize position informations. In other words, only force control is implemented in the master side. Therefore, the acceleration references of master and slave are expressed as follows.

$$\ddot{x}_m^{ref} = C_f(\hat{F}_s e^{-T_2 s} + \hat{F}_m) \tag{61.10}$$

$$\ddot{x}_s^{ref} = C_p(x_m e^{-T_1 s} - x_s) + C_f(\hat{F}_m e^{-T_1 s} + \hat{F}_s) \tag{61.11}$$

In this controller, models of a slave robot and communication delay are not required and 3ch controller with time delay gives better performance than 4ch controller [11].

### 61.2.4 Proposed Force Feedback Based Virtual Impedance Controller

Figure 61.2 shows the block diagram of proposed controller. Here, RTOB means Reaction Torque Observer which can estimate the reaction torque and achieve force/torque sensorless master-slave control. Using the proposed approach, position and force controllers can be designed independently. The force difference of master and slave is an input to the virtual impedance controller to generate motion commands and they are utilized to achieve a perfect tracking performance in the position controller. In the virtual impedance controller, the desired impedance models $M_c$, $D_c$ and $K_c$ are defined and the difference between human input $F_m$ and reaction force $F_s$ is transformed to position, velocity and acceleration command $(x_c, \dot{x}_c, \ddot{x}_c)$. A motion model of virtual impedance is given as (61.12).

$$M_c\ddot{x}_c + D_c\dot{x}_c + K_c x_c = \hat{F}_m + \hat{F}_s e^{-T_2 s} \tag{61.12}$$

Using the generated motion commands $(x_c, \dot{x}_c, \ddot{x}_c)$, the acceleration reference is synthesized in (61.13).

$$\ddot{x}_m^{ref} = \ddot{x}_c + K_p(x_c - x_m) + K_v(\dot{x}_c - \dot{x}_m) \tag{61.13}$$

Here, in the DOB based controller, the acceleration control, that is, $\ddot{x}^{res} = \ddot{x}^{ref}$ is achieved from (61.12) and (61.13). In slave side, position controller is designed to track the response of master manipulator. As well as the master manipulator, acceleration reference $\ddot{x}_s^{ref}$ for the slave manipulator is given as follows.

$$\ddot{x}_s^{ref} = \ddot{x}_m e^{-T_1 s} + K_p(x_m e^{-T_1 s} - x_s) + K_v(\dot{x}_m e^{-T_1 s} - \dot{x}_s) \tag{611.14}$$

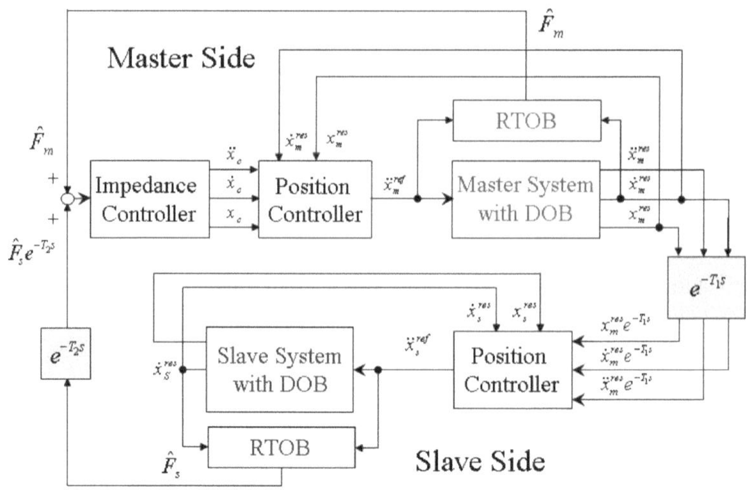


**Fig. 61.2** Block diagram of proposed controller with time delay

**Fig. 61.3** Network representation of master and slave system

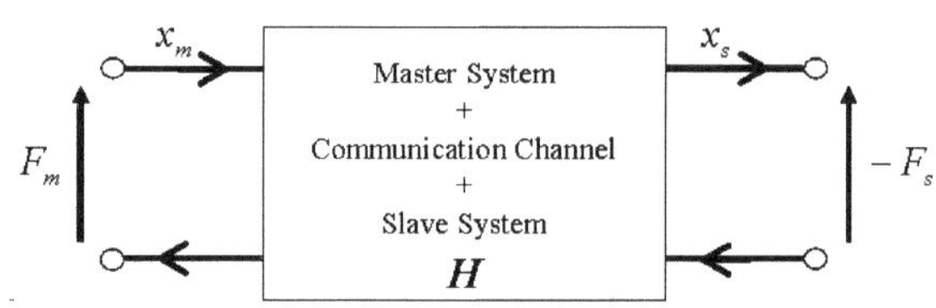


## 61.3 Performance Analysis

In this section, performance analysis is conducted. Here, the performance difference between proposed controller, 4ch and 3ch bilateral controller is analyzed by using "Reproducibility" and "Operationality" [8].

### *61.3.1 Hybrid Parameters*

Figure 61.3 shows the network representation of teleoperation systems. The relationship between master and slave can be formulated by independent variables $H$, called hybrid parameters [2] as follows.

$$\begin{bmatrix} F_m \\ -x_s \end{bmatrix} = \begin{bmatrix} H_{11} & H_{12} \\ H_{21} & H_{22} \end{bmatrix} \begin{bmatrix} x_m \\ -F_s \end{bmatrix} \tag{61.15}$$

In order to realize ideal condition, following hybrid parameters should be selected.

$$\begin{bmatrix} H_{11} & H_{12} \\ H_{21} & H_{22} \end{bmatrix} = \begin{bmatrix} 0 & 1 \\ -1 & 0 \end{bmatrix} \tag{61.16}$$

### 61.3.2 Reproducibility and Operationality

"Reproducibility" and "Operationality" are used as performance indices for Operability. First, slave force is treated as follows with impedance of environment $Z_e$.

$$F_s = Z_e x_s \tag{61.17}$$

From (61.15) and (61.17), force in master side is represented as follows.

$$F_m = \left( \frac{H_{12}H_{21}}{1 - H_{22}Z_e} Z_e + H_{11} \right) x_m \tag{61.18}$$

Here, $P_r$ and $P_o$ are defined as follows.

$$P_r = \frac{H_{12}H_{21}}{1 - H_{22}Z_e} \tag{61.19}$$

$$P_o = H_{11} \tag{61.20}$$

Hence, (61.18) is represented as follows.

$$F_m = (P_r Z_e + P_o) x_m \tag{61.21}$$

$P_r$ and $P_o$ are defined as "Reproducibility" and "Operationality". Because the reproduction of environmental impedance in master side is the most important condition in teleoperation, $P_r = 1$ should be satisfied. Additionaly, when $P_o = 0$ is realized, human operator feels real environmental impedance naturally. The ideal condition that satisfies perfect "Reproducibility" and "Operationality" is called Transparency [2].

### 61.3.3 Analysis of 4ch Controller

At first, 4ch controller is analyzed. In this analysis, it is assumed that cut-off frequency of LPF approaches infinity and ideal robust acceleration control is achieved. Hybrid parameters of 4ch controller can be calculated as follows. Here, $T$ indicates round-trip delay time $(T = T_1 + T_2)$.

$$\begin{bmatrix} H_{11} & H_{12} \\ H_{21} & H_{22} \end{bmatrix} = \begin{bmatrix} \dfrac{s^2\left(s^2 + 2C_p(s)\right) + Cp^2(1 - e^{-Ts})}{C_f s^2 + C_p(s)(1 + e^{-Ts})} & \dfrac{C_f\left(s^2 + 2C_p\right)e^{-T_2 s}}{C_f s^2 + C_p(s)(1 + e^{-Ts})} \\ -\dfrac{C_f\left(s^2 + 2C_p\right)e^{-T_1 s}}{C_f s^2 + C_p(s)(1 + e^{-Ts})} & \dfrac{C_f^2(1 - e^{-Ts})}{C_f s^2 + C_p(s)(1 + e^{-Ts})} \end{bmatrix} \tag{61.22}$$

If time delay is not included in the system, hybrid matrix can be shown as follows.

$$\begin{bmatrix} H_{11} & H_{12} \\ H_{21} & H_{22} \end{bmatrix} = \begin{bmatrix} \dfrac{s^2}{C_f} & 1 \\ -1 & 0 \end{bmatrix} \tag{61.23}$$

From (61.23), perfect transparency is achieved in 4ch controller without time delay, if force gain $C_f$ is large enough.

### 61.3.4 Analysis of 3ch Controller

Next, 3ch control is explained. This controller is designed based on acceleration control [11]. Hybrid parameters of 3ch controller can be shown as follows.

$$\begin{bmatrix} H_{11} & H_{12} \\ H_{21} & H_{22} \end{bmatrix} = \begin{bmatrix} \frac{s^2}{C_f} & e^{-T_2 s} \\ -e^{-T_1 s} & \dfrac{C_f(1 - e^{-Ts})}{s^2 + C_p} \end{bmatrix} \tag{61.24}$$

"Reproducibility" and "Operationality" of 3ch controller are shown as follows.

$$P_r = \frac{e^{-Ts}}{1 - \dfrac{C_f(1 - e^{-Ts})}{s^2 + C_p} Z_e} \tag{61.25}$$

$$P_o = \frac{s^2}{C_f} \tag{61.26}$$

In "Reproducibility", perfect performance cannot be achieved especially in contact motion. Moreover, "Reproducibility" degrades if force gain $C_f$ is large enough in contact motion. In "Operationality", perfect performance is achieved, if force gain $C_f$ is large enough. Therefore, the relationship between force gain $C_f$ and transparency is trade-off in contact motion with time delay in 3ch controller. If the system does not include time delay, hybrid matrix of 3ch controller is shown as follows.

**Table 61.1** Analysis parameters

| | Definitions | Value |
|---|---|---|
| $K_p$ | Position gain | 400.0 |
| $K_v$ | Velocity gain | 40.0 |
| $K_f$ | Force gain | 5.0 |
| $M_c$ | Virtual mass gain of master impedance | 0.2 |
| $D_c$ | Virtual viscosity gain of master impedance | 0.2 |
| $K_c$ | Virtual spring gain of master impedance | 0.0 |
| $Z_e$ | Environmental impedance | 10,000 + 20s |
| $T_1, T_2$ | Time delay [ms] | 1,000 |

$$\begin{bmatrix} H_{11} & H_{12} \\ H_{21} & H_{22} \end{bmatrix} = \begin{bmatrix} \frac{s^2}{C_f} & 1 \\ -1 & 0 \end{bmatrix} \tag{61.27}$$

As you can see from (61.23) and (61.27), hybrid parameters of 4ch controller and 3ch controller are same values if cut off frequency is infinity and the system does not include time delay ($T_1 = T_2 = 0\,\text{ms}$).

### 61.3.5 Analysis of Proposed Controller

Hybrid parameters of proposed controller can be calculated as follows.

$$\begin{bmatrix} H_{11} & H_{12} \\ H_{21} & H_{22} \end{bmatrix} = \begin{bmatrix} M_c s^2 + D_c s + K_c & e^{-T_2 s} \\ -e^{-T_1 s} & 0 \end{bmatrix} \tag{61.28}$$

The difference of performance between 3ch controller and proposed controller is $H_{11}$ and $H_{22}$. In $H_{11}$, since the ideal value is $H_{11} = 0$, how to design the impedance models $M_c$, $D_c$, $K_c$ is key factor about "Operationality". In $H_{22}$, since the ideal value is $H_{22} = 0$, proposed controller with time delay satisfies the ideal relationship constantly. Here, "Reproducibility" and "Operationality" of proposed controller are shown as follows.

$$P_r = e^{-Ts} \tag{61.29}$$

$$P_o = M_c s^2 + D_c s + K_c \tag{61.30}$$

In "Reproducibility", the effect of time delay is only transmitted to the slave side. In "Operationality", perfect "Operationality" is also achieved in the proposed controller, if the virtual impedance gain $(M_c, D_c, K_c)$ is small enough. Compared with 3ch controller, the relationship of $P_r$ and $P_o$ is simple and this makes clear parameter design of the controller with time delay in proposed controller. This is an important point in this paper.

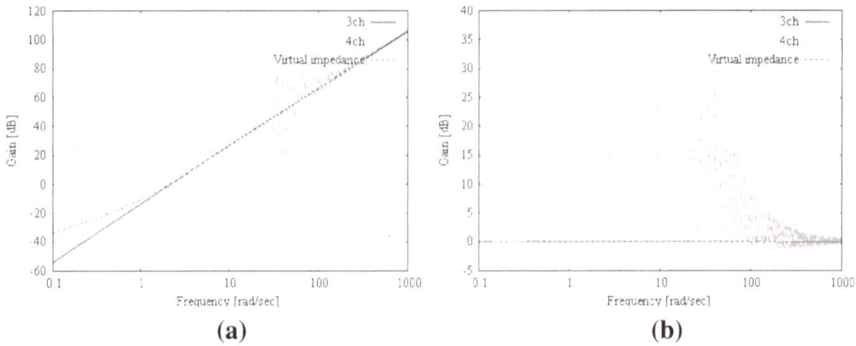


**Fig. 61.4** Comparison of controller characteristics by using hybrid parameters. **a** Gain characteristics of H11 **b** Gain characteristics of H12

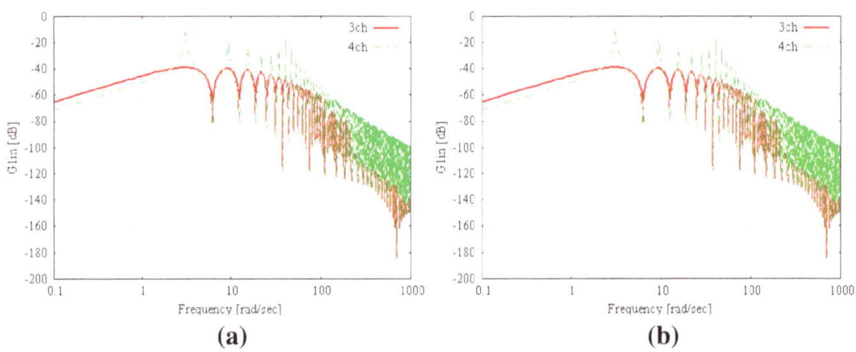


**Fig. 61.5** Comparison of controller characteristics by using hybrid parameters. **a** Gain characteristics of H21 **b** Gain characteristics of H22

## *61.3.6 Analysis by Using Bode Diagram*

Performance analysis is conducted by using Bode diagram. 4ch controller, 3ch controller and proposed controller are compared. Parameters in analysis are listed in Table 61.1. Figures 61.4 and 61.5 show the gain characteristics of Hybrid parameters. From Fig. 61.4a, 3ch controller gives the best performance in low frequency area. However, if virtual impedance gains $D_c$ and $K_c$ are zero, $H_{11}$ of 3ch controller and proposed controller become same values. As shown in Figs. 61.4b and 61.5a, both 3ch controller and proposed controller are same characteristics. These controllers are better than 4ch controller in $H_{12}$ and $H_{21}$. Figure 61.5b shows the gain diagram of $H_{22}$. In $H_{22}$, since $H_{22}$ of proposed controller is zero, the gain diagram cannot be indicated. However, proposed controller is better than 3ch controller in $H_{22}$.

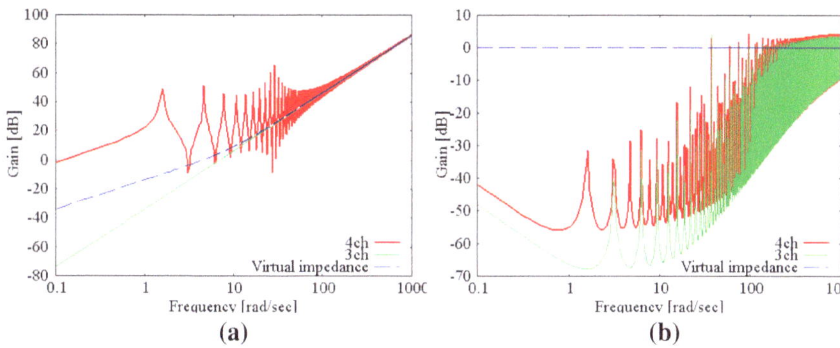


**Fig. 61.6** Comparison of controllers by using "Operationality" and "Reproducibility". **a** "Operationality" **b** "Reproducibility"

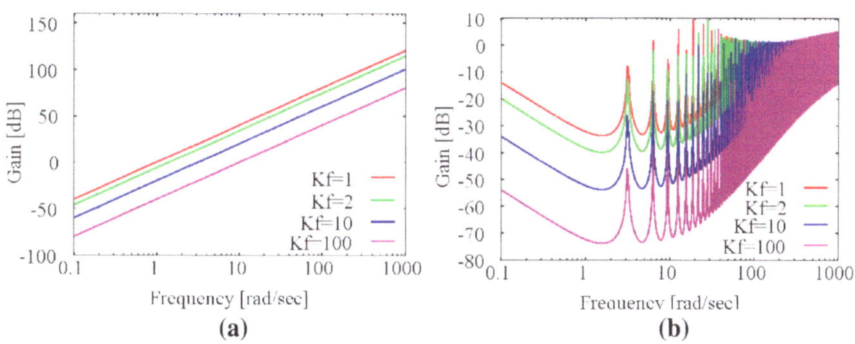


**Fig. 61.7** The relationship between force gain and transparency in 3ch controller. **a** Po (3ch controller) **b** Pr (3ch controller)

Next, discussions about "Reproducibility" and "Operationality" are conducted. Since "Operationality" and $H_{11}$ are same meaning, 3ch controller gives the best performance in low frequency area as you can see from Fig. 61.6a. However, if virtual impedance gains $D_c$ and $K_c$ are zero, "Operationality" of 3ch controller and proposed controller become same characteristics. On the other hand, proposed controller is the best performance in "Reproducibility" as you can see from Fig. 61.6b.

Thirdly, the relationship between force gain $C_f$(3ch controller), $M_c$(proposed controller) and transparency is discussed. Here, $C_f$ and $K_f$ are same values. As shown in Fig. 61.7a, b, the relationship between force gain $C_f$ and transparency is trade-off in contact motion with time delay in 3ch controller. In other words, if force gain $C_f$ is large enough, perfect "Operationality" can be achieved, but "Reproducibility" degrades in contact motion. On the other hand, Fig. 61.7a, b denote that if virtual mass gain $M_c$ is small enough, perfect "Operationality" can

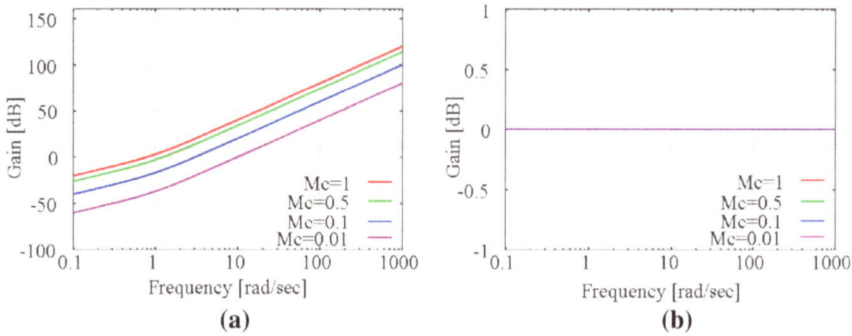


**Fig. 61.8** The relationship between force gain and transparency in proposed controller. **a** Po (Proposed controller) **b** Pr (Proposed controller)

**Table 61.2** Simulation parameters

| | Definitions | Value |
|---|---|---|
| $K_p$ | Position gain | 400.0 |
| $K_v$ | Velocity gain | 40.0 |
| $K_f$ | Force gain | 50.0 |
| $M_c$ | Virtual mass gain of master impedance | 0.02 |
| $D_c$ | Virtual viscosity gain of master impedance | 0.2 |
| $K_c$ | Virtual spring gain of master impedance | 0.0 |
| $Z_e$ | Environmental impedance | 10,000 + 20s |

be achieved and "Reproducibility" is constant with the value of $M_c$. Consequently, the relationship of $P_r$ and $P_o$ is simple and this makes clear parameter design of the controller with time delay in proposed controller (Fig. 61.8).

## 61.4 Simulation

In this section, numerical results are shown to confirm the validity of the proposed controller. Parameters in simulations are listed in Table 61.2.

### *61.4.1 Simulation Results Without Time Delay*

At first, simulation is conducted in the case that the teleoperation system does not include communication delay ($T_1 = T_2 = 0\,\text{ms}$). Figures 61.9 and 61.10 show the results of torque response and position response respectively. As shown in Figs. 61.9 and 61.10, the results of three types of controllers are almost the same.

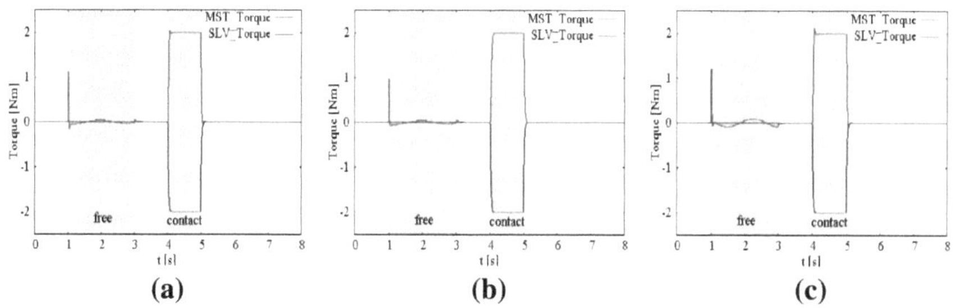


**Fig. 61.9** Torque response without time delay ($T_1 = T_2 = 0\,\text{ms}$) **a** 4ch controller **b** 3ch controller **c** Proposed controller

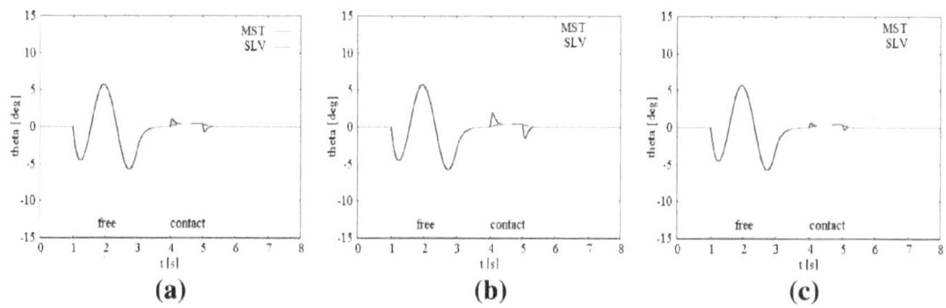


**Fig. 61.10** Position response without time delay ($T_1 = T_2 = 0\,\text{ms}$) **a** 4ch controller **b** 3ch controller **c** Proposed controller

Therefore, good transparency is achieved without time delay in proposed controller as well as 4ch controller and 3ch controller.

## *61.4.2 Simulation Results with Time Delay*

Secondly, simulation is conducted in the case that the teleoperation system includes the time-varying delay ($50\,\text{ms} < T_1, T_2 < 150\,\text{ms}$). Figures 61.11 and 61.12 show the results of torque response and position response respectively. From Fig. 61.11a and 61.12a, both "law of action and reaction" and position tracking and are not achieved in 4ch controller. In other words, ideal performance cannot be achieved in 4ch controller if the teleoperation system includes the time delay. As shown in Fig. 61.11b, c, good torque responses can be achieved in both 3ch controller and proposed controller. However, the results of position response are different especially in contact motion as you can see from Fig. 61.12b, c. In 3ch controller, position tracking is not achieved in contact motion. On the other hand, the position tracking can be almost achieved in proposed controller. Therefore, the

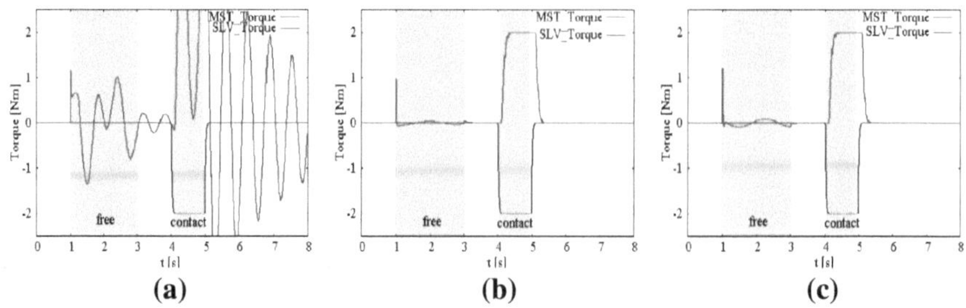


**Fig. 61.11** Torque response with time delay ($50\,\text{ms} < T_1, T_2 < 150\,\text{ms}$) **a** 4ch controller **b** 3ch controller **c** Proposed controller

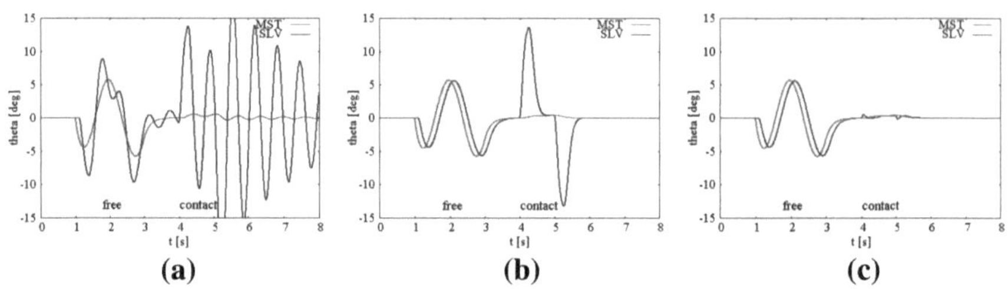


**Fig. 61.12** Position response with time delay ($50\,\text{ms} < T_1, T_2 < 150\,\text{ms}$) **a** 4ch controller **b** 3ch controller **c** Proposed controller

communication delay models are not required in proposed controller. As a result, the validity of proposed controller is confirmed as well as performance analysis in some simulations.

## 61.5 Conclusion and Future Works

In this paper, an analysis of a master-slave control by force feedback based virtual impedance controller with time delay is conducted. Compared with conventional controllers, the relationship of $P_r$ and $P_o$ is simple and this makes clear parameter design of the controller with time delay. Moreover, the proposed controller can be flexibly applied to many kinds of time-delayed teleoperation systems because models of a slave robot and communication delay are not required. The validity of proposed controller is confirmed by some simulations. As the future work, the stability analysis and experimental validation should be conducted.

**Acknowledgements** This work was supported in part by a Grant-in-Aid for Scientific Research(C)(22560280).

## References


1. Hannaford B (1989) A design framework for teleoperators with kinesthetic feedback. IEEE Trans Robot Autom 5(4):426–434
2. Lawrence DA (1993) Stability and transparency in bilateral teleoperation. IEEE Trans Robot Autom 9(5):624–637
3. Ohnishi K (1993) Robust motion control by disturbance observer. J Robot Soc Jpn 11(4):486–493 (in Japanese)
4. Matsumoto Y, Katsura S, Ohnishi K (2003) An analysis and design of bilateral control based on disturbance observer. Proceedings of the 10th IEEE international conference on industrial technology, pp 802–807
5. Murakami T, Yu F, Ohnishi K (1993) Torque sensorless control in multidegree-of-freedom manipulator. IEEE Trans Ind Electron 40(2):259–265
6. Horie R, Murakami, T (2010) Workspace based controller design and performance evaluation for MDOF bilatetal system. The 8th edition of France–Japan and 6th Europe–Asia congress on mechatronics, Yokohama (Japan), 22–24th Nov
7. Lasnier A, Murakami T (2010) Hybrid sensorless bilateral teleoperation of two-wheel mobile manipulator with underactuated joint. Int AIM2010, pp 347–352
8. Iida W, Ohnishi K (2004) Reproducibility and operationality in bilateral teleoperation. Proceedings of the 8th International workshop on AMC '04, pp 217–222, 25–28 March
9. Smith OJM (1959) A controller to overcome dead time. ISA J 6(2):28–33
10. Natori K., Tsuji T., Ohnishi K., Hace A., and Jezernik K. (2010) Time-delay compensation by communication disturbance observer for bilateral teleoperation under time-varying delay. IEEE Trans Ind Electron 57(3):1050–1062
11. Kubo R, Iiyama N, Natori K, Ohnishi K, Furukawa H (2007) Performance analysis of a three-channel control architecture for bilateral teleoperation with time delay. IEEJ Trans Ind Appl 127(12):1224–1230
12. Horie R, Murakami T (2011) An analysis of a novel master-slave control and its application to MDOF motion systems. The 37th annual conference of IEEE industrial electronics society (IECON'11), Melbourne (Australia), 7–10th Nov 2011

# Chapter 62
# Friction Compensation and Control Strategy for the Dexterous Robotic Hands

**Vahid Aminzadeh, Rich Walker, Ugo Cupcic, Hugo Elias and Jian S. Dai**

**Abstract** Friction in tendon based robot hands has been an unavoidable problem. With introduction of more dexterous and metamorphic hands this intrinsically non-linear and highly time-varying force has become an obstacle to achieve high performances and required accuracy which the robot designers are aiming for. This paper proposes a method to tackle the problem by measuring friction and applying it in a feed forward loop. The novelty of the method is in the possibility to perform the task automatically and on a regular basis to compensate the time varying phenomenon.

**Keywords** Robotic · Robot hand · Friction measurement · Friction compensation · Control

## 62.1 Introduction

Since Dr. Engelberger [1] invented the first industrial robot in the early 1960s, there is a growing interest in robotic hand research and in the past three decades a great variety of robotic hands have been developed. The industrial robotic hands were strong and robust with a simple design of two or three fingers and are referred as industrial grippers.

V. Aminzadeh (✉) · J. S. Dai
King's College London, Strand, London
e-mail: vahid.aminzadeh@kcl.ac.uk

R. Walker · U. Cupcic · H. Elias
Shadow Robot Company, 251 Liverpool Road, London, UK
e-mail: ugo@shadowrobot.com

J. S. Dai et al. (eds.), *Advances in Reconfigurable Mechanisms and Robots I*,
DOI: 10.1007/978-1-4471-4141-9_62, 

With the growing interest in humanoid robotics research, anthropomorphism of a robotic hand becomes a popular design goal, and has been proposed by researchers in the form of NASA Robonaut Hand [2], Stanford/JPL hand [3], Utah/MIT hand [4] and many more. These hands have been continuously developed in the past two to three decades but most of them have a common feature of having a fixed/rigid palm. Thus, mechanisms with more functionality are needed for secure grasping and manipulation of different and complex objects. This then leads to DLR/HIT Hand II [5], Fluidic Hand [6], IIT hand [7], Bi-on-ics [8] and Shadow Hand [9]. Also King's College London has developed a new metamorphic robotic hand which consists of a metamorphic mechanism for the palm [10].

As it can be seen there are a variety of robotic hands developed by different institutions and companies and it can be claimed that the hardware is already available. Nevertheless, in all the cases the control architecture fails to provide the required autonomous manipulation for the different objects and to the best of our knowledge there is little work regarding dexterous manipulation control strategies ([11], [12]) and these are generally theoretical studies on trajectory generation. The majority of the control strategy literature is based on the assumption of the existence of a relatively accurate dynamic model of the hand.

With the efforts for development of a prosthetic hand and also hands which are supposed to be the same size of the human hand it is very difficult to place actuators on the joints unless the fingers have low degree of freedom or the hand is larger than the human hand. In such cases the motors are usually very constrained due to the size and are either weak, very expensive or the fingers have very low degrees of freedom. As a result most of the existing robotic hands need to transfer the power from the motors (or pumps, air muscle, etc.) to the joints with the aid of a set of tendons.

Tendons on the other hand have their own disadvantages. Most of the good implementations of tendon operated robots require two tendons per joint. In this case spacing problems arise for the tendons. Furthermore the two tendons are being pulled in different directions. Furthermore tendons are in contact with surfaces in an almost random manner. This is particularly important when there is an element of metamorphism which increases the complexity of the kinematic chain.

Depending on the posture of the joints and the fingers the tendons are being deformed in different ways. Therefore the contact surfaces and the acting forces on the tendons vary which changes the friction forces. As the hands become more dexterous the friction forces are becoming almost impossible to predict. This varying friction results in large inaccuracies in the dynamic model of the system and as a result the dexterous hands do not have the same performance over the whole range of the motion of the joints.

One might argue that by using gain scheduling and derivation of an accurate enough model at a number of setpoints it will be possible to utilize model based control. This might be true however this requires an offline estimation process and will not be accurate when the model changes due to decay, etc. Furthermore it is almost impossible to guarantee that the tendons are always taut (even in the presence of tensioners). This will create a random delay in between the moment that the command for motion is being sent and the actual motion of the joint. Such a delay makes the system identification process even more complicated and almost impossible to perform autonomously.

The inaccuracy in control can have drastic consequences as an overshoot in the joint motion can be strong enough to damage the grasped object and a small error can cause a failure to hold an object. It can be argued that such errors can be avoided by using force feedback loops, however it is still desirable to have an extremely accurate position control.

We have proposed and implemented a method to measure friction as the largest contributor to the non-linearity of the system. The method can be performed autonomously in the form of an online upgrade of the friction map. This can also compensate the time-varying of the mechanism, particularly in the harsh environments and can be used as a measure for fault detection.

## 62.2 Review of Friction Models

Friction forces are created between two surfaces which are in contact due to the irregularities and asperities at the microscopical level. The friction force depends on a large number of factors including the surface material, temperature, etc. A variety of models have been proposed to formulate the friction which although very different in nature, are capable of more or less accurately estimate the friction parameters. A summary of the classic friction models can be found in [13], [14] and [15] which cover the basic ideas of the friction as a force opposing the motion to the more advanced static and dynamic models.

The simplest model proposed for friction has the form $F(v) = F_c\text{sgn}(v)$ where $F$ is the friction force; $v$ is the relative velocity of the contacted surfaces and $F_c$ is the coulomb friction. This model has widespread application in friction compensation due to its simplicity and in many cases ([16], [17]) it is enough for ad hoc purposes. The friction force when velocity is zero is not defined by this equation. The main problem with this model is the fact that it does not consider the drop in the friction force after motion starts and assumes that the friction is constant throughout the course of motion. In order to overcome this problem a better model can be constructed by considering stiction which is a larger friction force which acting on the system when it is at rest. Also the change in the friction is continuous [15]. There are a number of formulas describing this phenomenon and the most commonly used one is

$$F(v) = F_c + (F_s - F_c)e^{-\left|\frac{v}{v_S}\right|^{\delta_s}}\text{sgn}(v) + F_v v \qquad (1)$$

Where $v_s$ is the Stribeck velocity and $\delta_s$ is a free parameter usually assumed to be 2. During the development in the field of hydrodynamics expressions were derived the viscous frictions which are generally in the form of $F(v) = F_v v$. It is also possible to combine the viscous friction into the same equation either by $F(v) = F_c\text{sgn}(v) + F_v v$ or $F(v) = F_v|v|^{\delta_v}\text{sgn}(v)$ ([15], [18]).

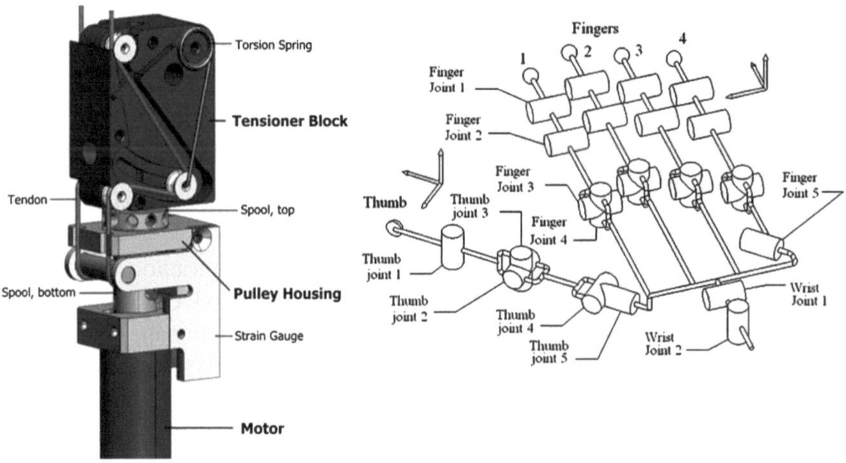


**Fig. 62.1** Power transmission (*left*) and kinematics of the finger of Shadow hand (*right*)

## 62.3 Apparatus Description

Figure 62.1 demonstrates the power transmission of the Shadow hand. A Motor is connected to a spool which the two tendons are connected to. The tendons are passing through the tensioner which tries to keep the tendons as taut as possible with the aid of a torsion spring. The tendons are directed through predefined paths to the respective finger joints.

## 62.4 Friction Measurement

In order to measure the friction the following control loop was implemented. By adjusting the PI controller values it is possible to obtain a relatively slow and continuous motion throughout the range of the motion of the finger. To do so, set the P value to the smallest value that can create a motion in the beginning of the range (or point with lowest friction). At the next step the value of I should be set such that it creates at the end of the range (or the point with highest friction).

The velocity setpoint has been chosen as a very low speed to guarantee that the robot is moving in the correct part of the friction-velocity curve. In order to make sure that the output velocity has the required specifications the velocity values throughout the motion has been recorded and the measurements of the motor input have begun afterwards.

The implementation of the PI controller is to guarantee the generation of motion in the joint i.e. if the friction force is very high the value of the controller output keeps increasing until the joint moves (Fig. 62.2).

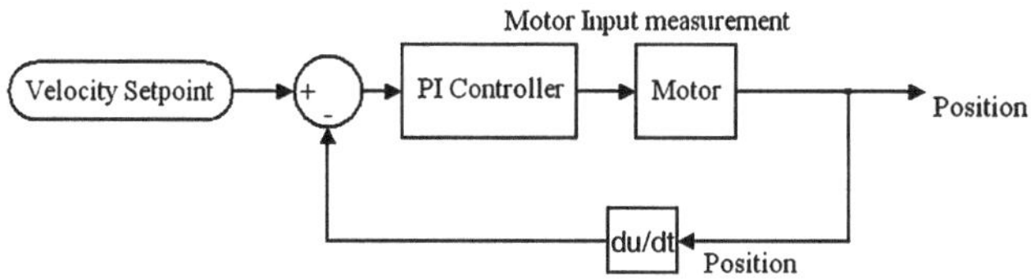


**Fig. 62.2** Control loop for friction measurement

One might argue that the velocity can be derived directly if the motor is equipped with shaft encoder or tachometer. However in such a case the velocity obtained from the shaft encoder does not necessarily reflect the actual velocity of the joint and it is possible that the joint has motion which is not reflected in the measurements and/or the shaft of the motor has motions (due to backlash, etc.) which cannot be seen in the velocity measurement. For this reason it is recommended to measure the velocity by differentiating the position.

On the other hand an appropriate technique should be employed for the differentiation of the position as it can be a noisy signal. A number of methods are proposed in the literature for differentiation of noisy signals. We employed Kalman filter to do so, however any other method can be utilized depending on the situation.

As it is mentioned before the tendons responsible for forward and backward motion are different. This resulted in different friction on the forward and backward motion and the corresponding frictions should be measured separately. One might argue that by arranging the tendons to take close paths it should be possible to record only one measurement and adjust it for both motions. This is not necessarily correct as the tendon path would be different when they are being pulled in different directions unless the tendons paths are also symmetrical. In any other case formation of a hysteresis loop can impose inaccuracies into the model.

As majority of the tendons utilized in the hands (including the apparatus used for the data in this paper) are PTFE, the friction is mainly a coulomb friction. As a result the second term of the equation above can be omitted. If the tendon is lubricated then the extra term needs to be estimated and considered in the feed forward loop.

The next question was to decide which friction model best suits our purpose. In order to do so we performed a set of tests with different velocities which are demonstrated in Fig. 62.3.

As it can be seen in Fig. 62.3 the friction generally decreases as the velocity increases. On the other hand any more specific relation between friction and velocity is almost impossible to derive. One solution might be to derive a friction map for a number of velocity set points. With this information at hand at each point the controller can measure (or estimate) the velocity and calculate the correct friction map. This will increase the overhead calculation and the storage space required for the friction tables can be a problem. On the other hand if only one of the friction maps is supposed to be chosen then the question is which one of the

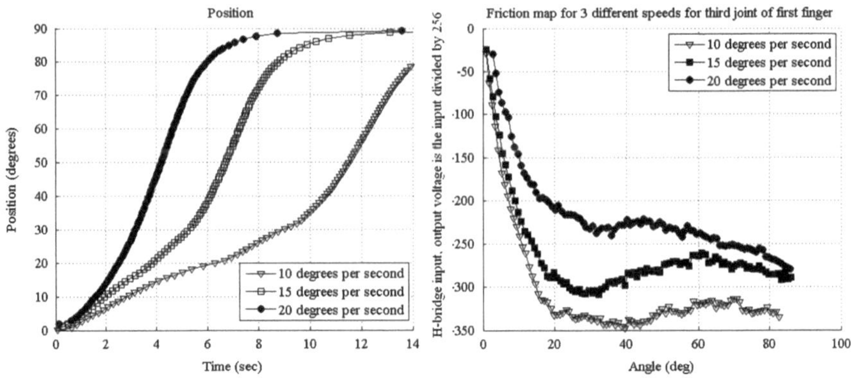


**Fig. 62.3** Control loop for friction measurement

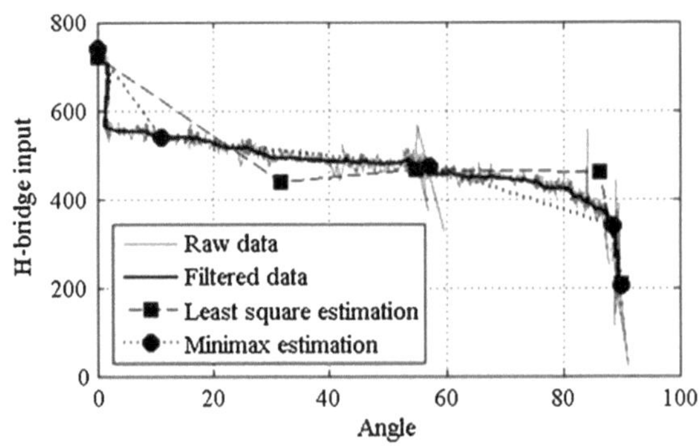


**Fig. 62.4** Raw data, filtered data and the points obtained by least square and minimax algorithms

maps should be chosen. We tested all the friction maps and based on our experience we observed that the friction maps with low velocities provide better accuracy for all the motions. On the other hand if required, it is possible to store more than one friction map. This is particularly useful if the mechanism demonstrate different friction maps for different velocities or configuration of hand changes the friction values. Also it is possible to form two friction maps for very low velocities (almost start the motion from steady state) to have an exact measurement for the stiction and one with an average velocity (can be mid range of the velocity). At the next step the velocity should again be calculated (or estimated) and the interpolation (or extrapolation) should be employed which can increase the overhead again.

The choice of the method depends on the system at hand. We decided to go with only one friction map as we observed that it can slow the sampling frequency. The obtained data has been further filtered to derive a better friction map. Derived friction map can be saved in its entirety in the robot framework, however as it can be seen from the Fig. 62.4 the data can be constructed with a limited number of points. Higher number of points results in a more accurate friction map; however it

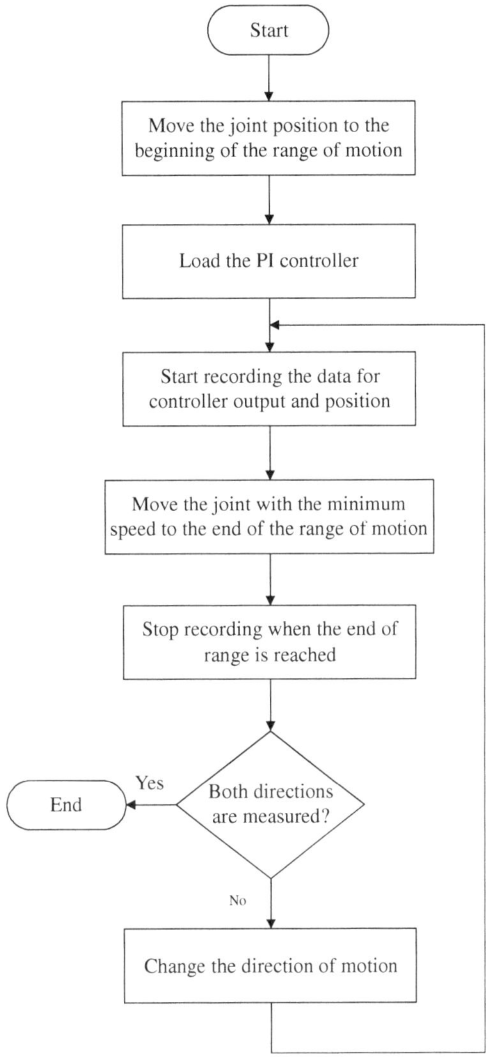


**Fig. 62.5** Flowchart of the algorithm

increases the memory usage. It was observed that 5 points can fully incorporate the trend of the function and interpolation between the 5 points results in very small error (less than 0.1 percent in majority of the cases).

A variety of the algorithms for deriving the points were considered however the two most suitable ones were mini-max and least square algorithm. Figure 62.4 demonstrates the results of the applying above algorithms to the sets. If the data is smooth enough it is even possible to choose the points which are spaced at equal intervals. Figure 62.5 demonstrates flowchart of the algorithm for friction measurement.

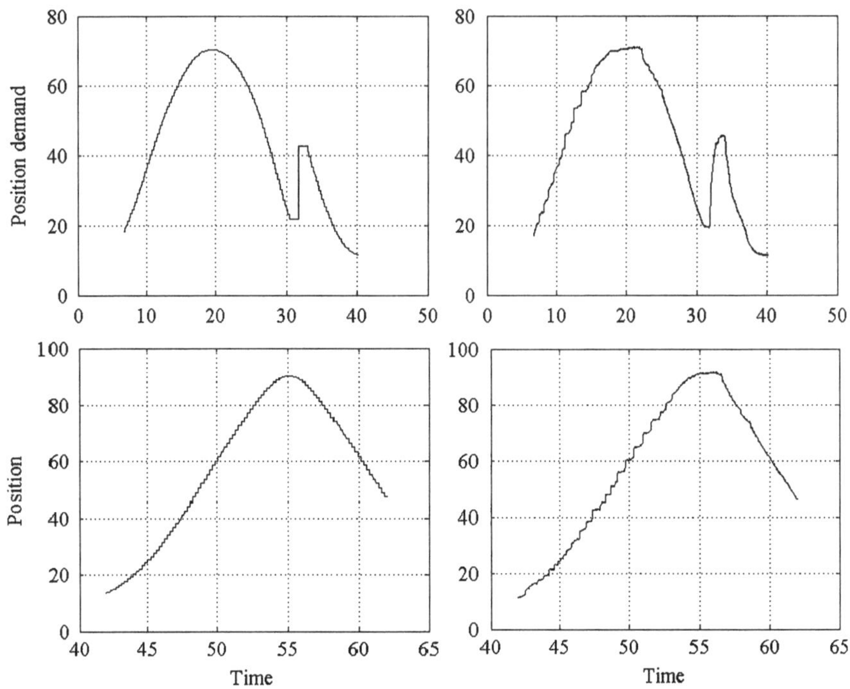


**Fig. 62.6** Position demand (*left*) and actual position (*right*) of the joint

## 62.5 Results

In order to check the performance of the friction compensation a simple PID loop with the friction feedforward has been arranged. To check the performance of the controller over the possible range of the motion of the joint a series of steps with different amplitudes have been applied to the system.

As it can be seen in Fig. 62.6 the system has followed the input without a significant overshoot. The system is capable of accurately reaching any position demand in the whole range of the motion of the joint.

One major source of error which contributes to the difference between the input and the output is the backlash of the tendon. The backlash detection and compensation methods can be of use in further improving the performance of the controller, however care should be taken into account when both friction and backlash compensation are being in use.

## 62.6 Conclusion

Various friction models were discussed. A new method for friction measurement and compensation the friction for tendon based robot hands was proposed. The proposed method was applied to Shadow robot hand and the results were presented.

## References


1. Shimon YN (1985) Handbook of industrial robot. Wiley, New York
2. Ambrose R, Aldridge H, Askew R, Burridge R, Bluethmann W, Diftler M, Lovchik C, Magruder D, Rehnmark F (2000) Robonaut: Nasa's space humanoid. IEEE Intell Syst 15(4):57–63
3. Salisbury JK, Craig JJ (1982) Articulated hands: force control and kinematic issues. Int J Rob Res 1:4–17
4. Jacobasen SC, Iversen EK, Knutti DF, Johnson RT, Biggers KB (1986) Design of the Utah/M.I.T. dexterous hand. In: IEEE international conference robotics and automation, pp 1520–1532
5. Liu H, Wu K, Meusel P, Hirzinger G, Jin MH, Liu YW, Fan SW, Lan T, Chen ZP (2008) A dexterous humanoid five-fingered robotic hand. In: IEEE international symposium on robot and human interactive communication, pp 371–376
6. Kargov A, Pylatiuk C, Oberle R, Klosek H, Werner T, Roessler W, Schulz S (2007) Development of a multifunctional cosmetic prosthetic hand. In: IEEE 10th International Conference Rehabilitation Robotics, pp 550–553
7. New scientist (2010) http://www.newscientist.com/article/mg20627566.800-robots-with-skin-enter-our-touchyfeely-world.html
8. Fischman J (2010) Bionics. Natl Geogr 217(1):35–53
9. Shadow_Robot_Company (2010) http://www.shadowrobot.com/
10. Wei G, Dai JS, Wang S, Luo H (2011) Kinematic analysis and prototype of a metamorphic anthropomorphic hand with a reconfigurable palm. Int J Humanoid Robotics 8(3):795–798
11. Rovetta A, Wen X (1993) Telemanipulation control of a robotic hand with cooperating fingers by means of telepresence with a hybrid virtual-real structure. Lect Notes Control Inf Sci 187:411–417
12. NASA (2011) Robonauts http://robonaut.jsc.nasa.gov/index.asp. Accessed on 28/3/2011
13. Armstrong-Hélouvry B (1991) Control of Machines with Friction. Kluwer Academic Publisher, Boston
14. Armstrong-Hélouvry B, Dupont P, Canudas de Wit C (1994) A survey of models, analysis tools and compensation methods for the control of machines with friction. Automatica 30(7):1083–1138
15. Olsson H, Åström K, Canudas de Wit C, Gäfvert M, Lischinsky P (1998) Friction models and friction compensation. Eur J Control 4:176–195
16. Baril CB (1993) Control of mechanical systems affected by friction and other nondifferentiable nonlinearities. PhD thesis, Israel Institute of Technology, Haifa, Israel
17. Friedland B, Park YJ (1991) On adaptive friction compensation. In: Proceedings of the IEEE conference on decision and control, pp 2899–2902
18. Bona B, Indri M (2005) Friction compensation in robotics: an overview. In: Proceedings of the 44th IEEE conference on decision and control, and the European control conference

# Chapter 63
# On the Task Specific Evaluation and Optimisation of Cable-Driven Manipulators

**Darwin Lau, Kishor Bhalerao, Denny Oetomo and Saman K. Halgamuge**

**Abstract** Cable-driven manipulators are traditionally designed for general performance objectives, such as maximisation of workspace. To take advantage of the reconfigurability of cable-driven mechanisms, the optimisation of cable-configurations for specific tasks is presented. Specifically, two types of task specific objectives are explored, the minimisation of cable forces over a desired trajectory and the maximisation of workspace about a desired pose. The formulation and incorporation to the optimisation problem for both task specific objectives are presented. Illustrated using a 3-DoF manipulator example, the results clearly demonstrated the advantages of optimising cable configurations for specific tasks. The potential ease of relocation in cable attachments makes task dependent reconfiguration feasible.

**Keywords** Cable-driven manipulators · Task-specific objectives · Optimisation

D. Lau · K. Bhalerao · D. Oetomo (✉) · S. K. Halgamuge
Department of Mechanical Engineering, The University of Melbourne, Melbourne, Australia
e-mail: doetomo@unimelb.edu.au

D. Lau
e-mail: kishorb@unimelb.edu.au

K. Bhalerao
e-mail: dtlau@student.unimelb.edu.au

S. K. Halgamuge
e-mail: saman@unimelb.edu.au

J. S. Dai et al. (eds.), *Advances in Reconfigurable Mechanisms and Robots I*,
DOI: 10.1007/978-1-4471-4141-9_63, © Springer-Verlag London 2012

## 63.1 Introduction

Cable-driven parallel manipulators refer to mechanisms where the end-effector is controlled through cables. Cables are attached to the end effector on one end and to the actuator located at the base platform on the other. Desirable characteristics over traditional parallel mechanisms include: reduced weight and inertia, simplified dynamics modelling and the ease of transportation. The unique property of cable driven mechanisms is that its cables can only be actuated unilaterally through tension and not compression (*positive cable force*).

Due to the potential ease in relocating the cable attachment locations (*cable configuration*) on the end effector and base platform, cable-driven manipulators can be regarded as a reconfigurable mechanism. A wide range of applications exist for cable-driven mechanisms, such as such as manufacturing [2], rehabilitation [13] and exoskeletons [14]. For a given application, cable configuration can be either determined by the designer or through an optimisation process [1, 9, 2].

Naturally, the selection of an appropriate objective function is crucial in the determination of a desired cable configuration. In [1], the workspace volume was maximised for a two-link upper arm exoskeleton. Maximisation of tension-closure workspace volume and Global Conditioning Index (*GCI*) was performed on a cable-driven universal joint module [9]. In [12], the configuration for a locomotion interface actuated by 16 cables was determined. The objectives involved the maximisation of mechanism workspace volume while minimising cable interference.

Objective functions such as workspace volume and GCI are similar to those employed in the design of rigid link mechanisms [5]. The goal is to achieve desirable general performance, as opposed to the efficiency of specific tasks. To take advantage of the potential reconfigurability of cable-driven manipulators, optimisation for specific tasks should be considered. For example, manipulators designed with a large workspace may not be energy efficient for a prescribed trajectory. This saving is particularly significant for highly repetitive tasks, such as pick-and-place in manufacturing or rehabilitation treatment for stroke patients.

In this paper, objective functions to evaluate the desirability of cable-driven manipulator configurations for specific tasks are investigated. Two classes of objective functions are presented, one based on the minimisation of cable forces for a desired trajectory, and the second is based on desired regions of the manipulator workspace. It is shown how workspace properties can be efficiently evaluated by extending the workspace analysis technique proposed in [8]. The impact of the proposed objective functions is illustrated through the optimisation of a 3-DoF manipulator, using a standard Particle Swarm Optimisation (*PSO*) algorithm. The results highlight the potential of task specific cable arrangement over a manipulator that is optimised for general or global performance.

The remainder of this paper is organised as follows: Sect. 63.2 presents the manipulator model, inverse dynamics problem and workspace analysis. The task specific objective functions and optimisation problem are formulated in Sect. 63.3. The results

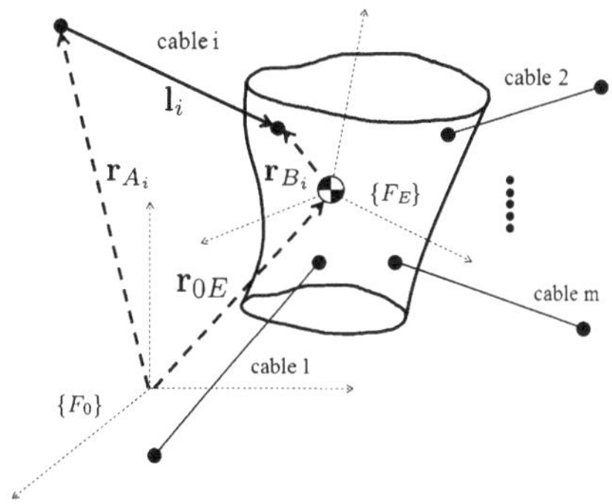


**Fig. 63.1** General model for cable manipulator

for the optimisation of an example 3-DOF manipulator are presented and discussed in Sect. 63.4. Finally, Sect. 63.5 concludes the paper and presents areas of future work.

## 63.2 Manipulator Model and Background

Consider the model shown in Fig. 63.1, where $\mathbf{r}_{A_i}$ and $\mathbf{r}_{B_i}$ represents the cable attachments at the base and the end effector for cable $i$, respectively. The vector $\mathbf{r}_{0E}$ represents the translation from the origin of the inertial frame $\{F_0\}$ to the origin of the end effector frame $\{F_E\}$.

The equations of motion a general $n$-DOF manipulator actuated by $m$-cables can be expressed in the form

$$M(\mathbf{q})\ddot{\mathbf{q}} + \mathbf{C}(\mathbf{q}, \dot{\mathbf{q}}) + \mathbf{G}(\mathbf{q}) = -J^T(\mathbf{q})\mathbf{f}, \tag{63.1}$$

where $\mathbf{q} = [q_1 \quad q_2 \quad \dots \quad q_n]^T \in \mathbb{R}^n$ represents the manipulator pose, $M$, $\mathbf{C}$, and $\mathbf{G}$ are the mass inertia matrix, centrifugal and Coriolis force vector, and gravitational vector, respectively. The resultant of the cable wrenches is denoted by $-J^T\mathbf{f}$, where $J^T(\mathbf{q}) \in \mathbb{R}^{n\times m}$ is the transpose of the Jacobian matrix. For a 6-DOF spatial manipulator

$$J^T = \begin{bmatrix} \hat{\mathbf{l}}_1 & \hat{\mathbf{l}}_2 & \dots & \hat{\mathbf{l}}_m \\ \mathbf{r}_{B_1} \times \hat{\mathbf{l}}_1 & \mathbf{r}_{B_2} \times \hat{\mathbf{l}}_2 & \dots & \mathbf{r}_{B_m} \times \hat{\mathbf{l}}_m \end{bmatrix}. \tag{63.2}$$

where $\mathbf{l}_i$ is the cable vector of cable $i$, defined as $\mathbf{l}_i = \mathbf{r}_{0E} + \mathbf{r}_{B_i} - \mathbf{r}_{A_i}$. The cable force vector is denoted by $\mathbf{f} = [f_1 \quad f_2 \quad \dots \quad f_m]^T \in \mathbb{R}^m$, where $f_i$ is the cable force in cable $i$. The allowable cable force range in cable $i$ can be defined as

$$0 < f_{i,min} \leq f_i \leq f_{i,max}, \tag{63.3}$$

where $f_{i,min}$ ensures positive cable force and prevents cable slackness, while $f_{i,max}$ provides an upper limit on allowable actuation through the cable.

### 63.2.1 Inverse Dynamics Problem

For a desired manipulator trajectory $\mathbf{q}_r(t), \dot{\mathbf{q}}_r(t)$ and $\ddot{\mathbf{q}}_r(t)$, the system dynamics from (63.1) at time $t$ can be expressed as a linear equation

$$-J^T\mathbf{f} = \mathbf{w}, \tag{63.4}$$

where $\mathbf{w} = M(\mathbf{q})\ddot{\mathbf{q}} + C(\mathbf{q}, \dot{\mathbf{q}}) + G(\mathbf{q})$. The inverse dynamics problem refers to the determination of cable forces, $\mathbf{f}(t)$, subject to the constraints from (63.3) to satisfy the system dynamics in (63.4). For completely and redundantly restrained cable systems ($m \geq n + 1$) an infinite number of solutions exist [10]. One approach to resolve the cable force redundancy is to introduce an objective function for cable forces [3, 7, 11]. The minimisation of the sum of the squared cable forces results in a Quadratic Programming problem

$$\begin{aligned} \mathbf{f}^* &= \arg\min_{\mathbf{f}} \; \mathbf{f}^T H \mathbf{f} \\ &\text{s.t} - J^T\mathbf{f} = \mathbf{w} \\ &\quad \mathbf{f}_{min} \leq \mathbf{f} \leq \mathbf{f}_{max}, \end{aligned} \tag{63.5}$$

where $H$ is a positive definite weighting matrix and $\mathbf{f}^*$ is the cable force solution. The cable force constraint refers to the allowable cable forces described in (63.3).

### 63.2.2 Wrench-Closure Workspace

The Wrench-Closure Workspace (*WCW*) is defined as the poses in which the manipulator can sustain any arbitrary wrench without any upper cable force bounds

$$W = \{\mathbf{q} : \forall \mathbf{q} \in \mathbb{R}^n, \exists \mathbf{f} > \mathbf{0}, -J(\mathbf{q})^T\mathbf{f} = \mathbf{w}\}, \tag{63.6}$$

where $\mathbf{f} > \mathbf{0}$ denotes the positive cable force constraint. Point-based evaluation techniques to determine the WCW have been studied in [4, 6]. In [8], a hybrid numerical–analytical approach was proposed to generate the WCW with increased accuracy and efficiency compared to point-wise methods. In this approach, the WCW from (63.6) is reduced to a set of univariate polynomial inequalities and analytically solved. From this approach, the workspace is represented by a set of linear regions defined by $\mathbf{q}_r \in \mathbb{R}^{n-1}$, the constant pose for $n - 1$ variables, and the continuous region $(q_l, q_u)$, where $q_l$ and $q_u$ are the lower and upper bounds for variable $q$, respectively. This approach is extended to evaluate the workspace in Sect. 63.3.2.

## 63.3 Optimisation Problem and Evaluation Methods

The cable configuration optimisation problem can be expressed as

$$\begin{aligned} \mathbf{x}^* = \{\mathbf{r}_A^*, \mathbf{r}_B^*\} &= \arg \min_{\mathbf{r}_A, \mathbf{r}_B} C(\mathbf{r}_A, \mathbf{r}_B) \\ \text{s.t}\mathbf{r}_A &\in \mathscr{A}, \mathbf{r}_B \in \mathscr{B} \end{aligned} \tag{63.7}$$

where $\mathbf{r}_A^*$ and $\mathbf{r}_B^*$ denote the optimal cable attachment locations at the base and end effector, respectively. The sets $\mathscr{A}$ and $\mathscr{B}$ represent the possible attachment locations at the base and end effector, respectively, and form the constraints on the optimisation variables. The objective function, $C(\mathbf{r}_A, \mathbf{r}_B)$, represents desired properties of the manipulator and is dependent on cable configuration. Two classes of task specific objectives appropriate to cable-driven manipulators are presented in this section.

### 63.3.1 Cable Force Characteristics

The evaluation of cable force is a direct and meaningful measure on the configuration's performance in executing a desired trajectory, $\mathbf{q}_r(t)$, $\dot{\mathbf{q}}_r(t)$ and $\ddot{\mathbf{q}}_r(t)$, for a time period of $0 \leq t \leq t_{max}$. Denoting a penalty function for cable forces at time $t$ as $h(\mathbf{f}^*, t)$, the objective function to minimise the penalty over the entire trajectory can be expressed as

$$C(\mathbf{r}_A, \mathbf{r}_B) = \int_0^{t_{max}} h(\mathbf{f}^*, t)dt \tag{63.8}$$

where $\mathbf{f}^*$ represents the cable forces from the inverse dynamics problem in (63.5). One possible cable force penalty function at time $t$ is

$$h(\mathbf{f}^*, t) = \begin{cases} Q(\mathbf{f}^*), & \mathbf{f}_{min} \leq \mathbf{f}^* \leq \mathbf{f}_{max} \\ Q(\mathbf{f}_b), & \text{otherwise} \end{cases}, \tag{63.9}$$

where $Q$ is the function from (63.5), and $\mathbf{f}_b > \mathbf{f}_{max}$ represents penalty cable forces when there are no solutions for (63.5), penalising configurations in which the trajectory cannot be executed.

### 63.3.2 Workspace Evaluation

Upon determination of the workspace, for example, WCW, existence of desired regions can be evaluated. This is particularly to ensure that the designed manipulator is able to operate within a specifically prescribed region for a desired task.

By prescribing a desirability of manipulator pose $\mathbf{q}$ within the workspace $W$ as $v(\mathbf{q})$, the objective function to achieve maximum workspace desirability can be expressed as

$$C(\mathbf{r}_A, \mathbf{r}_B) = -\int_W v(\mathbf{q})\, dW. \tag{63.10}$$

One of the simplest measures of workspace is its volume, where the entire workspace is equally weighted $v_1(\mathbf{q}) = 1$. The disadvantage in such a general performance measure is that the maximisation of volume does not imply that the manipulator can operate in a desired region. To accommodate for this, $v(\mathbf{q})$ can be constructed to favour desired regions. For example, if the desired workspace region is the region about a manipulator pose $\mathbf{q}_s = \{q_{s_1}, q_{s_2}, , q_{s_n}\}$, where $q_{s_i}$ corresponds to the pose variable $q_i$, respectively, one weighting function can be

$$v_2(\mathbf{q}) = \frac{1}{1 + a_1(q_1 - q_{s_1})^2 + \ldots + a_{n-1}(q_{n-1} - q_{s_{n-1}})^2 + a_n(q_n - q_{s_n})^2}. \tag{63.11}$$

The properties of (63.11) are $v_2(\mathbf{q}_s) = 1$, $v_2(\mathbf{q}) \leq 1$, and $v_2(\mathbf{q}) \to 0, ||\mathbf{q} - \mathbf{q}_s|| \to \infty$. The constant $a_i$ corresponds weighting in dimension $i$.

Taking advantage of the workspace generation approach proposed in [8], the workspace desirability from (63.10) can be expressed as

$$C = -\sum_{q_{r_1}} \sum_{q_{r_2}} \cdots \sum_{q_{r_{n-1}}} \int_{q_l}^{q_u} v(\mathbf{q}_r, q) dq \cdot \Delta q_{r_{n-1}} \ldots \Delta q_{r_2} \cdot \Delta q_{r_1}, \tag{63.12}$$

where $\mathbf{q}_r = \{q_{r_1}, q_{r_2}, \ldots, q_{r_{n-1}}\}$ represents the constant variables and $q$ is the analytical variable. It is important to note that if the definite integral of $v(\mathbf{q}_r, q)$ with respect to $q$ can be determined analytically, then (63.12) becomes

$$C = -\sum_{q_{r_1}} \sum_{q_{r_2}} \cdots \sum_{q_{r_{n-1}}} (V(q_u) - V(q_l)) \cdot \Delta q_{r_{n-1}} \ldots \Delta q_{r_2} \cdot \Delta q_{r_1}, \tag{63.13}$$

where $V$ is the definite integral of $v(\mathbf{q}_r, q)$ with respect to $q$. It is worth noting that for both the volume and proposed task specific workspace region function from (63.11), $V(q) = \int v(\mathbf{q}) dq$ can be determined. Evaluation of workspace in such a manner preserves the advantages of the workspace generation from [8], increased efficiency and accuracy.

## 63.4 Simulation and Results

To illustrate and compare the different task specific objective functions, the optimisation of a completely restrained 3-DoF manipulator driven by 4-cables is presented. As shown in Fig. 63.2, the manipulator is constrained at the base

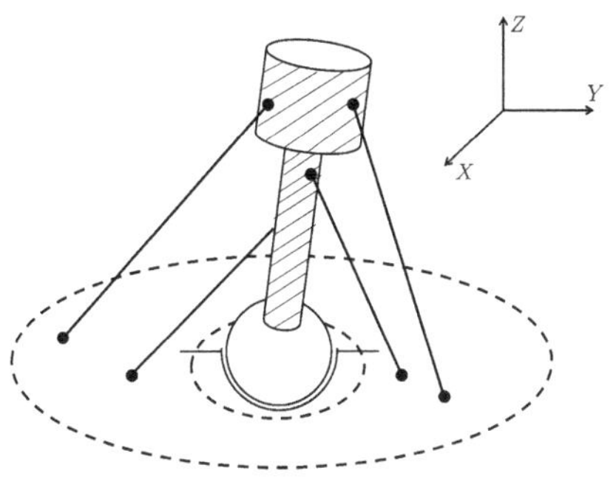


**Fig. 63.2** 3-DOF ball joint manipulator

**Table 63.1** Evaluation Function Descriptions

| Function | Type | Description |
|---|---|---|
| $C_1$ | *WCW region* | Function from (63.14) with weight values of $a_\alpha = a_\beta = a_\gamma = 100$ about pose $\alpha_s = -1, \beta_s = 0.5, \gamma_s = \frac{2\pi}{5}$ |
| $C_2$ | *Trajectory* | $\mathbf{q}_s = \left[-\frac{\pi}{10}\ \frac{\pi}{10}\ 0\right]^T$, $\mathbf{q}_e = \left[\frac{\pi}{10}\ -\frac{\pi}{10}\ 0\right]^T$ and $t_{max} = 1$ |
| $C_3$ | *Trajectory* | $\mathbf{q}_s = \left[\frac{3\pi}{10}\ \frac{2\pi}{5}\ -\frac{\pi}{6}\right]^T$, $\mathbf{q}_e = \left[\frac{\pi}{3}\ \frac{\pi}{5}\ -\frac{\pi}{3}\right]^T$ and $t_{max} = 1$ |
| $C_4$ | *WCW* | WCW volume (general performance measure as a baseline) |

through a ball joint, and the pose of the manipulator can be defined as $\mathbf{q} = [\alpha, \beta, \gamma]^{\mathrm{T}}$, where $\alpha$, $\beta$, and $\gamma$ are the *xyz*-Euler angles of rotation, respectively.

For the presented manipulator, three task specific objective functions $C_1$ to $C_3$ and one general measure $C_4$ have been selected for optimisation. These objective functions are described in Table 63.1. Trajectories are defined by a starting pose $\mathbf{q}_s$ and ending pose $\mathbf{q}_e$, assuming zero starting and ending velocities and accelerations $\dot{\mathbf{q}}_s = \dot{\mathbf{q}}_e = \ddot{\mathbf{q}}_s = \ddot{\mathbf{q}}_e = \mathbf{0}$. The trajectory $\mathbf{q}(t)$ is then generated from $t = 0$ to $t = t_{max}$ by quaternion interpolation. The evaluation function for a desired workspace region from (63.11) for the 3-DoF manipulator is

$$v_2(\mathbf{q}) = \frac{1}{1 + a_\alpha(\alpha - \alpha_s)^2 + a_\beta(\beta - \beta_s)^2 + a_\gamma(\gamma - \gamma_s)^2}. \tag{63.14}$$

For evaluation function $C_i$, the optimal cable configuration $\mathbf{x}_i^*$ was determined through a standard Particle Swarm Optimisation (*PSO*) algorithm. To allow comparison between task specific evaluation functions, each optimal configuration was also evaluated for $C_1$ to $C_4$. The results for these evaluations are shown in Table 63.2, where the minimum value representing the optimal cost is highlighted.

The optimal cable configuration for the WCW volume $\mathbf{x}_4^*$ is a robust configuration that allows the manipulator to satisfy a wide range of tasks. The disadvantage of such a configuration is that efficiency for specific tasks are not considered. For example, the $\alpha - \beta$ cross section of the WCW at $\gamma_s = \frac{2\pi}{5}$ is shown

**Table 63.2** Comparison of Evaluation Functions

| | $\mathbf{x}_1^*$ | $\mathbf{x}_2^*$ | $\mathbf{x}_3^*$ | $\mathbf{x}_4^*$ |
|---|---|---|---|---|
| $C_1$ | **−282.3** | −161 | −11.05 | −216.7 |
| $C_2$ | 116.7 | **3.8** | 8080006.5 | 13.9 |
| $C_3$ | 2333536.2 | 1864.8 | **282.6** | 6973 |
| $C_4$ | −35543 | −29966 | −2884 | **−36321** |

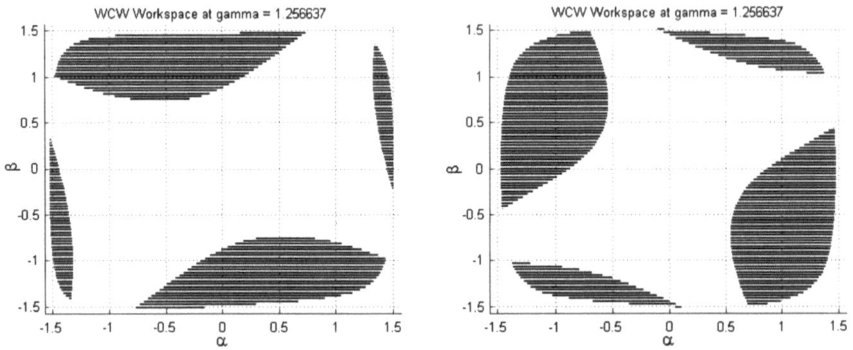


**Fig. 63.3** WCW for optimal configurations $\mathbf{x}_4^*$ and $\mathbf{x}_1^*$ **a** $\gamma = \frac{2\pi}{5}$ for $\mathbf{x}_4^*$ **b** $\gamma = \frac{2\pi}{5}$ for $\mathbf{x}_1^*$

in Fig. 63.3a. It can be observed that workspace does not exist about pose $\alpha = -1$ and $\beta = 0.5$. In comparison, the workspace from configuration $\mathbf{x}_1^*$, optimised for the workspace region about $(\alpha, \beta, \gamma) = (-1, 0.5, \frac{2\pi}{5})$, is shown in Fig. 63.3b, where the workspace about the desired location is satisfied. This is particularly useful to ensure that a manipulator is more robust to operate at a desired workspace region.

The benefits of task specific evaluations can be further observed for the minimisation of trajectory cable forces. It can be observed that $\mathbf{x}_4^*$ can perform the trajectories $C_2$ and $C_3$, but much less efficiently than the optimal configurations $\mathbf{x}_2^*$ and $\mathbf{x}_3^*$, respectively. The comparison in cable forces required to generate trajectory $C_3$ between configurations $\mathbf{x}_4^*$ and $\mathbf{x}_3^*$ is shown in Fig. 63.4. It is clear that $\mathbf{x}_3^*$ from Fig. 63.4b performs the trajectory much more efficiently, with a maximum force of approximately $20\,N$, compared to the maximum force of 100 N from $\mathbf{x}_4^*$.

From the comparison, the benefits in task specific evaluation to determine optimal cable configurations can be observed. This study suggests that it is difficult to obtain a robust cable configuration to satisfy a wide range of tasks, while performing them efficiently. In addition, the optimal configuration for one task specific objective typically performs poorly for another. Due to the potential reconfigurability of cable-driven systems, the adaptation of task specific objectives become feasible.

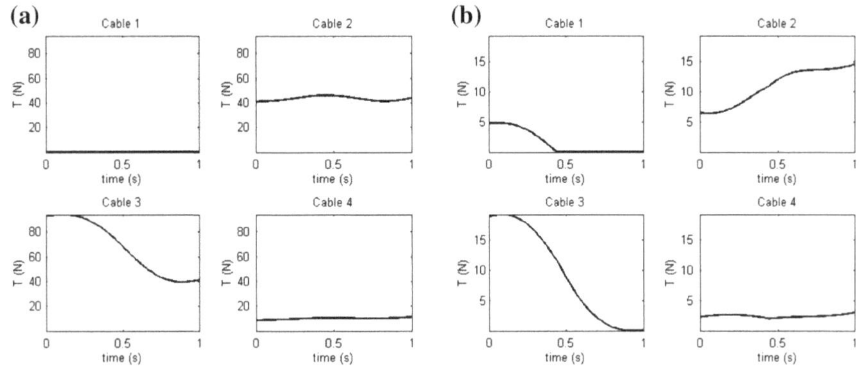


**Fig. 63.4** Cable forces for trajectory $C_3$ for optimal configurations $\mathbf{x}_4^*$ and $\mathbf{x}_3^*$ **a** Cable sources for $\mathbf{x}_4^*$ **b** Cable forces for $\mathbf{x}_3^*$

## 63.5 Conclusion

A study to compare the effectiveness of different task specific evaluation functions for optimising cable-driven manipulators was presented. Two classes of tasks were considered, minimisation of cable force to perform a trajectory, and maximisation of a workspace region. In addition, an efficient method to evaluate the quality of WCW was demonstrated extending from [8]. Performing optimisation on a 3-DoF cable-driven manipulator, the results indicated the improvement and savings of task specific objectives, as compared to a robust performance measure such as workspace volume. Future work should consider other task specific measures, such as cable interference, and to extend this to the optimisation of multilink manipulators.

## References

1. Agrawal SK, Dubey VN, Gangloff JJ Jr, Brackbill E, Mao Y, Sangwan V (2009) Design and optimization of a cable driven upper arm exoskeleton. J Med Dev 3:031, 004
2. Albus JS, Bostelman RV, Dagalakis N (1993) The NIST robocrane. J Rob Sys 10(5):709–724
3. Borgstrom PH, Jordan BL, Sukhatme GS, Batalin MA, Kaiser WJ (2009) Rapid computation of optimally safe tension distributions for parallel cable-driven robots. IEEE Trans Rob 25(6):1271–1281
4. Diao X, Ma O (2009) Force-closure analysis of 6-dof cable manipulators with seven or more cables. Robotica 27:209–215
5. Gosselin C, Angeles J (1991) Global performance index for the kinematic optimization of robotic manipulators. J Mech Des 113(3):220–226
6. Gouttefarde M, Gosselin CM (2006) Analysis of the wrench-closure workspace of planar parallel cable-driven mechanisms. IEEE Trans Robot 22(3):434–445
7. Hassan M, Khajepour A (2008) Optimization of actuator forces in cable-based parallel manipulators using convex analysis. IEEE Trans Robot 24(3):736–740

8. Lau D, Oetomo D, Halgamuge SK (2011) Wrench-closure workspace generation for cable driven parallel manipulators using a hybrid analytical-numerical approach. J Mech Des 133:071–004
9. Lim WB, Yeo SH, Yang G, Mustafa SK (2009) Kinematic analysis and design optimisation of a cable-driven universal joint module. Proceedings of IEEE/ASME international conference advanced intelligent mechatronics, pp 1933–1938
10. Ming A, Higuchi T (1994) Study on multiple degrees-of-freedom positioning mechanism using wires (part 1)—concept, design and control. Int J Jpn Soc Eng 28(2):131–138
11. Oh SR, Agrawal SK (2005) Cable suspended planar robots with redundant cables: controllers with positive tension. IEEE Trans Robot 21(3):457–464
12. Perreault S, Gosselin CM (2008) Cable-driven parallel mechanisms: application to a locomotion interface. J Mech Des 130:102–301
13. Surdilovic D, Bernhardt R (2004) STRING-MAN: A new wire robot for gait rehabilitation. Proceedings of the IEEE international conference on robotics and automation, pp 2031–2036
14. Yang G, Lin W, Kurbanhusen MS, Pham CB, Yeo SH (2005) Kinematic design of a 7-DOF cable-driven humanoid arm: a solution-in-nature approach. Proceedings of IEEE/ASME international conference advanced intelligent mechatronics, pp 444–449

# Chapter 64
# Motion Coordination Strategy of Quadruped Robot Inspired by the Locomotion of Animal

**Hongkai Li and Zhendong Dai**

**Abstract** Motion coordination is one of the key technologies that affect the stability and efficiency of robot. Animals' motion must have the advantages reserved from long time's evolution, which offers good prototypes for the control of robot. In this paper, the locomotion of quadruped animal, *gecko*, was observed when trotting on the level, and the trajectories of feet relative to the hip on the same side were analyzed. Results showed that the speed of feet on stance phase relative to the hips were the same, feet accomplished the change of moving direction and speed during transition process between stance and swing phases. This approach was simulated in a robot mode and used in trajectory plan of a quadruped sprawling robot.

**Keywords** Motion coordination · Robot · Kinematical chain · Trajectory

## 64.1 Introduction

Potential solutions to engineering problems can be found in investigating how these problems are solved in nature [1]. Over the long evolutionary periods of species, morphological characteristics of body and limbs have been optimized to improve the species' motion performance, which offers good bionic inspirations for the solution of practical motion control.

H. Li (✉) · Z. Dai
Institute of Bio-inspired Structure and Surface Engineering,
Nanjing University of Aeronautics and Astronautics, 210016 Nanjing, China
e-mail: hklee@nuaa.edu.cn

J. S. Dai et al. (eds.), *Advances in Reconfigurable Mechanisms and Robots I*,
DOI: 10.1007/978-1-4471-4141-9_64, © Springer-Verlag London 2012

The motion mechanism, composed of animal's body, limbs and the contact substrate, is variable and reconfigurable. Limb on swing phase is an open kinematic chain relative to the body. The limbs on stance phase together with body and the ground form close kinematic chains. During the motion, the previous motion mechanism is dissolved with the limb transfers from stance to swing phase, and the new is reconfigured with the limb transfer from swing to stance phase. The inter-coordination of limbs on stance phase and the inner-coordination of limb transfer between swing and stance phase are significant. These also exist in the control of robots. The motion coordination of robot should only be realized by man-made control system. But animals can achieve that inherently. So it will be an efficient way for the coordination control of robot to learn from animal.

Locomotion involves the harmonious activity of the entire animal, depending especially on the coordinated action of muscles, bones, nervous system and sense organs [2]. The movement coupling of the hindlimbs and forelimbs about intact and decerebrate cats under different types of locomotion in several preparations was analyzed. In pacing and trotting gaits, two basic patterns of coupling have been found, which simplify considerably the problem of neural control of the limbs in locomotion [3]. By training four cats locomotion on a treadmill at a variety of speeds, *Vilensky* et al. filmed at weekly intervals for three weeks to characterize the locomotor behavior variation exhibited by the cats in total cycle and swing/stance durations with the variation of speed [4]. *English* recorded the simultaneous electromyography of each limb of intact cats during repeated overground stepping trials to determine the temporal spacing of step cycles. Analysis of the coordination of step cycles of both homologous limb pairs, both homolateral limb pairs and both sets of diagonal limbs, suggest that the step cycles of the four limbs are coordinated according to a few frequently occurring patterns [5]. Symmetry is an typical pattern for the robot control [6, 7]. By coordinating the limbs pairs move simultaneously, symmetrical gaits such as trotting, pacing, bouncing was studied and used to the control of robot [8–12]. But this method requires the control system to calculate the accurate foot points of the next step.

Motion coordination of legged robot includes inter-coordination of legs at stance phase and inner-coordination of single leg at different motion state. The two kinds of coordination are simultaneous with new closed kinematic chain reconfigured and leg enters and exits the kinematic chain during the motion. Existing work mainly concentrated on the inter-coordination of legs and rarely on the inner-coordination and the relationship of the two kinds of coordination. As an outstanding climber, gecko could move in three dimensional spaces, and obtained desirable focus on behalf of its outstanding locomotion performance and biologic characters [13–15]. In this paper, the locomotion of gecko was observed, and the trajectories of feet tips relative to the hip were obtained by picking up the key points from the filmed records. The motion coordination strategies of gecko were concluded by analyzing the trajectories. At last a motion coordination method was proposed and simulated in a gecko-like robot model.

## 64.2 Trajectory of Gecko's Ankle Joint During Trotting

Trajectories of feet relative to the body were planned for the control of quadruped robot. Normally, the body of robot was rigid, and the relative position between the center of body and each hip was fixed. So the trajectory of foot could also be planned relative hip. In this part, trajectories of the two limbs on right side were observed to study the coordination pattern when gecko trotted on the level.

For the experiment, a three-dimensional locomotion observation system was built [16, 17]. Locomotion was filmed by the high speed camera when gecko trotted along the channel. The spatial position of limbs could be obtained from each recorded frames.

One of the recorded frames and the global coordinate system ($\Sigma o$) which build based on right hind rule were showed in Fig. 64.1. Four hip coordinate systems were built on each hip paralleling to the global coordinate system.

Homolateral limbs were selected because locomotion of two limb pairs on both sides was nearly same except for time phase when on trot gait. Here the limbs on right side were selected. The trajectories of feet relative to the corresponding hip, showed in Fig. 64.2, were smoothed by Savitzky-Golay method in Origin Pro 8 (Origin Lab Corp., Northampton, MA, USA) when gecko trotted straightly at 337.1 mm/s. The difference between smoothed and original data was considered to be insignificant if significance levels of 0.05 were reached by variance analysis.

In Fig. 64.2, the solid lines are trajectories of front foot and the dashes are that of rear foot. The duty cycle of front foot, determined by the locomotion at Z direction in reference coordinate system, is showed at the bottom.

The flexible body of gecko can swing during motion, even at the stance phase the position of hip joints still vary in the lateral and up-down directions. And the curves shape of front and rear feet in each plot are nearly the same except the attitudes. But along the moving direction (Y direction), two curves are nearly the same even though the length difference of front and rear legs.

The curves in Y direction were divided into several segments based on duty cycle for further analysis. $S_1$, $S_4$ are trajectories of rear foot at stance phase, and $S_2$ and $S_6$ are that of front foot. $S_3$ and $S_5$ are the trajectories of rear and front foot at swing and stance phase respectively. Linear-regression was analyzed about segments $S_1$, $S_2$, $S_4$ and $S_6$, and the result was showed in Table 64.1. It could be found that difference of slopes is not large, which means that the curves of each segments should be about parallel.

It implied that feet at stance phase kept the same speed though the ecologic difference when trotting at certain speed from the results of Y-direction. And the distance between front and rear hip of legs on stance phase kept constant along motion direction.

The trajectories of feet on swing phase in Y direction could be divided to three stages. At the beginning of swing phase, foot lift off and kept moving backward at the same speed as before. When feet totally departed from the ground, foot change motion direction smoothly and then move forward. In the second state the foot

**Fig. 64.1** Coordinate system

moves forward at a nearly constant speed. In the last phase, foot changed motion direction again to move backward, and ready to touch down. The foot reached the same speed with the other feet on stance before touching down the substrate. By this way, gecko satisfied the geometric constrains along the motion direction. In this process, the inner-coordination of each motion stage coordinated smoothly, and the inter-coordination of legs on stance phase was realized efficiently.

## 64.3 Motion Coordination Analysis of Robot

Animals have adapted to the environment strongly in morphology and motion style after long time's evolution. It has been a popular method for designing and controlling robot to learning from nature. Generally, the structure of quadruped robot is symmetric. Base on the mechanism of gecko's hind limb [18], a sprawling quadruped robot mode was brought forward (Fig. 64.3). There were four joints on one leg. $J_1$ and $J_2$ composed the compound hip joint. $J_3$ was knee joint, and $J_4$ is ankle joint. $J_1$, $J_2$ and $J_3$ were active revolution joints and were actuated by servo motors. Ankle joint was a passive spherical joint ($J_4$) to improve the adaptability. Extension springs were used to link the foot and the leg articulation to adapt the motion passively.

In swing phase, leg is an open kinematic chain, and foot could reach any position in its motion space freely. Legs in stance phase together with body and contact substrate compose a closed kinematic chain, and the motion of each leg on stance phase should be satisfied with the constraints among them to avoid the interference among them. The transition between two motion phases is also the moment that interference would emerge. The following would analyze the coordination from kinematics.

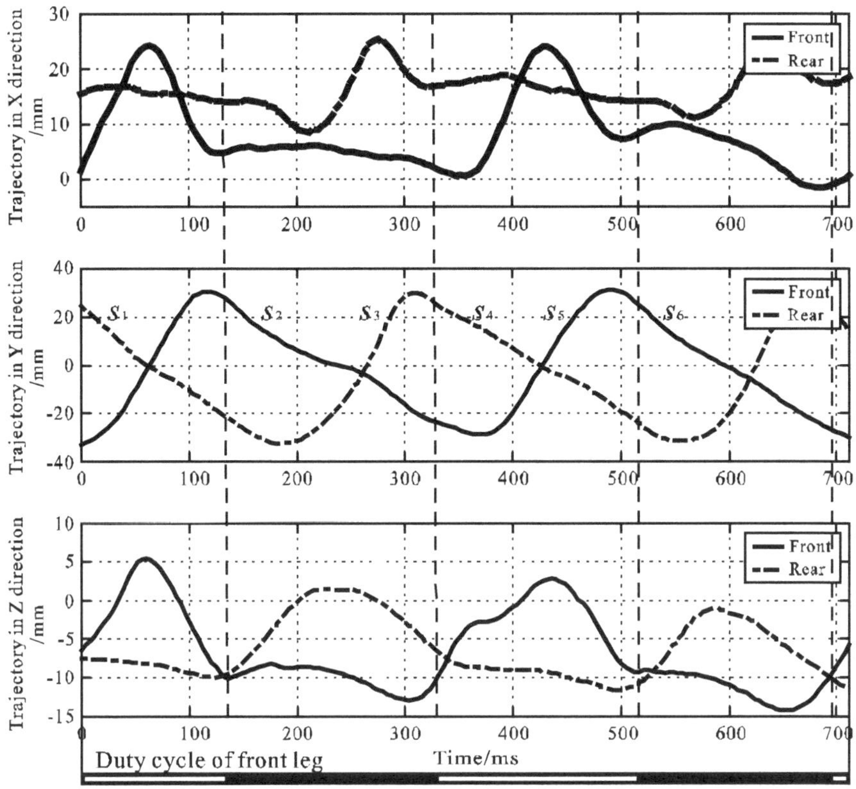


**Fig. 64.2** Curves of ankle joints relative corresponding hip joints

**Table 64.1** Results of linear-regression analysis

| Segment | Slope | $R^2$ |
|---|---|---|
| $S_1$ | −1.1971 | 0.9981 |
| $S_2$ | −1.1202 | 0.9911 |
| $S_4$ | −1.0542 | 0.9956 |
| $S_6$ | −1.0838 | 0.9971 |
| Avg. | −1.1138 | |

In Fig. 64.3, $\Sigma_O$ is the global coordinate system fixed on the ground. $\Sigma_{Hi}$ is the coordinate system of hip fixed on the hip joint, and the transition between the coordinate systems of hips is constant. $V_i$ (i = 1, 2, 3, 4) is the velocity of foot relative to hop joint ($\Sigma_{Hi}$), $A_i$ (i = 1, 2, 3, 4) is the position of foot in body coordinates. $\boldsymbol{r}_{hi}$ is the vector from hip joint to foot. Define $\mathbf{r}_{31}$ as the vector from $A_3$ to $A_1$, and $\mathbf{r}_{h31}$ is the vector between the two hips. So,

$$r_{31} = r_{h31} + r_{h1} - r_{h3} \tag{64.1}$$

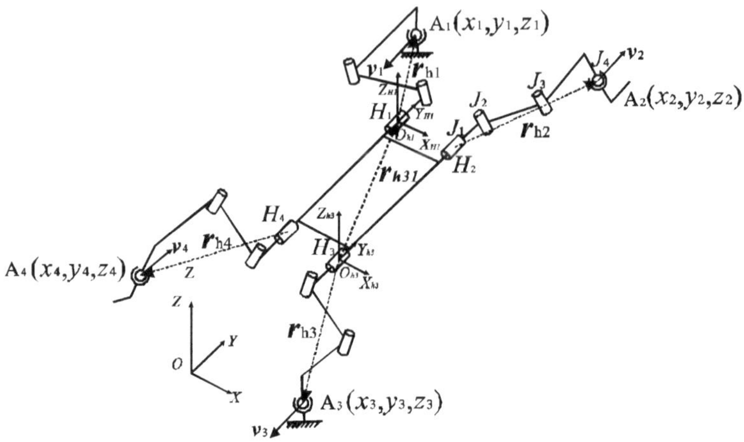


**Fig. 64.3** Mechanism of sprawling quadruped robot

The derivation of Eq. (64.1) respect to time is

$$\frac{dr_{31}}{dt} = \frac{dr_{h31}}{dt} + \frac{dr_{h1}}{dt} - \frac{dr_{h3}}{dt} \tag{64.2}$$

Here assumed that: a) the body of robot keeps parallel translation, and b) there's no sliding between feet and the substrate.

In Fig. 64.2, foot 1 and 3 are on stance phase. According to the assumption, $\mathbf{r}_{31}$ is a constant vector, and the positions of feet are fixed on the ground. Define $v_{\text{xhi}}$, $v_{\text{yhi}}$ and $v_{\text{zhi}}$ are the absolute speeds of hip along each arises, the following could be deduced from Eq. (64.22),

$$v_{xh1}i + v_{yh1}j + v_{zh1}k = v_{xh3}i + v_{yh3}j + v_{zh3}k \tag{64.3}$$

It could be considered that the body is fixed, and the feet move relative to the body according to the rule of relative motion, then,

$$v_{xA1}i + v_{yA1}j + v_{zA1}k = v_{xA3}i + v_{yA3}j + v_{zA3}k \tag{64.4}$$

where, $v_{\text{xAi}}$, $v_{\text{yAi}}$, $v_{\text{zAi}}$ are speeds along axis x, y and z axis in hip coordinate systems.

From Eq. (64.4) it could be concluded that when the body of robot keeps parallel translation, motion coordination could be realized kinematically among legs on stance phase by equalizing the relative velocity of foot on stance phase to the hip, and is regardless of the initial positions of contact point.

The topological structure of motion mechanism of robot was changed and reconfigured in motion cycle. When one leg switched into stance phase, there may be more than two legs on stance phase. With the number of the closed loop motion chains increasing, the coordination among legs will be more complicated. If the leg converts to stance phase without the same velocity with the others legs in stance phase, there interference will happed in the new closed kinematical chains.

Relative velocities of feet to hip should be equal to satisfy the motion coordination. And this is also the requirement for the legs that is going to switch into stance phase. This will be reached by adjusting the velocity of foot to the same with the feet on stance phase convert switching into stance phase. When to convert to swing phase, the foot keeps the same speed as on stance phase along forward and lateral directions and lift off at the same time. When departing from the substrate totally, it turns to move forward. And this strategy is also adopted by gecko. When the motion of body is translational, the general method for inter-coordination among feet on stance phase could be depicted as below:

$$\begin{cases} \left.\begin{aligned} v_{xi} &= v_{xj} \\ v_{yi} &= v_{yj} \\ v_{zi} &= v_{zj} \end{aligned}\right\} & \begin{aligned}&\text{(velocity of feet in stance phase relative to hip.} \\ &\qquad i,j = 1,\ 2,\ 3,\ 4;\ i \neq j) \end{aligned} \\ \left.\begin{aligned} x_i &= free \\ y_i &= free \\ z_i &= free \end{aligned}\right\} & \begin{aligned}&\text{(stance coordinates in hip coordinate system.} \\ &\qquad i = 1,\ 2,\ 3,\ 4) \end{aligned} \end{cases} \tag{64.5}$$

Interference might occur in the moment of touching. It's difficult to satisfy all the coordination conditions in advance for the foot that transfer to stance phase because at least one freedom should be left touch the substrate. Normally the condition in up-down direction would be the last to match, and impact always happed during the touch process. Sensors and energy storage mechanisms were always used to reduce the influence.

Motion of leg was actuated by the motors on the joints. The joints angles could be presented as the function of coordinates of foot in hip coordination system by inverse kinematics. After the motion of foot was schemed, the control parameters of motors could be deduced.

## 64.4 Simulation and Results

According the motion robot mechanism, a robot model was built and simulated in software ADAMS (Mechanical Dynamics Inc., Santa Ana, CA, USA). The motion was schemed to move forward straightly based on the proposed method.

The projections of trajectory at each axis in hip coordinate system were showed in Fig. 64.4. Foot moved to the initial position from the original position during the first 4 s, and then transferred into stance phase. From the 4th to 12th second the foot was on stance phase, it moved backward at a constant speed relative to body along motion direction, and kept the same distance with hip at up-down and lateral direction. At the 12th second, foot switched into swing phase. In the first a few

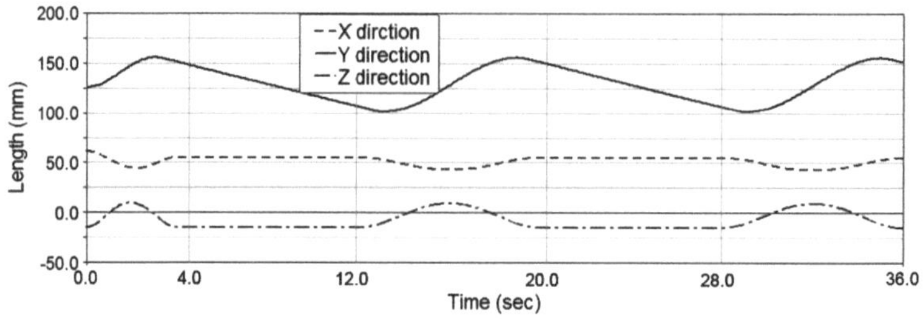


**Fig. 64.4** Displacement of front right foot relative to body center

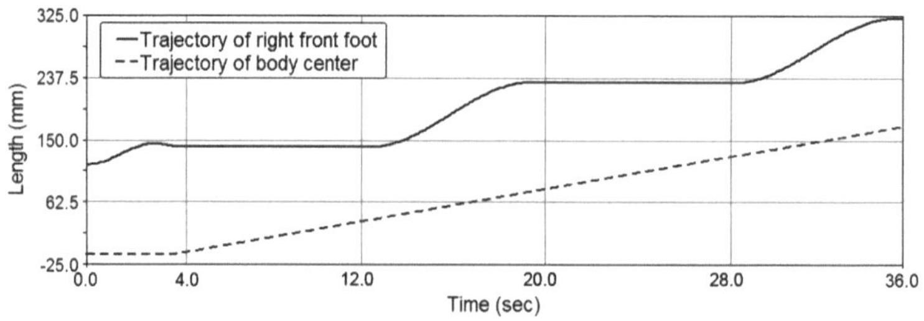


**Fig. 64.5** Displacement of front right foot and body relative to ground

seconds of this stage, it kept the same speed in each direction as in stance phase. Then motion direction changed, and the foot began to move forward relative to body. When moved to the forefront, it changed direction and moved backward, and accelerated to the same velocity with the others feet on stance phase. The swing phase ended at the 20th second. It entranced to stance phase after touching the substrate, and the whole motion cycle finished.

The displacements of foot and body relative to ground were showed in Fig. 64.5. According to the method of trajectory plan, the foot transferred between stance and swing phased smoothly. And the body moved forward at an even speed in the whole motion cycle.

Motion coordination of quadruped robot includes the inter-coordinate of legs on stance phase and the inner-coordinate of single leg switching between the stance and swing phases. The simulation results showed that the motion coordination of legs was realized and the body could move steady based on the method proposed in this paper.

**Acknowledgments** The project is supported by the National Program on Key Basic Research Project (2011CB302106), Foundation of Nanjing University of Aeronautics and Astronautics (4015-56YAH11017), Projects of International Cooperation and Exchanges NSFC (60910007) and Chinese NSFC (51175249, 61175105, 51105201).

## References


1. Espenschied KS, Chiel HJ, Quinn RD, Beer RD (1993) Leg coordination mechanisms in the stick insect applied to hexapod robot locomotion. Adapt Behav 1:455–468
2. Manter JT (1938) The dynamics of quadrupedal walking. J Exp Biol 15:522–540
3. Millera S, Burga JVD, Mechéa FGAVD (1975) Coordination of movements of the hindlimbs and forelimbs in different forms of locomotion in normal and decerebrate cats. Brain Res 91:217–237
4. JA, Vilensky, Patrick, MC (1984) Inter and intratrial variation in cat locomotor behavior. Physiol Behav 33:733–43
5. English AW (1979) Interlimb coordination during stepping in the cat: an electromyographic analysis. J Neurophysiol 42:229–243
6. Raibert M (1986) Symmetry in running. Sci. 231:1292–1294
7. Raibert MH (1986) Running with symmetry. IJRR 5:3–19
8. Raibert, MH(1986) Legged robots that balance. MIT Press, Cambridge
9. Raibert, M, Chepponis, M, Brown, HBJ (1986) Running on four legs as though they were one. IJRA. RA-2, 70–82
10. Won, M, Kang, T, Chung, WK (2008) Dynamical balancing control for stable walking of quadruped robot. In: 5th international conference on ubiquitous robots and ambient intelligence, 769–773
11. Raibert, M, Blankespoor, K, Nelson, G, Playter, R,Team, tB (2008) BigDog, the rough-terrain quadruped robot. In: Proceedings of the 17th world congress the international federation of automatic control, 10822–10825
12. Buehler, M, Playter, R, Raibert, M (2005) Robots step outside. In: Internal symposium on adaptive motion of animals and machines
13. Autumn K, Liang TA, Flsieh ST et al (2000) Adhesive force of a single gecko foot-flair. Nature 405:681–685
14. Bhushan, B, Sayer, RA (2008) gecko feet: natural attachment systems for smart adhesion-mechanism, modeling, and development of bio-inspired materials. In: Bhushan, B, Tomitori, M, Fuchs, H (eds) Applied scanning probe methods X. Springer, Berlin, 1–61
15. Santos D, Heyneman B, Kim S, Esparza N, Cutkosky MR (2007) Directional adhesion for climbing: theoretical and practical considerations. J Adhes Sci Technol 21:1317–1341
16. Li H, Dai Z, Shi A, Zhang H, Sun J (2009) Angular observation of joints of geckos moving on horizontal and vertical surfaces. ChSBu 54:592–598
17. Dai Z, Wang Z, Ji A (2011) Dynamics of gecko locomotion: a force-measuring array to measure 3D reaction forces. J Exp Biol 214:703–708
18. Zaaf A, Herrel A, Aerts P, Vree FD (1999) Morphology and morphometrics of the appendicular musculature in geckoes with different locomotor habits (Lepidosauria). Zoomorphology 119:9–22

# Chapter 65
# Robotics Studies in Europe

**Valentina Resaz, Fabrice Meriaudeau, David Fofi and Matteo Zoppi**

**Abstract** This paper describes the organization and teaching methodologies for mechanics and robotics related subjects in the European Erasmus Mundus master programmes EMARO (European master in Advanced Robotics), and VIBOT (master courses in VIsion & roBOTics). The structure of these masters is overviewed, the experience in designing and managing them is outlined and we point out the common effort for a global reach of these programmes and to build a transnational teaching architecture.

**Keywords** European higher education · Robotics · Computer vision · University networking

## 65.1 The Erasmus Mundus Programme

In these years we are facing unprecedented global challenges that affect all aspects of our lives and that are no more constrained by geographical borders. The European University Association (EUA) strongly believes that European

V. Resaz (✉) · M. Zoppi
School of Engineering and Architecture, University of Genoa, Genoa, Italy
e-mail: valentina.resaz@unige.it

M. Zoppi
e-mail: zoppi@dimec.unige.it

F. Meriaudeau · D. Fofi
Centre Universitaire Condorcet, Université de Bourgogne, Le Creusot, France
e-mail: fabrice.meriaudeau@u-bourgogne.fr

D. Fofi
e-mail: david.fofi@gmail.com

J. S. Dai et al. (eds.), *Advances in Reconfigurable Mechanisms and Robots I*,
DOI: 10.1007/978-1-4471-4141-9_65, 

universities are called to face such challenges by working within a global research community: Universities need to actively contribute to the creation of new knowledge by educating talented individuals in being creative and capable to find innovative solutions both to current and to new problems [1].

In this framework, Erasmus Mundus represents a truly effective and successful answer that is able to promote the circulation of talented students and scholars in order not only to enhance research capacity and bring new knowledge but also to foster language/cultural knowledge and, in a wider sense, interpersonal skills. Erasmus Mundus is a cooperation and mobility programme in the field of higher education whose major goals are to enhance the quality of European higher education and to promote dialogue and understanding between people and cultures through cooperation with third-countries [2].

The Erasmus Mundus programme supports selected European master's courses by providing very attractive scholarships to non-European and European students to engage in graduate study in the institutes involved in the programmes. It also provides grants to scholars worldwide (European and non European), to conduct teaching and research work related to master's topics in the selected consortium institutes.

Robotics related studies greatly benefit from the Erasmus Mundus programme since, to effectively teach robotics, different disciplines must be integrated such as mechanics, computer engineering, electronics, automatic control, sensors, signal and image processing, shape recognition and 3D vision. A project able to design, implement and manage scholar and student mobility, from and to different partner institutions, enables a synergistic combination of several excellent research activities, histories and expertise which significantly enrich both partners and the consortium as a whole.

In the Erasmus Mundus framework, EMARO and VIBOT represent two European curricula of excellence in the field of robotics.

## 65.2 EMARO

EMARO (http://EMARO.irccyn.ec-nantes.fr) is an integrated master course designed to promote high-quality education in the area of advanced and intelligent robotics [3]. The EMARO programme is driven by a consortium with multidisciplinary and synergistic competences, representing excellence in robotics research and education. The partners are: École Centrale de Nantes (ECN)–France (coordinator), Warsaw University of Technology (WUT)–Poland, University of Genova (UG)–Italy, Asian Institute of Technology (AIT)–Thailand, Faculty of Science and Technology, Keio University (KEIO)–Japan, Shanghai Jiao Tong University (SJTU)–China.

The EMARO consortium has implemented a solid academic program by a continuous transfer of research expertise and teaching methods together with a rich offer of specialization courses in advanced robotics. The Asian participation further extends EMARO visibility worldwide, particularly in the far-east. A clear evidence of that is the large number of applications from Asian countries.

### 65.2.1 Curriculum Description

The interdisciplinary EMARO programme comprises Mechanics, Electronics, Automatic Control, Computer Engineering, and Mathematics, as well as nanotechnology and neuroscience. The duration of the EMARO master is two academic years corresponding to 120 ECTS credits (European credit transfer system). The language of instruction and examination of all courses is English. Each student spends the first year in one institution and the second year in a second, different, institution. Successful students obtain the master degrees from both institutions.

The first year (semester 1 and 2) is offered simultaneously in the three European institutions with the same courses (with the same programs). Its aim is to provide the students with a solid background across the main areas of robotics: perception (vision, sensors), cognition (computer science, artificial intelligence, neural networks), action (actuators, motion control, cooperative control), mechanics (kinematics, dynamics mechatronic and sustainable design), Industrial and service applications (robot programming and group project), mathematical foundations (modeling, simulation, and optimization). Local language and culture modules are provided as well in order to strengthen the students' international mindset. Participation to summer internships and summer schools on robotics related subjects is encouraged during the summer break (July and August).

In the second semester, particular attention is dedicated to the courses providing skills in the mechanical sides of mechatronics and in physical integration of robotic and automatic systems. The aim is to form engineers with both solid practical ability and strong theoretical skills, overcoming a weakness typical of robotics curricula. Screw theory is thought and applied to the analysis and synthesis of mechanisms. Fundamentals of mechanical design and related tools and methods are addressed. A group project is assigned to all students, in small groups of two–three, in the second semester: the researchers of the hosting institution provide the subjects and involve the students in their current research; each assignment comprises parts of implementation, it is scheduled and considered complete only when planned results have been achieved. Each group prepares a report and presents their results to a committee for their evaluation.

During the third semester, the EMARO programme delivers more specialized courses, different in each partner institution, based on the specific research strengths available locally (Table 65.1). This customization of the program to local excellences in the partner universities helps students preparing to the research subjects they develop in the same institutions in the final thesis.

As the students move in a different country in the second year, all universities organize local language and culture modules as well as a course of research methodology, which is preparatory to the actual thesis work.

**Table 65.1** Semesters 3 courses

| ECN | | WUT | | UG | |
|---|---|---|---|---|---|
| Modules | ECTS | Modules | ECTS | Modules | ECTS |
| Advanced modeling of robots | 4 | Biomechanics * | 5 | Flexible automation,* | 6 |
| Identification and control of robots | 4 | Bio-robotics * | 5 | Systems identification* | 5 |
| Humanoid and walking robots | 4 | Dynamics of multi-body systems | 5 | Ambient intelligence | 5 |
| Optimal kinematic design of robots | 4 | Motion synthesis in robotics | 5 | Computer graphics | 5 |
| Biologically inspired robots | 4 | Walking robots | 5 | MEMS design (micro-robotics) | 6 |
| Ambient intelligence | 4 | Advanced mechanical design | 5 | Software architectures for robotics | 5 |
| | | Ambient intelligence | 5 | Hardware design languages | 5 |

Team works and research activities are offered in the two years of studies to provide students with the opportunity to apply their specialized knowledge to real problems. Practical *hands-on* experience on design and testing of a robotic system teaches the students skills and tools that are necessary to investigate a research topic.

During the fourth semester, the students carry out their Master Thesis work under the joint supervision of tutors from two different consortium partners. Because of the strong industrial *liasons* of the EMARO partners, the master research topic can be supervised and located either in the research labs of the partner institutions or in R&D department of an industrial company. A relevant part of the thesis topics are related to ongoing research projects funded by the European Commission under FP7 (NMP, ICT, Transport, Health). The students involved in these topics are involved in the projects, participate in the advancement meetings and spend periods in the partner organization further strengthening their skills in high level research organization and management.

The official master degrees delivered are the partner national degree as listed in Table 65.2: they are officially recognized by each local ministry and give full, worldwide access to PhD study programs.

To exploit the full potential of the complementary areas of expertise available in the consortium, more than 30 % of the modules in each institution are taught by staff from the other partners. Furthermore, the EMARO academic staff every year is enriched by scholars from outside of the consortium that are financed by the Erasmus Mundus programme to perform teaching and research activities in the partner institutions.

**Table 65.2** EMARO official master degrees

| Name of institution | Local degree awarded | Local degree awarded (translated in English) |
|---|---|---|
| Ecole Centrale de Nantes | Master automatique et systèmes de production, spécialité automatique, robotique, signal et images | Master in control engineering and production systems. Speciality: automatic control, robotics, signals and images |
| Warsaw University of | Technology | Magister, specjalność automatyka i robotyka |
| Master of | engineering in robotics and control | |
| University of Genova | Laurea magistrale in robotics engineering | Master in robotics engineering |

## *65.2.2 EMARO Students*

EMARO is open to worldwide talented students who already hold a first university degree in a field such as: Mechatronics, Automatic Control, Computer Science, Electrical Engineering, Mechanical Engineering, and Applied Mathematics. The applicants have to be fluent in English.

Since its first edition in 2008, the number of applications to EMARO has been very high compared to the number of accepted students: 400 to 600 candidates yearly for about 30 positions.

Even if the financial support provided by the scholarships has been every year reduced, the number of selected students has kept constant as the number of self-funding students has increased. The largest part of the application that EMARO received were from India, China and Pakistan, yet the consortium has strived to keep the presence of students of different nationality as balanced as possible by assigning no more than 2 scholarships for each country. Figure 65.1 shows the students distributions by continents.

Beside a strong technical skills portfolio, the EMARO graduate has an international profile, skill in foreign languages, a knack for mobility and an open attitude toward multicultural environments. These traits are highly appreciated by multinational companies spread around the world. The career prospects for EMARO graduates are therefore excellent in many industrial and economical fields as well as in the research field.

## 65.3 VIBOT

The VIBOT master is divided into four semesters designed and developed to ensure a clear and structured educational progress in the field of computer vision and robotics, fully integrated among the three universities of the consortium:

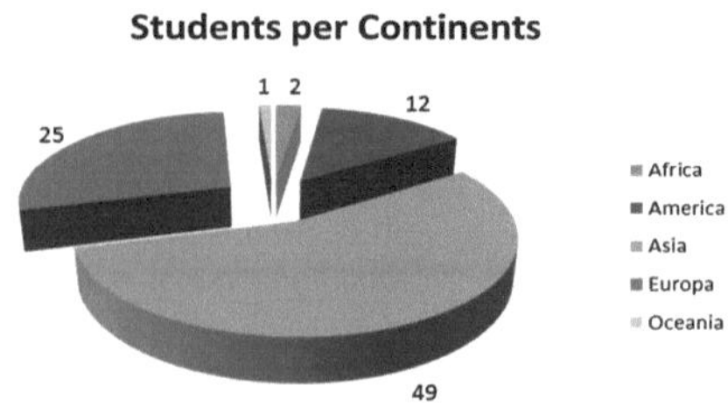


**Fig. 65.1** Student distribution per continent in the first three editions of EMARO

- Semester 1 University of Burgundy (UB): fundamentals on image processing, sensors and acquisition, mathematics and computer sciences (*basis*);
- Semester 2 University of Girona (UDG): improvements on image processing and pattern recognition, complements on computer vision and medical imaging and an introduction on robotics (*reinforcement*);
- Semester 3 Heriot–Watt University (HWU): high-level processing and methods for real-time implementation and integration of vision and robotics (*capitalization*);
- Semester 4 (research lab or company): research training in an applied or academic context (going further).

The overall objective is to start from science fundamentals and low-level image processing and then introduce mid-level image processing, computer vision and applications in medical imaging, ending with robotics and high-level integration so that the main methods and algorithms in the state-of-art of computer vision and robotics (through both a theoretical and applied point of view) are covered.

An introduction week is organized each year in September during which the enrolled students are informed about the programme, the assessment and the rules. The introduction week also permits new students to meet second-year students or former VIBOT students.

### *65.3.1 Curriculum Description*

The modules are mainly organized into lectures, tutorial work, lab work/projects and seminars:

- Lectures are taught to give the students the basic concepts, the state-of-art, the mathematical background and to foster exchanges and discussions with the professor permitting hindsight and reflection;
- Tutorial work are sessions of exercises based on lectures and permitting the students to solve theoretical problems with the help of a professor;
- Lab work/projects are practical problems given to the students to be solved autonomously (some are supervised by invited scholars)—note that robotics projects are prepared in France and Spain and finalized in Scotland by using the same robotic platform;

**Table 65.3** VIBOT courses and credits

| Semester 1: University of Burgundy | ECTS credits |
| --- | --- |
| B31XA digital signal processing | 5.5 |
| B31XB introduction to image processing | 5.5 |
| B31XC sensors and digitization | 5.5 |
| B31XD software engineering | 5.5 |
| B31XE applied mathematics | 5.5 |
| B31XF french culture | 2.5 |
| **Semester 2: University of Girona** | ***ECTS credits*** |
| B31XG probabilistic robotics | 6 |
| B31XH autonomous robots | 5 |
| B31XI scene segmentation and interpretation | 6 |
| B31XJ visual perception | 6 |
| B31XK medical image analysis | 5 |
| B31XL local culture | 2 |
| **Semester 3: Heriot–Watt University** | ***ECTS credits*** |
| B31XM advanced image analysis | 7.5 |
| B31XN multi-sensor fusion and tracking | 7.5 |
| B31XO real time imaging and control | 7.5 |
| B31XP robotics project | 5 |
| B31XQ local culture | 2.5 |
| **Semester 4 (at one partner, industry or research facility)** | ***ECTS credits*** |
| MSc project (M) | 30 |

- Seminars are research presentations given by renowned researchers and/or invited scholars permitting the students to go further into vision and robotics and to be aware of the novel methods and algorithms published in the field (Table 65.3).

### *65.3.2 VIBOT Students*

The masters in VIsion & roBOTics (VIBOT) has been selected by the European Commission since September 2006 in the Erasmus Mundus programme and is regularly running since then.

The following graphs demonstrate the attractiveness of the programme. It should be noticed that from the beginning of the programme, VIBOT enrolls more students than Erasmus Mundus grants, which illustrates the capacity to attract students without EM grants (sustainability) through industrial or institutional partnerships (Fig. 65.2).

The originality and a strong point of VIBOT is that students move in cohort from one university to the other, semester after semester. This group mobility, in contrast with option-based mobility, permits to create an *esprit de corps* and a true

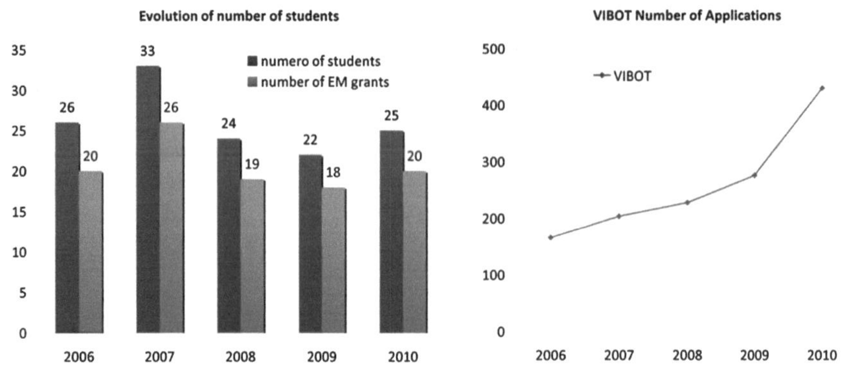


**Fig. 65.2** VIBOT students and number of applications

class which facilitates the belonging and integration to VIBOT and leads to an authentic VIBOT identity. Furthermore academic staff has built up valuable experience in multi-cultural group teaching, not only through the presence of international students but also through inclusion of international invited scholars involved in teaching or in research activities.

## 65.4 EACOVIROE

The consortia of VIBOT and EMARO have joined their forces to promote a new extended consortium, EACOVIROE (eacoviroe.org), whose aim is to enhance the attractiveness of European master programmes related to the fields of computer vision and robotics [4]. The specific targets of EACOVIROE are Asian students, as the applications rate from that area in VIBOT and EMARO is particularly high. A broad and general survey in Europe to identify masters in the field of computer vision and robotics has been done; quality assurance procedures have been investigated with particular attention to the hosting of Asian students (accommodation packs, visa procedures, residency cards, tutoring).

A web portal has been realized allowing potential students to obtain relevant information regarding the master degree programmes in computer vision and robotics. A web portal where companies and laboratories in the field of computer vision and robotics can post their job opportunities has been also created. A quality handbook has been edited which includes the best practices in Europe on Asian students hosting. The handbook serves also as a template for other European universities and can be downloaded from our website.

## 65.5 Conclusion

EMARO and VIBOT are examples of fully interdisciplinary integrated masters, the realization of which has been possible only through the impulse coming from the Erasmus Mundus programme of the European Commission; they are having a pioneering role in the construction of the future higher education offer in the fields of advanced technology in Europe. The number of applications to these programs is remarkably high compared to the number of positions currently available and it is constantly growing together with the number of self-supporting students.

The EACOVIROE project, originated from the experience of the two masters, has shown that the high quality international teaching offer in the new technology areas in Europe is fast increasing.

Worldwide research breakthroughs and technological improvements are generated from global cross-fertilization between traditionally different disciplines. The goal of the European university system is to become an active part of the global interaction, promoting the transformation of innovation systems and technologies in new jobs, for the promotion of local and global wellness, but also increasing the reciprocal knowledge and understanding in the international relations to fully achieve inclusivity and equity.

A European PhD program is in preparation on the topics of EMARO. It will give graduate students the opportunity to start a PhD with consistent continuity with the topics of their master studies and master thesis, in the same transnational and interdisciplinary framework and with similar strong links with China, Japan, Thailand and the whole East.

**Acknowledgments** The Erasmus Mundus program of the European Commission is here acknowledged for the co-funding. All the EMARO and VIBOT staff are also gratefully acknowledged.

## References

1. EUA Aarhus Declaration (2011) Investing today in talent for tomorrow. http://www.eua.be/Home.aspx. Accessed May 2011
2. Erasmus Mundus official site. http://eacea.ec.europa.eu/erasmus_mundus/index_en.php
3. Khalil, W, Molfino, R, Resaz, V (2011) EMARO: a European contribution in shaping the future world roboticists. Education of Robotics & Mechatronics 2011. J Robotics Mechatron 23(5)
4. Meriaudeau, F, Fofi, D, Meriaudeau, A, Adema-Labille, H, Torres, V, Tremeau, A, Zaccaria, R, Resaz, V (2010) European programs and their extension in the field of computer vision, color and robotics. J Electron Electrl Eng 6(102) (ISSN 1392–1215)

# Part VII
# Deployable Mechanisms and Applications of Reconfigurable Mechanisms

# Chapter 66
# Deployable Masts Based on the Bennett Linkage

**Hongwei Guo and Zhong You**

**Abstract** A concept of building a deployable mast using the Bennett linkage as the basic bay is proposed in this paper. The deployable mast is assembled by placing in series a number of modular bays constructed using the Bennett linkages. It has one degree of mobility. Geometrical analysis of the deployable mast has been conducted to determine the deployment length and cross section of the mast. The effects of the Bennett linkage parameters on the overall dimension of the mast are considered. The results show that the concept is valid and can be easily tailored to suit practical requirements.

**Keywords** Bennett linkage · Deployable mast · Geometry · Parametrical analysis

## 66.1 Introduction

Deployable masts are one of the most crucial structural components in space applications. Various concepts have been proposed in the past, some of which are constructed by assembling known mechanisms, e.g., the articulated mast and VG truss [1]. In these assemblies beam-like components are used to construct a basic module which has a single degree of freedom, and these modules are placed one on top of the other by repetition. They have been used in space widely for deploying and supporting space operations for large-area space structures such as space station,

H. Guo (✉)
Department of Engineering Science, School of Mechatronics Engineering,
Harbin Institute of Technology, University of Oxford, Oxford, UK
e-mail: guohw@hit.edu.cn

Z. You
Department of Engineering Science, University of Oxford, Oxford, UK
e-mail: zhong.you@eng.ox.ac.uk

J. S. Dai et al. (eds.), *Advances in Reconfigurable Mechanisms and Robots I*,
DOI: 10.1007/978-1-4471-4141-9_66, © Springer-Verlag London 2012 

large deployable antennae, large space mirrors, space telescope, space solar power station and power systems due to their low cost, light weight, high stiffness, high packaging efficiency, ease of transporting and assembling in space. Most of existing deployable masts use scissor-like mechanisms (also known as pantographic elements) or telescopic mechanisms. Some even involve mechanisms using spherical hinges. The mast with scissor-like mechanisms is simple in structural configuration but its members may experience excessive bending, whereas telescopic masts can be very heavy. The spherical joints lead to complex manufacturing process.

In this paper we explore a concept of building a mast using the Bennett linkage, a $4R$ spatial linkage where four bars are connected by revolute joints whose axes are neither parallel nor concurrent [2]. In a century after Bennett linkage was first reported, it has attracted huge research interests. More importantly, it was found that the Bennett linkage could be used as the basic elements to obtain other mechanisms. For instance, Goldberg combined a pair of Bennett linkages and improved it to produce a 5R linkage, he also found a set of 6R linkage by combining Bennett linkages [3]. Some of the other linkages with only revolute joints, e.g., the Myard linkage, the Altmann linkage, the Bennett 6R hybrid linkage, the Bennett-joint 6R linkage, the Dietmaier 6R linkage, the Double-Hooke's-joint linkage, the Schatz linkage and the Wohlhart double-Goldberg linkage are found to be or constructed by combinations of the basic Bennett linkages [4].

In a peering paper published in 2005, Chen and You [5] outlined the possibilities of using the Bennett linkages to form deployable surface grid-like structures as well as towers. Here we explore the tower structures further by providing a detailed analysis on the geometrical behavior of the tower structures with the aim to use them as deployable masts. The effect of geometrical parameters of the Bennett linkage on mast's configuration is given.

The layout of this article is as follows. First, we outline the basic geometrical relations for the Bennett linkage, including its closure equations. Then the key geometrical parameters which define the height and expanded shape of the Bennett linkage are introduced. In Sect. 66.3, a deployable mast is designed using a basic modular bay consisting of a number of nested Bennett linkages. The deployment sequence of a four-bay deployable mast is described. Section 66.4 is about the geometrical parameters of the mast. Section 66.5 concludes the paper with proposed future works.

## 66.2 The Bennett Linkage

### *66.2.1 Basics*

The Bennett linkage is a spatial $4R$ over constrained mobile loop using a minimum number of links with revolute connections, as shown in Fig. 66.1, the rotation axes of the revolute joints are neither parallel nor concurrent.

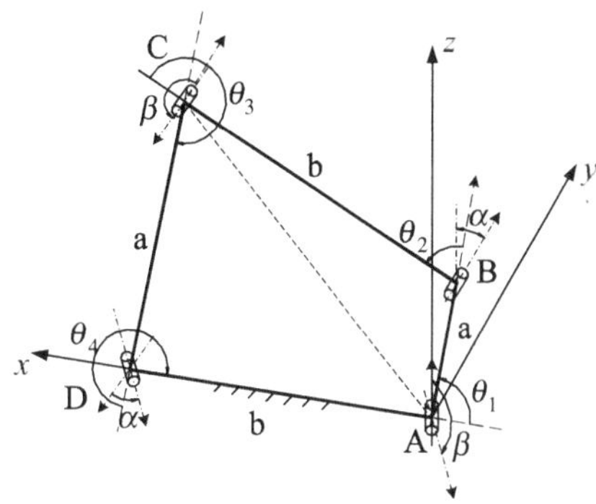


**Fig. 66.1** The Bennett linkage

Bennett identified the conditions for the linkage to have one degree of mobility, which are listed as follows.

Denote by $l_{AB}, l_{BC}, l_{CD}$ and $l_{DA}$ the distances and by $\alpha_{AB}, \alpha_{BC}, \alpha_{CD}$ and $\alpha_{DA}$ the skew angles between revolute joints A and B, B and C, C and D, as well as D and A, respectively. There are

$$l_{AB} = l_{CD} = a, \quad l_{BC} = l_{DA} = b \tag{66.1a}$$

$$\alpha_{AB} = \alpha_{CD} = \alpha, \quad \alpha_{AD} = \alpha_{BC} = \beta \tag{66.1b}$$

The link lengths and skew angles must satisfy

$$\frac{\sin\alpha}{\sin\beta} = \frac{a}{b} = k \tag{66.2}$$

The angle values between adjacent links, $\theta_1$, $\theta_2$, $\theta_3$, and $\theta_4$, vary when the linkage moves but there are

$$\theta_1 + \theta_3 = 2\pi, \; \theta_2 + \theta_4 = 2\pi \tag{66.3a}$$

$$\tan\frac{\theta}{2}\tan\frac{\varphi}{2} = \frac{\sin\frac{1}{2}(\alpha+\beta)}{\sin\frac{1}{2}(\alpha-\beta)} \tag{66.3b}$$

where $\theta = \theta_1$ and $\varphi = \theta_2$. Equations (66.3a, b) are the closure equations for the Bennett linkage.

## *66.2.2 Angle Output of the Bennett Linkage*

If link AD is taken as a reference, $\theta$ as the input angle and $\theta_4$ as the output angle, $\theta_4$ is only dependent on $\theta$ when $a$, $b$, $\alpha$ and $\beta$ are given. As shown in Fig. 66.2, output angle $\theta_4$ varies with the link length ratio $k$ and angle $\alpha$. In Fig. 66.2a the $\alpha$ are 0, $\pi/6$, $\pi/4$, $\pi/3$ and infinitely close to $\pi/2$ respectively. It is found that when $k$ equals to 1, the non-linearity of $\theta_4$ *vs* $\theta$ curves becomes more pronounced as the $\alpha$ increases, and the

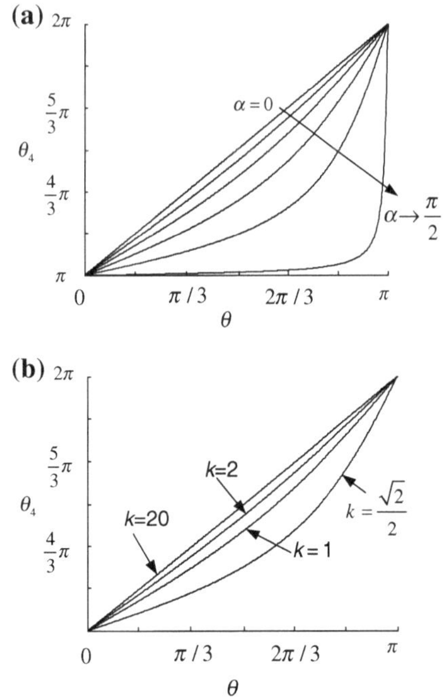


**Fig. 66.2** Angle input and output relations for the Bennett linkage with various $k$ and $\alpha$. **a** $\theta_4$ *vs* $\theta$ as $\alpha$ varies when $k = 1$. **b** $\theta_4$ *vs* $\theta$ as $k$ varies when $\alpha = \pi/4$

change rates of output angle $\theta_4$ increases sharply when $\theta$ is closer to $\pi$ if $\alpha$ is close to $\pi/2$. So this characteristic of Bennett linkage can be used to design some special mechanisms. On the other hand, for a given $\alpha$ ($\alpha = \pi/4$), the non-linearity of $\theta_4$ *vs* $\theta$ curves increases when $k$ is less than 1.

## *66.2.3 Dimensional Output of the Bennett Linkage*

The dimension of a Bennett linkage can be defined by the relative position of the corner hinges. If we take member AD shown in Fig. 66.1 as reference, position C defines the size of a Bennett linkage when it is mobile. Take link DA as $x$ axis and revolute axis at A as $z$ axis. Length $l_{AC}$ is calculated by considering $\Delta ACD$. The coordinates of point C can be obtained as

$$\begin{cases} x_C = (1 - k\cos(\pi - \varphi))b \\ y_C = \dfrac{k(1 + \cos\theta\cos(\pi - \varphi)) - (\cos\theta + \cos(\pi - \varphi))}{\sin\theta} b \\ z_C = b\sqrt{1 + k^2 - 2k\cos(\pi - \varphi) - (\frac{x_C}{b})^2 - (\frac{y_C}{b})^2} \end{cases} \tag{66.4}$$

The output trajectories of point C for various combination of $k$ and $\alpha$ are drawn in Fig. 66.3. For convenience $b$ is assumed to be equal to 1. The point C moves in

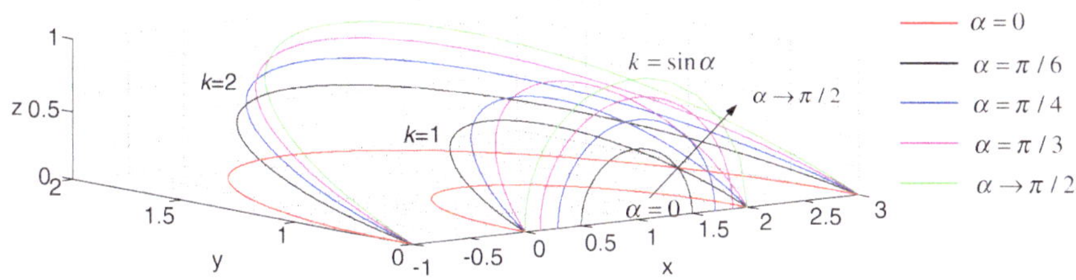


**Fig. 66.3** Output trajectories of Bennett linkages with different $k$ and $\alpha$

the planes from whose angles to plane $xoy$ are equal to corresponding values of $\beta$, and the trajectories of point C are a series of ellipses with different longer axis and shorter axis. It will move in the plane $xoz$ or $xoy$ when $\beta = \pi/2$ or $\beta = \pi$ respectively. It is found that the coordinates $y_c$ and $z_c$ reach their maximum synchronously when $\varphi = \pi/2$, where

$$y_{c\max} = b\sin\alpha, \quad z_{c\max} = a\cos(\pi - \beta) \tag{66.5}$$

So in some special cases, trajectory of point C becomes a circle in plane $xoz$ when $\alpha$ tends to $\pi/2$ or it is a circle in plane $xoy$ when $k = 1$ and $\beta = \pi$. If $\theta_1$ can achieve the scale from 0 to $2\pi$, then the trajectories of point C form a series of integrate ellipses.

## 66.3 The Deployable Mast

A number of Bennett linkages can be assembled together to construct a single or multiple layered curved grid-like deployable structures including arches or helices [4]. All of them can be folded into a small volume or even a bundle but they cover a scale of areas after deployment. The multiple layered grid concept can also lead to the construction of a tower structure. The latter is the focus of this paper and we adopt the concept for the deployable masts.

The deployable mast is constructed with a number of identical mobile bays connected one on the top of the other to obtain required the length. A bay is shown in Fig. 66.4, it contains eight links which in fact can be identified as two types of links marked in the red and green colors, respectively. The crossing points between links are the locations of the revolute joints. In the bay, four Bennett linkages can be found, which are ABCD, EFGH, AMEN and CPGQ, respectively, All of them are *equilateral* Bennett linkages, the skew angles between adjacent revolute axes in the four linkages are related, i.e., Bennett linkages ABCD, AMEN and CPGQ have the same angle $\alpha$ and $\beta$, where $\beta = \pi - \alpha$, whereas the angle in EFGH is $-\alpha$ and $-\beta$. Note that the Bennett linkages at each of the corners, i.e., AMEN and CPGO, have much smaller lengths. In fact, they can be made as connection junctions with a Bennett linkage concealed within.

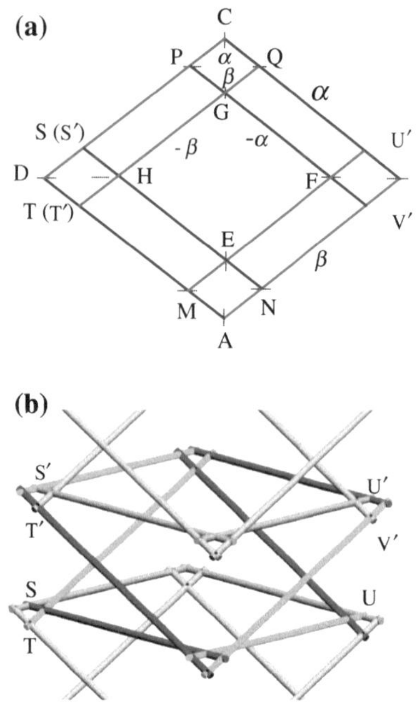


**Fig. 66.4** Deployable mast bay constructed by Bennett linkages. **a** The projection of a bay of the deployable mast; **b** Three-dimensional view

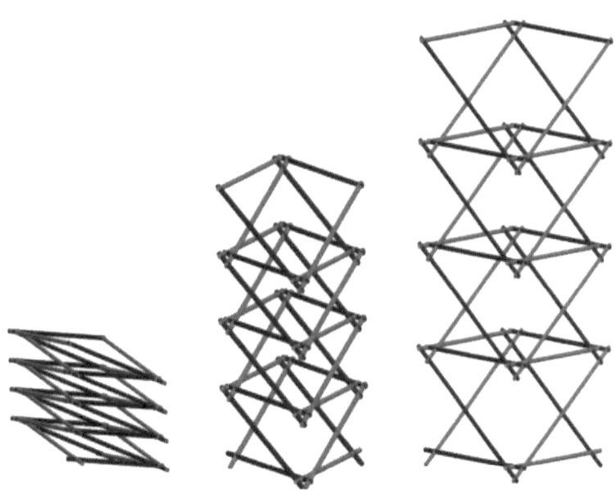

**Fig. 66.5** Deployment sequence of a four-bay deployable mast

The basic bay retains mobility one, so its deployment can be control conveniently. In Fig. 66.4b, the basic bay has eight connection points, S, T, U, V, S′, T′, U′ and V′, distributing on the upper and lower links, respectively, allowing the bay to be connected with its neighboring bays. For example, at points S, T, U, V the upper bay can be connected with the points S′, T′, U′, V′ of the lower bay. Two new Bennett linkages are formed after the connection with adjacent bays. The connections, once they are done, form Bennett linkages with smaller lengths. These Bennett linkages are made identical to those at AMEN and CPGO. A completed deployable mast with four bays and its deployment sequence are shown in Fig. 66.5.

## 66.4 Geometrical Analysis of the Deployable Mast

The primary geometrical parameter is the length of the mast, determined by number of the bays $N$ and the deployment height of the basic bay $H_{Bay}$. When the mast is folded, the distance between A(E) and C(G) is the smallest, whereas that between B(F) and D(H) is the largest. With the deployment of the mast, the distance between A(E) and C(G) increases while that that between B(F) and D(H) decreases, the height of the mast increases gradually. There is a moment that the mast reaches its highest height, after that the mast height decreases, and the mast tends to fold to the other direction where distance between A(E) and C(G) becomes the largest, meanwhile that between B(F) and D(H) becomes the smallest. According to [6] and [7], the distance between the diagonal points and the height of the equilateral Bennett linkage along its symmetrical axis, $H_{BL}$, can be expressed as

$$\overline{AC}^2 = 2l^2(1+\cos\varphi), \quad \overline{BD}^2 = 2l^2(1+\cos\theta), \tag{66.6a}$$

$$H_{BL} = \sqrt{-\frac{1}{2}(\cos\theta + \cos\varphi)}l. \tag{66.6b}$$

And for equilateral Bennett linkage, the $\theta$ and $\varphi$ satisfy the equation

$$\tan\frac{\theta}{2}\tan\frac{\varphi}{2} = \frac{1}{\cos\alpha}. \tag{66.7}$$

So $H_{BL}$ achieves its maxima when the Bennett linkage deploys completely and corresponding angles between links are $\theta_d$ and $\varphi_d$. It is found that $\theta_d = \varphi_d$ and $\overline{AC} = \overline{BD}$ when the Bennett linkage is fully deployed. When the equilateral Bennett linkage fully deploys, the highest height of the Bennett linkage is

$$\theta_d = \varphi_d = 2\arctan\sqrt{1/\cos\alpha}, \tag{66.8a}$$

$$H_{BL\max} = l\sqrt{-\cos(2\arctan\sqrt{1/\cos\alpha})}. \tag{66.8b}$$

After that, the dimension of the linkage shrinks. The height and the distance between diagonal points are dependent on length of the links $l$ and skew angle $\alpha$ as shown in Fig. 66.6, the peak heights of the Bennett linkage are denoted by the solid points. The angle $\theta$ or distance $\overline{BD}$ can been taken as input of the lowest mast bay to deploy the mast.

The overall height of the mast bay is the sum of heights of Bennett linkages ABCD and EFGH due to the fact that the skew angles of between adjacent revolute axes are $\alpha$ and $-\alpha$. Take the lengths of links in Bennett linkages ABCD and EFGH being $l_1$ and $l_2$. The height of a basic bay height can be expressed as

$$H_{Bay} = (l_1 + l_2)\sqrt{-\frac{1}{2}(\cos\theta + \cos\varphi)} \tag{66.9}$$

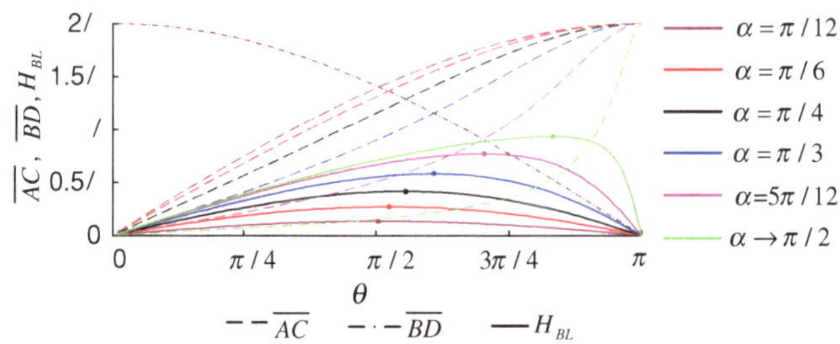


**Fig. 66.6** $\overline{AC}$, $\overline{BD}$ and $H_{BL}$vary with different α

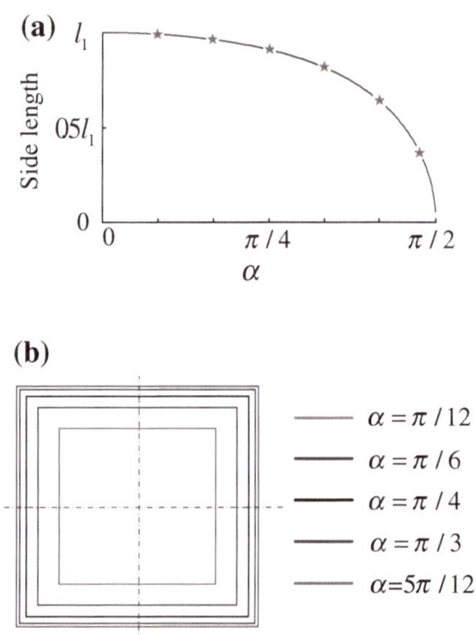


**Fig. 66.7** Projected cross-sections of the mast with different α. **a** Side length varies with α; **b** Projected cross-section for determined α

Then the height of the entire mast consisting of $N$ bays is $NH_{Bay}$. It is also interesting to note that the projected cross-section of the mast is a square with side length of $l_1\sqrt{1+\cos\theta_d}$ when the mast completely deploys, as shown in Fig. 66.7. The projected square side lengths vary as α increasing.

The mast height increase with α increasing while the side length of the mast projected cross-section decreases, so α must be considered carefully. As we can see that the small linkages in the mast bay do not affect its height and size, their function is purely for connecting adjacent bays and maintaining the mobility of the assembly.

## 66.5 Conclusion

The Bennett linkage is a spatial mobile loop contains four links and revolute joints, making is one of the most reliable mechanisms in comparison with those mechanisms consisting sliders or spherical joints. It was found in the past this linkage can be used as a building block to form many deployable assemblies such as grid-like surfaces and towers. In this paper, the focus is on its utilization of creating deployable masts. Geometrical analysis has been conducted to determine the deployment length of the mast and the basic parameters of the Bennett linkages which are used as basic units to form such mast. The relationships among these parameters, which are presented in charts, can be used to select a suitable design depending on practical requirements. The concept is modular so it provides ample freedom to obtain desirable length for the mast.

We proposed to have smaller Bennett linkages at the corners of each bay to facilitate connections between adjacent bays. These smaller Bennett linkages can be treated as a connection box in which a Bennett linkage is housed. Work is still on-going in designing such box.

In short, the proposed concept involves only revolute joints, for which lubrication can be carried out and their performance is proven to be robust. It has one mobility and therefore can be controlled easily. To explore the concept's full potential, further analysis such as the load-bearing capacity and its dynamic behavior should be carried out. We are continuing this work and will report in due course any further progress.

**Acknowledgments** H. W. Guo acknowledges support from the Program of Young Leadership in Universities (Grant No.B07018) and University of Oxford where he is currently a visiting scholar.

## References

1. AEC-Able Engineering Company, Inc. http://www.aec-able.com
2. Bennett GT (1903) A new mechanism. Eng-Lond 76:777–778
3. Goldberg M (1943) New five-bar and six-bar linkages in three dimensions. Trans ASME 65:649–663
4. You Z, Chen Y (2011) Motion structures, deployable assemblies of mechanisms. Spon Press, London
5. Chen Y, You Z (2005) Mobile assemblies based on the Bennett linkage. Proc R Soc A (Math Phys Eng Sci) 461:1229–1245
6. Chen Y, You Z (2006) Square deployable frame for space application: Part I: Theory. Proceedings of the Institution of Mechanical Engineers, Part G, J Aerospace Eng, 220(4):347–354
7. Chen Y, You Z (2007) Square deployable frame for space application: Part II: Realization. Proceedings of the Institution of Mechanical Engineers, Part G: J Aerospace Eng, 221(1):37–45

# Chapter 67
# Structural Synthesis of Ancient Chinese Chu State Repeating Crossbow

**Kuo-Hung Hsiao and Hong-Sen Yan**

**Abstract** Chu State repeating crossbow is a reconfigurable mechanism which was found in the tomb and dated back to the fourth century BC in ancient China. Since crossbows have diversified types and were widely used in ancient China after the Spring and Autumn Period (770–476 BC), Chu State repeating crossbow should have many designs in different eras and regions. The aim of this work is to synthesize mechanism structures of Chu State repeating crossbow. Based on the analyses of crossbows, the structural characteristics and design constraints of this device are concluded. Then, according to the concepts of generalization and specialization subject to the concluded design constraints, seven feasible structures of mechanism that meet the technological standards of the subject's time period are reconstructed.

**Keywords** Repeating crossbow · Structure of mechanism · Reconstruction design · History of machinery · Chu State

K.-H. Hsiao (✉)
Collections and Research Division, National Science and Technology Museum, No. 720, Jiouru 1st Road, Kaohsiung 80765 Taiwan
e-mail: khhsiao@mail.nstm.gov.tw

H.-S. Yan
Department of Mechanical Engineering, National Cheng Kung University, No.1, University Road, Tainan 70101, Taiwan
e-mail: hsyan@mail.ncku.edu.tw

J. S. Dai et al. (eds.), *Advances in Reconfigurable Mechanisms and Robots I*,
DOI: 10.1007/978-1-4471-4141-9_67,  

## 67.1 Introduction

The shooting process of crossbow includes three steps: drawing the bowstring, placing the bolt and shooting the bolt. According to the literatures and the excavations, ancient Chinese crossbows have three types including the original crossbow, Chu State repeating crossbow and Zhuge repeating crossbow. All of them are reconfigurable mechanisms which encounter a certain changes in its topological structure during shooting process. An original crossbow is a weapon bow mounted on a stock with a trigger mechanism for holding the drawn bowstring as shown in Fig. 67.1a [1]. It generally consists of six members including the frame $K_F(1)$, the bow $K_{CB}(2)$, the bowstring $K_T(3)$, input link $K_I(4)$, the percussion link $K_{PL}(5)$ and the connecting link $K_L(6)$. The trigger mechanism was designed in such a way that it was able to store a large amount of energy within the bow when drawn, but was easily fired with little recoil when the input link was pulled. It also allowed the shooter to aim the target for precision shooting. A repeating crossbow is a crossbow where the separate actions of three steps can be accomplished with a simple one-handed movement while keeping the crossbow stationary. This allows a higher rate of fire than an original crossbow. Based on the archeological finds, the repeating crossbow consists of six members and first appeared around 400 BC in the tomb in Hubei Province as shown in Fig. 67.1b, [2]. The magazine fixes on the top of the stock and contains twenty bolts, and the device is worked by moving the input link forward and backward. Two bolts can be shot at once in the reciprocating motion of the input link. Since the tomb belonged to Chu State during the War States Period (475–221 BC), the device was named Chu State repeating crossbow. The Records of the Three Kingdoms mentions another type of repeating crossbow named “Zhuge repeating crossbow” and describes the strategist Zhuge Liang (181–234 AD) of the Three Kingdoms Period (220–280 AD) as the inventor [3]. Figure 67.1c is a Zhuge repeating crossbow with five members in the book of Treatise on Armament in the Ming Dynasty (1368–1644 AD) [1]. The magazine contains a number of bolts and functions as a percussion link. The function of the percussion link with the magazine is to draw and release the bowstring to shoot the bolt. When the device is worked by moving the input link, one or several bolts can be shot at once in the oscillating motion of the input link. Since the types of crossbows were diversified and widely used in ancient China, the topological structures of each crossbow should have many designs in different eras and regions. In order to realize the development of ancient Chinese crossbows, it is necessary to study further for accessing and understanding the real ancient technology.

The purpose of this work is to synthesize all feasible mechanism structures of Chu State repeating crossbow which was dated back to the fourth century BC. In what follows, the representation of joints of ancient mechanisms is presented first. Then, an approach for synthesizing the structures of ancient mechanisms with diversified types is introduced. Finally, the structural synthesis of Chu State repeating crossbow is provided.

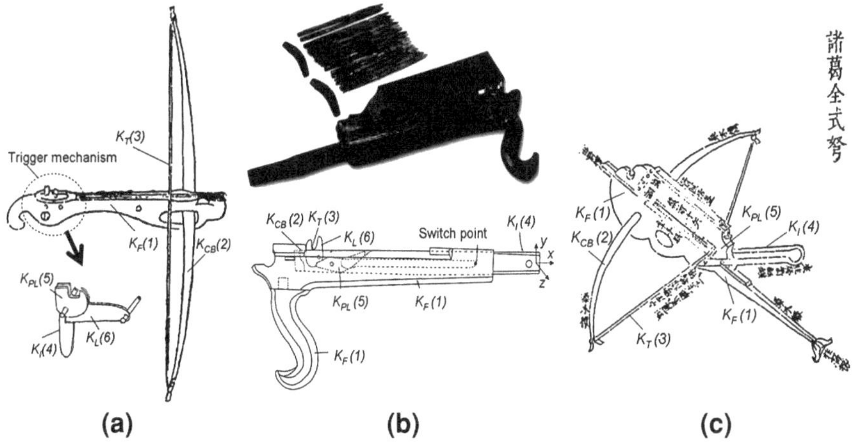


**Fig. 67.1** Ancient Chinese crossbows [1, 2]. **a** An original crossbow, **b** Chu State repeating crossbow, **c** Zhuge repeating crossbow

## 67.2 Representation of Joints

Since some special joints in ancient mechanisms can not be represented by modern schematic representations, a novel representation for such joints in ancient mechanisms is introduced here [4, 5].

The number of degrees of freedom is the number of independent parameters needed to specify the relative positions of the pairing elements of a joint. An unconstrained pairing element has six degrees of freedom including three translational and three rotational degrees of freedom. And, a joint is represented as:

$$J^{P_{xyz}}_{R_{xyz}}$$

in which the superscript $P_{xyz}$ denotes that it can translate as a prismatic joint ($J_P$) in x, y and z axes; and the subscript $R_{xyz}$ denotes that it can rotate as a revolute joint ($J_R$) in x, y and z axes. When a pairing element connects to another pairing element and forms a joint, a constraint is imposed and the motion of the original member is reduced by one or more degrees of freedom. For example, joint $J_{Rx}$ represents that a pairing element of a joint can rotate in x-axis with respect to the other pairing element. And, joint $J^{P_x}_{R_{yz}}$ represents that a pairing element of a joint not only translates in x-axis and also rotates in y and z axes with respect to the other pairing element.

Figure 67.2a shows a two-member mechanism with a joint from ancient Chinese book [6]. Since the drawing is not clear, link $K_L$ is adjacent to frame $K_F$ with an uncertain joint. Considering the type and the direction of motion of the link, the joint has two possible types: the link rotates in z-axis only, denoted as $J_{Rz}$;

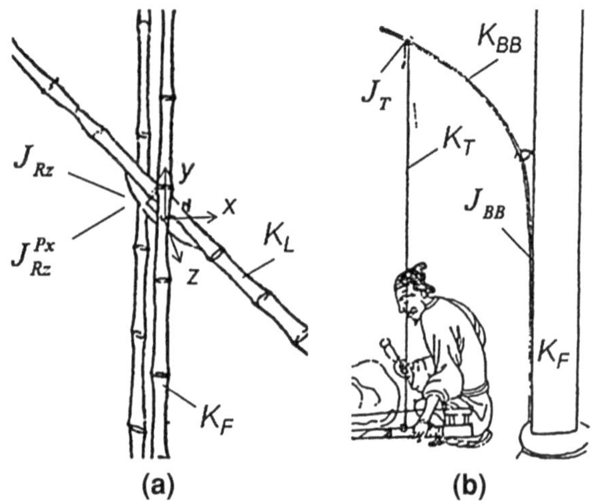


**Fig. 67.2** Special joints in ancient mechanisms [6]

and, the link rotates not only in z-axis but also translates in x-axis, denoted as $J_{Rz}^{Px}$. The x and y axes are defined as the horizontal and vertical directions in the drawing, respectively. The z-axis is determined by the right-hand rule.

Thread and bamboo often appear as members in ancient mechanisms especially in textile and agricultural devices. Figure 67.2b shows that bamboo $K_{BB}$ is adjacent to frame $K_F$ and thread $K_T$ from ancient Chinese book [6]. The bamboo is firmly fixed to the frame and the thread ties the bamboo directly. The joint incident to the bamboo and the frame is defined as a bamboo joint, denoted as $J_{BB}$. Using a thread tying a member to form a joint is rare in modern mechanisms but popular in ancient mechanisms. The joint incident to the thread and a member is defined as a thread joint, denoted as $J_T$.

## 67.3 Design Approach

Figure 67.3 shows the process for synthesizing all feasible structures of ancient mechanisms with diversified types [4, 7–9]. Each step of the process is explained as follows:

***Step 1. Analyses of the structures of mechanisms***

By analyzing the structure of a target mechanism, its structural characteristics are concluded including the possible number of members and incidences among members and joints of the device. Since the development of the similar mechanisms should influence each other, the structures of different devices with the same function are usually alike. Therefore, the number of members and the structures of joints of the relevant mechanisms can be considered. For example, ancient Chinese crossbows have three types. One's structural characteristics may be the other one's design constraints.

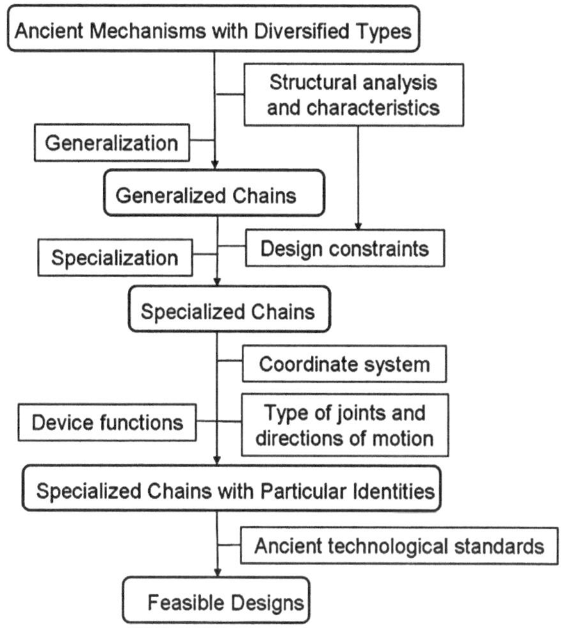


**Fig. 67.3** Process for structural synthesis of ancient mechanisms with diversified types

***Step 2. Generalized chains***

The second step is to identify the atlas of generalized chains with the same numbers of members and joints subject to concluded structural characteristics based on the concept of number synthesis [10, 11]. Generalization is to transform the mechanism, that involves various types of members and joints, into a generalized chain with only generalized links and joints. A generalized link is a link with generalized joints; it can be a binary link, ternary link,etc. A generalized joint is a joint in general; it can be a revolute joint, prismatic joint, or others. Graphically, a joint is symbolized by a small circle, and a member with $i$ incident joints is symbolized by an $i$-side shaded polygon with small circles as vertices. Some of the important atlases of generalized chains have already been built in references [10, 11].

***Step 3. Specialized chains***

Specialization is to assign specific types of members and joints in the available atlas of generalized chains, subject to required design constraints to have atlas of specialized chains [10, 11]. Design constraints are defined based on the concluded structural characteristics. And, a generalized chain after specialization is a specialized chain. Based on the process of specialization, all possible specialized chains can be identified after assigning the types of members and joints subject to the concluded design constraints for each generalized chain obtained in step 2.

***Step 4. Specialized chains with particular identities***

In order to represent the structure of mechanism, a right-hand rectangular Cartesian coordinate system is defined to describe each motion axis of the joints. By studying the target device, the function of the device is determined. In view of

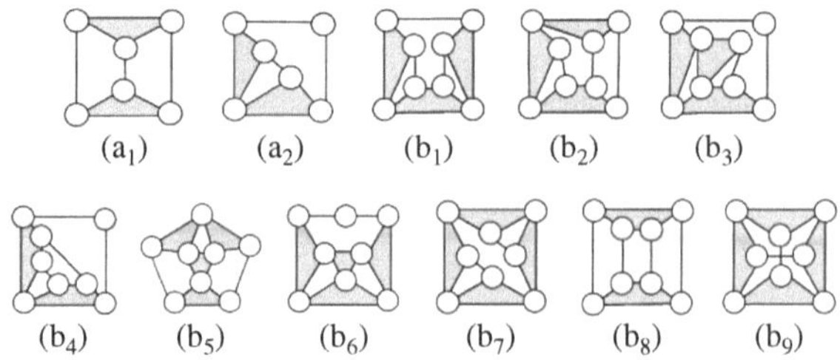


**Fig. 67.4** Atlas of some generalized chains with five and six members

the device function, these uncertain joints can be represented by different types of joints to achieve the equivalent function. Considering the types and the motion directions of the uncertain joints, all possible types of the uncertain joints are generated. By assigning the possible types of the uncertain joints to the specialized chains obtained in step 3, specialized chains with particular identities are obtained.

***Step 5. Feasible designs***

According to the motion and function requirements of the device, all feasible designs that meet the technological standards of the subject's time period can be obtained from the specialized chains with particular identities.

## 67.4 Structural Synthesis

According to the actual excavation, Chu State repeating crossbow comprises six members including a frame, a bow, a bowstring, an input link, a percussion link and a connecting link. The function of the connecting link is to draw the bowstring and release the bowstring by the driving of the percussion link. However, in Zhuge repeating crossbow, the percussion link is, instead of the connecting link, to draw the bowstring, and the number of members of the device becomes five. Using the same concept, the number of members of Chu State repeating crossbow should be five or six. All feasible structures of Chu State repeating crossbow are synthesized through the following steps to illustrate the introduced approach shown in Fig. 67.3:

***Step 1.*** The structural characteristics of the device are:

1. It is a cam mechanism with five-member or six-member.
2. The frame ($K_F$) is a ternary link and adjacent to a pair of binary links.
3. The bow ($K_{BC}$) is a binary link and adjacent to the frame ($K_F$) with a bamboo joint ($J_{BB}$).
4. The bowstring ($K_T$) is a binary link and adjacent to the bow ($K_{CB}$) and a percussion link ($K_{PL}$) or a connecting link ($K_L$) with thread joints ($J_T$).
5. The input link ($K_I$) is adjacent to the frame ($K_F$) with a prismatic joint ($J^{Px}$).
6. The percussion link ($K_{PL}$) is adjacent to the frame ($K_F$) with a cam joint ($J_A$).

***Step 2.*** Since this is a device with five-member and six-joint or six-member and eight-joint, the corresponding generalized chains are shown in Fig. 67.4.

***Step 3***. There must be a pair of binary links as a bow and a bowstring. And, the pair of binary links must be adjacent to a ternary link as the frame. Therefore, only three generalized chains shown in Fig. 67.4a$_2$, b$_3$ and b$_6$ are qualified for the process of specialization. And, all possible specialized chains are identified as follows:

*Frame, bow and bowstring*—link $K_F$, link $K_{CB}$ and link $K_T$

There must be a ternary link as the frame ($K_F$) and a pair of binary links as the bow ($K_{CB}$) and the bowstring ($K_T$) is adjacent to the frame. The frame, the bow and the bowstring are identified as follows:

1. For the generalized chain shown in Fig. 67.4a$_2$, the assignment of the frame ($K_F$), the bow ($K_{CB}$) and the bowstring ($K_T$) generates one non-isomorphic result, Fig. 67.5a$_1$.
2. For the generalized chain shown in Fig. 67.4b$_3$, the assignment of the frame ($K_F$), the bow ($K_{CB}$) and the bowstring ($K_T$) generates one result, Fig. 67.5a$_2$.
3. For the generalized chain shown in Fig. 67.4b$_6$, the assignment of the frame ($K_F$), the bow ($K_{CB}$) and the bowstring ($K_T$) generates one non-isomorphic result, Fig. 67.5a$_3$.

Therefore, three specialized chains with identified frame ($K_F$), bow ($K_{CB}$) and bowstring ($K_T$) are available as shown in Fig. 67.5a$_1$–a$_3$.

*Input link, percussion link and connecting link*—link $K_I$, link $K_{PL}$ and link $K_L$

There must be an input link ($K_I$) that is adjacent to the frame ($K_F$) with a prismatic joint ($J^{Px}$) and not adjacent to the bowstring ($K_T$). The percussion link ($K_{PL}$) must be adjacent to the frame ($K_F$) with a cam joint ($J_A$) and the remaining one is the connecting link. Therefore, the input link, the percussion link and the connecting link are identified as follows:

1. For the case shown in Fig. 67.5a$_1$, the assignment of the input link ($K_I$), the percussion link ($K_{PL}$) and an uncertain joint ($J_I$) generates one result, Fig. 67.5b$_1$.
2. For the case shown in Fig. 67.5a$_2$, the assignment of the input link ($K_I$), the percussion link ($K_{PL}$), the connecting link ($K_L$) and the uncertain joints generates one result, Fig. 67.5b$_2$.
3. For the case shown in Fig. 67.5a$_3$, the assignment of the input link ($K_I$), the percussion link ($K_{PL}$), the connecting link ($K_L$) and the uncertain joints generates one non-isomorphic result, Fig. 67.5b$_3$.

Therefore, three specialized chains with identified frame ($K_F$), bow ($K_{CB}$), bowstring ($K_T$), input link ($K_I$), percussion link ($K_{PL}$) and connecting link ($K_L$) are available as shown in Fig. 67.5b$_1$–b$_3$.

***Step 4.*** The coordinate system is defined as shown in Fig. 67.1b. The function of Chu State repeating crossbow is to complete the shooting process through the reciprocation of the input link. The uncertain joints may have multiple types to achieve the equivalent function. Considering the uncertain joint $J_I$, it has one possible type: the input link rotates in z-axis with respect to the percussion link, denoted as $J_{Rz}$. Considering the uncertain joints $J_2$, $J_3$ and $J_4$, each of joint has two

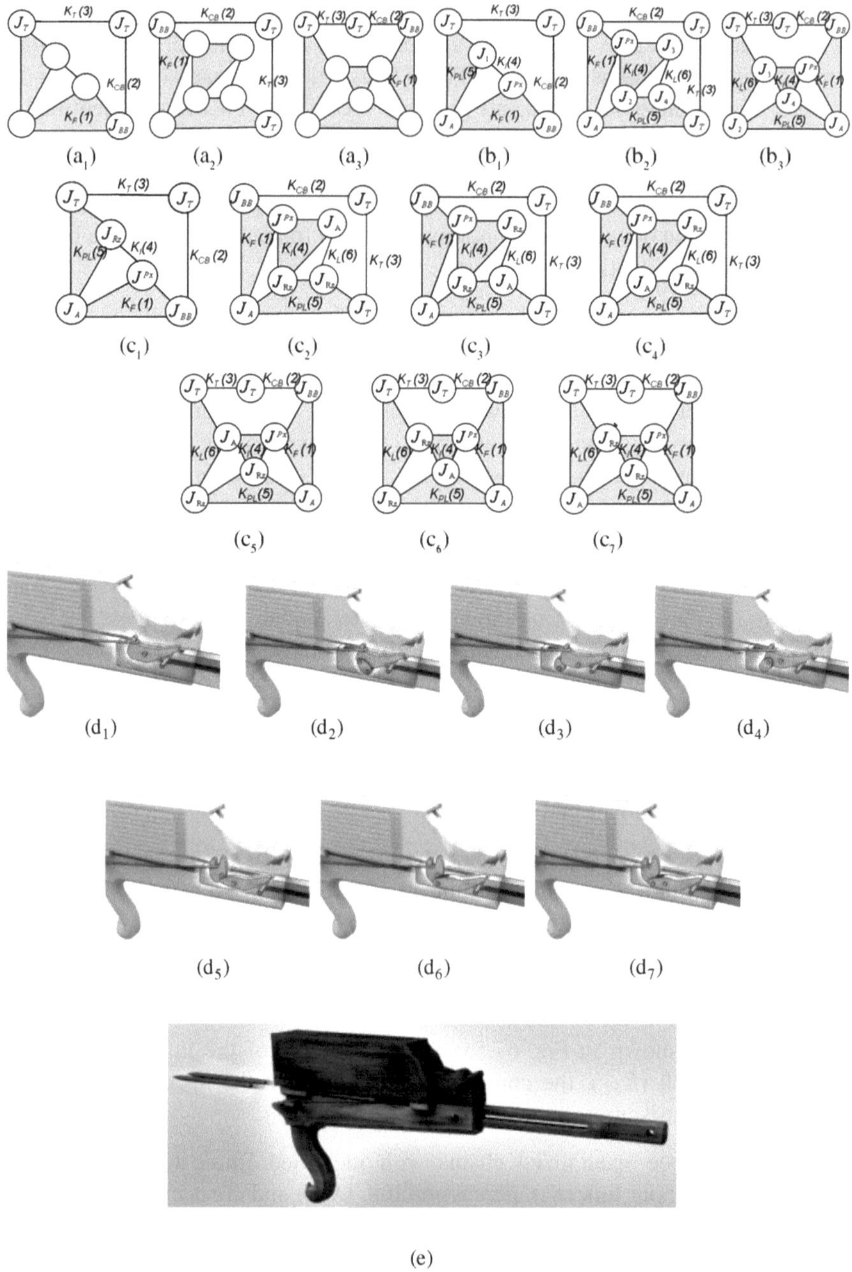


**Fig. 67.5** Structural synthesis of Chu State repeating crossbow

possible types. When any one of the joints is a cam joint, denoted as $J_A$, the others rotate in z-axis, denoted as $J_{Rz}$. By assigning the possible types of the uncertain joints $J_1$ ($J_{Rz}$), $J_2$ ($J_{Rz}$ and $J_A$), $J_3$ ($J_{Rz}$ and $J_A$) and $J_4$ ($J_{Rz}$ and $J_A$) to the specialized chains shown in Fig. 67.5$b_1$–$b_3$, seven specialized chains with particular identities as shown in Fig. 67.5$c_1$–$c_7$ are obtained.

***Step 5.*** Fig. 67.5$d_1$–$d_7$ show the corresponding 3D solid models of the feasible designs. Fig. 67.5e shows the imitation of Chu State repeating crossbow.

## 67.5 Conclusions

According to the existing records, Chu State repeating crossbow with reconfiguration mechanism is the earliest repeater weapon in ancient China. Since crossbows were popular and widely used in ancient China, the structures of Chu State repeating crossbow should have many designs. Through the study of mechanisms and applying the concepts of generalization and specialization of mechanisms subject to the concluded design constraints, all feasible specialized chains are generated. By assigning the possible joints to the feasible specialized chains, all possible structures of Chu State repeating crossbow are systematically synthesized. Via such an approach, the structures of ancient mechanisms with diversified types can be reasonably obtained. And, the proposed approach not only can be applied to Chu State repeating crossbow but also to other ancient machinery.

**Acknowledgments** The authors are grateful to the National Science Council (Taipei, Taiwan) under Grant NSC 97-2221-E-006-070-MY3 for the financial support of this work.

## References

1. Mao TY (2001) Treatise on armament. Hainan Press, Hainan (in Chinese)
2. Museum J (ed) (2009) Important archaeological discoveries. Cultural Relics Press, Beijing (in Chinese)
3. Chen S (1968) History of the three kingdoms. Taiwan Commercial Press, Taipei (in Chinese)
4. Hsiao KH, Yan HS (2010) Structural identification of the uncertain joints in the drawings of Tain Gong Kai Wu. J Chin Soc Mech Eng 31(5):383–392
5. Hsiao KH, Chen YH, Tsai PY, Yan HS (2011) Structural synthesis of ancient Chinese foot-operated slanting loom. Proc Ins Mech eng C J Mech Eng Sci 225:2685–2699
6. Song YX (1966) Chinese technology in the seventeen century. Dover Publications, New York (in Chinese, trans. Sun EZ, Sun SC)
7. Yan HS, Hsiao KH (2010) Structural synthesis of the uncertain joints in the drawings of Tain Gong Kai Wu. J Adv Mech Des, Syst Manuf Jpn Soc Mech Eng 4(4):773–784
8. Hsiao KH, Chen YH, Yan HS (2011) Structural synthesis of ancient Chinese foot-operated silk-reeling mechanism. Front Mech Eng China 5(3):279–288
9. Hsiao KH, Yan HS (2012) Structural synthesis of ancient Chinese Zhuge repeating crossbow. In: HMM2012 international symposium on the history of machines and mechanisms (accepted on 15 Jan 2012)

10. Yan HS (1998) Creative design of mechanical devices. Springer, Singapore
11. Yan HS (2007) Reconstruction designs of lost ancient Chinese machinery. Springer, Dordrecht

# Chapter 68
# Duality of the Platonic Polyhedrons and Isomorphism of the Regular Deployable Polyhedral Mechanisms (DPMs)

**Guowu Wei and Jian S. Dai**

**Abstract** This paper investigates the duality of the five Platonic polyhedrons via the ray and axis coordinates of Line geometry. Then, based on the five Platonic polyhedrons, regular deployable polyhedral mechanisms are constructed by implanting the plane-symmetric eight-bar linkages into the edges, faces and vertices of the polyhedrons. Due to the duality of the five regular Platonic polyhedrons, it is revealed in this paper that the dual regular deployable polyhedral mechanisms constructed based on the dual Platonic polyhedrons are isomorphic and the isomorphism of the dual regular deployable polyhedral mechanisms is then for the first time presented via topology graphs and their corresponding adjacency matrices.

**Keywords** Duality · Isomorphism · Deployable polyhedral mechanisms (DPMs) · Line geometry

## 68.1 Introduction

It is well known that there are five regular convex polyhedrons named after the great Greece philosopher and mathematician Plato as Platonic solids [1]. The five regular convex polyhedrons are, namely, the tetrahedron, octahedron, hexahedron (cube), dodecahedron and icosahedron. There are dualities in the five regular

G. Wei (✉) · J. S. Dai
Centre for Robotics Research, King's College London, Strand, London, WC2R 2LS, UK
e-mail: guowu.wei@kcl.ac.uk

J. S. Dai
e-mail: jian.dai@kcl.ac.uk

J. S. Dai et al. (eds.), *Advances in Reconfigurable Mechanisms and Robots I*,
DOI: 10.1007/978-1-4471-4141-9_68, 

Platonic polyhedrons [2] such that by connecting the centres of the adjacent faces of one Platonic polyhedron, a new dual polyhedron can be constructed turning out to be a Platonic polyhedron each vertex and face of which correspond respectively to a face and vertex of the original one. The tetrahedron is self-dual, the hexahedron and octahedron are dual to each other and so are the icosahedron and dodecahedron. These attractive geometrical objects have a universal appeal and their popularity has endured for centuries. The five Platonic polyhedrons have been widely explored and based on the regular polyhedrons, different types of expandable polyhedral mechanisms have been presented [3–5].

In this paper, the duality of the five Platonic polyhedrons is to be interpreted with the ray and axis coordinates representation of Line geometry. Subsequently, by implanting a group of plane-symmetric eight-bar linkages [6] into the Platonic polyhedrons, a family of regular deployable polyhedral mechanisms are to be constructed. According to the duality of the Platonic polyhedrons, the isomorphism of the dual deployable polyhedral mechanisms is to be presented through the topology graph and their corresponding adjacency matrices of the mechanisms.

## 68.2 Duality of Platonic Polyhedrons Via Line Geometry

From the kinematics point of view, the dualities of the five Platonic polyhedrons can be represented by the duality of ray and axis coordinates of the lines along the edges of the given Platonic polyhedrons. Herein, the duality of the hexahedron and the octahedron is presented via Line geometry. Using the same approach the self-dual of the tetrahedron and the duality between the dodecahedron and the icosahedron can be investigated.

### *68.2.1 Lines in Terms of Ray and Axis Coordinates*

From line geometry, the six homogeneous ray coordinates $(l, m, n; p, q, r)$ for a line jointing two points $\boldsymbol{r}_1(w_1, x_1, y_1, z_1)$ and $\boldsymbol{r}_2(w_2, x_2, y_2, z_2)$ can be given by the six $2 \times 2$ Grassmann determinants [7] of the array as

$$\begin{bmatrix} 1 & x_1 & y_1 & z_1 \\ 1 & x_2 & y_2 & z_2 \end{bmatrix}. \tag{68.1}$$

Based on Eq. (68.1), the three direction components $(l, m, n)$ of the line can be given as

$$l = \begin{vmatrix} 1 & x_1 \\ 1 & x_2 \end{vmatrix} = x_2 - x_1, \quad m = \begin{vmatrix} 1 & y_1 \\ 1 & y_2 \end{vmatrix} = y_2 - y_1 \quad \text{and} \quad n = \begin{vmatrix} 1 & z_1 \\ 1 & z_2 \end{vmatrix} = z_2 - z_1 \tag{68.2}$$

And the three moments of the three direction vectors about the origin are given as

$$p = \begin{vmatrix} y_1 & z_1 \\ y_2 & z_2 \end{vmatrix} = y_1 z_2 - y_2 z_1, \quad q = \begin{vmatrix} z_1 & x_1 \\ z_2 & x_2 \end{vmatrix} = x_2 z_1 - x_1 z_2, \\ r = \begin{vmatrix} x_1 & y_1 \\ x_2 & y_2 \end{vmatrix} = x_1 y_2 - x_2 y_1 \tag{68.3}$$

The line is said to be normalized when $l^2 + m^2 + n^2 = 1$.

Alternatively, the six homogeneous axis coordinates $(P, Q, R; L, M, N)$ for a line can be presented as the intersection of two planes $\pi_1(\mathrm{A}_1, \mathrm{B}_1, \mathrm{C}_1, \mathrm{D}_1)$ and $\pi_2(\mathrm{A}_2, \mathrm{B}_2, \mathrm{C}_2, \mathrm{D}_2)$ given by the $2 \times 2$ Grassmann determinants of the array as

$$\begin{bmatrix} D_1 & A_1 & B_1 & C_1 \\ D_2 & A_2 & B_2 & C_2 \end{bmatrix}, \tag{68.4}$$

And above gives the following axis coordinates of the line as,

$$P = \begin{vmatrix} D_1 & A_1 \\ D_2 & A_2 \end{vmatrix}, \quad Q = \begin{vmatrix} D_1 & B_1 \\ D_2 & B_2 \end{vmatrix}, \quad R = \begin{vmatrix} D_1 & C_1 \\ D_2 & C_2 \end{vmatrix}, \tag{68.5}$$

and

$$L = \begin{vmatrix} B_1 & C_1 \\ B_2 & C_2 \end{vmatrix}, \quad M = \begin{vmatrix} C_1 & A_1 \\ C_2 & A_2 \end{vmatrix}, \quad N = \begin{vmatrix} A_1 & B_1 \\ A_2 & B_2 \end{vmatrix}. \tag{68.6}$$

### *68.2.2 Line Geometry Representation of the Duality of the Platonic Polyhedrons*

The duality of the hexahedron and the octahedron can be presented by the ray and axis coordinates representations of the lines of the corresponding edges of a hexahedron and its dual octahedron. Figure 68.1 shows a hexahedron and its dual octahedron. If they are put together, the edges of the original hexahedron and the dual octahedron form pairs of perpendicular bisectors. The six faces of the original hexahedron are denoted as $f_\mathrm{a}, f_\mathrm{b}, f_\mathrm{c}, f_\mathrm{d}, f_\mathrm{e}$ and $f_\mathrm{f}$, the six vertices of the dual octahedron are indicated as A, B, C, D, E and F, and the edges of the original hexahedron and the dual octahedron are respectively denoted as $e_1$ to $e_{12}$ and $e'_1$ to $e'_{12}$.

Assuming that the edges of the hexahedron are of unit lengths, establishing a reference frame with the origin located at the centre O of the hexahedron (dual octahedron), $x$-axis passing through centers of faces $f_\mathrm{b}$ and $f_\mathrm{d}$ (points O and D), $y$-axis passing through centers of faces $f_\mathrm{a}$ and $f_\mathrm{c}$ (points O and A), and $z$-axis passing through centers of faces $f_\mathrm{e}$ and $f_\mathrm{f}$ (points O and E), the coordinates of the six faces of the original hexahedron can be obtained in the reference frame as

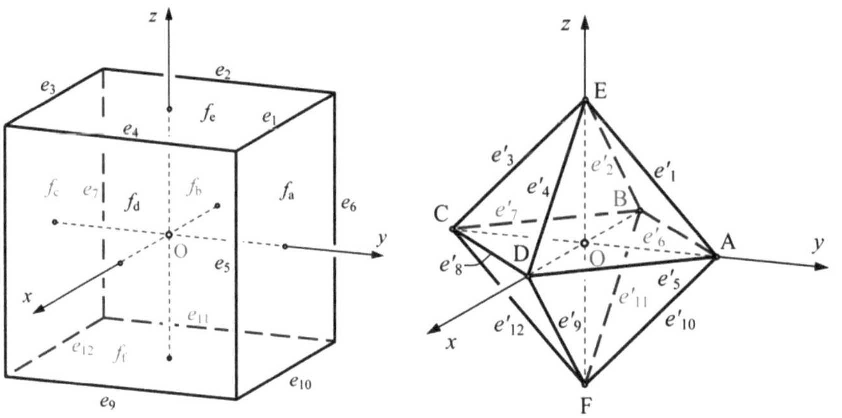


**Fig. 68.1** A hexahedron and its dual octahedron

$$\left\{\begin{array}{lll} \pi_{\rm a}\left(\frac{1}{2},\,0,\,1,\,0\right), & \pi_{\rm b}\left(\frac{1}{2},\,-1,0,\,0\right), & \pi_{\rm c}\left(\frac{1}{2},\,0,\,-1,\,0\right) \\ \pi_{\rm d}\left(\frac{1}{2},1,0,\,0\right), & \pi_{\rm e}\left(\frac{1}{2},\,0,0,1\right), & \pi_{\rm f}\left(\frac{1}{2},\,0,0,\,-1\right) \end{array}\right., \tag{68.7}$$

and if the lengths of the edges of the dual octahedral are $\sqrt{2}$, coordinates of the six vertices of the dual octahedron are

$$\left\{\begin{array}{lll} \boldsymbol{r}_{\rm A}(1,\,0,\,1,\,0), & \boldsymbol{r}_{\rm B}(1,\,-1,\,0,\,0), & \boldsymbol{r}_{\rm A}(1,\,0,\,-1,\,0) \\ \boldsymbol{r}_{\rm D}(1,\,1,\,0,\,0), & \boldsymbol{r}_{\rm E}(1,\,0,\,0,\,1), & \boldsymbol{r}_{\rm F}(1,\,0,\,0,\,-1) \end{array}\right.. \tag{68.8}$$

From Eq. (68.7), the normalized matrix presenting the 12 lines of the edges of the hexahedron in axis coordinates can be derived as

$$\mathbf{M}_{o2} = \left[\begin{array}{cccccc} 0 & -1/2 & 1/2 & 1 & 0 & 0 \\ 1/2 & 0 & 1/2 & 0 & 1 & 0 \\ 0 & 1/2 & 1/2 & -1 & 0 & 0 \\ -1/2 & 0 & 1/2 & 0 & -1 & 0 \\ 1/2 & -1/2 & 0 & 0 & 0 & -1 \\ -1/2 & -1/2 & 0 & 0 & 0 & 1 \\ 1/2 & -1/2 & 0 & 0 & 0 & 1 \\ 1/2 & 1/2 & 0 & 0 & 0 & 1 \\ -1/2 & 0 & -1/2 & 0 & 1 & 0 \\ 0 & -1/2 & -1/2 & -1 & 0 & 0 \\ 1/2 & 0 & -1/2 & 0 & -1 & 0 \\ 0 & 1/2 & -1/2 & 1 & 0 & 0 \end{array}\right] \begin{array}{l} e_1 \\ e_2 \\ e_3 \\ e_4 \\ e_5 \\ e_6 \\ e_7 \\ e_8 \\ e_9 \\ e_{10} \\ e_{11} \\ e_{12} \end{array}, \tag{68.9}$$

where each row of the matrix gives the normalized axis coordinates of a line of an edge in the hexahedron.

Similarly, based on the vertex coordinates in Eq. (68.8), the normalized matrix expressing the 12 lines of the dual octahedron in ray coordinates can be given as

$$\mathbf{M}_{\text{d2}} = \begin{array}{c} \begin{matrix} e_1' & e_2' & e_3' & e_4' & e_5' & e_6' & e_7' & e_8' & e_9' & e_{10}' & e_{11}' & e_{12}' \end{matrix} \\ \begin{bmatrix} 0 & \frac{\sqrt{2}}{2} & 0 & -\frac{\sqrt{2}}{2} & \frac{\sqrt{2}}{2} & -\frac{\sqrt{2}}{2} & \frac{\sqrt{2}}{2} & \frac{\sqrt{2}}{2} & -\frac{\sqrt{2}}{2} & 0 & \frac{\sqrt{2}}{2} & 0 \\ -\frac{\sqrt{2}}{2} & 0 & \frac{\sqrt{2}}{2} & 0 & -\frac{\sqrt{2}}{2} & -\frac{\sqrt{2}}{2} & -\frac{\sqrt{2}}{2} & \frac{\sqrt{2}}{2} & 0 & -\frac{\sqrt{2}}{2} & 0 & \frac{\sqrt{2}}{2} \\ \frac{\sqrt{2}}{2} & \frac{\sqrt{2}}{2} & \frac{\sqrt{2}}{2} & \frac{\sqrt{2}}{2} & 0 & 0 & 0 & 0 & -\frac{\sqrt{2}}{2} & -\frac{\sqrt{2}}{2} & -\frac{\sqrt{2}}{2} & -\frac{\sqrt{2}}{2} \\ \frac{\sqrt{2}}{2} & 0 & -\frac{\sqrt{2}}{2} & 0 & 0 & 0 & 0 & 0 & 0 & -\frac{\sqrt{2}}{2} & 0 & \frac{\sqrt{2}}{2} \\ 0 & \frac{\sqrt{2}}{2} & 0 & -\frac{\sqrt{2}}{2} & 0 & 0 & 0 & 0 & \frac{\sqrt{2}}{2} & 0 & -\frac{\sqrt{2}}{2} & 0 \\ 0 & 0 & 0 & 0 & -\frac{\sqrt{2}}{2} & \frac{\sqrt{2}}{2} & \frac{\sqrt{2}}{2} & \frac{\sqrt{2}}{2} & 0 & 0 & 0 & 0 \end{bmatrix} \end{array}. \tag{68.10}$$

where each column of the matrix gives the normalized ray coordinates of a line of an edge in the dual octahedron.

The matrices in Eqs. (68.9) and (68.10) are not square matrices, it is impossible to obtain their inversed matrices directly. In this case, the pseudo inverse, i.e. the Moore–Penrose inverse is employed and the normalized pseudo inverse of Eq. (68.9) is obtained as

$$\mathbf{M}_{\text{o2}}{}^{+} = \begin{bmatrix} 0 & \frac{\sqrt{2}}{2} & 0 & -\frac{\sqrt{2}}{2} & \frac{\sqrt{2}}{2} & -\frac{\sqrt{2}}{2} & \frac{\sqrt{2}}{2} & \frac{\sqrt{2}}{2} & -\frac{\sqrt{2}}{2} & 0 & \frac{\sqrt{2}}{2} & 0 \\ -\frac{\sqrt{2}}{2} & 0 & \frac{\sqrt{2}}{2} & 0 & -\frac{\sqrt{2}}{2} & -\frac{\sqrt{2}}{2} & -\frac{\sqrt{2}}{2} & \frac{\sqrt{2}}{2} & 0 & -\frac{\sqrt{2}}{2} & 0 & \frac{\sqrt{2}}{2} \\ \frac{\sqrt{2}}{2} & \frac{\sqrt{2}}{2} & \frac{\sqrt{2}}{2} & \frac{\sqrt{2}}{2} & 0 & 0 & 0 & 0 & -\frac{\sqrt{2}}{2} & -\frac{\sqrt{2}}{2} & -\frac{\sqrt{2}}{2} & -\frac{\sqrt{2}}{2} \\ \frac{\sqrt{2}}{2} & 0 & -\frac{\sqrt{2}}{2} & 0 & 0 & 0 & 0 & 0 & 0 & -\frac{\sqrt{2}}{2} & 0 & \frac{\sqrt{2}}{2} \\ 0 & \frac{\sqrt{2}}{2} & 0 & -\frac{\sqrt{2}}{2} & 0 & 0 & 0 & 0 & \frac{\sqrt{2}}{2} & 0 & -\frac{\sqrt{2}}{2} & 0 \\ 0 & 0 & 0 & 0 & -\frac{\sqrt{2}}{2} & \frac{\sqrt{2}}{2} & \frac{\sqrt{2}}{2} & \frac{\sqrt{2}}{2} & 0 & 0 & 0 & 0 \end{bmatrix}. \tag{68.11}$$

This is exactly the same as Eq. (68.10) that the columns of the matrix represent the ray coordinates of the lines of the edges in the dual octahedron.

Similarly, from Eq. (68.10), the normalized pseudo inverse of the matrix is

$$\mathbf{M}_{\text{d2}}{}^{+} = \begin{bmatrix} 0 & -1/2 & 1/2 & 1 & 0 & 0 \\ 1/2 & 0 & 1/2 & 0 & 1 & 0 \\ 0 & 1/2 & 1/2 & -1 & 0 & 0 \\ -1/2 & 0 & 1/2 & 0 & -1 & 0 \\ 1/2 & -1/2 & 0 & 0 & 0 & -1 \\ -1/2 & -1/2 & 0 & 0 & 0 & 1 \\ 1/2 & -1/2 & 0 & 0 & 0 & 1 \\ 1/2 & 1/2 & 0 & 0 & 0 & 1 \\ -1/2 & 0 & -1/2 & 0 & 1 & 0 \\ 0 & -1/2 & -1/2 & -1 & 0 & 0 \\ 1/2 & 0 & -1/2 & 0 & -1 & 0 \\ 0 & 1/2 & -1/2 & 1 & 0 & 0 \end{bmatrix}, \tag{68.12}$$

it is the same as Eq. (68.9), each row of which gives the axis coordinates of a line of the edge in the original hexahedron.

From the above analysis, it can be seen that the normalized pseudo inverse of the matrix in Eq. (68.9), each row of which represents a line of an edge of the hexahedron in the axis coordinates form, gives a matrix, each column of which represents a line of an edge of the dual octahedron in the ray coordinates form. In the same while, the normalized pseudo inverse of the matrix in Eq. (68.10), each column of which represents a line of an edge of the dual octahedron in the ray coordinates form, gives a matrix, each row of which represents a line of an edge of the original hexahedron in the axis coordinates form.

Further, from Eq. (68.9) the lines of the edges in the hexahedron can be obtained in the axis coordinates as

$$\boldsymbol{l}_{2i} = \boldsymbol{e}_{2i}^{\mathrm{T}}\mathbf{M}_{o2}\ (i = 1, 2, \ldots, 12), \tag{68.13}$$

where $e_{2i}$ is an $12 \times 1$ elementary vector with the $i$th element being 1 and the rest being 0.

Similarly, the lines of the edges in the dual octahedron can be obtained from Eq. (68.10), in the ray coordinates as

$$\boldsymbol{l'}_{2i} = \mathbf{M}_{d2}\boldsymbol{e}_{2i}(i = 1, 2, \ldots, 12), \tag{68.14}$$

Reciprocal multiplication of Eqs. (68.13) and (68.14) gives

$$\boldsymbol{l}_{2i}\boldsymbol{\Delta l'}_{2i} = \boldsymbol{e}_{2i}^{\mathrm{T}}\mathbf{M}_{o2}\Delta\mathbf{M}_{d2}\boldsymbol{e}_{2i} = 0, \tag{68.15}$$

where $\boldsymbol{\Delta}$ is give as $\boldsymbol{\Delta} = \begin{bmatrix} \mathbf{0} & \mathbf{I} \\ \mathbf{I} & \mathbf{0} \end{bmatrix}$ with $\mathbf{0}$ being $3 \times 3$ null matrix and $\mathbf{I}$ being $3 \times 3$ identity matrix, and Eq. (68.15) implies that each of the edges $e'_1$ to $e'_{12}$ in the dual octahedron is respectively perpendicular to one of the corresponding edges $e_1$ to $e_{12}$ in the hexahedron.

Thus, from above analysis, it is found that the duality between the hexahedron and octahedron can be interpreted by the ray and axis coordinates representation of the lines of the edges of the dual polyhedrons.

Similarly, using the same procedure, the duality of the tetrahedron itself, and the dodecahedron and icosahedron can be presented utilizing the same approach. It is also found in this research that the duality of the five Platonic polyhedrons leads to the isomorphism of the dual deployable polyhedral mechanisms constructed based on the five regular polyhedrons.

## 68.3 Construction of the Regular Deployable Polyhedral Mechanisms (DPMs)

Figure 68.2 gives the plane-symmetric eight-bar linkage [7] that contains four links 1, 3, 5 and 7 with two parallel joint axes at each end and four plates $V_0$, $V_2$, $V_4$ and $V_6$ in isosceles triangle shape with two revolute joint axes being spread at both ends of the plates by $\varphi_1$ and $\varphi_2$. The lengths of the four links 1, 3, 5 and 7 are the same, and plates $V_0$ and $V_4$ are identical and so are plates $V_2$ and $V_6$. The linkage is symmetric to two planes, one is the plane $\pi$ in Fig. 68.2 that passes the

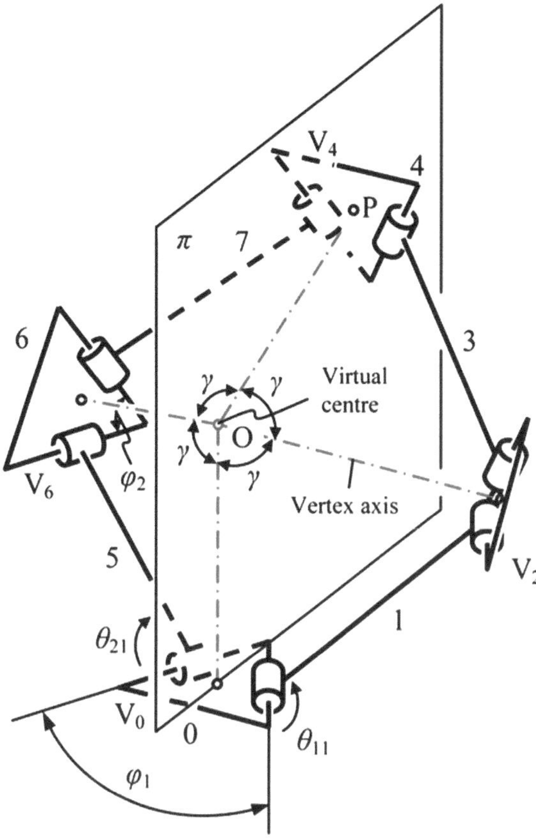


**Fig. 68.2** A plane-symmetric eight-bar linkage

centre points of $V_0$ and $V_4$ and is perpendicular to plates $V_0$ and $V_4$, and the other one is $\pi_1$ that passes the centre points of plates $V_2$ and $V_6$ and is perpendicular to plates $V_2$ and $V_6$. Given symmetric inputs at any pair of joints in any of the four vertices, e.g. $\theta_{11} = \theta_{21}$ in Fig. 68.2, and imaging that the linkage is fixed at point O, the four plates perform radially reciprocating motions along their vertex axes (see Fig. 68.2) in such a way that plates $V_0$ and $V_4$ move towards O, plates $V_2$ and $V_6$ move away from point O, and vice versa. The centre point O is therefore referred to as the virtual centre of the eight-bar linkage.

In this work, it is found that by implanting a group of the above plane-symmetric eight-bar linkages into the five Platonic polyhedral bases, a family of regular deployable polyhedral mechanisms (DPMs) can be constructed. It is also found that the dual deployable polyhedral mechanisms (DPMs) are isomorphic. Herein, in order to illustrate the construction of the DPMs, the hexahedron and its dual octahedron are picked up and based on them the deployable hexahedral mechanisms and deployable octahedral mechanism are constructed in this section. The deployable hexahedral mechanism and the deployable octahedral mechanism are isomorphic and the isomorphism of the DPMs is to be investigated in the next section.

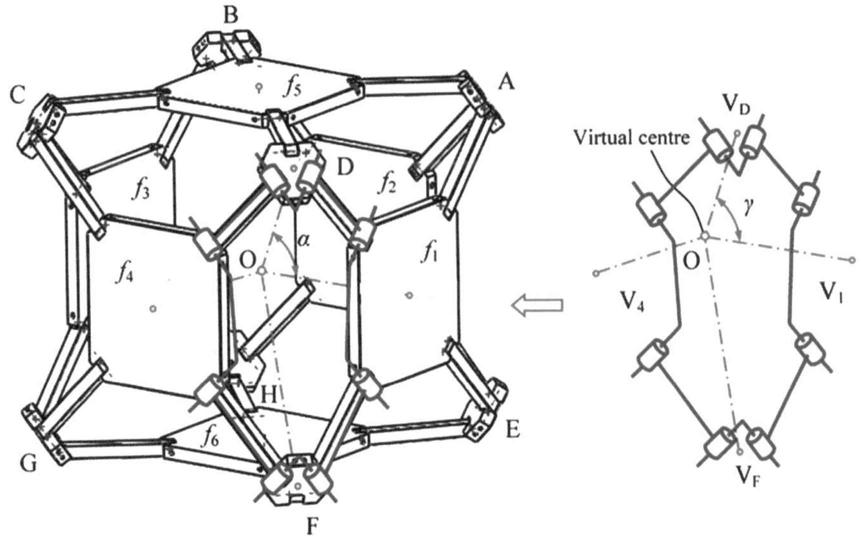


**Fig. 68.3** A deployable hexahedral mechanism and the eight-bar linkage

As illustrated in Fig. 68.3, a hexahedron has six identical square faces with 12 edges of the same length and each of the eight identical vertices contains three edges involving three faces. Based on this hexahedron, introduce 11 identical plane-symmetric eight-bar linkages in the faces, vertices and edges of the hexahedron and ensure that the angle $\gamma$ of the adjacent vertex axes equal the central angle $\alpha = 54.74°$ of the hexahedron, a deployable hexahedral mechanism can be generated in Fig. 68.3. It should be pointed out that during the constructing process, all the joint axes located at the faces of the hexahedron must be perpendicular to the corresponding diagonals of the faces. Figure 68.4 shows two of the typical phases of the mechanism.

Further, an octahedron consists of eight identical equilateral triangular faces, twelve edges of the same length and six identical vertices each of which contains four faces and four edges. Ensure that the vertex angle $\gamma$ of the eight-bar linkage equal the central angle $\alpha = 54.74°$ of the octahedron and then implant eleven identical plane-symmetric eight-bar linkages in a octahedral basis, a highly overconstrained deployable octahedral mechanism of mobility one can be constructed in Fig. 68.5.

One can find that, in every configuration, the F-plates that have the same shape as the faces of the polyhedron always form a polyhedron which is herein referred to as virtual polyhedron that is in the same shape as the polyhedral basis. Further, as aforementioned, the tetrahedron is self-dual, the hexahedron and the octahedron, and the dodecahedron and the icosahedron are respectively said to be dual to each other. Therefore from the above construction, one can find that the deployable hexahedral and octahedral mechanisms contain the same number of plane-symmetric eight-bar linkages. Due to the duality of the Platonic polyhedrons, one can also find that the deployable hexahedral mechanism is dual to the deployable octahedral mechanism

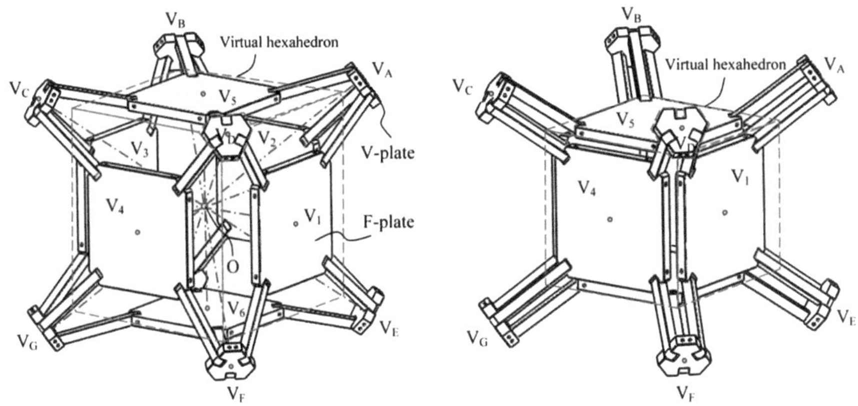


**Fig. 68.4** Two of the typical phases of the deployable hexahedral mechanism

because if we increase the sizes of the V-plates and reduce the sizes of the F-plates of the deployable hexahedral mechanism it becomes a deployable octahedral mechanism in which the V-plates in the original hexahedral mechanism become F-plates of the octahedral mechanism and the F-plates of the original hexahedral mechanism become V-plates of the octahedral mechanism, and vice versa. From mechanism point of view, the topologies of the deployable hexahedral and octahedral mechanisms are said to be isomorphic and so are that of the deployable dodecahedral and icosahedral mechanisms, the isomorphism of the Platonic-polyhedron-based deployable mechanisms is to be presented in the following section through the adjacency matrices of their topology graphs.

## 68.4 Isomorphism of the Regular Deployable Polyhedral Mechanisms (DPMs)

According to Fig. 68.4 the topology graphs of the deployable hexahedral mechanism are given in Fig. 68.6. In the figures, the links are denoted as dots labeled by the letters and numbers of the corresponding links in Fig. 68.4, joints are denoted as lines connecting the links. It should be noted that in Fig. 68.6, links 1–3 are three links around plate $V_A$ of the deployable hexahedral mechanism in Fig. 68.4, links 4–6 are around $V_B$, links 7–9 around $V_C$, links 10–12 around $V_D$, links 13–15 around $V_E$, links 16–18 around $V_F$, links 19–21 around $V_G$, and links 22–24 around $V_H$.

Based on the Fig. 68.6, there are totally $6 + 8 + 8 \times 3 = 38$ links in the mechanism and following the sequence as $V_A$, $V_B$, $V_C$, $V_D$, $V_E$, $V_F$, $V_G$, $V_H$, $V_1$, $V_2$, $V_3$, $V_4$, $V_5$, $V_6$, and 1–24, the $38 \times 38$ adjacency matrix of the deployable hexahedral mechanism can be written as

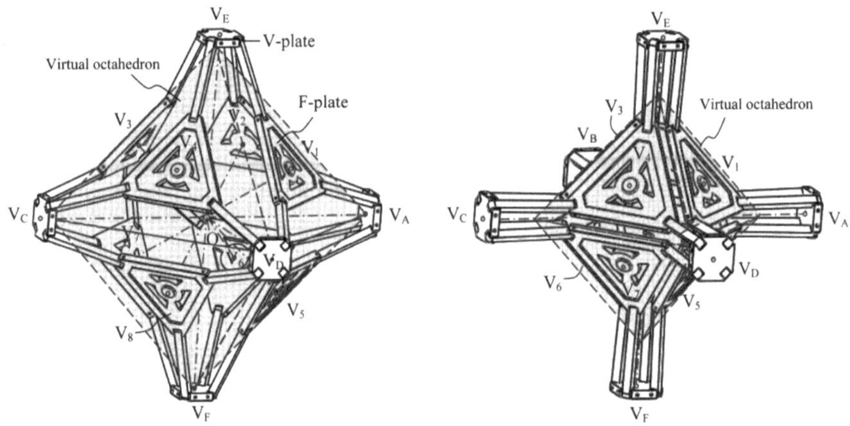


**Fig. 68.5** A deployable octahedral mechanism and two of the typical phases

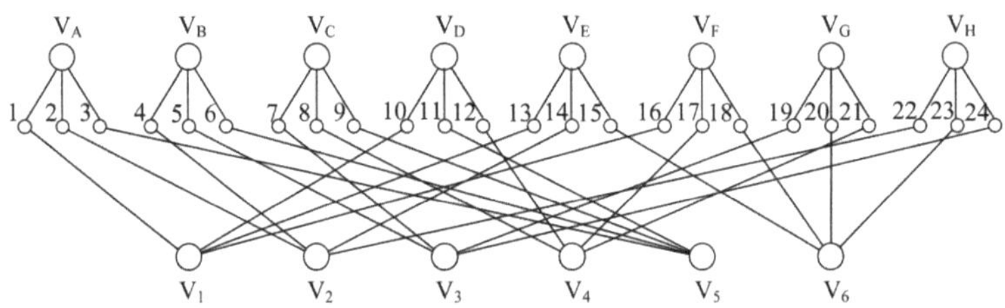


**Fig. 68.6** Topology graph of the deployable hexahedral mechanism

$$
\mathbf{A}_h = \begin{pmatrix}
0 & 0 & 0 & 0 & 0 & 0 & 0 & 0 & 0 & 0 & 0 & 0 & 0 & 0 & 1 & 1 & 1 & 0 & 0 & 0 & 0 & 0 & 0 & 0 & 0 & 0 & 0 & 0 & 0 & 0 & 0 & 0 & 0 & 0 & 0 & 0 & 0 & 0 \\
0 & 0 & 0 & 0 & 0 & 0 & 0 & 0 & 0 & 0 & 0 & 0 & 0 & 0 & 0 & 0 & 0 & 1 & 1 & 1 & 0 & 0 & 0 & 0 & 0 & 0 & 0 & 0 & 0 & 0 & 0 & 0 & 0 & 0 & 0 & 0 & 0 & 0 \\
0 & 0 & 0 & 0 & 0 & 0 & 0 & 0 & 0 & 0 & 0 & 0 & 0 & 0 & 0 & 0 & 0 & 0 & 0 & 0 & 1 & 1 & 1 & 0 & 0 & 0 & 0 & 0 & 0 & 0 & 0 & 0 & 0 & 0 & 0 & 0 & 0 & 0 \\
0 & 0 & 0 & 0 & 0 & 0 & 0 & 0 & 0 & 0 & 0 & 0 & 0 & 0 & 0 & 0 & 0 & 0 & 0 & 0 & 0 & 0 & 0 & 1 & 1 & 1 & 0 & 0 & 0 & 0 & 0 & 0 & 0 & 0 & 0 & 0 & 0 & 0 \\
0 & 0 & 0 & 0 & 0 & 0 & 0 & 0 & 0 & 0 & 0 & 0 & 0 & 0 & 0 & 0 & 0 & 0 & 0 & 0 & 0 & 0 & 0 & 0 & 0 & 0 & 1 & 1 & 1 & 0 & 0 & 0 & 0 & 0 & 0 & 0 & 0 & 0 \\
0 & 0 & 0 & 0 & 0 & 0 & 0 & 0 & 0 & 0 & 0 & 0 & 0 & 0 & 0 & 0 & 0 & 0 & 0 & 0 & 0 & 0 & 0 & 0 & 0 & 0 & 0 & 0 & 0 & 1 & 1 & 1 & 0 & 0 & 0 & 0 & 0 & 0 \\
0 & 0 & 0 & 0 & 0 & 0 & 0 & 0 & 0 & 0 & 0 & 0 & 0 & 0 & 0 & 0 & 0 & 0 & 0 & 0 & 0 & 0 & 0 & 0 & 0 & 0 & 0 & 0 & 0 & 0 & 0 & 0 & 1 & 1 & 1 & 0 & 0 & 0 \\
0 & 0 & 0 & 0 & 0 & 0 & 0 & 0 & 0 & 0 & 0 & 0 & 0 & 0 & 0 & 0 & 0 & 0 & 0 & 0 & 0 & 0 & 0 & 0 & 0 & 0 & 0 & 0 & 0 & 0 & 0 & 0 & 0 & 0 & 0 & 1 & 1 & 1 \\
0 & 0 & 0 & 0 & 0 & 0 & 0 & 0 & 0 & 0 & 0 & 0 & 0 & 0 & 1 & 0 & 0 & 0 & 0 & 0 & 0 & 0 & 0 & 1 & 0 & 0 & 1 & 0 & 0 & 1 & 0 & 0 & 0 & 0 & 0 & 0 & 0 & 0 \\
0 & 0 & 0 & 0 & 0 & 0 & 0 & 0 & 0 & 0 & 0 & 0 & 0 & 0 & 0 & 1 & 0 & 1 & 0 & 0 & 0 & 0 & 0 & 0 & 0 & 0 & 0 & 1 & 0 & 0 & 0 & 0 & 0 & 0 & 0 & 1 & 0 & 0 \\
0 & 0 & 0 & 0 & 0 & 0 & 0 & 0 & 0 & 0 & 0 & 0 & 0 & 0 & 0 & 0 & 0 & 0 & 1 & 0 & 1 & 0 & 0 & 0 & 0 & 0 & 0 & 0 & 0 & 0 & 0 & 0 & 1 & 0 & 0 & 0 & 0 & 1 \\
0 & 0 & 0 & 0 & 0 & 0 & 0 & 0 & 0 & 0 & 0 & 0 & 0 & 0 & 0 & 0 & 0 & 0 & 0 & 0 & 0 & 1 & 0 & 0 & 0 & 1 & 0 & 0 & 0 & 0 & 1 & 0 & 0 & 0 & 1 & 0 & 0 & 0 \\
0 & 0 & 0 & 0 & 0 & 0 & 0 & 0 & 0 & 0 & 0 & 0 & 0 & 0 & 0 & 0 & 1 & 0 & 0 & 1 & 0 & 0 & 1 & 0 & 1 & 0 & 0 & 0 & 0 & 0 & 0 & 0 & 0 & 0 & 0 & 0 & 0 & 0 \\
0 & 0 & 0 & 0 & 0 & 0 & 0 & 0 & 0 & 0 & 0 & 0 & 0 & 0 & 0 & 0 & 0 & 0 & 0 & 0 & 0 & 0 & 0 & 0 & 0 & 0 & 0 & 0 & 1 & 0 & 0 & 1 & 0 & 1 & 0 & 0 & 1 & 0 \\
1 & 0 & 0 & 0 & 0 & 0 & 0 & 0 & 1 & 0 & 0 & 0 & 0 & 0 & 0 & 0 & 0 & 0 & 0 & 0 & 0 & 0 & 0 & 0 & 0 & 0 & 0 & 0 & 0 & 0 & 0 & 0 & 0 & 0 & 0 & 0 & 0 & 0 \\
1 & 0 & 0 & 0 & 0 & 0 & 0 & 0 & 0 & 1 & 0 & 0 & 0 & 0 & 0 & 0 & 0 & 0 & 0 & 0 & 0 & 0 & 0 & 0 & 0 & 0 & 0 & 0 & 0 & 0 & 0 & 0 & 0 & 0 & 0 & 0 & 0 & 0 \\
1 & 0 & 0 & 0 & 0 & 0 & 0 & 0 & 0 & 0 & 0 & 1 & 0 & 0 & 0 & 0 & 0 & 0 & 0 & 0 & 0 & 0 & 0 & 0 & 0 & 0 & 0 & 0 & 0 & 0 & 0 & 0 & 0 & 0 & 0 & 0 & 0 & 0 \\
0 & 1 & 0 & 0 & 0 & 0 & 0 & 0 & 0 & 1 & 0 & 0 & 0 & 0 & 0 & 0 & 0 & 0 & 0 & 0 & 0 & 0 & 0 & 0 & 0 & 0 & 0 & 0 & 0 & 0 & 0 & 0 & 0 & 0 & 0 & 0 & 0 & 0 \\
0 & 1 & 0 & 0 & 0 & 0 & 0 & 0 & 0 & 0 & 1 & 0 & 0 & 0 & 0 & 0 & 0 & 0 & 0 & 0 & 0 & 0 & 0 & 0 & 0 & 0 & 0 & 0 & 0 & 0 & 0 & 0 & 0 & 0 & 0 & 0 & 0 & 0 \\
0 & 1 & 0 & 0 & 0 & 0 & 0 & 0 & 0 & 0 & 0 & 1 & 0 & 0 & 0 & 0 & 0 & 0 & 0 & 0 & 0 & 0 & 0 & 0 & 0 & 0 & 0 & 0 & 0 & 0 & 0 & 0 & 0 & 0 & 0 & 0 & 0 & 0 \\
0 & 0 & 1 & 0 & 0 & 0 & 0 & 0 & 0 & 0 & 1 & 0 & 0 & 0 & 0 & 0 & 0 & 0 & 0 & 0 & 0 & 0 & 0 & 0 & 0 & 0 & 0 & 0 & 0 & 0 & 0 & 0 & 0 & 0 & 0 & 0 & 0 & 0 \\
0 & 0 & 1 & 0 & 0 & 0 & 0 & 0 & 0 & 0 & 0 & 1 & 0 & 0 & 0 & 0 & 0 & 0 & 0 & 0 & 0 & 0 & 0 & 0 & 0 & 0 & 0 & 0 & 0 & 0 & 0 & 0 & 0 & 0 & 0 & 0 & 0 & 0 \\
0 & 0 & 1 & 0 & 0 & 0 & 0 & 0 & 0 & 0 & 0 & 0 & 1 & 0 & 0 & 0 & 0 & 0 & 0 & 0 & 0 & 0 & 0 & 0 & 0 & 0 & 0 & 0 & 0 & 0 & 0 & 0 & 0 & 0 & 0 & 0 & 0 & 0 \\
0 & 0 & 0 & 1 & 0 & 0 & 0 & 0 & 1 & 0 & 0 & 0 & 0 & 0 & 0 & 0 & 0 & 0 & 0 & 0 & 0 & 0 & 0 & 0 & 0 & 0 & 0 & 0 & 0 & 0 & 0 & 0 & 0 & 0 & 0 & 0 & 0 & 0 \\
0 & 0 & 0 & 1 & 0 & 0 & 0 & 0 & 0 & 0 & 0 & 0 & 1 & 0 & 0 & 0 & 0 & 0 & 0 & 0 & 0 & 0 & 0 & 0 & 0 & 0 & 0 & 0 & 0 & 0 & 0 & 0 & 0 & 0 & 0 & 0 & 0 & 0 \\
0 & 0 & 0 & 1 & 0 & 0 & 0 & 0 & 0 & 0 & 0 & 1 & 0 & 0 & 0 & 0 & 0 & 0 & 0 & 0 & 0 & 0 & 0 & 0 & 0 & 0 & 0 & 0 & 0 & 0 & 0 & 0 & 0 & 0 & 0 & 0 & 0 & 0 \\
0 & 0 & 0 & 0 & 1 & 0 & 0 & 0 & 1 & 0 & 0 & 0 & 0 & 0 & 0 & 0 & 0 & 0 & 0 & 0 & 0 & 0 & 0 & 0 & 0 & 0 & 0 & 0 & 0 & 0 & 0 & 0 & 0 & 0 & 0 & 0 & 0 & 0 \\
0 & 0 & 0 & 0 & 1 & 0 & 0 & 0 & 0 & 1 & 0 & 0 & 0 & 0 & 0 & 0 & 0 & 0 & 0 & 0 & 0 & 0 & 0 & 0 & 0 & 0 & 0 & 0 & 0 & 0 & 0 & 0 & 0 & 0 & 0 & 0 & 0 & 0 \\
0 & 0 & 0 & 0 & 1 & 0 & 0 & 0 & 0 & 0 & 0 & 0 & 0 & 1 & 0 & 0 & 0 & 0 & 0 & 0 & 0 & 0 & 0 & 0 & 0 & 0 & 0 & 0 & 0 & 0 & 0 & 0 & 0 & 0 & 0 & 0 & 0 & 0 \\
0 & 0 & 0 & 0 & 0 & 1 & 0 & 0 & 1 & 0 & 0 & 0 & 0 & 0 & 0 & 0 & 0 & 0 & 0 & 0 & 0 & 0 & 0 & 0 & 0 & 0 & 0 & 0 & 0 & 0 & 0 & 0 & 0 & 0 & 0 & 0 & 0 & 0 \\
0 & 0 & 0 & 0 & 0 & 1 & 0 & 0 & 0 & 0 & 0 & 1 & 0 & 0 & 0 & 0 & 0 & 0 & 0 & 0 & 0 & 0 & 0 & 0 & 0 & 0 & 0 & 0 & 0 & 0 & 0 & 0 & 0 & 0 & 0 & 0 & 0 & 0 \\
0 & 0 & 0 & 0 & 0 & 1 & 0 & 0 & 0 & 0 & 0 & 0 & 0 & 1 & 0 & 0 & 0 & 0 & 0 & 0 & 0 & 0 & 0 & 0 & 0 & 0 & 0 & 0 & 0 & 0 & 0 & 0 & 0 & 0 & 0 & 0 & 0 & 0 \\
0 & 0 & 0 & 0 & 0 & 0 & 1 & 0 & 0 & 0 & 1 & 0 & 0 & 0 & 0 & 0 & 0 & 0 & 0 & 0 & 0 & 0 & 0 & 0 & 0 & 0 & 0 & 0 & 0 & 0 & 0 & 0 & 0 & 0 & 0 & 0 & 0 & 0 \\
0 & 0 & 0 & 0 & 0 & 0 & 1 & 0 & 0 & 0 & 0 & 0 & 0 & 1 & 0 & 0 & 0 & 0 & 0 & 0 & 0 & 0 & 0 & 0 & 0 & 0 & 0 & 0 & 0 & 0 & 0 & 0 & 0 & 0 & 0 & 0 & 0 & 0 \\
0 & 0 & 0 & 0 & 0 & 0 & 1 & 0 & 0 & 0 & 0 & 1 & 0 & 0 & 0 & 0 & 0 & 0 & 0 & 0 & 0 & 0 & 0 & 0 & 0 & 0 & 0 & 0 & 0 & 0 & 0 & 0 & 0 & 0 & 0 & 0 & 0 & 0 \\
0 & 0 & 0 & 0 & 0 & 0 & 0 & 1 & 0 & 1 & 0 & 0 & 0 & 0 & 0 & 0 & 0 & 0 & 0 & 0 & 0 & 0 & 0 & 0 & 0 & 0 & 0 & 0 & 0 & 0 & 0 & 0 & 0 & 0 & 0 & 0 & 0 & 0 \\
0 & 0 & 0 & 0 & 0 & 0 & 0 & 1 & 0 & 0 & 0 & 0 & 0 & 1 & 0 & 0 & 0 & 0 & 0 & 0 & 0 & 0 & 0 & 0 & 0 & 0 & 0 & 0 & 0 & 0 & 0 & 0 & 0 & 0 & 0 & 0 & 0 & 0 \\
0 & 0 & 0 & 0 & 0 & 0 & 0 & 1 & 0 & 0 & 1 & 0 & 0 & 0 & 0 & 0 & 0 & 0 & 0 & 0 & 0 & 0 & 0 & 0 & 0 & 0 & 0 & 0 & 0 & 0 & 0 & 0 & 0 & 0 & 0 & 0 & 0 & 0
\end{pmatrix} \quad (68.16)
$$

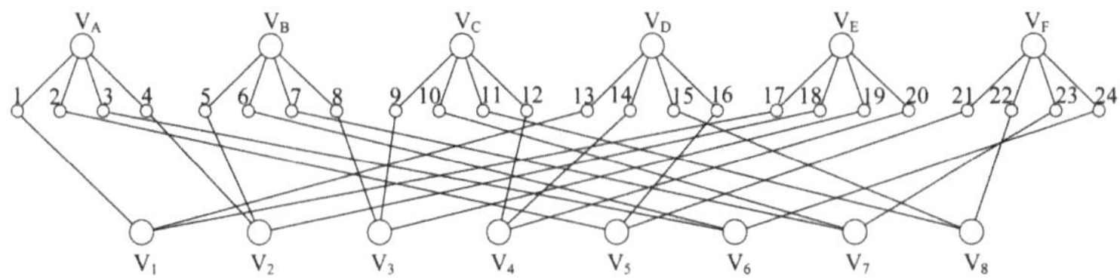


**Fig. 68.7** Topology graph of the deployable octahedral mechanism

Solving Eq. (68.16), the eigenvalues of adjacency matrix are $\{-\sqrt{7},\ \sqrt{7},\ -\sqrt{(7+\sqrt{17})/2},\ -\sqrt{(7+\sqrt{17})/2},\ -\sqrt{(7+\sqrt{17})/2},\ \sqrt{(7+\sqrt{17})/2},\ -\sqrt{(7+\sqrt{17})/2},\ -\sqrt{(7+\sqrt{17})/2},\ -2,\ -2,\ -2,\ -2,\ -\sqrt{3},\ -\sqrt{3},\ -\sqrt{3},\ -\sqrt{3},\ \sqrt{3},\ \sqrt{3},\ \sqrt{3},\ \sqrt{3},\ -\sqrt{(7-\sqrt{17})/2},\ -\sqrt{(7-\sqrt{17})/2},\ -\sqrt{(7-\sqrt{17})/2},\ \sqrt{(7-\sqrt{17})/2},\ \sqrt{(7-\sqrt{17})/2},\ \sqrt{(7-\sqrt{17})/2},\ 0,\ 0,\ 0,\ 0,\ 0,\ 0,\ 0,\ 0,\ 0,\ 0,\ 0,\ 0\}$.

Similarly, from Fig. 68.5, topology graph of the deployable octahedral mechanism can be obtained in Fig. 68.7. In Fig. 68.7, links 1–4 are four links arranged around plate $V_A$ of the deployable octahedral mechanism in Fig. 68.5, links 5–8 are around $V_B$, links 9–12 around $V_C$, links 13–16 around $V_D$, links 17–20 around $V_E$, and links 21–24 around $V_F$.

According to Fig. 68.7, there are totally $8 + 6 + 6 \times 4 = 38$ links in the mechanism and following the sequence as $V_A$, $V_B$, $V_C$, $V_D$, $V_E$, $V_F$, $V_1$, $V_2$, $V_3$, $V_4$, $V_5$, $V_6$, $V_7$, $V_8$, and 1–24, the $38 \times 38$ adjacency matrix of the deployable hexahedral mechanism can be written as

$$
\mathbf{A}_o = \begin{pmatrix}
0 & 0 & 0 & 0 & 0 & 0 & 0 & 0 & 0 & 0 & 0 & 0 & 0 & 0 & 1 & 1 & 1 & 1 & 0 & 0 & 0 & 0 & 0 & 0 & 0 & 0 & 0 & 0 & 0 & 0 & 0 & 0 & 0 & 0 & 0 & 0 & 0 & 0 \\
0 & 0 & 0 & 0 & 0 & 0 & 0 & 0 & 0 & 0 & 0 & 0 & 0 & 0 & 0 & 0 & 0 & 0 & 1 & 1 & 1 & 1 & 0 & 0 & 0 & 0 & 0 & 0 & 0 & 0 & 0 & 0 & 0 & 0 & 0 & 0 & 0 & 0 \\
0 & 0 & 0 & 0 & 0 & 0 & 0 & 0 & 0 & 0 & 0 & 0 & 0 & 0 & 0 & 0 & 0 & 0 & 0 & 0 & 0 & 0 & 1 & 1 & 1 & 1 & 0 & 0 & 0 & 0 & 0 & 0 & 0 & 0 & 0 & 0 & 0 & 0 \\
0 & 0 & 0 & 0 & 0 & 0 & 0 & 0 & 0 & 0 & 0 & 0 & 0 & 0 & 0 & 0 & 0 & 0 & 0 & 0 & 0 & 0 & 0 & 0 & 0 & 0 & 1 & 1 & 1 & 1 & 0 & 0 & 0 & 0 & 0 & 0 & 0 & 0 \\
0 & 0 & 0 & 0 & 0 & 0 & 0 & 0 & 0 & 0 & 0 & 0 & 0 & 0 & 0 & 0 & 0 & 0 & 0 & 0 & 0 & 0 & 0 & 0 & 0 & 0 & 0 & 0 & 0 & 0 & 1 & 1 & 1 & 1 & 0 & 0 & 0 & 0 \\
0 & 0 & 0 & 0 & 0 & 0 & 0 & 0 & 0 & 0 & 0 & 0 & 0 & 0 & 0 & 0 & 0 & 0 & 0 & 0 & 0 & 0 & 0 & 0 & 0 & 0 & 0 & 0 & 0 & 0 & 0 & 0 & 0 & 0 & 1 & 1 & 1 & 1 \\
0 & 0 & 0 & 0 & 0 & 0 & 0 & 0 & 0 & 0 & 0 & 0 & 0 & 0 & 1 & 0 & 0 & 0 & 0 & 0 & 0 & 0 & 0 & 0 & 0 & 0 & 1 & 0 & 0 & 0 & 1 & 0 & 0 & 0 & 0 & 0 & 0 & 0 \\
0 & 0 & 0 & 0 & 0 & 0 & 0 & 0 & 0 & 0 & 0 & 0 & 0 & 0 & 0 & 0 & 0 & 1 & 1 & 0 & 0 & 0 & 0 & 0 & 0 & 0 & 0 & 0 & 0 & 0 & 0 & 1 & 0 & 0 & 0 & 0 & 0 & 0 \\
0 & 0 & 0 & 0 & 0 & 0 & 0 & 0 & 0 & 0 & 0 & 0 & 0 & 0 & 0 & 0 & 0 & 0 & 0 & 0 & 0 & 1 & 1 & 0 & 0 & 0 & 0 & 0 & 0 & 0 & 0 & 0 & 1 & 0 & 0 & 0 & 0 & 0 \\
0 & 0 & 0 & 0 & 0 & 0 & 0 & 0 & 0 & 0 & 0 & 0 & 0 & 0 & 0 & 0 & 0 & 0 & 0 & 0 & 0 & 0 & 0 & 0 & 0 & 1 & 0 & 1 & 0 & 0 & 0 & 0 & 0 & 1 & 0 & 0 & 0 & 0 \\
0 & 0 & 0 & 0 & 0 & 0 & 0 & 0 & 0 & 0 & 0 & 0 & 0 & 0 & 0 & 1 & 0 & 0 & 0 & 0 & 0 & 0 & 0 & 0 & 0 & 0 & 0 & 0 & 0 & 1 & 0 & 0 & 0 & 0 & 1 & 0 & 0 & 0 \\
0 & 0 & 0 & 0 & 0 & 0 & 0 & 0 & 0 & 0 & 0 & 0 & 0 & 0 & 0 & 0 & 1 & 0 & 0 & 1 & 0 & 0 & 0 & 0 & 0 & 0 & 0 & 0 & 0 & 0 & 0 & 0 & 0 & 0 & 0 & 0 & 0 & 1 \\
0 & 0 & 0 & 0 & 0 & 0 & 0 & 0 & 0 & 0 & 0 & 0 & 0 & 0 & 0 & 0 & 0 & 0 & 0 & 0 & 1 & 0 & 0 & 1 & 0 & 0 & 0 & 0 & 0 & 0 & 0 & 0 & 0 & 0 & 0 & 0 & 1 & 0 \\
0 & 0 & 0 & 0 & 0 & 0 & 0 & 0 & 0 & 0 & 0 & 0 & 0 & 0 & 0 & 0 & 0 & 0 & 0 & 0 & 0 & 0 & 0 & 0 & 1 & 0 & 0 & 0 & 1 & 0 & 0 & 0 & 0 & 0 & 0 & 1 & 0 & 0 \\
1 & 0 & 0 & 0 & 0 & 0 & 1 & 0 & 0 & 0 & 0 & 0 & 0 & 0 & 0 & 0 & 0 & 0 & 0 & 0 & 0 & 0 & 0 & 0 & 0 & 0 & 0 & 0 & 0 & 0 & 0 & 0 & 0 & 0 & 0 & 0 & 0 & 0 \\
1 & 0 & 0 & 0 & 0 & 0 & 0 & 0 & 0 & 0 & 1 & 0 & 0 & 0 & 0 & 0 & 0 & 0 & 0 & 0 & 0 & 0 & 0 & 0 & 0 & 0 & 0 & 0 & 0 & 0 & 0 & 0 & 0 & 0 & 0 & 0 & 0 & 0 \\
1 & 0 & 0 & 0 & 0 & 0 & 0 & 0 & 0 & 0 & 0 & 1 & 0 & 0 & 0 & 0 & 0 & 0 & 0 & 0 & 0 & 0 & 0 & 0 & 0 & 0 & 0 & 0 & 0 & 0 & 0 & 0 & 0 & 0 & 0 & 0 & 0 & 0 \\
1 & 0 & 0 & 0 & 0 & 0 & 0 & 1 & 0 & 0 & 0 & 0 & 0 & 0 & 0 & 0 & 0 & 0 & 0 & 0 & 0 & 0 & 0 & 0 & 0 & 0 & 0 & 0 & 0 & 0 & 0 & 0 & 0 & 0 & 0 & 0 & 0 & 0 \\
0 & 1 & 0 & 0 & 0 & 0 & 0 & 1 & 0 & 0 & 0 & 0 & 0 & 0 & 0 & 0 & 0 & 0 & 0 & 0 & 0 & 0 & 0 & 0 & 0 & 0 & 0 & 0 & 0 & 0 & 0 & 0 & 0 & 0 & 0 & 0 & 0 & 0 \\
0 & 1 & 0 & 0 & 0 & 0 & 0 & 0 & 0 & 0 & 0 & 1 & 0 & 0 & 0 & 0 & 0 & 0 & 0 & 0 & 0 & 0 & 0 & 0 & 0 & 0 & 0 & 0 & 0 & 0 & 0 & 0 & 0 & 0 & 0 & 0 & 0 & 0 \\
0 & 1 & 0 & 0 & 0 & 0 & 0 & 0 & 0 & 0 & 0 & 0 & 1 & 0 & 0 & 0 & 0 & 0 & 0 & 0 & 0 & 0 & 0 & 0 & 0 & 0 & 0 & 0 & 0 & 0 & 0 & 0 & 0 & 0 & 0 & 0 & 0 & 0 \\
0 & 1 & 0 & 0 & 0 & 0 & 0 & 0 & 1 & 0 & 0 & 0 & 0 & 0 & 0 & 0 & 0 & 0 & 0 & 0 & 0 & 0 & 0 & 0 & 0 & 0 & 0 & 0 & 0 & 0 & 0 & 0 & 0 & 0 & 0 & 0 & 0 & 0 \\
0 & 0 & 1 & 0 & 0 & 0 & 0 & 0 & 1 & 0 & 0 & 0 & 0 & 0 & 0 & 0 & 0 & 0 & 0 & 0 & 0 & 0 & 0 & 0 & 0 & 0 & 0 & 0 & 0 & 0 & 0 & 0 & 0 & 0 & 0 & 0 & 0 & 0 \\
0 & 0 & 1 & 0 & 0 & 0 & 0 & 0 & 0 & 0 & 0 & 0 & 1 & 0 & 0 & 0 & 0 & 0 & 0 & 0 & 0 & 0 & 0 & 0 & 0 & 0 & 0 & 0 & 0 & 0 & 0 & 0 & 0 & 0 & 0 & 0 & 0 & 0 \\
0 & 0 & 1 & 0 & 0 & 0 & 0 & 0 & 0 & 0 & 0 & 0 & 0 & 1 & 0 & 0 & 0 & 0 & 0 & 0 & 0 & 0 & 0 & 0 & 0 & 0 & 0 & 0 & 0 & 0 & 0 & 0 & 0 & 0 & 0 & 0 & 0 & 0 \\
0 & 0 & 1 & 0 & 0 & 0 & 0 & 0 & 0 & 1 & 0 & 0 & 0 & 0 & 0 & 0 & 0 & 0 & 0 & 0 & 0 & 0 & 0 & 0 & 0 & 0 & 0 & 0 & 0 & 0 & 0 & 0 & 0 & 0 & 0 & 0 & 0 & 0 \\
0 & 0 & 0 & 1 & 0 & 0 & 1 & 0 & 0 & 0 & 0 & 0 & 0 & 0 & 0 & 0 & 0 & 0 & 0 & 0 & 0 & 0 & 0 & 0 & 0 & 0 & 0 & 0 & 0 & 0 & 0 & 0 & 0 & 0 & 0 & 0 & 0 & 0 \\
0 & 0 & 0 & 1 & 0 & 0 & 0 & 0 & 0 & 1 & 0 & 0 & 0 & 0 & 0 & 0 & 0 & 0 & 0 & 0 & 0 & 0 & 0 & 0 & 0 & 0 & 0 & 0 & 0 & 0 & 0 & 0 & 0 & 0 & 0 & 0 & 0 & 0 \\
0 & 0 & 0 & 1 & 0 & 0 & 0 & 0 & 0 & 0 & 0 & 0 & 0 & 1 & 0 & 0 & 0 & 0 & 0 & 0 & 0 & 0 & 0 & 0 & 0 & 0 & 0 & 0 & 0 & 0 & 0 & 0 & 0 & 0 & 0 & 0 & 0 & 0 \\
0 & 0 & 0 & 1 & 0 & 0 & 0 & 0 & 0 & 0 & 1 & 0 & 0 & 0 & 0 & 0 & 0 & 0 & 0 & 0 & 0 & 0 & 0 & 0 & 0 & 0 & 0 & 0 & 0 & 0 & 0 & 0 & 0 & 0 & 0 & 0 & 0 & 0 \\
0 & 0 & 0 & 0 & 1 & 0 & 1 & 0 & 0 & 0 & 0 & 0 & 0 & 0 & 0 & 0 & 0 & 0 & 0 & 0 & 0 & 0 & 0 & 0 & 0 & 0 & 0 & 0 & 0 & 0 & 0 & 0 & 0 & 0 & 0 & 0 & 0 & 0 \\
0 & 0 & 0 & 0 & 1 & 0 & 0 & 1 & 0 & 0 & 0 & 0 & 0 & 0 & 0 & 0 & 0 & 0 & 0 & 0 & 0 & 0 & 0 & 0 & 0 & 0 & 0 & 0 & 0 & 0 & 0 & 0 & 0 & 0 & 0 & 0 & 0 & 0 \\
0 & 0 & 0 & 0 & 1 & 0 & 0 & 0 & 1 & 0 & 0 & 0 & 0 & 0 & 0 & 0 & 0 & 0 & 0 & 0 & 0 & 0 & 0 & 0 & 0 & 0 & 0 & 0 & 0 & 0 & 0 & 0 & 0 & 0 & 0 & 0 & 0 & 0 \\
0 & 0 & 0 & 0 & 1 & 0 & 0 & 0 & 0 & 1 & 0 & 0 & 0 & 0 & 0 & 0 & 0 & 0 & 0 & 0 & 0 & 0 & 0 & 0 & 0 & 0 & 0 & 0 & 0 & 0 & 0 & 0 & 0 & 0 & 0 & 0 & 0 & 0 \\
0 & 0 & 0 & 0 & 0 & 1 & 0 & 0 & 0 & 0 & 1 & 0 & 0 & 0 & 0 & 0 & 0 & 0 & 0 & 0 & 0 & 0 & 0 & 0 & 0 & 0 & 0 & 0 & 0 & 0 & 0 & 0 & 0 & 0 & 0 & 0 & 0 & 0 \\
0 & 0 & 0 & 0 & 0 & 1 & 0 & 0 & 0 & 0 & 0 & 0 & 0 & 1 & 0 & 0 & 0 & 0 & 0 & 0 & 0 & 0 & 0 & 0 & 0 & 0 & 0 & 0 & 0 & 0 & 0 & 0 & 0 & 0 & 0 & 0 & 0 & 0 \\
0 & 0 & 0 & 0 & 0 & 1 & 0 & 0 & 0 & 0 & 0 & 0 & 1 & 0 & 0 & 0 & 0 & 0 & 0 & 0 & 0 & 0 & 0 & 0 & 0 & 0 & 0 & 0 & 0 & 0 & 0 & 0 & 0 & 0 & 0 & 0 & 0 & 0 \\
0 & 0 & 0 & 0 & 0 & 1 & 0 & 0 & 0 & 0 & 0 & 1 & 0 & 0 & 0 & 0 & 0 & 0 & 0 & 0 & 0 & 0 & 0 & 0 & 0 & 0 & 0 & 0 & 0 & 0 & 0 & 0 & 0 & 0 & 0 & 0 & 0 & 0
\end{pmatrix} \tag{68.17}
$$

From Eq. (68.17), the eigenvalues of the matrix can be computed as $\{-\sqrt{7}, \sqrt{7}, -\sqrt{(7+\sqrt{17})/2}, -\sqrt{(7+\sqrt{17})/2}, -\sqrt{(7+\sqrt{17})/2}, \sqrt{(7+\sqrt{17})/2}, -\sqrt{(7+\sqrt{17})/2}, -\sqrt{(7+\sqrt{17})/2}, -2, -2, -2, -2, -\sqrt{3}, -\sqrt{3}, -\sqrt{3}, -\sqrt{3}, \sqrt{3}, \sqrt{3}, \sqrt{3}, \sqrt{3}, -\sqrt{(7-\sqrt{17})/2}, -\sqrt{(7-\sqrt{17})/2}, -\sqrt{(7-\sqrt{17})/2}, \sqrt{(7-\sqrt{17})/2}, \sqrt{(7-\sqrt{17})/2}, \sqrt{(7-\sqrt{17})/2}, 0, 0, 0, 0, 0, 0, 0, 0, 0, 0, 0, 0\}$.

Comparing the eigenvalues of adjacency matrices $\mathbf{A}_h$ and $\mathbf{A}_o$ of the deployable hexahedral mechanism and the deployable octahedral mechanism, it is can be found that they are the same, so that matrices $\mathbf{A}_h$ and $\mathbf{A}_o$ have the same eigenvalues and eigenvectors, therefore, the deployable hexahedral mechanism and the deployable octahedral mechanism constructed based on the dual polyhedrons—the hexahedron and the octahedron are isomorphic [8]. Similarly, drawing the topology graphs of the deployable dodecahedral mechanism and the deployable icosahedral mechanism, the $92 \times 92$ adjacency matrices of the mechanisms and their corresponding eigenvalues and eigenvectors can be obtained. It can be found that the deployable dodecahedral mechanism and the deployable icosahedral mechanism are also isomorphic.

## 68.5 Conclusions

This paper investigated the duality of the five Platonic polyhedrons utilizing the ray and axis coordinates representation of Ling geometry, it was revealed that the inverse or pseudo inverse of the matrix each row of which representing the axis coordinates of the original polyhedral gives a matrix each column of which exactly represent the ray coordinates of the corresponding dual polyhedron, and vice versa. Then, based on the five Platonic polyhedrons, employing the plane-symmetric eight-bar linkage, deployable polyhedral mechanisms (DPMs) were constructed and the dual deployable polyhedral mechanisms are isomorphic. The isomorphism of the deployable polyhedral mechanisms was for the first time represented by the topology graphs and their corresponding adjacency matrices of the mechanisms. The synthesis method used in the paper for the construction of the five regular deployable polyhedral mechanisms based on the five Platonic polyhedrons can be extended to the synthesis of more deployable polyhedral mechanisms based on the much larger range of semi-regular of irregular polyhedrons.

**Acknowledgments** The authors gratefully acknowledge the support from the EU FP7 project TOMSY under Grant No. 270436, the EU FP7 project ECHORD DEXDEB under Grant No. 231143 and the Engineering and Physical Science Research Council (EPSRC) of the United Kingdom under grant number of EP/F031394/1.

## References

1. Coxeter HSM (1963) Regular polytopes. Macmillan, New York (Dover reprint, 1973)
2. Holden A (1991) Shapes, space and symmetry. Dover, New York
3. Wohlhart K (2008) New polyhedral star linkages. In: Proceedings of the 10th international conference on the theory of machines and mechanisms, Liberec, Czech Republic
4. Agrawal SK, Kumar S, Yim M (2002) Polyhedral single degree-of freedom expanding structures: design and prototypes. ASME J Mech Des 124(9):473–478
5. Gosselin CM, Gagnon-Lachance D (2006) Expandable polyhedral mechanisms based on polygonal one-degree-of-freedom faces. J Mech Eng Sci 220:1011–1018
6. Wei G, Dai JS (2011) Geometry and kinematics of a plane-symmetric spatial eight-bar linkage with exact straight-line motion. DETC2011-48281. In: Proceedings on ASME 2011 international design engineering and technical conferences, Washington, USA, 28–31 Aug 2011
7. Klein F (1924) Elementary mathematics from an advanced standpoint: geometry. Dover, New York (Reprinted in 1939)
8. He PR, Zhang WJ, Li Q (2002) Eigenvalues and eigenvectors information of graph and their validity in identification of graph isomorphism. In: Proceedings of ASME IDETC 2002, Montreal, Canada

# Chapter 69
# A Brief Survey on Inflatable Deployment Space Structures' Research and Development

**Jinguo Liu and Shufeng Sun**

**Abstract** Deployable structures are widely used in a variety of space spacecrafts and planetary detectors. Under the size and cost limitations of space transportation system, deployable structure usually requires large contraction ratio and high reliability, therefore, development of deployable structures faces a series of theoretical and technical challenges. Space inflatable structure is a key area in space mechanism's research. Compared to mechanically deployable structures, space inflatable structures have several unique advantages such as much lighter weight, higher packaging efficiency, lower life-cycle costs, simpler design with fewer parts, and higher deployment reliability. Inflatable structures with its irreplaceable advantages had attracted a lot of attentions during the past decades. This paper presents of a brief review of inflatable deployment space structures with stresses on their recent development and applications.

**Keywords** Space structure · Deployable structure · Inflatable

## 69.1 Introduction

With the spacecraft technology's unceasing development and wider application, the spacecraft structure has been more and more complex. Therefore, researchers and engineers must take all sorts of structure and mechanism to complete designated

J. Liu (✉) · S. Sun
State Key Laboratory of Robotics, Shenyang Institute of Automation,
Chinese Academy of Sciences, 110016 Shenyang, China
e-mail: liujinguo@sia.cn

S. Sun
School of Mechanical Engineering, Shenyang Ligong University,
110000 Shenyang, China
e-mail: xiaofeng1988918@126.com

J. S. Dai et al. (eds.), *Advances in Reconfigurable Mechanisms and Robots I*,
DOI: 10.1007/978-1-4471-4141-9_69, © Springer-Verlag London 2012

space missions. Deployable structure, being a kind of important structure, has become an essential component in the spacecraft structure. Deployable mechanism can change the main structure, secondary structure or part of the spacecraft structure from the folded state to the unfolded state. It has been widely used for variety of spacecraft and the planetary detector in aerospace, such as, satellite antenna, large optical mirror, spacecraft solar array, large support structure of space station, planetary detector, and the other space structures [1–8]. Because of dimensional constraints dictated by the finite size of the launch vehicle, large spacecraft appendages have to be stowed during launch and then developed once on-orbit.

Currently, development tendency of the space deployable structure mainly focus on two issues, one is to improve the traditional space deployable structure reliability and environmental adaptability, and the other is to explore new conceptual space deployable structure. Inflatable structures are expected to be widely applied in satellite antenna, radar array, solar array deployment structures as well as other spacecraft structures [9]. This paper will briefly review the inflatable structure's space application and its latest research progress with the intention to be a beneficial reference for future studies.

## 69.2 Representative Inflatable Deployment Space Structures

In the history, the most representative inflatable deployment space structures may be the inflatable antenna experiment (IAE). To demonstrate the potential of inflatable structure technology for large antennas, JPL initiated the IAE, as shown in Fig. 69.1.

The inflatable structure comprised two basic elements, the inflatable reflector assembly and the torus/strut supporting structure. The reflector film was stressed to approximately 1200 psi by the inflation pressure of $3 \times 10^4$ psi. The torus provides the rim support for the reflector assembly, and the reflector assembly will take a spherical shape after being inflated without the torus [11] Table 69.1.

Inflatable truss structures were an ideal for providing high specific stiffness with minimal stowed volume. Figure 69.2 shows how the mirror segments are supported from this truss. Figure 69.3 illustrates how a large space telescope employing an inflatable truss structure can be stowed to achieve very high packing efficiency and Fig. 69.4 illustrates the truss structure deployment and mirror assembly sequence.

## 69.3 Deployment Principles of the Inflatable Space Structures

A key technology of developing robust deployable inflatable space structures is selecting the proper material to meet the system requirements. In most cases, we choose the "rigidizable materials." Rigidizable materials, as described in terms of

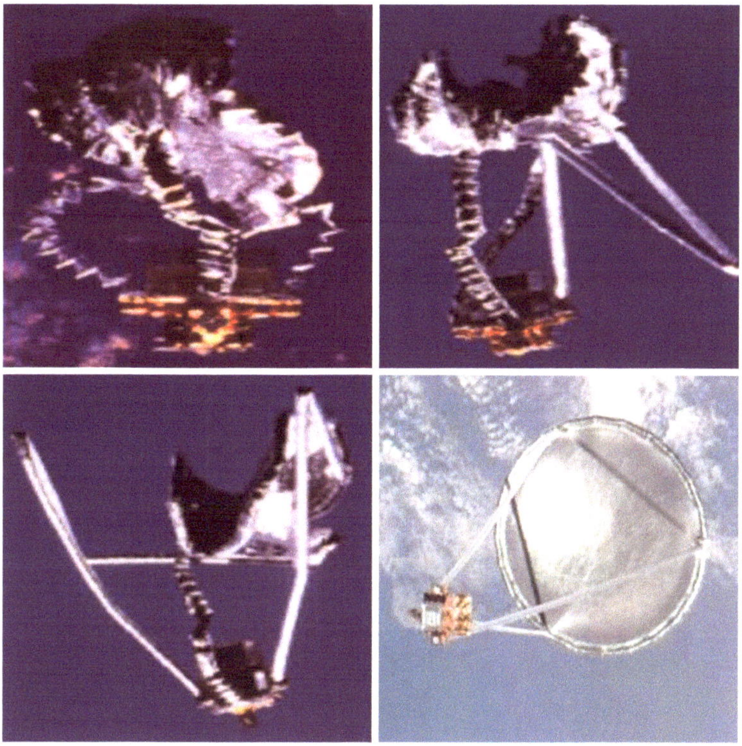

Fig. 69.1 The deployment process of IAE [10]

Table 69.1 Main specifications of IAE

| Surface precision | Inflation pressure | Torus/strut diameter | Material | The container volume |
|---|---|---|---|---|
| $RMS = 1$ mm | $3 \times 10^4$ psi | 24/18 inches | Made of 12 mil thick neoprene coated Kevlar | $14 \times 28$ m |

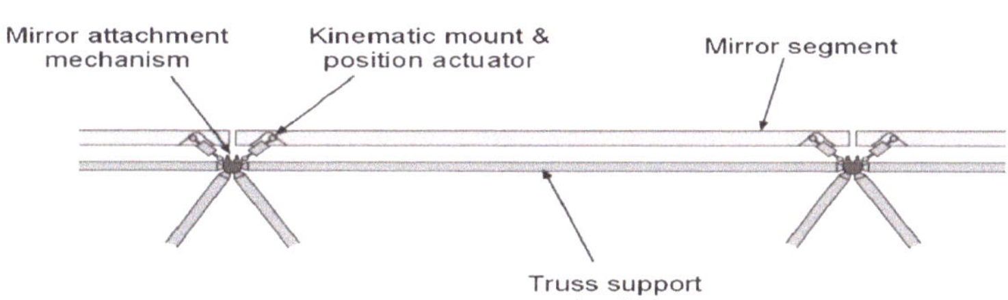


Fig. 69.2 Mirror segments mount to truss nodes through kinematic mounts [12]

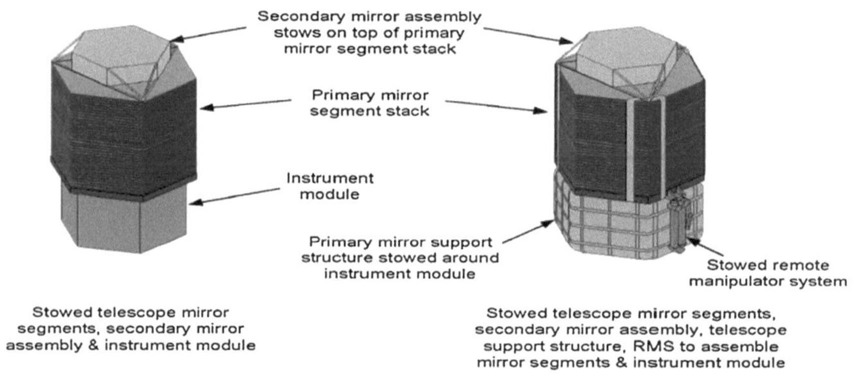


**Fig. 69.3** Illustration of an inflatable truss structure simply stowing around outside of instrument module [12]

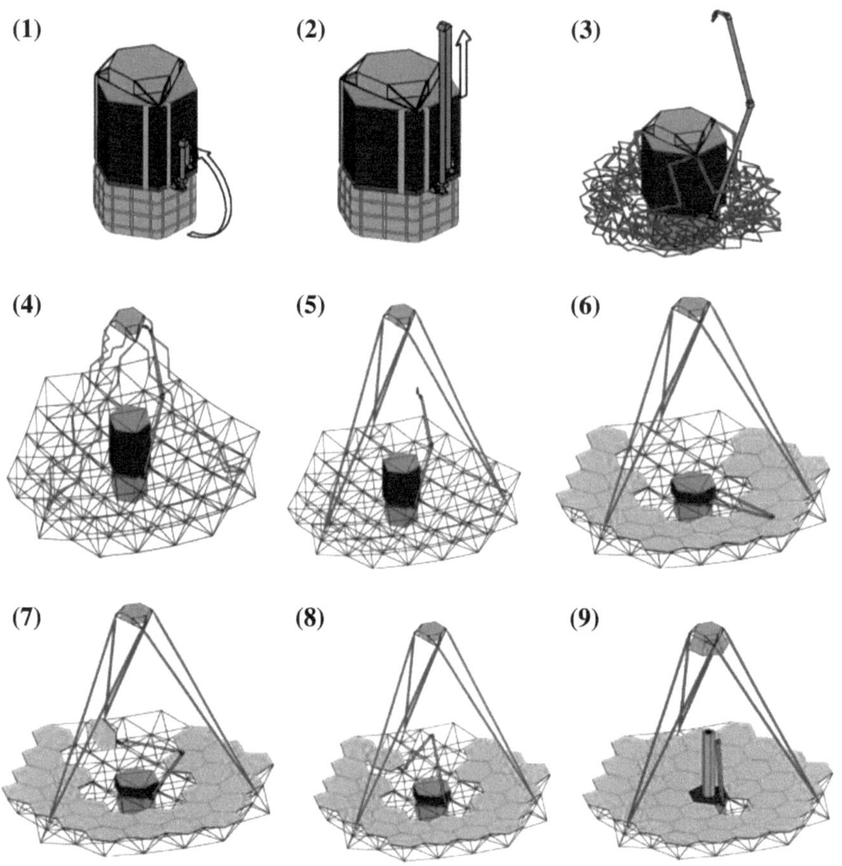


**Fig. 69.4** Deployment and assembly sequence of a large space telescope utilizing an inflatable truss support structure [12]

**Fig. 69.5** The 12 m diameter model of the inflatable space rigidised reflector antenna [18]

gossamer structures, can be defined as materials that are initially flexible to facilitate inflation or deployment, and become rigid when exposed to an external influence [13]. These external influences include heat, cold, ultra-violet radiation, and even the inflation gas itself. Rigidizable materials can be classified into several categories according to their base materials properties [14].

### *69.3.1 Thermally Cured Thermoset Composites*

Thermally cured thermoset composites are a kind of rigidizable materials that offer excellent structural performance and flexibility in design. The composite consists of a fibrous reinforcement that is impregnated with a thermoset polymer resin. The resin can be chemically hardened, or cross-linked, when exposed to heat. The cure cycle is dependent on the matrix material selected and can range from one to several hours. Several heat methods have been proposed and tested, including solar illumination of the material and resistive heating by embedded heater elements [15–17].

A construction of the antennas was made by using two parabolic membranes, and its periphery supported by a toroidal structure. The load carrying fibers in the gores become rigid on orbit from solar heating, after deployment by inflation, as shown in Fig. 69.5.

### *69.3.2 Foam Rigidization*

Foam can be not only evaluated as a method of rigidizing structures by filling their interior cavities, but has also be used in the deployment event as the inflation medium. Structures have been proposed to be reinforced with foam that is injected, or with walls that are coated with a film that reacts to an outside influence once deployed and thickened by foaming. Foam has been used as the structural material itself, and also as a strength and stiffness augmentation device when combined with composite laminate materials [19].

**Fig. 69.6** Echo I and II [9]

**Table 69.2** Main specifications of Echo

| Type | Diameter (m) | Orbit altitude (km) | Life time (years) | Material |
|---|---|---|---|---|
| Echo I | 30 | 1600 | 8 | Made of 12 um thick Mylar coated with 2000 angstroms of vapor-deposited aluminum. |
| Echo II | 40 | 1600 | 5 | Made of Mylar coated with aluminum on the inside. |

## *69.3.3 Aluminum and Film Laminates*

Aluminum laminates are a unique approach to providing a deployable rigidizable structure. The structures can be folded into small volumes and deployed via inflation pressure to yield a predetermined shape. The inflation pressure stresses the wall of the structure to smooth out the wrinkles. The structure is then pressurized to the aluminum's approximate work hardening stress, in its plastic deformation range, to slightly elevate the structural properties of the laminate and permanently remove wrinkles in the laminate. At this stress the polymer film is in its elastic range and thus when pressure is removed, the aluminum is put in compression. NASA Langley Research Center first developed aluminum laminates in the late 1950s in support of the Echo program [20], as shown in Fig. 69.6. The sphere with 100-ft diameter weighted 136 lbs. and was stowed in a 26 in. diameter's spherical container for launch. Table 69.2.

The current deployment techniques are directed at members such as tubes and struts. Normally these members are used to move the remainder of the inflatable system into position for inflation. Deployment methods must keep the deploying inflatable structure within a predictable envelope, and provide a well-defined deployment rate that is slow enough to prevent significant loads on the spacecraft. They must also provide restraint for the structure during launch, as well as large well-defined passages to vent entrapped gas during ascent to orbit. There are several techniques available to satisfy these requirements [9].

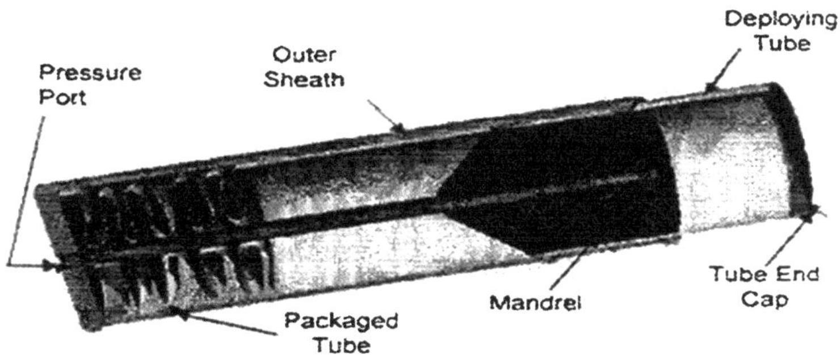


**Fig. 69.7** Mandrel method [9]

### *69.3.4 Roll-out Method*

This method is similar to the well known party favor. The primary difference is that the coiled spring that provides the deployment resistance in the party favor is replaced with Velcro@ on the top and bottom of the tube. By varying the area and location of the Velcro@ it is possible to vary the pressure necessary to deploy the tube, hence, its rigidity during deployment [9]. Depending on the flexibility of the strut material, it may be necessary to provide a method to assure the vent path.

### *69.3.5 Mandrel Method*

The feature of Mandrel method is with some venting paths to the packaged strut. The packaged strut is beneath a mandrel which with a conical shaped lead in. Application of pressure causes the packaged tube to be pulled over the mandrel. The mandrel provides the directional control of the deploying strut. In addition, friction between the mandrel and interior surface of the tube provides the resistance to deployment, which in turn controls the rate of deployment and strut stiffness during deployment [9], as shown in Fig. 69.7.

### *69.3.6 Fan Folded Method*

Directional control for this example is provided by folding the rigidizable tube 90° to the solar array folds. The deployment resistance is provided by the bending strength of the tube itself which, in this case, is made of Kapon–Aluminum–Kapton laminate [9]. The vent paths were inherent in the folded tubes by virtue of the stiffness of the tube material.

## 69.4 Key Issues in Inflatable Deployment Space Structures' Development

To develop an inflatable deployment space structure, there are many key technologies which can be generalized as follows:

### 69.4.1 Reflector Error Sources

Inflatable structures require unique manufacturing methods techniques because of the thin flexible materials that are used. A desirable method is to make use of an automated cutting system that can produce accurately cut gores with greater precision and lower cost [9].

### 69.4.2 Reflector Error Sources

As discussed above, the dimensional requirements for precision space reflectors involve accuracy beyond the customary tolerances of structural engineering. Thus, to achieve a high precision reflector, consideration must be given to all possible error sources. A listing of possible error sources are as follows: Material stiffness properties and real variation, Material thickness and real variation, creep, moisture effects, material "wrinkling" or creasing due to handling and packaging, fabrication, analytical shape prediction, and so on [9].

### 69.4.3 Dynamic Characteristics and Parameter Analysis

Currently, inflation deployable structure dynamic analysis methods are mainly in two aspects: experimental investigation and numerical simulation. Experimental research is mainly respect for the inflatable structure characteristics, from the contacting test means to the non-contact test means, but experimental studies was facing a big problem in vacuum and microgravity. Numerical simulation can effectively simulate structure dynamic behavior in vacuum and microgravity environment [21].

### 69.4.4 Potential Shape Adjustment

The major approaches for providing shape adjustment of a pressurized membrane reflector are as follows: adjust feed position, adjust edge radial displacements, change internal pressure, provide thermal gradient over reflector surface.

### 69.4.5 Inflation Subsystem

Numerous types of inflation systems have been used for inflatable space structures. The most used systems are using nitrogen gas and subliming powders. High vapor pressure fluids have been used where the pressures from sublimating powders are not adequate and using nitrogen is impractical [9]. Hydrazine systems are now being evaluated because of their capability to give medium level inflation pressures with low weight and volume when compared to nitrogen systems. Hydrazine application issues are handing, safety and cost.

## 69.5 Conclusion

Space inflatable structures and related technologies are hot topics in recent space structure researches world-widely. This field has been strongly sponsored by USA, Russia, and ESA with a large amount of funds. A series of research results have been achieved in USA. The Chinese scholars also have paid a lot of attentions to this area. However, there is still a big gap for inflatable structure to be used in aerospace in case of safety and reliability. With enough technology readiness or technology maturity and unique benefits, space inflatable structure will be a milestone in aerospace engineering.

**Acknowledgments** This work was supported by the Natural Science Foundation of China (Grant Nos. 51175494, 60705029).

## References

1. Kiper G, Söylemez E (2009) Deployable space structures. 4th international conference on recent advances in space technologies, 131–138 2009
2. Binkley, JF, Cheng PG., Smith PL, Tosney WF (2005) From data collection to lessons learned—space failure information exploitation at the aerospace corporation. Proceedings of the first international forum on integrated system health engineering and management in aerospace, 2005
3. Ma XR, Yu DY, Sun J, Hu CY (2006) The spacecraft deployment and driving mechanism research progress. J Astronaut 27(6):1123–1131
4. Deng ZQ (2009) Space folding and unfolding mechanism design research. Chinese J Sp Sci Seventh Acad Annu Conf Proc
5. Xu Y, Guan FL (2008) The deployable membrane structure of folding and unfolding process research. Eng Mech 25(5):176–181
6. Chen WJ, Zhang SJ (2006) The space deployable structure system and analysis. China Astronaut Publ H, Beijing
7. Qu GJ, Yu DY, Li FN, Zhang SJ (1997) Solar array deployment locking process flexible multi body impact dynamics finite element analysis. Spacecr Eng 6(1,2):10–17

8. Wang TS, Kong XR, Wang BL, Ma XR (2001) With hinge gap analysis of deployment in spacecraft appendages. J Harbin Inst Technol 33(3):283–286
9. Freeland, RE, Bilyeu, GD, Veal, GR (1998) Inflatable deployable space structures technology summary. IAF-98-1.5.01:1–16, 1998
10. Gunnar, T (2002) Deployable tensegrity structures for space applications. Doctoral Thesis, Stockholm
11. Freeland, RE, Bilyeu, GD (1992) IN-STEP inflatable antenna experiment. In: 43rd international astronautical federation congress. IAF-92-0301:29–40, 1992
12. Inflatable Truss Support Structures for Future Large Space Telescopes. http://www.ilcdover.com/products/aerospace_defense/supportfiles/teltruss.pdf
13. Billy, D (1992) Case studies in inflatable rigidizable structural concepts for space power. Proceedings of the 43th international astronautical congress. Washington: International Astronautical Federation, 1–16 1992
14. David, PC, Stephen, ES, ILC Dover, Inc (2001) Rigidizable materials for use in Gossamer space inflatable structures. AIAA 1417:1–14, 2001
15. Cadogan, D, Grahne, M, Mikulas, M (1998) Inflatable space structures: a new paradigm for space structure design, 49th international astronautical congress, 1998
16. Reibaldi GG, Bernasconi MC (1987) QUASAT programme: the ESA reflector paper IAF-85-400 presented at the 36th international astronautical congress, Stockholm, 7–12 Oct. Acta Astronaut 15:181–187
17. Sandy, C, Cadogan, D, Grahne, M (2000) Next generation space telescope inflatable sunshield development. IEEE Conference, 3–15 2000
18. Crone G (2000) Large deployable reflector antenna for advanced mobile communications. ESA/Ind Br Meet 9:4–20
19. Schnell, AR, Leigh, LM, Tinker, ML (2002) Deployment, foam rigidization, and structural characterization of inflatable thin-film booms. Proceedings of the 43th AIAA/ASME/ASCE/AHS/ASC structures, structural dynamics, and materials conference, Denver: American Institute of Aeronautics and Astronautics, 1–7 2002
20. Wilson, A (1981) A history of balloon satellites. British Interplanet Soc, 34:10–22
21. Tan HF, Li YL, Miao CQ (2007) Dynamic analysis of space inflatable structure progress. Adv Mechn 37(2):214–224

# Chapter 70
# A Novel Surface Deployable Antenna Structure Based on Special Form of Bricard Linkages

**Ji Cui, Hailin Huang, Bing Li and Zongquan Deng**

**Abstract** In this paper, a novel form of general line-symmetric Bricard linkage (GLSBL) that can be deployed onto arbitrary non-equilateral triangular profile and can be folded onto a bundle compact form with all the six links being parallel and contact to each other is presented. The mobility and detailed kinematic of the mechanism is studied. Using the novel GLSBL, a tripod mechanism using two similar deployable special form Bricard linkages as its bases connecting to three limbs is proposed. The tripod mechanism can be deployed onto triangular prism profile and can also be folded onto a bundle compact form so that it can be used as the basic building modules for the construction of large deployable trussed surface structures. Using this module, the deployable structure that can be deployed onto a spherical surface is presented.

**Keywords** Modular deployable mechanisms · Surface deployable trussed structure

J. Cui · H. Huang · B. Li (✉)
Shenzhen Graduated School, Harbin Institute of Technology, Shenzhen, People's Republic of China
e-mail: libing.sgs@hit.edu.cn

Z. Deng
School of Mechatronics Engineering, Harbin Institute of Technology, Harbin, People's Republic of China
e-mail: denzq@hit.edu.cn

J. S. Dai et al. (eds.), *Advances in Reconfigurable Mechanisms and Robots I*,
DOI: 10.1007/978-1-4471-4141-9_70, 

## 70.1 Introduction

Deployable structures are widely used in both aerospace applications and general civil engineering applications [1, 2]. In most design, planar mechanisms such as scissor shape mechanisms are selected as basic building elements for the construction of large deployable truss structures. Spatial mechanisms are rarely used in practical applications, the reason is probably due to the mathematical difficulty in finding solutions which ensure that the mobility of each building element is retained [3]. The spatial single loop mechanisms are good candidates for aerospace applications, as they can provide very good stiffness, low packaging/expansion ratio and easily be extended to large scale deployable networks [4]. In this paper, a novel deployable mechanism that can be deployed from a bundle compact configuration onto a large volume spherical surface structure is presented. The mechanism is constructed by a set of tripod basic deployable modules that is using two special forms of general line-symmetric Bricard mechanisms as its bases. The basic module can be deployed onto triangular prism profile with its bases can be designed into arbitrary non-equilateral triangular profile so that it can be used for constructing arbitrary surface structure. The rest of this paper is organized as follows: in Sect. 70.2, the novel deployable triangular prism mechanism is proposed; in Sect. 70.3, the kinematic of the proposed mechanism is studied; In Sect. 70.4, the construction of large surface deployable network using the triangular prism mechanism is presented and a conclusion to this paper is given in the last section.

## 70.2 Proposed of the Triangular Prism Deployable Mechanism

As reported by Bricard [5], there are six different types of mobile linkages containing six revolute joints. Of the six types, two types, i.e., the trihedral case and the general plane-symmetric case, can be used for the deployable mechanisms that can be folded onto a bundle compact form [6]. In this paper, we show that a third type mobile 6R linkage, the general line-symmetric case can also be applied to design the deployable mechanisms.

The general D–H model of 6R linkage is as shown in Fig. 70.1, for the general line-symmetric case, the geometric constraints are given as

$$\begin{cases} X_1 = X_2,\ X_3 = X_4,\ X_5 = X_6 \\ \alpha_{12} = \alpha_{45},\ \alpha_{23} = \alpha_{56},\ \alpha_{34} = \alpha_{61} \\ R_1 = R_4,\ R_2 = R_5,\ R_3 = R_6 \end{cases} \tag{70.1}$$

where $X_i$ is the common perpendicular of $z_i$ and $z_{i+1}$, $\alpha_{ij}$ is the rotation angle of axes $z_i$ and $z_{i+1}$ around $X_i$, $R_i$ is the distance of link $i-1$ and link $i$ along $z_i$, refers

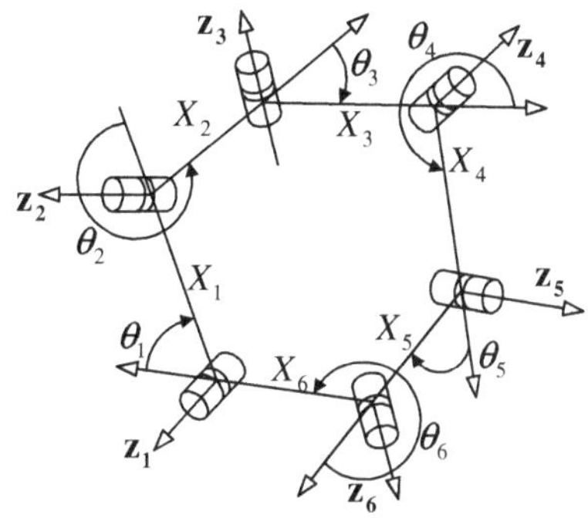


**Fig. 70.1** A general D–H model of 6R linkage

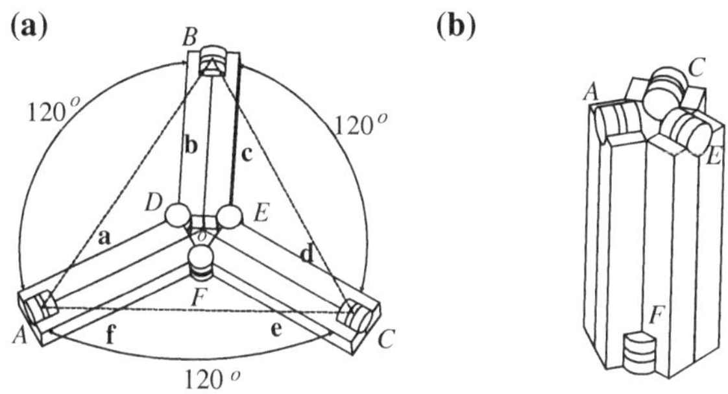


**Fig. 70.2** The Bricard mechanism that can be deployed onto "Y" shape profile. **a** Deployed configuration, **b** folded configuration

as the offset of joint $i$. Ref. [6] has presented the possibility of using a trihedral case Bricard mechanism to design deployable mechanism that can be deployed onto 'Y' shape profile. As shown in Fig. 70.2, it can be folded onto a bundle compact form as shown in Fig. 70.2b. For this mechanism, one can see that the link pairs $(a, f)$, $(b, c)$, $(d, e)$ of the mechanism can be designed with arbitrary length without changing the mobility of the mechanism, in the deployed configuration, three angles $\angle AOB$, $\angle BOC$ and $\angle COA$ are identical.

The GLSBL can also be designed to be deployable onto such "Y" shape deployed profile by using the joint axis position determination method as presented in [4]. Suppose that ABC is the required triangular deployed profile, for this case, however, the three angles $\angle AOB$, $\angle BOC$ and $\angle COA$ are no longer identical because the D–H lengths of the six links are not identical, so that the mechanism can be designed with non-equilateral triangular profiles, but the physical link of the six links for GLSBL can be designed with identical lengths. This is an important property for the construction of tripod deployable module.

As shown in Fig. 70.3, $AO$, $BO$ and $CO$ are the three joint axes of GLSBL connecting to the adjacent two parallel links. Let$AG = BH = CI$, i.e., all the lengths of the physical links are identical. $DG$, $GF$, $FI$, $IE$, $EH$, $HD$ are the

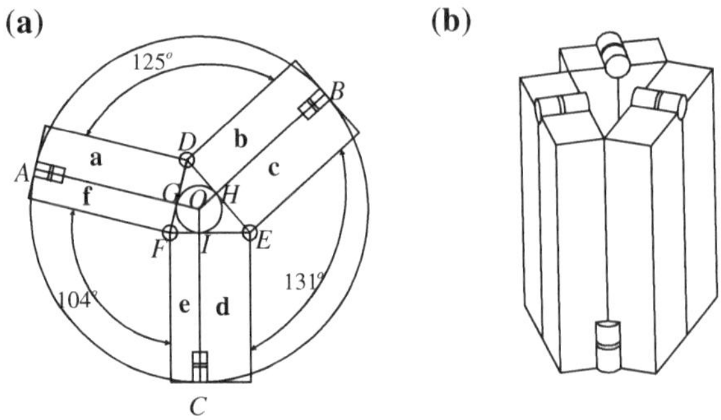


**Fig. 70.3** Top view of GLSBL in deployed configuration. **a** Top view of deployed configuration, **b** bundle compact form

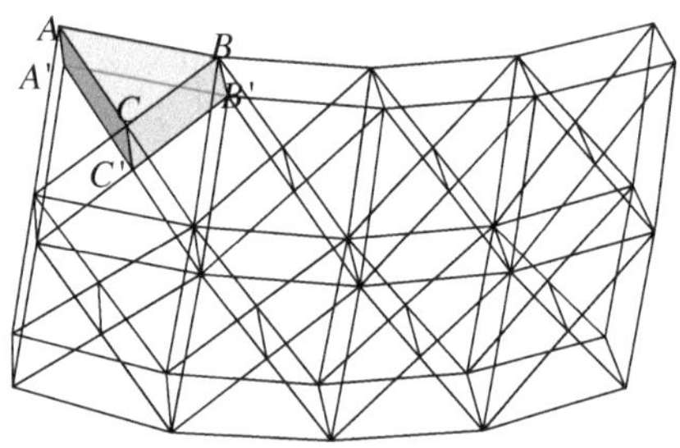


**Fig. 70.4** A conceptual model of a surface truss consisted of triangular prism

widths of the six physical links, $DG = HD$, $GF = FI$, $IE = EH$. As all the links are designed with rectangular shape in the top view, then $AG \perp DF$, $BH \perp DE$, $CI \perp EF$ and extension lines of $AG$, $BH$, $CI$ are concurrent at the point $O$ so that the point $O$ is the center of the inscribed circle of $\Delta DEF$. By changing the widths of the physical links, one can obtains different values for the angles $\angle AOB$, $\angle BOC$ and $\angle COA$ so as to realize different profile of the triangular $ABC$.

In order to build deployable mechanisms that can be deployed onto a surface structures, such as spherical surface or parabolic surface, the triangular prism mechanical modules with non-equilateral triangular base profiles are required, because any surface can be approximated by a set of triangular profiles [1]. Fig. 70.4 shows a conceptual model of a trussed surface structure that is consisted of triangular prism, $ABC$ and $A'B'C'$ are the two bases for one module. In the complicated surface, however, the parameters for the modules in the assembled structure are different from each other, this fact makes the mobile assembly of large deployable mechanism very difficult. Using GLSBL as the bases of the prism, one can easily construct a tripod mechanism as shown in Fig. 70.5. GLSBL and three long links forms the bases of the tripod and joints $\mathbf{k}_i (i = 1, 2, 3)$ connect GLSBL and the long links. For simplicity, the direction of joint $\mathbf{k}_i (i = 1, 2, 3)$ is along $\mathbf{z}_i (i = 1, 3, 5)$, respectively, as shown in Fig. 70.5. It can be

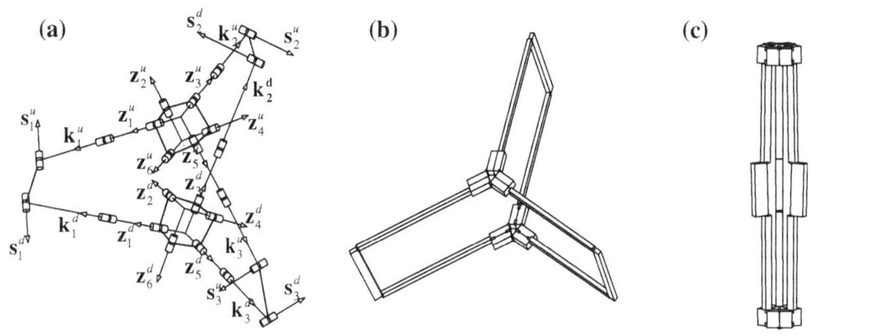


**Fig. 70.5** The CAD model of the proposed triangular prism mechanism module. **a** Theoretic model, **b** deployed configuration, **c** folded configuration

deployed onto a triangular prism deployed profile from a bundle compact form with all the links being parallel and contact to each other. As we can also use the trihedral Bricard linkage to design a non-equilateral triangular base profile with different physical link lengths, but in the tripod mechanism, it is required that the lengths of all physical links being identical, therefore, only the GLSBL case can be applied to this deployable tripod mechanism.

## 70.3 Kinematic Analysis of the Tripod Mechanism

As shown in Fig. 70.6, $DG$, $GF$, $FI$, $IE$, $EH$, $HD$ are the DH length of links, $Q$ is the concurrent point of the axes $\mathbf{z}_2$, $\mathbf{z}_4$, $\mathbf{z}_6$, $P$ is another concurrent point of axes $\mathbf{z}_1$, $\mathbf{z}_3$, $\mathbf{z}_5$. Based on Eq. (70.1), there are three pairs of the DH links are identical, i.e., $DG = HD$, $GF = FI$, $IE = EH$. Based on the geometric relations, one can prove that the three points $P$, $O$ and $Q$ are collinear and perpendicular to $\triangle GHI$. Then we have $QG = QH = QI$, $PG = PH = PI$.

From Fig 70.1 and Fig. 70.6, one can conclude that $\theta_1 + \angle DGF = \pi$, $\theta_3 + \angle DHE = \pi$, $\theta_5 + \angle EIF = \pi$, then we have

$$-\cos\theta_1 = \cos\angle DGF = \left(DG \cdot FG + \sqrt{QG^2 - DG^2} \cdot \sqrt{QG^2 - FG^2}\right)/QG^2 \tag{70.2}$$

$$\sin\theta_1 = \sin\angle DGF = \left(DG \cdot \sqrt{QG^2 - FG^2} + FG \cdot \sqrt{QG^2 - DG^2}\right)/QG^2 \tag{70.3}$$

Similarly

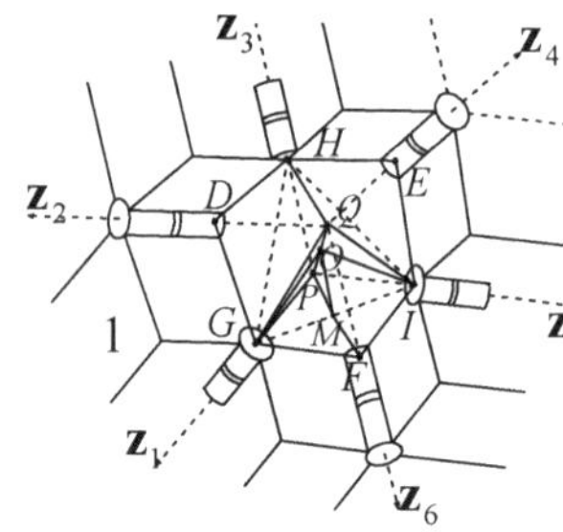


**Fig. 70.6** Geometry of general line-symmetric 6R mechanism

$$-\cos\theta_3 = \cos\angle DHE = \left(DH\cdot EH + \sqrt{QH^2 - DH^2}\cdot\sqrt{QH^2 - EH^2}\right)/QH^2 \tag{70.4}$$

$$\sin\theta_3 = \sin\angle DHE = \left(DH\cdot\sqrt{QH^2 - EH^2} + EH\cdot\sqrt{QH^2 - DH^2}\right)/QH^2 \tag{70.5}$$

$$-\cos\theta_5 = \cos\angle EIF = \left(EI\cdot FI + \sqrt{QI^2 - EI^2}\cdot\sqrt{QI^2 - FI^2}\right)/QI^2 \tag{70.6}$$

$$\sin\theta_5 = \sin\angle EIF = \left(EI\cdot\sqrt{QI^2 - FI^2} + FI\cdot\sqrt{QI^2 - EI^2}\right)/QI^2 \tag{70.7}$$

From Eq. (70.2) we have

$$QG^2 = QH^2 = QI^2 = (2DG\cdot FG\cdot\cos\theta_1 + DG^2 + FG^2)/\left(1-\cos^2\theta_1\right) \tag{70.8}$$

Substituting Eq. (70.8) into Eqs. (70.2–70.7) yields the relations of $\theta_3$, $\theta_5$ and $\theta_1$. Similarly, $\theta_2 = \angle GDH + \pi$, $\theta_4 = \angle HEI + \pi$, $\theta_6 = \angle IFG + \pi$, then

$$\tan(\theta_2/2) = -DH/PH, \quad \tan(\theta_4/2) = -EI/PI, \quad \tan(\theta_6/2) = -FG/PG \tag{70.9}$$

From Eq. (70.9) we have

$$\tan(\theta_4/2) = (EI/DH)\tan(\theta_2/2), \quad \tan(\theta_6/2) = (FG/DH)\tan(\theta_2/2) \tag{70.10}$$

Equation (70.10) expresses the relations of $\theta_2$, $\theta_4$, $\theta_6$. Then from D–H loop equation, we have the closure equation:

$$\cos\theta_2\cos\theta_4 - \cos\theta_3\sin\theta_2\sin\theta_4 = \cos\theta_6 \tag{70.11}$$

In Eqs. (70.9) and (70.10), there are three equations what include four variables, that can also prove that the general line-symmetric 6R mechanism has only one DOF.

In this research, it is assumed that the lateral profile of the triangular prism is isosceles trapezoid, i.e., all the $L_s$ s in one module are identical.

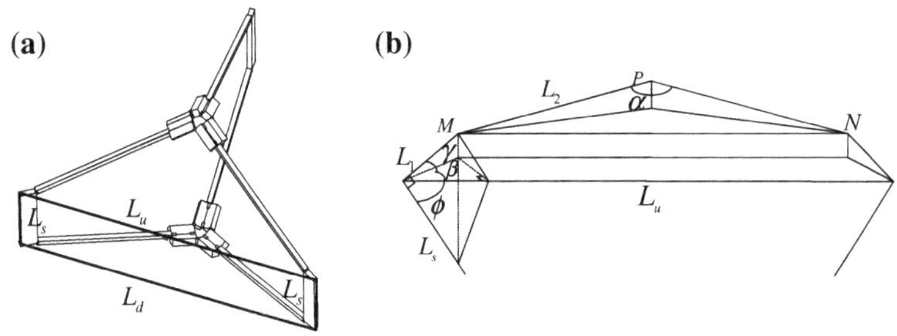


**Fig. 70.7** Parameters of the tripod mechanism. **a** The triangular prism profile, **b** lateral view of the prism

From Fig. 70.7, one can see that

$$L_u = 2L_2 \sin(\alpha/2) + 2L_1 \cos\gamma \cos\beta \tag{70.12}$$

$P$ is the concurrent point of the three axes $\mathbf{z}_1$, $\mathbf{z}_3$, $\mathbf{z}_5$, $\overrightarrow{PM}$ and $\overrightarrow{PN}$ are the two vectors along $\mathbf{z}_3$ and $\mathbf{z}_1$ respectively, $\alpha = \angle GPH$,

$$\alpha + \theta_2 = 2\pi \tag{70.13}$$

$L_2$ contains two parts, link length of Bricard $L_{c1}$, second part is length of $PG$, $PH$, $PI$

$$L_{c2} = -d_i \tan(\theta_2/2) \tag{70.14}$$

$L_1$ is the width of the lateral link, $\gamma$ is the obliquity of the lateral link, $\beta$ is the angle of $GH$ with respect to the radius of circumcircle of $\triangle GHI$. The edges of $\triangle GHI$ are $2L' \sin(\theta_2/2)L$, $2L' \sin(\theta_4/2)$ and $2L' \sin(\theta_6/2)$ respectively. $L'$ is length of $PG$. $\theta_2$, $\theta_4$, $\theta_6$ can be derived from Eq. (70.10). Given edges $l_1$, $l_2$, $l_3$ for one triangle, the radius of its circumcircle can be given as

$$r^2 = (l_1 l_2 l_3)^2 / \left[2\left(l_1^2 l_2^2 + l_2^2 l_3^2 + l_1^2 l_3^2\right) - \left(l_1^4 + l_2^4 + l_3^4\right)\right] \tag{70.15}$$

Then $\beta$ can be calculated by

$$\cos\beta = l_i/2r \tag{70.16}$$

where $l_i$ is edge of triangle, using Eqs. (70.12–70.16), one can obtain the relation of $\beta$ and $\theta_2$. From Fig. 70.7b, we can obtain the relation:

$$\cos\gamma = \sqrt{1 - (\cos\phi/\cos\beta)^2} \tag{70.17}$$

where $\phi$ is the angle of $L_u$ and $L_s$, then we can derive the relation of $L_u$ and $\theta_2$. The relation for the other base and the whole tripod mechanism can be derived using similar approach.

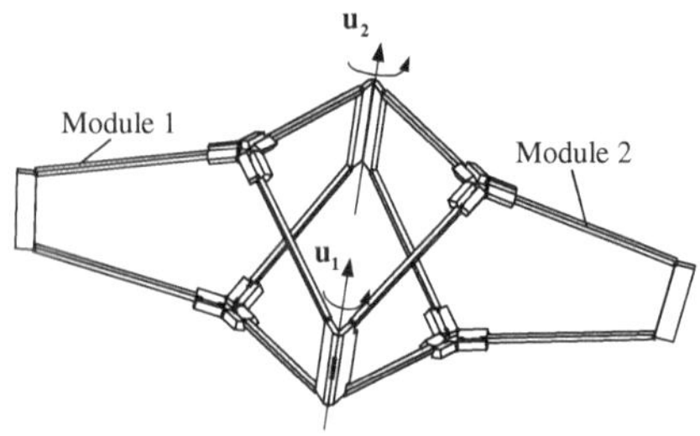


Fig. 70.8 Mobile connection of the adjacent two modules

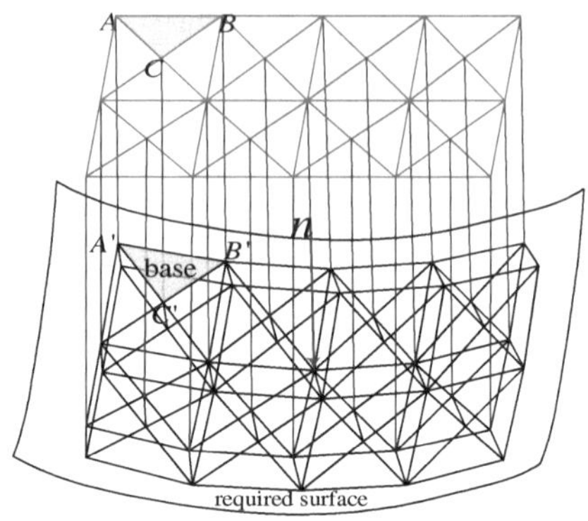


Fig. 70.9 Projections onto required surface

## 70.4 Mobile Assembly of the Large Trussed Surface Structure

Based on the conceptual trussed surface structure given in Fig. 70.4, one can see that in order to construct a modular trussed structure, every two adjacent triangular prisms are sharing one common lateral profiles, in order to build a "continuous and smooth" surface, the lateral profile of one module that is used to connected to the adjacent one must be identical, therefore, as soon as the parameters of one module, i.e., module 1 as given in Fig. 70.9 are determined, the parameters of the two lateral links of module 2 can be determined too. Based on the kinematic analysis as given in Sect. 70.3, the three lateral links of each tripod mechanism are not parallel to each other, therefore, every adjacent two modules can be connected via the common revolute joints, as shown in Fig. 70.8, $\mathbf{u}_1$ and $\mathbf{u}_2$ are the two axes of the revolute joints used for the mobile connections. The mechanism includes two modules still has two DOFs because no more constraints into the mechanism after connection.

Given a required surface for the deployed configuration, one has first to employ an efficient way for the description of the surface structure. In this paper, the projection method is used to determine the parameters of the discretized surface. As the basic building modules are triangular prisms, to design a module in the structure, the first step is to design the parameters of one base, as shown in

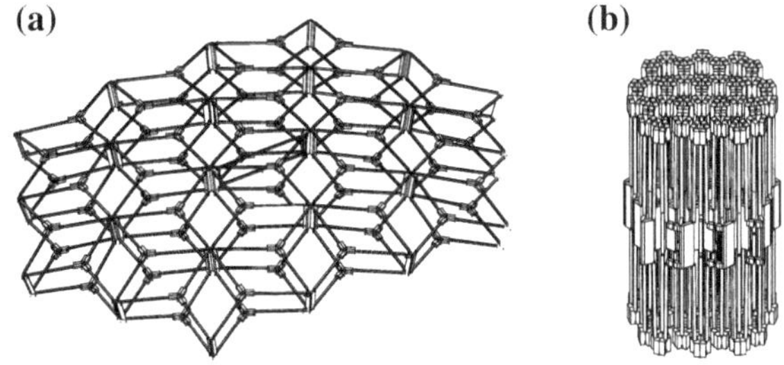


**Fig. 70.10** Spherical deployed configuration of the mechanism. **a** Deployed configuration, **b** folded configuration

Fig. 70.9, $A'B'C'$ is the triangular profile that one has to design first, in order to determine the parameters, it is assumed that there is a planar model with a set of identical triangular profiles, then the planar model is projected onto the required surface, every node of the planar model will intersect with the surface in one point, connected every two points if they are connected in the planar model, then one obtains the discretized surface model. In order to reduce the manufacturing cost, one has to make maximum number of identical modules in the structure, therefore, the projection direction can be set to be along the center normal direction of the surface, for example, $M$ is the center of the required surface, then the normal direction $\mathbf{n}$ of the surface in point $M$ is the projection direction. This projection makes that all the bases symmetric around the direction $\mathbf{n}$ are identical.

Using this method, a spherical surface structure based on the proposed tripod mechanism as shown in Fig. 70.10a, it can be folded onto a bundle compact form with all the links being parallel and contact to each others as shown in Fig. 70.10b. In the deployed configuration, the bases of the tripod mechanism form two discretized surfaces, the inner one is determined via the method as shown in Fig. 70.9. The directions of the lateral links are set to along the directions toward the spherical centre.

## 70.5 Conclusions

In this paper, a novel tripod deployable mechanism for constructing large surface deployable antenna structure has been presented. The tripod modules of the deployable mechanism uses two similar general line-symmetric Bricard mechanisms as its bases, both of which are connected by three limbs so that the mechanism can be deployed onto a triangular prism deployed profile and can be folded onto a bundle compact form with all the links being parallel and contact to each other. The projection method for constructing the surface structure has also presented in this paper, which assumed that there is a planar model with a set of identical triangular profiles, then the planar model is projected onto the required surface, every node of the planar model will intersect with the surface in one point, connected every two points if they are connected in the planar model, then one obtains the discretized surface model. Using this method, a surface structure was assembled by CAD model to show its feasibility.

**Acknowledgment** This work is financially supported by the National Science Foundation of China (Project No. 50935002 and 51175105).

## References

1. Gantes CJ (2001) Deployable structures: analysis and design. WIT Press, Boston
2. Puiga L, Barton A, Rando N (2010) Review: a review on large deployable structures for astrophysics missions. In Acta Astronautica 67(1–2):12–26
3. Chen, Y (2003) Design of structural mechanism. PhD dissertation, University of Oxford
4. Deng, Z, Huang, H, Li, B, Liu, R (2011) Synthesis of deployable/foldable single loop mechanisms with revolute joints. ASME J Mech Robotics, 3/031006, August
5. Bricard, R (1927) Leçons de cinématique. In: Tome II cinématique appliquée, Gauthier-Villars, Paris, 7–12 1927
6. Chen Y, You Z, Tarnai T (2005) Three fold-symmetric Bricard linkages for deployable structures. In: International journal of solids and structures 42(8):2287–2301

# Chapter 71
# Geometric and Combined Analysis of the Color-Flipping Ball

**Duanling Li, Guochao Bai, Qizheng Liao and Zhanjiang Du**

**Abstract** A color-flipping ball can be considered as a kind of spatial mechanism. In this paper, the geometric characteristics and combined patterns of the color-flipping ball are presented. The equivalent schematics of the color-flipping ball both in its initial state and its flip state are illustrated. The relationships between the rotational angle and the scaling triangle are analyzed. From the scaling range equation, the velocity and acceleration of the motion are analyzed, and the corresponding kinematic curves are obtained using Matlab. The combined analysis of the basic element extracted from this mechanism is conducted. At last based on the result, an example is presented to show the application of the mechanism.

**Keywords** Color-flipping ball · Geometric characteristic · Combined analysis · Articulated link

D. Li (✉) · G. Bai · Q. Liao · Z. Du
Automation School, Beijing University of Posts and Telecommunications, Beijing, China
e-mail: liduanling@163.com

G. Bai
e-mail: bgch2006@163.com

Q. Liao
e-mail: qzliao@bupt.edu.cn

Z. Du
e-mail: dzjtianjin@126.com

J. S. Dai et al. (eds.), *Advances in Reconfigurable Mechanisms and Robots I*,
DOI: 10.1007/978-1-4471-4141-9_71, 

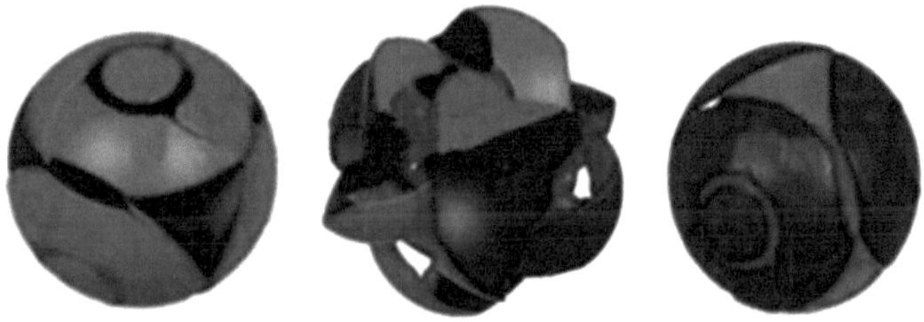

**Fig. 71.1** Three states of the color-flipping ball

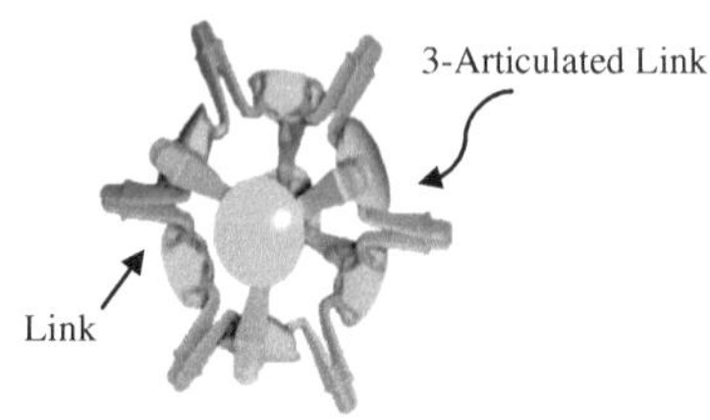


**Fig. 71.2** Flip state of the color-flipping ball

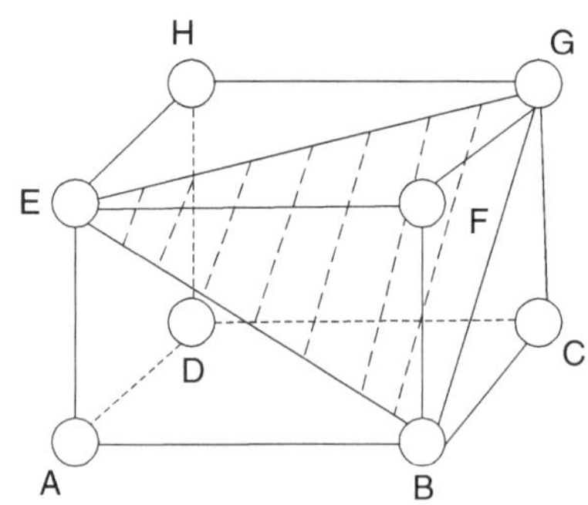


**Fig. 71.3** Schematic of the flip state

## 71.1 Introduction

A color-flipping ball is a transforming ball that can change colors when flipped [1]. It can be considered as a kind of spatial mechanism. Ding et al. [2] investigated typology of the mechanism and analyzed its structure and configurations during motion. Wei et al. [3, 4] presented a 20-bar evolved deployable ball mechanism, then reveals the geometric constraint and radio motion characteristics and investigates the mobility and the kinematics analysis of the mechanism. The mechanism of color-flipping ball also exhibits some characters of metamorphic mechanism [5].

Different scaling mechanisms have been studied so far. Liao [6] used interconnected multi-parallelograms to enlarge or shrink a planar graph by driving a revolute pair. Most of the scaling structures with scissor-like element (SLE) have

Fig. 71.4 The color-flipping ball in its final

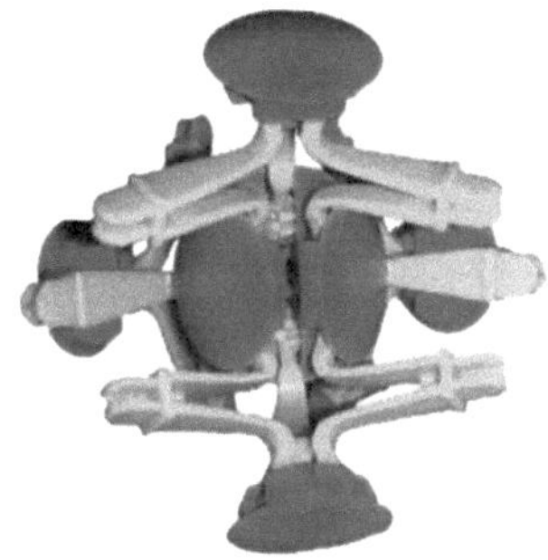

been studied by Zhao et al. [7]. Li et al. [8] studied the spatial deployable mechanisms based on the magic ball. Approaches to combine arbitrary special deployable mechanism were also proposed. Kiper [9] analyzed the geometric parameters of the scaling triangle and proposed a family of polyhedron.

In this paper, the scaling features of the color-flipping ball are presented during its motion. Figure 71.1 shows three states of color-flipping ball: initial state, flip state and final state.

## 71.2 The Geometric Principle of the Color-Flipping Ball

Figure 71.2 shows the flip state of a color-flipping ball. The mechanism is made up of 8 identical 3-articulated links and 12 equal links. Its skeleton is shown in Fig. 71.3 [10].

In Fig. 71.3, BE, EG and GB are connected in order to get a triangle EBG. Each loop in the mechanism is a square. By the formula of degree of freedom of planar mechanisms [11], the degree of freedom of each loop is one [12]. When one of the links is driven, the mechanism will have a certain motion.

Each face of the hexahedron has four links and four vertices in Figs. 71.4 and 71.5. When two opposite vertices are close to each other, the other two will be far away from each other. Each vertex of the hexahedron connects three links. When three links connected to a 3-articulated link are close to the center, the opposite 3-articulated links will leave apart from it. Because each link of the mechanism has equal length, four 3-articulated links join together and the others leave apart. The material object is shown in Fig. 71.4. Figure 71.5 is the schematic at the final state.

The hatched triangle in Fig. 71.6 is regarded as a random position during the scaling movement. The section *ACGE* is selected to be fixed in the coordinate system, as shown in Fig. 71.7.

We denote each length of the 12 links as $l$. When *AE* moves to its final position, *AE* and *OE* coincide with each other, as well as *CG* and *OG*. $\alpha$ represents the rotational angle between *AE* and *AO*.

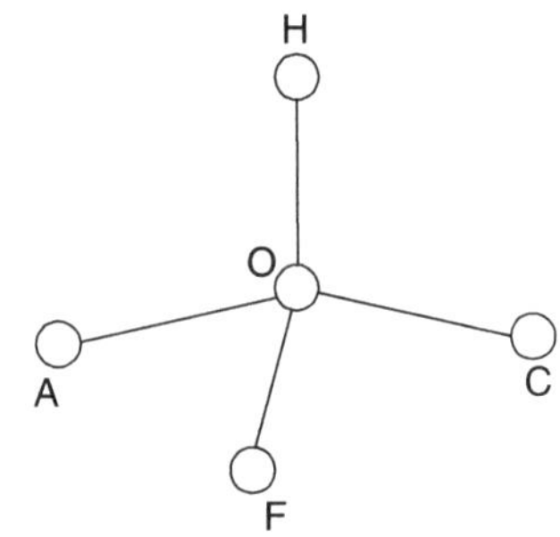


**Fig. 71.5** Schematic of the mechanism in its final position

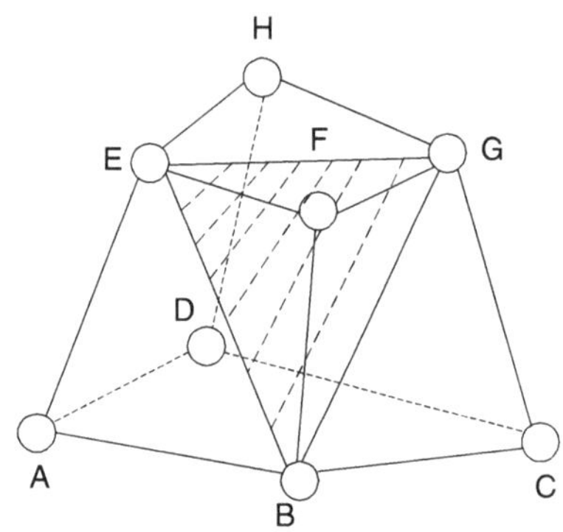


**Fig. 71.6** Schematic at a random position

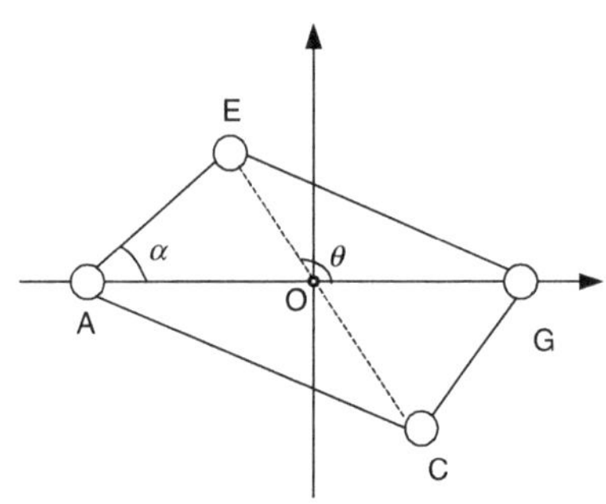


**Fig. 71.7** The section ACGE of the mechanism

$OE$ is equal to $OG$ and $\angle AOE = 180^\circ - \theta$. From Fig. 71.6, we can see triangle $EBG$ in the changing process is similar to itself in a flip state, i.e., $\theta$ has a fixed value.

For triangle $AOE$, we have:

$$l/\sin \angle AOE = OE/\sin \alpha \tag{71.1}$$

Similarly,

$$EG/\sin \theta = OE/\sin \angle EGO \tag{71.2}$$

Comparing Eqs. (71.1) and (71.2) yields:

$$EG/\sin\theta = l\sin\theta/(\sin\angle AOE \times \sin\angle EGO) \tag{71.3}$$

Suppose $EG = x$, then

$$x = l\sin\alpha\sin\theta \bigg/ \left(\sin(\pi-\theta)\sin\left(\frac{\pi-\theta}{2}\right)\right) \tag{71.4}$$

Suppose $k = l\sin\theta/\left(\sin(\pi-\theta)\sin\left(\frac{\pi-\theta}{2}\right)\right)$, we have

$$x = k\sin\alpha \tag{71.5}$$

$\alpha$ is angle between links $AE$ and $AO$ and $x$ is the side length of the scaling equilateral triangle.

Because of the symmetry of the mechanism, we know the shape will keep the similarity during motion.

## 71.3 Scaling Kinematic Analysis of the Color-Flipping Ball

### *71.3.1 The Scaling Range Analysis of the Color-Flipping Ball*

When $\alpha = 0$ and $x = 0$, if ignoring the overall dimensions of the links and their interference, the triangle will shrink to a point. That is to say, the minimum area of the triangle is 0.

The area of the scaling triangle reaches its maximum when the mechanism is in its final position shown in Fig. 71.8, and the corresponding side length is $AH$. The maximum area of the triangle is

$$s = \frac{\sqrt{3}}{4}m^2 = \frac{2\sqrt{3}}{3}l^2 \tag{71.6}$$

where $m$ the side length of tetrahedron $H\text{-}AFC$ and

$$m = \frac{2}{3}\sqrt{6}l \tag{71.7}$$

### *71.3.2 The Velocity and Acceleration Analysis of the Color-Flipping Ball*

The scaling velocity of the mechanism is the differential of the side length to the rotational angle. The side length of the scaling triangle can be obtained from Eq. (71.5). Differentiating Eq. (71.5) yields,

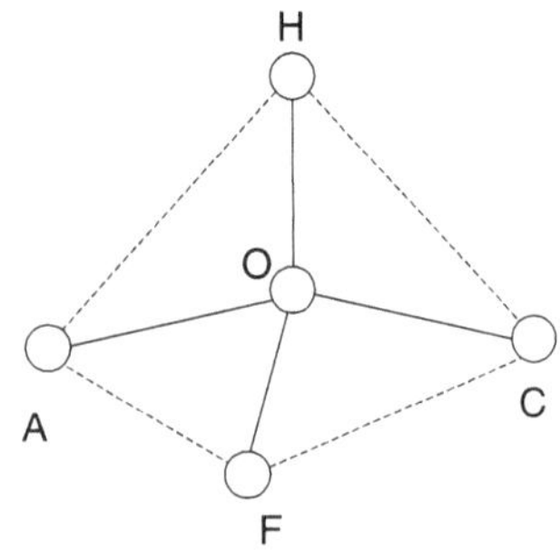


Fig. 71.8 The skeleton of tetrahedron H-AFC

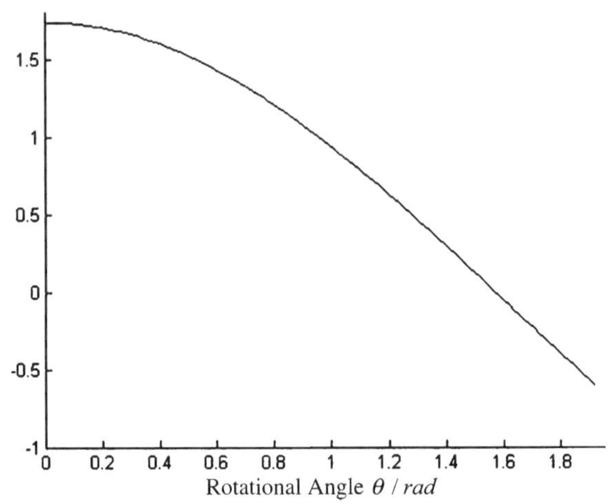


Fig. 71.9 Relationship between the scaling velocity and the rotational angle

$$v = \frac{dx}{d\alpha} = k\cos\alpha \tag{71.8}$$

where $v$ is the scaling velocity between the side length of scaling triangle and the rotational angle.

To get the acceleration, we differentiate Eq. (71.8) again with respect to $\alpha$

$$a = \frac{d^2x}{d\alpha^2} = \frac{d}{d\alpha}\left(\frac{dx}{d\alpha}\right) = -k\sin\alpha \tag{71.9}$$

where $a$ is the scaling acceleration between the side length of scaling equilateral triangle and the rotational angle.

The relationships between the scaling velocity and the rotational angle are shown in Fig. 71.9. It shows that the scaling velocity between the side length of scaling equilateral triangle and the rotational angle decreases as the angle increases.

Figure 71.10 shows the relationship between the scaling acceleration and the rotational angle. We can see from Fig. 71.10 that the scaling acceleration between the side length of the scaling equilateral triangle and the rotational angle also decreases as the angle increases.

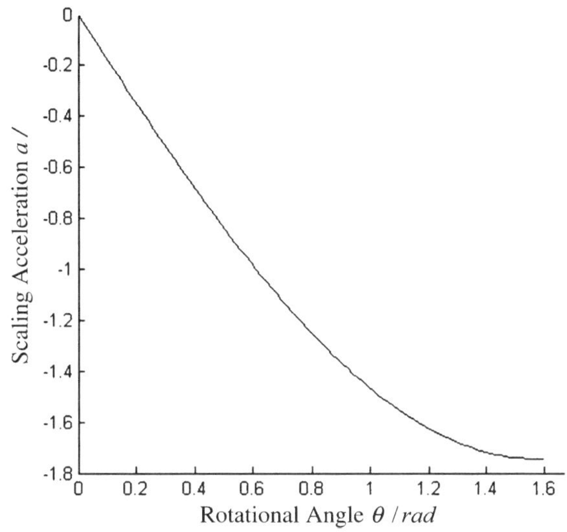


**Fig. 71.10** Relationship between the scaling acceleration and the rotational angle

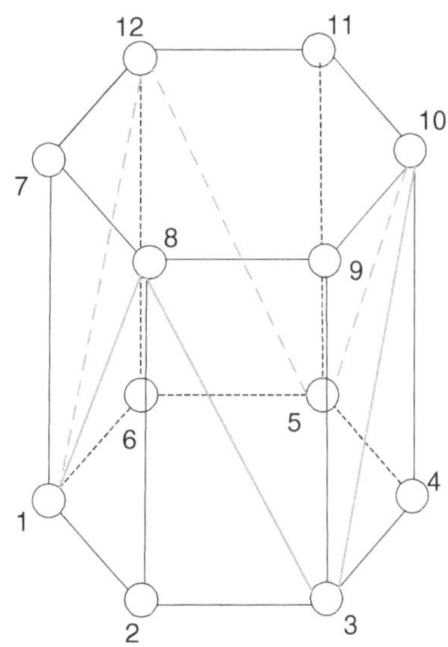


**Fig. 71.11** Schematic of octahedral mechanism

## 71.4 Combinational Analysis of the Color-Flipping Ball

From Sect. 71.2, we know that when two opposite vertices get close to each other, the other two will be far away from each other. An octahedral mechanism (shown in Fig. 71.11) can also be analyzed with the above method. We can connect the opposite vertices with lines and finally get a closed loop. All the connected vertices will coincide with or depart from each other at the same time. The number of the faces should be even, which is the necessary condition to form this kind of mechanism.

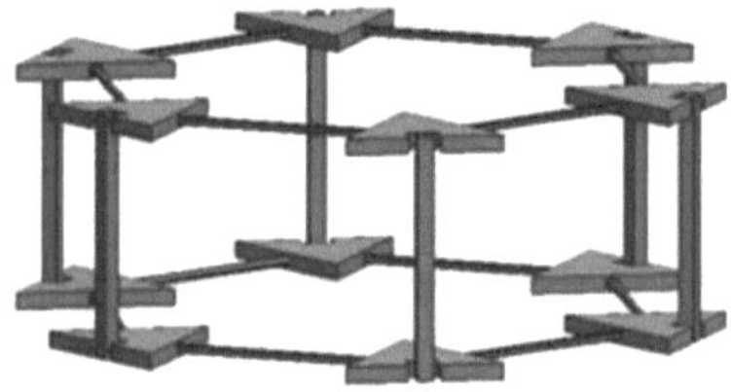

Fig. 71.12 A schematic model of octahedral mechanism

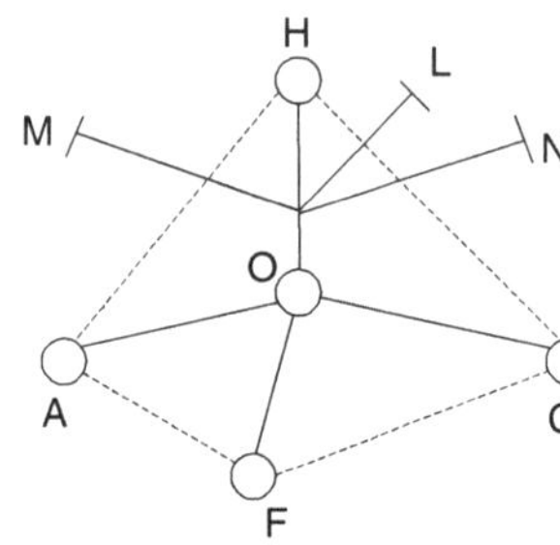


Fig. 71.13 Contract state of metamorphic lifting machine

Figure 71.12 shows a schematic of the mechanism with all kinematic pairs are revolute pairs. Although the octahedral mechanism accords with the condition of scaling, the revolute joints cannot yield the aim. Changing the revolute pairs to spherical pairs can make it possible.

All these come to the conclusion that, if the mechanism can make a reversal movement, the number of its faces (loop numbers) should be even. For a specific mechanism, the kinematic pairs should be changed to ensure the rotation.

## 71.5 An Application in a Lifting Machine

Currently, available lifting machines can be grouped into two categories: hydraulic lifting machine and mechanical scissor lifting machine. They all have their application areas. This paper presents a new lifting machine which has the kinematic characteristics of the color-flipping ball.

### *71.5.1 Structure of the Lifting Machine*

The mechanism has twelve links with equal length (denoted as $l$). Figure 71.13 shows the contract state of the machine. The three strut links $M$, $N$ and $L$ are of equal length denoted as $n$. They are positioned at one-third position of $OH$ from $O$. Three bearing links are perpendicular to the corresponding strut links. The actuator is positioned on point $O$. Connecting the 3-articulated links yields a tetrahedron

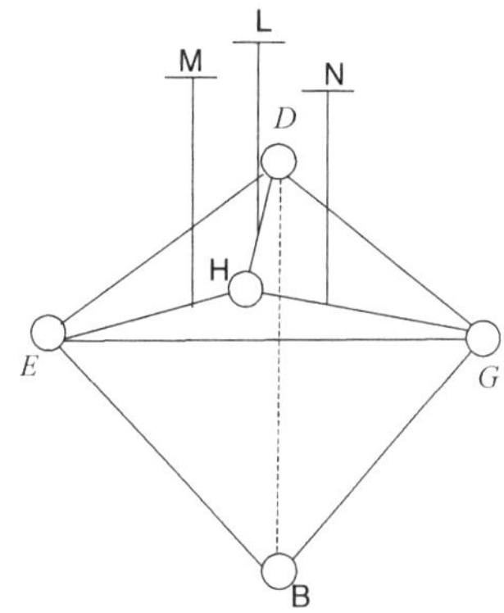


**Fig. 71.14** The structure of the lifting machine at its peak lifting position

*H-AFC*, whose side length is $m$. The height from $H$ to the bottom surface $AFC$ can be calculated by:

$$d = \sqrt{6}m\Big/3 \tag{71.10}$$

The relationships between the tetrahedral length and the link length are:

$$m = \frac{2}{3}\sqrt{6}l \tag{71.11}$$

Comparing Eqs. (71.10) and (71.11) yields:

$$d = \frac{4}{3}l \tag{71.12}$$

Equation (71.12) gives the minimum height $d_{\min}$ of the lifting machine in the contract state. The lifting machine is driven from 3-articulated link $B$ who coincided to a point $O$ with the other three 3-articulated links $E$, $G$, $D$. The mechanism will make a reverse movement. The three coincided links will separate from each other and the driving link will move upward.

When the three bearing links are parallel, the lifting machine reaches to its limit. The lifting distance is the length of the strut links (shown in Fig. 71.14). At this position, hinge $H$ is at the center of the face $EGD$ of the tetrahedron $O$—$EGD$ which is a regular tetrahedron. The distance from $B$ to face $EGD$ is: $h = \sqrt{2}l$.

The maximum lifting distance is calculated as:

$$d_{\max} = h + n = \sqrt{2}l + n \tag{71.13}$$

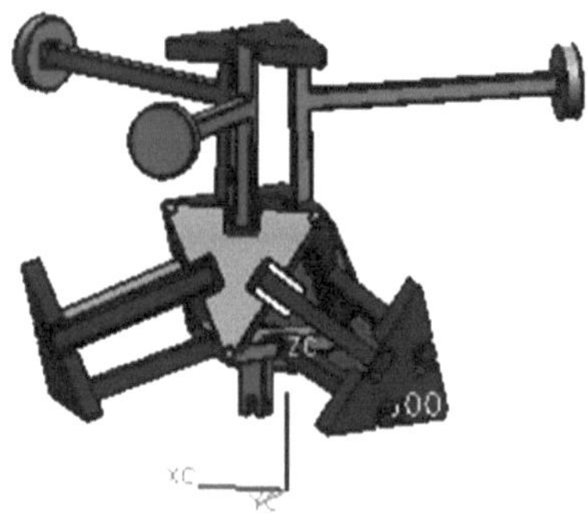

Fig. 71.15 The contract state

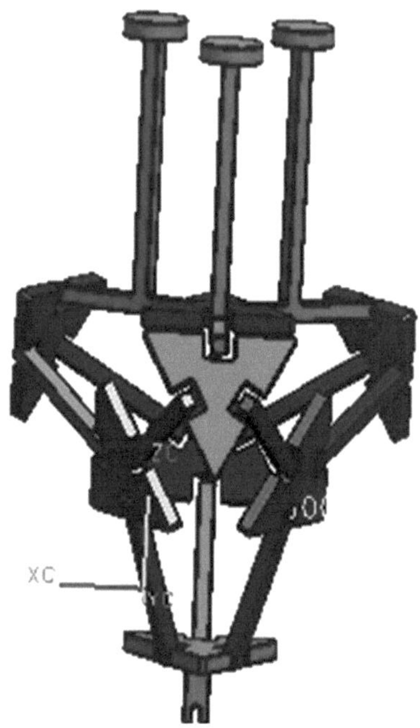

Fig. 71.16 The lifting state

### 71.5.2 Simulation Model of Lifting Machine

The 3-articulated links and the links of a lifting machine are modeled. The 3-articulated links are equilateral triangle with three revolute joints on each side. There are four parts of the model: general links, bearing links, driving 3-articulated links and coupling pins. All these parts are built and assembled.

When the links connected to the 3-articulated links are driven, the configuration of the mechanism changes. Figures 71.15 and 71.16 present the results of the simulation.

## 71.6 Conclusions

The schematics of the color-flipping ball both in its initial state and its flip state are presented. The geometric characteristics of the color-flipping ball are analyzed in this paper. The velocity and the acceleration related to the rotational angle are analyzed and the corresponding curves are obtained using Matlab. The method of the combined analysis is useful for searching for new mechanisms that have the same geometric characteristics as the color-flipping ball. At last, based on the color-flipping ball, a lifting machine is designed.

**Acknowledgments** The authors gratefully acknowledge the support of the National Natural Science Foundation of China (Granted No. 51075039).

## References

1. Switch kick, http://hoberman.com/fold/Switchkick/switchkick.htm
2. Ding X, Yang Y, Dai JS (2011) Topology and kinematic analysis of color-changing ball. Mech Mach Theory 46:67–81
3. Wei G, Ding X, Dai JS (2010) Mobility and geometric analysis of the Hoberman switch-pitch ball and its variant. Trans ASME J Mech Robotics 2:031010
4. Wei G, Ding X, Dai JS (2011) Geometric constraint of an evolved deployable ball mechanism. JSME J Adv Mech Des Syst Manuf 5(4):302–314
5. Dai JS, Jones RJ (1999) Mobility in metamorphic mechanisms of foldable/erectable kinds. J Mech Des Trans ASME 121(3):375–382 (25th ASME biennial mechanisms and robotics conference, Atlanta)
6. Liao Q, Li D (2005) Mechanisms for scaling planar graphs. Chin J Mech Eng 41:140–143
7. Zhao J-S, Chu F, Feng Z-J (2009) The mechanism theory and application of deployable structures based on SLE. Mech Mach Theory 44(2):324–335
8. Li D, Liao Q, Yin HX (2008) On construction of spatial deployable mechanisms based on the magic ball combination method. In: The proceeding of international mechanisms and machine science conference, Dalian, China, pp 71–73
9. Kiper G, Soylemez E, Ozgur-Kisisel AU (2008) A family of deployable polygons and polyhedral. Mech Mach Theory 43:627–640
10. Li D, Zhang Z (2011) A schematic representation of metamorphic links. Mech Mach Theory 46:228–238
11. Li D (2003) The mechanical analysis and application of metamorphic mechanism. Dissertation, Beihang University
12. Li D, Dai JS (2002) Loop characteristic analysis of parallelogram combined metamorphic mechanism. Mach Des Res 18:120–121

# Chapter 72
# Forward Displacement Analysis of two Foldable 3US Parallel Mechanisms

**Yun Qin and Jian S. Dai**

**Abstract** This paper investigates the foldability and the forward kinematics of the 3US parallel mechanisms. Based on the opportune arrangements of the Hooke joints axes, two foldable configurations with anticlockwise folding and clockwise deploying motion are presented. The forward kinematics of two configurations are analyzed which lead to the position expressions of the platform center and the orientation of the platform. The displacement expressions of two configurations are further compared. Moreover, the foldable configurations with clockwise folding and anticlockwise deploying motion are indicated to derive by mirroring the revolution axes of the Hooke joints associated with the base in each limb.

**Keywords** Foldability · Parallel mechanisms · Kinematics · Hooke joint arrangement

## 72.1 Introduction

The foldable mechanisms are the kind of mechanisms which can be transferred between a folded configuration or closed compact configuration and an expanded form or unfolded configuration which is capable of supporting loads [1–3]. The foldable mechanisms have wide applications in the space industry as well as

Y. Qin (✉) · J. S. Dai
Department of Mechanical Engineering, King's College London,
University of London, Strand Campus, London WC2R 2LS, UK
e-mail: yun.qin@kcl.ac.uk

J. S. Dai
e-mail: jian.dai@kcl.ac.uk

J. S. Dai et al. (eds.), *Advances in Reconfigurable Mechanisms and Robots I*,
DOI: 10.1007/978-1-4471-4141-9_72, © Springer-Verlag London 2012 

mechanical devices like instrument holders, strollers, construction site facilities and portable exhibition structures. They may be preferred over conventional unfoldable mechanisms due to their large-span, compact storage and high portability [1–5].

The foldable mechanisms attract a lot attention from many researchers [6, 7]. In 1962, Pinero proposed the expandable framing for space applications [8]. In recent years, the closed-form kinematics equations of the pantograph mast, which essentially can be seemed as a deployable structure, were presented and the global degree-of-freedom for the mast was investigated by Nagaraj et al. [4]. Kinematics singularity and actuation of a foldable 3-DOF parallel mechanism were discussed by Chung et al. in 2009 [9]. Kinematic path and bifurcation were analyzed for the 6R foldable frame and the frames without bifurcation were also designed by Chen et al. in 2009 [10].

Kinematics of the tripods (three-limbs-parallel mechanisms) have been widely studied in the past few years [11–16] since the tripods have great advantages compared with parallel mechanisms with full mobility in terms of less complex structure, lower cost and easier mechanical design. Both the parallel mechanisms with the different limb lengths [17] and the equal limb lengths [18, 19], which may have the folding potential [20], were studied in recent years. Kiper and Söylemez have proposed the modified Wren platforms with the ability to fold and deploy by replacing spherical joints with dihedral angle preserving joints in 2011 [21].

Two foldable configurations of 3US parallel mechanisms are proposed in this paper. In Sect. 72.2, the structures of two foldable configurations are presented and the geometrical conditions of the Hooke joints axes in two configurations are revealed. In Sect. 72.3, the positions of the extremities in three limbs are presented followed by the forward kinematics of the first configuration. In Sect. 72.4, the displacement analysis of the second configuration is demonstrated and the comparison of the forward kinematics expressions of two configurations is addressed.

## 72.2 The Foldable 3US Parallel Mechanisms

The 3US parallel mechanisms investigated in this paper have the full foldability, which means, they are able to fold from the deployed configuration, as shown in Fig. 72.1, to the fully folded configuration, as illustrated in Fig. 72.2, and deploy from the fully folded configuration to the deployed configuration. The configurations with the anticlockwise folding and clockwise deploying are addressed in detailed in this paper.

The foldable 3US parallel mechanisms are composed of a moving platform with the thickness $a$, a fixed base with the thickness $b$ and three identical limbs with the identical length $d$. Each limb connects the base to the platform by a lower Hooke joint and an upper spherical joint. The distance between the base center $O$ and the Hooke joint center $A_i$ ($i = 1, 2, 3$), where the subscript $i$ refers to limb sequence number, is denoted as $r_b$ and the distance between the platform center $P$ and the

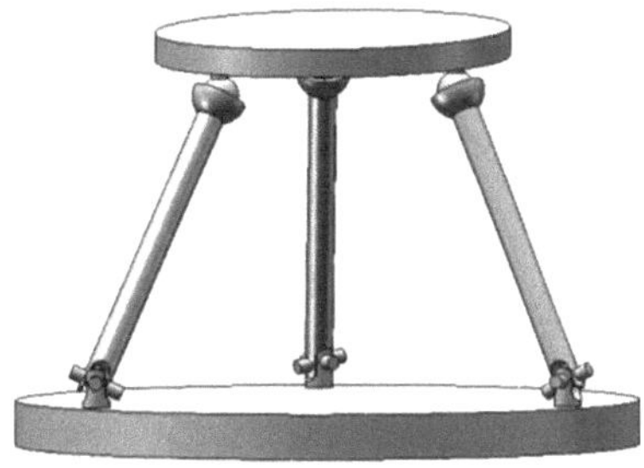

**Fig. 72.1** Unfolded configuration

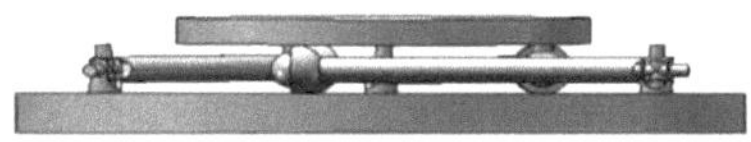

**Fig. 72.2** Folded configuration

spherical joint center $B_i$ $(i = 1, 2, 3)$ is $r_p$. The deployable ratio of the mechanism with the full foldability is defined as the ratio of the height of the mechanism when it is unfolded and the height when it is fully folded, as illustrated in Eq. (72.1).

$$q = \frac{\sqrt{d^2 - \left(r_b - r_p\right)^2} + a + b}{a + b} \tag{72.1}$$

Since the limb length of the parallel mechanism is always much bigger than the thickness of the base and the platform, the deployable radio can achieve a high ratio. A numerical evaluation is given as an example. If we choose the thickness $a$ and $b$ of the platform and the base as 5 and 10 mm, the limb length $d$ as 250 mm, the radius $r_p$ and $r_b$ of the platform and base as 65 and 200 mm, the folding radio will achieve as 15.027, which is a high ratio in industry. Besides the high deployable ratio, the 3US foldable parallel mechanism provides the high stiffness as well.

Aimed at the full foldability, the Hooke joint axes associated with the base in three limbs are opportunely placed so that they maintain relations with the limb length $d$, the distances $r_b$ and $r_p$. The expression of the angle $\alpha$, which is an interior angle of the triangle $A_1B_1O$ formed by $d$, $r_b$ and $r_p$, as shown in Fig. 72.3, is revealed in Eq. (72.2).

$$\alpha = \arccos\left(\frac{d^2 + r_b^2 - r_p^2}{2r_b d}\right) \tag{72.2}$$

Based on the axes directions of the Hooke joints, which can be essentially considered as the combination of two revolute joints, two configurations of 3US foldable parallel mechanisms with the anticlockwise folding and clockwise deploying motions are proposed. Figure 72.3 offers the observations of two configurations in the fully folded positions where the Hooke joint and spherical joint centers are coplanar.

In the first configuration of the 3US foldable parallel mechanism, the revolution axis of the Hooke joint associated with the base in the first limb is parallel to the

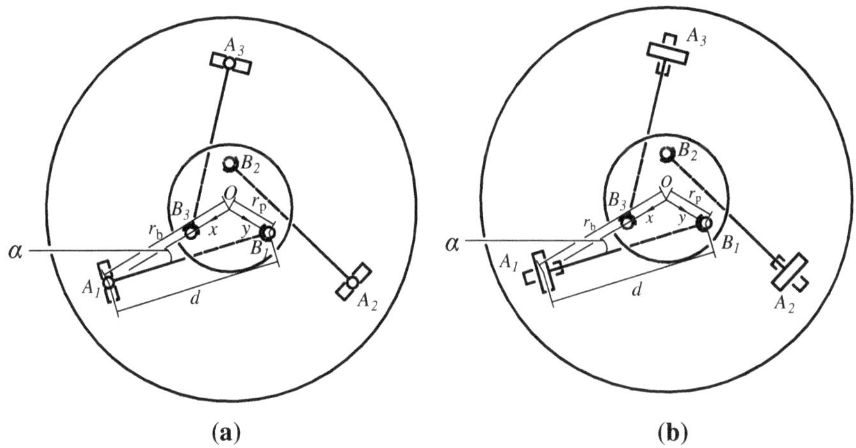


**Fig. 72.3** Fully folded position of two configurations. **a** The first configuration. **b** The second configuration

base and the angle from $OA_1$ to the axis is $\left(\frac{\pi}{2} - \alpha\right)$. The revolution axes of the Hooke joints associated with the base in three limbs form an equilateral triangle. During the folding process, the revolution axis of the Hooke joint associated with each limb changes its position in the plane which is perpendicular to that associated with the base. When the mechanism reaches the fully folded position, revolution axis of the Hooke joint associated with each limb is perpendicular to the base.

When it comes to the second foldable configuration, the axes of the Hooke joints in three limbs are arranged as follows: the revolution axis of the Hooke joint associated with the base in the first limb is fixed at an angle of $-\alpha$ with the line $OA_1$, where negative means the clockwise rotation; the revolution axes of the Hooke joints associated with the base in the other two limbs and that in the first limb form an equilateral triangle; the revolution axis of the Hooke joint associated with each limb changes its position in the plane which is perpendicular to that associated with the base until achieving the folded position where the revolution axis of the Hooke joint associated with each limb is parallel to the base and perpendicular to that associated with the base in each limb.

The Cartesian coordinate system $O(x, y, z)$ attached to the fixed base can also be seen in Fig. 72.3. The origin point $O$ is located at the base center, the $x$-axis points in the direction of $OA_1$, the $z$-axis is perpendicular to the base and points in the direction from the fixed base to the mobile platform, and $y$-axis is determined by the right hand rule.

If the 3US parallel mechanisms are arranged so that revolution axes of the Hooke joints associated with the base in three limb are mirrored with respect to the line $OA_i$ $(i = 1, 2, 3)$, the platform will undergo the anticlockwise deploying motion and clockwise folding motion.

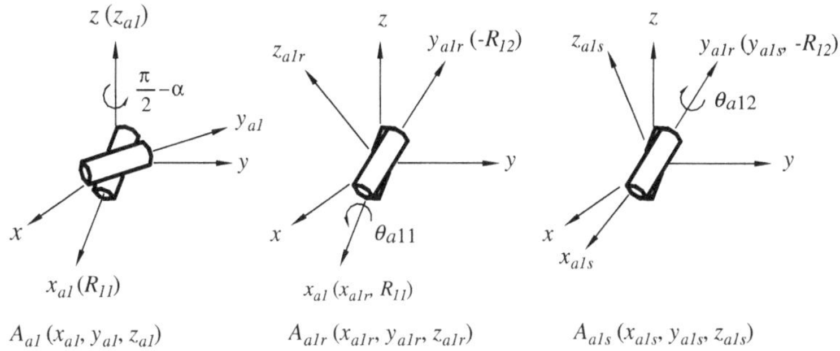


**Fig. 72.4** Rotations of the Hooke joints in the first limb

## 72.3 Forward Kinematics of the First Configuration

In the first foldable 3US configuration, the angle between line $OA_1$ and the revolution axis of the Hooke joint associated with the base in the first limb is $\left(\frac{\pi}{2}-\alpha\right)$.

The upper spherical joint center in each limb can be described with respect to the fixed coordinate system, as shown in Eq. (72.3).

$$\boldsymbol{p}_{Bi} = \boldsymbol{p}_{Ai} + \boldsymbol{d}_i \quad (i = 1, 2, 3) \tag{72.3}$$

where, $\boldsymbol{d}_i = \overline{A_iB_i}$ and $\boldsymbol{p}_{Ai}$ is the position vector of the lower Hooke joint center $A_i$. In order to obtain the direction of the first limb in the first configuration, the following rotations are applied to the fixed coordinate system $O(x, y, z)$ to get the moving coordinate systems $A_{a1}(x_{a1}, y_{a1}, z_{a1})$, $A_{a1r}(x_{a1r}, y_{a1r}, z_{a1r})$ and $A_{a1s}(x_{a1s}, y_{a1s}, z_{a1s})$ successively, where the subscribe $a$ indicates the first configuration, as shown in Fig. 72.4. By rotating the fixed coordinate system an angle $\left(\frac{\pi}{2}-\alpha\right)$ about $z$-axis, the revolution axis of the Hooke joint associated with the base in the first limb is gained. Followed by a second rotation of $\theta_{11}$, the revolution axis of the Hooke joint associated with the first limb achieves as the inverse direction of $y_{1r}$. Then a third rotation of angle $\theta_{12}$ about the revolution axis of the Hooke joint associated with the first limb occurs to get the coordinate system $A_{1s}(x_{1s}, y_{1s}, z_{1s})$. Positive rotation senses are selected such that the directed axes of the rotations point away from the centroid of the base according to the right hand rule.

In Fig. 72.4, $R_{11}$ and $R_{12}$ denote the revolution axes of the Hooke joint associated with the base and the first limb. $s\alpha$, $c\alpha$, $s\theta_{i1}$, $c\theta_{i1}$, $s\theta_{i2}$, $c\theta_{i2}$, $s\left(\frac{\pi}{3}+\alpha\right)$, $c\left(\frac{\pi}{3}+\alpha\right)$, $s\left(\frac{\pi}{6}+\alpha\right)$ and $c\left(\frac{\pi}{6}+\alpha\right)$ are short hands of $\sin\alpha$, $\cos\alpha$, $\sin\theta_{i1}$, $\cos\theta_{i1}$, $\sin\theta_{i2}$, $\cos\theta_{i2}$, $\sin\left(\frac{\pi}{3}+\alpha\right)$, $\cos\left(\frac{\pi}{3}+\alpha\right)$, $\sin\left(\frac{\pi}{6}+\alpha\right)$ and $\cos\left(\frac{\pi}{6}+\alpha\right)$, respectively. The rotation matrix of the first limb is provided as

$${}^{A}\mathbf{R}_{A_{1b}} = \begin{bmatrix} s\alpha c\theta_{12} + c\alpha s\theta_{11} s\theta_{12} & -c\alpha c\theta_{11} & -s\alpha s\theta_{12} + c\alpha s\theta_{11} c\theta_{12} \\ c\alpha c\theta_{12} - s\alpha s\theta_{11} s\theta_{12} & s\alpha c\theta_{11} & -c\alpha s\theta_{12} - s\alpha s\theta_{11} c\theta_{12} \\ c\theta_{11} s\theta_{12} & s\theta_{11} & c\theta_{11} c\theta_{12} \end{bmatrix} \tag{72.4}$$

Transferring the $\overline{A_1B_1}$ from the local coordinate system $A_{a1s}(x_{a1s}, y_{a1s}, z_{a1s})$ to the fixed coordinate system $O(x, y, z)$ gives

$$\boldsymbol{d}_1 = {}^{O}\mathbf{R}_{A_{a1s}}{}^{A_{a1s}}\boldsymbol{d}_1. \tag{72.5}$$

Substituting Eqs. (72.4 and 72.5) into Eq. (72.3), the position vector $\boldsymbol{p}_{aB1}$ of the spherical joint center $B_1$ in the first limb with respect to the fixed coordinate system $O(x, y, z)$ yields as illustrated in the following equation

$$\boldsymbol{p}_{aB1} = \begin{bmatrix} r_b + ds\alpha s\theta_{12} - dc\alpha s\theta_{11}c\theta_{12} & dc\alpha s\theta_{12} + ds\alpha s\theta_{11}c\theta_{12} & -dc\theta_{11}c\theta_{12} \end{bmatrix}^{\mathrm{T}}. \tag{72.6}$$

By the same approaches, we can get the position vectors of the spherical joints centers of the other two limbs with respect to the fixed coordinate system $O(x, y, z)$ as

$$\begin{aligned}\boldsymbol{p}_{aB2} &= \left[ r_b - ds\left(\tfrac{\pi}{3}+\alpha\right)s\theta_{22} + dc\left(\tfrac{\pi}{3}+\alpha\right)s\theta_{21}c\theta_{22} \quad -dc\left(\tfrac{\pi}{3}+\alpha\right)s\theta_{22} - ds\left(\tfrac{\pi}{3}+\alpha\right)s\theta_{21}c\theta_{22} \quad -dc\theta_{21}c\theta_{22} \right]^{\mathrm{T}}\end{aligned} \tag{72.7}$$

and

$$\begin{aligned}\boldsymbol{p}_{aB3} &= \left[ r_b + dc\left(\tfrac{\pi}{6}+\alpha\right)s\theta_{32} + ds\left(\tfrac{\pi}{6}+\alpha\right)s\theta_{31}c\theta_{32} \quad -ds\left(\tfrac{\pi}{6}+\alpha\right)s\theta_{32} + dc\left(\tfrac{\pi}{6}+\alpha\right)s\theta_{31}c\theta_{32} \quad -dc\theta_{31}c\theta_{32} \right]^{\mathrm{T}}.\end{aligned} \tag{72.8}$$

Since the moving platform center locates on the centroid of the triangle $\Delta B_1B_2B_3$, the position vector of the moving platform center $P$ with respect to the fixed coordinate system $O(x, y, z)$ can be obtained as

$$\boldsymbol{p}_{aP} = \frac{1}{3}(\boldsymbol{p}_{aB1} + \boldsymbol{p}_{aB2} + \boldsymbol{p}_{aB3}). \tag{72.9}$$

It should be noticed that Eq. (72.9) includes all the possible positions of the platform center but not all the points described in Eq. (72.9) are the positions of platform center. This is because that position vectors $\boldsymbol{p}_{aB1}$, $\boldsymbol{p}_{aB2}$ and $\boldsymbol{p}_{aB3}$ of extremities $B_1$, $B_2$ and $B_3$ are obtained by three limbs separately, which may lead to extremities $B_1$, $B_2$ and $B_3$ cannot form a triangle with the identical side length $\sqrt{3}r_p$ in some postures. Another three distance conditions, as shown in Eqs. (72.10–72.12), are added to remove the positions in Eq. (72.9) which disobey the parallel arrangements of three limbs.

$$\sqrt{(p_{aB1x} - p_{aB2x})^2 + (p_{aB1y} - p_{aB2y})^2 + (p_{aB1z} - p_{aB2z})^2} = \sqrt{3}r_p \tag{72.10}$$

$$\sqrt{(p_{aB1x} - p_{aB3x})^2 + (p_{aB1y} - p_{aB3y})^2 + (p_{aB1z} - p_{aB3z})^2} = \sqrt{3}r_p \tag{72.11}$$

$$\sqrt{\left(p_{aB2x} - p_{aB3x}\right)^2 + \left(p_{aB2y} - p_{aB3y}\right)^2 + \left(p_{aB2z} - p_{aB3z}\right)^2} = \sqrt{3}r_p \tag{72.12}$$

where, $p_{aBix}, p_{aBiy}, p_{aBiz}$ $(i = 1, 2, 3)$ are three components of vector $\boldsymbol{p}_{aBi}$ $(i = 1, 2, 3)$ in Eqs. (72.6–72.8), angles $\theta_{i1}$ $(i = 1, 2, 3)$ serve as input angles of the three degrees of freedom mechanism and angles $\theta_{i2}$ $(i = 1, 2, 3)$ are the negative angles. There are three Eqs. (72.10–72.12) and three unknowns $\theta_{12}, \theta_{22}, \theta_{32}$, hence angles $\theta_{12}, \theta_{22}, \theta_{32}$ can be solved.

For the forward displacement analysis, the input angles $\theta_{11}$, $\theta_{21}$ and $\theta_{31}$ are given, and the problem is to find the position of the platform center as well as the orientation of the platform. It is possible to describe the orientation of the platform by a vector $\boldsymbol{n}$ which is perpendicular to the platform plane, as illustrated in Eq. (72.13).

$$\boldsymbol{n} = \left(\boldsymbol{p}_{aB1} - \boldsymbol{p}_{aB2}\right) \times \left(\boldsymbol{p}_{aB1} - \boldsymbol{p}_{aB3}\right) \tag{72.13}$$

If the platform of the foldable mechanism is aimed at having anticlockwise deploying motion and clockwise folding motion, the revolution axis of the Hooke joint associated with the base in the first limb needs to be placed that the angle from $x$-axis to the revolution axis is $\left(\alpha - \frac{\pi}{2}\right)$ in the first configuration. An equilateral triangle is built up by revolution axes of the Hooke joints associated with the base in three limbs.

## 72.4 Displacement Analysis of Another Configuration and Their Comparison

The revolution axes of the Hooke joints associated with the base in the second configuration are placed in different directions compared to the first configuration. The angle between $OA_1$ and the axis in the first limb is $-\alpha$. By the similar approach, the position vectors of extremities $B_i$ $(i = 1, 2, 3)$ in the second configuration are derived as

$$\boldsymbol{p}_{bB1} = \left[r_b + dc\alpha s\theta_{12} + ds\alpha s\theta_{11}c\theta_{12} \quad -ds\alpha s\theta_{12} + dc\alpha s\theta_{11}c\theta_{12} \quad -dc\theta_{11}c\theta_{12}\right]^{\mathrm{T}} \tag{72.14}$$

$$\boldsymbol{p}_{bB2} = \left[r_b - dc\left(\tfrac{\pi}{3}+\alpha\right)s\theta_{22} - ds\left(\tfrac{\pi}{3}+\alpha\right)s\theta_{21}c\theta_{22} \quad ds\left(\tfrac{\pi}{3}+\alpha\right)s\theta_{22} - dc\left(\tfrac{\pi}{3}+\alpha\right)s\theta_{21}c\theta_{22} \quad -dc\theta_{21}c\theta_{22}\right]^{\mathrm{T}} \tag{72.15}$$

$$\boldsymbol{p}_{bB3} = \left[r_b - ds\left(\tfrac{\pi}{6}+\alpha\right)s\theta_{32} + dc\left(\tfrac{\pi}{6}+\alpha\right)s\theta_{31}c\theta_{32} \quad -dc\left(\tfrac{\pi}{6}+\alpha\right)s\theta_{32} - ds\left(\tfrac{\pi}{6}+\alpha\right)s\theta_{31}c\theta_{32} \quad -dc\theta_{31}c\theta_{32}\right]^{\mathrm{T}} \tag{72.16}$$

Subscript $b$ means the second configuration. It is possible to derive numerical evaluations of three surfaces made by all the points where extremities $B_i$ $(i = 1, 2, 3)$ in three limbs can achieve based on Eqs. (72.14–72.16). Without lose generality, the

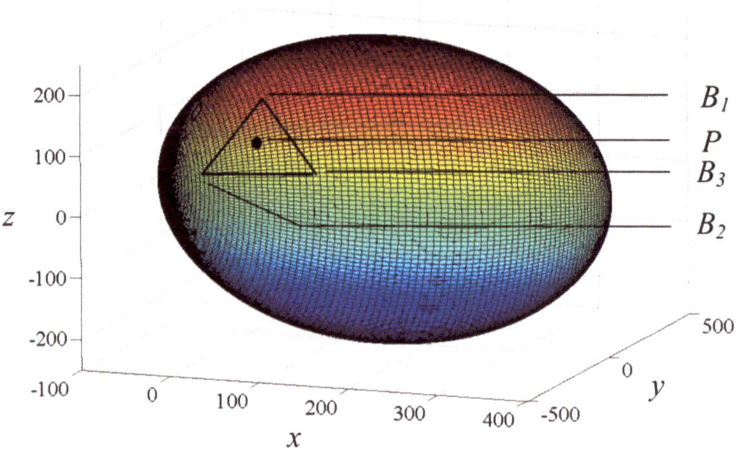


**Fig. 72.5** Possible positions of extremities $B_i$ in the second configuration

lengths $d$, $r_p$ and $r_b$ are chosen the same as the values mentioned in Sect. 72.2. When the Hooke joints in three limbs rotate freely, particularly angles $\theta_{i1}$ and $\theta_{i2}$ vary from $-2\pi$ to $2\pi$, three surfaces containing all the possible positions of extremities $B_i$ ($i = 1$, 2, 3) with respect to the fixed coordinate system $O(x, y, z)$ are presented in Fig. 72.5.

Figure 72.5 provides all the possible positions of three limbs extremities based on their free movements. After three limbs are assembled as a close chain mechanism, the parallel arrangement exerts constraints to three limbs, so they cannot reach all the positions in Fig. 72.5. However, all the actual reached position of extremities $B_i$ ($i = 1$, 2, 3) are included in three surfaces.

It can be seen that three surfaces of extremities $B_i$ ($i = 1$, 2, 3) in three limbs are coincident as one surface in Fig. 72.5. Since three limbs connected the platform at points $B_1$, $B_2$ and $B_3$, the three points build up an equilateral triangle with the side length $\sqrt{3}r_p$. Therefore, in Fig. 72.5, any three points on the surface with the identical distance $\sqrt{3}r_p$ are the positions of extremities $B_i$ ($i = 1$, 2, 3) after assembly. An example of extremities positions is marked as a triangle in Fig. 72.5. The center of the triangle in Fig. 72.5 is the position of the platform center $P$.

The analytic solutions from the input angles $\theta_{11}$, $\theta_{21}$ and $\theta_{31}$ to the platform center position and the platform orientation of the second configuration can be derived by the same method of Eqs. (72.9–72.13).

The above analysis shows the position of the platform when the foldable mechanisms anticlockwise folds and clockwise deploys. If the foldable 3US parallel mechanism is aimed at the inverse motion as clockwise folding and anti-clockwise deploying, the angle from the $x$-axis to the revolution axis of the Hooke joint associated with the base in the first limb is $\alpha$. The axes of the Hooke joints associated with the base in other two limbs and that in the first limb form an equilateral triangle.

Two foldable configurations have the same compositions as the base, the platform and three limbs with the Hooke joints and the spherical joints. The difference between two configurations only presents as how to place the Hooke joint axes. As far as the relative positions between the moving platform and the fixed base are concerned, the two folding configurations have the similar expressions in Eqs. (72.6–72.8) and (72.14–72.16). If we changes sin$\alpha$ to cos$\alpha$ and cos$\alpha$ to $-$sin$\alpha$ in the first limb, changes $\sin\left(\alpha+\frac{\pi}{3}\right)$ to $\cos\left(\alpha+\frac{\pi}{3}\right)$, $\cos\left(\alpha+\frac{\pi}{3}\right)$ to $-\sin\left(\alpha+\frac{\pi}{3}\right)$ in the second limb, and $\sin\left(\alpha+\frac{\pi}{6}\right)$ to $\cos\left(\alpha+\frac{\pi}{6}\right)$ and $\cos\left(\alpha+\frac{\pi}{6}\right)$ to $-\sin\left(\alpha+\frac{\pi}{6}\right)$ in the third limb, the displacement expressions for the first configuration will be the same as the second configuration.

## 72.5 Conclusions

Two new foldable configurations of 3US parallel mechanisms were proposed in this paper. Aimed at achieving full foldability, the Hooke joints axes at the base in two configurations were inclinably placed along opportune directions. The forward displacement analyses have been addressed lead to the position expressions of the platform center, and the orientation expressions of the platform. More precisely, the position vectors of three extremities were analyzed at first. Considering the geometrical conditions of three extremities, the position expressions of the platform center and the vector perpendicular to the platform were obtained. In the end, the comparison of the forward kinematics expressions in two configurations was addressed.

Two foldable configurations with the anticlockwise folding motions and clockwise deploying motions have been discussed in this paper. It was further indicated that, by mirror symmetrical arrangements of the current revolution axes associated with the base with the line connected the base centroid and the Hooke joints center, the inverse motion as clockwise folding and anticlockwise deploying undergoes.

Due to the advantage of compact folding and high stiffness, two configurations proposed in this paper may have the potential applications in the aerospace industry.

## References

1. Gantes CJ (2001) Deployable structures: analysis and design. WIT Press, Boston
2. Huang H, Li B, Liu R, Deng Z (2010) Type synthesis of deployable/foldable articulated mechanisms. Proceedings of the 2010 IEEE international conference on mechatronics and automation, Xi'an, China, pp 991–996
3. Gantes CJ, Konitopoulou E (2004) Geometric design of arbitrarily curved bi-stable deployable arches with discrete joint size. Int J Solids Struct 41:5517–5540

4. Nagaraj BP, Pandiyan R, Ghosal A (2009) Kinematics of pantograph masts. Mech Mach Theory 44:822–834
5. Zhang K, Dai JS, Fang Y (2010) Topology and constraint analysis of phase change in the metamorphic chain and its evolved mechanism. J Mech Des 132(12):121001-1-11
6. Hanaor A, Levy R (2001) Evaluation of deployable structures for space enclosures. Int J Space Struct 16:211–229
7. Kiper G, Söylemez E (2009) Deployable space structures. In: 4th international conference on recent advances in space technologies, pp 131–138
8. Pinero EP (1962) Expandable space framing. Prog Archit 43(6):154–155
9. Chung J, Yi B, Oh S (2009) A foldable 3-DOF parallel mechanism with application to a flat-panel TV mounting device. IEEE Trans Robotics 25(5):1214–1221
10. Chen Y, You Z, Tarnai T (2009) Two-fold symmetrical 6R foldable frame and its bifurcations. Int J Solids Struct 46(25–26):4504–4514
11. Tsai L, Joshi S (2000) Kinematics and optimization of a spatial 3-UPU parallel manipulator. J Mech Des 122(4):439–446
12. Gregorio RD, Parenti-Castelli V (2002) Mobility analysis of the 3-UPU parallel mechanism assembled for a pure translational motion. J Mech Des 124(2):259–264
13. Wang J, Gosselin CM (2004) Kinematic analysis and design of kinematically redundant parallel mechanisms. J Mech Des 126(109–118):109–118
14. Ji P, Wu H (2003) Kinematics analysis of an offset 3UPU translational parallel robtic manipulator. Robotics Auton Syst 42:117–123
15. Merlet JP, Gosselin CM, Mouly N (1998) Workspaces of planar parallel manipulators. Mech Mach Theory 33(1/2):7–20
16. Lu Y, Shi Y, Hu B (2008) Kinematic analysis of two novel 3UPU1 and 3UPU2 PKMs. Robotics Auton Syst 56:296–305
17. Walter DR, Husty ML, Pfurner M (2009) A complete kinematic analysis of the SNU 3-UPU parallel robot. Interactions of classical and numerical algebraic geometry, volume. Contemp Math 496:331–346
18. Tsai LW (1996) Kinematics of a three DOF platform with three extensible limbs. recent advances in robot kinematics. Kluwer Academic Publishers, Dordrecht, pp 401–410
19. Husty ML, Zsombor-Murray P (1994) A special type of singular Stewart Gough platform. In advances in robot kinematics and computational geometry. Kluwer Academic Publishers, Dordrecht, pp 449–458
20. Husty ML, Borel's E, Bricard's R (2000) Papers on displacements with spherical paths and their relevance to self-motions of parallel manipulators. Proceedings of international symposium on history of machines and mechanisms. Kluwer Academic Publishers, Dordrecht, pp 163–172
21. Kiper G, Söylemez E (2011) Modified Wren platforms. 13th IFToMM world congress, Guanojuato, Mexico

# Chapter 73
# A Reconfigurable Linkage and Its Applications in Lift Mechanism

**Jing-Shan Zhao, Zheng-Fang Yan and Fu-Lei Chu**

**Abstract** This paper proposes a reconfigurable lift mechanism which is composed of many planar revolute jointed linkages. Theoretically all of these linkages could be compacted to one line and then be further folded to a bundle. When working, the bundle first spreads out to a line and then connects two ends of the line to form a triangular closed loop so that the loop could ultimately deploy to act as a lift mechanism. Therefore, this kind of reconfigurable lift mechanism is portable and will be particularly used in some situations. The reconfigurable unit is first synthesized from the mobility requirement and the release of rigid joint constraint of the strengthened triangular frame is discussed to prove the possibility of completely folding after utilization for the whole mechanism. Prototype test shows that this kind of reconfigurable mechanism has very good stiffness and strength while keeping the portable characteristics.

## 73.1 Introduction

Reconfiguration and deployment are new critical requirements in the modern space engineering [1, 2]. Reconfigurable mechanisms are receiving more and more concerns in both theory and engineering applications because of their high-flexibility

J.-S. Zhao (✉) · Z.-F. Yan · F.-L. Chu
Department of Precision Instruments and Mechanology, Tsinghua University, Beijing 100084, People's Republic of China
e-mail: jingshanzhao@mail.tsinghua.edu.cn

Z.-F. Yan
e-mail: mryaoce@qq.com

F.-L. Chu
e-mail: chufl@mail.tsinghua.edu.cn

J. S. Dai et al. (eds.), *Advances in Reconfigurable Mechanisms and Robots I*,
DOI: 10.1007/978-1-4471-4141-9_73, © Springer-Verlag London 2012

in configuration, redundant constraints in structure and subsequently high stiffness and strength [3]. Structures whose shapes are formed by folding or deploying processes are thus called foldable/deployable structures. Research and development of such structures mainly took place from the 1960s to 1980s [4, 5]. These structures are also called metamorphic mechanisms [3] which initially occurred in the 1990s in a study of artifacts and their equivalent mechanisms and have now been widely investigated from the structure representation [3, 5–10], mechanism analysis [11] and synthesis [12, 13], kinematics [14–17], to the statics [18] and stiffness analysis [19]. A major problem in the study of kinematic structure is that of detecting possible isomorphism between any given chains [20].

Currently, reconfigurable polygonal frames are widely used in aerospace structures to support antenna reflectors and solar blankets. Generally, the frame consists of a closed loop of bars connected by revolute joints. They can have various forms, from squares, rectangles to hexagons [21]. Recently, a comprehensive review of the applications was given by Gan and Pellegrino [22]. The foldable systems for structural applications are often composed of a number of primary units such as truss module, which might include other additional elements to increase its stability [23]. Certain viruses having the shape of a truncated icosahedron expand under the effect of pH change in biological media. Therefore, biological modeling and evolution [24] was investigated for the synthesis and configuration design of metamorphic mechanisms. The expandable polyhedral structures consisting of prismatic faces linked along edges by hinged plates, provides a model for the swelling of viruses [25] to simulate their finite breathing modes. Symmetry extended mobility criteria show that the breathing motion is part of a set of face mechanisms for many expandable polyhedral structures.

Recent needs for space structures and nonconventional structures have propelled extensive researches on the mechanics and design of large flexible space structures. Space structures are usually made to be lightweight and flexible [26]. Foldable plate structures often consist of units of plate joined together along their edges by continuous hinges. These hinges allow the structure to be folded according to an origami pattern formed by intersecting mountain folds and valley folds [12]. Design of such linkages is usually complicated because that they are kinematically over constraint in most cases. It is their particular arrangement of specially designed and suitably proportioned rigid links that renders them mobile often with a single degree of freedom [14]. The elemental module that creates the structural typology is made up of two bars connected with each other by an inner joint. In order to form a foldable framework, these modules are juxtaposed by connecting the end of one with that of another. When the truss has a flat covering, all links are absolutely equal, in a way that the inner connection in each link is situated exactly at the midpoint of its length [23]. A simple, two-dimension foldable structure can be made from two sets of parallel, straight rods connected by pivots, or scissor-like hinges, at all intersection points. A scissor-like hinge is a revolute joint whose axis is perpendicular to the plane of the structure [27].

As a modern challenging structure, the reconfigurable and portable requirements raise a very difficult problem in engineering. However, the merits of the scissor-like hinged linkage that is overconstraint in structure but has mobile flexibility of a mechanism provide a possible frame for the particular goals. This paper dedicates to the synthesis of a reconfigurable mechanism unit which is composed of three planar scissor-like hinged linkages. Each mechanism unit could be contracted to one line and then be compacted to a concrete bundle. In applications, the bundle of links should first spread out to a line and then connect the two ends to form a closed loop and ultimately the loop stretches out to act as a lift mechanism. Because the reconfigurable unit is a three-plane overconstraint linkage, the lift mechanism has very good strength and stiffness.

## 73.2 Synthesis of a Unit Reconfigurable Mechanism

The planar linkage, $A_1B_{11}C_{11}$ shown in Fig. 73.1 (left), is an *RRR* open chain. The relative free motions at this instant of the revolute joint $C_{11}$ with respect to the fixed revolute joint, $A_1$, can be first examined as follows.

### *73.2.1 Deployable Unit of the Mechanism*

A coordinate system is established by setting the origin be at the fixed joint, $A_1$, $x$-axis be rightward along the horizontal direction and the $y$-axis be upward and pass through joint $C_{11}$. Suppose that the subtended angle of the minus $x$-axis and link $A_1B_{11}$ is $\theta_1$.

The screw matrix of the kinematic chain, $A_1B_{11}$ $C_{11}$ in Fig. 73.1, can be expressed as

$$\mathbb{S}_{A_1B_{11}C_{11}} = \begin{bmatrix} \mathbf{S}_{A_1} & \mathbf{S}_{B_{11}} & \mathbf{S}_{C_{11}} \end{bmatrix} \tag{73.1}$$

where $S_{A_1}$, $S_{B_{11}}$ and $S_{C_{11}}$ denote the kinematic screws of the joints $A_1$, $B_{11}$ and $C_{11}$, respectively, and $S_{A_1} = (0 \quad 0 \quad 1 \quad 0 \quad 0 \quad 0)^{\mathrm{T}}$ which represents a pure rotation around joint $A_1$, $S_{B_{11}} = (0 \quad 0 \quad 1 \quad -l_1 \sin\theta_1 \quad -l_1\cos\theta_1 \quad 0)^{\mathrm{T}}$ which indicates a rotation around joint $B_{11}$, and $S_{C_{11}} = (0 \quad 0 \quad 1 \quad -l_1 \sin\theta_1 - l_1' \sin\theta_2 \quad l_1\cos\theta_1 - l_1'\cos\theta_2 0)^T$ which stands for a pure rotation around joint $C_{11}$, and where $l_1$ denotes the length of $A_1$ $B_{11}$ and $l_1{}'$ denotes the length of $B_{11}C_{11}$. Therefore, the terminal constraint screw matrix of kinematic chain, $A_1B_{11}C_{11}$, can be obtained by solving the reciprocal screw equation

$$\mathbb{S}^T\mathbf{E}\mathbf{S}^{\tau} = \mathbf{0} \tag{73.2}$$

where $\mathbb{S}$ is a kinematic screw matrix, $\mathbf{S}^{\tau}$ is the terminal constraint and $\mathbf{E} = \begin{bmatrix} 0 & I \\ I & 0 \end{bmatrix}$, in which 0 and $I$ are zero and unit matrices, respectively.

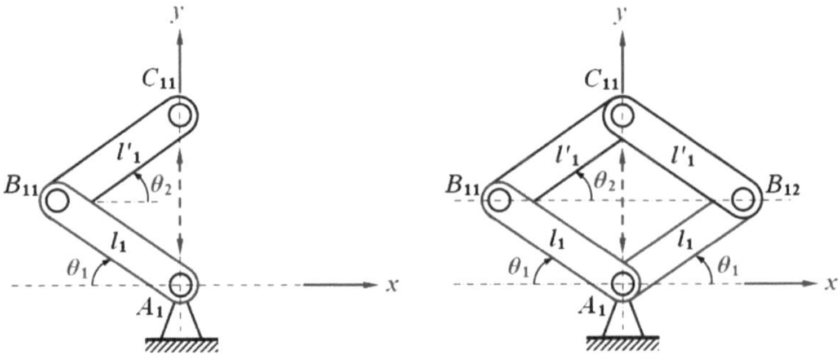


**Fig. 1** A planar quadrilateral linkage

According to Eq. (73.2), one can obtain the terminal constraint screw matrix for the chain $A_1B_{11}C_{11}$.

$$\mathbb{S}^{\tau}_{A_1B_{11}C_{11}} = [\mathbf{S}^{\tau}_1 \quad \mathbf{S}^{\tau}_2 \quad \mathbf{S}^{\tau}_3] \tag{73.3}$$

where $\mathbf{S}^{\tau}_1 = (0 \quad 0 \quad 1 \quad 0 \quad 0 \quad 0)^{\mathrm{T}}$ which denotes force constraint passing through the origin and being along the $z$-direction, $\mathbf{S}^{\tau}_2 = (0 \quad 0 \quad 0 \quad 1 \quad 0 \quad 0)^{\mathrm{T}}$ which represents a torque constraint being along the $x$-direction and $\mathbf{S}^{\tau}_3 = (0 \quad 0 \quad 0 \quad 0 \quad 1 \quad 0)^{\mathrm{T}}$ which is a torque constraint being along the $y$-direction.

The constraint screws represented by Eq. (73.3) are therefore called planar constraints that are restricted by the $xA_1y$-plane shown in Fig. 73.1. Similarly, when another identical *RRR* kinematic chain is connected $A_1$ with $C_{11}$ in parallel shown in Fig. 73.1 (right), the screw matrix of the chain can be also expressed as

$$\mathbb{S}_{A_1B_{12}C_{11}} = [\mathbf{S}_{A_1} \quad \mathbf{S}_{B_{12}} \quad \mathbf{S}_{C_{11}}] \tag{73.4}$$

where $\mathbf{S}_{B_{12}} = (0 \quad 0 \quad 1 \quad -l_1 \sin\theta_1 \quad l_1 \cos\theta_1 \quad 0)^T$ which denotes the kinematic screw of joint $B_{12}$. The terminal constraint screw matrix of kinematic chain, $A_1B_{12}C_{11}$, can be similarly obtained:

$$\mathbb{S}^{\tau}_{A_1B_{12}C_{11}} = \begin{bmatrix} 0 & 0 & 1 & 0 & 0 & 0 \\ 0 & 0 & 0 & 1 & 0 & 0 \\ 0 & 0 & 0 & 0 & 1 & 0 \end{bmatrix}^T \tag{73.5}$$

The constraint screw system exerted on the pin $C_{11}$ depends entirely on the union of Eqs. (73.3) and (73.5). Consequently, the constraint screw matrix for pin $C_{11}$ is expressed as:

$$\mathbb{S}^{\tau}_{C_{11}} = \begin{bmatrix} 0 & 0 & 1 & 0 & 0 & 0 \\ 0 & 0 & 0 & 1 & 0 & 0 \\ 0 & 0 & 0 & 0 & 1 & 0 \end{bmatrix}^T \tag{73.6}$$

Comparing Eq. (73.6) with Eqs. (73.3) and (73.5), one can find that the two chains, $A_1B_{11}C_{11}$ and $A_1B_{12}C_{11}$, provide the same constraints for pin $C_{11}$ and therefore they are two overconstraint kinematic chains. According to the reciprocal screw theory, the free motions the pin $C_{11}$ has and the constraint screw matrix for pin $C_{11}$ should satisfy that

$$\left[\mathbb{S}^{\tau}\right]^{T}\mathbf{E}\mathbf{S}^{F}=0 \tag{73.7}$$

where $\mathbf{S}^F$ denotes the screw of a free motion of an object.

The free motion screw matrix of pin $C_{11}$ can be obtained by substituting Eq. (73.6) into Eq. (73.7):

$$\mathbb{S}^{F}_{C_{11}}=\begin{bmatrix}\mathbf{S}^F_1 & \mathbf{S}^F_2 & \mathbf{S}^F_3\end{bmatrix} \tag{73.8}$$

where $\mathbf{S}^F_1=(0\quad 0\quad 1\quad 0\quad 0\quad 0)^{\mathrm{T}}$ which indicates a pure rotation around the joint $A_1$, $\mathbf{S}^F_2=(0\quad 0\quad 0\quad 1\quad 0\quad 0)^{\mathrm{T}}$ which represents a translation along the $x$-direction and $\mathbf{S}^F_3=(0\quad 0\quad 0\quad 0\quad 1\quad 0)^{\mathrm{T}}$ which stands for a translation along the $y$-direction. The real motion of point $C_{11}$ will be any combination of screws $\mathbf{S}^F_1$, $\mathbf{S}^F_2$ and $\mathbf{S}^F_3$ which can be expressed as

$$\boldsymbol{S}^{F}_{C_{11}}=\mathbb{S}^{F}_{C_{11}}k \tag{73.9}$$

where $\boldsymbol{S}^F_{C_{11}}$ represents the instantaneous screw of point $C_{11}$, $k=(k_1\ k_2\ k_3)^{\mathrm{T}}$ and $k_i=(i=1,2,3)$ could be any real numbers the range of which is determined by the sizes of the links shown in Fig. 73.1.

Equations (73.8) and (73.9) indicate that the lengths of link $A_1B_{11}$ and $B_{11}C_{11}$ do not provide any effect on the motions of the distal link connected by joint $C_{11}$, but only determine the range of its movements. Therefore the assumption $l_1'=l_1$ is adopted in the following analysis.

Now it is safe to draw a conclusion that the tip pin of a planar equilateral quadrangle can make full degrees of freedom motions within the plane it locates. In other words, the equivalent screw matrix for the tip pin of the planar equilateral quadrangle shown in Fig. 73.1 (right) has the form of Eq. (73.8). The motion of an equilateral triangle frame when its three side centers are connected with three such planar quadrangles will be discussed next. The equilateral triangle, $C_{12}C_{22}C_{32}$ shown in Fig. 73.2, is a platform which outputs the motion of the three complex chains.

To facilitate the analysis of the free motion of the triangle frame coupled by three planar equilateral quadrangle chains that are jointed at the middle of every side, a coordinate system is set as Fig. 73.2 shows. The constraints exerted to the triangular frame by the first equilateral quadrangle are expressed by formula (73.6) in the coordinate system shown in Fig. 73.2. Assume that the side length of the equilateral triangle is denoted by $2l_2$, the constraints exerted to the triangular frame by the other two equilateral quadrangles can be similarly obtained:

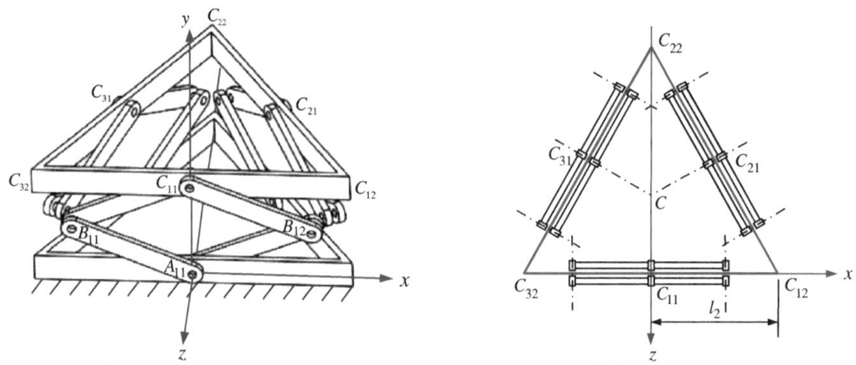


**(a)** Isometric view **(b)** Top view

**Fig. 2** A triangular platform with three planar equilateral quadrangle chains (a) Isometric view (b) Topview

$$\mathbb{S}^{\tau}_{C_{21}} = \begin{bmatrix} \cos\alpha & 0 & -\sin\alpha & 2l_1\sin\theta\sin\alpha & -l_2\sin\alpha & 2l_1\sin\theta\cos\alpha \\ 0 & 0 & 0 & -\sin\alpha & 0 & -\cos\alpha \\ 0 & 0 & 0 & 0 & 1 & 0 \end{bmatrix}^T \tag{73.10}$$

and

$$\mathbb{S}^{\tau}_{C_{31}} = \begin{bmatrix} -\cos\alpha & 0 & -\sin\alpha & 2l_1\sin\theta\sin\alpha & l_2\sin\alpha & -2l_1\sin\theta\cos\alpha \\ 0 & 0 & 0 & -\sin\alpha & 0 & \cos\alpha \\ 0 & 0 & 0 & 0 & 1 & 0 \end{bmatrix}^T \tag{73.11}$$

where $\alpha = \pi/6$ for the equilateral triangle and the geometry center of the triangular frame in the coordinate system shown in Fig. 73.2 are $C(0 \quad 2l_1\sin\theta \quad -l_2\tan\alpha)$.

Associating Eqs. (73.6) (73.10) and (73.11) and substituting them into Eq. (73.7) results in

$$\mathbb{S}^{F}_{C_{11}C_{21}C_{31}} = [0 \quad 0 \quad 0 \quad 0 \quad 1 \quad 0]^T \tag{73.12}$$

where $\mathbb{S}^{F}_{C_{11}C_{21}C_{31}}$ represents the free motion of the triangular frame connected by the three equilateral quadrangles at points, $C_{11}$, $C_{21}$ and $C_{31}$. The instantaneous motion of the triangular frame can be represented by

$$\mathbb{S}^{F}_{C_{11}C_{21}C_{31}} = k[0 \quad 0 \quad 0 \quad 0 \quad 1 \quad 0]^T \tag{73.13}$$

where $k$ denotes the instantaneous amplitude of the translation.

Equation (73.13) indicates that the triangular frame $C_{12}C_{22}C_{32}$ can only make a translation along the $x$-axis. When the two links, $B_{11}C_{11}$ and $B_{12}C_{11}$, of parallelogram linkage shown in Fig. 73.1 extend in their own central line directions to form

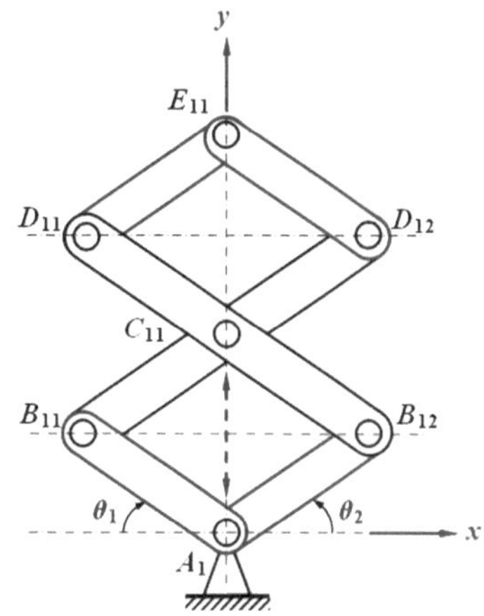


**Fig. 3** A two-level mechanism

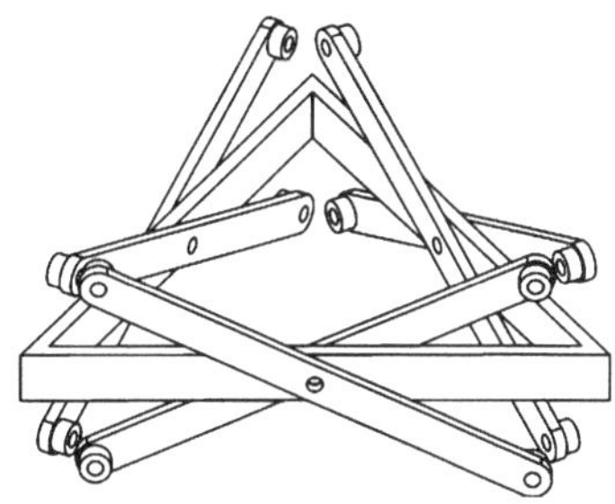

**Fig. 4** A deployable unit

a pair of scissor-like links with the common center at joint $C_{11}$, another unit $C_{11}D_{12}E_{11}D_{11}$ comes into being. As shown in Fig. 73.3, $C_{11}D_{12}E_{11}D_{11}$ is also an identical rhombus to $A_1B_{12}C_{11}B_{11}$. The free motions of joint $E_{11}$ turn out to be expressed by Eq. (73.8) with a similar analysis process above. So is true to the triangular frame $E_{12}E_{22}E_{32}$ the free motion of which is also depicted by Eq. (73.12).

The three scissor-like linkages and the triangular frame connected by them shown in Fig. 73.4 are called one deployable unit of the lift mechanism.

## 73.2.2 Constraint Release and the Generalized Unit

This section will discuss the case when the three vertexes of a triangular frame $C_{12}D_{22}C_{32}$ shown in Fig. 73.2 are split and connected by vertically placed pin joints. The overview of the mechanism is shown in Fig. 73.5.

To investigate the motion of the triangular frame pin jointed vertically at the vertexes, one can first analyze the motion of any side of the frame. For the sake of simplicity, the side $C_{32}C_{11}C_{12}$ of the frame is discussed here. As is shown in Fig. 73.6, $C_{32}C_{11}C_{12}$ has three complex kinematic chains, $A_3\left\langle \begin{matrix} B_{31} \\ B_{32} \end{matrix} \right\rangle C_{31}C_{32}$,

Fig. 5 Triangular frame vertically pin jointed

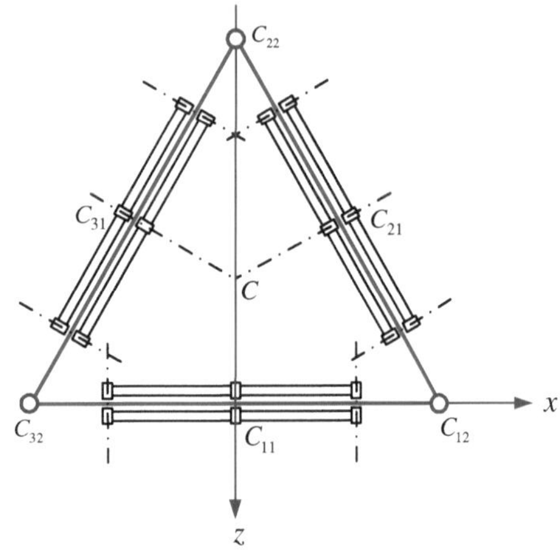


Fig. 6 A branch of the quadrilateral chain

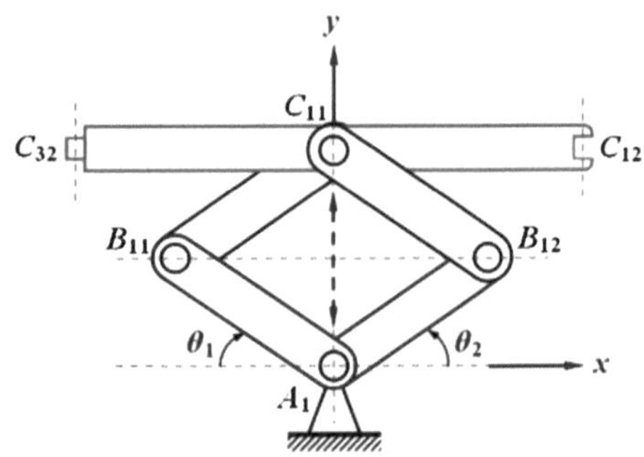


$A_1\left\langle\begin{matrix}B_{11}\\B_{12}\end{matrix}\right\rangle C_{11}$ and $A_2\left\langle\begin{matrix}B_{21}\\B_{22}\end{matrix}\right\rangle C_{21}C_{12}$ in which $\left\langle\begin{matrix}B_{21}\\B_{22}\end{matrix}\right\rangle$ represents that $B_{21}$ and $B_{22}$ are parallel in the hybrid chain $A_2\left\langle\begin{matrix}B_{21}\\B_{22}\end{matrix}\right\rangle C_{21}C_{12}$, the other symbols can be similarly deduced.

Similar to Eq. (73.8), the screw matrix of the complex chain $A_3\left\langle\begin{matrix}B_{31}\\B_{32}\end{matrix}\right\rangle C_{31}C_{32}$ can be expressed as

$$\mathbb{S}^F_{C_{32}} = \left[\boldsymbol{S}^{C_{32}}_1 \quad \boldsymbol{S}^{C_{32}}_2 \quad \boldsymbol{S}^{C_{32}}_3 \quad \boldsymbol{S}^{C_{32}}_4\right] \tag{73.14}$$

where $\mathbb{S}^F_{C_{32}}$ represents the kinematic screw matrix of the chain $A_3\left\langle\begin{matrix}B_{31}\\B_{32}\end{matrix}\right\rangle C_{31}C_{32}$, $S^{C_{32}}_1 = (-\cos\alpha \quad 0 \quad -\sin\alpha \quad 2l_1\sin\theta\sin\alpha \quad l_2\sin\alpha \quad -2l_1\sin\theta\cos\alpha)^{\mathrm{T}}$, which indicates a pure rotation around the joint $C_{31}$, $\boldsymbol{S}^{C_{32}}_2 = (0 \quad 0 \quad 0 \quad -\sin\alpha \quad 0\cos\alpha)^{\mathrm{T}}$ which represents a translation along the $C_{22}C_{32}$-direction, $\boldsymbol{S}^{C_{32}}_3 =$

$(0 \quad 0 \quad 0 \quad 0 \quad 1 \quad 0)^{\mathrm{T}}$ which stands for a translation along the $y$-direction and $\boldsymbol{S}_4^{C_{32}} = (0 \quad 1 \quad 0 \quad 0 \quad 0 \quad l_2)^{\mathrm{T}}$.

Substituting Eq. (73.14) into Eq. (73.2) produces the terminal constraints of the complex chain $A_3\left\langle\begin{matrix} B_{31} \\ B_{32} \end{matrix}\right\rangle C_{31}C_{32}$:

$$\mathbb{S}^{\tau}_{C_{32}} = \left[\boldsymbol{S}^{C_{32}}_{\tau 1} \quad \boldsymbol{S}^{C_{32}}_{\tau 2}\right] \tag{73.15}$$

where $\mathbb{S}^{\tau}_{C_{32}}$ represents the terminal constraint matrix of the chain $A_3\left\langle\begin{matrix} B_{31} \\ B_{32} \end{matrix}\right\rangle C_{31}C_{32}$, $\boldsymbol{S}^{C_{32}}_{\tau 1} = (-\cos\alpha \quad 0 \quad -\sin\alpha \quad 2l_1\sin\theta\sin\alpha \quad l_2\sin\alpha \quad -2l_1\sin\theta\cos\alpha)^{\mathrm{T}}$ which denotes a force constraint perpendicular to the plane of equilateral quadrangle $A_3B_{31}B_{32}C_{31}$, $\boldsymbol{S}^{C_{32}}_{\tau 2} = (0 \quad 0 \quad 0 \quad -\sin\alpha \quad 0 \quad \cos\alpha)^{\mathrm{T}}$ which denotes a torque constraint about the direction of side $C_{31}C_{32}$.

Similarly, the terminal constraints of the complex chain $A_2\left\langle\begin{matrix} B_{21} \\ B_{22} \end{matrix}\right\rangle C_{21}C_{12}$ can also be obtained with a process above:

$$\mathbb{S}^{\tau}_{C_{12}} = \left[\boldsymbol{S}^{C_{12}}_{\tau 1}\boldsymbol{S}^{C_{12}}_{\tau 2}\right] \tag{73.16}$$

where $\mathbb{S}^{\tau}_{C_{12}}$ represents the terminal constraint matrix of the chain $A_2\left\langle\begin{matrix} B_{21} \\ B_{22} \end{matrix}\right\rangle C_{21}C_{12}$, $\boldsymbol{S}^{C_{12}}_{\tau 1} = (\cos\alpha \quad 0 \quad -\sin\alpha \quad 2l_1\sin\theta\sin\alpha \quad -l_2\sin\alpha \quad 2l_1\sin\theta\cos\alpha)^{\mathrm{T}}$ which denotes a force constraint perpendicular to the plane of equilateral quadrangle $A_2B_{21}B_{22}C_{12}$, $\boldsymbol{S}^{C_{12}}_{\tau 2} = (0 \quad 0 \quad 0 \quad -\sin\alpha \quad 0 \quad -\cos\alpha)^{\mathrm{T}}$ which denotes a torque constraint about the direction of side $C_{21}C_{12}$.

The terminal constraints of $C_{32}C_{11}C_{12}$ can be expressed in the matrix form by associating Eqs. (73.6), (73.15) and (73.6):

$$\mathbb{S}^{\tau}_{C_{32}C_{11}C_{12}} = \begin{bmatrix} 0 & 0 & 1 & 0 & 0 & 0 \\ 0 & 0 & 0 & 1 & 0 & 0 \\ 0 & 0 & 0 & 0 & 1 & 0 \\ -\cos\alpha & 0 & -\sin\alpha & 2l_1\sin\theta\sin\alpha & l_2\sin\alpha & -2l_1\sin\theta\cos\alpha \\ 0 & 0 & 0 & -\sin\alpha & 0 & \cos\alpha \\ \cos\alpha & 0 & -\sin\alpha & 2l_1\sin\theta\sin\alpha & -l_2\sin\alpha & 2l_1\sin\theta\cos\alpha \\ 0 & 0 & 0 & -\sin\alpha & 0 & -\cos\alpha \end{bmatrix}^{T} \tag{73.17}$$

Substituting Eq. (73.17) into Eq. (73.7) yields

$$\mathbb{S}^{F}_{C_{32}C_{11}C_{12}} = [0 \quad 0 \quad 0 \quad 0 \quad 1 \quad 0]^{T} \tag{73.18}$$

Equation (73.18) indicates that the frame side $C_{32}C_{11}C_{12}$ still has one translational degree of freedom along the $y$-axis shown in Fig. 73.6. In addition, it can be found from Eqs. (73.17) and (73.18) that the frame side, $C_{32}C_{11}C_{12}$, is subjected to over constraints the redundant number of which is three.

In a similar way, the conclusion that the other two sides of the triangular frame also only have the same one translational degree of freedom along the $y$-axis shown in Fig. 73.6 can be drawn. Therefore, the rigid connection of the strengthened triangular frame can be released to vertically pin jointed while not changing the motion of the frame which is still subjected to the over constraints of three nonparallel planes. The conclusion will also be true when two or more deployable units are connected in series. This provides an outstanding advantage for the portable design of the mechanism.

## 73.3 Structure of Completely Foldable Lift Mechanism

When the unit mechanism shown in Fig. 73.6 compacts, any vertically pined joint of the triangular frame can be unlocked and therefore, the three equilateral quadrangles should be stretched out to form one planar mechanism. Figure 73.7 illustrates the case when the three complex chains are untied at joint $C_{22}$ and then deployed to one plane.

If the free motions of the outspread frame $C_{22}C_{31}C_{32}C_{11}C_{12}C_{21}C_{22}$ which forms one beam within the $xA_1y$-plane are further investigated, the terminal constraints contributed to this beam by the three complex planar kinematic chains, $A_1\left\langle\begin{matrix}B_{11}\\B_{12}\end{matrix}\right\rangle C_{11}$, $A_2\left\langle\begin{matrix}B_{21}\\B_{22}\end{matrix}\right\rangle C_{21}$ and $A_3\left\langle\begin{matrix}B_{31}\\B_{32}\end{matrix}\right\rangle C_{31}$, can be analyzed in a like manner above:

$$\mathbb{S}^{\tau}_{C_{22}-C_{22}} = \begin{bmatrix} 0 & 0 & 1 & 0 & 0 & 0 \\ 0 & 0 & 0 & 1 & 0 & 0 \\ 0 & 0 & 0 & 0 & 1 & 0 \end{bmatrix}^T \tag{73.19}$$

where $\mathbb{S}^{\tau}_{C_{22}-C_{22}}$ represents the terminal constraint matrix of the planar beam

$$C_{22}C_{31}C_{32}C_{11}C_{12}C_{21}C_{22}.$$

Substituting Eq. (73.19) into Eq. (73.7) results

$$\mathbb{S}^{F}_{C_{22}-C_{22}} = \begin{bmatrix} 0 & 0 & 1 & 0 & 0 & 0 \\ 0 & 0 & 0 & 1 & 0 & 0 \\ 0 & 0 & 0 & 0 & 1 & 0 \end{bmatrix}^T \tag{73.20}$$

here $\mathbb{S}^{F}_{C_{22}-C_{22}}$ represents the screw matrix of the full three planar free motions of the planar beam $C_{22}C_{31}C_{32}C_{11}C_{12}C_{21}C_{22}$. The instantaneous free motion of the outspread frame can be expressed as

$$\boldsymbol{S}^{F}_{C_{22}-C_{22}} = \mathbb{S}^{F}_{C_{22}-C_{22}}\boldsymbol{k}_{C_{22}-C_{22}} \tag{73.21}$$

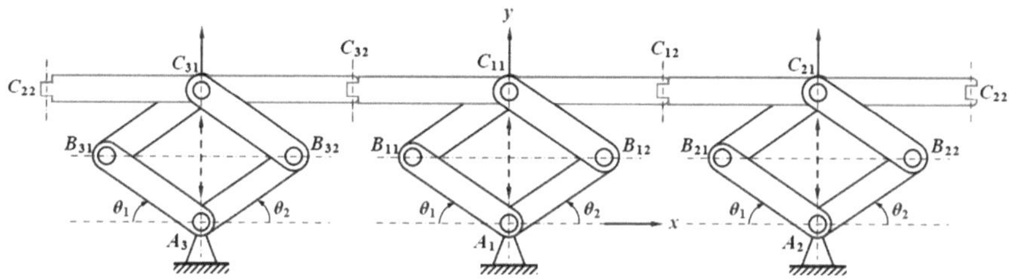


**Fig. 7** One level of the mechanism deployed to one planar linkage

where $\boldsymbol{k}_{C_{22}-C_{22}} = (k_1 \quad k_2 \quad k_3)^{\mathrm{T}}$ and $k_i (i = 1, 2, 3)$ can be any real numbers the range of which are specified by the sizes of the links.

Comparing Eq. (73.20) with Eq. (73.18), one can find that the free motions of the frame $C_{32}C_{11}C_{12}$ has changed from one translational degree of freedom along the $y$-axis shown in Fig. 73.6 to the full three planar degrees of freedom. This indicates that the free motions of a mechanism might be shifted with the change of configurations.

To completely fold the structure shown in Fig. 73.7, the triangle frame can be modified as Fig. 73.8 shows. The final folding sequence is shown in the left of Fig. 73.8 and the completely folded case is shown in the right of Fig. 73.8.

The least length of the zigzag segment of each side should be

$$\begin{cases} s_{1\,\min} = 2t\cot\dfrac{\pi}{6} = 2\sqrt{3}t \\ s_{2\,\min} = 2t\cot\dfrac{\pi}{6} = 2\sqrt{3}t \\ s_{3\,\min} = 0 \end{cases} \tag{73.22}$$

Therefore, to keep the possibility of completely folding of the structure shown in the right of Fig. 73.8, the minimum length of the zigzag segment should be $2\sqrt{3}t$. Corresponding to Fig. 73.4, the deployable unit mechanism connected by the pin jointed triangular frame, which can also be untied to spread out in one plane, is shown in Fig. 73.9.

When more deployable units shown in Fig. 73.9 are jointed one by one, a lift mechanism comes into existence. The outspread, folded and deployed structures of the reconfigurable lift mechanism are presented in Fig. 73.10.

The deployed lift mechanism is a three-plane overconstraint rectilinear moving platform. Therefore, the stiffness and stability of the platform should be higher than the existing counterparts. For the portable and reconfigurable requirements, the actuation unit should be particularly designed. Because every level of the unit shown in Fig. 73.10c only has linear relative motion, a hydraulic cylinder spherically jointed at both ends and some accessory parts are enough to keep the linear movement. The completely folded bundle shown in Fig. 73.10b can be taken away in a case or in the car boot.

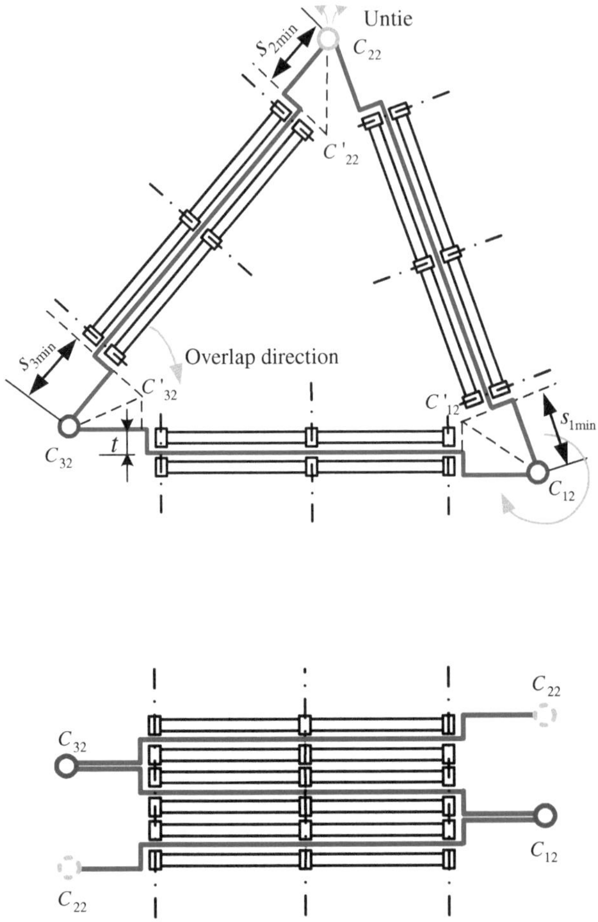


**Fig. 8** The final folding sequence

**Fig. 9** A unit mechanism jointed by a vertically pinned frame

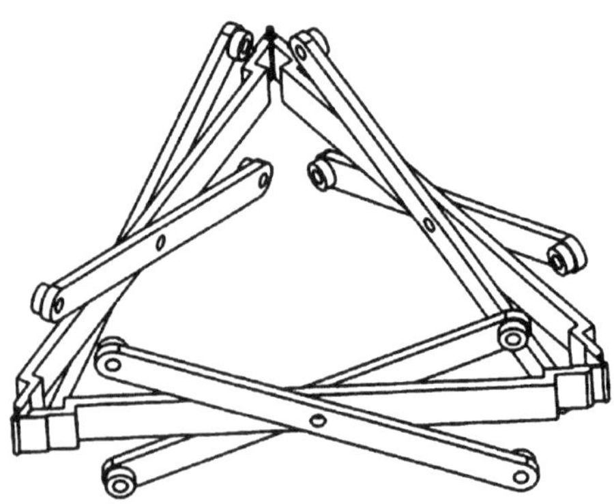

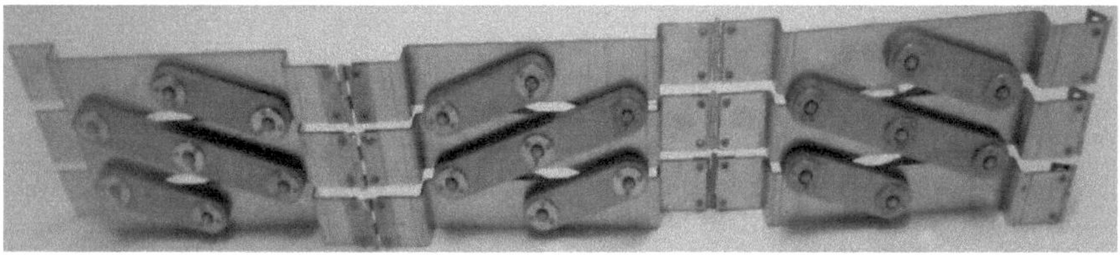

**(a)** Outspread structure in one plane

**(b)** A bundle of folded mechanism

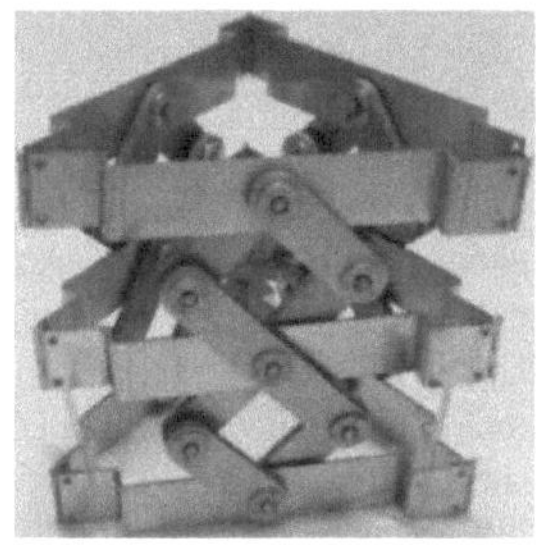

**(c)** Deployed mechanism

**Fig. 10** A prototype for the lift mechanism (a) Outspread structure in one plane (b) A bundle of folded mechanism (c) Deployed mechanism

## 73.4 Conclusions

This paper investigated a reconfigurable linkage that can be used as the unit mechanism of a portable lift platform. The reconfigurable linkage is a planar revolute jointed mechanism which could be compacted to one line and then be further folded to a bundle. This provides the possibility for the portable target. When using, the bundle of linkages should be first spread out and tied the two ends to form a closed triangular loop, and under the actuating of the actuation unit, the folded loop can be deployed to a lift mechanism. This kind of mechanism has a wide adaptation in engineering and can be surely used in the situations when a column structure must be enclosed by a lift mechanism for work. Because every reconfigurable unit is three-plane overconstraint mechanism connected by a strengthened triangular frame, the portable lift platform has very high stiffness and strength in applications. The fact that all the mechanism units only have linear relative movement provides the facility of actuator's setting. It can be mounted between any two adjacent levels.

**Acknowledgments** This research was supported by the National Natural Science Foundation of China under Grant 51175277, the Natural Science Foundation of Beijing under Grant 3112014 and the Program for New Century Excellent Talents in Tsinghua University. The authors gratefully acknowledge these support agencies. In addition, the authors are grateful to Dr. Ketao Zhang and Dr. Guowu Wei from King's College London, University of London for their kind help in improving the manuscript.

# References


1. Chen W-J, Luo Y-Z, Fu G-Y, Gong J-H, Dong S-L (2001) A study on space masts based on octahedral truss family. Int J Space Struct 16(1):75–82
2. Zhao J-S, Chu F, Feng Z-J (2009) The mechanism theory and application of deployable structures based on SLE. Mech Mach Theory 44(2):324–335
3. Dai JS, Jones JR (1999) Mobility in metamorphic mechanisms of foldable/erectable kinds. Trans ASME J Mech Des 121(3):375–382
4. Jaksch S, Sedlak V (2011) A foldable umbrella structure-developments and experiences, Int J Space Struct 26(1):1–18
5. Wei G-W, Ding X-L, Dai JS (2010) Mobility and geometric analysis of the hoberman switch-pitch ball and its variant. Trans ASME J Mech Robotics 2: 031010
6. Pinero EP (1962) Expandable space framing. Prog Architect 43(6):154–155
7. Gantes C, Giakoumakis A, Vousvounis P (1997) Symbolic manipulation as a tool for design of deployable domes. Comput Struct 64(1–4):865–878
8. Dai JS, Jones JR (2005) Matrix representation of topological configuration transformation of metamorphic mechanisms. Trans ASME J Mech Des 127(4):837–840
9. Lan ZH, Du R (2008) Representation of topological changes in metamorphic mechanisms with matrices of the same dimension. Trans ASME J Mech Des 130(7):074501
10. Li D-L, Zhang Z-H, McCarthy JM (2011) A constraint graph representation of metamorphic linkages. Mech Mach Theory 46(2):228–238
11. Dai JS, Li D, Zhang QX, Jin GG (2004) Mobility analysis of a complex structured ball based on mechanism decomposition and equivalent screw system analysis. Mech Mach Theory 39(4):445–458
12. Baker J (2006) Eddie: on generating a class of foldable six-bar spatial linkages. Trans ASME J of Mech Des 128(2):374–383
13. De Temmerman N, Mollaert M, Van Mele T, De Laet L (2007) Design and analysis of a foldable mobile shelter system. Int J Space Struct 22(3):161–168
14. Warnaar D.B., Chew M (1995) Kinematic synthesis of deployable-foldable truss structures using graph theory, part 1: graph generation. ASME Trans J Mech Des 117(1):112–116
15. Warnaar D.B., Chew M (1995) Kinematic synthesis of deployable-foldable truss structures using graph theory. part 2: generation of deployable truss module design concepts. ASME Trans J Mech Des 117(1):117–122
16. Patel J, Ananthasuresh GK (2007) A kinematic theory for radially foldable planar linkages. Int J Solids Struct 44(18–19):6279–6298
17. Wei G-W, Dai JS (2010) Geometric and kinematic analysis of a seven-bar three-fixed-pivoted compound-joint mechanism. Mech Mach Theory 45 (2):170–184
18. Zhao J-S, Wang J-Y, Chu F, Feng Z-J, Dai JS (2011) Structure synthesis and statics analysis of a foldable stair. Mech Mach Theory 46(7):998–1015
19. Kaveh A, Davarant A (1996) Analysis of pantograph foldable structures. Comput Struct 59(1):131–140
20. Mruthyunjaya TS (2003) Kinematic structure of mechanisms revisited. Mech Mach Theory 38(4):279–320
21. Chen Y, You Z (2009) Two-fold symmetrical 6r foldable frame and its bifurcations\. Int J Solids Struct 46(25–26):4504–4514
22. Gan WW, Pellegrino S (2006) A numerical approach to the kinematic analysis of deployable structures forming a closed loop. J Mech Eng Sci 220(7):1045–1056
23. Gutiérrez EM, Valcárcel JP (2004) Foldable systems based on bundle modules with quadrangular base. Int J Solids Struct 19(3):155–165
24. Zhang L-P, Wang D-L, Dai J S (2008) Biological modeling and evolution based synthesis of metamorphic mechanisms. J Mech Design 130(7):072303
25. Kovács F, Tarnai T, Fowler PW, Guest SD (2004) A class of expandable polyhedral structures. Int J Solids Struct 41(3–4):1119–1137

26. Ragavan V, Amde MA (2001) An algorithm for nonlinear stability analysis of an expandable self-erecting structure. Comput Struct 79(29–30):2587–2893
27. You Z, Pellegrino S (1997) Foldable bar structures. Int J Solids Struct 34(15):1825–1847

# Chapter 74
# Dynamic Analysis on Crank-Connecting Rod Mechanism of Reciprocating Pumps with Crankshaft–Bushing Clearance

**Lixin Xu, Yuhu Yang, Yonggang Li and Chongning Li**

**Abstract** The clearance appears in crankshaft pin and connecting rod bushing by manufacturing tolerance or wear of components is the most important factor that influences the dynamic performance and reliability of reciprocating pumps. In this paper, a parametric dynamic model of the crank-connecting rod mechanism of reciprocating pumps considering the crankshaft–bushing clearance is proposed based on the multibody dynamics theory. By simulation, the dynamic response of the mechanism with and without clearance is obtained and discussed. It is observed that the free flight motion between the pin and bushing occurs followed by the impact and rebound phenomenon when the mechanism moves from discharging period into sucking period. The simulation result of the joint force can be used for the strength and stiffness checking of the crankshaft.

**Keywords** Reciprocating pumps · Clearance · Crank-connecting rod mechanism · Dynamic analysis

L. Xu (✉) · Y. Li · C. Li
School of Mechanical Engineering, Tianjin University of Technology and Education, Tianjin, China
e-mail: xulixin_tju@yahoo.cn

Y. Li
e-mail: lygang1975@163.com

C. Li
e-mail: licn126@126.com

Y. Yang
School of Mechanical Engineering, Tianjin University, Tianjin, China
e-mail: yangyuhu@tju.edu.cn

J. S. Dai et al. (eds.), *Advances in Reconfigurable Mechanisms and Robots I*,
DOI: 10.1007/978-1-4471-4141-9_74, © Springer-Verlag London 2012

## 74.1 Introduction

The crank-connecting rod mechanism (as shown in Fig. 74.1), which can transform the rotary motion of motor into the reciprocating motion of plunger piston, is a vital component in the structure of reciprocating pumps. The dynamic response of the mechanism directly determines the working performance and reliability of the reciprocating pumps, which motivated researchers to investigate the dynamic behavior of them [1–4]. In these works, the detailed dynamic models of the crank-connecting rod mechanism were developed.

However, in the above researches no clearance between the crankshaft pin and connecting rod bushing is considered. Actually, the clearance appears in crankshaft–bushing joint induced by manufacturing tolerance or wear of components is the most important factor that influences the dynamic performance of reciprocating pumps. Due to the existence of the clearance, the serious impact, noise and vibration phenomenon would be caused in the motion of crank-connecting rod mechanism.

The subject of the modeling and dynamic analysis of mechanical systems with joint clearance, especially the representation of real joints, draw the attention of a large mount of researchers that produced many theoretical and experimental works [5–21]. For these works, the revolute clearance joints were comprehensively defined. In general, there are three main modeling strategies for mechanical systems with revolute clearance joints, namely, the massless link approach [5–7], the spring-damper approach [8, 9], and the momentum exchange approach [10–21]. In the massless link approach, the presence of clearance at a joint is modeled by adding a link of zero mass that has a constant length equal to the radial clearance. This model assumes that there is contact between the journal and the bearing all the time, hence unable to reveal the dynamic characteristics of separation and collision between the two contact elements. In the spring-damper approach, the clearance is modeled by introducing a spring-damper element, which simulates the surface elasticity. This model does not represent the physical nature of energy transfer during the impact process. Moreover, there is a real difficulty in quantifying the parameters of the spring and damper elements. In the momentum exchange model, the contact elements are considered as two colliding bodies and the contact forces control the dynamics of the clearance joint. This model is more realistic since the impact force model allows, with high level of approximation, to simulate the elasticity of the contacting surfaces as well as the energy dissipation during the impact.

In this paper, a parametric dynamic model of the crank-connecting rod mechanism of reciprocating pumps considering the crankshaft–bushing clearance is proposed based on the multibody dynamics theory. By simulation, the dynamic response of the mechanism with and without clearance is obtained and discussed.

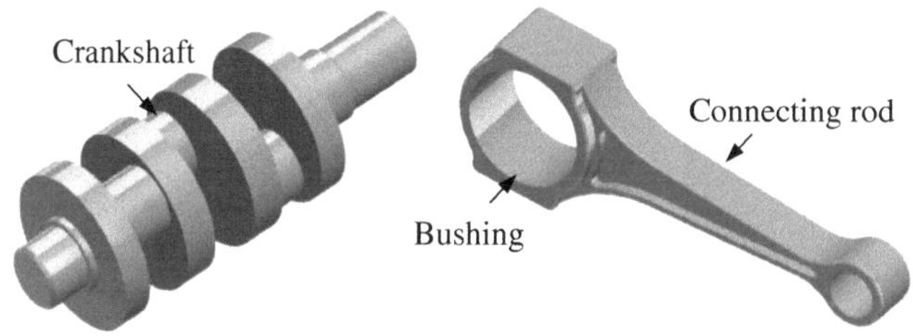


**Fig. 74.1** Crankshaft and connecting rod of the reciprocating pumps

## 74.2 Modeling of Revolute Joint with Clearance in a Multibody System

Two bodies $i$ and $j$ connected by a revolute joint with clearance [22] can be depicted in Fig. 74.2. Part of body $i$ is the bushing and part of body $j$ is the journal. The center of mass of bodies $i$ and $j$ are $O_i$ and $O_j$, respectively. Body-fixed coordinate systems $\xi\eta$ are attached to the center of mass of each body, while the XOY coordinate frame represents the global coordinate system. Point $P_i$ indicates the center of the bushing, and the center of the journal is denoted by point $P_j$.

The eccentricity vector **e** connecting the centers of the bushing and the journal is calculated as

$$\mathbf{e} = \mathbf{r}_j^P - \mathbf{r}_i^P = \left(\mathbf{r}_j + \mathbf{A}_j\mathbf{s}_j'^P\right) - \left(\mathbf{r}_i + \mathbf{A}_i\mathbf{s}_i'^P\right) \tag{74.1}$$

where $\mathbf{r}_i$ and $\mathbf{r}_j$ are the vectors linking the global origin and the center or masses of the bodies, $\mathbf{s}_i^P$ and $\mathbf{s}_j^P$ are vectors in the local coordinate system that link the center of masses to the journal and bushing centers, respectively, and $\mathbf{A}_i$ and $\mathbf{A}_j$ are matrices that transform the vectors $\mathbf{s}_i^P$ and $\mathbf{s}_j^P$ from the local coordinate system to the global system.

The magnitude of the eccentricity vector is evaluated as

$$e = \sqrt{\mathbf{e}^{\mathrm{T}}\mathbf{e}} \tag{74.2}$$

The unit normal vector **n** to the surfaces in collision between the bushing and the journal, is aligned with the eccentricity vector,

$$\mathbf{n} = \mathbf{e}/e \tag{74.3}$$

The penetration depth caused by the impact between the journal and the bushing, as shown in Fig. 74.3, is evaluated as

$$\delta = e - c \tag{74.4}$$

where $c$ is the radial clearance, defined as the difference between the radius of the bushing and the radius of the journal.

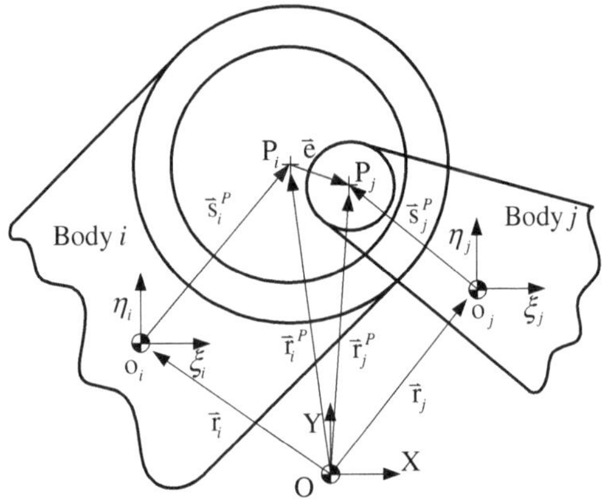


**Fig. 74.2** Revolute joint with clearance in a multibody system [22]

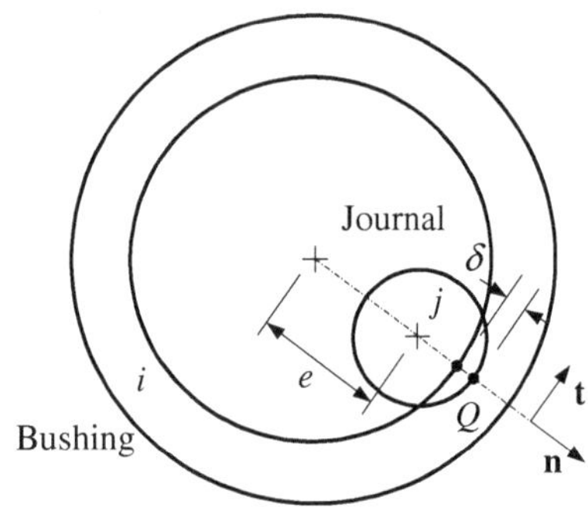


**Fig. 74.3** Penetration during contact between the pin and the bushing

When the journal is not in contact with the bushing, the eccentricity is smaller than the clearance and the penetration has a negative value. When the penetration has a value equal or greater than zero, contact is established. Thus, when it is greater than zero, a contact force is applied between the bodies. The contact force vanishes when it is equal to or less than zero.

The location of contact point Q with respect to the journal and bushing can then be expressed as

$$\mathbf{r}_k^Q = \mathbf{r}_k + \mathbf{A}_k \mathbf{s}_k'^Q + R_k \mathbf{n}, \ \ (k = i, j) \tag{74.5}$$

where $R_k, (k = i, j)$ are the bushing and journal radius, respectively. To determine the contact force, the relative penetration velocity is also required. This is the difference between the velocities of the contact point. The velocity of the contact points $Q^i$ and $Q^j$ in the global coordinate system is found by differentiating Eq. (74.5) with respect to time t

$$\dot{\mathbf{r}}_k^Q = \dot{\mathbf{r}}_k + \dot{\mathbf{A}}_k \mathbf{s}_k'^Q + R_k \dot{\mathbf{n}}, \ \ (k = i, j) \tag{74.6}$$

The relative velocities in the normal and tangential direction can then be computed as

$$\upsilon_N = \left(\dot{\mathbf{r}}_j^Q - \dot{\mathbf{r}}_i^Q\right)^{\mathrm{T}} \mathbf{n} \tag{74.7}$$

$$\upsilon_T = \left(\dot{\mathbf{r}}_j^Q - \dot{\mathbf{r}}_i^Q\right)^{\mathrm{T}} \mathbf{t} \tag{74.8}$$

where $t$ is obtained by rotating the vector $\mathbf{n}$, calculated using Eq. (74.7), in the counter clockwise direction by 90°.

After the relative penetration and velocity have been determined, the normal contact and friction forces can be computed. The normal contact force can be calculated by the model with hysteresis damping discussed by Lankarani and Nikravesh [23]. The model is expressed as

$$F_N = K\delta^{1.5}\left(1 + \frac{3\left(1 - e_r^2\right)}{4}\frac{\dot{\delta}}{\dot{\delta}^{(-)}}\right) \tag{74.9}$$

where $\dot{\delta}$ is the relative normal penetration velocity, $\dot{\delta}^{(-)}$ is the initial normal impact velocity, $e_r$ is the restitution coefficient, and $K$ is a constant that is dependent on the material properties of the components and their geometry. The constant $K$ is expressed as

$$K = \frac{4}{3\pi(h_1 + h_2)}\left(\frac{R_i R_j}{R_i + R_j}\right)^{1/2} \tag{74.10}$$

and

$$h_k = \frac{1 - v_k^2}{\pi E_k}, \ (k = i, j) \tag{74.11}$$

where $v_k$, $(k = i, j)$ is Poisson's ratio of the components and $E_k$ is the elastic modulus of the components.

In addition to the reaction forces, a friction force can be included to enhance the model. In this work, coulomb friction is applied. The friction force is

$$F_T = \mu F_N \tag{74.12}$$

where $\mu$ is the coefficient of friction which can be determined through experiments as discussed by Schmitz [24] and $F_N$ is the normal contact force.

After calculation of the normal contact force $F_N$ and the friction force $F_T$. The contributions to the generalized vector of forces and moments, $\mathbf{Q}^{\mathrm{A}}$ in the equations of multibody system (74.13), are found.

$$\begin{bmatrix} \mathbf{M} & \mathbf{\Phi}_q^{\mathrm{T}} \\ \mathbf{\Phi}_q & 0 \end{bmatrix}\begin{bmatrix} \ddot{\mathbf{q}} \\ \boldsymbol{\lambda} \end{bmatrix} = \begin{bmatrix} \mathbf{Q}^{\mathrm{A}} \\ \gamma \end{bmatrix} \tag{74.13}$$

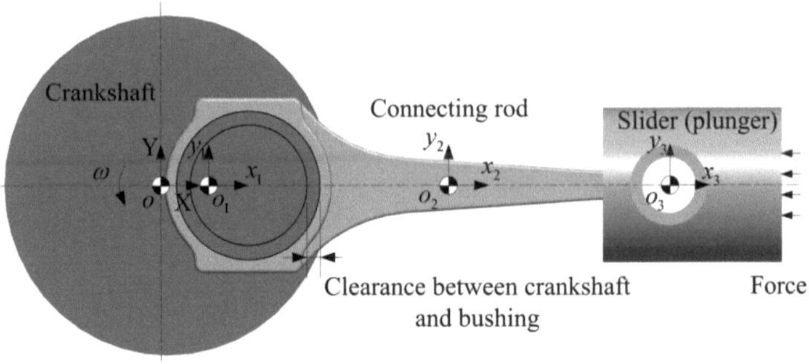


**Fig. 74.4** Model of the crank-connecting rod mechanism of reciprocating pumps with crankshaft–bushing clearance

These forces that act on the contact points of bodies $i$ and $j$ are transferred to the center of mass of bodies and an equivalent transport moment is applied to the rigid body. As discussed by Flores [22], the forces $\mathbf{f}_i$ and moment $T_i$ that act on the center of mass of body $i$ due to the clearance joint contact can be expressed as follows

$$\mathbf{f}_i = \mathbf{f}_N + \mathbf{f}_T \tag{74.14}$$

$$T_i = -(y_i^Q - y_i)f_i^x + (x_i^Q - x_i)f_i^y \tag{74.15}$$

The corresponding forces and moments applied to the body j are

$$\mathbf{f}_j = -\mathbf{f}_i \tag{74.16}$$

$$T_j = \left(x_j^Q - x_j\right)f_j^y - \left(y_j^Q - y_j\right)f_j^x \tag{74.17}$$

## 74.3 Crank-Connecting Rod Mechanism of Reciprocating Pump with Crankshaft–Bushing Clearance

Figure 74.4 shows a diagram of the crank-connecting rod mechanism of reciprocating pumps which consists of three moving components (crankshaft, connecting rod, and plunger). The components are connected to each other by two revolute joints and a translational joint. For this example, the revolute joint between the crankshaft pin and the connecting rod bushing is modeled as a clearance joint (with a clearance size c = 0.5 mm). The dimension and mass properties for the mechanism are shown in Table 74.1. Also, the radius and material properties for the joint components (the pin of crankshaft and bushing of the connecting rod) are shown in Table 74.2. In this case a steel pin and a steel bushing are used. For the

**Table 74.1** Dimensions and mass parameters for the crank-connecting rod mechanism

| Bodies | Length (m) | Mass (kg) | Moment of inertia (kg•m$^2$) |
|---|---|---|---|
| Crank | 0.27 | 32 | 1.81 |
| Connecting rod | 0.7 | 60.5 | 3.12 |
| Slider (plunger) | – | 12 | – |

**Table 74.2** Material properties for the clearance joint components

| | Crankshaft (pin) | Bushing |
|---|---|---|
| Young's modulus | 207GPa | 207GPa |
| Poisson's ratio | 0.3 | 0.3 |
| Radius | 12 mm | 12.5 mm |

contact force model, a value of 0.15 and 0.95 were used for the friction coefficient and the coefficient of restitution, respectively.

In the model, the fluid force acts on the plunger can be expressed as follows

$$F = \begin{cases} AP & \text{discharging} \\ 0 & \text{sucking} \end{cases} \tag{74.18}$$

Where A is the cross-sectional area of the plunger, P is the fluid pressure. Variation of the fluid force acts on the plunger is shown in Fig. 74.5. In the first half cycle of the crankshaft rotation (sucking period), the fluid force acts on the plunger is zero. Contrarily, in the second half cycle of the crankshaft rotation (discharging period), the fluid force reaches the maximum value. In simulation, the crankshaft is assumed to rotate at a constant angular velocity of 300 rpm. It is supposed that the radius of the plunger is 20 mm, and the maximum fluid pressure on the plunger is 40 MPa.

## 74.4 Results and Discussion

By simulation, the dynamic response of the crank-connecting rod mechanism of reciprocating pumps with crankshaft–bushing clearance is obtained. Figure 74.6 shows the relative motion between crankshaft pin and connecting rod bushing centers in a complete motion cycle of crank rotation. Different types of motion between the joint elements can be observed, namely, free flight, impact and rebound, and continuous contact. It is interesting to observe that the crankshaft pin and connecting rod bushing are in continuous contact condition in most of time. When the mechanism moves from discharging period into sucking period, the free flight motion between pin and bushing occurs as depicted from A to B in Fig. 74.6. In this process, the crank rotation position changes from 0.21° to 4.4°. Then, the impact and rebound phenomenon accompanied by the penetration and impact force between joint elements is caused.

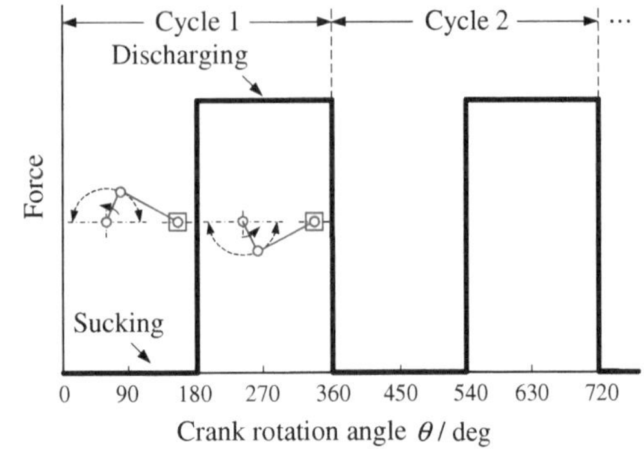


**Fig. 74.5** Variation of the fluid force on the plunger

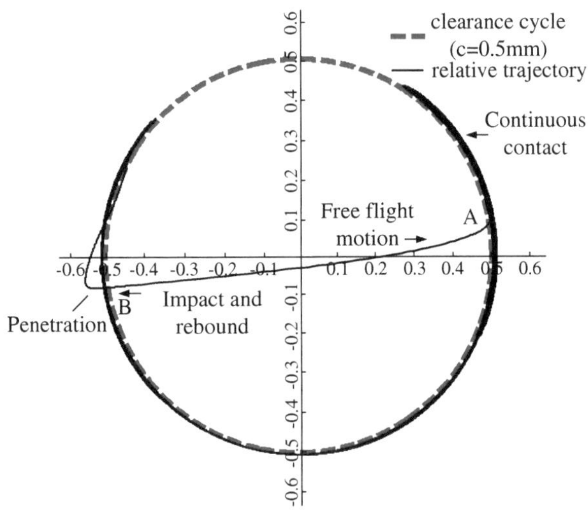


**Fig. 74.6** Crankshaft center trajectory with respect to the connecting rod bushing center

Figure 74.7 shows the constraint force in crankshaft–bushing clearance joint. It is found that the magnitude of the impact force due to collision between the joint elements reaches 8.6968e5N. This force is far greater than the ideal value without clearance. The simulation result of the joint force can be used for the strength and stiffness checking of the crankshaft.

## 74.5 Conclusions

In the present work, a parametric dynamic model of the crank-connecting rod mechanism of reciprocating pumps considering the crankshaft–bushing clearance is proposed based on the multibody dynamics theory. A detailed numerical simulation is carried out to compare the dynamic response of the mechanism with and

**Fig. 74.7** Constraint force in crankshaft–bushing joint with and without clearance

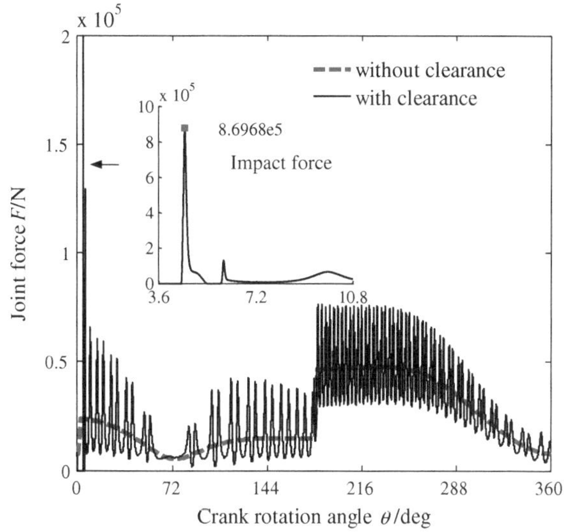


without clearance. According to the numerical solutions of the theoretical model, effect of the clearance is investigated. Some conclusions can be drawn from the investigation.

It is interesting to observe that different types of motion between the joint elements (crankshaft pin and connecting rod bushing) can be observed, namely, free flight, impact and rebound, and continuous contact. The free flight motion occurs when the mechanism moves from discharging period into sucking period. Then, the impact and rebound phenomenon accompanied by the penetration and impact force between joint elements is caused.

Due to collision between the joint elements, a large impact force can be got in the joint. The simulation result of the joint force can be used for the strength and stiffness checking of the crankshaft of reciprocating pumps.

**Acknowledgments** This project is supported by the National High Technology Research and Development Program of China ("863" Program), (Grant No. 2011AA04A102) and the Key Project of Chinese Ministry of Education, (Grant No. 210005).

## References

1. Han XM, Du CL, Zhang YZ (2008) Wear fault diagnosis of an emulsion pump crank bearing. J China Univ Min Technol 18(3):470–474
2. Yuan DH, Hua SB, Gu ZH (2009) Analysis of reciprocating pump vibration and noise mechanism and reduction measures (in Chinese). Chin J Ship Res 4(5):75–80
3. Li CM (2010) Dynamics simulation of the main mechanism for triplex reciprocating pump (in Chinese). Mach Tool Hydraul 38(15):73–76

4. Peng B, Zhang HS, Zhao RZ (2010) FEM of high pressure reciprocating pump crankshaft intensity (in Chinese). Chin Agric Mech 227:70–74
5. Wu CLS, Earles SWE (1977) A determination of contact-loss at a bearing of a linkage mechanism. J Eng Ind 99(2):375–380
6. Haines RS (1980) A theory of contact loss at resolute joints with clearance. J Mech Eng Sci 22(3):129–136
7. Earles SWE, Seneviratne LD (1990) Design guidelines for predicting contact loss in revolute joints of planar mechanisms. Proc Inst Mech Engrs Part C 204(1):9–18
8. Dubowsky S, Freudenstein F (1971) Dynamic analysis of mechanical systems with clearances, part 1: formulation of dynamic model. J Eng. Ind 93(1):305–309
9. Dubowsky S, Freudenstein F (1971) Dynamic analysis of mechanical systems with clearances, part 2: dynamic response. J Eng Ind 93(1):310–316
10. Soong K, Thompson BS (1990) A theoretical and experimental investigation of the dynamic response of a slider-crank mechanism with radial clearance in the gudgeon-pin joint. J Mech Des 112(2):183–189
11. Ravn P (1998) A continuous analysis method for planar multibody systems with joint clearance. Multibody Syst Dyn 2(1):1–24
12. Ravn P, Shivaswamy S, Alshaer BJ (2000) Joint clearances with lubricated long bearings in multibody mechanical systems. J Mech Des 122(4):484–488
13. Schwab AL, Meijaard JP, Meijers P (2002) A comparison of revolute joint clearance model in the dynamic analysis of rigid and elastic mechanical systems. Mech Mach Theory 37(9):895–913
14. Flores P, Ambrósio J, Claro JP (2004) Dynamic analysis for planar multibody mechanical systems with lubricated joints. Multibody Syst Dyn 12(1):47–74
15. Alshaer BJ, Nagarajan H, Beheshti HK (2005) Dynamics of a multibody mechanical system with lubricated long journal bearings. J Mech Des 127(3):493–498
16. Flores P, Ambrósio J, Claro JP (2006) A study on dynamics of mechanical systems including joints with clearance and lubrication. Mech Mach Theory 41(3):247–261
17. Flores P, Ambrósio J, Claro JP et al (2007) Dynamic behavior of planar rigid multi-body systems including revolute joints with clearance. Proc IMechE Part K J Multi-body Dyn 221(K2) 161–174
18. Imed K, Lotfi R (2008) Dynamic analysis of a flexible slider–crank mechanism with clearance. Eur J Mech A Solids 27(5):882–898
19. Shi B, Jin Y (2008) Dynamic analysis of the reheat-stop-valve mechanism with revolute clearance joint in consideration of thermal effect. Mech Mach Theory 43(12):1625–1638
20. Tian Q, Zhang YQ, Chen LP (2010) Simulation of planar flexible multibody systems with clearance and lubricated revolute joints. Nonlinear Dyn 60(4):489–511
21. Flores P (2010) A parametric study on the dynamic response of planar multibody systems with multiple clearance joints. Nonlinear Dyn 61(4):633–653
22. Flores, P, Ambrósio J, Claro JP et al (2008) Kinematics and dynamics of multibody systems with imperfect joints: models and case studies. Springer, Berlin, Heidelberg
23. Lankarani HM, Nikravesh PE (1990) A contact force model with hysteresis damping for impact analysis of multibody systems. J Mech Des 112(3):369–376
24. Schmitz TL, Action JE, Ziegert JC et al (2005) The difficulty of measuring low friction: uncertainty analysis for friction coefficient measurement. J Tribol 127(3):673–678

# Chapter 75
# Design and Analysis of a Biphasic Variable Impedance Actuator

**Bo Han, Matteo Zoppi and Rezia Molfino**

**Abstract** Variable impedance actuators guarantee control performance for robots and robotic peripherals that are inherently safe to humans in their environment especially in cooperative tasks involving human-robot interaction. This paper presents a concept of variable impedance actuator using a combination of fluid and gas to realize the adaptation of the impedance. For this actuator, the ranges of variation of the impedance are studied with different initial conditions and positions of the piston in the actuator; The dynamic response characteristic is analyzed in presence of a variable external load. A method of control of the actuator is proposed and the performance shown while performing a testing cycle.

**Keywords** Variable impedance actuation · Biphasic fluid

## 75.1 Introduction

Impedance control can be used in a robotic system for safer interaction between robot and environment, including unexpected impacts. Variable impedance actuators (VIAs) are actuators whose impedance can be actively controlled together with the output motion. These actuators allow the robot to behave more or less

B. Han (✉) · M. Zoppi · R. Molfino
PMAR Robotics, University of Genoa, Genoa, Italy
e-mail: zoppi@dimec.unige.it

M. Zoppi
e-mail: molfino@dimec.unige.it

R. Molfino
e-mail: hanbo@dimec.unige.it

J. S. Dai et al. (eds.), *Advances in Reconfigurable Mechanisms and Robots I*,
DOI: 10.1007/978-1-4471-4141-9_75, © Springer-Verlag London 2012

compliant, depending on the task, and thus allow accurate motion under loads when required and safe human-robot interaction when expected [1–5].

A constant compliance in the transmissions between an actuator and a robot link can be introduced to attenuate the reflected inertia of the motor and mechanics on the driven link [3, 5]. The drawback is a reduction of performance for part of the operative conditions since there is no optimal compliance for all tasks.

Several designs of variable impedance actuators have been proposed in the literature, generally using systems of mechanical springs [6, 7] or other similar elastic elements. This paper discusses a new type of VIA [8, 9] using a biphasic medium with a gas fraction with the function of nonlinear elastic element [10, 11]. The variation of impedance comes from the variation of pressure of the gas. This leads to a simple and robust design realizing a significant range of stiffness variation. A further point of interest is that, compared to the state of the art design principles, the use of this biphasic medium is simpler and easier to apply since there is no need for motors and complex mechanical transmissions [12, 13].

The paper is organized as follows: Sect. 75.2 details the concept of the new biphasic adjustable stiffness mechanism and its modeling. Section 75.3 discusses the stiffness scope in different initial conditions of gas and also the stiffness dynamic response characteristic. Section 75.4 proposes the control method of the actuator and studies two cases according to specific task. Section 75.5 draws conclusions and ongoing development.

## 75.2 Physical Modeling of the Actuator

A schematic of the biphasic fluid actuator is shown in Fig. 75.1. In this schematic, the actuator comprises two chambers each containing a mass of gas segregated in a compliant bag and a volume of liquid.

$q^A$ and $q^B$ indicate the volumes of liquid moving in/out the chambers (positive if inlet and negative if outlet); $p^A$ and $p^B$ are the pressures in the chambers. The liquid is used to modify the volume of the gas in the chambers and consequently its stiffness.The volume of fluid in the chambers can be controlled using any type of servo controlled pump and system of valves [14–16]. Sensors can be used to transduce the values of relevant functional parameters such as pressures in the chambers and output motion. The schematic in Fig. 75.1 represents the functioning principle used in the actuator to achieve the variation of stiffness. The gas behaves like a nonlinear spring. Two hydraulic servo pumps are used to move liquid inside-outside the chambers and so drive the mobile part of the actuator, represented by the center beam. Two pressure sensors are mounted on the chambers to measure the instantaneous pressures of the fluid inside.

The actuator is modeled referring to the schematic in Fig. 75.1. Assumptions are listed below. The following differential equations hold, linking the output motion (described in this case by the displacement $x$ of the piston) to the fluid

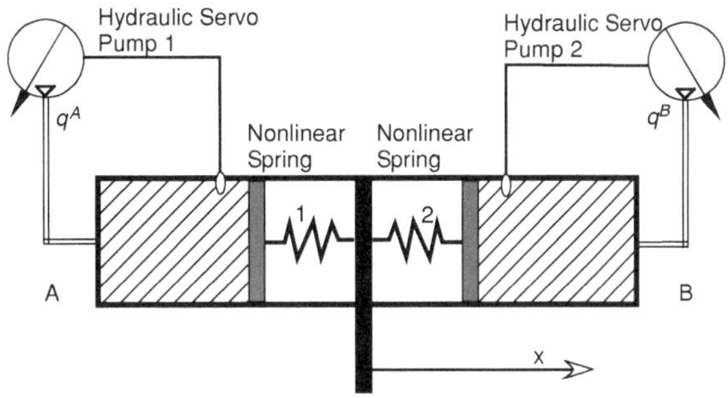


**Fig. 75.1** The biphasic gas-fluid variable impedance actuator schematic

pressures, $p^A$ and $p^B$, and gas volumes, $V^A$ and $V^B$, and environment temperature $T$ in the chambers, according to the equation of state of a hypothetical ideal gas:

$$p^A V^A = n^A RT \tag{75.1}$$

$$p^B V^B = n^B RT \tag{75.2}$$

with $m$ the mass of the piston (meaning by piston the output link of the actuator); $S$ the net area of the piston (on which fluid pressures acts), being identical for both chambers; $x$ the displacement of the piston and $\dot{x}$ and $\ddot{x}$ the first and second derivative of $x$ in the time; $\mu$ a damping coefficient related to the viscosity and dissipative properties of the fluid; $R$ the ideal gas constant; $n^A$ and $n^B$ the numbers of moles for the gas fractions in the chambers. Using dynamic equation and combine with the expressions in Eqs. (75.1) and (75.2), we obtain:

$$m\ddot{x} + \mu\dot{x} - \left(\frac{n^A RT}{V^A} - \frac{n^B RT}{V^B}\right)S = 0 \tag{75.3}$$

where $q^A$, $q^B$, $T$, $x$, are functions of the time $t$; $V^A = V^A(0) - q^A + Sx$; $V^B = V^B(0) - q^B - Sx$. The structure of this nonlinear partial differential equation is $U(t, q^A, q^B, T, x, \dot{x}, \ddot{x}) = 0$. We consider $q^A$ and $q^B$ assigned and used to command motion and stiffness of the actuator.

The gas is ideal and sealed in compliant bags respectively. The temperature depends on the environment and should be constant if operating at low frequency.

### 75.2.1 Reference Case Studies

The model of the actuator is non linear and the behavior depends on its design parameters and initial conditions. We proceed in the analysis numerically and consider the reasonable design parameters as $m = 1\,\text{kg}$, $\mu = 20\,\text{Ns/m}$, $S = 9.62 \times 10^{-4}\,\text{m}^2$; initial conditions as $V^A(0) = V^B(0) = 2.405 \times 10^{-5}\,\text{m}^3$, $p^A(0) = p^B(0) = 2.5\,\text{MPa}$, $n^A RT = n^B RT = 60.125\,\text{m}^3$ Pa. The model has been implemented in Simulink.

## 75.3 Analysis of Stiffness Variation Under Load Induced Displacements

When the actuator is in equilibrium under an external force, the difference between the pressures in the two chambers is related to the value of the external force by the geometry and the design of the actuator. In the design in Fig. 75.1, the external force is proportional to the difference of pressure through the surface area of the piston: $F = (p^A - p^B)S$. In statics, for any given external force, the values of the pressures in the two chambers may be any provided that their difference is the one required to balance the external force.

The instantaneous stiffness $K$ of the actuator (of the piston) is the derivative of the external force applied to it with respect to the displacement. In our case the stiffness is a function of the volumes of liquid moved $q^A$ and $q^B$; Considering ideal the gas, its expression is:

$$K = \frac{\partial F}{\partial x} = -\left(\frac{(p^A S)^2}{n^A RT} + \frac{(p^B S)^2}{n^B RT}\right) \tag{75.4}$$

According to Eq. (75.4), it is possible and easy to adapt the stiffness to a desired value or to specific actuation requirements using $q^A$ and $q^B$ as control variables.

### *75.3.1 Stiffness Range for Different Initial Gas Constant*

We study the variation of stiffness with the displacement of the piston without in/out flows of liquid in the chambers. Figure 75.2a shows the stiffness variation with the displacement of the piston for different initial values of the gas pressure in the chambers. The lines are symmetrical with respect to the position of static equilibrium of the piston with no external load applied. The stiffness is larger if the piston is closer to the extreme position while it tends to infinite when the volume of the gas in the left or right chamber tends to zero. As shown in Fig. 75.2 a, using the parameters described in Sect. 75.2.1, the piston stiffness ranges from $10^{3.28}$ N/m to $10^{6.28}$ N/m in stable position when the gas initial pressure is from 0.025 MPa to 25 MPa while in the position $X = -0.024$ m or 0.024 m, the stiffness is from $10^{5.78}$ N/m to $10^{8.78}$ N/m.

### *75.3.2 Dynamic Response to Step External Load for Different Gas Masses*

The dynamic responses with different gas constant for same initial stiffness 192.5 N/mm and actuator in two initial conditions (represented by a solid and dashed line) are shown in Fig. 75.2b: The initial gas constant and gas mass are

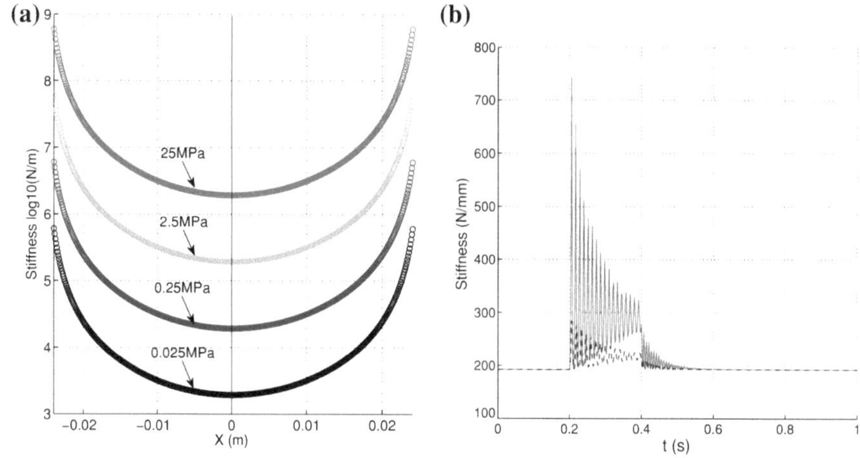


**Fig. 75.2** *Stiffness scope* of different initial gas pressures corresponding piston displacement, *left* **a**; *Dynamic response* in different initial gas masses, *right* **b**

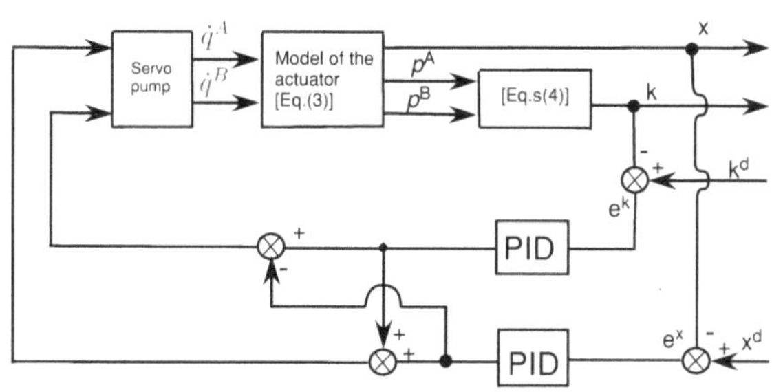


**Fig. 75.3** Schematic of the control loop

respectively four and two times larger for the dashed line than for the solid line. With external force 2,000 N applied to the actuator between 0.2 and 0.4 s, the change in stiffness is larger for the actuator represented by the solid line (maximum 743 N/mm) than for the actuator represented by the dashed line (maximum is 284 N/mm). Hence the initial gas constant and volume have a significant impact on stiffness and dynamic response. For same external force applied, the maximum transient stiffness is smaller if the initial gas volume is larger for same initial gas pressure and actuator.

## 75.4 Controllability of the Actuator

As illustrated in the schematic and mathematic model, the stiffness and displacement of the piston can be controlled through the flows of liquid in the chambers. At this preliminary stage of the investigation of the actuator, we have

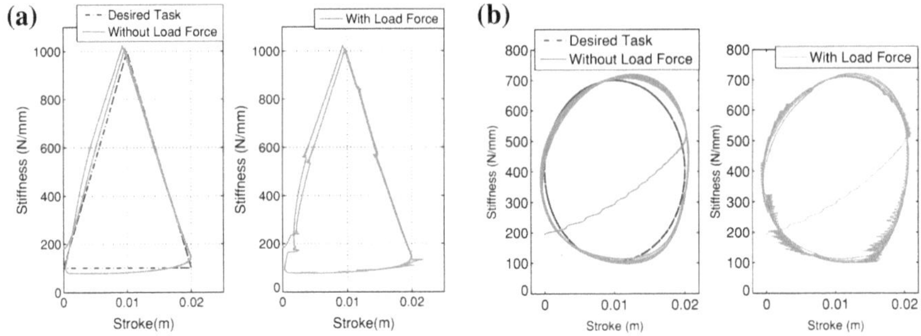


**Fig. 75.4** Testing tasks in the *position-stiffness* plane *without (left)* and *with (right)* external force applied, *left* **a** *Triangle task*; *right* **b** *Ellipse task*

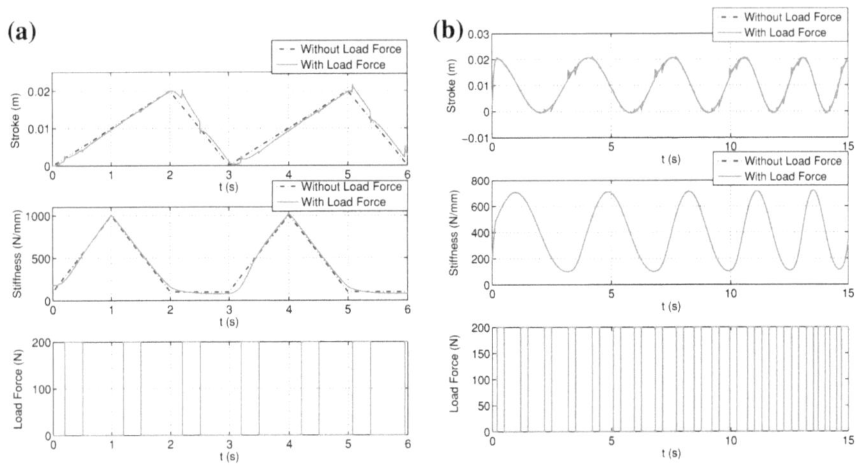


**Fig. 75.5** Position and stiffness in the time without and with external force for the testing loops *(top and center)*; the external force applied *(bottom)*, *left* **a** *Triangle task*; *right* **b** *Ellipse task*

applied a simple PID controller [17] with controlled variables $q^A$ and $q^B$ with the aim to investigate through examples the controllability of the VIA concept proposed. The measured quantities are the position of the piston and the pressures in the two chambers. The pressures are used to calculate the current stiffness of the actuator using Eqs. (75.1), (75.2), (75.4). Since the position of the piston depends on $q^A - q^B$ and the stiffness on $q^A + q^B$, two PID controllers are used relating the control flows $\dot{q}^A$ and $\dot{q}^B$ to the position error $e^x$ and stiffness error $e^K$ as follows:

$$\begin{aligned}\dot{q}^A + \dot{q}^B &= K_{px}e^x + K_{dx}e^x + K_{ix}e^x \\ \dot{q}^A - \dot{q}^B &= K_{pK}e^K + K_{dK}e^K + K_{iK}e^K\end{aligned} \tag{75.5}$$

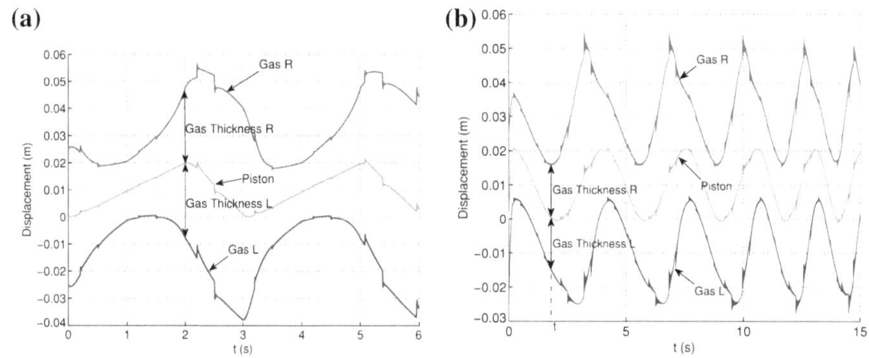


**Fig. 75.6** Piston position and gas volume in the time with external force for the testing loops, *left* **a** *Triangle task*; *right* **b** *Ellipse task*

From Eq. (75.5), $\dot{q}^A$ and $\dot{q}^B$ are easily expressed as functions of $e^x$, $e^K$ and the proportional, derivative and integral gains.

The schematic of the control is shown in Fig. 75.3. The response of the actuator is simulated using all sets of equations presented above, grouped in a block having $\dot{q}^A$ and $\dot{q}^B$ as input and, as output, the position $x$ and the pressures $p^A$ and $p^B$. A separate block calculates $K$ from $p^A$ and $p^B$. $x$ and $K$ are subtracted to the task-desired values at each control loop, generating the errors $e^x$ and $e^K$ used in the two PIDs for the generation of $\dot{q}^A$ and $\dot{q}^B$.

The simulated tasks are a triangle and an ellipse in the position-stiffness plane of the actuator respectively. The triangular task is performed by the actuator in simulation two times at constant velocity while the elliptical task is performed by the actuator in simulation five times at increasing velocity (the last triangle/ellipse at twice the velocity of the first triangle/ellipse) and both tasks (two consecutive loops for the triangular and five consecutive loops for the elliptical) are repeated twice: without and with an external force applied. The external force is applied instantaneously 0.2 s after each corner of the triangle and each quadrant of the ellipse, and removed instantaneously after 0.3 s, as clarified in Fig. 75.5a, b where the value of the force is shown in the time Fig. 75.4.

The deformation of the gas at both sides of the piston can be visualized in the time. Assume that the volume of gas $V^A(t) = S \cdot L_{gas}(t)$ where $S$ (constant) is the area of the piston and $L_{gas}(t)$ the height of the gas at the left side of the piston supposed in a cylindrical container with base area $S$. $L_{gas}(t) \propto V^A(t)$; $R_{gas}(t) \propto V^B(t)$ with $R_{gas}(t)$ the height of gas at the right side of the piston supposes in a cylindrical container of base area $S$. The deformation of the gas can be described as the changes of these heights in the time, as shown by the three curves in Fig. 75.6a, b where the two heights are represented together with the displacement in the time of the piston (central line). According to the time axis of Figs. 75.5a and 75.6a, Figs. 75.5b and 75.6b appear well physically consistent.

## 75.5 Conclusions

The paper discusses stiffness characteristics and performance of a new type of variable impedance actuator using a combination of liquid and gas. The actuator is modeled under the hypothesis of ideal gas and incompressible liquid. Its operation is simulated numerically and covers a big stiffness range implementation which gets the advantage over mechanical springs or other similar elastic mechanism structure. A PID control method has fulfilled the possibility to get specified stiffness-stroke tasks.

The variable impedance actuator design proposed can replace a standard hydraulic actuator. Thanks to its simple mechanic structure, the actuator can be used in daily life robotics where the requirement of cost effectiveness is relevant; and also, it's easy to realize at micro scale. The surgical application is the one for which the development of this actuator started [18].

## References

1. Bicchi A, Tonietti G (2004) Fast and soft-arm tactics:dealing with the safety-performance tradeoff in robot arms design and control. IEEE Robot Autom Mag 11(2):22–33
2. Bicchi A, Lodi Rizzini S, Tonietti G (2001) Compliant design for intrinsic safety: general issues and preliminary design. Proceedings of the IEEE/RSJ international conference on intelligent robots and systems, Maui, Hawaii
3. Filippini R, Sen S, Bicchi A. (2007) Variable impedance actuation for physical human cooperating robots: a comparative analysis of safety, performance and dependability. IARP-IEEE/RAS-EURON joint workshop on technical challenges for dependable robots in human environments, Rome, Italy
4. Ikuta K, Ishii H, Nokata M (2003) Safety evaluation method of design and control for human-care robots. Int J Robot Res 22(5):281–297
5. De Santis A, Siciliano B, De Luca A, Bicchi A (2008) An atlas of physical human-robot interaction. Mech Mach Theory 43(3):253–270
6. English C, Russell D (1999) Implementation of variable joint stiffness through antagonistic actuation using rolamite springs. Mech Mach Theory 34(1):27–40
7. Choi JH, Hong SH, Lee WS, Kang SC (2009) A variable stiffness joint using leaf springs for robot manipulators. Proceedings of the IEEE international conference on robotics and automation ICRA, pp 4363–4368, NJ, USA
8. Hurst AJW, Chestnutt JE, Rizzi AA (2004) An actuator with physical variable stiffness for highly dynamic legged locomotion. Proceedings of the IEEE international conference on robotics and automation ICRA, pp 4662–4667, New Orleans, LA, USA, 26 April–1 May
9. Wolf S, Hirzinger G, (2008) A new variable stiffness design: matching requirements of the next robot generation. Proceedings of the IEEE international conference on robotics and automation ICRA, pp 1741–1746, Pasadena, California, USA, 19–23 May
10. Tonietti G, Schiavi R, Bicchi A (2006) Optimal mechanical/control design for safe and fast robotics. In: Proceedings of Experimental robotics IX: the 9th international symposium on experimental robotics, 21:311–320
11. Koganezawa K, Nakazawa T, Inaba T (2006) Antagonistic control of multi-dof joint by using the actuator with non-linear elasticity. Proceedings of the IEEE international conference on robotics and automation, pp 2201–2207, Florida, USA

12. Wassink M, Carloni R, Stramigioli S (2010) Port-hamiltonian analysis of a novel robotic finger concept for minimal actuation variable impedance grasping. Proceedings of the IEEE international conference on robotics and automation ICRA, pp 771–776, Anchorage, Alaska, USA, 3–8 May
13. Hyun D, Yang HS, Park J, Shim Y (2010) Variable stiffness mechanism for human-friendly robots. Mech Mach Theory 45(6):880–897
14. Jardim B, Siqueira AAG (2009) Development of series elastic actuators for impedance control of an active ankle foot orthosis. 20th international congress of mechanical engineering, Gramado, RS, Brazil, 15–20 Nov
15. Hogan N, Buerger SP (2005) Impedance and interaction control. Robotics and automation handbook, CRC Press, Boca. Raton, pp 19.1–19.24
16. Pratt G, Willison P, Bolton C, Hofman A (2004) Late motor processing in low-impedance robots: impedance control of serieselastics actuators.Proceedings of the American control conference, pp 3245–3251
17. Blaya JA, Herr H (2004) Adaptive control of a variable-impedance ankle-foot orthosis to assist drop-foot gait. IEEE Trans Neural Sys Rehab Eng 12(1):24–31
18. Zoppi M, Molfino R, Cerveri P (2010) Modular micro robotic instruments for transluminal endoscopic robotic surgery: new perspectives. In: International IEEE/ASME conference on mechatronic and embedded systems and applications MESA2010, Qingdao, ShanDong, China, July

# Chapter 76
# H-Beam Cutting System Based on Pro/E Model-Driven Technology

**Wenbin Duan, Juliang Xiao and Gang Wang**

**Abstract** In order to overcome the shortcomings of existing cutting methods, H-beam cutting system was developed. This system which based on three-dimensional Pro/E model is three-cutting-torch H-beam robot cutting machine, it can realize the position control and the motion control of some auxiliary operations. This paper mainly describes system's architecture, control method and key technologies used in this system. VBAPI, one of the redeveloping methods, is used in the model-driven technology and method of minimum delay-time is used in the path planning. The experiment results in practice show that: This system has well operability and applicability and can significantly improve the efficiency and quality of H-beam cutting.

**Keywords** Cutting system · Nested hierarchical architecture · Model-driven · Minimum delay-time

## 76.1 Introduction

Large-scale structural frameworks based on H-beam design are widely used in various fields, such as drilling platform, supermarkets, bridges, and so on. Due to the complexity and diversity of H-beam, for more accurately welding, H-beam cutting system was put forward higher requirements [1]. Currently, cutting H-beam mainly rely on manual. In order to overcome the shortcomings of manual cutting, such as low quality, low efficiency and so on, specialized cutting equipments appear. Existing cutting equipments in the domestic, however, mainly based

W. Duan · J. Xiao (✉) · G. Wang
Key Laboratory of Mechanism Theory and Equipment Design of Ministry of Education, Tianjin University, 92Weijin Road, Nankai District, Tianjin 300072, China
e-mail: tianjinxjl@163.com

J. S. Dai et al. (eds.), *Advances in Reconfigurable Mechanisms and Robots I*,
DOI: 10.1007/978-1-4471-4141-9_76, © Springer-Verlag London 2012

on CAD graph-driven [2] or parameter-driven [3]. In this case, it is invisible to increase workload of designers and operators. In addition, most of the cutting equipments belong to the Cartesian coordinate and single robot system. They are low efficiency in respect of cutting H-beams.

In response to these deficiencies mentioned above, this paper presents a cutting system. The system adopts three-cutting-torch robots. Each robot is responsible for cutting one side of H-beam. It can reduce the human–computer interaction greatly and improves the cutting efficiency, quality and accuracy highly. The system has been tested in the Offshore Oil Engineering (Qingdao) Co., Ltd. Practice shows that it can improve the efficiency and quality of cutting effectively.

## 76.2 Nested Hierarchical Architecture

Now, comparing with the traditional enclosed system, the open architecture system has more merits [2], such as interoperability, portability, interchangeability, extendibility, etc. So, in order to overcome these deficiencies, traditional enclosed system is changing to general open-type real-time dynamic control system gradually.

This paper, on the basis of traditional enclosed system, proposes a novel architecture to improve deficiencies of traditional enclosed system [4]. This system is a nested hierarchical architecture (shown in Fig. 76.1). The entire system is composed by three subsystems (as follows below). Three subsystems are closely linked which forming a large closed-loop control system. In the subsystem, in order to accurately complete the work, a small closed-loop control system is formed. They constitute double closed-loop control system. Besides, the whole system adopts the "PC + DMC" open architecture, which improves deficiencies of traditional enclosed system and makes the whole system have better interoperability and portability.

- *Control system.* This system is mainly in charge of receiving tasks to control corresponding hardware. In this subsystem, touch panel and control panel are mainly responsible for human–computer interaction. The role of PC is responsible for resource management, task scheduling and instructions issue. Programmable logic controllers (PLC) are utilized to control buttons on the control panel and motion control cards (DMC) to control field devices.
- *Execution system.* The *function* of this system is mainly executing the corresponding action in terms of the corresponding instructions or signals. Through analysis and calculation, digital pulse signals are sent to the servo motor drivers. Then the motor drivers drive motors rotation according to the received instructions. Speed signals and position signals are detected by the encoders and feedback to the DMC.
- *Detection system.* This *subsystem* adopts structured light detection technology. Structured light detector is composed by image collection card and structured light generator. Through industrial camera, image information of real object is real-time sent to image collection card. Through error measuring, error information feedback to control system for correcting cutting trajectory. Eventually, the whole system forms a correct cutting path.

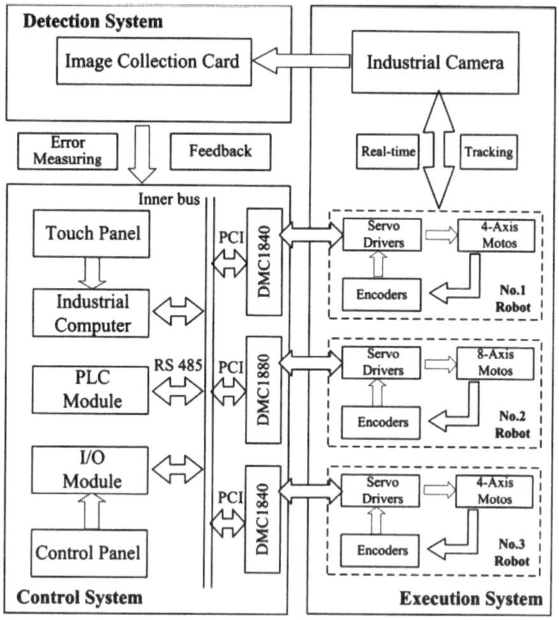


**Fig. 76.1** Nested hierarchical architecture of system

## 76.3 Control Algorithms

DMC [5] provided a variety of functions. The instructions of functions can be directly used for motion control of Cartesian robots. However, three-cutting-torch robots used in this system are planar multi-joint robots. Instructions of these functions mentioned above can't be used directly, they must be rewritten.

Industrial computer regularly sent the incremental pulse of the motor shafts to DMC so as to achieve coordinated movement of the motor shafts. Joint movement can be realized in the contour mode by send the following instructions to DMC:

CM X,Y,Z,W Set X, Y, Z and W axis to contour mode

DT 5 Time interval is 32 ms ($2^5 = 32$)

CD $d_x,d_y,d_z,d_w$ $d_x,d_y,d_z,d_w$ are incremental pulse of corresponding axis

WC Wait for motion to complete

As long as sending the last two instructions continuously, robots can realize continuous motion until a stop instruction.

## 76.4 Pro/E Model-Driven Technology

Now, graph-driven technology and parameter-driven technology are relatively common. Model-driven technology (MDT) was used relatively less than them. However, MDT has more merits, such as operability and intuitiveness. Aiming at the characteristic of model, this paper proposed a MDT based on software——Pro/ Engineer (Pro/E).

VBAPI [6, 7], one of the redeveloping methods, provides a lot of classes and methods, through these classes and methods [8] we can easily and accurately extract path data. This method is simple and practical for new beginners.

### *76.4.1 Extraction of Path Data*

Path extraction is a procedure of distinguishing, storing path data and preparing for the following path generation. The processes and methods used in this procedure are shown in the Fig. 76.2.

As shown in Fig. 76.2, extraction of path data is an important step. On the right we list the ideas of extraction and methods used in this step. Because path data we want to obtain are concentrated in three top surfaces of the flange plate and the web plate. Normal vector of these three surfaces are same and normal vector in this direction are only these three surfaces. So using the method of IpfcSurface.Eval3DData, we can accurately extract these three surfaces. Then in turn extract the corresponding contours and edges. Finally we can get data of all elements.

Set a reference point. According to the distance from this point, we can distinguish the data that we want from all. Then, these data are tagged and categorized. Store them so as to facilitate the next path generation.

In the path generation phase, through discrediting and translating points on the trajectory into pulse sequences which can be read by DMC (mentioned in the Sect. 76.3), path program are generated.

## 76.5 Path Planning

The path planning is based on experience-based path, namely a viable path scheme will be prepared based on experience. Quality and accuracy of this cutting system are mainly determined by hardware, followed by software. For quality, fixed preheating time and cutting speed are set to ensure the cutting quality of H-beam. For accuracy, structured light detector is adopted in this cutting system. This equipment can measure error and feedback to control system to correct cutting trajectory. Therefore, in the case that cutting path and auxiliary cutting parameters are predetermined, we, for multi-robots, can plan a collision-free, non-interference and minimum-time path

Fig. 76.2 Processes and methods for extraction of path data

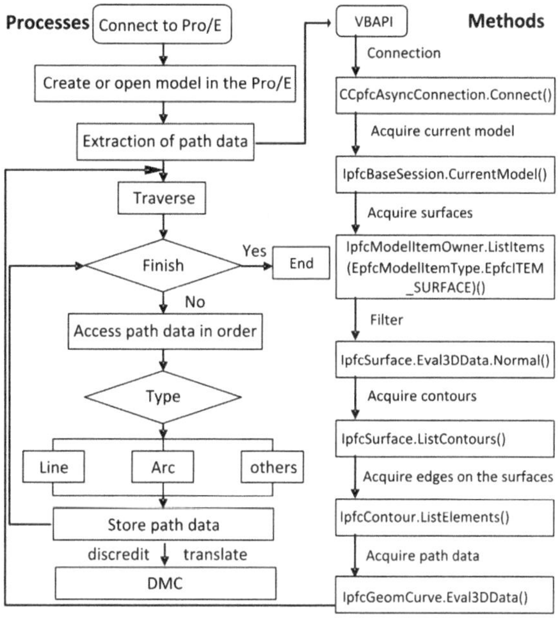


to improve the efficiency of the cutting system. In this paper, method of minimum delay-time [9–11] is adopted in the collision detection module. Through this method, the cutting path can be optimized. Path planning will increase participation of three robots, thereby reduce the whole cutting time and improve efficiency.

## 76.5.1 Experience-Based

Path planning is based on experience. It needs designers plan a feasible path based on the designers' experience. Considering the shape characteristics of H-beam, therefore, experience-based path should consist of two parts: cutting path of the flange plate and the web plate. In which, cutting path of flange plate is a line and cutting path of the web plate is composed by line and arc. Details of the path are shown in Fig. 76.3.

## 76.5.2 Collision Detection Module

To ensure efficiency and security of system, the system imported the collision detection module. The main purpose of the module is to ensure collision-free and to acquire a group of minimum delay-time. Mathematical model is as follows.

$$T_{\min} = F_1(t) + F_2(t) + F_3(t) + \cdots \quad (76.1)$$

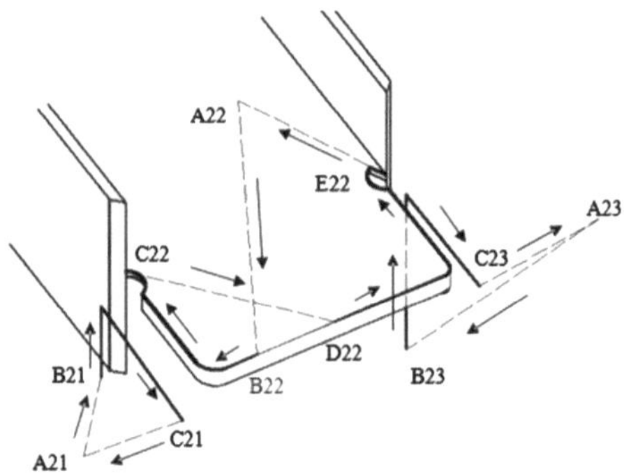


**Fig. 76.3** The second stage of cutting path

where the time vector $t = \{t_{11}, t_{12}, t_{21}, t_{22}, t_{31} \ldots\}$, $t_{ij}$ indicates *i*th robot required delay time in the *j*th path sequence. $T(t)$ indicates the time consumed of the whole path. The whole cutting process is divided into several cutting stages. Therefore, in order to optimize the cutting time, it should ensure that each stage consumes minimum time, just as Eq. (76.2).

$$\begin{cases} F(t) = \min\{\max(R_1(t),\ R_2(t),\ R_3(t)) + P(t)\} \\ s.t. C(t) = 0, \quad \forall t_{ij} \in t \quad \text{and} \quad t > 0 \end{cases} \tag{76.2}$$

where $F(t)$ indicates the time spent in each cutting stage. $R(t)$ indicates the time spent by each robot to complete their tasks. $C(t)$ is the optimum constraint function based on the collision detection results, if $C(t) = 0$ there are no collisions between robots. Otherwise, collision occurs. Besides, $P(t)$ is the preheat time given by Eqs. (76.3).

$$P(t) = m(n(t) - 1) \tag{76.3}$$

where *m*, a constant, indicates the time consumed by a once preheat. *n(t)* indicates the number of nonzero elements in the vector *t*.

A group of delay-time can be obtained by the above method. By adding these delay-times in corresponding stage, a highly efficient path with collision-free be obtained.

## 76.6 Experiment

The experiment [12] used the equipment of H-beam steel-cutting system (shown in Fig. 76.4). To ensure the cutting quality, cutting speed was set a fixed value, 3 mm/s. Take class A of HM588X300 for example, here, we selected the second stage of cutting path pre-designed by us to do optimization (shown in Fig. 76.3). The paths of robots in the second stage respectively are: No.1 robot: A21-B21-C21-A21; No.2 robot: A22-B22-C22-D22-E22-A22; No.3 robot: A23-B23-C23-A23. Each letter represents a potential pause point. Namely, these points are likely to insert a delay segment. Default delay-time is 0 s.

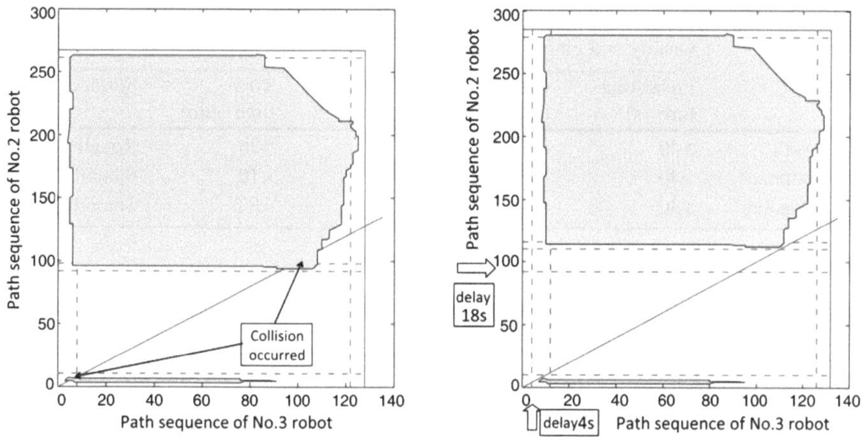


**Fig. 76.4** Interference graph before (*left*) and after (*right*) optimization

**Fig. 76.5** Cutting sample (*left*) and prototype of H-beam cutting system (*right*)

Three robots are designed in different rotational forms, and they are installed at different altitudes. Moreover, three robots owned their work space. Summing up the appeal, collision phenomenon will only appear between No.2 robot's arm and No.3 robot's screw.

From the Fig. 76.3 above, we can know that No.2 robot consumed the most time. The time consumed by each segment after initial planning are as follows. A22-B22:10 s; B22-C22:82 s; C22-D22:6 s; D22-E22:163 s; E22-A22:6 s; A23-B23:8 s; B23-C23:114 s; C23-B23:6 s. So the total time consumed in the second stage is 267 s. Simulate the collision of No.2 robot and No.3 robot. From the Fig. 76.5 (left) we can know that there are two collisions. The first collision is at 4 s, and the second is at 89 s. These results are as the constraints. Then, through Eqs. (76.2) we can obtain a group of time $t = \{0,0,0,0,0,18,0,0,4,0,0\}$. That means at A23 No.3 robot waits for 4.0 s and at C22 No.2 robot waits for 18 s. In this way, collision can be avoided after optimization [as shown in Fig. 76.5 (*right*)].

Experiment results are shown in Table 76.1. By adding 18 s delay-time, the consumed time of cutting process reduced from 10 to about 8.5 min, which save

**Table 76.1** Experimental results

| Class A | Quality and efficiency analysis | | | | |
|---|---|---|---|---|---|
| | Preheating time (s) | Cleaning time (min) | Cutting time (min) | Total time (min) | Surface quality |
| Handwork | >40 | >5 | >10 | >26 | Rough |
| Not optimized | >40 | <2 | <7 | <10 | Smooth |
| Path planning | >40 | <2 | <5.5 | <8.5 | Smooth |

23.3 % of time and improve the efficiency significantly. Figure 76.4 shows the prototype of H-beam cutting system and the sample cut by this system. Therefore we can know that the system has a high cutting efficiency and cutting quality.

## 76.7 Conclusion

According to some shortcomings of existing cutting system, this paper proposes a H-beam steel-cutting system. This system with nested hierarchical architecture is based on Pro/E MDT. Besides, the method of minimum delay-time is applied to optimize the cutting path in the machining process. The experiment results show that this H-beam steel-cutting system only needs few human–computer interactions to complete cutting tasks. The system can improve cutting quality and efficiency greatly, which has some practical significance for H-beam cutting system and even the actual intelligent CNC system.

## References

1. Li S, Deng H, Yang J et al (2011) Research on multi-robot H-beam steel cutting CNC system. Chin J Mach Tool Hydraul 21:98–100
2. Li P, Gao T, Wang J et al (2010) Open architecture of CNC system research based on CAD graph-driven technology. J Robotics Comput Integr Manuf 26(6):720–724
3. Song W, Wang G, Xiao J et al (2012) Research on multi-robot open architecture of an intelligent CNC system based on parameter-driven technology. J Robotics Comput Integr Manuf 28(3):326–330
4. Yang H (2010) Research of software for three-cutting-torch H-beam robot cutting machine. Tianjin University, China
5. Wang G, Wang G, Xiao J et al (2003) DMC and application in the development of robot controller with open architecture. Chin J Mach Tool Hydraul 6:206–209
6. Shang Y (2009) Discussion on the secondary development of Pro/E based on VB API. Chin J Mod Mach 4:19
7. Li S (2010) A product digital model serialized design method based on Pro/VBAPI. Chin J Mod Manuf Eng 8:22
8. Parametric technology corporation (2010) The VB API User Guide, USA
9. Bien Z, Lee J (1992) A minimum-time trajectory planning method for two robot. IEEE Transaction on Robotics and Automation, pp 413–418

10. Chang C, Chung MJ (1994) Collision avoidance of two general robot manipulators by minimum delay time. IEEE Trans Syst Man Cybern, pp 517–522.
11. Shen G (2010) Path planning for H-beam cutting system based on multi-robot. Tianjin University, China
12. Fan M, Xiao J, Wang G (2011) Optimum path planning for multi-robot H-beam cutting system. In: 3rd International conference on measuring technology and mechatronics automation, pp 823–826, IEEE Press, Shanghai

# Chapter 77
# The Typical Box-Taking Mechanism and Its Development Trend of Automatic Cartoning Machine

**Jujiang Cao, Lei Zhang and Long Li**

**Abstract** Introducing the history of automatic cartoner from seventies in our country; Introducing the typical box-taking mechanism of automatic cartoner. Analyzing the principle and advantages of mechanisms and proposing the future development trend of automatic cartoning machine. There is a small gap between our country and foreign country in high-level automatic cartoning machine, some machines have achieved the synchronization with the world's latest technology, but there are still gaps with foreign advanced level in the whole aspects of automatic cartoning machine.

**Keywords** Automatic cartoning machine · Box-taking machine · Development trend

## 77.1 Introduction

Automatic cartoning machine can put the medicine bottle, medicine board, creams etc. instruction into the folding carton and complete the cover action. Some full-feature automatic cartoning machines contain additional features, such as paste sealing label or thermal contraction wrapping.

J. Cao · L. Zhang (✉) · L. Li
Shaanxi University of Science and Technology, Xi'an, China
e-mail: laylay1121@126.com

J. Cao
e-mail: jjcao@sust.edu.cn

L. Li
e-mail: llabc125@163.com

J. S. Dai et al. (eds.), *Advances in Reconfigurable Mechanisms and Robots I*,
DOI: 10.1007/978-1-4471-4141-9_77, © Springer-Verlag London 2012

Early in our country, put the drugs production into box mainly by manual. With the development of production technology, the mechanized cartoning is used gradually. The earliest introduction of foreign automatic cartoning machine in our country is in the mid-seventies, but because of the quality of our packaging materials and medicine box production are lower than the automatic cartoning machine packaging requirements, so our pharmaceutical production have to continue to use inefficient manual packaging in a very long period of time. Until in the late 80 s, with the rapid development of the relevant technical field of pharmaceutical production in our country, the packaging materials and medicine box processing quality have a great step forward. Then the automatic cartoning machine has gradually been a wide range of applications in China's pharmaceutical industry.

With the continuous development of pharmaceutical industry in our country, the number of types of drugs is increasing; the pharmaceutical industries constantly improve the quality of their various pharmaceutical products. Demand for pharmaceutical packaging has also been enhanced, especially for automatic cartoning machine requirement. Automatic cartoner can be divided into three classes according its running speed. Category 1 for low-speed type, about 60–80 box/min; Category 2 for medium to high type, about 80–120 boxes/min; Category 3 for high-speed type, faster than 150 boxes/min. Each Category has its advantages, disadvantages and different purposes [1]. Although China has more and more types of automatic cartoning machine, and they can basically meet the demand of the domestic drug manufacturers produce a variety of different dosage forms of pharmaceutical packaging, well-known domestic pharmaceutical manufacturers of automatic cartoning machine, technical content of their products compared with foreign similar products, there are still a large gap. Competitiveness in the market, it only got the upper hand in the lower and medium t market, in market share, domestic automatic cartoning machine only have 30 % of market, the remaining markets are being carved up by imports. Therefore, most of large-scale pharmaceutical industries still use the import cartoner mainly; the cause of this situation is that the quality of domestic high-speed automatic cartoner is a huge problem for domestic pharmaceutical industry to solve.

Box-taking mechanism is an important mechanism in automatic cartoner, it takes drug box out from shelf and shove it off during the process of operation, at last, put the box on the conveyor belt smoothly in order that the medicine board can be put into it in next station. On the whole, box-taking mechanism is a "bottleneck" in the cartoner; its efficiency affects the whole efficiency. Therefore, developing a high-quality box-taking mechanism will be a sally port to solve the problem of the high-speed cartoner.

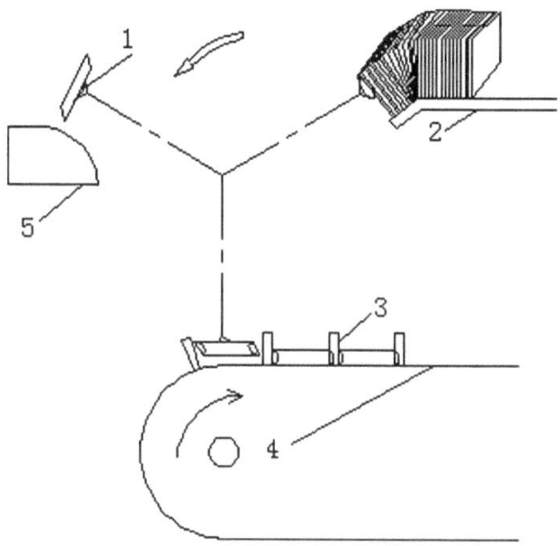


**Fig. 77.1** Movements of taking box. *1* vacuum suction head *2* shelf *3* block groove *4* conveyor belt *5* block

## 77.2 Typical Box-Taking Mechanism

### *77.2.1 Introduction of Box-Taking Movement*

In order to improve the efficiency of taking box, the present box-taking Mechanisms are using continuous rotary structure. Movements of taking box request actuators have a complicated trajectory. As shown in Fig. 77.1. Rotating vacuum suction head 1 suctioned the box out from the shelf 2, after rotate 120°, utilizing extrusion between medicine box and block 5 to shove box off slightly. Continue rotate 120°, and then put it in the block groove utilizing extrusion between boards to open the box completely.

The mainly mechanism of low-speed box-taking mechanism is gear. It has the compact structure, small size, good adaptability and low cost, the disadvantage is not easy to adjust and the speed is low. The main mechanism of high-speed box-taking is cam. The advantage of relatively stable operation, easy adjustment, wide field of applications, the disadvantage is typically narrowly targeted, so demanding for packing material and poor stability.

### *77.2.2 Planet Gear Style Box-Taking Mechanism*

It is a common type of low-speed box-taking mechanism [2]. As shown in Fig. 77.2 Which uses principles of planetary gear. Steering gear 7 and driven gear

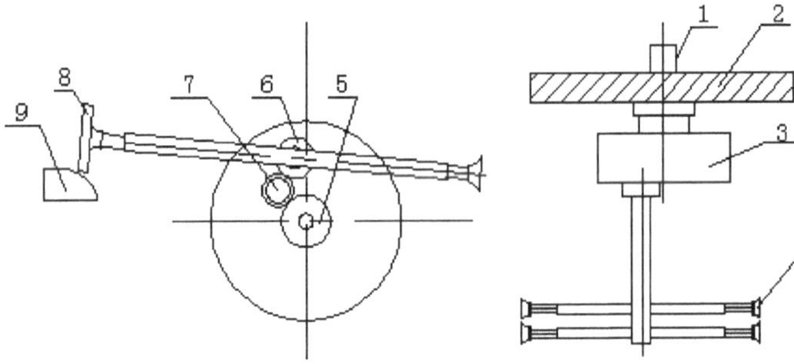


**Fig. 77.2** Principles of planetary gear type box-taking mechanism *1* fixed spindle *2* fixed board *3* eccentric *4* sucker *5* center gear *6* driven gear *7* turning gear *8* drug box *9* block

6 installed in the eccentric 3, the fixed gear 5, and the driven gear 6 and steering gear 7 form planetary gear train.

When the turntable 3 is rotated clockwise, the revolution of the driven gear 6 around turntable 3, meanwhile the driven gear 6 rotates in planetary gear train, Aspirate the driven gear 6 fixed to the sucker link, so the sucker both revolution and rotation. Revolution controls the sucker form a circular motion and rotation control the suction head at the specified location to achieve the specified action.

The planetary gear box-taking mechanism has simple structure and small volume, transmission with gear and has a smooth operation. In addition, the mechanism is easy to make and have no complicated parts, but the mechanism can only get one box every round. When the spindle speed is faster, it will affect the quality of the suction box, so the mechanism is only applicable to low-speed box-taking.

### *77.2.3 Multi-heads Planet Gear Style Box-Taking Mechanism*

Planetary gear box-taking mechanism only applies to low-speed box- taking. With the rapid development of automatic cartoning machine, low-speed ones no longer meet the needs of the market. In order to improve the efficiency of box-taking, the new planetary gear box-taking mechanism was born at the moment. The new mechanism also uses planetary theory, and develop to three suckers based on the old planetary gear, so the mechanism can get three boxes every round, the new planetary gear mechanism increases twofold efficiency to the old one in theory.

As shown in Fig. 77.3. When big turntable 1 rotating, shaft gear 4 and turntable gear 5 rotates around the fixed gear I and fixed gear II respectively. Small turntable 7 revolution with the big turntable 1. Meanwhile, it rotates with the controls of

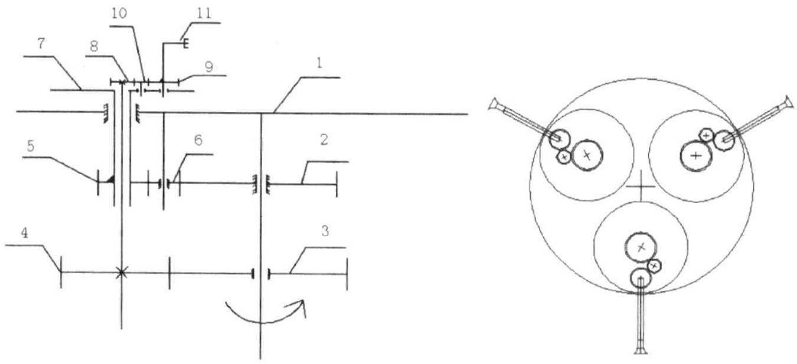


**Fig. 77.3** Principles of multi-head planetary gear type box-taking mechanism. *1* big turntable *2* fixed gear I *3* fixed gear II *4* shaft gear *5* turntable gear *6* steer gear I *7* small turntable *8* driving gear *9* driven gear *10* steer gear II *11* vacuum sucker

turntable gear 5. In like manner, driving gear 8 revolution with shaft gear 4 and it also drives driven gear 9 rotation. The vacuum sucker 11 fixed with driven gear 9, so the sucker revolute and rotate at the same time. So long as the control the ratio transmission among gears, can control the sucker to achieve the specified action at the specified location.

The new planetary gear mechanism with compact structure can run smoothly and is easy to make. Which is suit for medium–high speed automatic cartoning machine, but its disadvantage is that the quality of the mechanism is too heavy, and it is no longer fit the requirement when we want to improve the speed of box-taking.

### *77.2.4 Cam-Link Box-Taking Mechanism*

The mechanism uses rotary execution path in vertical plane. It uses double cam groove driven five-bar mechanism. As Fig. 77.4 shows, the Mechanism has four suction heads, only analyze one of them, the rest of them have same trajectory [3].

Working principle: Cam plate 3 fixed in pedestal 2, connect Spindle 1 with turntable 7. When spindle 1 revolving with turntable 7, five-bar mechanism rotates with turntable 7 together. Five-bar mechanism has two degrees of freedom and needs two input power to form precise trajectory. Rollers 6 in cam external groove 4 and cam internal groove 5 drive shaft A 11 and shaft B 12 respectively through the linkages, and then drive two input power linkages of five-bar mechanism, making C end point in five-bar mechanism execute the designated action in designated position. Suction head pole 8 fixed in the C end point of five-bar mechanism, therefore, trajectory of C point is the trajectory of suction head.

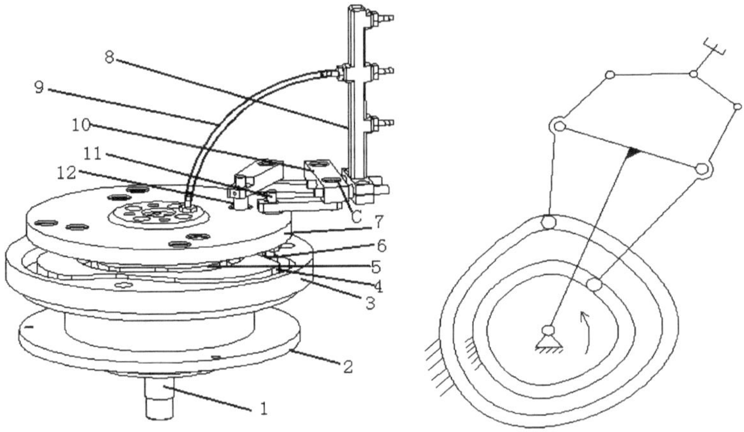


**Fig. 77.4** Cam-link box-taking mechanism

**Fig. 77.5** Five-head box-taking mechanism

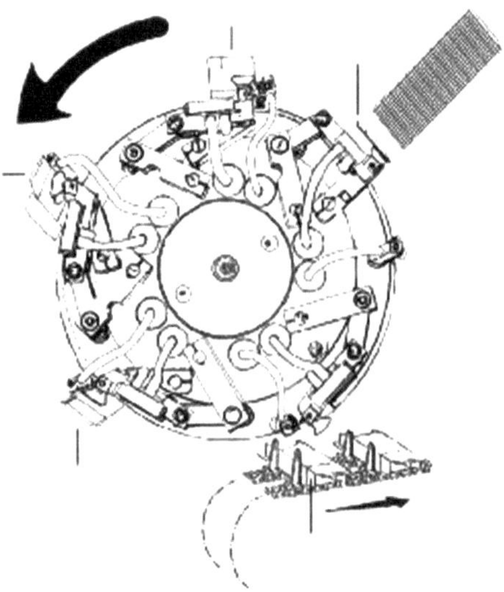

Vacuum pump transfer gas to the turntable 7 through the spindle 1, the hose 9 connects turntable with vacuum suction head, the situation of box depends on the gas on or off.

In order to improve the efficiency of box-taking, we must first solve the problem of high-speed box-taking, multi-head sucker will be the development trends of high-speed box-taking mechanism. A pharmaceutical factory of China newly developed five-head high-speed box-taking mechanism shown as Fig. 77.5, which also used a combination of cam-link to achieve high-speed box-taking action.

## 77.3 Development Trend of Automatic Cartoning Machine

With the growing scale of China's pharmaceutical market, the increasing of product categories, number of pharmaceutical manufacturers and degree of automation of production, market demand for pharmaceutical cartoning machine is continually expanding; Automatic cartoning machine market is very promising.

In the high-end level of pharmaceutical packing machinery, it is not a large gap between our country and foreign country, some china's equipments have been achieved the synchronization with the current progress of the world's latest technology, but there are still gaps with foreign advanced level in the whole aspects. So when we develop pharmaceutical packaging machinery and improve our overall level, especially needs further efforts in the following aspects.

### *77.3.1 Automated Design*

With the deepening of GMP certification work, automatic packaging has become an inevitable trend [4]. We need to solve testing, feedback, automatic adjustment and a series of questions. For non-standard forms of drugs, they have different physical forms, it also would encounter more problems when detect them, system requirements for post-test data processing are also increasing. Therefore, to achieve mechanical–electrical integration of the packaging machinery, it needs electronic, mechanical, packaging industry to work together and apply the advanced electronic technology up to the packaging industry in a timely manner.

### *77.3.2 Modular Design*

Because there are many types of drugs, the same drug product have different packaging forms because of different dosage, if only one type of packaging machinery can pack one model drug, it would formed a waste of resources. If we can make the equipments of similar packaging process become one or more standard equipment, replacement parts need to be made to operate independently of the basic unit of the body, that is "modular design", then the basic unit connects host and become a community through the interface, and combinate the basic units according to their own requirements in the practical application, so the device can complete the packaging of different drugs, and reflect the good applicability and flexibility. This "modular design" is an advanced design methods. Through a combination of different modules in the system, achieve multi-functional, serial machine. This is the future development of the trend of packaging machinery.

### *77.3.3 Humanization Design*

Modern packaging design, emphasized environmental protection, resource recycling, emphasizing humane, people-oriented in any field, Automatic cartoning machine as a product which reduces the strength of human labor should take into account the human factor, this idea should through the mechanical design of every detail, such as the machine's overall aesthetics, environmental protection, work surface height, the visual interface, operational safety, ease of maintenance, etc.

## 77.4 Conclusion

For automatic cartoning machine, the depth direction of development is faster, more robust, higher levels of automation and control, operating more humane, more convenient, etc. Automatic packaging machinery in China is only slightly dominant in the low-end market, but it is at the initial position in the high-end market, there are still gap between ours and international advanced level in R & D funding sources, test conditions, product stability and automatic control, etc. [5]. In addition to the use of PLC control technology in domestic cartoning machine, we must improve materials, design and all the relevant aspects of the package, which is possible to achieve intelligent control.

## References

1. Huang L (2006) The present situation and development research of leechdom installing box machine in our country. J Pack Food Mach 27:36–38
2. Li L, Tian X, Cao J (2011) Design and research of motion path for high-speed box-taking mechanism. J Pack Food Mach 29:28–30
3. Cao J, Long Li C (2011) Research and design of box-taking mechanism in automatic cartoner. In: 2011 International conference on mechatronics and materials processing, pp 743–746, Trans Tech Publications, Switzerland
4. Sun H, Liao Y (2009) The typical stucture of medicine box-taking machinery and its development trend. J Mech Electr Inf 20:30–33
5. Zhu YJ, Liang Y (2010) Observation and discussion of the latest high speed cartoner. J Mech Electr Inf 23:53–54

# Chapter 78
# A 'Multilink Spatial Hyper Redundant' Manipulator

**Praveen Jagadeesan, Sabarish Sivaprakasham, Dinesh Kumar and M. Madhu**

**Abstract** Conventionally, industrial robots manipulate their end effector in a pre defined workspace. Structures are made rigid to cater the pay load requirements. Navigation of these arm inspired robots in cluttered environment pose a challenging task, hence the focus on the continuum robots becomes a cynosure. This precisely defines the requirement of a redundant manipulator where in the controllable degrees of freedom is more than the total degree of freedom in the work space. Present work is focused on design of one such manipulator that can function in a hassle milieu. This serial manipulator briefly named as multilink spatial hyper redundant robot comprises of sixteen links and fifteen joints. With each joint providing a prescribed degree of freedom a thirty degree of freedom is realized on the system. The degree of freedom of the system is precisely controlled using ropes which are driven by a rotary actuator. Kinematic approach to the configuration is proposed. A prototype of this robot was developed for analyzing the functionality and for future study.

**Keywords** Spatial · Hyper · Continuum robots · Rope robots · Serpentine · Redundant · Backbone robot · Metamorphic robot

P. Jagadeesan (✉) · S. Sivaprakasham · D. Kumar · M. Madhu
PSG College of Technology, Coimbatore, India
e-mail: jparasu@gmail.com

S. Sivaprakasham
e-mail: sabarish1990psg@gmail.com

D. Kumar
e-mail: psgtech.dinesh@gmail.com

M. Madhu
e-mail: mmu@pro.psgtech.ac.in

J. S. Dai et al. (eds.), *Advances in Reconfigurable Mechanisms and Robots I*,
DOI: 10.1007/978-1-4471-4141-9_78, © Springer-Verlag London 2012 

## 78.1 Introduction

The slave arm robot, a simulacrum of human arm has rudimental control and little perception about the environment. Structures of these robots are designed as rigid modules considering the payload. Advent of technologies replacing man power with machines and demand for human safety in work environment sparked the study of field robotics [1]. These robots necessitate a high degree of flexibility to man oeuvre in hazardous work place which is not met by the conventional robots. Traditional approach of developing structures that would simulate a human action needs to be considered critically for this purpose. Several researches pertaining to the locomotive mechanisms (reptiles), prehensile nature (Monkeys tail), grasping ability (elephant trunk) and movement of invertebrate organs (octopus legs) were successfully studied and these ideas translated into robotic application. For an example, a rope drive controlled serpentine robot used in rescue operations that navigates with gait motion has been developed previously [2]. Narrowing down the study, this paper deals with development of a robot which has a base mounted structure precisely controlled by actuators and exhibit a non locomotive nature unlike the above described bio inspired robots. A non holonomic robot where in the controllable degree of freedom less than the total degree of freedom limits the application of robots in complex environment. For example, a continuum intrinsic type (actuator integrated with mechanism) robot described in literature [3] operates by fluid. It relies on elastic deformation of actuator plates that offers a multiple degree of freedom but has less controllable degree of freedom. The current work is an inspiration of the back bone structure that would produce a continuum motion in human. The following are the criteria that would aptly name the robot as Multi link spatial hyper redundant robot, identical links are manipulated in a three dimensional space with a constrained base link. Each Joint offering prescribed degree of freedom between links which cumulatively offers a high degree of freedom for the entire robot. A redundancy system where in there is more controllable degree of freedom than degree of freedom in work space is preferred over a holonomic (controllable degree of freedom equals the total degree of freedom) and non holonomic systems(Fig. 78.1).

The robot possesses 30 degree of freedom comprising of 15 joints and 16 links. Links are manipulated by ropes. A rotary actuator can be used to control the rope. The robot finds application in diverse fields where in the redundancy provides an advantage. This manipulator can be employed in production purposes, surveillance tasks and systems that demands redundancy in operation.

## 78.2 Design of Manipulator Segments

The manipulator is a serial type that comprises of links, spider joints and springs. The end effector is fitted to the last link. Link is a cylindrical disk of diameter 60 mm with a thickness of 15 mm and as the robot is rope driven, each link is

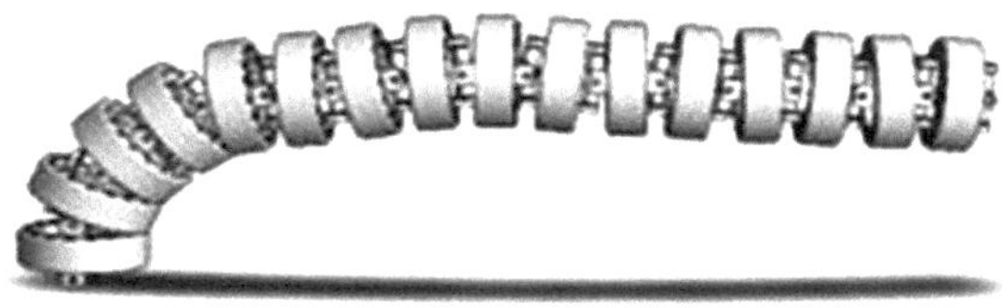

**Fig. 78.1** CAD simulated model of multilink spatial hyper redundant robot

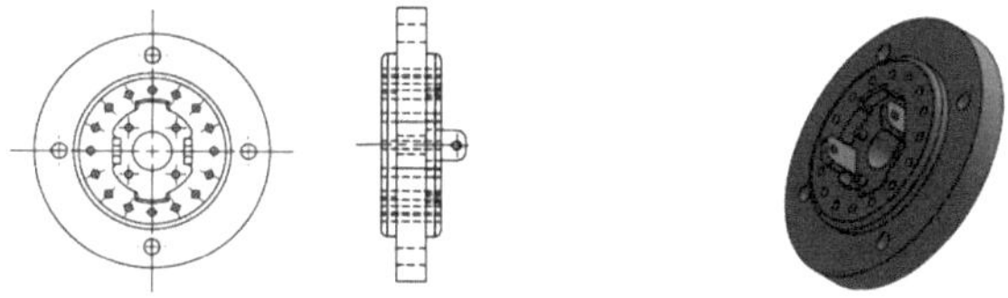

**Fig. 78.2** Manipulators dummy link geometry

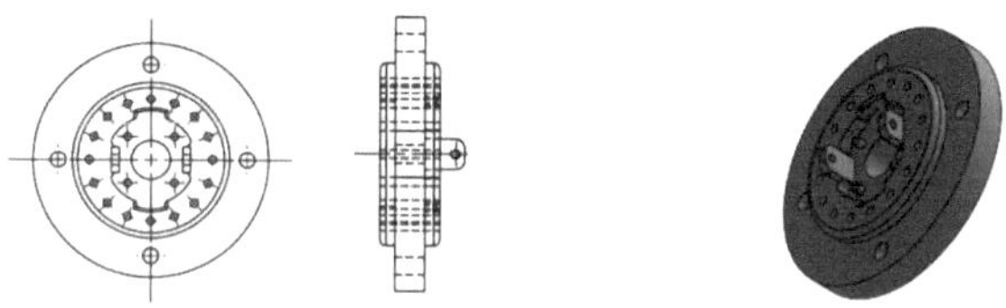

**Fig. 78.3** Manipulators first link geometry

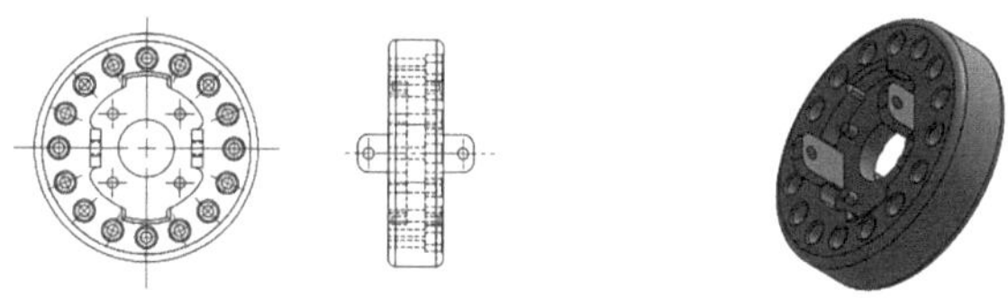

**Fig. 78.4** Manipulators targeted link geometry

provided with 16 through holes, see Fig. 78.2. These holes are made on the cylinder laid over a pitch circle diameter of 47 mm. Holes serve as a guiding element for the ropes. The first link is provided with provisions for rigid fastening to a plate, see Fig. 78.3. This plate enables the manipulators parts to be fitted to the actuator pack. Every fourth link serves as anchorage for four ropes. The ropes are placed orthogonal to one another such that two ropes control the motion of the robot in one plane and the remaining in the other plane. Hence eight individually controllable degrees of freedom are attained. Dummy links in between serves to increase the length as well to provide more degree of freedom. One end of the rope is tied to the actuator and the other end to the specifically targeted link (every fourth link), see Fig. 78.4. The targeted links are provided with pockets for anchorage of the ropes. Pulling a rope makes the targeted link to rest over the bottom link making an operating angle of 20° in a plane.

Projections made on the links serve to hinge the joints. Flexibility to link is provided with help of universal spider joints, see Fig. 78.5. Joints are housed between two identical links and each universal joint provides two degree of freedom.

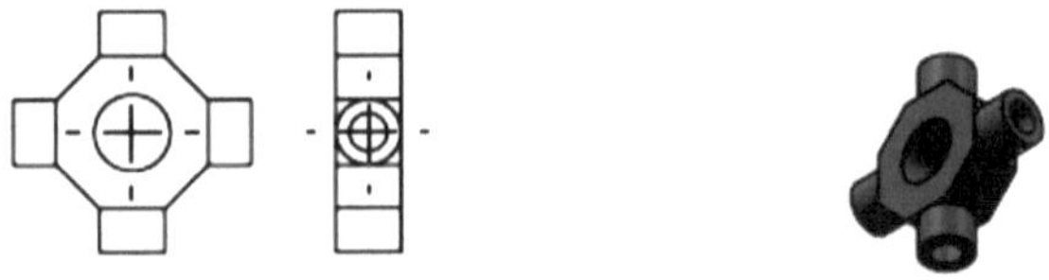

**Fig. 78.5** Spider joint

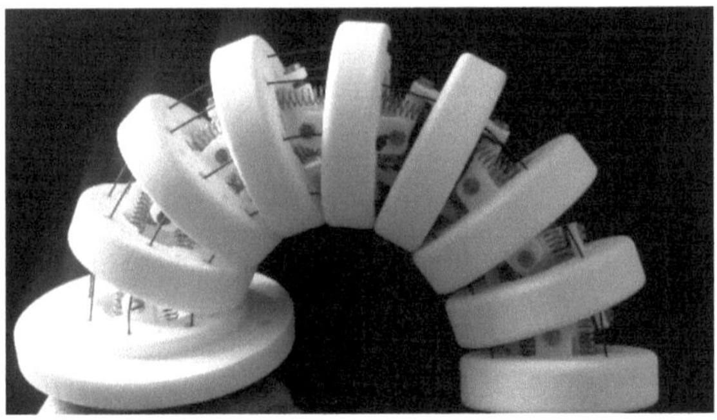

**Fig. 78.6** Rapid prototyped model

A centre through hole is provided on links and joints for bypassing the end effector cables to the last link. An axial play is provided between joint and link housing to prevent jam of the joints during motion. Four compression springs are provided between each link assembly which aid in restoration of links to the original position.

A prototype was developed for the purpose of concept realization using rapid prototyping technique, see Fig. 78.6. The assembly displayed in the figure consists of nine links and eight joints. Material chosen for prototyping was nylon as its abrasion resistance up to 5 mg/1000 cycles [4] and could be used for further experimentation purposes. The prototype is presently available at Department of production engineering, PSG College of technology.

## 78.3 Selection of Actuator and Rope

A silk braided rope of 0.5 mm diameter is used for actuating the manipulator. The premise of selecting a silk rope is it is free from springing nature unlike steel ropes and the prototype is of nylon. For heavy load application steel wires can be considered. The ropes were tested for its load carrying capacity and it was found out that the ropes are capable of carrying a 2 kg load (inclusive of factor of safety). A rotary geared stepper motor is used for the actuation purpose. The holding torque requirement depends upon the weight of the links and joints.

The springs are loaded perpendicular to its axis which tries to restore the moment produced by the links. Based on the spring stiffness and the applied load, springs yields differently and hence the $x_d$ alters proportionately. The following equations give the holding torque and the results are considered for selection of actuators (Fig. 78.7).

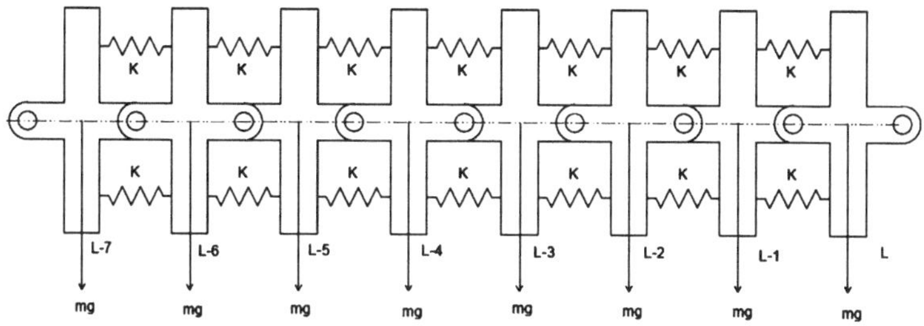


**Fig. 78.7** Forces acting on the system, K spring constant; L, total number of links; $P_c$, holding torque required at the targeted link; c, targeted link (t = 4,8,12,16,..,n); d, c + i; g, acceleration due to gravity; m, mass of a link; $x_d$, perpendicular distance between the targeted link and link d

**Fig. 78.8** Kinematic representation

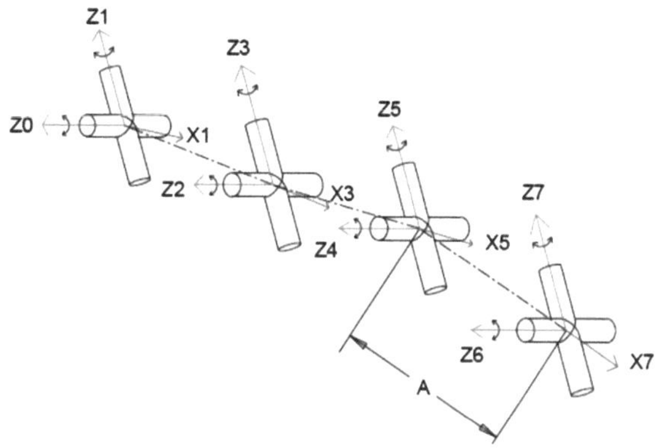


$$P_4 = x_d \tag{78.1}$$

$$P_8 = x_d \tag{78.2}$$

$$P_{12} = x_d \tag{78.3}$$

Actuators driving the fourth link are found to have a maximum holding torque requirement compared to the subsequent targeted links. This is because the fourth link bears the weight of the forthcoming links.

## 78.4 Kinematic Analyses

A kinematic analysis can be conveniently carried out for the manipulator using Denavit–Hartenberg. Since the total degree of freedom in the system is thirty, equal number of position variables is to be specified in order to locate all parts of the mechanism. A kinematic representation of the robot is shown in Fig. 78.8.

**Table 78.1** Danavit–Hartenberg for the manipulator

| # | $\alpha$ | A | $\theta$ | d |
|---|---|---|---|---|
| L1 | −90 | 0 | $\Theta_1$ | 0 |
| L2 | 90 | A | $\Theta_2$ | 0 |
| L3 | −90 | 0 | $\Theta_3$ | 0 |
| L4 | 90 | A | $\Theta_4$ | 0 |
| L5 | −90 | 0 | $\Theta_5$ | 0 |
| L6 | 90 | A | $\Theta_6$ | 0 |
| ⋮ | ⋮ | ⋮ | ⋮ | ⋮ |
| L28 | 90 | A | $\Theta_{28}$ | 0 |
| L29 | −90 | 0 | $\Theta_{29}$ | 0 |
| L30 | 90 | A | $\Theta_{30}$ | 0 |

A, length of common normal (*joint offset*) ; $\alpha$, angle of rotation along X axis (*joint twist*); $\theta$, rotation along Z axis; d, distance along Z axis between two common normal

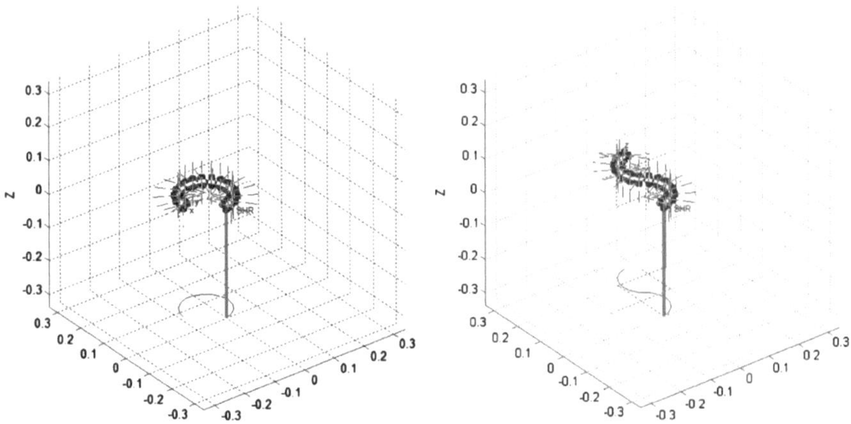


**Fig. 78.9** Motion of robot on specifying spatial coordinates

It can be inferred from Table 78.1 that a repeating set appears since the joints are identical in nature. The robot is tested for its manipulation in workspace using MATLAB, see Fig. 78.9. Joint variables are assigned to find the end effectors position.

There exists more than one kinematic approach for this manipulator. The robot is of continuum nature and hence the use of curvature to describe the manipulator is an alternative approach. A position vector expressed in terms of magnitude and angle of rotation can be taken as the DH parameters [5]. This can be arrived using Fernet Serret formula which is a differential geometry approach. F-S parameterizes a spatial curve by its arc length. The arc length and angle of rotation can also be found out using simple geometrical approach.

## 78.5 Conclusions

A novel approach for hyper redundancy has been presented in this paper. By manufacturing the prototype, it was confirmed that the design adhered to the functionality of the manipulator. Kinematic realization by DH approach aid for further studies. Future works involves practical testing of the robot with actuators and arriving at the inverse kinematic for the proposed manipulator.

## References

1. Siciliano B, Khatib O (2008) Springer handbook of robotics. ISBN: 978-3-540-23957-4.
2. Junyao G, Xueshan G, Wei Z, Jianguo Z, Boyu W (2008) Design and research of a new structure rescue snake robot with all body drive system. In: Proceedings of IEEE international conference on mechatronics and automation, pp 119–124.
3. Robinson G, Davies JBC (1999) Continuum robots—A state of the art. In Proceedings IEEE international conference on robotics and automation, Detroit, Michigan, vol 4, pp 2849–2854.
4. James ED, Shackelford F, Alexander W (2001) Material science and engineering handbook, 3rd edn., CRC press LLC, Washington, pp 763–764
5. Gravange IA, Walker ID (2000) On the kinematics of remotely actuated continuum robots. In: Proceedings IEEE international conference on robotics and automation, San Francisco, CA, vol 3, pp 2544–2550.

# Author Index



J. S. Dai et al. (eds.), *Advances in Reconfigurable Mechanisms and Robots I*,
DOI: 10.1007/978-1-4471-4141-9, © Springer-Verlag London 2012

FSC
www.fsc.org
MIX
Papier aus verantwortungsvollen Quellen
Paper from responsible sources
FSC® C105338

If you have any concerns about our products,
you can contact us on
**ProductSafety@springernature.com**

In case Publisher is established outside the EU,
the EU authorized representative is:
**Springer Nature Customer Service Center GmbH**
**Europaplatz 3, 69115 Heidelberg, Germany**

Printed by Libri Plureos GmbH
in Hamburg, Germany